Differentiable Measures and the Malliavin Calculus

Mathematical Surveys and Monographs
Volume 164

Differentiable Measures and the Malliavin Calculus

Vladimir I. Bogachev

American Mathematical Society
Providence, Rhode Island

EDITORIAL COMMITTEE

Ralph L. Cohen, Chair Michael A. Singer
Eric M. Friedlander Benjamin Sudakov
Michael I. Weinstein

2010 *Mathematics Subject Classification.* Primary 28Cxx, 46Gxx, 58Bxx, 60Bxx, 60Hxx.

For additional information and updates on this book, visit
www.ams.org/bookpages/surv-164

Library of Congress Cataloging-in-Publication Data
Bogachev, V. I. (Vladimir Igorevich), 1961–
  Differentiable measures and the Malliavin calculus / Vladimir I. Bogachev
    p. cm. – (Mathematical surveys and monographs ; v. 164)
  Includes bibliographical references and index.
  ISBN 978-0-8218-4993-4 (alk. paper)
  1. Measure theory. 2. Theory of distributions (Functional analysis) 3. Sobolev spaces.
4. Malliavin calculus. I. Title.

QA312.B638 2010
515'.42–dc22
                                                                            2010005829

Copying and reprinting. Individual readers of this publication, and nonprofit libraries acting for them, are permitted to make fair use of the material, such as to copy a chapter for use in teaching or research. Permission is granted to quote brief passages from this publication in reviews, provided the customary acknowledgment of the source is given.
  Republication, systematic copying, or multiple reproduction of any material in this publication is permitted only under license from the American Mathematical Society. Requests for such permission should be addressed to the Acquisitions Department, American Mathematical Society, 201 Charles Street, Providence, Rhode Island 02904-2294 USA. Requests can also be made by e-mail to **reprint-permission@ams.org**.

© 2010 by the American Mathematical Society. All rights reserved.
The American Mathematical Society retains all rights
except those granted to the United States Government.
Printed in the United States of America.

∞ The paper used in this book is acid-free and falls within the guidelines
established to ensure permanence and durability.
Visit the AMS home page at **http://www.ams.org/**

10 9 8 7 6 5 4 3 2 1   15 14 13 12 11 10

Contents

Preface		ix
Chapter 1.	**Background material**	1
1.1.	Functional analysis	1
1.2.	Measures on topological spaces	6
1.3.	Conditional measures	19
1.4.	Gaussian measures	23
1.5.	Stochastic integrals	29
1.6.	Comments and exercises	35
Chapter 2.	**Sobolev spaces on $\mathbb{R}^n$**	39
2.1.	The Sobolev classes $W^{p,k}$	39
2.2.	Embedding theorems for Sobolev classes	45
2.3.	The classes BV	51
2.4.	Approximate differentiability and Jacobians	52
2.5.	Restrictions and extensions	56
2.6.	Weighted Sobolev classes	58
2.7.	Fractional Sobolev classes	65
2.8.	Comments and exercises	66
Chapter 3.	**Differentiable measures on linear spaces**	69
3.1.	Directional differentiability	69
3.2.	Properties of continuous measures	73
3.3.	Properties of differentiable measures	76
3.4.	Differentiable measures on $\mathbb{R}^n$	82
3.5.	Characterization by conditional measures	88
3.6.	Skorohod differentiability	91
3.7.	Higher order differentiability	100
3.8.	Convergence of differentiable measures	101
3.9.	Comments and exercises	103
Chapter 4.	**Some classes of differentiable measures**	105
4.1.	Product measures	105
4.2.	Gaussian and stable measures	107
4.3.	Convex measures	112
4.4.	Distributions of random processes	122

4.5.	Gibbs measures and mixtures of measures	127
4.6.	Comments and exercises	131

Chapter 5. Subspaces of differentiability of measures — 133
- 5.1. Geometry of subspaces of differentiability — 133
- 5.2. Examples — 136
- 5.3. Disposition of subspaces of differentiability — 141
- 5.4. Differentiability along subspaces — 149
- 5.5. Comments and exercises — 155

Chapter 6. Integration by parts and logarithmic derivatives — 157
- 6.1. Integration by parts formulae — 157
- 6.2. Integrability of logarithmic derivatives — 161
- 6.3. Differentiability of logarithmic derivatives — 168
- 6.4. Quasi-invariance and differentiability — 170
- 6.5. Convex functions — 173
- 6.6. Derivatives along vector fields — 180
- 6.7. Local logarithmic derivatives — 183
- 6.8. Comments and exercises — 187

Chapter 7. Logarithmic gradients — 189
- 7.1. Rigged Hilbert spaces — 189
- 7.2. Definition of logarithmic gradient — 190
- 7.3. Connections with vector measures — 194
- 7.4. Existence of logarithmic gradients — 198
- 7.5. Measures with given logarithmic gradients — 204
- 7.6. Uniqueness problems — 209
- 7.7. Symmetries of measures and logarithmic gradients — 217
- 7.8. Mappings and equations connected with logarithmic gradients — 222
- 7.9. Comments and exercises — 223

Chapter 8. Sobolev classes on infinite dimensional spaces — 227
- 8.1. The classes $W^{p,r}$ — 227
- 8.2. The classes $D^{p,r}$ — 230
- 8.3. Generalized derivatives and the classes $G^{p,r}$ — 233
- 8.4. The semigroup approach — 235
- 8.5. The Gaussian case — 239
- 8.6. The interpolation approach — 242
- 8.7. Connections between different definitions — 246
- 8.8. The logarithmic Sobolev inequality — 250
- 8.9. Compactness in Sobolev classes — 253
- 8.10. Divergence — 254
- 8.11. An approach via stochastic integrals — 257
- 8.12. Some identities of the Malliavin calculus — 263
- 8.13. Sobolev capacities — 265
- 8.14. Comments and exercises — 274

Contents   vii

Chapter 9.   The Malliavin calculus   279
   9.1.   General scheme   279
   9.2.   Absolute continuity of images of measures   282
   9.3.   Smoothness of induced measures   288
   9.4.   Infinite dimensional oscillatory integrals   297
   9.5.   Surface measures   299
   9.6.   Convergence of nonlinear images of measures   307
   9.7.   Supports of induced measures   319
   9.8.   Comments and exercises   323

Chapter 10. Infinite dimensional transformations   329
   10.1.   Linear transformations of Gaussian measures   329
   10.2.   Nonlinear transformations of Gaussian measures   334
   10.3.   Transformations of smooth measures   338
   10.4.   Absolutely continuous flows   340
   10.5.   Negligible sets   342
   10.6.   Infinite dimensional Rademacher's theorem   350
   10.7.   Triangular and optimal transformations   358
   10.8.   Comments and exercises   365

Chapter 11. Measures on manifolds   369
   11.1.   Measurable manifolds and Malliavin's method   370
   11.2.   Differentiable families of measures   379
   11.3.   Current and loop groups   390
   11.4.   Poisson spaces   393
   11.5.   Diffeomorphism groups   394
   11.6.   Comments and exercises   398

Chapter 12. Applications   401
   12.1.   A probabilistic approach to hypoellipticity   401
   12.2.   Equations for measures   407
   12.3.   Logarithmic gradients and symmetric diffusions   414
   12.4.   Dirichlet forms and differentiable measures   416
   12.5.   The uniqueness problem for invariant measures   420
   12.6.   Existence of Gibbs measures   422
   12.7.   Comments and exercises   424

References   427

Subject Index   483

Preface

Among the most notable events in nonlinear functional analysis for the last three decades one should mention the development of the theory of differentiable measures and the Malliavin calculus. These two closely related theories can be regarded at the same time as infinite dimensional analogues of such classical fields as geometric measure theory, the theory of Sobolev spaces, and the theory of generalized functions.

The theory of differentiable measures was suggested by S.V. Fomin in his address at the International Congress of Mathematicians in Moscow in 1966 as a candidate for an infinite dimensional substitute of the Sobolev–Schwartz theory of distributions (see [**440**]–[**443**], [**82**], [**83**]). It was Fomin who realized that in the infinite dimensional case instead of a pair of spaces (test functions – generalized functions) it is natural to consider four spaces: a certain space of functions S, its dual S', a certain space of measures M, and its dual M'. In the finite dimensional case, M is identified with S by means of Lebesgue measure which enables us to represent measures via their densities. The absence of translationally invariant nonzero measures destroys this identification in the infinite dimensional case. Fomin found a convenient definition of a smooth measure on an infinite dimensional space corresponding to a measure with a smooth density on $\mathbb{R}^n$. By means of this definition one introduces the space M of smooth measures. The Fourier transform then acts between S and M and between S' and M'. In this way a theory of pseudo-differential operators can be developed: the initial objective of Fomin was a theory of infinite dimensional partial differential equations. However, as it often happens with fruitful ideas, the theory of differentiable measures overgrew the initial framework. This theory has become an efficient tool in a wide variety of the most diverse applications such as stochastic analysis, quantum field theory, and nonlinear analysis. At present it is a rapidly developing field abundant with many challenging problems of great importance for better understanding of the nature of infinite dimensional phenomena.

It is worth mentioning that before the pioneering works of Fomin related ideas appeared in several papers by T. Pitcher on the distributions of random processes in functional spaces. T. Pitcher investigated an even more general situation, namely he studied the differentiability of a family of measures $\mu_t = \mu \circ T_t^{-1}$ generated by a family of transformations T_t of a fixed measure μ. Fomin's differentiability corresponds to the case where the measures μ_t are the shifts μ_{th} of a measure μ on a linear space. However, the theory constructed for this more special case is much richer. Let us also mention similar ideas of L. Gross [**322**] developed in the analysis on Wiener space.

In the mid 1970s P. Malliavin [**751**] suggested an elegant method for proving the smoothness of the transition probabilities of finite dimensional diffusions. The essence of the method is to consider the transition probabilities P_t as images of the Wiener measure under some nonlinear transformations (generated by stochastic differential equations) and then apply an integration by parts formula on the Wiener space leading to certain estimates of the generalized derivatives of P_t which ensure the membership in C^∞. As an application, a probabilistic proof of Hörmander's theorem on hypoelliptic second order operators was given. Malliavin's method (now known as *the Malliavin calculus* or *the stochastic variational calculus*) attracted considerable attention and was refined and developed by many authors from different countries, including P. Malliavin himself. It was soon realized that the Malliavin calculus is a powerful tool for proving the differentiability of the densities of measures induced by finite dimensional nonlinear mappings of measures on infinite dimensional spaces (under suitable assumptions of regularity of the mappings and measures). Moreover, the ideas of the Malliavin calculus have proved to be very efficient in many applications such as stochastic differential equations, filtering of stochastic processes, analysis on infinite dimensional manifolds, asymptotic behavior of heat kernels on finite dimensional Riemannian manifolds (including applications to the index theorems), and financial mathematics.

Although the theory of differentiable measures and the Malliavin calculus have a lot in common and complement each other in many respects, their deep connection was not immediately realized and explored. The results of this exploration have turned out to be exciting and promising.

Let us consider several examples illuminating, on the one hand, the basic ideas of the theory of differentiable measures and the Malliavin calculus, and, on the other hand, explaining why the analysis on the Wiener space (or the Gaussian analysis) is not sufficient for dealing with smooth measures in infinite dimensions. First of all, which measures, say, on a separable Hilbert space X, should be regarded as infinite dimensional analogues of absolutely continuous measures on $\mathbb{R}^n$? It is well-known that there is no canonical infinite dimensional substitute for Lebesgue measure. For example, no nonzero locally finite Borel measure μ on l^2 is invariant or quasi-invariant under translations, that is, if $\mu \sim \mu(\cdot + h)$ for every h, then μ is identically zero. Similarly, if μ is not the Dirac measure at the origin, it cannot be invariant under all orthogonal transformations of l^2. In the finite dimensional case, instead of referring to Lebesgue measure one can fix a nondegenerate Gaussian measure γ. Then such properties as the existence of densities or regularity of densities can be described via densities with respect to γ. For example, it is possible to introduce Sobolev classes over γ as completions of the class of smooth compactly supported functions with respect to certain natural Sobolev norms. The Malliavin calculus pursues this possibility and enables us to study measures whose Radon–Nikodym densities with respect to γ are elements of such Sobolev classes. However, in the infinite dimensional case, there is a continuum of mutually singular nondegenerate Gaussian measures (e.g., all measures $\gamma_t(A) = \gamma(tA)$, $t > 0$, are mutually singular). Therefore, by fixing one of them, one restricts significantly the class of measures which can be investigated. Moreover, even if we do not fix a particular Gaussian measure, but consider measures for which there exist dominating Gaussian measures, we exclude from our considerations wide classes of smooth measures arising in applications. Let us consider an example of such a measure with an especially simple structure.

Example 1. Let μ be the countable product of measures m_n on the real line given by densities $p_n(t) = 2^n p(2^n t)$, where p is a smooth probability density with bounded support and finite Fisher's information, i.e., the integral of $(p')^2 p^{-1}$ is finite. It is clear that μ is a probability measure on l^2. The measure μ is mutually singular with *every* Gaussian measure (which will be proven in §5.3).

Certainly, this measure μ can be regarded and has the same property on $\mathbb{R}^\infty$ (the countable power of the real line). Note that μ is an invariant probability for the diffusion process ξ_t governed by the following stochastic differential equation (on the space $\mathbb{R}^\infty$):

$$d\xi_t = dW_t + \frac{1}{2}\beta(\xi_t)dt, \tag{1}$$

where the Wiener process W_t in $\mathbb{R}^\infty$ is a sequence (w_t^n) of independent real Wiener processes, and the vector field β is given by

$$\beta(x) = \bigl(p_n'(x_n)/p_n(x_n)\bigr)_{n=1}^\infty.$$

Therefore, in order to study differential properties of the transition probabilities and invariant measures of infinite dimensional diffusions we have to admit measures which cannot be characterized via Gaussian measures. Recent investigations of infinite dimensional diffusions have attracted considerable attention to the theory of differentiable measures and stimulated further progress in this field. One of the key achievements in this direction is the characterization of the drifts of symmetrizable diffusions as logarithmic gradients of measures. In the finite dimensional case, the logarithmic gradient of a measure μ with density p in the Sobolev class $W^{1,1}(\mathbb{R}^n)$ is given by $\beta = \nabla p/p$. It is known that in this case there is a diffusion ξ_t determined by equation (1) for which μ is a symmetric invariant measure. Conversely, if μ is a symmetric invariant measure for a diffusion ξ_t generated by equation (1), where β is μ-square integrable, then μ has a density $p \in W^{1,1}(\mathbb{R}^n)$ and $\beta = \nabla p/p$ μ-a.e. As we already know, in infinite dimensions, neither p nor ∇p are meaningful. Surprisingly enough, it is possible to give a rigorous meaning to the expression $\beta = \nabla p/p$ without defining its entries separately in such a way that in the finite dimensional case one obtains the usual interpretation. We shall discuss logarithmic gradients in Chapters 7 and 12 and see there that in the situation of Example 1 one gets exactly the mapping $\beta(x) = \bigl(p'_n(x_n)/p_n(x_n)\bigr)$. Moreover, we shall also see that in the infinite dimensional case logarithmic gradients of measures coincide with mappings β in equation (1) giving rise to symmetrizable diffusions.

Another related important problem, which both the theory of differentiable measures and the Malliavin calculus deal with, is the smoothness of the transition probabilities $P(t, x, \,\cdot\,)$ of the diffusion process ξ_t as set functions. In the finite dimensional case this corresponds to the differentiability of the transition density $p(t, x, y)$ with respect to y. In infinite dimensions such a problem in principal cannot be reduced to the study of densities because typically all measures $P(t, x, \,\cdot\,)$ are mutually singular for different t. For example, the transition probabilities $P(t, 0, \,\cdot\,)$ of an infinite dimensional Wiener process W_t are given by $P(t, 0, A) = \gamma(A/t)$, where γ is a Gaussian measure with infinite dimensional support. One can show that there is no σ-finite measure with respect to which all of these transition probabilities would be absolutely continuous. However, the theory of differentiable measures provides adequate tools for dealing with such examples.

Now we address the following problem. Let μ be a measure on a measurable space X and let F be a real function on X or an $\mathbb{R}^n$-valued mapping. Does the induced measure $\mu \circ F^{-1}$ have a density? Is this density smooth? Questions of this kind arise both in probability theory and analysis. The following reasoning in the spirit of the theory of differentiable measures and the Malliavin calculus illuminates one of the links between the two subjects.

Example 2. Let μ be a measure on $\mathbb{R}^n$ given by a density p from the Schwartz class $\mathcal{S}(\mathbb{R}^n)$ and let F be a polynomial on $\mathbb{R}^n$. Suppose that p and all of its derivatives vanish on the set $Z = \{x \colon \nabla F(x) = 0\}$. For example, let μ be the standard Gaussian measure and $Z = \varnothing$. Then the measure $\nu = \mu \circ F^{-1}$ admits an infinitely differentiable density q.

For the proof we denote by D_v the differentiation along the vector field $v = \nabla F$. For any $\varphi \in C_0^\infty(\mathbb{R}^1)$ the integration by parts formula yields

$$\int \varphi'(y)\, \nu(dy) = \int \varphi'\bigl(F(x)\bigr)\, \mu(dx)$$
$$= \int D_v(\varphi \circ F)\frac{1}{D_v F}\, p\, dx = -\int \varphi \circ F\, D_v \frac{p}{D_v F}\, dx.$$

The right-hand side is estimated by $\sup_t |\varphi(t)| \, \|D_v(p/D_v F)\|_{L^1(\mathbb{R}^1)}$, whence it follows that the derivative of ν in the sense of generalized functions is a measure and so ν has a density of bounded variation. For justification of integration by parts we observe that $D_v F = |\nabla F|^2$ and the function
$$D_v(p/D_v F) = (\nabla p, \nabla F)|\nabla F|^{-2} - p|\nabla F|^{-4}\bigl(\nabla F, \nabla(|\nabla F|^2)\bigr)$$
has the form $|\nabla F|^{-2}(\nabla p, \nabla F) + |\nabla F|^{-4} g p$, where g is a polynomial. The principal problem is to show that for every partial derivative $\partial^r p$ of the function p and every $m \geqslant 1$ the function $\partial^r p / |\nabla F|^m$ belongs to $L^1(\mathbb{R}^n)$. This can be done by means of the well-known Seidenberg–Tarski theorem (see Trèves [**1122**, Lemma 5.7, p. 315]). For example, if the set Z is empty, then this theorem gives two positive numbers a and b for which one has the inequality
$$|\nabla F(x)| \geqslant a(1+|x|)^{-b}.$$
In this case,
$$f/|\nabla F|^m \in L^1(\mathbb{R}^n) \quad \text{for all } f \in \mathcal{S}(\mathbb{R}^n) \quad \text{and} \quad D_v(p/D_v F) \in L^1(\mathbb{R}^n).$$
In the general case,
$$|\nabla F(x)| \geqslant a \operatorname{dist}(x, Z)^{b_1}(1+|x|)^{-b_2},$$
which also ensures the required integrability due to our condition on p. Repeating this procedure we obtain that the integral of $\varphi^{(r)}$ against the measure ν is estimated by $C_r \sup_t |\varphi(t)|$, where C_r is independent of φ. Hence ν is an element of the Sobolev class $W^\infty(\mathbb{R}^n)$, which yields the existence of a smooth density q of the measure ν.

Trying to follow the same plan in the infinite dimensional case, we at once encounter the obvious difficulty that the last equality in the integration by parts formula makes no sense due to the lack of infinite dimensional analogues of Lebesgue measure. This difficulty can be overcome if we define the action of differential operators directly on measures. This is, in fact, the essence of Fomin's theory of differentiable measures and the Malliavin calculus. Then a more delicate problem arises of finding vector fields for which the integration by parts holds. Most of the other problems discussed below are connected in one way or another to this central one.

The goal of this book is to present systematically the basic facts of the theory of differentiable measures, to give an introduction to the Malliavin calculus with emphasizing Malliavin's method of studying nonlinear images of measures and its connections with the theory of differentiable measures, and to discuss applications of both theories in stochastic analysis. The applications we consider include the study of some problems in the theory of diffusion processes (such as the existence and symmetrizability of diffusions, regularity of invariant measures and transition probabilities), and some standard applications of the Malliavin calculus. As compared to the finite dimensional case, one can say that we deal with geometric measure theory and Sobolev classes. Naturally, not all aspects of infinite dimensional analysis are touched upon, in particular, we do not discuss at all or just briefly mention such directions as harmonic analysis, white noise analysis, generalized functions, and differential equations in infinite dimensional spaces.

Chapter 1 provides some background material which may be useful for reading the subsequent chapters. Chapter 2 gives a concise introduction into the theory of classical Sobolev spaces. Almost all results in this chapter are given with proofs in view of the fact that very similar problems are studied further for measures on infinite dimensional spaces and hence it is useful to have at hand not only the formulations, but also the details of proofs from the finite dimensional case. In Chapters 3–7 we discuss the properties of measures on infinite dimensional spaces that are analogous to such properties as the existence and smoothness of densities of measures with respect to Lebesgue measure on $\mathbb{R}^n$. Chapter 8 is concerned with Sobolev classes over infinite dimensional spaces. Chapters 9–11 are devoted to the Malliavin calculus and nonlinear transformations of smooth measures, in particular, to the regularity of measures induced by smooth functionals (which is the central problem in the Malliavin calculus). In addition, we discuss measurable manifolds (measurable spaces with some differential structure). Finally, in Chapter 12 diverse applications are considered.

Dependence between the chapters. Chapters 1 and 2 are auxiliary, one can consult their sections when needed. Chapter 3 is one of the key chapters; the concepts introduced there are used throughout. Adjacent are Chapter 4 and Chapter 5, which are not essentially used in the subsequent exposition. Chapters 6 and 7 are used in Chapters 8, 9, and 12. Chapter 8 is important for Chapters 9, 10, and 12. Finally, Chapter 11 is relatively independent and self-contained; it employs only some general ideas from the previous chapters along with a few concrete results from them.

For all statements and formulas we use the triple enumeration: the chapter number, section number, and assertion number (all statements are numbered independently of their type within each section); numbers of formulas are given in parenthesis. For every work in the references all page numbers are indicated in square brackets where the corresponding work is cited.

The presentation is based on the courses taught by the author at the Department of Mechanics and Mathematics at Moscow Lomonosov State University, Scuola Normale Superiore di Pisa (an extended exposition of that course was published in 1995 in Pisa, and in 1997 an abridged journal version was published, see Bogachev [**189**]), and at the University of Minnesota in Minneapolis, as well as on lectures at many other universities and mathematical institutes, in particular, in St. Petersburg, Kiev, Paris, Orsay, Strasbourg, Marseille, Bordeaux, Bielefeld, Berlin, Bonn, Kaiserslautern, Munich, Göttingen, Heidelberg, Passau, Warwick, Cambridge,

London, Edinburgh, Rome, Torino, Lecce, Delft, Liège, Luxembourg, Stockholm, Copenhagen, Vienna, Madrid, Barcelona, Lisbon, Coimbra, Athens, Zürich, Haifa, Berkeley, Boston, Edmonton, Vancouver, Montreal, Kyoto, Tokyo, Osaka, Nagoya, Fukui, Peking, Sydney, Canberra, Melbourne, Santiago.

Acknowledgements. I am grateful to L. Accardi, H. Airault, S. Albeverio, L. Ambrosio, G. Ben Arous, V. Bentkus, S.G. Bobkov, A.-B. Cruzeiro, G. Da Prato, A.A. Dorogovtsev, D. Elworthy, H. Föllmer, M. Fukushima, B. Goldys, M. Gordina, F. Götze, L. Gross, P. Imkeller, A.I. Kirillov, N.V. Krylov, A.M. Kulik, R. Léandre, P. Lescot, P. Malliavin, E. Mayer-Wolf, P. Meyer, S.A. Molchanov, J. van Neerven, D. Nualart, E. Pardoux, A.Yu. Pilipenko, D. Preiss, E. Priola, Yu.V. Prohorov, M. Röckner, B. Schmuland, E.T. Shavgulidze, I. Shigekawa, H. Shimomura, A.N. Shiryaev, A.V. Skorohod, O.G. Smolyanov, V.N. Sudakov, A. Thalmaier, A.V. Uglanov, F.Y. Wang, H. von Weizsäcker, N. Yoshida, M. Zakai, and V.V. Zhikov for valuable discussions. I am especially indebted to A.I. Kirillov for careful reading of the whole text and valuable suggestions and E.P. Krugova, A.A. Lipchius, O.V. Pugachev, S.V. Shaposhnikov, N.A. Tolmachev, and R.A. Troupianskii for their help in preparation of the manuscript.

CHAPTER 1

Background material

This chapter contains the material from functional analysis, measure theory and stochastic analysis used throughout but not belonging to standard university courses. The role of this chapter is auxiliary, it should be consulted for references in the subsequent chapters; however, a quick look at this chapter may be useful at the first acquaintance with the book. For a reader not familiar with the principal concepts of measure theory on infinite dimensional spaces and the theory of Gaussian measures, the necessary minimum is presented in §§1.3–1.4.

1.1. Functional analysis

Here we recall some concepts and facts from functional analysis used throughout. We assume acquaintance with such notions as metric space, Hilbert space, Banach space, topological space, the weak topology of a Banach space at the level of a standard university course and knowledge of the fundamentals of the theory of the Lebesgue integral (see, e.g., Kolmogorov, Fomin [**626**], Rudin [**976**]). Some acquaintance with the concept of a locally convex space might also be useful, although not necessary: one can assume that everywhere normed spaces are considered in place of locally convex spaces.

A *locally convex space* is a real linear space X equipped with a collection of seminorms $\{p_\alpha\}$ such that for every $x \neq 0$ there exists α with $p_\alpha(x) > 0$. It is clear that any normed space is a locally convex space. Throughout, the following example of a locally convex space, which is not normable, will be important: the countable power $\mathbb{R}^\infty$ of the real line, i.e., the set of all infinite real sequences $x = (x_1, x_2, \ldots)$ with the seminorms $p_n(x) := |x_n|$, where $n \in \mathbb{N}$.

A locally convex space X with the corresponding collection of seminorms $\{p_\alpha\}$ is equipped with the topology in which the family of open sets consists of the empty set and arbitrary unions of the so-called basis neighborhoods of the form

$$U_{\alpha_1,\ldots,\alpha_n,\varepsilon_1,\ldots,\varepsilon_n}(a) := \{x \in X\colon p_{\alpha_i}(x-a) < \varepsilon_i, i = 1, \ldots, n\}, \ a \in X, \ \varepsilon_i > 0.$$

A locally convex space is called *sequentially complete* if in this space every fundamental (or Cauchy) sequence converges ("fundamental" means fundamental with respect to every seminorm from a family defining the topology). Similarly one defines sequentially complete sets. A complete metrizable locally convex space is called a *Fréchet space*.

Let X^* denote the topological dual to a locally convex space X (i.e., the space of all continuous linear functionals on X).

We shall often use the following classical Banach spaces: the space l^p of real sequences with finite norm $(x_n) \mapsto \left(\sum_{n=1}^{\infty} |x_n|^p\right)^{1/p}$, where $p \in [1, +\infty)$, the space c_0 of sequences convergent to zero with the norm $(x_n) \mapsto \sup_n |x_n|$, the space of bounded continuous functions $C_b(T)$ on a topological space T, equipped with the norm $x \mapsto \sup_{t \in T} |x(t)|$, and the space $L^p(\mu)$ (see below). The norm and inner product in an abstract Hilbert space H are denoted by $|\cdot|$ and $(\cdot, \cdot)$, respectively, or by $|\cdot|_H$ and $(\cdot, \cdot)_H$ if it is necessary to indicate the space explicitly. Similarly the norm in an abstract Banach space X is denoted by $\|\cdot\|$ or $\|\cdot\|_X$.

We shall also use the following classical locally convex spaces of functions: the space $\mathcal{D}(\mathbb{R}^n) := C_0^\infty(\mathbb{R}^n)$ consisting of all infinitely differentiable functions on $\mathbb{R}^n$ with compact support, the classes $\mathcal{D}(U) := C_0^\infty(U)$ defined similarly for domains $U \subset \mathbb{R}^n$, and the space $\mathcal{S}(\mathbb{R}^n)$ of rapidly decreasing smooth functions on $\mathbb{R}^n$, i.e., infinitely differentiable functions f such that for all integer nonnegative numbers k and m one has an estimate $|f^{(k)}(x)| \leqslant C_{k,m}(1+|x|)^{-m}$. The dual spaces to $\mathcal{D}(\mathbb{R}^n)$ and $\mathcal{S}(\mathbb{R}^n)$, called spaces of generalized functions or distributions, are denoted by the symbols $\mathcal{D}'(\mathbb{R}^n)$ and $\mathcal{S}'(\mathbb{R}^n)$.

For the study of measures the following two infinite dimensional spaces are particularly important: the space $\mathbb{R}^\infty$ of all infinite real sequences $x = (x_1, x_2, \ldots)$ and the Hilbert space l^2. The space $\mathbb{R}^\infty$ has the natural coordinatewise convergence; this convergence is generated by the metric

$$d(x,y) := \sum_{n=1}^{\infty} 2^{-n} \frac{|x_n - y_n|}{|x_n - y_n| + 1},$$

with respect to which $\mathbb{R}^\infty$ is a complete separable metric space. The space l^2 is a separable Hilbert space with respect to the inner product

$$(x,y) := \sum_{n=1}^{\infty} x_n y_n.$$

Each of these two spaces contains the union of all spaces $\mathbb{R}^n$ (the vectors of which become infinite sequences by adding zero components) and presents an infinite dimensional version of one of the two natural and equivalent convergences on $\mathbb{R}^n$: coordinatewise convergence and norm convergence.

A set V is called *convex* if $\lambda x + (1-\lambda)y \in V$ whenever $x, y \in V$, $\lambda \in [0,1]$. If, in addition, $\lambda x \in V$ for all $x \in V$ and all λ with $|\lambda| \leqslant 1$, then V is called *absolutely convex*. The *convex hull* of A is the intersection of all convex sets $V \supset A$; similarly one defines the *absolutely convex hull*.

1.1. Functional analysis

If V is an absolutely convex bounded set in a locally convex space X, then its linear span E_V is a normed space with respect to the norm q_V called the *Minkowski functional* of V and defined by the formula

$$q_V(x) := \inf\{t > 0 \colon t^{-1}x \in V\}.$$

If V is sequentially complete, then E_V is a Banach space and V is the closed unit ball in E_V (see Bogachev, Smolyanov [**234**, §8.6(v)] or Edwards [**379**, Lemma 6.5.2, p. 609]). Let us remark that if X is sequentially complete, then the closed absolutely convex hull of any metrizable compact set is metrizable and compact.

Let $\mathcal{L}(X, Y)$ be the space of all continuous linear operators between normed (or locally convex) spaces X and Y (and $\mathcal{L}(X) := \mathcal{L}(X, X)$), which in the case of normed spaces is equipped with the operator norm $\|\cdot\|_{\mathcal{L}(X,Y)}$, where

$$\|A\|_{\mathcal{L}(X,Y)} := \sup_{\|x\|_X \leq 1} \|Ax\|_Y.$$

A bounded linear operator A between two separable Hilbert spaces H and E is called a *Hilbert–Schmidt operator* if for some orthonormal basis $\{e_n\}$ in H we have

$$\|A\|_{\mathcal{H}(H,E)} := \Big(\sum_{n=1}^{\infty} |Ae_n|_E^2\Big)^{1/2} < \infty.$$

It is known that in this case such a sum is finite for every orthonormal basis in H and does not depend on the basis. The quantity $\|A\|_{\mathcal{H}(H,E)}$ is called the Hilbert–Schmidt norm of the operator A. The space of Hilbert–Schmidt operators between H and E is denoted by $\mathcal{H}(H, E)$. Equipped with the Hilbert–Schmidt norm it becomes a separable Hilbert space. The inner product in it is given by the formula

$$(A, B)_{\mathcal{H}(H,E)} := \sum_{n=1}^{\infty} (Ae_n, Be_n)_E,$$

where $\{e_n\}$ is an arbitrary orthonormal basis in H. Let $\mathcal{H}(H) = \mathcal{H}(H, H)$.

The class $\mathcal{H}_n(H, E)$ of n-linear Hilbert–Schmidt mappings with values in E is defined analogously and equipped with the inner product

$$(A, B)_{\mathcal{H}_n} := \sum_{i_1,\ldots,i_n \geq 1} \big(A(e_{i_1}, \ldots, e_{i_n}), B(e_{i_1}, \ldots, e_{i_n})\big)_E.$$

Letting $\mathcal{H}_n := \mathcal{H}(H, \mathcal{H}_{n-1})$, where $\mathcal{H}_1 := H$, $\mathcal{H}_0 := \mathbb{R}^1$, we can naturally identify the class $\mathcal{H}_n(H, \mathbb{R}^1)$ with the class $\mathcal{H}_n$ since every mapping $\Lambda \in \mathcal{H}_n$ defines an n-linear function

$$(h_1, \ldots, h_n) \mapsto \Lambda[h_1, \ldots, h_n] := \big(\Lambda(h_1) \cdots (h_{n-1}), h_n\big)_H.$$

This convention will be in force throughout.

An operator $A \in \mathcal{L}(H)$ in a separable Hilbert space H is called *nuclear* (or an *operator with trace*) if, for every orthonormal basis $\{e_n\}$ in H, the series

$$\sum_{n=1}^{\infty}(Ae_n, e_n)$$

converges. Let $\mathcal{L}_{(1)}(H)$ be the class of all nuclear operators. It is known that $A \in \mathcal{L}_{(1)}(H)$ if and only if $A = BC$, where B and C are Hilbert–Schmidt operators. An alternative description of nuclear operators is this: there exists an orthonormal basis $\{e_n\}$ in H such that $\sum_{n=1}^{\infty}|Ae_n|_H < \infty$. However, *not every* orthonormal basis has this property. By the Hilbert–Schmidt theorem, a compact selfajoint operator A in H has an orthonormal eigenbasis $\{e_n\}$, i.e., $Ae_n = \alpha_n e_n$. Such an operator is a Hilbert–Schmidt operator precisely when $\sum_{n=1}^{\infty}|\alpha_n|^2 < \infty$. A compact selfadjoint operator A is nuclear if and only if the series $\sum_{n=1}^{\infty}|\alpha_n|$ convergence.

If $A = U|A|$ is the *polar decomposition* of an operator A, i.e., $|A|$ is a selfadjoint operator such that $(|A|h, h) \geq 0$ and $(U|A|h, U|A|h) = (|A|h, |A|h)$ for all $h \in H$, then membership of A in the classes of Hilbert–Schmidt operators or nuclear operators is equivalent to the respective inclusion for $|A|$. An operator A between normed spaces X and Y is called *absolutely summing* if $\sum_n \|Ax_n\| < \infty$ for every unconditionally convergent series $\sum_n x_n$ (i.e., the series whose arbitrary rearrangements converge). It is known that absolutely summing operators between Hilbert space are precisely Hilbert–Schmidt operators.

A locally convex space X is called *continuously embedded* into a locally convex space Y if X is a linear subspace in Y and the identical embedding $X \to Y$ is continuous (i.e., the topology of X is stronger than the topology induced from Y). If some neighborhood of the origin in X has compact closure in Y, then the embedding is called *compact*. For example, the space $C[0,1]$ with its usual norm is continuously embedded into the Hilbert space $L^2[0,1]$, but this embedding is not compact; the space $C^1[0,1]$ of continuously differentiable functions with the norm $x \mapsto \max_t |x(t)| + \max_t |x'(t)|$ is compactly embedded into $C[0,1]$.

A mapping f between metric spaces (X, d_X) and (Y, d_Y) is called *Lipschitzian* if $d_Y(f(x), f(y)) \leq L d_X(x, y)$, where L is a number called the *Lipschitz constant* of f. For example, the distance function to a set E in X defined by the formula

$$\operatorname{dist}(x, E) := \inf\{d(x, y) \colon y \in E\}$$

is Lipschitzian. A mapping f from $[a, b]$ to a normed space E is called *absolutely continuous* if, for every $\varepsilon > 0$, there exists $\delta > 0$ such that $\sum_{i=1}^{n}\|f(b_i) - f(a_i)\| \leq \varepsilon$ for all pairwise disjoint intervals $[a_1, b_1], \ldots, [a_n, b_n]$ with $\sum_{i=1}^{n}|b_i - a_i| \leq \delta$.

Let us recall basic concepts related to the differentiability in infinite dimensional spaces. Let X be a locally convex space. Denote by $\mathcal{FC}_b^{\infty}(X)$

1.1. Functional analysis

the class of all functions f on X of the form
$$f(x) = \varphi\bigl(l_1(x),\ldots,l_n(x)\bigr), \quad \varphi \in C_b^\infty(\mathrm{I\!R}^n),\ l_i \in X^*.$$
Such functions will be called *smooth cylindrical*.

A mapping $F\colon X \to Y$ is called *differentiable* at a point x if there exists a continuous linear operator $DF(x)\colon X \to Y$, called the *derivative* of F at the point x, such that
$$F(x + th) - F(x) = tDF(x)(h) + r(x, th),$$
where the mapping $r(x, \cdot)\colon X \to Y$ is small in a certain sense (our choice of this sense corresponds to the type of differentiability). One often uses the differentiability with respect to a class $\mathcal{K}$ of subsets of X, where the smallness of r means by definition that uniformly in h from every fixed $K \in \mathcal{K}$ one has
$$\lim_{t \to 0} t^{-1} r(x, th) = 0.$$

A typical choice of $\mathcal{K}$ is one of the following classes:

a) all finite sets (*Gâteaux differentiability*),

b) all compact sets (*Hadamard differentiability* or compact differentiability),

c) all bounded sets (*Fréchet differentiability* or bounded differentiability).

For example, if X and Y are normed spaces, then Gâteaux differentiability means that for every $h \in X$ we have $\|r(x, th)/t\| \to 0$, while Fréchet differentiability is the relationship $\|r(x, h)\|/\|h\| \to 0$ as $\|h\| \to 0$.

The derivative of the mapping $F\colon X \to Y$ is denoted by DF or F'. If E is a locally convex space embedded linearly and continuously into X, then we denote by $D_E F$ the derivative of F along E, which is defined as the derivative at the origin of the mapping $h \mapsto F(x + h)$, $E \to Y$. We shall be mainly concerned with the case where $E = H$ is a separable Hilbert space. Then the symbol $\nabla_H F$ will be used as well. Below in Chapters 7–12 we shall also deal with certain generalized derivatives along a subspace.

Let us recall that a Banach space X is called a space with the *Radon–Nikodym property* if every Lipschitzian mapping f from $[0, 1]$ to X is almost everywhere differentiable. There is a number of equivalent definitions, for example, the property that every absolutely continuous mapping f from $[0, 1]$ to X is almost everywhere differentiable (see Benyamini, Lindenstrauss [**138**, Ch. 5] or Diestel, Uhl [**351**]). It is known that for any $p \in (1, \infty)$ the space $L^p(\mu)$ has the Radon–Nikodym property.

1.1.1. Lemma. *Let X be a Banach space having the Radon–Nikodym property and let $\{T_t\}$ be a one-parameter group of linear operators on X with $\|T_t\| = 1$ such that for some $h \in X$ one has $\|T_t h - h\| \leqslant C|t|$ for all $t \in [-1, 1]$. Then the mapping $F\colon t \mapsto T_t h$ is continuously differentiable on the real line and Lipschitzian with constant $\|F'(0)\| \leqslant C$.*

PROOF. Since $T_{t+s} = T_t \circ T_s$ and $\|T_t\| = 1$, the mapping F is Lipschitzian with constant C and hence differentiable at some point τ. Then the mapping F is differentiable everywhere because $F(t+\delta)-F(t) = T_{t-\tau}[F(\tau+\delta)-F(\tau)]$. In addition, one has the equality $F'(t) = T_t F'(0)$. By the Baire theorem, the mapping $t \mapsto F'(t)$ has a point of continuity being a pointwise limit of continuous mappings. It is readily seen that this yields the continuity of this mapping everywhere. $\square$

1.2. Measures on topological spaces

Here we give a brief outline of measure theory on topological spaces. For the purposes of this book, a rather modest background from this theory is needed. Here is the list of the most useful things:

(1) Radon measures (i.e., approximated from within by compact sets) and the fact that all Borel measures on reasonable spaces (e.g., complete separable metric) are Radon, (2) Souslin sets and their measurability, (3) Kolmogorov's theorem on constructing measures by means of consistent finite dimensional distributions, (4) conditional measures, (5) weak convergence of measures and the theorems of A.D. Alexandroff and Yu.V. Prohorov.

All the facts from the theory of measure and integral used below can be found in the author's book [**193**].

Let I_A denote the indicator function of a set A: $I_A(x) = 1$ if $x \in A$ and $I_A(x) = 0$ if $x \notin A$.

Throughout the term "measure on a measurable space $(X, \mathcal{A})$" means a real (finite) countably additive measure on a σ-algebra $\mathcal{A}$ of subsets of some space X. Mostly we are concerned with Borel measures on topological spaces, i.e., measures on Borel σ-algebras. Let us recall that the *Borel* σ-algebra $\mathcal{B}(X)$ of a topological space X is the smallest σ-algebra containing all open sets. Sets in $\mathcal{B}(X)$ are called Borel. Sometimes the *Baire* σ-algebra $\mathcal{B}a(X)$ of the topological space X is used, i.e., the smallest σ-algebra with respect to which all continuous functions on X are measurable. If X is a locally convex space, then we also use the σ-algebra $\sigma(X)$ generated by all sets of the form $\{x \colon l(x) \leqslant c\}$, where $l \in X^*$ and $c \in \mathbb{R}^1$, i.e., the smallest σ-algebra with respect to which all functionals $l \in X^*$ are measurable.

Any measure μ on a σ-algebra $\mathcal{A}$ in a space X can be decomposed into a difference of two nonnegative measures, i.e., there exists a decomposition $\mu = \mu^+ - \mu^-$, called the Hahn–Jordan decomposition, in which the nonnegative measures μ^+ and μ^- are concentrated on disjoint sets $X^+, X^- \in \mathcal{A}$ with $X = X^+ \cup X^-$. It is clear that by the latter condition the measures μ^+ and μ^- are uniquely determined. Let $|\mu|$ denote the variation of μ, i.e., the sum $|\mu| := \mu^+ + \mu^-$. Set $\|\mu\| := |\mu|(X)$. The space of all bounded measures on $\mathcal{A}$ is Banach with the norm $\mu \mapsto \|\mu\|$.

For a nonnegative measure μ, let $\mathcal{A}_\mu$ denote the Lebesgue completion of $\mathcal{A}$ with respect to μ, i.e., the class of sets of the form $A \cup C$, where $A \in \mathcal{A}$ and C has outer measure zero. The sets from $\mathcal{A}_\mu$ are called μ-measurable.

1.2. Measures on topological spaces

For a signed measure μ, we set $\mathcal{A}_\mu := \mathcal{A}_{|\mu|}$. A set E is called a *set of full μ-measure* if $|\mu|(X\backslash E) = 0$. A measure $\mu \geq 0$ is called *atomless* if every set of positive measure has a subset of a strictly smaller nonzero measure. A Radon measure (see the definition below) is atomless precisely when it vanishes at each point.

Let $(X, \mathcal{A})$ and $(Y, \mathcal{B})$ be two measurable spaces, i.e., spaces equipped with σ-algebras of subsets. A mapping $f \colon X \to Y$ is measurable with respect to the pair of σ-algebras $(\mathcal{A}, \mathcal{B})$ (or $(\mathcal{A}, \mathcal{B})$-measurable) if

$$f^{-1}(B) \in \mathcal{A} \quad \text{for all} \quad B \in \mathcal{B}.$$

If $(Y, \mathcal{B})$ is the real line $\mathbb{R}^1$ with its Borel σ-algebra $\mathcal{B} = \mathcal{B}(\mathbb{R}^1)$, then $(\mathcal{A}, \mathcal{B})$-measurable functions are called $\mathcal{A}$-measurable. Thus, a real function f on X is $\mathcal{A}$-measurable if $f^{-1}(B) \in \mathcal{A}$ for every Borel set $B \subset \mathbb{R}^1$. This is equivalent to the inclusion $\{x \colon f(x) < c\} \in \mathcal{A}$ for every $c \in \mathbb{R}^1$.

If a measure μ is given on a measurable space $(X, \mathcal{A})$, then we obtain a wider σ-algebra $\mathcal{A}_\mu$ of all sets measurable with respect to μ. In this case, we call μ-measurable also such functions f on X that f is defined and finite on a set $E \subset X$ of full μ-measure and is $\mathcal{A}_\mu$-measurable on it, i.e., $\{x \in E \colon f(x) < c\}$ belongs to $\mathcal{A}_\mu$ for all $c \in \mathbb{R}^1$. Outside of E the function f may be undefined at all or can assume infinite values.

The class of all μ-measurable functions is denoted by $\mathcal{L}^0(\mu)$, and its subclass of functions integrable to power $p \in (0, +\infty)$ is denoted by $\mathcal{L}^p(\mu)$. The class $\mathcal{L}^\infty(\mu)$ consists of bounded *everywhere defined* μ-measurable functions. If $f = g$ almost everywhere, then f and g are called equivalent and also *versions* or *modifications* of each other; notation: $f \sim g$. The space of equivalence classes of functions from $\mathcal{L}^p(\mu)$ is denoted by $L^p(\mu)$, $p \in (0, +\infty]$. In the case of a signed measure μ the integrability of a function f with respect to μ is understood as the integrability with respect to $|\mu|$. By definition we set

$$\int_X f(x)\,\mu(dx) := \int_{X^+} f(x)\,\mu^+(dx) - \int_{X^-} f(x)\,\mu^-(dx).$$

For $p \in [1, +\infty)$ the space $L^p(\mu)$ is equipped with the norm

$$\|f\|_{L^p(\mu)} := \|f\|_p := \left(\int_X |f(x)|^p\,|\mu|(dx) \right)^{1/p}.$$

The standard norm on $L^\infty(\mu)$ is defined as

$$\|f\|_\infty := \inf_{g \sim f} \sup_{x \in X} |g(x)|,$$

where inf is taken over all functions g equivalent to f.

In the case of a signed measure μ the symbol $(f, g)_{L^2(\mu)}$ is understood as $(f, g)_{L^2(|\mu|)}$.

If E is a Banach space, then $\mathcal{L}^p(\mu, E)$ denotes the space of μ-measurable mappings f with values in E such that f has values in a separable subspace and $\|f(\cdot)\| \in L^p(\mu)$, where $p \in [1, +\infty)$. The space $L^p(\mu, E)$ of equivalence classes in $\mathcal{L}^p(\mu, E)$ is equipped with the norm

$$\|f\|_p = \left(\int \|f(x)\|^p \, |\mu|(dx) \right)^{1/p},$$

with respect to which it is a Banach space.

Let two measures μ and ν be given on a σ-algebra $\mathcal{A}$ in a space X. The measure ν is called absolutely continuous with respect to μ, which is denoted by $\nu \ll \mu$, if $\nu(A) = 0$ for all sets $A \in \mathcal{A}$ with $|\mu|(A) = 0$. According to the Radon–Nikodym theorem, this is equivalent to the fact that ν has the form $\nu = \varrho \cdot \mu$, where $\varrho \in \mathcal{L}^1(\mu)$, i.e.,

$$\nu(A) = \int_A \varrho(x) \, \mu(dx), \quad A \in \mathcal{A}.$$

The function ϱ is called a density of ν with respect to μ (or a Radon–Nikodym density); it is denoted by $d\nu/d\mu$. If $\nu \ll \mu$ and $\mu \ll \nu$, then the measures μ and ν are called equivalent; notation: $\mu \sim \nu$. This is equivalent to that $\nu \ll \mu$ and $|\mu|$-a.e. one has $d\nu/d\mu \neq 0$.

The measures μ and ν are called mutually singular, which is denoted by $\mu \perp \nu$, if there exists a set $A_0 \in \mathcal{A}$ such that one has $|\mu|(A_0) = |\mu|(X)$ and $|\nu|(X \backslash A_0) = |\nu|(X)$. In the general case, one has the decomposition $\nu = \nu_1 + \nu_2$, where $\nu_1 \ll \mu$ and $\nu_2 \perp \mu$.

Let $\lambda \geqslant 0$ be a measure on $\mathcal{A}$ such that $\mu = f \cdot \lambda$ and $\nu = g \cdot \lambda$ (for example, one can take $\lambda = |\mu| + |\nu|$). Then one has the equality

$$\|\mu - \nu\| = \int_X |f - g| \, d\lambda.$$

It is useful to introduce Hellinger's integral

$$H(\mu, \nu) := \int_X |fg|^{1/2} \, d\lambda.$$

It is easily verified that $H(\mu, \nu)$ does not depend on a measure λ with respect to which μ and ν are absolutely continuous. An easy justification of the following assertion from Steerneman [**1077**] (which generalizes a well-known result for probability measures) is delegated to Exercise 1.6.11. Set

$$d_H(\mu, \nu) := \left[\|\mu\| + \|\nu\| - 2H(\mu, \nu) \right]^{1/2}.$$

1.2.1. Proposition. *The function d_H is a metric on the space of bounded measures on $\mathcal{A}$ and*

$$d_H(\mu, \nu)^2 \leqslant \|\mu - \nu\| \leqslant \left[(\|\mu\| + \|\nu\|)^2 - 4H(\mu, \nu)^2 \right]^{1/2}$$
$$\leqslant (\|\mu\|^{1/2} + \|\nu\|^{1/2}) d_H(\mu, \nu).$$

This estimate shows that d_H defines the same topology as the one generated by the variation norm.

1.2. Measures on topological spaces

1.2.2. Corollary. *For any measures* $\mu_1, \ldots, \mu_n$ *and* $\nu_1, \ldots, \nu_n$ *we have*

$$\prod_{k=1}^{n} \|\mu_k\| + \prod_{k=1}^{n} \|\nu_k\| - 2^{1-n} \prod_{k=1}^{n} \Big[(\|\mu_k\| + \|\nu_k\|)^2 - \big| \|\mu_k\| - |\nu_k| \big|^2 \Big]^{1/2}$$

$$\leqslant \Big\| \bigotimes_{k=1}^{n} \mu_k - \bigotimes_{k=1}^{n} \nu_k \Big\| \leqslant \sum_{k=1}^{n} \|\mu_k - \nu_k\| \prod_{j=1}^{k-1} \|\mu_j\| \prod_{j=k+1}^{n} \|\mu_j\|.$$

If μ_k *and* ν_k *are probability measures, then*

$$\Big\| \bigotimes_{k=1}^{n} \mu_k - \bigotimes_{k=1}^{n} \nu_k \Big\| \geqslant 2 - 2\exp\Big(-8^{-1} \sum_{k=1}^{n} \|\mu_k - \nu_k\|^2\Big). \tag{1.2.1}$$

The image of a measure μ on a measurable space $(X, \mathcal{A})$ under a measurable mapping f from X to a measurable space $(Y, \mathcal{B})$ is denoted by $\mu \circ f^{-1}$ and defined by the equality

$$\mu \circ f^{-1}(B) = \mu\big(f^{-1}(B)\big).$$

This definition yields the following change of variables formula:

$$\int_X \varphi(f(x)) \, \mu(dx) = \int_{\mathbb{R}^1} \varphi(t) \, \mu \circ f^{-1}(dt), \tag{1.2.2}$$

valid for every bounded $\mathcal{B}$-measurable function φ; more generally, this formula remains valid for every $\mathcal{B}$-measurable function φ integrable with respect to the image measure $|\mu| \circ f^{-1}$.

A Borel measure μ is called *Radon* if, for every Borel set B and every $\varepsilon > 0$, there exists a compact subset K in B such that $|\mu|(B \backslash K) \leqslant \varepsilon$.

Every Radon measure has a *topological support*, i.e., the smallest closed set of full measure.

The set of all Radon measures on X is denoted by $\mathcal{M}_r(X)$, and its subset consisting of probability measures is denoted by $\mathcal{P}_r(X)$.

Any Radon measure μ is *regular*: for every $B \in \mathcal{B}(X)$ and every $\varepsilon > 0$ there exist a closed set $Z \subset B$ and an open set $U \supset B$ with $|\mu|(U \backslash Z) < \varepsilon$. However, the regularity is not equivalent to being Radon.

1.2.3. Proposition. *Every Borel measure μ on a metric space X is regular.*

PROOF. We can assume that $\mu \geqslant 0$. Let $\mathcal{E}$ be the class of all sets $B \in \mathcal{B}(X)$ such that for each $\varepsilon > 0$ there exist a closed set Z and an open set U with $Z \subset B \subset U$ and $\mu(U \backslash Z) < \varepsilon$. All closed sets belong to $\mathcal{E}$ since in the case of closed B one can take $Z = B$ and an open set $U_n := \{x \colon \mathrm{dist}(x, B) < 1/n\}$ with some n because the sets U_n decrease to B. The class $\mathcal{E}$ is obviously closed with respect to complementation. In addition, $B = \bigcup_{n=1}^{\infty} B_n \in \mathcal{E}$ if $B_n \in \mathcal{E}$. Indeed, given $\varepsilon > 0$ we find open sets $U_n \supset B_n$ with $\mu(U_n \backslash B_n) < \varepsilon 4^{-n}$ and take $U = \bigcup_{n=1}^{\infty} U_n$. Next we find N such that $\mu(B \backslash \bigcup_{n=1}^{N} B_n) < \varepsilon/4$ and take closed sets $Z_n \subset B_n$

with $\mu(B_n\backslash Z_n) < \varepsilon 4^{-n}$. The set $Z = \bigcup_{n=1}^N Z_n \subset B$ is closed, and we have $\mu(U\backslash Z) < \varepsilon$. Hence $\mathcal{E}$ is a σ-algebra containing all closed sets, i.e., we obtain the equality $\mathcal{E} = \mathcal{B}(X)$. □

1.2.4. Remark. It is seen from the proof that the class $\mathcal{E}$ is a σ-algebra also in the case of a topological space. The metrizability of X is needed to conclude that $\mathcal{E}$ contains all closed sets.

Let us take a nonmeasurable (with respect to Lebesgue measure λ) set $X \subset [0,1]$ with $\lambda^*(X) = 1$. Then the formula $\mu(B \cap X) := \lambda(B)$ correctly defines a Borel probability measure on X. Indeed, it is readily verified that any Borel set in X has the form $B \cap X$, where B is a Borel set in $[0,1]$. In addition, $\lambda(B_1) = \lambda(B_2)$ if $B_1 \cap X = B_2 \cap X$ since $\lambda^*(X) = 1$. It follows that the measure μ is regular, but is not Radon, since X is not measurable in $[0,1]$ and compact sets from X are compact also in $[0,1]$.

A Radon measure μ is *tight*: for every $\varepsilon > 0$ there is a compact set $K_\varepsilon \subset X$ with $|\mu|(X\backslash K_\varepsilon) < \varepsilon$.

However, this is also not equivalent to being Radon: there are examples of nonregular Borel measures on nonmetrizable compact sets. Only the combination of regularity and tightness is equivalent to the Radon property. On most of the spaces encountered in applications all Borel measures are Radon (although there are exceptions, for example, the product of the continuum of intervals). In particular, the following theorem of Ulam is valid.

1.2.5. Theorem. *On every complete separable metric space all Borel measures are Radon.*

PROOF. Let μ be a nonnegative Borel measure on a complete separable metric space X. We already know that this measure is regular. By using the separability and completeness, we establish its tightness. Let $\varepsilon > 0$ and let $\{x_n\}$ be a countable everywhere dense set in X. For every $k \in \mathbb{N}$ the union of the open balls $B(x_n, 1/k)$ of radius $1/k$ is X. Hence there exists N_k such that $\mu\bigl(X\backslash \bigcup_{n=1}^{N_k} B(x_n, 1/k)\bigr) < \varepsilon 2^{-k}$. Let $B := \bigcap_{k=1}^\infty \bigcup_{n=1}^{N_k} B(x_n, 1/k)$. Then $\mu(X\backslash B) \leqslant \varepsilon$. The set B is completely bounded since for every k it is covered by finitely many balls of radius $1/k$. Hence B has compact closure (due to the completeness of X) with the desired property. □

For complete metrizable locally convex spaces this theorem can be refined as follows (see, for example, [**193**, Theorem 7.12.4]).

1.2.6. Theorem. *Let μ be a Radon measure on a Fréchet space X. Then there exists a continuously embedded reflexive separable Banach space $E \subset X$ such that $|\mu|(X\backslash E) = 0$ and the closed unit ball from E is compact in X.*

In the general case one cannot choose E to be Hilbert; however, for measures on $\mathbb{R}^\infty$ this can be easily done.

A measure μ on a Banach space has *moment of order m* if $\|\cdot\|^m \in L^1(\mu)$.

1.2. Measures on topological spaces

One of the most often applied results in measure theory on topological space is the theorem on the measurability of Souslin sets, going back to Lusin and formulated below. Let us recall that a *Souslin set* in a topological space X is the image of a complete separable metric space M under a continuous mapping from M to X. It is known that Borel sets in complete separable metric spaces are Souslin. However, every uncountable complete separable metric space contains a Souslin set that is not Borel.

Souslin sets in complete separable metric spaces can be described also by means of the Souslin operation over closed (or open) sets.

Let $\mathcal{E}$ be some class of subsets of a space X. If to every finite sequence of natural numbers $(n_1, \ldots, n_k)$ there corresponds a set $A_{n_1,\ldots,n_k} \in \mathcal{E}$, then we say that a Souslin scheme (or table) of sets $\{A_{n_1,\ldots,n_k}\}$ with values in $\mathcal{E}$ is defined. The Souslin operation (also known as the A-operation) on the class $\mathcal{E}$ is a mapping which associates to every Souslin scheme $\{A_{n_1,\ldots,n_k}\}$ with values in $\mathcal{E}$ the set

$$A = \bigcup_{(n_i) \in \mathbb{N}^\infty} \bigcap_{k=1}^\infty A_{n_1,\ldots,n_k}.$$

Sets of this form along with the empty set are called $\mathcal{E}$-Souslin or $\mathcal{E}$-analytic. The class of such sets is denoted by $\mathcal{S}(\mathcal{E})$. It turns out that by applying the Souslin operation to the class of closed (or open) sets in a complete separable metric space, we obtain precisely that class of Souslin sets. In the case of the real line it suffices to apply the Souslin operation to the class of intervals. Yet another equivalent description of $\mathcal{E}$-Souslin sets is this: such sets are exactly the projections to X of sets in $X \times [0,1]$ representable as countable intersections of countable unions of products of the form $E \times [a,b]$, where $E \in \mathcal{E}$ and $[a,b] \subset [0,1]$. The same class coincides with the projections of sets in $\mathcal{E} \otimes \mathcal{B}([0,1])$. For example, a set $A \subset \mathbb{R}^1$ is Souslin precisely when it is the projection of a Borel set in the plane. It is important to remember that such projections need not belong to the class of Borel sets.

The role of these objects in measure theory is clear from the following assertions.

1.2.7. Theorem. (i) *Let μ be a measure on a measurable space $(X, \mathcal{A})$. The Souslin operation on sets in $\mathcal{A}_\mu$ yields sets from $\mathcal{A}_\mu$, i.e., the Souslin operation preserves the measurability.*

(ii) *All Souslin sets in a topological space X are measurable with respect to every Radon measure on X.*

It is clear that assertion (i) is nontrivial: in the formation of sets by means of the Souslin operation, uncountable unions over all infinite sequences (n_i) of natural numbers are used! Let us mention yet another useful result in this area (see [**193**, Corollary 6.10.10]).

1.2.8. Theorem. *Let $(X, \mathcal{E})$ be a measurable space and let Y be a Souslin space. Then the projection to X of any set $M \in \mathcal{S}\big(\mathcal{E} \otimes \mathcal{B}(Y)\big)$ belongs to $\mathcal{S}(\mathcal{E})$ and hence is measurable with respect to every measure on $(X, \mathcal{E})$.*

For example, if X is a Hausdorff topological space and $\mathcal{E} = \mathcal{B}(X)$, then for any Souslin space Y we have the equality (see [**193**, Lemma 6.4.2])

$$\mathcal{B}(X \times Y) = \mathcal{B}(X) \otimes \mathcal{B}(Y). \tag{1.2.3}$$

Hence, for every set B in $\mathcal{B}(X \times Y)$, its projection

$$\{x \in X \colon (x, y) \in B\}$$

is measurable with respect to all Borel measures on X; although even in the case $X = Y = [0, 1]$ this projection need not be Borel.

1.2.9. Example. Let X be a Hausdorff topological space, let Y be a Souslin space, and let f be a bounded Borel function on $X \times Y$. Then the function

$$g(x) := \sup_{y \in Y} f(x, y)$$

is measurable with respect to every Borel measure on X. Indeed, we see that the set $\{x \colon g(x) \leqslant c\}$ is the projection of the set $\{(x, y) \colon f(x, y) \leqslant c\}$. In the case of an unbounded function f the same is true for the mapping g with values in $(-\infty, +\infty]$.

It is known that if μ and ν are two Radon measures on a completely regular space X such that

$$\int_X f(x)\, \mu(dx) = \int_X f(x)\, \nu(dx)$$

for all functions $f \in C_b(X)$, then $\mu = \nu$. A bit stronger assertion is true.

1.2.10. Lemma. *If the above equality is true for all functions from some algebra $\mathcal{F} \subset C_b(X)$ with the property that for every two distinct points $x, y \in X$ there exists a function $f \in \mathcal{F}$ with $f(x) \neq f(y)$, then $\mu = \nu$. The same is true in the complex case if $\overline{f} \in \mathcal{F}$ for all $f \in \mathcal{F}$.*

PROOF. By the Stone–Weierstrass theorem, given a compact set K, a function $\varphi \in C_b(X)$ with $\|\varphi\|_\infty \leqslant 1$, and $\varepsilon > 0$, one can find a function $f \in \mathcal{F}$ with $|f(x) - \varphi(x)| \leqslant \varepsilon$ for all $x \in K$ and $\|f\|_\infty \leqslant 1$. Now it is readily seen that the integrals of φ with respect to μ and ν coincide: it suffices to take a compact set K with $[|\mu| + |\nu|](X \backslash K) \leqslant \varepsilon$, find f as above and observe that the absolute value of the integral of φ with respect to $\mu - \nu$ is estimated by $2\varepsilon + \varepsilon \|\mu - \nu\|$. □

In the study of measures on infinite dimensional spaces it is useful to consider as model examples the spaces $\mathbb{R}^\infty$ and l^2 introduced above, for which many abstract constructions acquire a simple and transparent form. In order to uniquely determine a probability measure μ on the Borel σ-algebra in $\mathbb{R}^\infty$ it suffices to know its finite dimensional projections μ_n, i.e., the images of the measure μ under the linear projections $\pi_n\colon \mathbb{R}^\infty \to \mathbb{R}^n$, $x \mapsto (x_1,\ldots,x_n)$. These projections are consistent: the projection of μ_{n+1} to $\mathbb{R}^n$ is μ_n. The celebrated theorem of Kolmogorov says that the converse is also true: if, for every n, we are given a Borel probability measure μ_n on $\mathbb{R}^n$ and the consistency condition mentioned above is fulfilled, then there exists a unique Borel probability measure μ on $\mathbb{R}^\infty$ for which the measures μ_n serve as the finite dimensional projections. To know the measures μ_n means to know the measures of all cylindrical sets of the form

$$C = \bigl\{x\colon (x_1,\ldots,x_n) \in B\bigr\}, \quad B \in \mathcal{B}(\mathbb{R}^n).$$

Such sets form an algebra, but not a σ-algebra. So one should not think that the measure of a general Borel set can be explicitly expressed by means of projections. Apart from a simple coordinate representation, the importance of $\mathbb{R}^\infty$ is explained by the fact that this space arises in the consideration of countable products of measures on the real line. Let us recall the corresponding construction (details can be found in [**193**, Ch. 3]). Given a sequence of probability measures ν_n on measurable spaces $(X_n, \mathcal{A}_n)$, on the countable product $X = \prod_{n=1}^\infty X_n$ of all sequences of the form $x = (x_1, x_2, \ldots)$, where $x_i \in X_i$, we obtain the σ-algebra $\mathcal{A} = \bigotimes_{n=1}^\infty \mathcal{A}_n$ generated by products of sets of the form $A_1 \times A_2 \times A_n \times X_{n+1} \times X_{n+2} \times \cdots$, where $A_i \in \mathcal{A}_i$, called measurable rectangles. The measure of this set is defined to be $\prod_{i=1}^n \nu_i(A_i)$. Then one proves that there exists a measure on $\mathcal{A}$, denoted by $\bigotimes_{n=1}^\infty \nu_n$, having on all measurable rectangles the indicated values. In the case where we multiply the real lines, the existence of countable products of measures follows at once by the existence of finite products and Kolmogorov's theorem on consistent probabilities. In the general case this requires a justification.

Some features of the space $\mathbb{R}^\infty$ with respect to constructing measures is inherited by its subset l^2. Here finite dimensional projections μ_n of the measure μ are useful as well. Yet, now for the existence of a measure with given projections μ_n their consistency is not sufficient. For example, if we take the countable power of the measure with some probability density ϱ on the real line, then we obtain consistent measures μ_n with densities $\varrho(x_1)\cdots\varrho(x_n)$ on $\mathbb{R}^n$. However, one can verify that on l^2 there is no measure with projections μ_n (Exercise 1.6.13). In order to guarantee that the measures μ_n generate a measure on l^2 and not on the larger space $\mathbb{R}^\infty$, one more condition is required.

1.2.11. Proposition. *Suppose we are given a consistent sequence of Borel probability measures μ_n on the spaces $\mathbb{R}^n$. For the existence of a Borel probability measure μ on l^2 with the projections μ_n it is necessary and sufficient that for every $\varepsilon > 0$ we could find a number $R_\varepsilon > 0$ such that $\mu_n\bigl(B_n(0, R_\varepsilon)\bigr) \geqslant 1 - \varepsilon$ for all n, where $B_n(0, R_\varepsilon)$ is the closed ball of radius R_ε centered at the origin in $\mathbb{R}^n$.*

PROOF. Indeed, if a measure μ on l^2 with the projections μ_n exists, then the indicated additional condition is fulfilled since for every $\varepsilon > 0$ one can find a ball B in l^2 of a sufficiently large radius R such that $\mu(B) > 1 - \varepsilon$. For projections we obtain analogous inequalities since for every n the ball B is contained in the cylinder $C_{n,R}$ with the base $B_n(0,R)$ in $\mathbb{R}^n$, whence we find $\mu_n(B_n(0,R)) = \mu(C_{n,R}) \geqslant \mu(B)$. On the other hand, the stated condition is sufficient. For a proof we observe that this condition yields that the measure μ on $\mathbb{R}^\infty$ with the finite dimensional projections μ_n is concentrated on l^2. Indeed, otherwise $\mu(l^2) = \alpha < 1$. Let us take $\varepsilon > 0$ such that $1 - \varepsilon > \alpha$. Then we find R_ε according to our condition. Let us observe that the intersection of the cylinders

$$C_n = \left\{ x \in \mathbb{R}^\infty \colon \sum_{i=1}^n x_i^2 \leqslant R_\varepsilon^2 \right\}$$

coincides with the closed ball B of radius R_ε centered at the origin in l^2, i.e., with the set $\{x \in \mathbb{R}^\infty \colon \sum_{i=1}^\infty x_i^2 \leqslant R_\varepsilon^2\}$. Since we have $C_{n+1} \subset C_n$ and the estimate $\mu(C_n) = \mu_n(B_n(0, R_\varepsilon)) \geqslant 1 - \varepsilon$ for all n, we obtain the estimate $\mu(B) \geqslant 1 - \varepsilon > \alpha$, which is a contradiction. $\square$

Kolmogorov's theorem on consistent probability distributions is valid in a more general case. For example, if we are given a nonempty set T, then one can take the space $\mathbb{R}^T$ of all functions $x \colon T \to \mathbb{R}^1$. The case of $\mathbb{R}^\infty$ considered above corresponds to $T = \mathbb{N}$. In a similar way one introduces cylindrical sets of the form

$$C_{t_1,\ldots,t_n,B} := \{x \colon (x(t_1), \ldots, x(t_n)) \in B\},$$

where $t_i \in T$ and $B \in \mathcal{B}(\mathbb{R}^n)$. Then we obtain the σ-algebra $\sigma(\mathbb{R}^T)$ generated by such cylinders. If $\mathbb{R}^T$ is regarded as a locally convex space with the topology of pointwise convergence, then this will be precisely the σ-algebra generated by its dual since all continuous linear functionals on the space $\mathbb{R}^T$ have the form $x \mapsto c_1 x(t_1) + \cdots + c_n x(t_n)$, where $c_i \in \mathbb{R}$, $t_i \in T$. Any fixed collection of points $t_1, \ldots, t_n \in T$ determines a finite dimensional operator

$$\pi_{t_1,\ldots,t_n} \colon x \mapsto (x(t_1), \ldots, x(t_n)), \quad \mathbb{R}^T \to \mathbb{R}^n.$$

Hence for any measure μ on $\sigma(\mathbb{R}^T)$ one can define finite dimensional distributions $\mu_{t_1,\ldots,t_n} := \mu \circ \pi_{t_1,\ldots,t_n}^{-1}$. These finite dimensional distributions are related by the following two consistency conditions:

(i) the projection of $\mu_{t_1,\ldots,t_{n+1}}$ on $\mathbb{R}^n$ is $\mu_{t_1,\ldots,t_n}$,

(ii) if the collection $s_1, \ldots, s_n$ is obtained from $t_1, \ldots, t_n$ by some permutation of indices, then $\mu_{s_1,\ldots,s_n}$ is the image of $\mu_{t_1,\ldots,t_n}$ under the linear transformation of $\mathbb{R}^n$ generated by this permutation of coordinates.

Kolmogorov's theorem in this more general situation states that every collection of probability measures $\mu_{t_1,\ldots,t_n}$ on the spaces $\mathbb{R}^n$ satisfying conditions (i) and (ii) is the family of finite dimensional distributions for some probability measure μ on $\sigma(\mathbb{R}^T)$.

In the case where $T \subset \mathbb{R}^1$, it suffices to consider only finite collections of the form $t_1 < \cdots < t_n$. Hence, in this case, only condition (i) must be verified. Yet, in the general case, in place of collections $t_1,\ldots,t_n$ one can deal with disordered finite subsets $\Lambda \subset T$ and define finite dimensional distributions not on $\mathbb{R}^n$, but on finite dimensional spaces $\mathbb{R}^\Lambda$ consisting of real functions on Λ. The projection $\pi_\Lambda\colon \mathbb{R}^T \to \mathbb{R}^\Lambda$ is merely the restriction of the function to Λ. In this representation only one consistency condition arises: whenever $\Lambda_1 \subset \Lambda_2$, it is required that the measure μ_{Λ_1} be the image of the measure μ_{Λ_2} under the natural projection from $\mathbb{R}^{\Lambda_2}$ to $\mathbb{R}^{\Lambda_1}$ given by the restriction of functions to Λ_1.

Most of the results in this book deal with abstract Hilbert, Banach or locally convex spaces, but for understanding the essence of the discussion it suffices (and is useful) to have in mind l^2 or $\mathbb{R}^\infty$ as model spaces. A general locally convex space X also has *cylindrical sets* (or *cylinders*) of the form

$$C_{l_1,\ldots,l_n,B} := \Big\{x \in X\colon\ \big(l_1(x),\ldots,l_n(x)\big) \in B\Big\},$$

where $l_1,\ldots,l_n \in X^*$ and $B \in \mathcal{B}(\mathbb{R}^n)$. The σ-algebra generated by all cylindrical sets is denoted by $\sigma(X)$. If X is a separable normed space (or a separable metrizable locally convex space), then $\sigma(X) = \mathcal{B}(X)$. The class of all cylindrical sets is an algebra. Hence for every measure μ on $\sigma(X)$, every set $B \in \sigma(X)$ and every $\varepsilon > 0$, there exists a cylindrical set C such that $|\mu|(B \bigtriangleup C) < \varepsilon$. If μ is a Radon measure on X, then the same is true for every set $B \in \mathcal{B}(X)$ (even in the case where $\sigma(X) \neq \mathcal{B}(X)$). If the measure μ is defined only on $\sigma(X)$ (for example, appeared as the distribution of some random process in the space of all paths $\mathbb{R}^T$), then the question arises of its extendibility to a Radon measure on $\mathcal{B}(X)$. The answer is given by the following theorem.

1.2.12. Theorem. *Let μ be a measure on $\sigma(X)$. A necessary and sufficient condition for the existence of an extension of μ to a Radon measure on X is the following: for every $\varepsilon > 0$ there exists a compact set K_ε in X such that $|\mu|(C) < \varepsilon$ for every cylindrical set C disjoint with K_ε. In other words, $|\mu|^*(K_\varepsilon) \geq \|\mu\| - \varepsilon$.*

In spite of a simple formulation, this theorem is rather subtle. For example, in general it is false that a Radon extension can be obtained by the Lebesgue completion. Though, for the purposes of this book such subtleties are not important. There is also a version of this theorem generalizing Proposition 1.2.11.

1.2.13. Theorem. *Suppose that on the algebra of all cylinders in a locally convex space X we are given a real additive function μ whose images under the finite dimensional mappings $(l_1,\ldots,l_n)\colon X \to \mathbb{R}^n$, where $l_i \in X^*$, are countably additive and have uniformly bounded variations. Then the condition indicated in the previous theorem is necessary and sufficient for the existence of a Radon extension of μ.*

For measures on locally convex spaces one defines the Fourier transform and the convolution by analogy with $\mathbb{R}^n$. The Fourier transform of a measure μ defined on the σ-algebra $\sigma(X)$ in a locally convex space X is defined by the formula

$$\widetilde{\mu}\colon X^* \to C, \quad \widetilde{\mu}(l) = \int_X \exp\bigl(il(x)\bigr)\,\mu(dx).$$

Since two measures on $\mathbb{R}^n$ with equal Fourier transforms coincide, one can see from the above results that two measures on $\sigma(X)$ with equal Fourier transforms are equal. Hence two Radon measures with equal Fourier transforms coincide also.

Given two Hausdorff spaces X and Y with Radon measures μ and ν, the product measure $\mu \otimes \nu$ is defined on the σ-algebra $\mathcal{B}(X) \otimes \mathcal{B}(Y)$, which in the general case can be strictly smaller than the Borel σ-algebra $X \times Y$ (if at least one of the spaces has a countable base, then both σ-algebras coincide). However, the measure $\mu \otimes \nu$ always extends to a Radon measure on $\mathcal{B}(X \times Y)$, denoted by the same symbol and satisfying the equality

$$\mu \otimes \nu(B) = \int_Y \mu(B_y)\,\nu(dy), \quad B \in \mathcal{B}(X \times Y), \qquad (1.2.4)$$

where $B_y := \{x \in X\colon (x,y) \in B\}$. The right-hand side is meaningful since the function $y \mapsto \mu(B_y)$ is Borel on Y. In what follows the product of measures will mean this extension. Certainly, all these refinements are not needed if we deal with separable metric spaces.

The convolution $\mu * \nu$ of Radon measures μ and ν on a locally convex space is the image of the Radon measure $\mu \otimes \nu$ on $X \times X$ under the mapping $(x,y) \mapsto x + y$. The convolution can be defined by an explicit formula.

1.2.14. Lemma. *Let μ be a Radon measure on a locally convex space X. Then, for every $B \in \mathcal{B}(X)$, the function $x \mapsto \mu(B - x)$ is Borel measurable.*

PROOF. The set $E := \{(x,y)\colon x + y \in B\}$ is Borel in $X \times Y$ being the preimage of a Borel set under a continuous mapping. Since $E_x = B - x$, our claim follows from what has been said above. □

Thus, the convolution is given by the formula

$$\mu * \nu(B) := \int_X \mu(B - x)\,\nu(dx).$$

It is verified directly that

$$\widetilde{\mu * \nu}(l) = \widetilde{\mu}(l)\widetilde{\nu}(l).$$

Therefore, one has the equality $\mu * \nu = \nu * \mu$.

1.2.15. Lemma. *Let X be a locally convex space and let Y be its linear subspace. Then the mapping*

$$(x,y) \mapsto x + y, \quad X \times Y \to X,$$

1.2. Measures on topological spaces

is measurable with respect to the pair of σ-algebras $\big(\sigma(X) \otimes \sigma(Y), \sigma(X)\big)$. If Y is metrizable and separable, then this mapping is measurable also with respect to the pair $\big(\mathcal{B}(X) \otimes \mathcal{B}(Y), \mathcal{B}(X)\big)$.

PROOF. For the proof of measurability in the case of the cylindrical σ-algebra it suffices to prove the measurability of the indicated mapping for $Y = X$ because by the Hahn–Banach theorem the elements of Y^* are restrictions to Y of functionals in X^*. In turn, it suffices to show that the preimage of every cylinder of the form $U + L$, where U is an open set in a finite dimensional subspace $E \subset X$ and L is a closed linear subspace with $X = E \oplus L$, is a cylinder as well. The set
$$W := \{(x, y) \colon x \in E, y \in E, x + y \in U\}$$
is open in $E \times E$ by the continuity of addition. The preimage of $U + L$ has the form $W + (L \times L)$, i.e., is a cylinder. The measurability in the case of separable metrizable Y and the Borel σ-algebra follows by the continuity of this mapping and the equality $\mathcal{B}(X \times Y) = \mathcal{B}(X) \otimes \mathcal{B}(Y)$, which is valid under the stated assumptions on Y (see [**193**, Lemma 6.4.2]). $\square$

One more basic concept of measure theory on topological spaces is weak convergence of measures. A sequence of Radon measures μ_n on a topological space X is called *weakly convergent* to a Radon measure μ if for every bounded continuous function f one has the equality
$$\lim_{n \to \infty} \int_X f(x) \, \mu_n(dx) = \int_X f(x) \, \mu(dx).$$
Similarly one defines convergence of a net of measures μ_α to μ (concerning nets of measures, see [**193**]).

For nonnegative measures, there is the following criterion of weak convergence due to A.D. Alexandroff.

1.2.16. Theorem. *Suppose that we are given a completely regular topological space X, a sequence of Radon probability measures $\{\mu_n\}$, and a Radon probability measure μ. Then the following conditions are equivalent:*
(i) *the sequence $\{\mu_n\}$ converges weakly to μ;*
(ii) *for every closed set F one has*
$$\limsup_{n \to \infty} \mu_n(F) \leqslant \mu(F);$$
(iii) *for every open set U one has*
$$\liminf_{n \to \infty} \mu_n(U) \geqslant \mu(U).$$

The same is true for nets.

We observe that in the sufficiency part in conditions (ii) and (iii), in place of arbitrary closed and open sets one can take functionally closed and functionally open sets, i.e., sets of the form $\{x \colon \varphi(x) = 0\}$ and $\{x \colon \varphi(x) \neq 0\}$, respectively, where $\varphi \in C_b(X)$.

In applications it is very important to have efficient criteria that a given sequence of measures contains a weakly convergent subsequence. The main general result is due to Yu.V. Prohorov.

1.2.17. Theorem. *Let X be a complete metric space and let $\{\mu_n\}$ be a sequence of Radon measures on X. A necessary and sufficient condition that every subsequence in $\{\mu_n\}$ contains a further subsequence weakly convergent to a Radon measure is the following: the sequence $\{\mu_n\}$ is uniformly bounded in variation and uniformly tight, i.e., for every $\varepsilon > 0$ there exists a compact set K_ε such that $|\mu_n|(X\backslash K_\varepsilon) < \varepsilon$ for all n.*

Below we often use a number of fundamental theorems from the theory of integration: Lebesgue's dominated convergence theorem, Fatou's theorem, Beppo Levi's monotone convergence theorem, and Fubini's theorem. From more special results we shall use the following theorem of Nikodym (see [193, Theorem 4.6.3]).

1.2.18. Theorem. *Let $\{\mu_n\}$ be a sequence of countably additive measures on a σ-algebra $\mathcal{A}$ in a space X such that, for every $A \in \mathcal{A}$, there exists a finite limit $\mu(A) = \lim\limits_{n\to\infty} \mu_n(A)$. Then μ is also a countably additive measure and the sequence $\{\mu_n\}$ is bounded in variation.*

1.2.19. Theorem. *Suppose that in the previous theorem we are given a uniformly bounded sequence of $\mathcal{A}$-measurable functions f_n such that for all x there exists a limit $f(x) = \lim\limits_{n\to\infty} f_n(x)$. Then*

$$\int_X f(x)\,\mu(dx) = \lim_{n\to\infty} \int_X f_n(x)\,\mu_n(dx).$$

PROOF. Let us consider the measure $\nu := \sum_{n=1}^\infty 2^{-n}|\mu_n|$. Then $\mu \ll \nu$ and $\mu_n \ll \nu$ for all n, whence we obtain $\mu_n = \varrho_n \cdot \nu$, $\mu = \varrho \cdot \nu$. Under our assumptions the integrals of the functions ϱ_n against the measure ν are uniformly absolutely continuous (see [193, Theorem 4.6.3]), i.e., for every $\varepsilon > 0$ there exists $\delta > 0$ such that the integrals of $|\varrho_n|I_A$ and $|\varrho|I_A$ with respect to ν are less than ε for all $A \in \mathcal{A}$ with $\nu(A) < \delta$. By Egoroff's theorem one can take a set A such that outside of A the functions f_n converge uniformly to f. Then it is clear that our assertion reduces to proving convergence of the integrals of $f\varrho_n$ against the measure ν to the integral of $f\varrho$. Since the function f is bounded, such convergence follows by setwise convergence of μ_n to μ. □

Let us give one more result on convergence of the images of measures.

1.2.20. Theorem. *Let μ be a probability measure defined on a measurable space $(X, \mathcal{A})$, let ν be a Radon probability measure on a completely regular topological space Y, and let $T, T_n\colon X \to Y$ be $(\mathcal{A}_\mu, \mathcal{B}(Y))$-measurable mappings such that the sequence $T_n(x)$ converges μ-a.e. to $T(x)$. Suppose that the measure $\mu\circ T^{-1}$ is Radon, the measures $\mu\circ T_n^{-1}$ are absolutely continuous with respect to ν and their Radon–Nikodym densities ϱ_n are uniformly*

1.3. Conditional measures

ν-integrable. Then the measure $\mu \circ T^{-1}$ is also absolutely continuous with respect to ν, and its Radon–Nikodym density ϱ is the limit of the sequence $\{\varrho_n\}$ in the weak topology of the space $L^1(\nu)$.

If, in addition, we are given ν-measurable functions f_n convergent in measure ν to f, then the functions $f_n \circ T_n$ converge in measure μ to $f \circ T$.

A proof can be found in [**193**, Proposition 9.9.10, Corollary 9.9.11].

1.3. Conditional measures

The following facts related to conditional measures will be important below. Let X and Y be topological spaces such that X is Souslin (for example, complete separable metrizable), let μ be a nonnegative Radon measure on $X \times Y$, and let μ_Y be the projection of μ on Y. Then there exist Radon probability measures μ^y, $y \in Y$, on $X \times \{y\}$ such that, for every set $A \in \mathcal{B}(X \times Y)$, the function $y \mapsto \mu^y\bigl(A \cap (X \times \{y\})\bigr)$ is measurable with respect to μ_Y and the following equality is fulfilled:

$$\mu(A) = \int_Y \mu^y\bigl(A \cap (X \times \{y\})\bigr)\,\mu_Y(dy). \tag{1.3.1}$$

An analogous assertion is true also for signed measures with the only difference that in place of μ_Y one should take the projection $|\mu|_Y$ of the measure $|\mu|$ to Y and the measures μ^y can be signed; in addition, one has $\|\mu^y\| = 1$. To this end we write μ as $\mu = \varrho \cdot |\mu|$, where $|\varrho| = 1$, and take the measures $\mu^y = \varrho \cdot |\mu|^y$.

The measures μ^y are called *conditional* or *regular conditional*. Conditional measures μ^y on $X \times \{y\}$ with the stated properties are determined μ_Y-uniquely: any two families of such measures coincide μ_Y-a.e.

If the space Y is also Souslin (which is fulfilled in most of applications and hence can be assumed throughout this book without considerable loss of generality), then the conditional measures μ^y can be chosen in such a way that the functions $y \mapsto \mu^y\bigl(A \cap (X \times \{y\})\bigr)$ will be Borel measurable for all Borel sets $A \in \mathcal{B}(X \times Y)$.

In place of measures μ^y on the fibers $X \times \{y\}$ one can define conditional measures μ_y on the space X. Then, in (1.3.1), in place of $A \cap (X \times \{y\})$ one should take the sections $A_y := \{x \in X \colon (x, y) \in A\}$.

For all sets $A \in \mathcal{B}(X) \otimes \mathcal{B}(Y)$ and $B \in \mathcal{B}(Y)$, equality (1.3.1) yields the equality

$$\mu\bigl(A \cap (X \times B)\bigr) = \int_B \mu^y\bigl(A \cap (X \times \{y\})\bigr)\,\mu_Y(dy). \tag{1.3.2}$$

It also follows from (1.3.1) that for every $|\mu|$-integrable Borel function f one has the equality

$$\int_{X \times Y} f\,d\mu = \int_Y \int_{X \times \{y\}} f(x,y)\,\mu^y(dx)\,|\mu|_Y(dy), \tag{1.3.3}$$

where the repeated integral exists in the following sense: for $|\mu|_Y$-a.e. $y \in Y$, the function $x \mapsto f(x,y)$ is integrable with respect to the conditional measure μ^y and the obtained integral is a $|\mu|_Y$-integrable function.

Let us note a useful corollary of the last formula. Let ν be a Radon measure on $X \times Y$ such that $\nu \ll \mu$. Then ν can be written in the form

$$\nu(A) = \int_Y \int_{X \times \{y\}} \sigma^y\bigl(A \cap (X \times \{y\})\bigr) \, |\mu|_Y(dy), \qquad (1.3.4)$$

where, for each $y \in Y$, σ^y is a Radon measure on the space $X \times \{y\}$, and the functions $y \mapsto \|\sigma^y\|$ and $y \mapsto \sigma^y(A)$ for any $A \in \mathcal{B}(X)$ are integrable with respect to $|\mu|_Y$. Indeed, we may assume that $\mu \geqslant 0$. By the Radon–Nikodym theorem $\nu = \varrho \cdot \mu$, where $\varrho \in \mathcal{L}^1(\mu)$ is a Borel function. According to formula (1.3.3) we find that

$$\nu(A) = \int_{X \times Y} I_A \varrho \, d\mu = \int_Y \int_{X \times \{y\}} I_A(x,y) \varrho(x,y) \, \mu^y(dx) \, \mu_Y(dy)$$
$$= \int_Y \sigma^y\bigl(A \cap (X \times \{y\})\bigr) \, \mu_Y(dy),$$

where $\sigma^y := \varrho(\,\cdot\,,y) \cdot \mu^y$. The measures σ^y with the indicated properties are also uniquely determined up to a redefinition of such measures for points y from some set of $|\mu|_Y$-measure zero.

If a sequence of functions f_n on the product $X \times Y$ converges in $L^1(\mu)$, then in general it is false that the restrictions of f_n to the slices $X \times \{y\}$ converge in $L^1(\mu^y)$ (Exercise 1.6.16). However, the following somewhat weaker assertion is true.

1.3.1. Proposition. *Let a sequence $\{f_n\}$ converge in $L^1(\mu)$. Then one can find a subsequence $\{n_k\}$ such that for μ_Y-a.e. y, the restrictions of the functions f_{n_k} to $X \times \{y\}$ converge in $L^1(\mu^y)$.*

PROOF. Take increasing numbers n_k such that $\|f_{n_{k+1}} - f_{n_k}\|_{L^1(\mu)} \leqslant 2^{-k}$. By the monotone convergence theorem the series $\sum_{k=1}^\infty \|f_{n_{k+1}} - f_{n_k}\|_{L^1(\mu^y)}$ converges for μ_Y-a.e. y, which shows that the sequence $\{f_{n_k}\}$ on $X \times \{y\}$ is norm Cauchy in $L^1(\mu^y)$. $\square$

For the proof of the following result on convergence of conditional measures see [**193**, Proposition 10.4.23].

1.3.2. Proposition. *Let X and Y be Souslin spaces and let Borel measures μ_n on the space $X \times Y$ converge in variation to a measure μ. Then one can choose a subsequence $\{n_i\}$ and conditional measures $\mu_{n_i}^y$ and μ^y on $X \times \{y\}$ for μ_{n_i} and μ in such a way that for $|\mu|_Y$-a.e. y the measures $\mu_{n_i}^y$ will converge in variation to μ^y.*

Sometimes it is useful to represent a measure μ on $X \times Y$ in the form

$$\mu(B) = \int_Y \mu^y(B_y) \, \sigma(dy), \quad B_y := \{x \in X \colon (x,y) \in B\}, \qquad (1.3.5)$$

where each μ^y is a Borel measure on X, σ is some nonnegative Radon measure on Y with $\mu_Y \ll \sigma$, and the functions $y \mapsto \mu^y(B_y)$ and $y \mapsto \|\mu^y\|$ are integrable with respect to the measure σ. The measures μ^y are uniquely determined up to a redefinition for points y from a set of σ-measure zero. If we are given two measures μ and ν on $X \times Y$ such that the measure $\zeta := |\mu| + |\nu|$ possesses conditional measures ζ^y on X, then one can take $\sigma = \zeta_Y$ and obtain a representation of the form (1.3.5) for μ and ν with a common measure σ. The same is possible for a countable family of measures.

Note also that if μ is a signed measure, then we have

$$|\mu|(B) = \int_Y |\mu^y|(B_y)\, \sigma(dy), \quad \|\mu\| = \int_Y \|\mu^y\|\, \sigma(dy).$$

1.3.3. Proposition. (i) *The relation $\mu \ll \nu$ is equivalent to the relation $\mu^y \ll \nu^y$ for σ-a.e. y.* (ii) *The relation $\mu \perp \nu$ is equivalent to the relation $\mu^y \perp \nu^y$ for σ-a.e. y.*

PROOF. Let μ, ν, σ be probability measures. If $\mu^y \ll \nu^y$ for σ-a.e. y, then $\mu \ll \nu$. Conversely, suppose that $\mu \ll \nu$ and $f = d\mu/d\nu$. Then the measure μ can be represented by means of the measures $f(\,\cdot\,, y) \cdot \nu^y$ as well as by means of the measures μ^y, whence $\mu^y = f(\,\cdot\,, y) \cdot \nu^y$ for σ-a.e. y due to the essential uniqueness of representing measures. Hence $\mu^y \ll \nu^y$. If $\mu \perp \nu$, then there exists a set B with $\mu(B) = 0$, $\nu(B) = 1$. Hence $\mu^y(B_y) = 0$ and $\nu^y(B_y) = 1$ σ-a.e., i.e., $\mu^y \perp \nu^y$. Conversely, suppose that $\mu^y \perp \nu^y$ σ-a.e. If it is not true that $\mu \perp \nu$, then $\mu = \mu_1 + \mu_2$, where $\mu_1 \ll \nu$, $\mu_2 \perp \nu$. Hence $\mu^y = \mu_1^y + \mu_2^y$ by the uniqueness, and $\mu_1^y \ll \nu^y$ and $\mu_2^y \perp \nu^y$ according to what has already been proven. Since $\mu^y \perp \nu^y$, one has $\mu_1^y = 0$ σ-a.e., which gives $\mu_1 = 0$. The case of signed measures reduces to the considered one by means of the Hahn decomposition. $\square$

Some results on conditional measures are valid in a more general form. For simplicity we suppose that $\pi \colon X \to Y$ is a Borel mapping between Souslin spaces. Let μ be a Borel probability measure on X and $\nu = \mu \circ \pi^{-1}$. Then, there exist Borel probability measures μ^y on the sets $\pi^{-1}(y)$, where $y \in Y$, such that, for every set $B \in \mathcal{B}(X)$, the function $y \mapsto \mu^y(B \cap \pi^{-1}(y))$ is measurable with respect to ν and one has

$$\mu(B) = \int_Y \mu^y(B \cap \pi^{-1}(y))\, \nu(dy). \tag{1.3.6}$$

For any μ-integrable Borel function f we have

$$\int_X f\, d\mu = \int_Y \int_X f(x)\, \mu^y(dx)\, \nu(dy).$$

One can make all functions $y \mapsto \mu^y(B)$ Borel measurable if we admit that the measure μ^y is concentrated on $\pi^{-1}(y)$ not for every y, but only for ν-a.e. y. In the general case it is impossible to combine the requirement of the Borel measurability of all such functions with the equality $\mu^y(\pi^{-1}(y)) = 1$ for all y even for Borel functions π on $[0,1]$. If X and Y are Borel sets in

Polish spaces, then one can combine both requirements only when $\pi(X)$ is a Borel set and π possesses a Borel selection, i.e., there exists a Borel mapping $g\colon \pi(X)\to X$ with $\pi\bigl(g(y)\bigr)=y$ (see the result of Blackwell and Ryll-Nardzewski described in [**193**, Exercise 10.10.55]). The existence of the required conditional measures in the presence of the indicated mapping g is established as follows. Let us take conditional measures μ^y such that the functions $y\mapsto \mu^y(B)$, $B\in\mathcal{B}(X)$, are Borel measurable and $\mu^y\bigl(\pi^{-1}(y)\bigr)=1$ for all $y\in Y_0$, where $Y_0\subset \pi(X)$ is a Borel set of full ν-measure. For $y\in \pi(X)\backslash Y_0$ we set $\mu^y:=\delta_{g(y)}$. It is readily verified that the functions $y\mapsto \delta_{g(y)}(B)$, $B\in\mathcal{B}(X)$, are Borel measurable. A Borel selection does not always exist (see [**193**, §6.9]). If X and Y are Polish spaces, it exists under the following condition: the sets $\pi^{-1}(y)$ are countable unions of compact sets.

The Hahn–Jordan decomposition gives (1.3.6) for any signed measure μ with $\nu=|\mu|\circ\pi^{-1}$, and the measures μ^y on $\pi^{-1}(y)$ are uniquely determined ν-a.e. (for the uniqueness a.e. it suffices that $\mathcal{B}(X)$ be countably generated).

The situation of a product space considered above is the partial case in which π is the projection from the product to a factor. In turn, from this partial case the general one can be easily deduced. To this end we consider the space $X\times Y$ with the measure μ_0 on the graph of π obtained as the image of the measure μ under the mapping $x\mapsto \bigl(x,\pi(x)\bigr)$ from X to $X\times Y$. The projection of this measure on Y equals $\nu=\mu\circ\pi^{-1}$. According to (1.3.1), on the slices $X\times\{y\}$ there are conditional measures μ_0^y. These measures are concentrated on the intersections of the slices with the graph of π, i.e., on the sets $\pi^{-1}(y)\times\{y\}$, which enables us to define the required measures μ^y on $\pi^{-1}(y)$. In the case of arbitrary spaces, these formulations are not equivalent; moreover, there might be no conditional measures at all.

Conditional measures are useful for defining conditional expectations. Let us recall that if $(X,\mathcal{A},\mu)$ is a space with a nonnegative measure, then every sub-σ-algebra $\mathcal{B}\subset\mathcal{A}$ generates the operator $\mathbb{E}^\mathcal{B}$ of conditional expectation with respect to $\mathcal{B}$ in the spaces $L^p(\mu)$. This means that for every element $f\in\mathcal{L}^p(\mu)$ one has a $\mathcal{B}$-measurable function $\mathbb{E}^\mathcal{B}f\in\mathcal{L}^p(\mu)$, the operator $\mathbb{E}^\mathcal{B}$ is linear and bounded on $L^p(\mu)$, and for every bounded $\mathcal{B}$-measurable function g the following equality holds:

$$\int_X f(x)g(x)\,\mu(dx) = \int_X [\mathbb{E}^\mathcal{B}f](x)g(x)\,\mu(dx).$$

In the case of $L^2(\mu)$ this operator is just the orthogonal projection on the closed linear subspace in $L^2(\mu)$ generated by $\mathcal{B}$-measurable functions. For $p\neq 2$ this operator can be constructed by using its action on $L^2(\mu)$, but it is even simpler to apply the Radon–Nikodym theorem to the restriction of the measure $f\cdot\mu$ to $\mathcal{B}$. If the σ-algebra $\mathcal{B}$ has the form $\mathcal{B}=f^{-1}(\mathcal{E})$, where f is a measurable mapping from X to a measurable space $(Y,\mathcal{E})$, and we have conditional measures μ^y on the slices $f^{-1}(y)$, then $\mathbb{E}^\mathcal{B}f$ is given by the

formula
$$\mathbb{E}^{\mathcal{B}} f(x) = g(f(x)), \quad g(y) := \int_X f(z)\, \mu^y(dz). \quad (1.3.7)$$

We observe that every sub-σ-algebra $\mathcal{B}$ has the indicated form: one can take $\mathcal{E} = \mathcal{B}$ and the identical mapping.

Conditional measures and conditional expectations are employed for approximation of functions on infinite dimensional spaces with measures by functions of "finite dimensional" argument. Suppose we are given measurable spaces $(X_n, \mathcal{A}_n)$ and let $X = \prod_{n=1}^{\infty} X_n$, $\mathcal{A} := \bigotimes_{n=1}^{\infty} \mathcal{A}_n$, be their product. Denote by $\mathcal{B}_n$ the sub-σ-algebra in $\mathcal{A}$ generated by the projection on $X_1 \times \cdots \times X_n$. Sets from this σ-algebra are cylinders of the form $A \times X_{n+1} \times X_{n+2} \times \cdots$, where $A \in \bigotimes_{i=1}^n \mathcal{A}_i$. Suppose that $(X, \mathcal{A})$ is equipped with a probability measure μ. Then, for any $p \in [1, +\infty)$ and every function $f \in \mathcal{L}^p(\mu)$, the conditional expectations $\mathbb{E}^{\mathcal{B}_n} f$ with respect to $\mathcal{B}_n$ converge to f in $L^p(\mu)$ and almost everywhere. In the simplest case of a product-measure $\mu = \bigotimes_{n=1}^{\infty} \mu_n$ one has a simple explicit formula

$$\mathbb{E}^{\mathcal{B}_n} f(x_1, \ldots, x_n) = \int_{\prod_{i \geq n+1} X_i} f(x_1, \ldots, x_n, y_{n+1}, y_{n+2}, \ldots) \bigotimes_{i=n+1}^{\infty} \mu_i(dy_i). \quad (1.3.8)$$

In the general case, having conditional measures $\mu^{x_1, \ldots, x_n}$ on the slices of the form $(x_1, \ldots, x_n) \times \prod_{i=n+1}^{\infty} X_i$, we can apply the formula

$$\mathbb{E}^{\mathcal{B}_n} f(x_1, \ldots, x_n) = \int_X f(y)\, \mu^{x_1, \ldots, x_n}(dy). \quad (1.3.9)$$

Certainly, this formula, unlike the previous one, cannot be called explicit since usually conditional measures are not given explicitly.

1.4. Gaussian measures

A detailed discussion of Gaussian measures with an extensive bibliography can be found in the author's book [**191**]. Here we recall only some basic notions and facts necessary for the following.

A *Gaussian measure* on the real line is a Borel probability measure which is either concentrated at some point a (i.e., is Dirac's measure δ_a at a) or has density $(2\pi\sigma)^{-1/2} \exp(-(2\sigma)^{-1}(x-a)^2)$ with respect to Lebesgue measure, where $a \in \mathbb{R}^1$ is its *mean* and $\sigma > 0$ is its *dispersion*. The measure for which $a = 0$ and $\sigma = 1$ is called *standard Gaussian*.

1.4.1. Definition. *A Radon probability measure γ on a locally convex space X (or a measure on $\sigma(X)$) is called Gaussian if for every $f \in X^*$ the induced measure $\gamma \circ f^{-1}$ is Gaussian on the real line.*

An important example of a Gaussian measure is the countable product of the standard Gaussian measures on the real line. This measure is defined on the space $\mathbb{R}^{\infty}$. Another key example is the Wiener measure P_W on the space $\mathbb{R}^{[0,T]}$ of all functions on $[0, T]$ or on the space of continuous functions

$C[0,T]$. On the space $\mathbb{R}^{[0,T]}$ the Wiener measure is defined by means of its finite dimensional projections $P_{t_1,\ldots,t_n}$, $0 < t_1 < \cdots < t_n \leqslant T$, whose densities $p_{t_1,\ldots,t_n}(x_1,\ldots,x_n)$ with respect to Lebesgue measures on $\mathbb{R}^n$ have the following form:

$$\frac{1}{\sqrt{2\pi t_1}}\exp\Bigl(-\frac{x_1^2}{2t_1}\Bigr) \times \frac{1}{\sqrt{2\pi(t_2-t_1)}}\exp\Bigl(-\frac{(x_2-x_1)^2}{2(t_2-t_1)}\Bigr) \times \cdots$$

$$\times \frac{1}{\sqrt{2\pi(t_n-t_{n-1})}}\exp\Bigl(-\frac{(x_n-x_{n-1})^2}{2(t_n-t_{n-1})}\Bigr).$$

In addition, it is required that $P_0 = \delta_0$. Kolmogorov's theorem on consistent finite dimensional distributions yields the existence of the Wiener measure on $\mathbb{R}^{[0,T]}$. Then it is verified that the set $C[0,T]$ of continuous functions has outer measure 1 and hence the measure P_W can be defined also on $C[0,T]$ or on $C_0[0,T] := \{x \in C[0,T] \colon x(0) = 0\}$, which is called the *classical Wiener measure*. Though, there are constructions of the Wiener measure directly on $C[0,T]$.

It is known that any Radon Gaussian measure γ has mean $m \in X$, i.e., m is a vector such that

$$f(m) = \int_X f(x)\,\gamma(dx), \quad \forall f \in X^*.$$

If $m = 0$, i.e., the measures $\gamma \circ f^{-1}$ for $f \in X^*$ have zero mean, then γ is called *centered*. Any Radon Gaussian measure γ is a shift of a centered Gaussian measure γ_m defined by the formula $\gamma_m(B) := \gamma(B+m)$. Hence for many purposes it suffices to consider only centered Gaussian measures.

For a centered Radon Gaussian measure γ we denote by X_γ^* the closure of X^* in $L^2(\gamma)$. The elements of X_γ^* are called γ-measurable linear functionals. There is an operator $R_\gamma \colon X_\gamma^* \to X$, called the *covariance operator* of the measure γ, such that

$$f(R_\gamma g) = \int_X f(x)g(x)\,\gamma(dx), \quad \forall f \in X^*,\, g \in X_\gamma^*.$$

Set $g := \widehat{h}$ if $h = R_\gamma g$. Then $\widehat{h}$ is called the γ-measurable linear functional generated by h. One has the vector equality (if X is a Banach space, then it holds in Bochner's sense)

$$R_\gamma g = \int_X g(x)x\,\gamma(dx), \quad \forall g \in X_\gamma^*.$$

For example, if γ is a centered Gaussian measure on a separable Hilbert space X, then there exists a nonnegative nuclear operator K on X for which $Ky = R_\gamma y$ for all $y \in X$, where we identify X^* with X. Then we obtain

$$(Ky,z) = (y,z)_{L^2(\gamma)} \quad \text{and} \quad \widetilde{\gamma}(y) = \exp\bigl(-(Ky,y)/2\bigr).$$

Let us take an orthonormal eigenbasis $\{e_n\}$ of the operator K with eigenvalues $\{k_n\}$. Then γ coincides with the image of the countable power γ_0 of

the standard Gaussian measure on $\mathbb{R}^1$ under the mapping
$$\mathbb{R}^\infty \to X, \quad (x_n) \mapsto \sum_{n=1}^\infty \sqrt{k_n} x_n e_n.$$
This series converges γ_0-a.e. in X by convergence of the series $\sum_{n=1}^\infty k_n x_n^2$, which follows by convergence of the series of k_n and the fact that the integral of x_n^2 against the measure γ_0 equals 1. Here X_γ^* can be identified with the completion of X with respect to the norm $x \mapsto |\sqrt{K}x|_X$, i.e., the embedding $X = X^* \to X_\gamma^*$ is a Hilbert–Schmidt operator.

The space $H(\gamma) = R_\gamma(X_\gamma^*)$ is called the *Cameron–Martin space* of the measure γ. It is a Hilbert space with respect to the inner product
$$(h, k)_H := \int_X \widehat{h}(x)\widehat{k}(x)\,\gamma(dx).$$
The corresponding norm is given by the formula
$$|h|_H := \|\widehat{h}\|_{L^2(\gamma)}.$$
Moreover, it is known that $H(\gamma)$ with the indicated norm is separable. Note that the same norm is given by the formula
$$|h|_H = \sup\{f(h)\colon f \in X^*, \|f\|_{L^2(\gamma)} \leq 1\}.$$
It should be noted that if $\dim H(\gamma) = \infty$, then $\gamma(H(\gamma)) = 0$.

In the above example of a Gaussian measure γ on a Hilbert space we have the equality $H(\gamma) = \sqrt{K}(X)$.

The following equality is also worth noting:
$$\widetilde{\gamma}(l) = \exp\bigl(-|R_\gamma(l)|_H^2/2\bigr), \quad l \in X^*.$$

Let us observe that $H(\gamma)$ coincides also with the set of all vectors of the form
$$h = \int_X f(x)x\,\gamma(dx), \quad f \in L^2(\gamma).$$
Indeed, letting f_0 be the orthogonal projection of f onto X_γ^* in $L^2(\gamma)$, we see that the integral of the difference $[f(x) - f_0(x)]x$ over X vanishes since the integral of $[f(x) - f_0(x)]l(x)$ vanishes for each $l \in X^*$.

The mapping $h \mapsto \widehat{h}$ establishes a linear isomorphism between $H(\gamma)$ and X_γ^* preserving the inner product. In addition, $R_\gamma \widehat{h} = h$. If $\{e_n\}$ is an orthonormal basis in $H(\gamma)$, then $\{\widehat{e_n}\}$ is an orthonormal basis in X_γ^* and $\widehat{e_n}$ are independent random variables. One can take an orthonormal basis in X_γ^* consisting of elements $\xi_n \in X^*$. The general form of an element $l \in X_\gamma^*$ is this: $l = \sum_{n=1}^\infty c_n \xi_n$, where the series converges in $L^2(\gamma)$. Since ξ_n are independent Gaussian random variables, this series converges also γ-a.e. The domain of its convergence is a Borel linear subspace L of full measure (we even have $L \in \sigma(X)$). One can take a version of l which is linear on all of X in the usual sense; it is called a *proper linear version*. For example, let l on L be defined as the sum of the indicated series; then we extend l to all of X by linearity in an arbitrary way (e.g., taking a linear subspace L_1 which algebraically complements L and setting $l(x+y) = l(x)$ whenever $x \in L$, $y \in L_1$). Such a version is not unique in the infinite dimensional case, but any two proper linear versions coincide on the subspace $H(\gamma)$ (although it has measure zero!). Thus, every γ-measurable linear functional f has a version linear on the whole space. Such a version is automatically continuous on $H(\gamma)$ with the norm $|\cdot|_H$. Conversely, any continuous linear functional l on the Hilbert space $H(\gamma)$ admits a unique extension to a γ-measurable proper linear functional $\widehat{l}$ such that $\widehat{l}$ coincides with l on $H(\gamma)$. For every $h \in H(\gamma)$, such an extension of the functional $x \mapsto (x, h)_H$ is exactly $\widehat{h}$. If $h = \sum_{n=1}^{\infty} c_n e_n$, then $\widehat{h} = \sum_{n=1}^{\infty} c_n \widehat{e}_n$. Two γ-measurable linear functionals are equal almost everywhere precisely when their proper linear versions coincide on $H(\gamma)$. It is known that every proper linear γ-measurable function belongs to X_γ^*.

If a measure γ on $X = \mathbb{R}^\infty$ is the countable power of the standard Gaussian measures on the real line, then X^* can be identified with the space of all sequences of the form $f = (f_1, \ldots, f_n, 0, 0, \ldots)$. Here we have

$$(f, g)_{L^2(\gamma)} = \sum_{i=1}^{\infty} f_i g_i.$$

Hence X_γ^* can be identified with l^2; any element $l = (c_n) \in l^2$ defines an element of $L^2(\gamma)$ by the formula $l(x) := \sum_{n=1}^{\infty} c_n x_n$, where the series converges in $L^2(\gamma)$. Therefore, the Cameron–Martin space $H(\gamma)$ coincides with the space l^2 with its natural inner product. An element l represents a continuous linear functional precisely when only finitely many numbers c_n are nonzero. For the Wiener measure on $C[0,1]$ the Cameron–Martin space coincides with the class $W_0^{2,1}[0,1]$ of all absolutely continuous functions h on $[0,1]$ such that $h(0) = 0$ and $h' \in L^2[0,1]$; the inner product is given by the formula

$$(h_1, h_2)_H := \int_0^1 h_1'(t) h_2'(t) \, dt.$$

The general form of a measurable linear functional with respect to the Wiener measure is given by the stochastic integral (see §1.5)

$$l(x) = \int_0^1 h'(t) \, dx(t).$$

This functional is continuous precisely when h' has bounded variation.

1.4.2. Theorem. *The set $H(\gamma)$ is the collection of all $h \in X$ such that $\gamma_h \sim \gamma$, where $\gamma_h(B) := \gamma(B + h)$, and the Radon–Nikodym density of the measure γ_h with respect to γ is given by the following Cameron–Martin formula:*

$$\frac{d\gamma_h}{d\gamma} = \exp\bigl(-\widehat{h} - |h|_H^2/2\bigr).$$

For every $h \notin H(\gamma)$ we have $\gamma \perp \gamma_h$.

1.4. Gaussian measures

A centered Radon Gaussian measure is uniquely determined by its Cameron–Martin space (with the indicated norm!): if μ and ν are centered Radon Gaussian measures such that $H(\mu) = H(\nu)$ and $|h|_{H(\mu)} = |h|_{H(\nu)}$ for all $h \in H(\mu) = H(\nu)$, then $\mu = \nu$. The Cameron–Martin space is also called the reproducing Hilbert space.

1.4.3. Definition. *A Radon Gaussian measure γ on a locally convex space X is called nondegenerate if for every nonzero functional $f \in X^*$ the measure $\gamma \circ f^{-1}$ is not concentrated at a point.*

The nondegeneracy of γ is equivalent to the fact that $\gamma(U) > 0$ for all nonempty open sets $U \subset X$. This is also equivalent to the Cameron–Martin space $H(\gamma)$ being dense in X. For every degenerate Radon Gaussian measure γ there exists the smallest closed linear subspace $L \subset X$ for which $\gamma(L + m) = 1$, where m is the mean of the measure γ. Moreover, $L + m$ coincides with the topological support of γ. If $m = 0$, then on the space L the measure γ is nondegenerate.

The role of the countable power of the standard Gaussian measure is clear from the following important theorem due to B.S. Tsirelson.

1.4.4. Theorem. *Let γ be a centered Radon Gaussian measure on a locally convex space X, let $\{e_n\}$ be an orthonormal basis in $H(\gamma)$, and let $\{\xi_n\}$ be independent standard Gaussian random variables (for example, the sequence of coordinate functions on $\mathbb{R}^\infty$ with the countable power of the standard Gaussian measure on the real line). Then the series $\sum_{n=1}^\infty \xi_n(\omega) e_n$ converges in X for a.e. ω and the distribution of its sum is γ. In particular, this is true if $\xi_n = \widehat{e_n}$.*

In addition, there exists a Souslin linear subspace $S \subset X$ such that one has $\gamma(S) = 1$.

This theorem shows that the countable power of the standard Gaussian measure on the real line is the main (and essentially unique) example of a centered Radon Gaussian measure since every centered Radon Gaussian measure γ is the image of this countable product under a measurable linear mapping T (however, T need not be continuous); the mapping T is given by the series indicated in the theorem, and its restriction to l^2 is an isometry between l^2 and $H(\gamma)$.

By analogy with functionals, a mapping T from X to a locally convex space Y is called a γ-measurable linear operator if it is measurable with respect to the pair of σ-algebras $\mathcal{B}(X)_\gamma$ and $\mathcal{B}(Y)$ and has a version linear in the usual sense (called proper linear).

1.4.5. Theorem. *Let γ be a centered Radon Gaussian measure on a locally convex space X with the Cameron–Martin space H. Then, for every operator $T \in \mathcal{L}(H)$, there exists a γ-measurable proper linear mapping $\widehat{T} \colon X \to X$ with the following properties:*

(i) *the mapping $\widehat{T}$ coincides with T on H,*

(ii) *the image of the measure γ under the mapping $\widehat{T}$ is a centered Radon Gaussian measure μ with the Cameron–Martin space $H(\mu) = T(H)$.*

Any two such mappings are equal γ-a.e. If the measure γ is the distribution of the series $\sum_{n=1}^{\infty} \xi_n(\omega) e_n$ from Theorem 1.4.4, then μ is the distribution of the series $\sum_{n=1}^{\infty} \xi_n(\omega) T e_n$, which converges a.e.

Let us observe that the Fourier transform of the measure μ has the form
$$\widetilde{\mu}(l) = \exp\bigl(-|TR_\gamma(l)|_H^2/2\bigr).$$
By using this theorem one can obtain a somewhat more general result.

1.4.6. Corollary. *Let γ and μ be centered Radon Gaussian measures on locally convex spaces X and Y, respectively, and let $A\colon H(\gamma) \to H(\mu)$ be a continuous linear operator. Then A extends to a γ-measurable linear mapping $\widehat{A}\colon X \to Y$ such that the image of γ under this mapping is a centered Radon Gaussian measure with the Cameron–Martin space $A\bigl(H(\gamma)\bigr)$.*

1.4.7. Corollary. *Let X and Y be locally convex spaces and let γ be a centered Radon Gaussian measure on X. A continuous linear operator $A\colon H(\gamma) \to Y$ is the restriction to $H(\gamma)$ of a measurable linear operator $\widehat{A}\colon X \to Y$ for which the measure $\gamma \circ \widehat{A}^{-1}$ is Radon precisely when there exists a Gaussian Radon measure ν on Y such that $A\bigl(H(\gamma)\bigr) \subset H(\nu)$. If Y is a Hilbert space, then this is equivalent to the inclusion $A \in \mathcal{H}\bigl(H(\gamma), Y\bigr)$, hence to the existence of a Hilbert–Schmidt operator T on Y satisfying the condition $A\bigl(H(\gamma)\bigr) \subset T(Y)$.*

An important property of Gaussian measures is the so-called 0-1 (or zero-one) law which asserts that certain sets of special form may have measure only 0 or 1.

1.4.8. Theorem. *Let γ be a Radon Gaussian measure on a locally convex space X.*

(i) *For every γ-measurable affine subspace $L \subset X$ we have either $\gamma(L) = 0$ or $\gamma(L) = 1$.*

(ii) *Let $\{e_n\}$ be an orthonormal basis in $H(\gamma)$ and let a γ-measurable set E be such that, for every n and every rational number r, the sets E and $E + re_n$ coincide up to a set of measure zero. Then either $\gamma(E) = 0$ or $\gamma(E) = 1$. In particular, this is true if a γ-measurable set E is invariant with respect to the shifts by the vectors re_n.*

Another classical alternative in the theory of Gaussian measure is the Hajek–Feldman theorem on equivalence and singularity.

1.4.9. Theorem. *Let μ and ν be Radon Gaussian measures on the same space. Then either $\mu \sim \nu$ or $\mu \perp \nu$.*

One more very important fact is the following theorem due to Fernique.

1.4.10. Theorem. *Let γ be a centered Radon Gaussian measure and let a γ-measurable function q be a seminorm on a γ-measurable linear subspace of full measure. Then $\exp(\varepsilon q^2) \in L^1(\gamma)$ for some $\varepsilon > 0$.*

In the theory of Gaussian measures an important role is played by the Hermite (or Chebyshev–Hermite) polynomials H_n defined by the equalities

$$H_0 = 1, \quad H_n(t) = \frac{(-1)^n}{\sqrt{n!}} e^{t^2/2} \frac{d^n}{dt^n}(e^{-t^2/2}), \quad n > 1.$$

They have the following properties:

$$H'_n(t) = \sqrt{n} H_{n-1}(t) = tH_n(t) - \sqrt{n+1} H_{n+1}(t).$$

In addition, the system of functions $\{H_n\}$ is an orthonormal basis in $L^2(\gamma)$, where γ is the standard Gaussian measure on the real line.

For the *standard Gaussian* measure γ_n on $\mathbb{R}^n$ (the product of n copies of the standard Gaussian measure on $\mathbb{R}^1$) an orthonormal basis in $L^2(\gamma_n)$ is formed by the polynomials of the form

$$H_{k_1,\ldots,k_n}(x_1, \ldots, x_n) = H_{k_1}(x_1) \cdots H_{k_n}(x_n), \quad k_i \geq 0.$$

If γ is a centered Radon Gaussian measure on a locally convex space X and $\{l_n\}$ is an orthonormal basis in X^*_γ, then a basis in $L^2(\gamma)$ is formed by the polynomials

$$H_{k_1,\ldots,k_n}(x) = H_{k_1}(l_1(x)) \cdots H_{k_n}(l_n(x)), \quad k_i \geq 0, n \in \mathbb{N}.$$

For example, for the countable power of the standard Gaussian measure on the real line such polynomials are $H_{k_1,\ldots,k_n}(x_1, \ldots, x_n)$. It is convenient to arrange polynomials $H_{k_1,\ldots,k_n}$ according to their degrees $k_1 + \cdots + k_n$. For $k = 0, 1, \ldots$ we denote by $\mathcal{X}_k$ the closed linear subspace generated by the functions $H_{k_1,\ldots,k_n}$ with $k_1 + \cdots + k_n = k$. The space $\mathcal{X}_0$ is one-dimensional and consists of constants and $\mathcal{X}_1 = X^*_\gamma$. One can show that every element $f \in \mathcal{X}_2$ can be written in the form

$$f = \sum_{n=1}^{\infty} \alpha_n (l_n^2 - 1),$$

where $\{l_n\}$ is an orthonormal basis in X^*_γ and $\sum_{n=1}^{\infty} \alpha_n^2 < \infty$ (i.e., the series for f converges in $L^2(\gamma)$).

In the Gaussian case Theorem 1.2.12 can be somewhat reinforced.

1.4.11. Theorem. *Let X be a locally convex space and let γ be a Gaussian measure on $\sigma(X)$ such that there is a compact set K with $\gamma^*(K) > 0$. Then γ is tight, hence has a Radon extension.*

PROOF. The linear span L of K is a union of compact sets K_n (Exercise 1.6.15), hence $\gamma^*(L) = 1$ by the zero-one law (see Talagrand [**1105**]), so $\gamma^*(K_n) \to 1$. Now Theorem 1.2.12 applies. □

1.5. Stochastic integrals

In this section we discuss abstract stochastic integrals. The most essential things here are Itô's integral, Itô's formula, and Girsanov's theorem. Let us recall that a random process w_t, $t \in [0, T]$, is called a *Wiener process* if for all times $t_0 < t_1 < \cdots < t_n$ the increments $w_{t_1} - w_{t_0}, \ldots, w_{t_n} - w_{t_{n-1}}$ are independent, w_t is a centered Gaussian random variable with dispersion t, and the trajectory $t \mapsto w_t(\omega)$ is continuous for a.e. ω. Any Wiener process generates the Wiener measure P_W on $C[0,T]$, which is concentrated on the hyperplane $C_0[0,T] = \{x\colon x(0) = 0\}$ and is nondegenerate there. Conversely, the process $w_t(\omega) = \omega(t)$ on the probability space $(C[0,T], P_W)$ is Wiener. Note that there is a more general concept of a Wiener process with respect to a filtration $\{\mathcal{F}_t\}$, i.e., a family of sub-σ-algebras of the main σ-algebra $\mathcal{F}$ such that $\mathcal{F}_s \subset \mathcal{F}_t$ if $s \leqslant t$. In this concept it is required that w_t be $\mathcal{F}_t$-measurable for each t and that $w_{t+h} - w_t$ be independent of $\mathcal{F}_t$ for all $h > 0$.

Vector Wiener processes in $\mathbb{R}^d$ are defined as $w_t = (w_t^1, \ldots, w_t^d)$, where $w_t^1, \ldots, w_t^d$ are independent real Wiener processes.

Let us introduce the classical integral of Itô, which is defined for every random process $u\colon [0,T] \times \Omega \to \mathbb{R}^1$, $(t,\omega) \mapsto u_t(\omega)$ satisfying the following conditions:

(i) u is an adapted (nonanticipating) process, i.e., for every t, the random variable u_t is measurable with respect to the σ-algebra $\mathcal{F}_t$ generated by the random variables w_s, $s \leqslant t$, where w_t is a Wiener process on $[0,T]$;

(ii) $u \in L^2([0,T] \times \Omega, \lambda \otimes P)$, where λ is Lebesgue measure on $[0,T]$.

If $u_t(\omega) = \sum_{i=1}^{n-1} u_{t_i}(\omega) I_{[t_{i+1},t_i]}(t)$, $0 = t_0 \leqslant \cdots \leqslant t_n = T$, then we set

$$\int_0^T u_t(\omega)\, dw_t := I(u) := \sum_{i=1}^{n-1} u_{t_i}(\omega)\big(w_{t_{i+1}}(\omega) - w_{t_i}(\omega)\big).$$

It is readily seen that $\mathbb{E}I(u)^2 = \|u\|_2^2$. Hence the stochastic integral I extends to an isometry of the space of all processes with properties (i) and (ii) to the space $L^2(P)$. In addition, $\mathbb{E}I(u) = 0$ and for every $t \in [0,T]$ the random variable

$$\int_0^t u_s\, dw_s$$

is $\mathcal{F}_t$-measurable. Similarly one defines the Itô integral in the case of a Wiener process with respect to a general filtration $\{\mathcal{F}_t\}$. Similarly one defines the stochastic integral of operator-valued processes with respect to vector-valued Wiener processes.

The stochastic integral is defined, in particular, for every nonrandom function $u \in L^2[0,T]$ and is a Gaussian random variable ξ with $\mathbb{E}\xi = 0$ and $\mathbb{E}\xi^2 = \|u\|_{L^2}^2$. In this special case the stochastic integral is called the *Wiener integral*. It is known that the trajectories of a Wiener process almost surely have unbounded variation. So the stochastic integral is not a Stieltjes integral. However, if a nonrandom function u on $[a,b]$ is absolutely continuous, then one has

$$\int_a^b u(t)\, dw_t = u(b)w_b - u(a)w_a - \int_a^b u'(t)w(t)\, dt.$$

1.5. Stochastic integrals

The main purpose of introducing the stochastic integral is that it enables us to consider stochastic differential equations of the form

$$d\xi_t = \sigma(t, \xi_t)dw_t + b(t, \xi_t)dt, \quad \xi_a = \eta.$$

This equation is understood as the integral equation

$$\xi_t = \eta + \int_a^t \sigma(s, \xi_s)\,dw_s + \int_a^t b(s, \xi_s)\,dt, \qquad (1.5.1)$$

where the second integral is the Lebesgue integral. However, the reader should be warned that there exist two *nonequivalent* ways of interpreting (1.5.1). The first one (*strong* solution) requires that $\{\xi_t\}$ be adapted and (1.5.1) be fulfilled for the initial Wiener process w_t and the filtration generated by it, and the second one (*weak* solution) allows us to choose another Wiener process w_t with respect to some filtration $\{\mathcal{F}_t\}$ (not necessarily generated by $\{w_t\}$) such that process ξ_t is adapted with respect to the filtration $\{\mathcal{F}_t\}$ and (1.5.1) is fulfilled. A strong solution ξ has the form $\xi_\bullet = \Phi(w_\bullet)$, where Φ is a Borel mapping on the space of continuous functions. Let a random variable η be $\mathcal{F}_a$-measurable. If the mappings σ and b are Borel and uniformly Lipschitzian in the second argument, then both concepts of solution coincide and (1.5.1) has a unique solution. In applications one has to deal with initial conditions η that are not $\mathcal{F}_a$-measurable (note that if $a = 0$, then only constants are $\mathcal{F}_0$-measurable). Moreover, it turns out to be necessary to consider more general stochastic equations

$$d\xi_t = \sigma(t, \omega)dw_t + b(t, \omega)dt$$

with nonadapted coefficients. For such purposes more general stochastic integrals were invented (see Ikeda, Watanabe [**571**] and Nualart [**847**], [**848**]). In Chapter 8 we shall encounter the Skorohod and Ogawa integrals. One of the simplest stochastic equations is the linear equation

$$d\xi_t = dw_t - \frac{1}{2}\xi_t\,dt.$$

Its solution is called the *Ornstein–Uhlenbeck process* and is expressed via w_t by the formula

$$\xi_t = e^{-t/2}\xi_0 + e^{-t/2}\int_0^t e^{s/2}\,dw_s.$$

If a one-dimensional process ξ_t has the stochastic Itô differential

$$\sigma(t, \omega)dw_t + b(t, \omega)dt,$$

then for every smooth function f the following *Itô formula* is valid:

$$df(\xi_t) = f'(\xi_t)\sigma(t, \omega)dw_t + \left[f'(\xi_t)b(t, \omega) + \frac{1}{2}f''(\xi_t)\sigma(t, \omega)^2\right]dt.$$

Given a process ξ_t with a stochastic differential $\sigma(t, \omega)dw_t + b(t, \omega)dt$, the stochastic integral of a process u with respect to ξ_t is defined (under

suitable measurability and integrability conditions) in a natural way as
$$\int_0^t u(s,\omega)\sigma(s,\omega)\,dw_s + \int_0^t u(s,\omega)b(s,\omega)\,ds.$$

1.5.1. Remark. Let u_t and v_t be adapted processes with trajectories in $L^2[0,T]$ and let $d\xi_t = u_t dw_t + v_t dt$. The Stratonovich integral of ξ_t is defined as an L^2-limit
$$\int_0^T \xi_t \circ dw_t = (L^2)\lim_{\delta\to 0} \sum_{i=1}^n \frac{1}{2}(\xi_{t_{i+1}} + \xi_{t_i})(w_{t_{i+1}} - w_{t_i}),$$
where $\{t_i\}$ is a finite partition of $[0,T]$ and $\delta = \max_i |t_{i+1} - t_i|$. The Stratonovich integral of the process ξ_t of the indicated form exists and is related with the Itô integral by the equality
$$\int_0^T \xi_t \circ dw_t = \int_0^T \xi_t \, dw_t + \frac{1}{2}\int_0^T u_t \, dt.$$
Sometimes the Stratonovich integral is called the symmetric integral. A nice feature of this integral is that it behaves like the ordinary differential under nonlinear mappings: in Itô's formula for the Stratonovich integral there is no last term.

The integral with respect to the Wiener process can be generalized.

Let $(T,\mathcal{B})$ be a measurable space with a finite positive atomless measure μ.

1.5.2. Definition. *Let $\{w(B), B \in \mathcal{B}\}$ be a centered Gaussian process on some probability space $(\Omega, \mathcal{F}, P)$ indexed by elements of $\mathcal{B}$ such that*
$$\mathbb{E}[w(B_1)w(B_2)] = \mu(B_1 \cap B_2), \quad \forall B_1, B_2 \in \mathcal{B}.$$
Then this process is called a Gaussian orthogonal measure on the space $(T, \mathcal{B}, \mu)$ or a Gaussian measure with orthogonal increments.

Suppose that the σ-algebra $\mathcal{F}$ is generated by random variables $w(B)$, where $B \in \mathcal{B}$, and that the space $H := L^2(\mu)$ is separable. Let us define the Wiener integral $w(h)$ for any element $h \in H$. If h is a simple function assigning values c_i to disjoint measurable sets B_i, $i = 1, \ldots, n$, then
$$w(h) := \sum_{i=1}^n c_i w(B_i).$$
It is clear that
$$\mathbb{E}|w(h)|^2 = \sum_{i=1}^n |c_i|^2 \mu(B_i) = |h|_H^2.$$
Hence the mapping $h \mapsto L^2(P)$ uniquely extends to a linear isometry of H to $L^2(P)$. For every $h \in H$ we denote by $w(h)$ the result of this extension. An alternative notation is $I_1(h) := w(h)$.

1.5. Stochastic integrals

1.5.3. Example. Let $T = [0, 1]$, let $\mathcal{B} = \mathcal{B}([0, 1])$, and let μ be Lebesgue measure. If w_t is a Wiener process and
$$w(B) = \int_0^1 I_B(t)\, dw_t,$$
where the integral is the standard Wiener one, then we obtain a Gaussian orthogonal measure on $[0, 1]$ with Lebesgue measure. In this case for every $h \in H = L^2[0, 1]$, the function $w(h)$ is the usual Wiener integral $\int_0^1 h(t)\, dw_t$.

Let μ^m be the product of m copies of μ defined on T^m. Let us define the multiple Itô integral $I_m(f_m)$ of functions $f_m \in L^2(\mu^m)$. If
$$f_m(t_1, \ldots, t_m) = \sum_{i_1, \ldots, i_m} c_{i_1, \ldots, i_m} I_{A_{i_1}}(t_1) \cdots I_{A_{i_m}}(t_m),$$
where $A_1, \ldots, A_n \in \mathcal{B}$ are pairwise disjoint and $c_{i_1, \ldots, i_m} = 0$ if some of $i_1, \ldots, i_m$ coincide, then we set
$$I_m(f_m) := \sum_{i_1, \ldots, i_m} c_{i_1, \ldots, i_m} w(A_{i_1}) \cdots w(A_{i_m}).$$

Note that $I_m(f_m)$ does not change under permutations of the variables of the function f_m. Let us consider the symmetrization of the function f_m defined by
$$\widetilde{f_m}(t_1, \ldots, t_m) = \frac{1}{m!} \sum_{\sigma} f_m(t_{\sigma(1)}, \ldots, t_{\sigma(m)}),$$
where the summation is taken over all permutations of $\{1, \ldots, m\}$. Then
$$I_m(f_m) = I_m(\widetilde{f_m})$$
and
$$\mathbb{E}\big[I_m(\varphi) I_k(\psi)\big] = \begin{cases} 0 & \text{if } m \neq k, \\ m!(\widetilde{\varphi}, \widetilde{\psi})_{L^2(\mu^m)} & \text{if } m = k. \end{cases}$$
This is verified directly for the functions f_m of the form indicated above by using the disjointness of A_i and the independence of $w(A_i)$. Therefore,
$$\mathbb{E} I_m(f_m) = 0, \quad \mathbb{E}[I_m(f_m)^2] \leqslant m! \|f_m\|_{L^2(\mu^m)}^2.$$
This enables us to extend I_m to a bounded operator from $L^2(\mu^m)$ to $L^2(P)$ satisfying the above relations since the functions f_m of the indicated form are dense in $L^2(\mu^m)$ (Exercise 1.6.21). The operator $(m!)^{-1/2} I_m$ is an isometry on the space of symmetric functions; a symmetric kernel f is uniquely determined by its multiple Itô integral $I_m(f)$.

1.5.4. Theorem. *For every element $F \in L^2(P)$, there exist symmetric kernels $f_n \in L^2(T^n, \mu^n)$ such that*
$$F = \mathbb{E} F + \sum_{n=1}^{\infty} I_n(f_n), \text{ where the series converges in } L^2(P). \tag{1.5.2}$$

For the standard Wiener process w_t on $[0,1]$ we obtain

$$I_m(f_m) = m! \int_0^1 \int_0^{t_m} \cdots \int_0^{t_2} \widetilde{f_m}(t_1,\ldots,t_m)\, dw_{t_1} \cdots dw_{t_m}. \quad (1.5.3)$$

For a proof it suffices to take f_m of the form $I_{A_1}(t_1) \cdots I_{A_m}(t_m)$, where the sets A_i are disjoint intervals.

1.5.5. Corollary. *Let w_t be a Wiener process on a probability space (Ω, P) and let ξ be a random variable in $L^2(P)$ with zero mean such that it is measurable with respect to the σ-algebra generated by $\{w_t, t \leqslant 1\}$. Then there exists an adapted process $u \in L^2(P \otimes \lambda)$ such that $\xi = \int_0^1 u_t\, dw_t$, $\|\xi\|_{L^2(P)} = \|u\|_{L^2(P \otimes \lambda)}$.*

PROOF. For $\xi = I_m(f_m)$ let u_t be the stochastic integral in (1.5.3) of order $m-1$. We have

$$\|u\|^2_{L^2(P \otimes \lambda)} = \|m!\widetilde{f_m} I_{\{t_1 \leqslant \cdots \leqslant t_m\}}\|^2_{L^2([0,1]^m)} = m!\, \|\widetilde{f_m}\|^2_{L^2([0,1]^m)},$$

which gives our claim. $\square$

One of the most important results in stochastic analysis is the *Girsanov theorem*. This theorem has several different formulations. The best known one is the theorem on an equivalent change of measure, which asserts that under certain conditions the solution of a stochastic integral equation is a Wiener process with respect to some measure equivalent to the initial measure of the given probability space. We shall need another formulation, which gives a condition of equivalence of the distributions of two diffusion processes. We consider a special case of a more general result.

Suppose that diffusion processes ξ_t and η_t in $\mathbb{R}^d$ are given by stochastic differential equations

$$d\xi_t = \sigma(t, \xi_t) dw_t + b_\xi(t, \xi_t) dt, \quad d\eta_t = \sigma(t, \eta_t) dw_t + b_\eta(t, \eta_t) dt$$

on $[0,T]$ with the same initial condition $\xi_0 = \eta_0$, where the operator function σ and the vector functions b_ξ, b_η are continuous, locally Lipschitzian in x uniformly in $t \in [0,T]$, and the operators $\sigma(t,x)^{-1}$ are uniformly bounded. Denote by μ^ξ and μ^η the distributions of these processes in the path space $\Omega := C([0,T], \mathbb{R}^d)$. Set

$$\zeta(t) := x(t) - \int_0^t b_\xi\bigl(s, x(s)\bigr) ds.$$

Then the process $\zeta(t)$ is a local martingale on a probability space (Ω, μ^ξ). Hence we obtain a well-defined process

$$z(t) := \int_0^t \sigma\bigl(s, x(s)\bigr)^{-1} d\zeta(s).$$

It turns out that this process is a Wiener process on the probability space (Ω, μ^ξ). Set
$$h\big(x(t)\big) := \sigma\big(t, x(t)\big)^{-1}\big[b_\eta\big(t, x(t)\big) - b_\xi\big(t, x(t)\big)\big], \quad x \in \Omega,$$
and define Girsanov's density by the formula
$$\varrho(x) := \exp\bigg\{\int_0^T \big(h(x(t)), dz(t)\big) - \frac{1}{2}\int_0^T |h(x(t))|^2\, dt\bigg\}.$$
The stochastic integral in this expression is a function of $x(\,\cdot\,)$ since $z(\,\cdot\,)$ is a transformation of the path $x(\,\cdot\,)$. Hence $\varrho(x)$ is indeed a function of $x(\,\cdot\,)$ on Ω.

1.5.6. Theorem. *Under the stated assumptions the measures μ^ξ and μ^η are equivalent and $d\mu^\eta/d\mu^\xi = \varrho$.*

1.6. Comments and exercises

In the author's books [191] and [193] one can find an extensive bibliography on general measure theory, measures on topological spaces, and the theory of Gaussian measures. In this chapter we present basic facts which will be needed later, but not all necessary technical results; references will be given at appropriate places. Concerning Gaussian measures, see also the books Fernique [423], Ibragimov, Rozanov [566], Kuo [666], Lifshits [723], and Rozanov [974]; on measures on linear spaces see also Vakhania, Tarieladze, Chobanyan [1167], Kruglov [644], and Mushtari [819]. Modern theory of weak convergence of measures goes back to the classical work Prohorov [927]. All necessary facts from the theory of random processes can be found in the excellent texts Kallenberg [590] and Wentzell [1191]. On stochastic differential equations, diffusion processes, and Girsanov's theorem see Ikeda, Watanabe [571] and Gikhman, Skorokhod [477, V. 3, Chap. III], [478].

Exercises

1.6.1. Let h be a nonzero vector in a locally convex space X. Prove that there exists a closed hyperplane Y in X such that $X = Y \oplus \mathbb{R}^1 h$ and the projection on Y is continuous.

1.6.2. (i) Let X be a separable metrizable locally convex space. Prove that there exists an injective continuous linear mapping of the space X into $\mathbb{R}^\infty$.
(ii) Prove that there is no surjective continuous linear mapping of the space l^2 onto $\mathbb{R}^\infty$.
(iii) Prove that there is no surjective continuous linear mapping of the space $\mathbb{R}^\infty$ onto l^2.
(iv) Prove that there is no injective continuous linear mapping of the space $\mathbb{R}^\infty$ into l^2.

HINT: (i) It suffices to find a countable family of continuous linear functionals separating the points in this space; let $\{p_n\}$ be a sequence of seminorms defining the topology and let $\{x_n\}$ be a countable everywhere dense set; let us fix n, distinct points x_k and x_m and $p \in \mathbb{N}$ with $2p < p_n(x_k - x_m)$; use the Hahn–Banach theorem to find a continuous linear functional l_{nkmp} separating the convex sets $\{x\colon p_n(x-x_k) < 1/p\}$ and $\{x\colon p_n(x-x_m) < 1/p\}$; verify that if $x \neq y$, then there are n, k, m, p such that $l_{nkmp}(x-y) \neq 0$. (ii) Observe that otherwise $\mathbb{R}^\infty$ would be linearly homeomorphic to a Hilbert space (the factor of l^2 by the kernel of the surjection). (iii) Observe that if $T\colon \mathbb{R}^\infty \to l^2$ is a continuous linear surjection, then $T^*\colon l^2 \to (\mathbb{R}^\infty)^*$ is a continuous linear injection, but $(\mathbb{R}^\infty)^*$ is the space of all finite sequences, i.e., a countable union of finite dimensional spaces E_n, which is impossible since otherwise some of the subspaces $T^{-1}(E_n)$ would coincide with l^2 by the Baire theorem. (iv) Show that there is no continuous norm on $\mathbb{R}^\infty$.

1.6.3. Prove that for every separable infinite dimensional Banach (or Fréchet) space X there is a continuous linear injective operator $T\colon l^2 \to X$ with dense range.

HINT: Take a bounded linear operator $l^2 \to X$ whose range contains a countable dense set and then restrict it to the orthogonal complement of the kernel.

1.6.4. Let E be a Hilbert space, let X be a separable Banach space, and let an operator $A \in \mathcal{L}(E, X)$ be injective. Prove that E is separable.

HINT: Using injective bounded linear operators $X \to C[0,1] \to L^2[0,1]$, reduce to the case where X is also Hilbert and then consider the range of the adjoint operator $A^*\colon X \to E$.

1.6.5. (i) Construct injective symmetric compact operators A and B on a separable Hilbert space H such that $(A(H) \cap B(H) = 0$.

(ii) Let A be a compact operator in l^2. Prove that there exists a nonnegative Hilbert–Schmidt operator B on l^2 with dense range such that $A(l^2) \cap B(l^2) = 0$.

1.6.6. Let A be a bounded nonnegative operator in a Hilbert space H. Prove that $\sqrt{A}(H)$ is the set of all vectors $v \in H$ such that $\sup\{(v,u)\colon (Au,u) \leqslant 1\} < \infty$.

1.6.7. Show that a set $K \subset l^2$ has compact closure if and only if it is norm bounded and
$$\lim_{n\to\infty} \sup_{x=(x_k)\in K} \sum_{k=n}^\infty x_k^2 = 0.$$

1.6.8. Show that a set $K \subset \mathbb{R}^\infty$ has compact closure if and only if there are numbers C_n such that $|x_n| \leqslant C_n$ for all $x = (x_n) \in K$ and all n.

1.6.9. A locally convex space X is called quasicomplete if every bounded closed set in X is complete. Show that such a space is sequentially complete.

1.6.10. Prove that a sequence of measures μ_n converges in variation precisely when for some probability measure ν the measures μ_n are given by densities with respect to ν being convergent in $L^1(\nu)$.

1.6.11. Prove Proposition 1.2.1 and Corollary 1.2.2.

1.6.12. Let μ be a Borel measure on the space $\mathbb{R}^\infty$ and let μ_n be its projection on the n-dimensional subspace X_n of all vectors with zero components x_i for $i > n$. Prove that the measures μ_n converge weakly to μ.

HINT: Prove that the sequence of measures $\{\mu_n\}$ is uniformly tight.

1.6. Comments and exercises

1.6.13. Let ν be a Borel probability measure on the real line different from Dirac's measure at the origin. Let ν_n be the measure on $\mathbb{R}^n$ equal to the product of n copies of the measure ν. Prove that there is no measure on l^2 with the projections ν_n.

HINT: Show that this product vanishes on every product of intervals $[-a_n, a_n]$ with $a_n \to 0$ and deduce that it vanishes on every compact set in l^2, hence vanishes on l^2.

1.6.14. Let X be an infinite dimensional Banach space and let $K \subset X$ be compact. Prove that every ball of radius $r > 0$ in X contains a closed ball of radius $r/4$ without common points with K.

HINT: Use that the unit sphere contains a sequence of vectors with mutual distances greater than $1/2$.

1.6.15. Prove that the absolutely convex hull of a compact set in a locally convex space is a countable union of compact sets, hence is a Borel set. The same is true for the linear span.

HINT: If K is compact, then $K^n \times [-C, C]^n$ is compact and the mapping $(x_1, \ldots, x_n, t_1, \ldots, t_n) \mapsto t_1 x_1 + \cdots + t_n x_n$ is continuous.

1.6.16. Give an example of a sequence of functions f_n on $[0,1]^2$ with Lebesgue measure such that $\|f_n\|_{L^1} \to 0$, but there is no $y \in [0,1]$ such that the functions $x \mapsto f_n(x,y)$ converge in $L^1[0,1]$.

1.6.17. Let μ be a Radon measure on an infinite dimensional Banach space X without an atom at the origin. Prove that the infimum of μ-measures of closed balls of radius 1 containing the origin is zero.

1.6.18. It is known (see Benyamini, Lindenstrauss [138, Chap. 4, §2]) that on any Banach space X with a separable dual there is an equivalent norm that is continuously Fréchet differentiable outside the origin. By using this fact and the fact that measures with equal Fourier transforms coincide, prove that the class $C^1_{0,b}(X)$ of continuously Fréchet differentiable Lipschitzian functions on X with bounded support separates Borel measures on X. Note that on $C[0,1]$ there are no nonzero functions of this type.

HINT: If $f \in C^1_{0,b}(X)$ and $l \in X^*$, then $f \exp(il) \in C^1_{0,b}(X)$, which gives the equality of the measures $f \cdot \mu$ and $f \cdot \nu$ for two measures μ and ν with equal integrals of functions from $C^1_{0,b}(X)$. This easily yields the equality $I_B \cdot \mu = I_B \cdot \nu$ for every ball B of radius R centered at the origin. Letting $R \to \infty$ we have $\mu = \nu$.

1.6.19. Let γ be a centered Radon Gaussian measure on a locally convex space X, $h \in H = H(\gamma)$ and $|h|_H = 1$. Show that the measure γ does not change under the mapping $x \mapsto x - 2\widehat{h}(x)h$.

1.6.20. In the notation of §1.5 let $A \in \mathcal{B}$, where $\mathcal{B}_A$ is the σ-algebra generated by all sets $A \cap B$, where $B \in \mathcal{B}$, and let F be given by series (1.5.2). Prove that

$$\mathbb{E}^{\mathcal{B}_A} F = \mathbb{E} F + \sum_{n=1}^\infty I_n(f_n I_A^{\otimes n}), \quad I_A^{\otimes n}(t_1, \ldots, t_n) := I_A(t_1) \cdots I_A(t_n).$$

1.6.21. Let μ be an atomless probability measure on $(T, \mathcal{B})$. Prove that a dense set in $L^2(\mu^m)$ is formed by functions of the form

$$f = \sum_{i_1, \ldots, i_m} c_{i_1, \ldots, i_m} I_{A_{i_1}}(t_1) \cdots I_{A_{i_m}}(t_m),$$

where $A_1, \ldots, A_n \in \mathcal{B}$ are disjoint and $c_{i_1,\ldots,i_m} = 0$ if some of the indices $i_1, \ldots, i_m$ coincide.

HINT: Since μ is atomless, every set A with $\mu(A) > 0$ can be partitioned in n pieces of measure $\mu(A)/n$; hence the function $I_B(t_1) \cdots I_B(t_m)$ can be approximated by functions of the indicated form with $A_{i,j} \subset B$.

1.6.22. (i) Let A and B be bounded linear operators on a Hilbert space H such that $A(H) \subset B(H)$. Prove that there exists a bounded linear operator C such that $A = BC$.

(ii) Let A and B be two compact symmetric operators on a separable Hilbert space H having eigenvalues $\alpha_n > 0$ and $\beta_n > 0$, respectively. Suppose that we have $A(H) \subset B(H)$. Prove that there exists a number $C > 0$ such that $\alpha_n \leqslant C\beta_n$.

HINT: Reduce (i) to the case of injective B and then consider the linear mapping $C := B^{-1}A$ and use the closed graph theorem to verify its continuity. Use (i) to prove (ii).

1.6.23. Let μ and ν be two equivalent measures. Suppose that a sequence of measurable functions f_n converges in measure μ. Show that it converges in measure ν as well.

HINT: Use that every subsequence in $\{f_n\}$ contains some further subsequence convergent almost μ-everywhere, hence ν-everywhere.

1.6.24. Let μ be a probability measure on $\mathbb{R}^2$ with a bounded density. Is it true that its projection on $\mathbb{R}^1$ has a bounded density?

1.6.25. Find a probability measure on $\mathbb{R}^2$ with a density $\varrho \in L^1(\mathbb{R}^2) \cap L^2(\mathbb{R}^2)$ such that the density of its projection on $\mathbb{R}^1$ is not in $L^2(\mathbb{R}^1)$.

1.6.26. (i) Show that l^2 with the weak topology is not complete, but is quasicomplete (hence sequentially complete).

(ii) Show that l^1 with the weak topology is not quasicomplete, but is sequentially complete.

(iii) Show that $C[0,1]$ with the weak topology is not sequentially complete.

HINT: (i) Take a discontinuous linear function on l^2 and show that it belongs to the closure of the space of continuous linear functions in the topology of the pointwise convergence. (ii) Use that $(l^1)^* = l^\infty$, $(l^\infty)^* \neq l^1$, and any element in $(l^\infty)^*$ belongs to the closure of a ball in l^1 in the topology of the pointwise convergence on l^∞. (iii) Use that $I_{[0,1/2]}$ is a pointwise limit of a bounded sequence of continuous functions.

1.6.27. Let T be an uncountable set and let γ be the product of T copies of the standard Gaussian measure on the real line. Show that $\gamma(K) = 0$ for every compact set K in $\mathbb{R}^T$.

HINT: Use that any compact set is contained in some product $\prod_{t \in T}[-C_t, C_t]$ and that for some k the set $\{t: C_t \leqslant k\}$ is infinite.

1.6.28. Let T be an uncountable set and let λ^T be the product of T copies of Lebesgue measure on $[0,1]$. Show that $L^2(\lambda^T)$ is not separable.

CHAPTER 2

Sobolev spaces on $\mathbb{R}^n$

This chapter gives an introduction to the theory of Sobolev spaces on $\mathbb{R}^n$. The ideas and methods of this chapter will be further developed in the infinite dimensional case, and many concrete results will be used throughout the book. First we consider the classes $W^{p,k}$ consisting of functions whose generalized derivatives up to order $k \in \mathbb{N}$ belong to L^p. In addition to several equivalent descriptions of such classes we shall obtain the so-called embedding theorems for them, i.e., assertions which give certain additional properties of integrability or continuity not obvious from the definitions. Some other interesting properties of the classes $W^{p,k}$ will be derived as well. Next we briefly discuss more general Sobolev classes, in particular, certain classes with the so-called fractional derivatives and certain classes obtained by interpolation.

2.1. The Sobolev classes $W^{p,k}$

A principal role in the theory of Sobolev spaces is played by the notion of generalized or Sobolev derivatives.

Let Ω be an open set in $\mathbb{R}^n$ and let

$$L^p_{\mathrm{loc}}(\Omega) = \{f \in L^0(\Omega) \colon \zeta f \in L^p(\Omega)\ \forall \zeta \in C_0^\infty(\Omega)\}.$$

2.1.1. Definition. *Let $f \in L^1_{\mathrm{loc}}(\Omega)$. A function $g \in L^1_{\mathrm{loc}}(\Omega)$ is called a generalized or Sobolev derivative of the function f with respect to the variable x_i if, for every function $\varphi \in C_0^\infty(\Omega)$, one has the equality*

$$\int_\Omega f(x) \partial_{x_i} \varphi(x)\, dx = -\int_\Omega g(x) \varphi(x)\, dx.$$

The function g is denoted by $\partial_{x_i} f$.

Thus, a locally integrable function f has a Sobolev derivative $\partial_{x_i} f$ if its derivative in the sense of generalized functions with respect to the variable x_i is a locally integrable function. Obviously, a Sobolev derivative is uniquely defined up to equivalence. Higher order generalized derivatives are defined inductively.

2.1.2. Definition. *Let $p \in [1, +\infty)$, $k \in \mathbb{N}$. The Sobolev class $W^{p,k}(\Omega)$ consists of all functions $f \in L^p(\Omega)$ such that, for each $m \leqslant k$ and each multi-index $i_1, \ldots, i_m$ with $i_j \in \{1, \ldots, n\}$, the generalized derivative $\partial_{x_{i_1}} \cdots \partial_{x_{i_m}} f$*

belongs to $L^p(\Omega)$. The space $W^{p,k}(\Omega)$ is equipped with the Sobolev norm

$$\|f\|_{p,k} := \|f\|_p + \sum_{m=1}^{k} \sum_{i_1,\ldots,i_m=1}^{n} \|\partial_{x_{i_1}} \cdots \partial_{x_{i_m}} f\|_p.$$

Let

$$W^{p,k}(\Omega, \mathbb{R}^d) := \{(f_1, \ldots, f_d) \colon \Omega \to \mathbb{R}^d, \ f_i \in W^{p,k}(\Omega)\};$$

$$W^{p,k}_{\text{loc}}(\Omega, \mathbb{R}^d) := \{f \colon \zeta f \in W^{p,k}(\Omega, \mathbb{R}^d), \ \forall \zeta \in C_0^\infty(\Omega)\}.$$

As in the case of the spaces $L^p(\Omega)$, we identify functions that are equal almost everywhere, i.e., $W^{p,k}(\Omega)$ is the space of equivalence classes. Hence $W^{p,k}(\Omega)$ is the subspace in $L^p(\Omega)$ consisting of all elements with finite norm $\|\cdot\|_{p,k}$. It is clear that $W^{p,k}(\Omega)$ is a linear subspace in $L^p(\Omega)$.

2.1.3. Proposition. *A function f belongs to the class $W^{1,1}(\mathbb{R}^1)$ precisely when it is integrable and has an absolutely continuous version whose derivative is integrable on $\mathbb{R}^1$.*

PROOF. If f is integrable and absolutely continuous with $f' \in L^1(\mathbb{R}^1)$, then the integration by parts formula for absolutely continuous functions shows that f' serves as a generalized derivative. Conversely, let f have a generalized derivative $g \in L^1(\mathbb{R}^1)$. Set

$$f_0(x) := \int_{-\infty}^{x} g(t)\, dt.$$

For every function $\varphi \in C_0^\infty(\mathbb{R}^1)$ we have

$$-\int_{-\infty}^{+\infty} \varphi'(t) f(t)\, dt = \int_{-\infty}^{+\infty} \varphi(t) g(t)\, dt = -\int_{-\infty}^{+\infty} \varphi'(t) f_0(t)\, dt.$$

It is known that this yields that the function $f - f_0$ coincides almost everywhere with some constant C. This constant equals zero since $f_0(x) \to 0$ as $x \to -\infty$ and by the integrability of f one can choose $x_n \to -\infty$ with $f(x_n) = f_0(x_n) + C$ such that $f(x_n) \to 0$. □

2.1.4. Lemma. *The space $W^{p,k}(\Omega)$ is complete with respect to the indicated norm.*

PROOF. If a sequence $\{f_j\}$ is fundamental with respect to the norm $\|\cdot\|_{p,k}$, then all partial derivatives of the functions f_j up to order k form fundamental sequences, hence converge in $L^p(\Omega)$. It is readily seen that the obtained limits are generalized partial derivatives of the corresponding order for the limit of f_j in $L^p(\Omega)$. □

2.1.5. Lemma. *Let $f \in W^{p,k}(\Omega)$ and $\zeta \in C_b^\infty(\Omega)$. Then $\zeta f \in W^{p,k}(\Omega)$ and the generalized partial derivatives of ζf up to order k are calculated according to the Leibniz rule. For example, $\partial_{x_i}(\zeta f) = \zeta \partial_{x_i} f + f \partial_{x_i} \zeta$.*

PROOF. The product of a smooth function and a generalized function is differentiated by the Leibniz rule. In addition, the corresponding derivatives are elements of $L^p(\Omega)$. □

In the theory of Sobolev spaces one often employs convolutions. We recall a simple result on approximations by convolutions (which is proved by using the estimate $\|f * g\|_p \leqslant \|f\|_p \|g\|_1$).

2.1.6. Theorem. *Let $f \in \mathcal{L}^p(\mathbb{R}^n)$ and let a function $\varrho \in C_0^\infty(\mathbb{R}^n)$ have the integral 1 over $\mathbb{R}^n$. Set $\varrho_\varepsilon(x) := \varepsilon^{-n}\varrho(x/\varepsilon)$, $\varepsilon > 0$. Then we have $f * \varrho_\varepsilon \in C_b^\infty(\mathbb{R}^n)$ and $f * \varrho_\varepsilon \to f$ in $L^p(\mathbb{R}^n)$ as $\varepsilon \to 0$.*

The next result characterizes $W^{p,k}(\mathbb{R}^n)$ as the completion with respect to the Sobolev norm.

2.1.7. Theorem. *The space $W^{p,k}(\mathbb{R}^n)$ coincides with the completion of the class $C_0^\infty(\mathbb{R}^n)$ with respect to the Sobolev norm $\|\cdot\|_{p,k}$.*

PROOF. By the completeness of $W^{p,k}(\mathbb{R}^n)$ the aforementioned completion is contained in $W^{p,k}(\mathbb{R}^n)$. We show that every function f from $W^{p,k}(\mathbb{R}^n)$ is approximated by C_0^∞-functions in the Sobolev norm. Let us observe that f is approximated in $W^{p,k}(\mathbb{R}^n)$ by functions with bounded support. To this end it suffices to take a sequence of smooth functions $\zeta_j(x) = \zeta(x/j)$, where $0 \leqslant \zeta \leqslant 1$, $\zeta(x) = 1$ if $|x| \leqslant 1$, $\zeta(x) = 0$ if $|x| \geqslant 2$. Then $\zeta_j f \in W^{p,k}(\mathbb{R}^n)$ and the derivatives are calculated by the Leibniz rule. One has

$$\partial_{x_{i_1}} \cdots \partial_{x_{i_m}} (\zeta_j f) \to \partial_{x_{i_1}} \cdots \partial_{x_{i_m}} f$$

in $L^p(\mathbb{R}^n)$, which follows by the Lebesgue dominated convergence theorem and the fact that $\zeta_j(x) = 1$ and $\partial_{x_i} \zeta_j(x) = 0$ whenever $|x| \leqslant j$. For example, in the equality

$$\partial_{x_i}(\zeta_j f) = (\partial_{x_i} \zeta_j) f + \zeta_j \partial_{x_i} f$$

the first term tends to zero and the second one tends to $\partial_{x_i} f$.

In the case where f has support in the ball of radius R centered at the origin, we take for approximations the convolutions $f_j := f * \varphi_j$, where $\varphi_j(x) = j^n \varphi(jx)$ and φ is a smooth probability density with support in the unit ball. It is clear that $f_j \in C_0^\infty(\mathbb{R}^n)$. We observe that $\partial_{x_i} f_j = \partial_{x_i} f * \varphi_j$. To this end it suffices to verify that the functions $\psi \partial_{x_i} f_j$ and $\psi \partial_{x_i} f * \varphi_j$ have equal integrals for every function $\psi \in C_0^\infty(\mathbb{R}^n)$. By the definition of a generalized derivative, for every fixed x we have

$$\int_{\mathbb{R}^n} \partial_{y_i} f(y) \varphi_j(x-y)\, dy = -\int_{\mathbb{R}^n} f(y) \partial_{y_i} \varphi_j(x-y)\, dy.$$

By multiplying both sides of this equality by $\psi(x)$ and integrating in x, on account of the equality $\partial_{y_i} \varphi_j(x-y) = -\partial_{x_i} \varphi_j(x-y)$ and Fubini's theorem we obtain

$$\int_{\mathbb{R}^n} \partial_{x_i} f * \varphi_j(x) \psi(x) \, dx = - \int_{\mathbb{R}^n} \int_{\mathbb{R}^n} f(y) \partial_{y_i} \varphi_j(x-y) \psi(x) \, dy \, dx$$

$$= \int_{\mathbb{R}^n} \int_{\mathbb{R}^n} f(y) \partial_{x_i} \varphi_j(x-y) \psi(x) \, dx \, dy$$

$$= - \int_{\mathbb{R}^n} \int_{\mathbb{R}^n} f(y) \varphi_j(x-y) \partial_{x_i} \psi(x) \, dx \, dy$$

$$= - \int_{\mathbb{R}^n} f * \varphi_j(x) \partial_{x_i} \psi(x) \, dx = \int_{\mathbb{R}^n} \partial_{x_i}(f * \varphi_j)(x) \psi(x) \, dx.$$

By induction we obtain $\partial_{x_{i_1}} \cdots \partial_{x_{i_m}} (f * \varphi_j) = \partial_{x_{i_1}} \cdots \partial_{x_{i_m}} f * \varphi_j$ for all $m \leqslant k$ and $i_j \leqslant n$. By Theorem 2.1.6 this gives convergence of $f * \varphi_j$ to f in the Sobolev norm. $\square$

Let us prove one more close result.

2.1.8. Theorem. *Let $1 < p < \infty$ and let a sequence $\{f_j\} \subset W^{p,k}(\mathbb{R}^n)$ be bounded in the Sobolev norm $\|\cdot\|_{p,k}$. Then, there is a subsequence $\{f_{j_l}\}$ weakly convergent in $L^p(\mathbb{R}^n)$ to a function $f \in W^{p,k}(\mathbb{R}^n)$ such that the generalized partial derivatives of the functions f_{j_l} up to order k converge weakly in $L^p(\mathbb{R}^n)$ to the corresponding partial derivatives of f.*

PROOF. It suffices to extract from $\{f_j\}$ a subsequence weakly convergent in $L^p(\mathbb{R}^n)$ such that its generalized partial derivatives up to order k also converge weakly in $L^p(\mathbb{R}^n)$. $\square$

For $p = 1$ the proven assertion is false. For example, the indicator of $[-1, 1]$ is a limit in L^1 of the bounded sequence $\{f_j\}$ in $W^{1,1}$ defined as follows: $f_j(t) = 1$ if $t \in [-1, 1]$, $f_j(t) = 0$ if $|t| > 1 + 1/j$ and f_j is linear on $[-1-1/j, -1]$ and $[1, 1+1/j]$.

If $n = 1$, then every function from the Sobolev class $W^{p,1}$ has a continuous modification. For $n > 1$ this is no longer true.

2.1.9. Example. (i) Let $\varphi \in C_0^\infty(\mathbb{R}^2)$ have support in the unit disc, let $\varphi \geqslant 0$, and let $\varphi = 1$ in a neighborhood of the origin. The function $f(x) = \varphi(x) \ln |x|$ belongs to $W^{p,1}(\mathbb{R}^2)$ whenever $1 \leqslant p < 2$, but has no bounded modification. Indeed, $f \in L^p(\mathbb{R}^2)$ for all $p < \infty$. The functions

$$\partial_{x_i} f(x) = \varphi(x) x_i |x|^{-2} + \partial_{x_i} \varphi(x) \ln |x|$$

are defined outside the origin and belong to $L^p(\mathbb{R}^2)$ if $p < 2$. By the one-dimensional integration by parts formula and Fubini's theorem it is verified directly that they serve as generalized partial derivatives of f.

(ii) Similarly, the function $f(x) = \varphi(x) \ln(-\ln|x|)$ belongs to $W^{2,1}(\mathbb{R}^2)$.

(iii) The function $f(x) = \varphi(x) |x|^{-1}$ on $\mathbb{R}^3$, where $\varphi \in C_0^\infty(\mathbb{R}^3)$ and $\varphi = 1$ in a neighborhood of the origin, belongs to $W^{p,1}(\mathbb{R}^3)$ if $1 \leqslant p < 3/2$, but has no bounded modification. This is verified similarly.

(iv) By using the indicated function f one obtains a function of the form

$$g(x) = \sum_{j=1}^{\infty} 2^{-j} f(x - x_j),$$

where $\{x_j\}$ is an everywhere dense set. This function belongs to the respective space $W^{p,1}(\mathbb{R}^n)$, but has no modification bounded at least in some neighborhood.

In spite of the existence of such examples, there are important positive results on additional regularity of Sobolev functions, which we consider below.

For $k = 1$ Sobolev functions can be described by their behavior on straight lines in the following way.

2.1.10. Theorem. *A function $f \in L^p(\mathbb{R}^n)$ belongs to $W^{p,1}(\mathbb{R}^n)$ precisely when f has a modification g with the following property: for each fixed i, for almost every point $(y_1, \ldots, y_{i-1}, y_{i+1}, \ldots, y_n) \in \mathbb{R}^{n-1}$ the function*

$$x_i \mapsto g(y_1, \ldots, y_{i-1}, x_i, y_{i+1}, \ldots, y_n)$$

is absolutely continuous on all intervals and the function $\partial_{x_i} g$ (which exists almost everywhere) belongs to $L^p(\mathbb{R}^n)$.

PROOF. To simplify formulas we consider the case $n = 2$, $p = 1$. If such a version g exists, then by using the one-dimensional integration by parts formula and Fubini's theorem one can easily show that the functions $\partial_{x_i} g$ serve as generalized partial derivatives of f. Let $f \in W^{1,1}(\mathbb{R}^2)$. Let us take functions $\varphi_j \in C_0^\infty(\mathbb{R}^2)$ convergent to f in $W^{1,1}(\mathbb{R}^2)$. Passing to a subsequence we may assume that one has $\|\varphi_{j+1} - \varphi_j\|_{W^{1,1}} \leqslant 2^{-j}$ and $\{\varphi_j\}$ converges almost everywhere. We define a version g of the function f as the limit of $\{\varphi_j(x)\}$, where it exists. For any fixed x_2 we set

$$h_j(x_2) := \int_{\mathbb{R}^1} \big[|\varphi_{j+1}(x_1, x_2) - \varphi_j(x_1, x_2)| + |\partial_{x_1} \varphi_{j+1}(x_1, x_2) - \partial_{x_1} \varphi_j(x_1, x_2)|\big] \, dx_1.$$

Then the series of the integrals of h_j converges, hence $\sum_{j=1}^\infty h_j(x_2) < \infty$ for almost all x_2. In addition, for almost every x_2 there is a point x_1 at which the functions $\psi_{x_2,j} \colon t \mapsto \varphi_j(t, x_2)$ converge. Therefore, for almost every x_2, the sequence of functions $\psi_{x_2,j}$ converges at some point and is fundamental in $W^{1,1}(\mathbb{R}^1)$. By the Newton–Leibniz formula this implies its convergence on the whole real line to a locally absolutely continuous function from $W^{1,1}(\mathbb{R}^1)$. Thus, the function g has the required property with respect to x_1. It is clear that we have the same for x_2. □

2.1.11. Remark. (i) Let $f \in W^{1,1}(\mathbb{R}^n)$, let L be a two-dimensional plane in $\mathbb{R}^n$, and let S be an orthogonal operator which is a rotation in L and identical on $L^\perp$. Then f has a version which is absolutely continuous on almost all orbits of S. This is seen from reasoning similar to that given above if we pass to the coordinates (r, θ, y), where (r, θ) are the polar coordinates in L and $y \in L^\perp$.

(ii) The version constructed in the theorem has the stated properties not only in the standard coordinates in $\mathbb{R}^n$, but also with respect to any other basis.

2.1.12. Theorem. *Let $f \in W^{p,1}(\mathbb{R}^n)$ and let a Lipschitzian function ψ on the real line satisfy the condition $\psi(0) = 0$. Then $\psi(f) \in W^{p,1}(\mathbb{R}^n)$.*

PROOF. We take a modification of f from the previous theorem and observe that $\psi(f) \in L^p(\mathbb{R}^n)$ because $|\psi(f)| \leqslant C|f|$. In addition, for every absolutely continuous function g on the interval $[a, b]$, the function $\psi(g)$ is also absolutely continuous and $\psi(g)'(t) = \psi'\bigl(g(t)\bigr)g'(t)$ almost everywhere on $[a, b]$ (Exercise 2.8.1). Hence the desired conclusion follows by the previous theorem. $\square$

The following characterization is useful.

2.1.13. Theorem. *Let $p > 1$. A function $f \in L^p(\mathbb{R}^n)$ belongs to $W^{p,1}(\mathbb{R}^n)$ precisely when for every $h \in \mathbb{R}^n$ the mapping $t \mapsto f(\,\cdot\, + th)$ with values in $L^p(\mathbb{R}^n)$ is differentiable (then this mapping is continuously differentiable). An equivalent condition: $\|f(\,\cdot\, + th) - f\|_{L^p} \leqslant C(h)|t|$ for all t.*

PROOF. Let $f \in W^{p,1}(\mathbb{R}^n)$. We show that

$$\|f(\,\cdot\, + h) - f\|_{L^p} \leqslant |h|\|\nabla f\|_{L^p}. \qquad (2.1.1)$$

It suffices to establish (2.1.1) for $f \in C_0^\infty(\mathbb{R}^n)$. In this case

$$|f(x+h) - f(x)| = \left|\int_0^1 \bigl(\nabla f(x+th), h\bigr)\,dt\right| \leqslant |h|\int_0^1 |\nabla f(x+th)|\,dt.$$

Integrating both sides to power p in x over $\mathbb{R}^n$ we obtain (2.1.1) by Hölder's inequality and the fact that the integral of $|\nabla f(x+th)|^p$ in x equals the integral of $|\nabla f(x)|$. Estimate (2.1.1) yields the Lipschitzness of the mapping $F\colon t \mapsto f(\,\cdot\, + th)$ from $\mathbb{R}^1$ to $L^p(\mathbb{R}^n)$. Here we have $F(t) = T_t(f)$, where T_t is the operator on $L^p(\mathbb{R}^n)$ defined by the formula $T_t(\varphi)(x) := \varphi(x+th)$. It is clear that $\|T_t\varphi\|_{L^p} = \|\varphi\|_{L^p}$. Therefore, we are in the situation of Lemma 1.1.1, which gives the continuous differentiability of the mapping F. The same conclusion holds if we have the estimate mentioned in the theorem. Now we have to verify that this estimate (or the differentiability in L^p implied by it) yields the inclusion $f \in W^{p,1}(\mathbb{R}^n)$. Let $g_i \in L^p(\mathbb{R}^n)$ be the derivative at zero of the mapping $t \mapsto f(\,\cdot\, + te_i)$ to $L^p(\mathbb{R}^n)$, $i = 1, \ldots, n$. Hence the differences $j[f(\,\cdot\,+j^{-1}e_i) - f]$ converge to g_i in $L^p(\mathbb{R}^n)$ as $j \to \infty$. It follows that g_i is a generalized derivative of f with respect to x_i since

$$\int_{\mathbb{R}^n} j[f(x+j^{-1}e_i) - f(x)]\varphi(x)\,dx \to \int_{\mathbb{R}^n} g_i(x)\varphi(x)\,dx$$

as $j \to \infty$ for all $\varphi \in C_0^\infty(\mathbb{R}^n)$. The left side equals the integral of the function $jf(x)[\varphi(x - j^{-1}e_i) - \varphi(x)]$, hence it converges to the integral of $-f(x)\partial_{x_i}\varphi(x)$ as $j \to \infty$. Therefore, $f \in W^{p,1}(\mathbb{R}^n)$. □

In the case $p = 2$ the Sobolev spaces $W^{p,k}$ are Hilbert spaces, which in many respects leads to considerable simplifications. As an example we give a description of the classes $W^{2,k}$ in terms of the Fourier transform.

2.1.14. Theorem. *The space $W^{2,k}(\mathbb{R}^n)$ consists of all $f \in L^2(\mathbb{R}^n)$ such that the function $y \mapsto |y|^k \widehat{f}(y)$ belongs to $L^2(\mathbb{R}^n)$.*

PROOF. Let $f \in W^{2,k}(\mathbb{R}^n)$. Then we have $\partial_{x_i}^k f \in L^2(\mathbb{R}^n)$ for generalized derivatives, which gives the square integrability of the functions $y_i^k \widehat{f}(y)$. Therefore, the function $|y|^k \widehat{f}(y)$ is in $L^2(\mathbb{R}^n)$. Conversely, let the functions f and $|y|^k \widehat{f}(y)$ be square integrable. Then, for every nonnegative integer numbers $k_1, \ldots, k_n$ with $k_1 + \cdots + k_n = m \leqslant k$, the function

$$g_{k_1, \ldots, k_n}(y) := i^m y_1^{k_1} \cdots y_n^{k_n} \widehat{f}(y)$$

belongs to $L^2(\mathbb{R}^n)$. Let $f_{k_1, \ldots, k_n}$ denote the inverse Fourier transform of the function $g_{k_1, \ldots, k_n}$. We show that $f_{k_1, \ldots, k_n}$ is the generalized derivative $\partial_{x_1}^{k_1} \cdots \partial_{x_n}^{k_n} f$. To this end, we fix a real function $\varphi \in \mathcal{D}(\mathbb{R}^n)$ and denote by ψ its Fourier transform. By the Parseval equality we have

$$\int_{\mathbb{R}^n} f(x) \partial_{x_1}^{k_1} \cdots \partial_{x_n}^{k_n} \varphi(x) \, dx = \int_{\mathbb{R}^n} \widehat{f}(y)(-i)^m y_1^{k_1} \cdots y_n^{k_n} \overline{\psi}(y) \, dy$$

$$= (-1)^m \int_{\mathbb{R}^n} g_{k_1, \ldots, k_n}(y) \overline{\psi}(y) \, dy = (-1)^m \int_{\mathbb{R}^n} f_{k_1, \ldots, k_n}(x) \varphi(x) \, dx,$$

which proves our assertion. □

2.1.15. Example. For every function $g \in L^2(\mathbb{R}^n)$ there exists a unique function $f \in W^{2,2}(\mathbb{R}^n)$ satisfying the equation $\Delta f - f = g$. This function is determined by the equality

$$\widehat{f}(y) = \frac{1}{1 + |y|^2} \widehat{g}(y).$$

By the aid of the Fourier transform one can introduce the fractional Sobolev classes $H^{2,r}(\mathbb{R}^n)$ with $r \in (-\infty, +\infty)$ consisting of all generalized functions $f \in \mathcal{S}'(\mathbb{R}^n)$ such that $(1 + |y|^r)\widehat{f}(y) \in L^2(\mathbb{R}^n)$. In applications the fractional classes $W^{p,r}$ as well as some other generalizations or analogs of Sobolev classes are also useful (for example, the Besov classes). Below we consider some of them.

2.2. Embedding theorems for Sobolev classes

In the theory of Sobolev spaces and its applications an important role is played by the so-called embedding theorems, which assert that, under appropriate conditions on p, k, and n, the space $W^{p,k}(\mathbb{R}^n)$ consists of continuous functions or is embedded into some $L^q(\mathbb{R}^n)$. The simplest example is the space $W^{1,1}(\mathbb{R}^1)$ is embedded into the space of bounded continuous functions. In the multidimensional case the situation is more complicated, but also here there are simply formulated embedding theorems. We prove the most typical and important result and formulate some other results.

2.2.1. Theorem. (i) *If $p > n$ or $p = n = 1$, then one has the embedding $W^{p,1}(\mathbb{R}^n) \subset C_b(\mathbb{R}^n) = C(\mathbb{R}^n) \cap L^\infty(\mathbb{R}^n)$. Moreover, there exists a number $C(p,n) > 0$ such that*

$$\|f\|_{L^\infty} \leqslant C(p,n)\|f\|_{W^{p,1}}, \quad f \in W^{p,1}(\mathbb{R}^n). \tag{2.2.1}$$

(ii) *If $p \in [1,n)$, then $W^{p,1}(\mathbb{R}^n) \subset L^{np/(n-p)}(\mathbb{R}^n)$. Moreover, there exists a number $C(p,n) > 0$ such that*

$$\|f\|_{L^{np/(n-p)}} \leqslant C(p,n)\|f\|_{W^{p,1}}, \quad f \in W^{p,1}(\mathbb{R}^n). \tag{2.2.2}$$

For any domain Ω with Lipschitzian boundary, analogous embeddings hold with some $C(p,n,\Omega)$.

PROOF. The case $n = 1$ is simple. Let $p > n > 1$. It suffices to establish an estimate

$$\sup_x |\varphi(x)| \leqslant C(p,n)\|\varphi\|_{W^{p,1}}, \quad \varphi \in C_0^\infty(\mathbb{R}^n).$$

In turn, it suffices to show that $|\varphi(0)| \leqslant C(p,n)\|\varphi\|_{W^{p,1}}$ for all $\varphi \in C_0^\infty(\mathbb{R}^n)$. This will be done once we find a number C such that $|\varphi(0)| \leqslant C\|\nabla\varphi\|_{L^p}$ for all φ with support in the unit ball U. Indeed, let us take $\zeta \in C^\infty(\mathbb{R}^n)$ with $\zeta(0) = 1$ and with support in U. Then, for any $\varphi \in C_0^\infty(\mathbb{R}^n)$, one has

$$|\zeta(0)\varphi(0)| \leqslant C\|\nabla(\zeta\varphi)\|_{L^p} \leqslant C\max_x\big[|\zeta(x)| + |\nabla\zeta(x)|\big]\|\varphi\|_{W^{p,1}}.$$

For every unit vector u the Newton–Leibniz formula gives

$$-\varphi(0) = \int_0^1 \big(\nabla\varphi(tu), u\big)\, dt.$$

We integrate over the unit sphere S with the standard surface measure σ_n and obtain

$$-\varphi(0) = \sigma_n(S)^{-1} \int_S \int_0^1 \big(\nabla\varphi(tu), u\big)\, dt\, \sigma_n(du).$$

On the right we easily recognize the integral over U written in the spherical coordinates. By the change of variables formula we find

$$-\varphi(0) = \sigma_n(S)^{-1} \int_U |x|^{1-n} \big(\nabla\varphi(x), |x|^{-1}x\big)\, dx. \tag{2.2.3}$$

Hölder's inequality yields

$$|\varphi(0)| \leqslant \sigma_n(S)^{-1}\|\nabla\varphi\|_{L^p}\left(\int_U |x|^{p(1-n)/(p-1)}\, dx\right)^{(p-1)/p}$$
$$= C(n,p)\|\nabla\varphi\|_{L^p}$$

2.2. Embedding theorems for Sobolev classes

since the integral of $|x|^{p(1-n)/(p-1)}$ over U is finite. This is verified by returning to the spherical coordinates, which leads, up to a factor, to a power of the integral of the function t^α over $[0,1]$, where $\alpha = (1-n)/(p-1) > -1$.

(ii) It suffices to establish estimate (2.2.2) for $\varphi \in C_0^\infty(\mathbb{R}^n)$. For every $j \leqslant n$ the Newton–Leibniz formula yields

$$|\varphi(x)| \leqslant \int_{-\infty}^{+\infty} |\partial_{y_j}\varphi(x_1,\ldots,y_j,\ldots,x_n)|\, dy_j.$$

Let $|\partial_{y_j}\varphi|$ denote the function under the integral. Then

$$|\varphi(x)|^{n/(n-1)} \leqslant \prod_{j=1}^{n}\left(\int_{-\infty}^{+\infty} |\partial_{y_j}\varphi|\, dy_j\right)^{1/(n-1)}.$$

Let us integrate this inequality successively in $x_1,\ldots,x_n$, applying after every integration the generalized Hölder inequality

$$\|\psi_1\cdots\psi_{n-1}\|_{L^1} \leqslant \|\psi_1\|_{L^{n-1}}\cdots\|\psi_{n-1}\|_{L^{n-1}}.$$

This gives the estimate

$$\|\varphi\|_{L^{n/(n-1)}} \leqslant \left\{\prod_{j=1}^{n}\int_{\mathbb{R}^n}|\partial_{y_j}\varphi|\,dx\right\}^{1/n} \leqslant C(n)\big\||\nabla\varphi|\big\|_{L^1}.$$

Therefore, we obtain inequality (2.2.2) for $p=1$. It is clear that it remains true also for $|\varphi|^\alpha$, where $\alpha = (n-1)p/(n-p) > 1$ in the case $p > 1$. Hence we have

$$\big\||\varphi|^\alpha\big\|_{L^{n/(n-1)}} \leqslant C(n)\alpha\big\||\varphi|^{\alpha-1}|\nabla\varphi|\big\|_{L^1}$$
$$\leqslant C(n)\alpha\big\||\varphi|^{\alpha-1}\big\|_{L^{p'}}\big\||\nabla\varphi|\big\|_{L^p}.$$

By the equalities $\alpha n/(n-1) = np/(n-p)$ and $(\alpha-1)p' = np/(n-p)$ we obtain

$$\big\||\varphi|^\alpha\big\|_{L^{n/(n-1)}} = \|\varphi\|_{L^{np/(n-p)}}^{(n-1)p/(n-p)}, \quad \big\||\varphi|^{\alpha-1}\big\|_{L^{p'}} = \|\varphi\|_{L^{np/(n-p)}}^{n(p-1)/(n-p)},$$

which completes our proof. $\square$

We remark that in place of (2.2.2) we obtained the inequality

$$\|f\|_{L^{np/(n-p)}} \leqslant C(p,n)\big\||\nabla f|\big\|_{L^p}, \quad f \in W^{p,1}(\mathbb{R}^n), \tag{2.2.4}$$

which for $p=1$ is called the *Galiardo–Nirenberg inequality*; it shows that an integrable function on $\mathbb{R}^n$ with an integrable gradient belongs in fact to the class $L^{n/(n-1)}(\mathbb{R}^n)$, hence also to all $L^p(\mathbb{R}^n)$ with $1 \leqslant p \leqslant n/(n-1)$.

A function from the class $W^{n,1}(\mathbb{R}^n)$ need not be even locally bounded (see Example 2.1.9(ii)), but on every ball U it belongs to all $L^r(U)$.

The following assertions can be easily derived from the above results.

2.2.2. Corollary. *One has the following embeddings.*
(i) *If $kp < n$, then $W^{p,k}(\mathbb{R}^n) \subset L^{np/(n-kp)}(\mathbb{R}^n)$.*
(ii) *If $kp > n$, then $W^{p,k}(\mathbb{R}^n) \subset C(\mathbb{R}^n) \cap L^\infty(\mathbb{R}^n)$.*
(iii) *$W^{1,n}(\mathbb{R}^n) \subset C(\mathbb{R}^n) \cap L^\infty(\mathbb{R}^n)$.*

A closer look at the proof leads to the following conclusion.

2.2.3. Theorem. *Let $rp > n$, let U be a ball of radius 1 in $\mathbb{R}^n$, and let $f \in W^{p,r}(U)$. Then f has a modification f_0 which satisfies Hölder's condition with exponent $\alpha = \min(1, r - n/p)$, and there exists $C(n,p,r) > 0$ such that for all $x, y \in U$ one has the inequality*

$$|f_0(x) - f_0(y)| \leqslant C(n,p,r)\|D^r f\|_{L^p(U)}|x - y|^\alpha, \qquad (2.2.5)$$

where $\|D^r f\|_{L^p(U)}$ denotes the $L^p(U)$-norm of the real function

$$x \mapsto \sup_{|v_i| \leqslant 1} |D^r f(x)(v_1, \ldots, v_r)|.$$

A similar assertion is true for domains with sufficiently regular boundaries, but the constants will depend also on the domains.

Unlike the whole space, for a bounded domain $\Omega \subset \mathbb{R}^n$, one has the inclusion $L^p(\Omega) \subset L^r(\Omega)$ whenever $p > r$. This yields a wider spectrum of embedding theorems. We formulate the main results for a ball $\Omega \subset \mathbb{R}^n$. Set $W^{q,0} := L^q$.

2.2.4. Theorem. (i) *Let $kp < n$. Then*

$$W^{p,j+k}(\Omega) \subset W^{q,j}(\Omega), \quad q \leqslant \frac{np}{n - kp}, \; j \in \{0, 1, \ldots\}.$$

(ii) *Let $kp = n$. Then*

$$W^{p,j+k}(\Omega) \subset W^{q,j}(\Omega), \quad q < \infty, \; j \in \{0, 1, \ldots\}.$$

If $p = 1$, then $W^{j+n,1}(\Omega) \subset C_b^j(\Omega)$.

(iii) *Let $kp > n$. Then*

$$W^{p,j+k}(\Omega) \subset C_b^j(\Omega), \quad j \in \{0, 1, \ldots\}.$$

In addition, these embeddings are compact operators, with the exception of case (i) *with $q = np/(n - kp)$.*

Detailed proofs can be found in Adams [**2**].

It is seen from the derivation of inequality (2.2.1) that for $p > n$ and any function $f \in W^{p,1}(\mathbb{R}^n)$ with support in a ball of radius R one has the estimate

$$\|f\|_{L^\infty} \leqslant C(p, n, R) \|\,|\nabla f|\,\|_{L^p}.$$

Neither this estimate nor (2.2.4) hold for functions on bounded domains (for example, for constant functions).

The next result, called the Sobolev inequality (for $q = p = 2$ it is called the Poincaré inequality), enables one to estimate the L^q-norm of a function on a domain through the norm of its gradient and the integral of the function. For any integrable function f on a measurable set U of finite Lebesgue measure $\lambda_n(U)$ we set

$$f_U := \frac{1}{\lambda_n(U)} \int_U f(x)\,dx.$$

2.2. Embedding theorems for Sobolev classes

2.2.5. Theorem. *Let $1 \leqslant p < n$, let $q = \frac{np}{n-p}$, and let $U \subset \mathbb{R}^n$ be a ball of radius 1. There exists a number $C(p,n) > 0$ such that*

$$\|f - f_U\|_{L^q(U)} \leqslant C(p,n) \||\nabla f|\|_{L^p(U)}, \quad f \in W^{p,1}(U). \tag{2.2.6}$$

PROOF. It is clear that it suffices to obtain this estimate for smooth functions f with zero integral over U. Whenever $x, y \in U$ we have

$$f(y) - f(x) = \int_0^1 \langle \nabla f(x + t(y-x)), y - x \rangle \, dt.$$

Integrating in y we obtain the estimate

$$\lambda_n(U)|f(x)| \leqslant \int_0^1 \int_U |y - x| |\nabla f(x + t(y-x))| \, dy \, dt.$$

Hölder's inequality gives

$$\lambda_n(U)^p |f(x)|^p \leqslant \lambda_n(U)^{(p-1)/p} \int_0^1 \int_U |y - x|^p |\nabla f(x + t(y-x))|^p \, dy \, dt.$$

Let us integrate this relationship in x, interchange the integrals in x and t and then, for fixed t, apply the change of variables $u = (1-t)x + ty$, $v = y - x$, which has the Jacobian 1 and takes $U \times U$ to a subdomain of $U \times 2U$. We obtain

$$\lambda_n(U)^p \int_U |f(x)|^p \, dx \leqslant \lambda_n(U)^{(p-1)/p} \int_0^1 \int_{U \times 2U} |v|^p |\nabla f(u)|^p \, du \, dv \, dt$$

$$\leqslant 2^p \lambda_n(U)^{(p-1)/p} \lambda_n(2U) \int_U |\nabla f(u)|^p \, du,$$

whence $\|f\|_{L^p(U)} \leqslant c(p,n) \||\nabla f|\|_{L^p(U)}$, where $c(p,n)$ depends only on p and n. Therefore, we have $\|f\|_{W^{p,1}(U)} \leqslant (c(p,n) + 1) \||\nabla f|\|_{L^p(U)}$. Now Theorem 2.2.1 (see estimate (2.2.2) for a ball) yields (2.2.6). □

2.2.6. Lemma. *Suppose that U is a ball in $\mathbb{R}^n$ and a nonnegative function $f \in W^{1,1}(U)$ is such that $|\nabla f|/f \in L^1(U)$, where we set $\nabla f(x)/f(x) := 0$ whenever $f(x) = 0$. Then either $f = 0$ a.e. in U or $f > 0$ a.e. in U, $\ln f \in W^{1,1}(U)$ and $\nabla \ln f = \nabla f / f$.*

PROOF. Set $g_k := \ln(f + k^{-1})$. As one can readily verify, $g_k \in W^{1,1}(U)$ and $\nabla g_k = \nabla f/(f + k^{-1})$. Almost everywhere on the set $\{f = 0\}$ we have $\nabla f = 0$, hence $\nabla g_k = 0$ almost everywhere on $\{f = 0\}$. Therefore, as $k \to \infty$ we have $\nabla g_k \to \nabla f/f$ in $L^1(U)$ and almost everywhere in U. We show that the functions g_k converge in $L^1(U)$ if f is not zero almost everywhere. This will yield that their limit $\ln f$ belongs to $W^{1,1}(U)$. Denote by I_k the integral of g_k over U. Let us show that the sequence $\{I_k\}$ is bounded. Otherwise, passing to a subsequence, we may assume that $I_k \to \infty$. The previous theorem gives the inequality

$$\|g_k - I_k\|_{L^1(U)} \leqslant C(U) \|\nabla g_k\|_{L^1(U)},$$

which along with the uniform boundedness of the norms $\|\nabla g_k\|_{L^1(U)}$ yields the boundedness of the sequence $\{g_k - I_k\}$ in $L^1(U)$. According to Fatou's theorem one has $\liminf\limits_{k\to\infty} |g_k(x) - I_k| < \infty$ a.e. This is only possible if $\lim\limits_{k\to\infty} g_k(x) = +\infty$ a.e. Hence $f(x) = 0$ a.e. In all other cases $\sup_k I_k < \infty$. Passing to a subsequence once again we may assume that the numbers I_k have a finite limit. Then the above inequality applied to the differences of the functions $g_k - I_k$ shows that the sequence $\{g_k - I_k\}$ is fundamental in $L^1(U)$, which shows that $\{g_k\}$ is fundamental as well. □

Let us state two more useful forms of the embedding theorem. In the estimates established above the L^q-norm of f is estimated via the L^p-norm of f and ∇f, which are involved in a symmetric way. However, it is often useful to make the coefficient at the derivative smaller at the expense of increasing the coefficient at the function. The next two theorems demonstrate how to do this.

2.2.7. Theorem. *Let U be a connected open set in $\mathbb{R}^n$ with Lipschitzian boundary, let $p < n$, and let $q \leqslant np/(n-p)$ (or $p = n$ and $q < \infty$). Then, there exists a number $C = C(U, n, p, q)$ such that, whenever $1 \leqslant r < q$, for all $f \in W^{p,1}(U)$ and $\varepsilon \in (0, 1]$ one has the inequality*

$$\|f\|_{L^q(U)} \leqslant \frac{C}{\varepsilon^{n/r - n/q}} \|f\|_{L^r(U)} + C\varepsilon^{1 + n/q - n/p} \|f\|_{W^{p,1}(U)}. \qquad (2.2.7)$$

For a proof see Gol'dshteĭn, Reshetnyak [**490**, §4.8]. Taking $r = p$, one can obtain an estimate of the form

$$\|f\|_{L^q(U)} \leqslant C(\varepsilon)\|f\|_{L^p(U)} + \varepsilon \|\nabla f\|_{L^p(U)}$$

with $\varepsilon > 0$ as small as we wish (certainly, $C(\varepsilon)$ can be very large).

There is also a multiplicative estimate, a proof of which can be found in Adams, Fournier [**3**, p. 139]. Let U satisfy the hypotheses of the previous theorem.

2.2.8. Theorem. *Let either $mp > n$ and $p \leqslant q \leqslant \infty$, or $mp = n$ and $p \leqslant q < \infty$, or $mp < n$ and $p \leqslant q \leqslant np/(n - mp)$. Then, there exists a number $C = C(U, n, m, p, q)$ such that for all $f \in W^{p,m}(U)$ one has the inequality*

$$\|f\|_{L^q(U)} \leqslant C\|f\|_{W^{p,m}(U)}^{\theta} \|f\|_{L^p(U)}^{1-\theta}, \qquad (2.2.8)$$

where $\theta := n \cdot m^{-1}(p^{-1} - q^{-1})$.

Let us give a version of this result for the whole space which fails for bounded domains (see Shigekawa [**1017**, Lemma 5.1, p. 104] for a proof).

2.2.9. Theorem. *For any $p > n$ there exists a number $C = C(n, p)$ such that for all $f \in W^{p,1}(\mathbb{R}^n)$ one has the inequality*

$$\|f\|_{L^\infty(\mathbb{R}^n)} \leqslant C\|\nabla f\|_{L^p(\mathbb{R}^n)}^{n/p} \|f\|_{L^p(\mathbb{R}^n)}^{1 - n/p}. \qquad (2.2.9)$$

Applications of these results to differentiable measures will be discussed in §3.4.

2.3. The classes BV

We have considered above certain classes of functions whose generalized derivatives are represented by functions, but one often encounters functions whose generalized derivatives are measures (possibly, singular). For example, the indicator function $I_{[0,1]}$ of the interval $[0,1]$ has the measure $\delta_0 - \delta_1$ as its generalized derivative. So it is natural to introduce the class $V^k(\mathbb{R}^n)$ of bounded measures ν on $\mathbb{R}^n$ such that, whenever $k_1 + \cdots + k_n \leqslant k$, the partial derivatives $\partial_{x_1}^{k_1} \cdots \partial_{x_n}^{k_n} \nu$ in the sense of generalized functions are also measures. For $k = 0$ we obtain the space of all bounded measures. For $k \geqslant 1$ all measures from $V^k(\mathbb{R}^n)$ are absolutely continuous and are identified with their densities. The space $V^1(\mathbb{R}^n)$ is denoted by the symbol $BV(\mathbb{R}^n)$ and called the space of functions of bounded variation. Let $\|\nu\|$ be the variation of the measure ν. The space $V^k(\mathbb{R}^n)$ is linear and, as one can easily verify, is Banach with the norm

$$\|\nu\|_{V^k} := \|\nu\| + \sum_{1 \leqslant k_1 + \cdots + k_n \leqslant k} \|\partial_{x_1}^{k_1} \cdots \partial_{x_n}^{k_n} \nu\|.$$

Similar classes are defined on domains.

2.3.1. Proposition. (i) *If $k \geqslant 1$ and $n \geqslant 1$, then one has the inclusion $V^k(\mathbb{R}^n) \subset L^1(\mathbb{R}^n) \cap L^{n/(n-1)}(\mathbb{R}^n)$.*

(ii) *If $n = k = 1$, then the class $BV(\mathbb{R}^1) = V^1(\mathbb{R}^1)$ consists of integrable functions of bounded variation.*

PROOF. (i) Let $\nu \in BV(\mathbb{R}^n)$, let $\varrho \in C_0^\infty(\mathbb{R}^n)$ be a probability density, and let $\varrho_m(x) = m^{-n} \varrho(mx)$ and $f_m = \varrho_m * \nu$, $m \in \mathbb{N}$. Then, for all m, we have $f_m \in C^\infty(\mathbb{R}^n) \cap L^1(\mathbb{R}^n)$, $\partial_{x_i} f_m \in L^1(\mathbb{R}^n)$ and $\|\partial_{x_i} f_m\|_{L^1} \leqslant \|\partial_{x_i} \nu\|$, i.e., the sequence $\{f_m\}$ is bounded in $W^{1,1}(\mathbb{R}^n)$. Therefore, it is bounded in the space $L^{n/(n-1)}(\mathbb{R}^n)$. Hence $\{f_m\}$ contains a subsequence weakly convergent in the space $L^{n/(n-1)}(\mathbb{R}^n)$ to some function $f \in L^{n/(n-1)}(\mathbb{R}^n)$. It is readily seen that ν coincides with the measure $f \, dx$.

(ii) Let σ be the measure which is the generalized derivative of ν. Let us set $f(x) := \sigma\bigl((-\infty, x)\bigr)$. Then f is a function of bounded variation. It defines a locally bounded measure $\nu_0 = f \, dx$. It is verified directly that

$$\int_{-\infty}^{+\infty} \varphi'(x) f(x) \, dx = -\int_{-\infty}^{+\infty} \varphi(x) \, \sigma(dx)$$

for all $\varphi \in C_0^\infty(\mathbb{R}^1)$. Therefore, the generalized derivative of the measure $\nu - \nu_0$ equals zero. Hence ν has a density $f + c$, where c is a constant. This constant can only be zero since f has a zero limit at $-\infty$ and the measure ν is finite. $\square$

It is seen from the proof that for $V^k(\mathbb{R}^n)$ we have the same embeddings to $L^q(\mathbb{R}^n)$ as for $W^{1,k}(\mathbb{R}^n)$.

A measurable set $E \subset \mathbb{R}^n$ is called a set of finite perimeter or a Caccioppoli set if its indicator function I_E belongs to $BV(\mathbb{R}^n)$. One can show that bounded convex sets are Caccioppoli sets.

Let us mention the following important theorem due to A.D. Alexandroff.

2.3.2. Theorem. *Let Ψ be a convex function on an open convex set $\mathrm{Dom}\,\Psi$ in $\mathbb{R}^n$. Then, for every open set U with compact closure in $\mathrm{Dom}\,\Psi$, one has $\Psi \in V^2(U)$ and, for every Borel set $B \subset U$, the matrix with entries $\partial_{x_i}\partial_{x_j}\Psi(B)$ is nonnegative.*

In addition, if $\partial_{x_i}\partial_{x_j}\Psi^{\mathrm{ac}}$ is a density of the absolutely continuous component of the measure $\partial_{x_i}\partial_{x_j}\Psi$ and $D^2_{\mathrm{ac}}\Psi(x)$ is the matrix with entries $\partial_{x_i}\partial_{x_j}\Psi^{\mathrm{ac}}(x)$, then almost everywhere in U one has

$$\Psi(x+h) - \Psi(x) - D\Psi(x)(h) - \frac{1}{2}\big(D^2_{\mathrm{ac}}\Psi(x)h, h\big) = o(|h|^2) \text{ as } |h| \to 0.$$

Conversely, if a locally integrable function f on an open convex set U is such that for all $\xi_1, \ldots, \xi_n$ from $\mathbb{R}^1$ the generalized function $\sum_{i,j\leqslant n} \xi_i\xi_j\partial_{x_i}\partial_{x_j}f$ is nonnegative, then f has a convex modification.

A proof can be found in Gol'dshteĭn, Reshetnyak [**490**, §4.10] and Evans, Gariepy [**400**, Ch. 6].

2.4. Approximate differentiability and Jacobians

Let $A \subset \mathbb{R}^n$ and $x \in A$. If

$$\lim_{r \to 0} \frac{\lambda\big(K(x,r) \cap A\big)}{\lambda\big(K(x,r)\big)} = 0,$$

where $K(x,r)$ is the closed ball of radius $r > 0$ centered at x, then the point x is called a point of density 0 for the set A. A point $y \in \mathbb{R}^d$ is called an approximate limit of the mapping $f\colon A \to \mathbb{R}^d$ at the point x if the density of the set $\mathbb{R}^n \setminus f^{-1}(U)$ at x equals 0 for every neighborhood U of the point y. It is clear that there might be at most one such y; we denote it by $\mathrm{ap}\lim_{z \to x} f(z)$. If f is a real function, then its upper approximate limit $\mathrm{ap}\limsup_{x \to a} f(x)$ at the point a is defined as the infimum of numbers t such that the point a is a point of density 0 for the set $\{z\colon f(z) > t\}$.

A mapping $f\colon A \to \mathbb{R}^d$ is called *approximately differentiable* at the point a if there exists a linear mapping $L\colon \mathbb{R}^n \to \mathbb{R}^d$ such that

$$\mathrm{ap}\lim_{x \to a} |f(x) - f(a) - L(x-a)|/|x-a| = 0.$$

It follows that $\mathbb{R}^n \setminus A$ has density 0 at the point a; in addition, such L is unique (see Federer [**421**, 3.1.2]). The operator L is denoted by $\mathrm{ap}Df(a)$ and called the *approximate derivative* of f at the point a.

The *approximate Jacobian* $J(f)$ of the mapping f is the determinant of the approximate derivative of f.

2.4. Approximate differentiability and Jacobians

We define the *approximate partial derivatives* of the mapping $f\colon A \to \mathbb{R}^d$ by the formula

$$\mathrm{apD}_i f(a) = \mathrm{ap}\lim_{t\to 0}\,[f(\ldots, a_{i-1}, a_i + t, a_{i+1}, \ldots) - f(a)]/t.$$

The existence almost everywhere of the approximate partial derivatives of a measurable mapping $f\colon A \to \mathbb{R}^d$ implies its approximate differentiability almost everywhere on A (see Federer [**421**, Theorem 3.1.4]), whence it follows in turn that

$$\mathrm{ap}\limsup_{x\to a} |f(x) - f(a)|/|x - a| < \infty \qquad (2.4.1)$$

for almost all $a \in A$. The usual differentiability of f at a point implies the approximate differentiability at this point and the usual derivative Df in this case serves as an approximate derivative. Unlike ordinary derivatives, the definition of an approximate derivative does not require that the mapping f be defined in a neighborhood of the point x. Let us also note that every mapping $f \in W^{1,1}_{\mathrm{loc}}(\mathbb{R}^n, \mathbb{R}^d)$ has a modification possessing partial derivatives almost everywhere, which gives the approximate differentiability almost everywhere. Any function $f \in BV(\mathbb{R}^n)$ is almost everywhere approximately differentiable as well (see Ambrosio, Fusco, Pallara [**65**]), and its approximate partial derivatives coincide with the absolutely continuous components of the generalized partial derivatives (which are measures).

One of the most important classical results on approximate derivatives is the following theorem due to Whitney. Its proof can be found in Federer [**421**, Theorem 3.1.8, Theorem 3.1.16].

2.4.1. Theorem. *Let $A \subset \mathbb{R}^n$ be a measurable set and let a measurable mapping $f\colon A \to \mathbb{R}^d$ have approximate partial derivatives $\mathrm{apD}_i f$ on A. Then, up to a set of measure zero, A is the union of countably many compact sets on which the restrictions of f are Lipschitzian, and f is approximately differentiable almost everywhere on A. In addition, for every $\varepsilon > 0$ there is a mapping $g \in C^1(\mathbb{R}^n, \mathbb{R}^d)$ such that $\lambda\bigl(x \in A\colon f(x) \neq g(x)\bigr) < \varepsilon$, where λ is Lebesgue measure.*

If a mapping f from an open set $U \subset \mathbb{R}^n$ to $\mathbb{R}^d$ has an approximate derivative L at a point $x \in U$ and there exists a sequence $r_j \to 0$ such that

$$\lim_{j\to\infty} \sup_{|h|=1} \left| \frac{f(x + r_j h) - f(x)}{r_j} - Lh \right| = 0,$$

then L is called the *regular approximate derivative* of f at the point x (the operator L is also called the *K-derivative* or the *total weak differential*). It is clear that the usual differentiability at a point implies the existence of a regular approximate derivative at this point and its coincidence with the usual derivative. It is known that every mapping $f \in W^{p,1}_{\mathrm{loc}}(\mathbb{R}^n, \mathbb{R}^n)$ with $p > n - 1$ has a modification $\widetilde{f}$ which almost everywhere has a regular approximate derivative that coincides almost everywhere with the Sobolev derivative of f (see Reshetnyak [**953**, §2], Ziemer [**1224**, p. 170]). If $p > n$, then the mapping f has a continuous modification which almost everywhere has the usual derivative (see Gol'dshteĭn, Reshetnyak [**490**, Ch. 2, §5.4, Theorem 5.2]).

In the study of multidimensional mappings the so-called Lusin's property (N) (or (N)-property) is useful. A mapping $F\colon \mathbb{R}^n \to \mathbb{R}^n$ is called a mapping with Lusin's property (N) if it takes all sets of Lebesgue measure zero to sets of measure zero. A simple but important example of such a mapping is any Lipschitzian mapping. The following fact established in Bojarski, Iwaniec [**235**, Lemma 8.1] is less obvious.

2.4.2. Theorem. *The continuous version of any mapping F of the class $W_{\mathrm{loc}}^{p,r}(\mathbb{R}^n, \mathbb{R}^n)$ with $pr > n$ possesses Lusin's property (N). In particular, a continuous version of every mapping of the class $W_{\mathrm{loc}}^{p,1}(\mathbb{R}^n, \mathbb{R}^n)$ with $p > n$ has this property.*

PROOF. Let $A \subset \mathbb{R}^n$ be a set of Lebesgue measure zero. In order to show that $\lambda\bigl(F(A)\bigr) = 0$, where λ is Lebesgue measure, we assume that A is contained in a ball U and that $F \in W^{p,r}(U, \mathbb{R}^n)$. By Theorem 2.2.3 the mapping F has a version which is Hölder continuous of order $\alpha = \min(1, r - n/p)$. We shall assume that $\alpha = r - n/p < 1$ since for Lipschitzian mappings our assertion is true. Let $\varepsilon > 0$. There is a sequence of cubes Q_i with disjoint interiors for which $A \subset \bigcup_{i=1}^{\infty} Q_i \subset U$ and $\lambda\bigl(\bigcup_{i=1}^{\infty} Q_i\bigr) < \varepsilon$. By estimate (2.2.5) the diameter of the set $F(Q_i)$ does not exceed $C r_i^\alpha \|D^r F\|_{L^p(Q_i)}$, where r_i is the edge length of Q_i and C is independent of i and F. Hence

$$\lambda\bigl(F(A)\bigr) \leqslant C^n \sum_{i=1}^{\infty} r_i^{n\alpha} \|D^r F\|_{L^p(Q_i)}^n.$$

By Hölder's inequality with exponents α^{-1} and $t = (1-\alpha)^{-1}$ the right-hand side is estimated from above by

$$C^n \Bigl(\sum_{i=1}^{\infty} r_i^n\Bigr)^\alpha \Bigl(\sum_{i=1}^{\infty} \|D^r F\|_{L^p(Q_i)}^{nt}\Bigr)^{1/t}$$

$$= C^n \Bigl(\sum_{i=1}^{\infty} r_i^n\Bigr)^\alpha \Bigl(\sum_{i=1}^{\infty} \bigl(\|D^r F\|_{L^p(Q_i)}/\|D^r F\|_{L^p(U)}\bigr)^{nt} \|D^r F\|_{L^p(U)}^{nt}\Bigr)^{1/t}.$$

Since $nt \geqslant p$, by the inequality $n \geqslant p(1-\alpha) = p - pr + n$, the right-hand side does not exceed

$$C^n \Bigl(\sum_{i=1}^{\infty} r_i^n\Bigr)^\alpha \Bigl(\sum_{i=1}^{\infty} \|D^r F\|_{L^p(Q_i)}^{p}\Bigr)^{1/t} \|D^r F\|_{L^p(U)}^{n-p/t}$$

$$\leqslant C^n \lambda\Bigl(\bigcup_{i=1}^{\infty} Q_i\Bigr)^\alpha \|D^r F\|_{L^p(U)}^{p/t} \|D^r F\|_{L^p(U)}^{n-p/t} \leqslant C^n \varepsilon^\alpha \|D^r F\|_{L^p(U)}^n,$$

whence our claim follows. □

2.4. Approximate differentiability and Jacobians

One should keep in mind that the proof has employed not only the Hölder continuity of the mapping. The Hölder continuity of order $\alpha < 1$ by itself is not enough even in the case $n = 1$ (Exercise 2.8.9). Ponomarev [**909**] proved that, for every $n > 1$, there exists a homeomorphism $h\colon K \to K$, $K = [0,1]^n$, such that $h \in W^{p,1}(K, \mathbb{R}^n)$ for all $p < n$, $h^{-1} \in W^{q,1}(K, \mathbb{R}^n)$ for all $q < \infty$, $h|_{\partial K} = I$, but h has no Lusin's property (N) (see also Kauhanen [**598**]). However, every homeomorphism $h \in W^{n,1}_{\mathrm{loc}}(\mathbb{R}^n, \mathbb{R}^n)$ has Lusin's property (N) (see Reshetnyak [**952**] or [**953**, Ch. II, §2, Corollary 1 of Theorem 6.2]). According to Malý, Martio [**775**], a mapping $f \in W^{n,1}_{\mathrm{loc}}(\mathbb{R}^n, \mathbb{R}^n)$ has Lusin's property (N) if it is locally Hölder continuous or if it is continuous and open; in the same work there is an example of a continuous mapping $g \in W^{n,1}_{\mathrm{loc}}(\mathbb{R}^n, \mathbb{R}^n)$ with $n > 1$ whose Jacobian vanishes almost everywhere, but the mapping takes the one-dimensional interval onto the cube $[0,1]^n$. Property (N) is also studied in Ponomarev [**911**]. It is also worth noting that in view of Theorem 2.4.1 any mapping $f \in W^{1,1}_{\mathrm{loc}}(\mathbb{R}^n, \mathbb{R}^n)$ has a version with property (N). However, it follows from what was said above that even for a continuous mapping this version may be discontinuous.

The objects introduced above are important for obtaining change of variables formulas under assumptions considerably weaker than the standard ones (when one deals with continuously differentiable mappings). Let us mention some typical results.

For a mapping $f\colon U \to \mathbb{R}^n$ and a set $E \subset U$ let $N(f, E, y)$ denote the cardinality of the set $\{x \in E\colon f(x) = y\}$, where $y \in \mathbb{R}^n$. The function $y \mapsto N(f, E, y)$ takes values in the set $\{0, 1, \ldots, +\infty\}$.

2.4.3. Theorem. *Let a set $U \subset \mathbb{R}^n$ be open and let a measurable mapping $f\colon U \to \mathbb{R}^n$ have Lusin's property (N) and possess approximate partial derivatives almost everywhere in U such that its approximate Jacobian $J(f)$ is locally integrable in U. Then, for every measurable set $E \subset U$, the function $N(f, E, y)$ is measurable and, for every measurable function g on $\mathbb{R}^n$ for which the function $g(y)N(f, E, y)$ is integrable on U, the following equality is fulfilled:*

$$\int_E g(f(x))|J(f)(x)|\,dx = \int_{\mathbb{R}^n} g(y)N(f, E, y)\,dy. \qquad (2.4.2)$$

2.4.4. Corollary. *The previous theorem is valid for the continuous version of any mapping $f \in W^{p,1}_{\mathrm{loc}}(U, \mathbb{R}^n)$ with $p > n$. In particular, it is valid for locally Lipschitzian mappings.*

Taking $E = \{x\colon J(f)(x) = 0\}$ we obtain from (2.4.2) that the set $f(E)$ has measure zero, which is a special case of Sard's theorem.

A similar but somewhat more complicated formula holds for mappings $f = (f_1, \ldots, f_k)\colon \mathbb{R}^n \to \mathbb{R}^k$ with $k < n$. In this case $|Jf(x)|$ is defined as the volume of the k-dimensional parallelepiped spanned by the approximate gradients $\operatorname{app} Df_i(x)$.

2.4.5. Theorem. *Let $f\colon \Omega \to \mathbb{R}^k$ satisfy the hypotheses of the previous theorem and let $k < n$. Then, for every measurable set $E \subset \Omega$, we have*

$$\int_E |Jf(x)|\, dx = \int_{\mathbb{R}^k} H^{n-k}\bigl(E \cap f^{-1}(y)\bigr)\, dy, \qquad (2.4.3)$$

where H^{n-k} is the $(n-k)$-dimensional Hausdorff measure.

Proofs of these results for Lipschitzian mappings can be found in Bogachev [**193**, §5.8(x)] and Federer [**421**, §3.2] (see also Gol'dshteĭn, Reshetnyak [**490**, Ch. 5]); the case of a Lipschitzian mapping (or even of a continuously differentiable mapping) yields the general case by Theorem 2.4.1 and property (N) (which was first observed in Hajłasz [**513**]).

2.5. Restrictions and extensions

Given a function from a Sobolev class on a domain, the question arises whether it can be extended to a function from a Sobolev class on the whole space. The following theorem gives a sufficient condition for the existence of extensions; moreover, the corresponding extension is given by means of a continuous linear operator.

2.5.1. Theorem. *Let U be a bounded domain in $\mathbb{R}^n$ with Lipschitzian boundary. Then, there exists a linear mapping $E\colon L^0(U) \to L^0(\mathbb{R}^n)$ with the following properties: $Ef(x) = f(x)$ for a.e. $x \in U$ and if $f \in W^{p,r}(U)$, then $Ef \in W^{p,r}(\mathbb{R}^n)$, where the operator $E\colon W^{p,r}(U) \to W^{p,r}(\mathbb{R}^n)$ is continuous.*

In many important partial cases a relatively difficult proof of this theorem considerably simplifies and can be explained shortly. For example, if U has smooth boundary, then by using a finite partition of unity the assertion on extensions of functions from fixed $W^{p,r}(U)$ reduces to the case of functions on the half-space $\{x_n > 0\}$. If $r = 1$, then an extension for the half-space is given by an explicit formula: for $x_n < 0$ we set $Ef(x_1, \ldots, x_{n-1}, x_n) := f(x_1, \ldots, x_{n-1}, -x_n)$, which defines a continuous linear extension operator. For a discussion of more general cases, see Stein [**1079**] and Adams, Fournier [**3**].

For an arbitrary domain, the extension theorem may fail. The following interesting connection between extensions and Sobolev embeddings was discovered in Koskela [**634**].

2.5.2. Theorem. *Let a domain $D \subset \mathbb{R}^n$ and $p > n$ be such that the class $W^{p,1}(D)$ is embedded into the space $C^{1-n/p}(\overline{D})$ of functions which are Hölder of order $1 - n/p$. Then, for every $q > p$, there exists a bounded extension operator from $W^{q,1}(D)$ to $W^{q,1}(\mathbb{R}^n)$.*

Another interesting question concerns restrictions of functions of the class $W^{p,r}(\mathbb{R}^n)$ to subspaces of lower dimension. It is seen from the proof of Theorem 2.1.10 how to find a version with nice restrictions to almost all subspaces. Let us give a precise formulation.

2.5. Restrictions and extensions

2.5.3. Theorem. *Let $f \in W^{p,r}(\mathbb{R}^n)$. Then, for every k-dimensional subspace $L \subset \mathbb{R}^n$, the function $y \mapsto f(x+y)$ on L belongs to $W^{p,r}(L)$ for a.e. $x \in L^\perp$.*

This theorem asserts nothing concerning restrictions to a fixed subspace. However, there are positive results also for such restrictions. For example, if an embedding theorem provides a continuous version, then one can take just the restriction of this version. Let us mention a more subtle result.

2.5.4. Theorem. *Suppose that a bounded domain $U \subset \mathbb{R}^n$ has Lipschitzian boundary, $mp < n$ and either $n - mp < k \leqslant n$ or $p = 1$ and $n - m \leqslant k \leqslant n$. Let $p \leqslant q \leqslant kp/(n-mp)$. Then there is a continuous linear operator*

$$\mathcal{R}_k \colon W^{p,m}(U) \to L^q(U \cap \mathbb{R}^k)$$

for which $\mathcal{R}_k f(x) = f(x)$ whenever $x \in U \cap \mathbb{R}^k$ for all $f \in C(U) \cap W^{p,m}(U)$. In particular, there is a continuous linear restriction operator

$$\mathcal{R}_{n-1} \colon W^{1,1}(U) \to L^1(U \cap \mathbb{R}^{n-1}).$$

The difference between this theorem and the previous one is the following: the section $U \cap \mathbb{R}^k$ is fixed and the operator depends on it. The main contents of this theorem is that for nice functions the L^q-norm on sections is estimated via the $W^{p,m}(U)$-norm. Then the operator extends by continuity. If $m = 1$ and $k = n-1$, then one can take $q = p(n-1)/(n-p)$, where $p > 1$. If $m = 1$ and $k = 1$, then p must be greater than $n-1$ and $q = p/(n-p)$. The problem of describing the collection of restrictions of functions from a given Sobolev class to a fixed hyperplane turns out to be more delicate. It involves scales of Besov spaces (see Stein [**1079**], Adams, Fournier [**3**]).

Similarly, one defines a continuous restriction operator $\mathcal{R}_S$ to a sufficiently regular surface S (say of class C^1). This is useful for extending integration by parts formulae which employ surface integrals. For example, if a bounded open set $\Omega \subset \mathbb{R}^n$ has C^1-boundary $\partial\Omega$ with the unit outer normal $n_{\partial\Omega}$, then, for all functions $f \in W^{1,1}(\mathbb{R}^n)$ and $\varphi \in C^2(\mathbb{R}^n)$, one has

$$\int_\Omega \Delta\varphi \, f \, dx = -\int_\Omega (\nabla\varphi, \nabla f) \, dx + \int_{\partial\Omega} (\nabla\varphi, n_{\partial\Omega}) \mathcal{R}_{\partial\Omega} f \, dS.$$

This follows from the classical formula and the continuity of $\mathcal{R}_{\partial\Omega}$.

The following interesting fact is related to the results on restrictions (see Evans, Gariepy [**400**, Ch. 6]).

2.5.5. Theorem. *Let $f \in W^{p,1}(\mathbb{R}^n)$ and $1 \leqslant p < \infty$. Then, for every $\varepsilon > 0$, there exists a function $g \in C^1(\mathbb{R}^n)$ such that $\|f - g\|_{W^{p,1}} \leqslant \varepsilon$ and*

$$\lambda\bigl(x \colon f(x) \neq g(x) \text{ or } \nabla f(x) \neq \nabla g(x)\bigr) < \varepsilon.$$

2.6. Weighted Sobolev classes

In the definition of Sobolev norms, Lebesgue measure was used. However, it is clear that analogous constructions are meaningful for every nonnegative locally finite Borel measure μ on $\mathbb{R}^n$. For example, one can set

$$\|f\|_{p,1,\mu} := \|f\|_{L^p(\mu)} + \sum_{i=1}^{n} \|\partial_{x_i} f\|_{L^p(\mu)}, \quad f \in C_0^\infty(\mathbb{R}^n),$$

and then complete $C_0^\infty(\mathbb{R}^n)$ with respect to this norm, which will give some Banach space $W^{p,1}(\mu)$. However, as in the case of $W^{p,1}(\mathbb{R}^n)$, we would like to have this space as a subset of $L^p(\mu)$: we complete a subspace of $L^p(\mu)$ with respect to a norm greater than the norm of $L^p(\mu)$, so that, at first glance, the completion must be automatically embedded into $L^p(\mu)$. In fact, this is false. The first (and trivial) reason is that μ-equivalent functions from C_0^∞ may have different norms in $W^{p,1}(\mu)$. For example, if μ is Dirac's measure δ at zero, then any function f such that $f(x) = x$ in a neighborhood of zero coincides with zero in $L^p(\mu)$, but $f' = 1$ as an element of $L^p(\mu)$. This trivial reason can be excluded if we require that the measure μ does not vanish on nonempty open sets. In this case, if $f, g \in C_0^\infty(\mathbb{R}^n)$ and $f(x) = g(x)$ for μ-a.e. x, then $\nabla f(x) = \nabla g(x)$ for μ-a.e. x. However, there might be deeper reasons to prevent $W^{p,1}(\mu)$ from being a subspace in $L^p(\mu)$.

2.6.1. Example. Let us take the measure $\mu = \sum_{k=1}^{\infty} 2^{-k} \delta_{r_k}$ on the real line, where $\{r_k\}$ is the set of all rational numbers. Let

$$\psi_j \in C_0^\infty(\mathbb{R}^1), \ |\psi_j(x)| \leqslant 1, \ |\psi_j'(x)| \leqslant 1,$$

$\psi_j(t_k) = 0$ if $k = 1, \ldots, j$, $\psi_j'(t_k) = 1$ if $k = 1, \ldots, j$. Then, as $j \to \infty$, we have $\psi_j \to 0$ and $\psi_j' \to 1$ in $L^2(\mu)$, i.e., the sequence $\{\psi_j\}$ is fundamental in norm $\|\cdot\|_{p,1,\mu}$ and converges to zero in $L^2(\mu)$, but does not converge to zero in $W^{2,1}(\mu)$. So the limit of this sequence in $W^{2,1}(\mu)$ cannot be identified with an element of $L^2(\mu)$.

2.6.2. Definition. *The norm $\|\cdot\|_{p,1,\mu}$ is called closable and the class $W^{p,1}(\mu)$ is called well defined if, for each sequence of functions $f_j \in C_0^\infty(\mathbb{R}^n)$ that is fundamental with respect to this norm and converges to zero in the space $L^p(\mu)$, one has the equality $\lim\limits_{j \to \infty} \|f_j\|_{p,1,\mu} = 0$.*

If the indicated norm is closable, then the space $W^{p,1}(\mu)$ is a Banach space continuously embedded into $L^p(\mu)$.

Similarly, one defines the space $W^{p,r}(\mu)$ as the completion of $C_0^\infty(\mathbb{R}^n)$ with respect to the Sobolev norm

$$\|f\|_{p,r,\mu} = \|f\|_{L^p(\mu)} + \sum_{1 \leqslant |\alpha| \leqslant r} \|\partial^{(\alpha)} f\|_{L^p(\mu)}.$$

The closability of this norm is defined analogously. We observe that the closability of the norm $\|\cdot\|_{p,1,\mu}$ implies the closability of all norms $\|\cdot\|_{p,r,\mu}$.

2.6. Weighted Sobolev classes

Moreover, all spaces $W^{p,r}(\mu)$ are continuously embedded into $L^p(\mu)$. Indeed, if $\|f_j\|_{L^p(\mu)} \to 0$ and $\partial^{(\alpha)} f_j \to g^{(\alpha)}$ whenever $|\alpha| \leqslant r$, then $g^{(\alpha)} = 0$ for $|\alpha| = 1$, whence by induction we obtain the equality $g^{(\alpha)} = 0$ for $|\alpha| \leqslant r$.

2.6.3. Example. (i) Let $n = 1$, $p \in (1, \infty)$ and let μ have a density ϱ with respect to Lebesgue measure such that the function $\varrho^{1/(1-p)}$ is locally integrable with respect to Lebesgue measure. Then the class $W^{p,1}(\mu)$ is well defined. Indeed, convergence in $L^p(\mu)$ along with Hölder's inequality

$$\int_a^b |f(x)|\, dx \leqslant \left(\int_a^b |f(x)|^p\, \varrho(dx)\right)^{1/p} \left(\int_a^b \varrho(x)^{-q/p}\, dx\right)^{1/q}$$

yields convergence in $L^1[a,b]$. Therefore, if smooth functions f_j converge to zero in $L^p(\mu)$ and the functions f_j' converge in $L^p(\mu)$ to a function g, then $f_j \to 0$ and $f_j' \to g$ in $L^1[a,b]$ for every $[a,b]$. This gives the equality $g = 0$ since for any $\psi \in C_0^\infty(\mathbb{R}^1)$ we have

$$\int g(x)\psi(x)\, dx = \lim_{j \to \infty} \int f_j'(x)\psi(x)\, dx = -\lim_{j \to \infty} \int f_j(x)\psi'(x)\, dx = 0.$$

(ii) If $n = 1$ and the class $W^{p,1}(\mu)$ is well-defined, then the measure μ is absolutely continuous. Indeed, otherwise there exists a compact set K of Lebesgue measure zero with $\mu(K) > 0$. We construct a sequence $\{f_i\}$ such that it is fundamental in norm $\|\cdot\|_{p,1,\mu}$ and converges to zero in $L^p(\mu)$, but the functions f_i' do not converge to zero in $L^p(\mu)$. We may assume that $K \subset (0,1)$. Let us take a sequence of functions $\varphi_i \in C_0^\infty((0,1))$ decreasing to I_K and set $\psi_i(x) := \varphi_i(x) - \varphi_i(x-1)$,

$$f_i(x) := \int_{-\infty}^x \psi_i(y)\, dy.$$

Then $f_i \in C_0^\infty(\mathbb{R}^1)$, $f_i = 0$ outside $(0,2)$ and $\sup_{i,x} |f_i(x)| < \infty$. Since

$$\lim_{i \to \infty} \psi_i(x) = g(x) := I_K(x) - I_K(x-1),$$

one has $f_i(x) \to 0$. Hence $\|f_i\|_{L^p(\mu)} \to 0$. In addition, $f_i'(x) = \psi_i(x) \to g(x)$ and $\sup_{i,x} |\psi_i(x)| < \infty$, which yields convergence $f_i' \to g$ in $L^p(\mu)$. Since $g(x) = 1$ whenever $x \in K$, one has $\|g\|_{L^p(\mu)} > 0$.

However, the absolute continuity of μ even combined with the positivity on all nonempty open sets still does not guarantee the closability of Sobolev norms.

For any measurable function ϱ let $S(\varrho)$ denote the set of all points x in no neighborhood of which the function $1/\varrho$ is integrable. It is readily seen that this set is closed.

2.6.4. Theorem. *Let μ be a nonnegative locally finite Borel measure on $\mathbb{R}^1$ and $p > 1$. The norm $\|\cdot\|_{p,1,\mu}$ is closable precisely when the measure μ is absolutely continuous and its density ϱ vanishes almost everywhere on $S(\varrho^{1/(p-1)})$.*

PROOF. Let $\mu = \varrho \, dx$, where the function ϱ vanishes almost everywhere on the set $S(\varrho^{1/(p-1)})$. Suppose that functions $f_n \in C_0^\infty(\mathbb{R}^1)$ converge to zero in $L^p(\mu)$ and the functions f_n' converge to g in $L^p(\mu)$. We show that $g = 0$ μ-a.e. Passing to a subsequence we may assume that $f_n(x) \to 0$ μ-a.e. Let $(a,b) \subset \mathbb{R}^1 \backslash S(\varrho^{1/(p-1)})$, where $f_n(a) \to 0$, $f_n(b) \to 0$. Then one has
$$\left| \int_a^b f_n'(x) \, dx \right| \leqslant |f_n(b) - f_n(a)| \to 0,$$
and the integral of the function $|f_n'(x) - g(x)|$ is estimated by
$$\left(\int_a^b |f_n'(x) - g(x)|^p \varrho(x) \, dx \right)^{1/p} \left(\int_a^b \varrho(x)^{1/(1-p)} \, dx \right)^{1-1/p} \to 0,$$
whence we obtain $\int_a^b g(x) \, dx = 0$. Hence $g(x) = 0$ a.e. outside $S(\varrho^{1/(p-1)})$. So $g(x) = 0$ μ-a.e. Let us prove the converse. The necessity of the absolute continuity of μ was already proven above. We show that ϱ must vanish on $S(\varrho^{1/(p-1)})$ by modifying the proof given in the case $p = 2$ in Fukushima, Oshima, Takeda [**466**, Theorem 3.1.6]. If it is false that $\varrho = 0$ a.e. on the set $S(\varrho^{1/(p-1)})$, then there exists a number $\alpha > 0$ such that the set $S(\varrho^{1/(p-1)}) \cap \{\varrho \geqslant \alpha\}$ is of positive Lebesgue measure. We may assume that the set $A := S(\varrho^{1/(p-1)}) \cap \{\varrho \geqslant \alpha\} \cap (0,1)$ is of positive Lebesgue measure. Let $B := (0,1) \backslash A$ and $I_{k,n} := [k/n, (k+1)/n]$, where $n \in \mathbb{N}$, $0 \leqslant k \leqslant n-1$. We choose a sequence $\{\varphi_n\} \subset L^\infty(R^1)$ as follows: $\varphi_n = 0$ on $\mathbb{R}^1 \backslash (0,1)$, $\varphi_n = 1$ on A and $\varphi_n = 0$ on $B \cap I_{k,n}$ if $|A \cap I_{k,n}| = 0$, where $|E|$ denotes Lebesgue measure of E. It is clear that the integral of $\varrho^{-\frac{1}{p-1}}$ over $B \cap I_{k,n}$ is infinite. Hence there exists a number $M_n > 0$ such that
$$\int_{B \cap I_{k,n}} M_n \wedge \varrho^{-\frac{1}{p-1}} \, dx \geqslant 1 \quad \text{whenever } k = 1, \ldots, n-1,$$
where $s \wedge t := \min(s,t)$. Let
$$\varphi_n := \beta_{n,k} \left(M_n \wedge \varrho^{-\frac{1}{p-1}} \right) \quad \text{on } B \cap I_{k,n} \text{ if } |A \cap I_{k,n}| > 0,$$
where a number $\beta_{n,k} \in (-1, 0)$ is chosen in a such a way that
$$|A \cap I_{k,n}| + \beta_{n,k} \int_{B \cap I_{k,n}} M_n \wedge \varrho^{-\frac{1}{p-1}} \, dx = 0.$$
Let us observe that
$$\int_{I_{k,n}} \varphi_n \, dx = 0 \quad \text{whenever } 0 \leq k \leqslant n-1.$$
In addition, we have the inequality
$$\int_{B \cap I_{k,n}} |\varphi_n|^p \varrho \, dx \leqslant \frac{1}{n^p} \quad \text{whenever } 0 \leqslant k \leqslant n-1,$$

2.6. Weighted Sobolev classes

since by our choice of $\beta_{n,k}$ and M_n we have

$$\int_{B\cap I_{k,n}} |\varphi_n|^p \varrho \, dx = |\beta_{n,k}|^p \int_{B\cap I_{k,n}} \left[M_n \wedge \varrho^{-\frac{1}{p-1}}\right]^p \varrho \, dx$$

$$\leqslant |\beta_{n,k}|^p \int_{B\cap I_{k,n}} M_n \wedge \varrho^{-\frac{1}{p-1}} \, dx = |\beta_{n,k}|^{p-1} |A \cap I_{k,n}| \leqslant |A \cap I_{k,n}|^p$$

due to the estimate $\left(M_n^p \wedge \varrho^{-\frac{p}{p-1}}\right)\varrho \leqslant M_n \wedge \varrho^{-\frac{1}{p-1}}$. Indeed, the latter is true if $M_n^p \geqslant \varrho^{-\frac{p}{p-1}}$ and if $M_n^p \leqslant \varrho^{-\frac{p}{p-1}}$. Now let

$$\psi_n(x) := \int_0^x \varphi_n(t) \, dt, \quad x \in \mathbb{R}^1.$$

The functions ψ_n are Lipschitzian and possess the following properties:
(a) $\|\psi_n\|_{L^p(\mu)} \to 0$;
(b) $\|\psi'_n - I_A\|_{L^p(\mu)} \to 0$, $\mu(A) \geqslant \alpha|A| > 0$.
This is seen from the following estimates:

$$|\psi_n(x)| \leqslant \int_{I_{k,n}} |\varphi_n| \, dx = 2 \int_{I_{k,n}} I_A \varphi_n \, dx = 2 \int_{I_{k,n}} I_A \, dx \leqslant \frac{2}{n},$$

$$\|\psi'_n - I_A\|_{L^p(\mu)}^p = \int_0^1 |\varphi_n - I_A|^p \varrho \, dx = \int_B |\varphi_n|^p \varrho \, dx \leqslant \frac{1}{n^{p-1}}.$$

Now we can replace ψ_n by functions from $C^\infty(\mathbb{R}^1)$ with analogous properties (for example, by means of convolutions). $\square$

The condition indicated in the theorem is called *Hamza condition* (H) in the case $p = 2$ (see Hamza [515]). For $p \neq 2$ we shall call it condition (H_p). The presented justification was suggested by A.S. Tregubov.

2.6.5. Corollary. *In the previous theorem, $W^{p,r}(\mu)$ is a Banach space continuously embedded in $L^p(\mu)$.*

In the case of an absolutely continuous measure $\mu = \varrho \, dx$ on $\mathbb{R}^n$ an obvious sufficient condition for the closability of norm $\|\cdot\|_{p,1,\mu}$ is the closability of the norms $N_i \colon f \mapsto \|f\|_{L^p(\mu)} + \|\partial_{x_i} f\|_{L^p(\mu)}$ generated by partial derivatives. For the closability of the norms N_i it suffices to have condition (H_p) for the functions $x_i \mapsto \varrho(x_1, \ldots, x_n)$ for almost all fixed $(x_1, \ldots, x_{i-1}, x_{i+1}, \ldots, x_n)$. Pugachev [932], [934] constructed an example showing that the closability of norm $\|\cdot\|_{2,1,\mu}$ for the measure $\mu = \varrho \, dx$ on $\mathbb{R}^2$ does not imply the closability of the partial norms N_i. The problem of the existence of such examples was open for a long time. In this relation note that the following problem raised by M. Fukushima more than 30 years ago is still open: Does there exist a singular measure μ on $\mathbb{R}^2$ for which the norm $\|\cdot\|_{2,1,\mu}$ is closable? In place of condition (H_p) with respect to all arguments one can use the following stronger condition: there exists a closed set $Z \subset \mathbb{R}^n$ of measure zero such that the function $\varrho^{-1/(p-1)}$ is locally integrable on $\mathbb{R}^n \backslash Z$. Sufficiency of this condition is verified in the same manner as in Example 2.6.3.

Even in the case of an absolutely continuous measure $\mu = \varrho\, dx$ with closable Sobolev norms other problems can arise. For example, apart from the class $W^{p,1}(\mu)$ obtained as the completion of $C_0^\infty(\mathbb{R}^n)$, there is the space $V^{p,1}(\mu)$ of all functions $u \in W^{1,1}_{\mathrm{loc}}(\mathbb{R}^n)$ such that $\|u\|_{p,1,\mu} < \infty$. In general, $W^{p,1}(\mu)$ is strictly contained in the class $V^{p,1}(\mu)$ (see Zhikov [**1222**]). A sufficient condition for the equality $W^{2,1}(\mu) = V^{2,1}(\mu)$ is membership of ϱ in the Muckenhoupt class A_2, i.e., the estimate

$$\sup_B \left(\frac{1}{\lambda_n(B)} \int_B \varrho\, dx \right) \left(\frac{1}{\lambda_n(B)} \int_B \varrho^{-1}\, dx \right) < \infty,$$

where sup is taken over all balls (see [**1222**]). However, this condition is very far from being necessary. According to De Giorgi's conjecture, which remains unproved, if $\exp(t\varrho), \exp(t\varrho^{-1}) \in L^1_{\mathrm{loc}}(\mathbb{R}^n)$ for all $t > 0$, then $W^{2,1}(\mu) = V^{2,1}(\mu)$. The question of completeness of $V^{p,1}(\mu)$ also arises. If we have $\varrho^{-1/(p-1)} \in L^1_{\mathrm{loc}}(\mathbb{R}^n)$, then the space $V^{p,1}(\mu)$ is complete (Exercise 2.8.10).

There is one more natural Sobolev class (probably, more natural than the class $V^{p,1}(\mu)$, which will be especially clear in the infinite dimensional case in Chapter 8). Let $e_1, \ldots, e_n$ be the standard basis in $\mathbb{R}^n$.

2.6.6. Definition. *Let $\mu = \varrho\, dx$, where $\varrho \in W^{1,1}_{\mathrm{loc}}(\mathbb{R}^n)$. Let $f \in L^1(\mu)$. A function $g_i \in L^1(\mu)$ will be called a generalized derivative of f along e_i if we have $f\partial_{x_i}\varrho \in L^1_{\mathrm{loc}}(\mathbb{R}^n)$ and*

$$\int_{\mathbb{R}^n} \partial_{x_i}\zeta f\, \varrho\, dx = -\int_{\mathbb{R}^n} \zeta g_i \varrho\, dx - \int_{\mathbb{R}^n} \zeta f \partial_{x_i}\varrho\, dx \qquad (2.6.1)$$

for all $\zeta \in C_0^\infty(\mathbb{R}^n)$. Let $\partial_{e_i} f := g_i$. Denote by $G^{p,1}(\mu)$ the class of all functions $f \in L^p(\mu)$ such that $\partial_{x_i} f \in L^p(\mu)$ whenever $i = 1, \ldots, n$.

It is clear that the function g_i is unique up to μ-equivalence. The space $G^{p,1}(\mu)$ is equipped with the Sobolev norm $\|\cdot\|_{p,1,\mu}$.

2.6.7. Lemma. *Let $|\nabla\varrho|/\varrho \in L^{p'}_{\mathrm{loc}}(\mu)$, where $p' = p/(p-1)$. Then $G^{p,1}(\mu)$ is a Banach space.*

PROOF. If $f_k \to f$ in $L^p(\mu)$ and $\partial_{e_i} f_k \to g_i$ in $L^p(\mu)$, then we have $f\partial_{x_i}\varrho \in L^1_{\mathrm{loc}}(\mathbb{R}^n)$ by Hölder's inequality and our assumption on ϱ. Hence equality (2.6.1) is true. $\square$

2.6.8. Theorem. *Let $\mu = \varrho\, dx$, where $\varrho \in W^{1,1}_{\mathrm{loc}}(\mathbb{R}^n)$. A function f from $L^p(\mu) \cap L^1_{\mathrm{loc}}(|\nabla\varrho|\, dx)$ belongs to $G^{p,1}(\mu)$ precisely when for every e_i it has a modification $\widetilde{f}$ with the following properties: for almost every $y \in e_i^\perp$, the function $t \mapsto \widetilde{f}(y + te_i)$ is locally absolutely continuous on the set $\{t: \varrho(y + te_i) > 0\}$, where we consider a version of ϱ which is absolutely continuous on the straight lines $y + \mathbb{R}^1 e_i$, and the function $\partial_{x_i}\widetilde{f}$, which exists μ-a.e., belongs to $L^p(\mu)$.*

PROOF. Let f have a modification with the stated properties, which we denote also by f. For verification of formula (2.6.1), by the existence of all integrals and Fubini's theorem, it suffices to observe that, for every interval $[a,b]$ in which the function $\varrho(y+te_i)$ equals zero only at the endpoints, we have

$$\int_a^b \partial_{x_i}\zeta(y+te_i)f(y+te_i)\varrho(y+te_i)\,dt$$

$$= -\int_a^b \zeta(y+te_i)\partial_{x_i}f(y+te_i)\varrho(y+te_i)\,dt$$

$$-\int_a^b \zeta(y+te_i)f(y+te_i)\partial_{x_i}\varrho(y+te_i)\,dt$$

for all y such that $f(y+te_i)$ is absolutely continuous on all inner intervals $[a+\varepsilon, b-\varepsilon]$ and the indicated integrals exist (which holds for almost all $y \in e_i^{\perp}$). This is seen from the integration by parts formula for $[a+\varepsilon, b-\varepsilon]$ and the fact that there exist two sequences $a_j \downarrow a$ and $b_j \uparrow b$ such that $f(y+a_je_i)\varrho(y+a_je_i) \to 0$ and $f(y+b_je_i)\varrho(y+b_je_i) \to 0$. Indeed, if we had $f(y+te_i)\varrho(y+te_i) \geqslant c > 0$ for $t \in (a, a+\varepsilon)$, then by the integrability of $f(y+te_i)\partial_{x_i}\varrho(y+te_i)$ on $[a, a+\varepsilon]$ the function $\partial_{x_i}\varrho(y+te_i)/\varrho(y+te_i)$ would be also integrable on that interval, which would yield that $\varrho(y+ae_i) > 0$. Let us prove the converse. Let $f \in G^{p,1}(\mu)$. Equality (2.6.1) means that $f\varrho \in W_{\mathrm{loc}}^{1,1}(\mathrm{I\!R}^n)$ and the corresponding Sobolev derivatives have the form $\partial_{x_i}(f\varrho) = g_i\varrho + f\partial_{x_i}\varrho$. For any fixed i we take versions of ϱ and $f\varrho$ that are locally absolutely continuous along e_i. For a desired version of f we take $(f\varrho)/\varrho$. On the set $\{\varrho > 0\}$ we have $\partial_{x_i}f = [\partial_{x_i}(f\varrho)\varrho - (f\varrho)\partial_{x_i}\varrho]\varrho^{-2} = g_i$, which completes the proof. □

Let us consider compositions and products of Sobolev functions with continuously differentiable functions.

2.6.9. Lemma. (i) If $f \in W^{p,1}(\mu)$, then we have $g \cdot f \in W^{p,1}(\mu)$ and $\varphi \circ f \in W^{p,1}(\mu)$ for all $g \in C_b^1(\mathrm{I\!R}^n)$ and $\varphi \in C_b^1(\mathrm{I\!R}^1)$.

(ii) If $f \in G^{p,1}(\mu)$, then $g \cdot f \in G^{p,1}(\mu)$ for all $g \in C_b^1(\mathrm{I\!R}^n)$. If $\mu = \varrho\,dx$, where $\varrho \in W_{\mathrm{loc}}^{1,1}(\mathrm{I\!R}^n)$, then $\varphi \circ f \in G^{p,1}(\mu)$ for all $\varphi \in C_b^1(\mathrm{I\!R}^1)$.

PROOF. Let $f \in G^{2,1}(\mu)$ and $g \in C_b^1(\mathrm{I\!R}^n)$. Then for any $\zeta \in C_0^\infty(\mathrm{I\!R}^n)$ we have

$$\int_{\mathrm{I\!R}^n} gf\partial_{x_i}\zeta\,\mu(dx) = \int_{\mathrm{I\!R}^n} f\partial_{x_i}(g\zeta)\,\mu(dx) - \int_{\mathrm{I\!R}^n} f\zeta\partial_{x_i}g\,\mu(dx)$$

$$= -\int_{\mathrm{I\!R}^n} \zeta[gf\beta_i + g\partial_{e_i}f + f\partial_{x_i}g]\,\mu(dx).$$

Since $gf, g\partial_{e_i}f + f\partial_{x_i}g \in L^p(\mu)$, one has $g \cdot f \in G^{p,1}(\mu)$. For the proof of an analogous fact for $W^{p,1}(\mu)$ it suffices to observe that if functions f_j from $C_0^\infty(\mathrm{I\!R}^n)$ converge to f in the Sobolev norm, then $g \cdot f_j \in W^{p,1}(\mu)$ and $g \cdot f_j \to g \cdot f$ in $W^{p,1}(\mu)$. Similarly, we verify the inclusion $\varphi \circ f \in W^{p,1}(\mu)$.

If μ is given by a density $\varrho \in W^{1,1}_{\text{loc}}(\mathbb{R}^n)$, then the inclusion $\varphi \circ f \in G^{p,1}(\mu)$ for $\varphi \in C^1_b(\mathbb{R}^1)$ is obvious from Theorem 2.6.8. □

2.6.10. Corollary. *The set of functions with bounded support is dense in the classes $W^{p,1}(\mu)$ and $G^{p,1}(\mu)$.*

PROOF. Let f be a function of one of these classes. It is readily seen that f is approximated in the Sobolev norm by the functions $\chi_k f$, where $\chi_k \in C_0^\infty(\mathbb{R}^n)$, $0 \leqslant \chi_k \leqslant 1$, $\chi_k(x) = 1$ if $|x| \leqslant k$, and $\sup_{k,x} |\nabla \chi_k(x)| < \infty$. In addition, the function f is a limit of functions $\theta_k \circ f \in G^{2,1}(\mu)$, where the functions $\theta_k \in C_b^\infty(\mathbb{R}^1)$ are odd, $\theta_k(t) = t$ if $|t| \leqslant k$, $\theta_k(t) = k+1$ if $t \geqslant k+1$, and $\sup_{k,t} |\theta'_k(t)| < \infty$. □

The next result in the case $p = 2$ was obtained in Röckner, Zhang [**963**] and later another proof was given in Cattiaux, Fradon [**270**]. We present a slightly simplified modification of the latter approach suitable for all p.

2.6.11. Theorem. *Let $\mu = \psi^p \, dx$, where $\psi \in W^{p,1}_{\text{loc}}(\mathbb{R}^n)$. The set $C_0^\infty(\mathbb{R}^n)$ is dense in the space $G^{p,1}(\mu)$, hence $G^{p,1}(\mu) \subset W^{p,1}(\mu)$. If one has $|\nabla \psi|/\psi \in L^{p'}_{\text{loc}}(\mu)$, where $p' = p/(p-1)$, then $G^{p,1}(\mu) = W^{p,1}(\mu)$.*

PROOF. It is clear from what has been proven above that it suffices to deal with a function f that is bounded and has bounded support. Multiplying μ by a smooth function with compact support we pass to the case where ψ vanishes outside the ball of radius 2 centered at the origin. Let us take a smooth probability density g with support in the unit ball and set $f_k := f * g_k$, where $g_k(x) = k^n g(kx)$. Since the functions f_k are uniformly bounded and converge a.e. to f, one has $\|f - f_k\|_{L^p(\mu)} \to 0$. Let $\partial_{x_i} f$ be derivatives in $G^{p,1}(\mu)$. We show that $\partial_{x_i} f_k \to \partial_{x_i} f$ in $L^p(\mu)$, which can be written as $\psi \partial_{x_i} f_k \to \psi \partial_{x_i} f$ in $L^p(\mathbb{R}^n)$. Let us denote the norm in $L^p(\mathbb{R}^n)$ by $\|\cdot\|_p$. The quantity $\|\psi \partial_{x_i} f - \psi \partial_{x_i} f_k\|_p$ does not exceed the sum of the quantities $\|\psi \partial_{x_i} f - g_k * (\psi \partial_{x_i} f)\|_p$ and

$$R_k(f) := \|g_k * (\psi \partial_{x_i} f) - \psi \partial_{x_i}(g_k * f)\|_p,$$

where the first one tends to zero as $k \to \infty$ because $\psi \partial_{x_i} f \in L^p(\mathbb{R}^n)$. Let us show that $R_k(f) \to 0$. By the definition of the convolution we have

$$R_k(f) = \int_{\mathbb{R}^n} \left| \int_{\mathbb{R}^n} \partial_{x_i} g_k(x-y) f(y) [\psi(x) - \psi(y)] \, dy \right.$$
$$\left. + \int_{\mathbb{R}^n} [\partial_{x_i} g_k(x-y) f(y) \psi(y) - g_k(x-y) \psi(y) \partial_{y_i} f(y) \, dy \right|^p dx.$$

Let us observe that for every function $\zeta \in C_0^\infty(\mathbb{R}^n)$ one has the equality

$$\int_{\mathbb{R}^n} [\zeta \psi \partial_{x_i} f + \partial_{x_i} \psi \zeta f + f \psi \partial_{x_i} \zeta] \, dx = 0.$$

Indeed, all three products are integrable. Evaluating the integral by Fubini's theorem we see that the integral in the variable x_i is zero. Indeed, the integral of $\zeta\psi\partial_{x_i}f$ over every interval $[a,b]$ in which ψ vanishes only at the end points equals the integral over $[a,b]$ of the function $\partial_{x_i}\psi\zeta f + f\psi\partial_{x_i}\zeta$, where we consider a version of f which is locally absolutely continuous with respect to the variable x_i on those intervals in the variable x_i where ψ is positive. Applying this equality to $\zeta(y) = g_k(x - y)$ we obtain

$$R_k(f) = \tag{2.6.2}$$
$$= \int_U \left| \int_U \big(\partial_{x_i} g_k(x-y) f(y)[\psi(x) - \psi(y)] - g_k(x-y) f(y) \partial_{y_i} \psi(y)\big) \, dy \right|^p dx$$
$$= \int_U \left| \int_U \big(\partial_{u_i} g(u) f(y) k[\psi(y + k^{-1}u) - \psi(y)] - g(u) f(y) \partial_{y_i} \psi(y)\big) \, dy \right|^p du,$$

where U is the ball of radius 3 centered at the origin. By Exercise 2.8.13 the functions $\eta_k(y,u) := k[\psi(y+k^{-1}u) - \psi(y)]$ converge in $L^p(U \times U)$. According to Exercise 2.8.14 we have $\lim_{j\to\infty} \sup_k R_k(f_j - f) = 0$. For any fixed j we have $f_j \in C_0^\infty(U)$, hence $\lim_{k\to\infty} R_k(f_j) = 0$, which is obvious from the determining formula for $R_k(f)$. Let $\varepsilon > 0$. Let us take j_1 such that $\sup_k R_k(f_{j_1} - f) < \varepsilon$. Since $R_k(f) \leqslant R_k(f - f_{j_1}) + R_k(f_{j_1})$, it remains to take a number k_1 such that $R_k(f_{j_1}) \leqslant \varepsilon$ for all numbers $k \geqslant k_1$. $\square$

2.6.12. Remark. The assumption on the measure μ in this theorem is equivalent to the following one: $\mu = \varrho \, dx$, where $\varrho \in W^{1,1}_{\text{loc}}(\mathbb{R}^n)$ and $|\nabla \varrho|/\varrho \in L^p_{\text{loc}}(\mu)$. Indeed, if $\psi \in W^{p,1}_{\text{loc}}(\mathbb{R}^n)$, then

$$\varrho := |\psi|^p \in W^{1,1}_{\text{loc}}(\mathbb{R}^n) \quad \text{and} \quad |\nabla \varrho| = p|\nabla\psi| |\psi|^{p-1} \in L^1_{\text{loc}}(\mathbb{R}^n)$$

by Hölder's inequality. In addition, we have $|\nabla \varrho|^p \varrho^{1-p} = p^p |\nabla \psi|^p$. For the proof of the converse let us set $\psi = \varrho^{1/p}$ and observe that we have the inclusions $\psi \in L^p_{\text{loc}}(\mathbb{R}^n)$ and $|\nabla \psi| \in L^p_{\text{loc}}(\mathbb{R}^n)$. In order to verify the latter we consider the function $(\varrho + \varepsilon)^{1/p}$ and pass to the limit as $\varepsilon \to 0$.

2.7. Fractional Sobolev classes

Here we briefly discuss continuous scales of Sobolev classes. Most of the methods of constructing such classes employ explicitly or implicitly the interpolation methods. Considering the Sobolev classes $W^{2,r}(\mathbb{R}^n)$ with $r \in \mathbb{N}$, which are characterized in terms of the Fourier transform by the inclusion $(1+|x|^2)^{r/2} \widehat{f} \in L^2(\mathbb{R}^n)$, we have observed that the latter condition is meaningful for all r and enables one to introduce the class of generalized functions $f \in \mathcal{S}'(\mathbb{R}^n)$ such that $\widehat{f}$ in the sense of generalized functions is given by a usual function satisfying the indicated condition. Here $r \in (-\infty, +\infty)$. In the case $p \neq 2$ the Fourier transform is a more complicated object, however, also here the described idea can be implemented. To this end for any $p \in (1, +\infty)$ and $r \in (-\infty, +\infty)$ we denote by $W^{p,r}(\mathbb{R}^n)$ the class of generalized functions $f \in \mathcal{S}'(\mathbb{R}^n)$ such that the Fourier transform of the function f in $\mathcal{S}'(\mathbb{R}^n)$ satisfies the following condition: $\mathcal{F}^{-1}[(1+|x|^2)^{r/2} \mathcal{F}(f)] \in L^p(\mathbb{R}^n)$, i.e., $(I-\Delta)^{r/2}f \in L^p(\mathbb{R}^n)$. If $r = 2k$ is natural and even, then $W^{p,r}(\mathbb{R}^n)$ consists of all functions $f \in L^p(\mathbb{R}^n)$ such that $(I-\Delta)^k f \in L^p(\mathbb{R}^n)$, where the left-hand side is understood in the sense of generalized functions. It is known that for any $r \in \mathbb{N}$ the class $W^{p,r}(\mathbb{R}^n)$ defined in this way coincides with the earlier defined Sobolev space (however, the proof of this fact is not simple at all, unlike the case $p = 2$, see Stein [**1079**, Ch. V]).

There are many other scales of Sobolev spaces (see a detailed account in Triebel [**1123**]), in particular, especially important Besov spaces $B^{p,q,s}(\mathbb{R}^n)$ defined as follows. Let $1 < p < \infty$, $1 \leq q < \infty$, $0 < s < m$, where $m \in \mathbb{N}$ (for example, if s is fractional, then let $m-1$ be the integer part of s). Set

$$\Delta_h f(x) := f(x) - f(x-h).$$

The class $B^{p,q,s}(\mathbb{R}^n)$ consists of all functions $f \in L^p(\mathbb{R}^n)$ such that the function $h \mapsto |h|^{-n-sq}\|(\Delta_h)^m f\|_{L^p}^q$ is integrable over $\mathbb{R}^n$. For example, if we have $0 < s < 1$ and $m = 1$, then the integrability of the function

$$h \mapsto |h|^{-n-sq}\|f(\,\cdot\,) - f(\,\cdot\,-h)\|_{L^p}^q$$

is required.

The Besov spaces enable one to describe completely the restrictions of Sobolev functions to surfaces. For example, if $r \in \mathbb{N}$, $rp < n+1$ and $p \leq q \leq np/(n+1-rp)$, then there is continuous linear restriction operator $R \colon W^{p,r}(\mathbb{R}^{n+1}) \to L^q(\mathbb{R}^n)$, which is the usual restriction on smooth functions and the image of this operator is exactly the Besov class $B^{p,p,r-1/p}(\mathbb{R}^n)$.

Some other constructions will be considered in Chapter 8.

2.8. Comments and exercises

The most suitable texts for the first acquaintance with Sobolev spaces are the comprehensible and well-written books, Evans, Gariepy [**400**] and Adams [**2**] (the second edition: Adams, Fournier [**3**]). Brief introductions to this theory are given in many texts on functional analysis and modern theory of partial differential equations. A thorough study of diverse aspects of the theory can be found in the books Besov, Il'in, Nikolskiĭ [**145**], Gol'dshteĭn, Reshetnyak [**490**], Maz'ja [**795**], Nikolskiĭ [**829**], Stein [**1079**], Stein, Weiss [**1080**], Triebel [**1123**], Kufner, John, Fučik [**650**], and Ziemer [**1224**]. Fractional Sobolev classes are discussed in Adams [**2**], Adams, Fournier [**3**], Stein [**1079**], Stein, Weiss [**1080**], and Triebel [**1123**]. Weighted Sobolev classes are studied in Kufner [**649**]. Concerning Sobolev classes on manifolds, see Hebey [**526**]. Multidimensional transformations by Sobolev class mappings and related problems are studied in Gol'dshteĭn, Reshetnyak [**490**] and Reshetnyak [**953**]. Functions of bounded variation are considered in Giusti [**482**], and Ambrosio, Fusco, Pallara [**65**].

2.8. Comments and exercises

Exercises

2.8.1. Prove that for every absolutely continuous function φ on $[a, b]$ and every Lipschitzian function ψ on the real line the function $\psi \circ \varphi$ is absolutely continuous as well and the equality $(\psi \circ \varphi)'(t) = \psi'(\varphi(t))\varphi'(t)$ holds almost everywhere on $[a, b]$.

2.8.2. Construct a function $f \in W^{1,1}(\mathbb{R}^2)$ which has no modification absolutely continuous in every variable separately.
HINT: First prove the following fact: If a function on $\mathbb{R}^2$ is continuous in every variable separately, then it has a point of continuity (use Baire's theorem).

2.8.3. Let ϱ be an absolutely continuous probability density on the real line. Prove the inequality
$$\int_{-\infty}^{+\infty} x^2 \varrho(x)\,dx \int_{-\infty}^{+\infty} \frac{|\varrho'(x)|^2}{\varrho(x)}\,dx \geqslant 1.$$
The integral of the function $x^2 \varrho(x)$ is called the second moment of ϱ, and the integral of the function $|\varrho'(x)|^2/\varrho(x)$ is called the *Fisher information* of ϱ.
HINT: If both integrals on the left are finite, then the function $x\varrho'(x)$ is integrable; use integration by parts to evaluate its integral and apply the Cauchy inequality with weight ϱ.

2.8.4. Let $f \in W^{1,1}(\mathbb{R}^n)$. Prove that the function f vanishes almost everywhere on the set $\{x\colon \nabla f(x) = 0\}$.
HINT: First show that this is true on the real line and then use a version that is absolutely continuous along almost all coordinate lines.

2.8.5. Prove that the space $W^{2,1}_{2\pi}[0,1]$ of all absolutely continuous complex functions f on $[0,1]$ with $f' \in L^2[0,1]$ and $f(0) = f(2\pi)$ is the range of the operator B on $L^2[0,1]$ with the eigenfunctions $\exp(ikt)$ corresponding to the eigenvalues $\beta_0 = 1$, $\beta_k = k^{-1}$, where $k \in \mathbb{Z}\setminus\{0\}$.
HINT: If $f \in W^{2,1}[0,1]$ and $f(t) = \sum_{k=-\infty}^{+\infty} c_k \exp(i2\pi kt)$, then
$$f'(t) = i2\pi \sum_{k=-\infty}^{+\infty} kc_k \exp(i2\pi kt);$$
conversely, if for the Fourier coefficients c_k of the function f series of $|kc_k|^2$ converges, then it is readily seen that $f \in W^{2,1}[0,1]$.

2.8.6. Determine whether the embedding $W^{1,1}\big((0,1)\big) \subset C(0,1)$ is compact.

2.8.7. Show that the spaces $W^{p,k}(\mathbb{R}^n)$, $1 \leqslant p < \infty$, are separable.

2.8.8. Let U be the half-space $\{x\colon x_1 > 0\}$ in $\mathbb{R}^n$. Prove that to every function $f \in W^{p,1}(U)$ one can associate a function $Tf \in W^{p,1}(\mathbb{R}^n)$ such that the operator T is linear and continuous and $Tf = f$ if f is the restriction to U of a smooth function which is even with respect to the variable x_1.

2.8.9. Let $\alpha \in (0,1)$. Construct an increasing function on $[0,1]$ which is Hölder of order $\alpha < 1$ and takes some measure zero set to a set of positive measure. Show that this holds for the distribution function of the random variable $\xi = \sum_{n=1}^{\infty} 2^{-n}\xi_n$, where $\{\xi_n\}$ is a sequence of independent equally distributed random variables taking the value 1 with probability p and the value 0 with probability $1-p$, where $p \in (0,1)$ and $p \neq 1/2$.

2.8.10. Let a nonnegative measure μ on $\mathbb{R}^n$ be given by a density ϱ such that
$$\varrho^{-1/(p-1)} \in L^1_{\mathrm{loc}}(\mathbb{R}^n).$$
Prove that the space $V^{p,1}(\mu)$ of all functions $u \in W^{1,1}_{\mathrm{loc}}(\mathbb{R}^n)$ with $\|u\|_{p,1,\mu} < \infty$ is complete with respect to the norm $\|\cdot\|_{p,1,\mu}$.

2.8.11. (Hajłasz [513]) Let $f \in W^{1,1}_{\mathrm{loc}}(\mathbb{R}^n)$. Then there exist increasing compact sets K_j and Lipschitzian functions g_j on $\mathbb{R}^n$ such that $f|_{K_j} = g_j|_{K_j}$ for all j and
$$\mathbb{R}^n \setminus \bigcup_{j=1}^\infty K_j \subset \{x\colon \mathcal{M}|\nabla f|(x) = \infty\},$$
where $\mathcal{M}\psi$ denotes the Hardy–Littlewood maximal function of ψ.

2.8.12. Let $f \in W^{1,1}_{\mathrm{loc}}(\mathbb{R}^n)$. Prove that the equality
$$f(y) - f(x) = \int_0^1 \Big(\nabla f(x + t(y-x)), y - x\Big)\, dt$$
holds for almost all pairs $(x, y) \in \mathbb{R}^n \times \mathbb{R}^n$.

HINT: Use that this is true for smooth functions and holds in the limit when one takes a sequence of smooth functions convergent to f in $W^{1,1}(U)$ and almost everywhere on U, where U is a fixed ball.

2.8.13. Let $f \in W^{p,1}(\mathbb{R}^n)$ and $\varepsilon_k \to 0$. Prove that the sequence of functions
$$\varphi_k(x, y) := \varepsilon_k^{-1}[f(x + \varepsilon_k y) - f(x)]$$
converges in $L^p(\mathbb{R}^n \times U)$ to $\big(\nabla f(x), y\big)$, where U is a ball in $\mathbb{R}^n$.

HINT: Use the previous exercise to represent $f(x + \varepsilon_k y) - f(x)$.

2.8.14. Let $\varphi_j \to \varphi$ in $L^p(\mu)$, $\sup_j \|\psi_j\|_{L^\infty(\mu)} < \infty$ and $\psi_j \to 0$ in measure μ. Prove that $\psi_j \varphi_j \to 0$ in $L^p(\mu)$. Deduce that, for every compact set $K \subset L^p(\mu)$, one has the equality $\lim\limits_{j \to \infty} \sup_{\varphi \in K} \|\psi_j \varphi\|_{L^p(\mu)} = 0$.

2.8.15. Suppose that $f \in W^{1,1}_{\mathrm{loc}}(\mathbb{R}^n)$ and $|\nabla f| \exp f \in L^1_{\mathrm{loc}}(\mathbb{R}^n)$. Prove that $\exp f \in L^1_{\mathrm{loc}}(\mathbb{R}^n)$.

HINT: Take the functions $f_k = \min(f, k)$, observe that the functions $|\nabla \exp f_k|$ are uniformly bounded in $L^1(U)$ for every ball U, and deduce that the same is true for the functions $\exp f_k$ using the fact that they have a finite pointwise limit.

2.8.16. Let a function f on $\mathbb{R}^d$ be such that f and all functions $x_i^k f$ belong to all classes $W^{1,r}(\mathbb{R}^d)$, $r \in \mathbb{N}$. Prove that $f \in \mathcal{S}(\mathbb{R}^d)$.

2.8.17. Give an example of a probability density on $\mathbb{R}^1$ which does not satisfy the Hamza condition.

2.8.18. Let $\varphi(t) = t^{1/3}$ if $t \geq 0$ and $\varphi(t) = -\varphi(-t)$ if $t < 0$. Let us consider the function $g(t) = \sum_{n=1}^\infty 2^{-n} \varphi(t - t_n)$, $t \in [0, 1]$, where $\{t_n\}$ is a dense set in $[0, 1]$. Show that the function g is continuous and strictly increasing and at every point has either a finite or infinite derivative. Prove that the inverse function $f = g^{-1}$ is Lipschitzian on $[g(0), g(1)]$ and differentiable at every point and that $f' = 0$ on a dense set.

2.8.19. Investigate weighted Besov classes on $\mathbb{R}^n$.

CHAPTER 3

Differentiable measures on linear spaces

In this chapter we consider the basic properties of differentiable measures on linear spaces. As everywhere in this book, the reader may assume that we are concerned with measures on Banach spaces or Souslin locally convex spaces or even on l^2 or $\mathbb{R}^\infty$, which simplifies some technical details. However, most of the definitions are given in their natural generality.

3.1. Directional differentiability

Let X be a linear space equipped with some σ-algebra $\mathcal{A}$ invariant with respect to the shifts along a given vector $h \in X$, i.e., $A + th \in \mathcal{A}$ for all $A \in \mathcal{A}$ and $t \in \mathbb{R}^1$. We are going to define certain properties of measures on $\mathcal{A}$ such as the differentiability, continuity, and analyticity along h. The most important example is the case where X is a locally convex space with one of our two standard σ-algebras $\mathcal{B}(X)$ or $\sigma(X)$. Most of the special results below will be obtained for these cases. The only purpose for our consideration in this generality is to emphasize that only linear and measurable structures are intrinsically connected with our basic concepts.

If μ is a measure on $\mathcal{A}$, then we set
$$\mu_h(A) := \mu(A + h).$$
The measure μ_h is the image of the measure μ under the mapping $x \mapsto x - h$. It is called the *shift of the measure* μ by h. The following identity is valid:
$$(f \cdot \mu)_h = f(\,\cdot\, + h) \cdot \mu_h, \quad f \in \mathcal{L}^1(\mu).$$
For the proof we use the relations
$$\int_X \varphi(x)\,(f \cdot \mu)_h(dx) = \int_X \varphi(x - h) f(x)\,\mu(dx) = \int_X \varphi(y) f(y + h)\,\mu_h(dy),$$
which are valid for every bounded $\mathcal{A}$-measurable function φ.

The following fundamental concept is due to S.V. Fomin [440]–[443].

3.1.1. Definition. *A measure μ on X is called differentiable along the vector h in Fomin's sense if, for every set $A \in \mathcal{A}$, there exists a finite limit*
$$d_h\mu(A) := \lim_{t \to 0} \frac{\mu(A + th) - \mu(A)}{t}. \tag{3.1.1}$$

Since in place of A in (3.1.1) one can substitute $A + t_0 h$, the function
$$t \mapsto \mu(A + th)$$
is differentiable at every point.

The set function $d_h\mu$ defined by (3.1.1) can be written as a pointwise limit of the sequence of measures $A \mapsto n\bigl(\mu(A + n^{-1}h) - \mu(A)\bigr)$. Therefore, by the Nikodym theorem recalled in §1.3 it is automatically a countably additive measure on $\mathcal{A}$ of bounded variation. Here $d_h\mu$ is always a signed measure: $d_h\mu(X) = 0$. Hence it is reasonable from the very beginning to admit signed measures.

The measure $d_h\mu$ defined by (3.1.1) is called the *derivative of the measure μ along the vector h* or *in the direction h* (it is also called *Fomin's derivative*).

Higher order derivatives $d_h^n\mu$ as well as mixed derivatives $d_{h_1}\cdots d_{h_n}\mu$ are defined inductively. This will be discussed below in §3.7.

3.1.2. Definition. *A measure μ on $\mathcal{A}$ is called continuous along the vector h if we have*
$$\lim_{t \to 0} \|\mu_{th} - \mu\| = 0. \tag{3.1.2}$$

It is clear that if the measure μ is continuous along the vector h, then for every set $A \in \mathcal{A}$ the function $t \mapsto \mu(A + th)$ is continuous. As we shall see below, the converse is also true.

3.1.3. Definition. *A measure μ on $\mathcal{A}$ is called quasi-invariant along the vector h if the measures μ_{th} and μ are equivalent for every real t.*

It is readily verified that a measure μ continuous or quasi-invariant along two vectors h and k has the same property also along their linear combinations.

Two other concepts of differentiability were introduced by Albeverio and Høegh-Krohn [41] and Skorohod [1046].

3.1.4. Definition. *We shall say that a nonnegative measure μ on $\mathcal{A}$ is differentiable along the vector h in the sense of Albeverio and Høegh-Krohn (or L^2-differentiable) if there exists a measure $\lambda \geq 0$ on $\mathcal{A}$ such that $\mu_{th} \ll \lambda$, $\mu_{th} = f_t \cdot \lambda$ and the mapping*
$$t \mapsto f_t^{1/2}$$
from $\mathbb{R}^1$ to $L^2(\lambda)$ is differentiable (with respect to the norm of $L^2(\lambda)$).

More generally, given $p \in [1, +\infty)$, let us define L^p-*differentiability* of the measure μ as the differentiability of the mapping $t \mapsto f_t^{1/p}$ from $\mathbb{R}^1$ to the space $L^p(\lambda)$, where we have $\mu_t = f_t \cdot \lambda$ for all t.

Let us note at once that L^2-differentiability implies Fomin's differentiability and the equality
$$d_h\mu = 2f_0^{1/2}\psi \cdot \lambda, \quad \text{where} \quad \psi := \frac{d}{dt}\sqrt{f_t}\Big|_{t=0}. \tag{3.1.3}$$

3.1. Directional differentiability

Indeed, for every $A \in \mathcal{A}$, the function

$$t \mapsto \mu_{th}(A) = \int_A f_t(x)\,\lambda(dx) = (I_A f_t^{1/2}, f_t^{1/2})_{L^2(\lambda)}$$

is differentiable and its derivative at zero is $2(I_A \psi, f_0^{1/2})_{L^2(\lambda)}$. It will be seen from the results in §3.3 that Fomin's differentiability of a nonnegative measure is equivalent to its L^1-differentiability.

3.1.5. Definition. *A Baire measure μ on a locally convex space X is called Skorohod differentiable or S-differentiable along the vector h if, for every function $f \in C_b(X)$, the function*

$$t \mapsto \int_X f(x - th)\,\mu(dx)$$

is differentiable. Skorohod differentiability of a Borel measure is understood as the differentiability of its restriction to the Baire σ-algebra.

It follows from this definition that, for every sequence $t_n \to 0$, the sequence of Baire measures $t_n^{-1}(\mu_{t_n h} - \mu)$ is fundamental in the weak topology in the space of measures. By a theorem due to A.D. Alexandroff there exists a Baire measure ν which is the weak limit of the indicated sequence (see [**193**, Theorem 8.7.1]). The measure ν is independent of our choice of $\{t_n\}$ since the integral of any bounded continuous function f with respect to the measure ν equals the derivative at zero of the function indicated in the definition. Thus, for every bounded continuous function f on X we have

$$\lim_{t \to 0} \int_X \frac{f(x - th) - f(x)}{t}\,\mu(dx) = \int_X f(x)\,\nu(dx). \quad (3.1.4)$$

3.1.6. Definition. *A Baire measure ν on X satisfying identity (3.1.4) is called the Skorohod derivative (or the weak derivative) of the measure μ along h.*

Thus, every Skorohod differentiable Baire measure has a weak derivative. However, the space of Radon measures need not be sequentially complete in the weak topology (certainly, for most of the spaces encountered in real applications such problems do not arise since in such spaces the Baire and Borel σ-algebras coincide and all measures are Radon). Nevertheless, it will be shown below that for any Radon measure μ the Skorohod derivative is Radon as well.

The natural concept of analyticity of a measure was introduced by Albeverio and Høegh-Krohn [**41**] and Bentkus [**131**], [**132**].

3.1.7. Definition. *A measure μ on $\mathcal{A}$ is called analytic along the vector h if, for every set $A \in \mathcal{A}$, the function $t \mapsto \mu(A + th)$ admits a holomorphic extension to a circle of the form $\{z\colon |z| < c\}$ independent of A.*

Replacing A by $A + t_0 h$ we see that for any analytic measure μ the function $t \mapsto \mu(A + th)$ has a holomorphic extension to the strip $\{z\colon |\operatorname{Im} z| < c\}$ independent of A.

3.1.8. Example. Let μ be a measure on $\mathbb{R}^1$ with a continuously differentiable density p and bounded support. Then the measure μ is continuous along every vector $h \in \mathbb{R}^1$ and is differentiable both in the sense of Fomin and Skorohod.

PROOF. Let $h = 1$. Then
$$\frac{\mu(A+t) - \mu(A)}{t} = \int_A \frac{p(x+t) - p(x)}{t}\, dx \to \int_A p'(x)\, dx \text{ as } t \to 0.$$

So the measure μ is Fomin differentiable. In addition, the measure $d_1\mu$ has density p'/p with respect to μ. Similarly, it is Skorohod differentiable. It is clear that
$$\lim_{t \to 0} \|\mu_t - \mu\| = \lim_{t \to 0} \int |p(x+t) - p(x)|\, dx = 0,$$
i.e., the measure μ is continuous. $\square$

Below we shall encounter less trivial examples (and also completely characterize continuous and differentiable measures on the real line). However, even this example exhibits two interesting properties of Fomin differentiable measures: the variation $|\mu|$ of a differentiable measure is again differentiable (in our example one can verify that the measure with density $|p|$ is differentiable) and the derivative $d_h\mu$ is absolutely continuous with respect to μ. It turns out that these two properties have a very general character.

Everywhere below the term "differentiable measure" will be used for the differentiability in the sense of Fomin and all other kinds of differentiability will be properly specified.

Now we give an important example of a differentiable measure on an infinite dimensional space.

3.1.9. Theorem. *Let γ be a centered Radon Gaussian measure on a locally convex space X. Then*

(i) *its Cameron–Martin space $H(\gamma)$ coincides with the collection of all vectors of continuity as well as with the collection of all vectors of differentiability;*

(ii) *for every $h \in H(\gamma)$ we have $d_h\gamma = -\widehat{h} \cdot \gamma$, where $\widehat{h}$ is the measurable linear functional generated by h (see Chapter 1).*

PROOF. By Theorem 1.4.2 for any $h \in H(\gamma)$ the measure γ_h is given by density $\exp(-\widehat{h} - |h|_H^2/2)$ with respect to the measure γ and for any other shift we have $\gamma_h \perp \gamma$. The ratio $t^{-1}\bigl[\exp(-t\widehat{h} - t^2|h|_H^2/2) - 1\bigr]$ has a limit $-\widehat{h}$ in $L^1(\gamma)$ as $t \to 0$. This is seen from the fact that for $|t| \leqslant 1$ this ratio is majorized in the absolute value by the integrable function $|\widehat{h}|\exp|\widehat{h}|$ and its pointwise limit is $-\widehat{h}$. Hence the measures $t^{-1}(\gamma_{th} - \gamma)$ converge in variation to the measure $-\widehat{h} \cdot \gamma$, which means that $d_h\gamma = -\widehat{h} \cdot \gamma$. $\square$

3.2. Properties of continuous measures

Here we establish a number of elementary properties of continuous measures. Let us observe that by the estimate from Proposition 1.2.1 the continuity of μ along h is characterized in terms of the Hellinger integral as the relationship
$$H(\mu, \mu_{th}) \to \|\mu\| \quad \text{as } t \to 0.$$

3.2.1. Proposition. *Suppose that a measure μ on $\mathcal{A}$, where $\mathcal{A}$ is one of our main σ-algebras $\mathcal{B}a(X)$, $\sigma(X)$ or $\mathcal{B}(X)$ (where in the case of $\mathcal{B}(X)$ the measure is Radon) is continuous along h. Then every measure ν that is absolutely continuous with respect to μ is continuous along h as well.*

PROOF. By the Radon–Nikodym theorem $\nu = f \cdot \mu$, where $f \in \mathcal{L}^1(\mu)$. Since $(f \cdot \mu)_{th} = f(\cdot + th) \cdot \mu_{th}$, as noted in §3.1, assuming that the function f is bounded and continuous along the vector h, we obtain
$$\|\nu_{th} - \nu\| \leqslant \|f(\cdot + th) \cdot (\mu_{th} - \mu)\| + \|(f(\cdot + th) - f) \cdot \mu\|$$
$$\leqslant \sup_x |f(x)| \|\mu_{th} - \mu\| + \|f(\cdot + th) - f)\|_{L^1(|\mu|)},$$
which, as $t \to 0$, tends to zero by the continuity of μ along h and the Lebesgue dominated convergence theorem. In the general case, given $\varepsilon > 0$, it suffices to find a bounded function g continuous along the vector h such that $\|f - g\|_{L^1(|\mu|)} \leqslant \varepsilon$. This is possible in all three cases mentioned in the theorem. Then it remains to employ the estimate
$$\|(f \cdot \mu)_{th} - f \cdot \mu\| \leqslant \|(f \cdot \mu)_{th} - (g \cdot \mu)_{th}\| + \|(g \cdot \mu)_{th} - g \cdot \mu\| + \|g \cdot \mu - f \cdot \mu\|$$
$$= 2\|g \cdot \mu - f \cdot \mu\| + \|(g \cdot \mu)_{th} - g \cdot \mu\|,$$
which follows from the invariance of the variation with respect to shifts. □

3.2.2. Corollary. *A measure μ on the σ-algebra $\mathcal{A}$ from the previous proposition is continuous along h if and only if the measure $|\mu|$ is also.*

Let ϱ be a fixed probability density on the real line with Lebesgue measure. Set $\sigma := \varrho \, dx$.

3.2.3. Corollary. *Suppose that for every $A \in \mathcal{A}$ the map $(x,t) \mapsto x - th$ from $X \times \mathbb{R}^1$ to X is measurable with respect to the pair of σ-algebras $\mathcal{A} \otimes \mathcal{B}(\mathbb{R}^1)$ and $\mathcal{A}$ (which is fulfilled if $\mathcal{A} = \sigma(X)$ or $\mathcal{A} = \mathcal{B}(X)$). A measure μ on $\mathcal{A}$ is continuous along h precisely when it is absolutely continuous with respect to the measure*
$$|\mu| * \sigma(A) := \int_{-\infty}^{+\infty} |\mu|(A - th) \varrho(t) \, dt.$$

PROOF. We observe that due to our assumption $|\mu| * \sigma$ is a nonnegative measure on $\mathcal{A}$. If the measure μ is continuous along h and $|\mu| * \sigma(A) = 0$, then $|\mu|(A - th) = 0$ for almost all t, which by the continuity of $|\mu|$ along h yields the equality $\mu(A) = 0$. So $\mu \ll |\mu| * \sigma$. For the proof of the converse it suffices to verify the continuity of $|\mu| * \sigma$ along h. Let ϱ be a smooth function with support in the interval $[-M, M]$. Then, for every $A \in \mathcal{A}$ and $|t| \leqslant 1$, we have

$$|\mu| * \sigma(A - th) - |\mu| * \sigma(A) = \int_{-\infty}^{+\infty} [|\mu|(A - th - sh) - |\mu|(A - sh)]\varrho(s)\, ds$$

$$= \int_{-\infty}^{+\infty} |\mu|(A - zh)[\varrho(z - t) - \varrho(z)]\, dz \leqslant 2(M + 1)|t|\|\mu\| \sup_z |\varrho'(z)|.$$

So $\|(|\mu| * \sigma)_{th} - |\mu| * \sigma\| \to 0$ as $t \to 0$. For completing the proof it remains to approximate the density ϱ by smooth densities with bounded support in the norm of $L^1(\mathbb{R}^1)$. Finally, the measurability of the mapping $(x, t) \mapsto x - th$ in the case $\mathcal{A} = \sigma(X)$ or $\mathcal{A} = \mathcal{B}(X)$ follows from Lemma 1.2.15. □

3.2.4. Corollary. *A measure μ on the σ-algebra $\sigma(X)$ in a locally convex space X is continuous along h precisely when it is absolutely continuous with respect to the measure $|\mu| * \gamma$, where γ is the image of the standard Gaussian measure on the real line under the mapping $t \mapsto th$. The same is true for Radon measures on $\mathcal{B}(X)$.*

3.2.5. Corollary. *The measure μ on $\sigma(X)$ is continuous along h if and only if the function $t \mapsto \mu(A + th)$ is continuous for every $A \in \sigma(X)$. In the case of Radon measures the same is true for $\mathcal{B}(X)$ in place of $\sigma(X)$.*

In addition, the continuity of a Radon measure μ along h follows from the continuity of all functions $t \mapsto \mu(A + th)$, where $A \in \sigma(X)$.

PROOF. The continuity of the measure μ along h implies the continuity of the functions $t \mapsto \mu(A + th)$ for all $A \in \sigma(X)$ (in the case of a Radon measure also for $A \in \mathcal{B}(X)$). Suppose that all these functions are continuous. Then $\mu \ll |\mu| * \sigma$ (where the measure σ is the same as above). Indeed, if $|\mu| * \sigma(A) = 0$, then there exists a sequence $t_n \to 0$ with $|\mu|(A - t_n h) = 0$, whence we find $\mu(A - t_n h) = 0$ and so $\mu(A) = 0$. Hence the measure μ is continuous along h. In the case of a Radon measure the continuity along h on $\sigma(X)$ yields at once the continuity on $\mathcal{B}(X)$. □

3.2.6. Corollary. *Let a σ-algebra $\mathcal{A}$ be the same as in the proposition above and let a sequence of measures μ_n continuous along h be such that for every $A \in \mathcal{A}$ there exists a finite limit $\mu(A) = \lim_{n \to \infty} \mu_n(A)$. Then the measure μ is continuous along h as well.*

PROOF. By the Nikodym theorem μ is indeed a measure and the measures μ_n are uniformly bounded in variation. The measure $\nu = \sum_{n=1}^{\infty} 2^{-n}|\mu_n|$ is continuous along h. It is clear that $\mu \ll \nu$. □

3.2.7. Remark. If a measure μ is continuous along h, then so is the measure $I_A \cdot |\mu|$ for every $A \in \mathcal{B}(X)$. Hence, for every set $A \in \mathcal{B}(X)$ with $|\mu|(A) > 0$, we obtain $(A + th) \cap A \neq \varnothing$ for sufficiently small t. In particular, h belongs to the linear span of every compact set K with $|\mu|(K) > 0$. It follows that the property of a measure to be continuous along h is invariant with respect to the space carrying the measure: if X is embedded by means

3.2. Properties of continuous measures

of an injective continuous linear operator to a locally convex space Y, then the measure μ considered on Y is continuous along the same vectors along which it is continuous on X. This is true, of course, for other differential properties introduced in §3.1.

Let us mention the following result from Yamasaki, Hora [**1198**].

3.2.8. Proposition. *There is no nonzero Borel measure on an infinite dimensional locally convex space that is continuous or quasi-invariant along all vectors.*

The proof follows from the next proposition, which is of interest in its own right and shows that in infinite dimensions a measure has no minimal linear subspace of full measure unless it is concentrated on a finite dimensional subspace.

3.2.9. Proposition. *Let X be an infinite dimensional locally convex space and let μ be a Borel probability measure on X. Then there is a Borel proper linear subspace $L \subset X$ with $\mu(L) > 0$.*

PROOF. Suppose that $\mu(L) = 0$ for every Borel proper linear subspace. Let us equip X^* with the following metric corresponding to convergence in measure μ:

$$d(f,g) := \int_X \min(|f(x) - g(x)|, 1)\, \mu(dx).$$

Note that d is indeed a metric since if $l \in X^*$ vanishes μ-a.e. we have $l = 0$ due to our assumption (otherwise $L = l^{-1}(0)$ is a Borel linear subspace of positive measure). Set

$$V_n := \{l \in X^*\colon d(l,0) \leqslant n^{-1}\}, \quad U_n := \{x \in X\colon \sup_{l \in V_n} |l(x)| \leqslant 1\}.$$

The sets U_n are closed. We observe that $X = \bigcup_{n=1}^\infty U_n$. Indeed, otherwise we would have an element $x_0 \in X$ and functionals l_n with $d(l_n, 0) \leqslant n^{-1}$ and $l_n(x_0) > 1$. The sequence $\{l_n\}$ converges to zero in measure, hence there is a subsequence $\{l_{n_i}\}$ convergent to zero μ-a.e. The linear subspace L consisting of all $x \in X$ such that $\lim_{i \to \infty} l_{n_i}(x) = 0$ is Borel and has positive measure, but $x_0 \notin L$, which is impossible. There is n such that $\mu(U_n) > 0$. This yields that the metric d on X^* generates the same convergence as the norm from $L^2(\mu|_{U_n})$. Indeed, let $l_i \to 0$ with respect to this norm. Then there is a subsequence $\{l_{i_j}\}$ convergent to zero μ-a.e. on U_n. As above, this shows that $\{l_{i_j}\}$ converges to zero pointwise, hence in metric d. It follows that the whole sequence converges to zero in metric d. Conversely, if $d(l_i, 0) \to 0$, then we may assume that $\{l_i\} \subset V_n$, hence $\sup_{x \in U_n} |l_i(x)| \leqslant 1$, which yields convergence $l_i \to 0$ in $L^2(\mu|_{U_n})$ by the dominated convergence theorem. Now we arrive at a contradiction by showing that X^* must be finite dimensional. Indeed, we have proved that there is $R > 0$ such that the unit ball B of X^* with the $L^2(\mu|_{U_n})$-norm is contained in RV_n. Hence, for any collection $l_1, \ldots, l_k \in X^*$ orthonormal in $L^2(\mu|_{U_n})$, we have

$$k = \sum_{i=1}^{k} \int_{U_n} |l_i|^2 \, d\mu \leqslant \sup_{x \in U_n} \sum_{i=1}^{k} |l_i(x)|^2$$
$$= \sup_{x \in U_n} \Big| \Big\langle \Big(\sum_{i=1}^{k} |l_i(x)|^2\Big)^{-1/2} \sum_{i=1}^{k} l_i(x) l_i, x \Big\rangle \Big|^2 \leqslant \sup_{x \in U_n} \sup_{l \in B} |l(x)|^2$$
$$\leqslant \sup_{x \in U_n} \sup_{l \in RV_n} |l(x)|^2 \leqslant R^2,$$

since for any fixed x, the $L^2(\mu|_{U_n})$-norm of $\sum_{i=1}^{k} l_i(x) l_i$ is $\big(\sum_{i=1}^{k} |l_i(x)|^2\big)^{1/2}$. This shows that $\dim X^* \leqslant R^2$. □

3.3. Properties of differentiable measures

We assume that $\mathcal{A}$ is one of our three main σ-algebras and that in the case of the Borel σ-algebra the considered measures are Radon, although some of the results below are also valid in more general cases. In particular, the following theorem is valid under the only assumption of invariance of $\mathcal{A}$ with respect to the shifts along the vectors th, which will be obvious from the proof.

3.3.1. Theorem. *Suppose that a measure μ on $\mathcal{A}$ is Fomin differentiable along the vector h. Then, its positive part μ^+, its negative part μ^-, and its variation $|\mu|$ are Fomin differentiable along h as well. In addition, if X^+ and X^- are the sets from the Hahn–Jordan decomposition for μ, then*

$$d_h(\mu^+) = d_h\mu(\cdot \cap X^+), \quad d_h(\mu^-) = d_h\mu(\cdot \cap X^-).$$

PROOF. Let $X = X^+ \cup X^-$ be the Hahn decomposition for the measure μ. Then the restriction of μ to X^+ is nonnegative and its restriction to X^- is nonpositive. The function $\varphi(t) = \mu(X^+ + th)$ is differentiable and attains its maximum at zero since $\mu(X^+)$ is the maximal value of μ. Hence its derivative at zero vanishes, i.e., as $t \to 0$ we have

$$\frac{\varphi(t) - \varphi(0)}{t} = t^{-1}\mu\big((X^+ + th)\backslash X^+\big) - t^{-1}\mu\big(X^+ \backslash (X^+ + th)\big) \to 0.$$

Since $\mu\big((X^+ + th)\backslash X^+\big) \leqslant 0$ and $\mu\big(X^+ \backslash (X^+ + th)\big) \geqslant 0$, we obtain as $t \to 0$ that

$$t^{-1}\mu\big((X^+ + th)\backslash X^+\big) \to 0, \quad t^{-1}\mu\big(X^+\backslash(X^+ + th)\big) \to 0.$$

Set $X_t := X^+ + th$. It is easy to verify the equality

$$X^+ = \big(X_t \cup (X^+\backslash X_t)\big)\backslash(X_t\backslash X^+).$$

3.3. Properties of differentiable measures

Therefore, for every $A \in \mathcal{A}$, as $t \to 0$ we have

$$\frac{\mu^+(A+th) - \mu^+(A)}{t} = \frac{\mu\big((A+th) \cap X^+\big) - \mu(A \cap X^+)}{t}$$
$$= \frac{\mu\big((A+th) \cap X_t\big) - \mu(A \cap X^+)}{t} + \frac{\mu\big(A \cap (X^+ \setminus X_t)\big) - \mu\big(A \cap (X_t \setminus X^+)\big)}{t}$$
$$\to d_h\mu(A \cap X^+)$$

since the second term in the second equality tends to zero according to what has been said above. The reasoning is similar for μ^-. □

3.3.2. Corollary. *If a measure μ is Fomin differentiable along h, then one has $d_h\mu \ll \mu$.*

PROOF. The previous theorem reduces the general case to the case of a nonnegative measure μ, in which our claim is obvious since if $\mu(A) = 0$, then the function $t \mapsto \mu(A + th)$ has minimum at zero, whence it follows that its derivative vanishes at zero. □

3.3.3. Corollary. *If a measure μ is Fomin differentiable along a vector h, then the measure $d_h\mu$ is continuous along h. In particular, all functions $t \mapsto \mu(A + th)$, where $A \in \mathcal{A}$, are continuously differentiable.*

3.3.4. Proposition. *A measure μ is Fomin differentiable along the vector h precisely when there exists a measure ν on $\mathcal{A}$ continuous along h such that*

$$\mu(A + th) = \mu(A) + \int_0^t \nu(A + sh)\, ds, \quad \forall A \in \mathcal{A}, t \in \mathbb{R}^1. \tag{3.3.1}$$

In this case $d_h\mu = \nu$. In addition, for every bounded $\mathcal{A}$-measurable function f for all $t \in \mathbb{R}^1$ the following equality holds:

$$\int_X f(x)\, (\mu_{th} - \mu)(dx) = \int_0^t \int_X f(x - sh)\, d_h\mu(dx)\, ds. \tag{3.3.2}$$

Finally, yet another necessary and sufficient condition for Fomin's differentiability of μ is the existence of a measure $\nu \ll \mu$ such that one has (3.3.1).

PROOF. Let the measure μ be Fomin differentiable along h. Then the derivative of the function $\varphi\colon t \mapsto \mu(A + th)$ is $d_h\mu(A + th)$. So we have $|\varphi'(t)| \leqslant \|d_h\mu\|$, i.e., φ is Lipschitzian, which yields equality (3.3.1) with $\nu = d_h\mu$. Then we have (3.3.2) for simple functions f, which gives (3.3.2) for bounded functions f by means of uniform approximations by simple functions. As we have proved earlier, $d_h\mu \ll \mu$ and the measure $d_h\mu$ is continuous along h. The converse is obvious. If we are given identity (3.3.1) with some measure $\nu \ll \mu$, then we obtain the continuity of the measure μ along h, which implies the continuity of ν along h. □

3.3.5. Corollary. *Suppose that in the situation of the previous proposition for μ-a.e. x there exists $\partial_h f(x) := \lim_{t \to 0} t^{-1}[f(x+th) - f(x)]$ and*

$$|t^{-1}[f(x+th) - f(x)]| \leqslant g(x), \quad \text{where } g \in \mathcal{L}^1(\mu).$$

Then

$$\int_X \partial_h f(x) \, \mu(dx) = -\int_X f(x) \, d_h\mu(dx). \quad (3.3.3)$$

PROOF. On the left in (3.3.2) we have the integral of $f(x - th) - f(x)$, which enables us to apply the Lebesgue dominated convergence theorem to the sequence of functions $n[f(x - h/n) - f(x)]$. □

3.3.6. Corollary. *If a measure μ is differentiable along two vectors h and k, then it is differentiable along every linear combination of these vectors and*

$$d_{\alpha h + \beta k}\mu = \alpha d_h\mu + \beta d_k\mu.$$

PROOF. The cases $k = 0$ and $h = 0$ are obvious. So it suffices to consider the case $\alpha = \beta = 1$. Let us recall that the measure $d_h\mu$ is continuous along h. In addition, it is continuous also along k since it is a limit of the sequence of measures $n(\mu_{n^{-1}h} - \mu)$ that are continuous along k by the differentiability along k. We have

$$\mu(A + th + tk) - \mu(A) = \mu(A + th + tk) - \mu(A + tk) + \mu(A + tk) - \mu(A)$$

$$= \int_0^t d_h\mu(A + tk + sh) \, ds + \int_0^t d_k\mu(A + sk) \, ds.$$

The right-hand side divided by t tends to $d_h\mu(A) + d_k\mu(A)$ as $t \to 0$. This follows from the continuity of the function $(t, s) \mapsto d_h(A + tk + sh)$ of two variables, which is clear from the estimate

$$|d_h\mu(A + tk + sh) - d_h\mu(A + t_0k + s_0h)|$$
$$\leqslant |d_h\mu(A + tk + sh) - d_h\mu(A + t_0k + sh)|$$
$$+ |d_h\mu(A + t_0k + sh) - d_h\mu(A + t_0k + s_0h)|$$
$$\leqslant \|(d_h\mu)_{tk} - (d_h\mu)_{t_0k}\| + \|(d_h\mu)_{sh} - (d_h\mu)_{s_0h}\|$$

and the continuity of $d_h\mu$ along h and k. □

3.3.7. Theorem. *Suppose that a measure μ is differentiable along the vector h. Then*
  (i) *one has $\|\mu_h - \mu\| \leqslant \|d_h\mu\|$;*
  (ii) *for every $t \in \mathbb{R}^1$ one has*

$$\|\mu_{th} - \mu - t d_h\mu\| \leqslant |t| \sup_{0 < \tau < t} \|(d_h\mu)_{\tau h} - d_h\mu\|;$$

  (iii) *one has $\|(\mu_{th} - \mu)/t - d_h\mu\| \to 0$ as $t \to 0$.*

3.3. Properties of differentiable measures

PROOF. Assertion (i) is readily seen from equality (3.3.2). The same relationship yields (ii). Finally, (iii) follows from (ii) by the continuity of $d_h\mu$ along h. □

3.3.8. Definition. *The Radon–Nikodym density of the measure $d_h\mu$ with respect to μ is denoted by β_h^μ and is called the logarithmic derivative of μ along h.*

This terminology is justified by Example 3.1.8. As we have seen in Theorem 3.1.9, if γ is a Gaussian measure and $h \in H(\gamma)$, then $\beta_h^\gamma = -\widehat{h}$ is a measurable linear function.

It follows from what has been said above that
$$d_h\mu = \beta_h^{\mu^+} \cdot \mu^+ - \beta_h^{\mu^-} \cdot \mu^-, \quad d_h|\mu| = \beta_h^{\mu^+} \cdot \mu^+ + \beta_h^{\mu^-} \cdot \mu^-.$$

Since the measures μ^+ and μ^- are mutually singular, $|\mu|$-a.e. we have
$$|\beta_h^\mu| = |\beta_h^{\mu^+}| + |\beta_h^{\mu^-}| = |\beta_h^{|\mu|}|. \tag{3.3.4}$$

This relationship is analogous to the equality $|\nabla f| = |\nabla |f||$ a.e. for functions from Sobolev classes.

If a nonnegative measure μ is L^2-differentiable along h, then equality (3.1.3) yields the following representation of the logarithmic derivative:
$$\beta_h^\mu = 2\psi f_0^{-1/2}, \tag{3.3.5}$$
where $\mu_{th} = f_t \cdot \lambda$ and ψ is the derivative at zero of the mapping $t \mapsto f_t^{1/2}$ to $L^2(\lambda)$. Since $\psi \in L^2(\lambda)$, one has $\beta_h^\mu \in L^2(\mu)$. The latter condition characterizes L^2-differentiability. An analogous assertion is true for all numbers $p \in (1, +\infty)$.

3.3.9. Theorem. *A measure $\mu \geqslant 0$ is L^p-differentiable along h precisely when it is Fomin differentiable along h and $\beta_h^\mu \in L^p(\mu)$. If $p > 1$, this is also equivalent to the estimate*
$$\|f_t^{1/p} - f_0^{1/p}\|_{L^p(\lambda)} \leqslant C|t|.$$
In this case one can take $C = \|\beta_h^\mu\|_{L^p(\mu)}/p$.

A short proof of a more general fact (for differentiable families of measures) is given in Theorem 11.2.13 in Chapter 11, so here we do not reproduce this proof.

For $p = 2$, by using the so-called Hellinger integral
$$H(\mu_{th}, \mu) := \int_X f_t^{1/2}(x) f_0^{1/2}(x) \, \lambda(dx)$$
this estimate can be written in the form
$$H(\mu_{th}, \mu) \geqslant 1 - t^2 \|\beta_h^\mu\|_{L^2(\mu)}^2 / 8.$$

Let us observe that the quantity $H(\mu_{th}, \mu)$ is independent of the measure λ with the property that $\mu_{th} \ll \lambda$ for all t.

3.3.10. Proposition. *A measure μ on $\mathcal{A}$ is analytic along h if and only if there exists a number $\delta > 0$ such that for all complex numbers z with $|z| < \delta$ the series*

$$\sum_{n=1}^{\infty} \frac{1}{n!} z^n d_h^n \mu$$

converges in variation in the Banach space of complex measures on X. In this case, for some $C, r > 0$ one has the estimate

$$\|d_h^n \mu\| \leqslant C r^n n!, \quad n \in \mathbb{N},$$

and μ is quasi-invariant along h.

PROOF. Convergence of the series obviously yields the analyticity of the measure. Conversely, if the measure is analytic, then the series with terms $z^n d_h^n \mu(A)/n!$ converges whenever $|z| < c$, $A \in \mathcal{A}$, which for any fixed z with $|z| < c$ implies the uniform boundedness of the quantities $|z|^n \|d_h^n \mu\|/n!$. This gives convergence of the series indicated in the theorem and the estimate $\|d_h^n \mu\| \leqslant C(c/2)^{-n} n!$ for some number $C > 0$. If $|\mu|(A) = 0$ for some $A \in \mathcal{A}$, then by the relationships $d_h^{k+1} \mu \ll d_h^k \mu$ we obtain $d_h^k \mu(A) = 0$ for all $k \geqslant 1$, whence $\mu(A + th) = 0$ for all $t \in \mathbb{R}^1$. Hence $\mu_{th} \ll \mu$. Therefore, we have the relationship $\mu_{th} \sim \mu$. □

3.3.11. Corollary. *If a Radon measure μ on a locally convex space X is analytic along a vector $h \in X$, then $d_h^n \mu \sim \mu$ for all n.*

PROOF. Let $|d_h \mu|(A) = 0$ for some set $A \in \mathcal{B}(X)$. Then $|d_h \mu|(K) = 0$ for every compact set $K \subset A$. According to what has been proved above, $|d_h^n \mu|(K) = 0$ for all $n \geqslant 1$, whence $\mu(K + th) = \mu(K)$ for all $t \in \mathbb{R}^1$. Since μ is Radon, we obtain the equality $\lim_{t \to \infty} \mu(K + th) = 0$. Thus, $\mu(K) = 0$ and so $|\mu|(A) = 0$, i.e., $\mu \sim d_h \mu$. Then $\mu \sim d_h^n \mu$ for all n. □

Let us discuss some elementary operations on differentiable measures.

3.3.12. Proposition. *Let μ be a measure differentiable along h and let f be a bounded measurable function possessing a uniformly bounded partial derivative $\partial_h f$. Then, the measure $\nu = f \cdot \mu$ is differentiable along h as well and one has*

$$d_h \nu = \partial_h f \cdot \mu + f \cdot d_h \mu.$$

PROOF. For every $A \in \mathcal{A}$ we have

$$\nu(A + th) - \nu(A) = \int_X [I_A(x - th) - I_A(x)] f(x) \, \mu(dx)$$
$$= \int_X I_A(y) f(y + th) \, \mu_{th}(dy) - \int_X I_A(x) f(x) \, \mu(dx)$$
$$= \int_X I_A(y) [f(y + th) - f(y)] \, \mu_{th}(dy) + \int_X I_A(x) f(x) \, (\mu_{th} - \mu)(dx).$$

Substituting in place of t points $t_n \to 0$ and dividing by t_n we obtain in the limit the sum of the integrals of $I_A \partial_h f$ and $I_A f \beta_h^\mu$ against the measure μ. This follows from our assumptions and Theorem 1.2.19. □

3.3.13. Proposition. *Let X be a locally convex space equipped with one of our three σ-algebras $\mathcal{A}$. Let μ be a measure on $\mathcal{A}$ possessing one of the properties along h described by our definitions from §3.1 and let ν be a measure on a locally convex space Y equipped with a σ-algebra of the same kind as $\mathcal{A}$. Then the following assertions are true.*

(i) *If $T\colon X \to Y$ is a continuous linear mapping (or, in the case of Fomin differentiability, a linear mapping measurable with respect to μ), then the measure $\mu \circ T^{-1}$ has the same property along Th that μ has along h. In the case of differentiability one has*

$$d_{Th}(\mu \circ T^{-1}) = (d_h \mu) \circ T^{-1}.$$

In the case of a Fomin differentiable probability measure μ, the logarithmic derivative $\beta_{Th}^{\mu \circ T^{-1}}$ is a Borel function β on Y such that the function $\beta \circ T$ coincides with the conditional expectation $\mathbb{E}^T \beta_h^\mu$ of the function β_h^μ with respect to the σ-algebra σ_T generated by the mapping T. Conversely, every Borel function β on Y for which $\beta \circ T = \mathbb{E}^T \beta_h^\mu$ μ-a.e. coincides with $\beta_{Th}^{\mu \circ T^{-1}}$ almost everywhere with respect to the measure $\mu \circ T^{-1}$.

(ii) *The measure $\mu \otimes \nu$ on $X \times Y$ has the same property along $(h, 0)$ that μ has along h.*

(iii) *If $X = Y$, then the measure $\mu * \nu$ has the corresponding property along h and in the case of differentiability*

$$d_h(\mu * \nu) = d_h \mu * \nu.$$

PROOF. The verification is direct by using the definitions and Fubini's theorem. Let us only explain the last claim in (i). For definiteness we shall assume that we deal with Borel measures. For every $B \in \mathcal{B}(Y)$ on account of the equality in (i) we have

$$\int_{T^{-1}(B)} \beta_h^\mu(x)\, \mu(dx) = \int_B \beta_{Th}^{\mu \circ T^{-1}}(y)\, \mu \circ T^{-1}(dy) = \int_{T^{-1}(B)} \beta_{Th}^{\mu \circ T^{-1}}(Tx)\, \mu(dx).$$

It follows that the function $\beta_{Th}^{\mu \circ T^{-1}} \circ T$ coincides with the conditional expectation of β_h^μ with respect to the σ-algebra σ_T since the latter is the class of all sets of the form $T^{-1}(B)$, where $B \in \mathcal{B}(Y)$. Conversely, if a Borel function β on Y is such that we have $\beta \circ T = \mathbb{E}^T \beta_h^\mu$ μ-a.e., then

$$\int_B \beta(y)\, \mu \circ T^{-1}(dy) = \int_{T^{-1}(B)} \beta(Tx)\, \mu(dx) = \int_{T^{-1}(B)} \mathbb{E}^T \beta_h^\mu(x)\, \mu(dx)$$

$$= \int_{T^{-1}(B)} \beta_h^\mu(x)\, \mu(dx) = d_h\mu(T^{-1}(B)) = d_{Th}(\mu \circ T^{-1}),$$

whence it follows that β coincides with $\beta_{Th}^{\mu \circ T^{-1}}$. □

3.3.14. Example. Let X and $\mathcal{A}$ be the same as in the proposition above, let μ be a measure on $\mathcal{A}$, and let f be a bounded $\mathcal{A}$-measurable function on X. If μ is Fomin differentiable along h, then the function
$$f * \mu(x) := \int_X f(x-y)\,\mu(dy)$$
has a partial derivative
$$\partial_h(f * \mu)(x) = \int_X f(x-y)\,d_h\mu(dy).$$

PROOF. If f is the indicator function of a measurable set, then this is the definition. Therefore, our claim is true for linear combinations g of indicator functions. Let x be fixed. According to Corollary 3.3.3, the functions $t \mapsto d_h\mu(A + x + th)$ are continuous. Hence so are the functions
$$t \mapsto \int_X g(x + th - y)\,d_h\mu(dy).$$
It remains to choose a sequence of functions g_n of the above type which converges to f uniformly on X. Then the function $t \mapsto f * \mu(x + th)$ is differentiable being a uniform limit of differentiable functions whose derivatives converge uniformly. $\square$

3.4. Differentiable measures on $\mathbb{R}^n$

In this section we discuss differentiable measures on the space $\mathbb{R}^n$. The term "measure on $\mathbb{R}^n$" will always mean a bounded Borel measure on $\mathbb{R}^n$ unless we deal with Lebesgue measure (which will always be explicitly explained). First we consider the case $n = 1$.

3.4.1. Proposition. *Let μ be a measure on $\mathbb{R}^1$.*

(i) *The measure μ is continuous along $h \neq 0$ if and only if it is absolutely continuous with respect to Lebesgue measure.*

(ii) *The measure μ is Skorohod differentiable along $h \neq 0$ if and only if it has a density of bounded variation. In particular, any Skorohod differentiable measure on $\mathbb{R}^1$ admits a bounded density.*

(iii) *The measure μ is Fomin differentiable along $h \neq 0$ if and only if it has an absolutely continuous density ϱ whose derivative ϱ' is integrable. In this case $d_1\mu = \varrho'dx$.*

(iv) *The measure μ is quasi-invariant along $h \neq 0$ if and only if it has a density ϱ such that $\varrho \neq 0$ almost everywhere.*

(v) *If $\mu \geqslant 0$, then μ is L^2-differentiable along $h \neq 0$ if and only if it admits an absolutely continuous density ϱ for which $|\varrho'|^2/\varrho \in L^1(\mathbb{R}^1)$, where we set $0/0 := 0$. The quantity*
$$I(\varrho) := \int_{-\infty}^{+\infty} \frac{|\varrho'(t)|^2}{\varrho(t)}\,dt$$
is called the Fisher information of the measure μ or the Fisher information of the density ϱ.

3.4. Differentiable measures on $\mathbb{R}^n$

PROOF. To prove (i) we observe that if μ is continuous along 1, then $|\mu|$ is continuous as well, which easily yields that $|\mu| \ll |\mu| * \sigma$, where σ is any probability measure with a smooth density (if $|\mu|(A-x) = 0$ for a.e. x, then $|\mu|(A) = 0$ by the continuity). Similar justification of (iv) is left as an easy exercise (Exercise 3.9.1). Suppose that the measure μ is Skorohod differentiable along $h = 1$. Then it has a density ϱ. Denote by $J(f)$ the derivative at zero of the function

$$t \mapsto \int_{\mathbb{R}^1} f(x-t)\varrho(x)\,dx, \quad \text{where } f \in C_b(\mathbb{R}^1).$$

Let $\|f\| = \sup_x |f(x)|$. By the Banach–Steinhaus theorem there exists a number C such that $|J(f)| \leqslant C\|f\|$ for all $f \in C_b(\mathbb{R}^1)$. Therefore, there exists a bounded measure ν on $\mathbb{R}^1$, for which

$$J(f) = \int_{\mathbb{R}^1} f(x)\,\nu(dx), \quad \forall f \in C_0^1(\mathbb{R}^1).$$

Thus, ν is the derivative of ϱ in the sense of generalized functions. Now it is readily verified that ϱ coincides almost everywhere with the distribution function of ν. Hence ϱ is of bounded variation. Conversely, suppose that μ admits a density ϱ of bounded variation. Then its derivative in the sense of generalized functions is a bounded measure ν. For every $f \in C_0^\infty(\mathbb{R}^1)$ this yields

$$\int_{\mathbb{R}^1} [f(x-t) - f(x)]\,\mu(dx) = \int_0^t \int_{\mathbb{R}^1} f(x-s)\,\nu(dx)\,ds$$

since the derivatives of both sides coincide. Therefore, this integral identity is true for all $f \in C_b(\mathbb{R}^1)$, which is Skorohod differentiability.

If the measure μ is differentiable, then according to what has been said above the measure $d_1\mu$ is absolutely continuous, hence has a density g. From our previous consideration we conclude that the indefinite integral of g (taken from $-\infty$) serves as a density for μ. Conversely, if the measure μ admits an absolutely continuous density ϱ, then it is Skorohod differentiable and ϱ' is a density of the measure $d_1\mu$. Therefore, μ is also Fomin differentiable (see Proposition 3.3.4).

Finally, if a nonnegative measure μ is L^2-differentiable, then, as noted above, it is Fomin differentiable and has an absolutely continuous density ϱ. It is clear from Theorem 3.3.9 that the function $|\varrho'|^2/\varrho$ belongs to $L^1(\mathbb{R}^1)$. The same theorem gives the converse. $\square$

In terms of the theory of generalized functions (distributions) the obtained result has the following interpretation: Skorohod differentiability of a measure μ is equivalent to that its derivative in the sense of distributions is a bounded measure, and Fomin differentiability is equivalent to that its generalized derivative is given by an integrable function.

3.4.2. Example. (i) Let $p(t) = 1$ on $[0,1]$ and $p(t) = 0$ at all other points. Then the measure μ with density p is Skorohod differentiable (but

not Fomin differentiable) and we have the equality
$$d_1\mu = \delta_0 - \delta_1.$$

(ii) There exists a probability measure μ with an infinitely differentiable density on $\mathbb{R}^1$ which is not quasi-invariant, but $\mu(U) > 0$ for every open set U.

PROOF. Assertion (i) is verified directly. For proving (ii) it suffices to find a smooth probability density whose zero set is nowhere dense, but is of positive Lebesgue measure. □

3.4.3. Proposition. (i) *A measure μ on $\mathbb{R}^n$ is continuous along n linearly independent directions precisely when μ is absolutely continuous.*

(ii) *A measure μ on $\mathbb{R}^n$ is Skorohod differentiable along n linearly independent vectors if and only if its generalized partial derivatives are bounded measures.*

(iii) *A measure μ on $\mathbb{R}^n$ is Fomin differentiable along n linearly independent vectors if and only if its generalized partial derivatives are integrable functions. This is equivalent to the following: $\mu = \varrho\, dx$, where $\varrho \in W^{1,1}(\mathbb{R}^n)$.*

PROOF. (i) If the measure μ is absolutely continuous, then it is absolutely continuous with respect to the standard Gaussian measure γ, the continuity of which along all vectors is obvious. If we are given the continuity of μ along n linearly independent vectors, then according to what has been said in §3.2 we see that $\mu \ll |\mu| * \gamma$, which yields the absolute continuity of the measure μ since the measure $|\mu| * \gamma$ is absolutely continuous. Justification of (ii) and (iii) is much the same as in the case $n = 1$ considered above. Yet, in the multidimensional case one cannot assert the existence of a bounded density. □

The following result is a direct corollary of Theorem 2.2.1 and Proposition 2.3.1.

3.4.4. Corollary. *Let a measure μ on $\mathbb{R}^n$ be Fomin or Skorohod differentiable along n linearly independent vectors. Then it has a density from the class $L^p(\mathbb{R}^n)$, where $p = n/(n-1)$.*

3.4.5. Remark. (i) In the language of Sobolev spaces Skorohod differentiability of a measure μ on $\mathbb{R}^n$ along n linearly independent vectors is equivalent to the representation $\mu = \varrho\, dx$, where $\varrho \in BV(\mathbb{R}^n)$ (the definition of the class BV is given in §2.3).

(ii) There exists an example of a function $\varrho \in W^{1,1}(\mathbb{R}^2)$ without locally bounded modifications (see Example 2.1.9). Thus, differentiable measures on $\mathbb{R}^n$ with $n \geqslant 2$ may have densities that are essentially locally unbounded.

The next result follows from the properties of functions of the class $W^{1,1}(\mathbb{R}^n)$ mentioned in Theorem 2.1.10 and Remark 2.1.11.

3.4.6. Proposition. *Let a measure μ on $\mathbb{R}^n$ be Fomin differentiable along all vectors. Then it has a density ϱ which is absolutely continuous on almost every straight line parallel to the ith coordinate line for each fixed index $i = 1, \ldots, n$. In addition, there is a version of ϱ which is absolutely continuous on the straight lines generated by almost all points in the sphere equipped with Lebesgue surface measure. Finally, for every two-dimensional plane L and every orthogonal operator S which is a rotation in L and the identity operator on the orthogonal complement of L one can find a version of ϱ which is absolutely continuous on almost all orbits of S.*

The theory of Sobolev spaces provides some additional information about the properties of differentiable measures on $\mathbb{R}^n$.

3.4.7. Theorem. *Let $\{e_i\}$ be the standard basis in $\mathbb{R}^n$, let $k \in \mathbb{N}$, and let μ be a measure on $\mathbb{R}^n$ such that, for every k vectors $h_1, \ldots, h_k$, there exists a derivative $d_{h_1} \cdots d_{h_k} \mu$ in the sense of Fomin. Then*

(i) if $k = n$, then μ has a bounded continuous density ϱ with respect to Lebesgue measure and $\|\varrho\|_{L^\infty} \leqslant \|d_{e_1} \cdots d_{e_n} \mu\|$;

(ii) if $k > n$, then μ has a $(k - n - 1)$-times continuously differentiable density ϱ with respect to Lebesgue measure such that the partial derivatives of ϱ up to order $k - n$ exist a.e. and are integrable and the derivatives up to order $k - n - 1$ are bounded;

(iii) the measure μ has derivatives of all orders along all vectors if and only if it admits an infinitely differentiable density ϱ such that all partial derivatives of ϱ belong to $L^1(\mathbb{R}^n)$.

Let us observe that a density ϱ of a measure μ on $\mathbb{R}^n$ which has partial derivatives up to order n (in the Fomin or Skorohod sense) can be written as

$$\varrho(x) = \frac{\partial^n}{\partial x_1 \cdots \partial x_n} \mu(y\colon\ y_i < x_i,\ i = 1, \ldots, n).$$

In the case of Skorohod differentiability ϱ may fail to have a continuous density, but the estimate $\|\varrho\|_{L^\infty} \leqslant \|d_{e_1} \cdots d_{e_n} \mu\|$ still holds.

3.4.8. Theorem. *Let $1 < p < \infty$. A Borel probability measure μ on $\mathbb{R}^n$ is L^p-differentiable along all vectors precisely when it is Fomin differentiable with $\beta_h^\mu \in L^p(\mu)$ for all $h \in \mathbb{R}^n$. This is also equivalent to that μ is given by a density ϱ with $\varrho^{1/p} \in W^{p,1}(\mathbb{R}^n)$.*

PROOF. Suppose that the measure μ is Fomin differentiable along all vectors and has logarithmic derivatives from $L^p(\mu)$. Hence it has a density $\varrho \in W^{1,1}(\mathbb{R}^n)$ and $\beta_h^\mu = (\nabla \varrho/\varrho, h)$, whence $|\nabla \varrho|/\varrho \in L^p(\mu)$. Set $\psi := \varrho^{1/p}$. Then $\psi \in L^p(\mathbb{R}^n)$. On every ball U the functions $\psi_\varepsilon := (\varrho + \varepsilon)^{1/p}$, where $\varepsilon > 0$, belong to $W^{p,1}(U)$ since

$$|\nabla \psi_\varepsilon|^p = p^{-p} |\nabla \varrho|^p (\varrho + \varepsilon)^{1-p} \leqslant p^{-p} |\nabla \varrho|^p \varrho^{1-p} \in L^p(U).$$

Letting $\varepsilon \to 0$ we get $\psi \in W^{p,1}(U)$ and $\nabla \psi = p^{-1}\varrho^{1/p-1}\nabla\varrho$. By the integrability of ψ^p and $|\nabla\psi|^p$ we obtain the inclusion $\psi \in W^{p,1}(\mathbb{R}^n)$. Conversely, if we know that $\psi = \varrho^{1/p} \in W^{p,1}(\mathbb{R}^n)$, then we obtain the equality $\nabla\varrho = p\psi^{p-1}\nabla\psi$, whence the equality $|\nabla\varrho/\varrho|^p = p^p|\nabla\psi|^p\varrho^{-1} \in L^p(\mu)$ follows. Let us show that the inclusion $\psi := \varrho^{1/p} \in W^{p,1}(\mathbb{R}^n)$ gives L^p-differentiability of μ along every $h \in \mathbb{R}^n$. It suffices to have the differentiability of the mapping $t \mapsto \psi(\,\cdot\, + th)$ with values in $L^p(\mathbb{R}^n)$. The latter holds by Theorem 2.1.13. Finally, if μ is L^p-differentiable, then $\beta_h^\mu \in L^p(\mu)$ by Theorem 3.3.9. □

3.4.9. Corollary. *Suppose that a measure μ on $\mathbb{R}^n$ is Fomin differentiable along linearly independent vectors $e_1, \ldots, e_n$ and $\beta_{e_i}^\mu \in L^p(|\mu|)$, where $p \in (1, +\infty)$. Then $|\mu|$ is L^p-differentiable along all vectors and $\mu = \varrho\, dx$, where $\varrho \in W^{1,1}(\mathbb{R}^n)$ is locally Hölder continuous and $|\varrho|^{1/p} \in W^{p,1}(\mathbb{R}^n)$.*

PROOF. We already know that $\mu = \varrho\, dx$, where $\varrho \in W^{1,1}(\mathbb{R}^n)$. Hence we have $|\mu| = |\varrho|\, dx$, $|\varrho| \in W^{1,1}(\mathbb{R}^n)$, $|\beta_{e_i}^{|\mu|}(x)| = |\beta_{e_i}^\mu(x)|$ for $|\mu|$-a.e. x, whence we obtain $\beta_{e_i}^{|\mu|} \in L^p(|\mu|)$. Now the previous theorem applies and gives $|\varrho|^{1/p} \in W^{p,1}(\mathbb{R}^n)$. Hence the function $|\varrho|^{1/p}$ has a Hölder continuous version. Therefore, $|\varrho|$ is locally Hölder continuous. Since $\varrho \in W^{1,1}(\mathbb{R}^n)$ and $|\nabla\varrho/\varrho|^p|\varrho| \in L^p(\mathbb{R}^n)$, this implies that $|\nabla\varrho| \in L^p_{\text{loc}}(\mathbb{R}^n)$, whence we obtain $\varrho \in W^{p,1}_{\text{loc}}(\mathbb{R}^n)$. Consequently, ϱ has a locally Hölder continuous version. □

3.4.10. Corollary. *Let μ be a measure on $\mathbb{R}^n$ such that $\beta_{e_i}^\mu \in L^p(|\mu|)$ for some $p > n$. Then μ admits a bounded locally Hölder continuous density ϱ satisfying the estimate*

$$\sup_x |\varrho(x)| \leqslant M(n,p)\|\mu\|^{1-n/p}\sum_{i=1}^n \|\beta_{e_i}^\mu\|_{L^p(|\mu|)}^n, \qquad (3.4.1)$$

where $M(n,p)$ is a constant that depends only on n and p.

PROOF. The previous corollary yields the existence of a density ϱ in the class $W^{1,1}(\mathbb{R}^n)$ such that $f := |\varrho|^{1/p} \in W^{p,1}(\mathbb{R}^n)$, whence by the Sobolev embedding theorem we obtain a bounded continuous modification of ϱ. Now Theorem 2.2.9 gives a constant $C(n,p)$ such that

$$\sup_x |\varrho(x)| \leqslant C(n,p)\|\nabla f\|_{L^p(\mathbb{R}^n)}^n \|f\|_{L^p(\mathbb{R}^n)}^{p-n}.$$

Here we have the equalities $\|f\|_{L^p(\mathbb{R}^n)} = \|\mu\|^{1/p}$, $|\nabla f| = p^{-1}|\varrho|^{1/p-1}|\nabla\varrho|$, so $\|\nabla f\|_{L^p(\mathbb{R}^n)} = p^{-1}\|\nabla\varrho/\varrho\|_{L^p(|\mu|)}$, which gives (3.4.1). □

3.4.11. Corollary. *Let μ be a measure on $\mathbb{R}^n$ with a bounded density ϱ and let a measure ν on $\mathbb{R}^n$ be such that $\nu = \psi \cdot \mu$ and $d_{e_i}\nu = g_i \cdot \mu$, where $\psi, g_i \in L^p(|\mu|)$ and $p > n$. Then ν admits a bounded locally Hölder continuous density ϱ_ν satisfying the estimate*

$$\sup_x |\varrho_\nu(x)| \leqslant C(n,p)\|\varrho\|_\infty^{1-1/p}\|\psi\|_{L^p(|\mu|)}^{1-n/p}\sum_{i=1}^n \|g_i\|_{L^p(|\mu|)}^{n/p}, \qquad (3.4.2)$$

where $C(n,p)$ is a constant that depends only on n and p. In addition, if ϱ is Hölder continuous in a neighborhood of some point x_0 and $\varrho(x_0) = 0$, then $\varrho_\nu(x_0) = 0$.

PROOF. Our hypotheses yield that ν is Fomin differentiable along all e_i, hence ν has a density $\varrho_\nu \in W^{1,1}(\mathbb{R}^n)$. Therefore, $\varrho_\nu = \psi\varrho$, $\partial_{x_i}\varrho_\nu = g_i\varrho$ and
$$\|\varrho_\nu\|_{L^p(\mathbb{R}^n)} \leqslant \|\varrho\|_\infty^{1-1/p}\|\psi\|_{L^p(|\mu|)}, \quad \|\partial_{x_i}\varrho_\nu\|_{L^p(\mathbb{R}^n)} \leqslant \|\varrho\|_\infty^{1-1/p}\|g_i\|_{L^p(|\mu|)}.$$
It remains to use again Theorem 2.2.9. Suppose now that ϱ is Hölder continuous of order $\kappa > 0$ in a ball centered at x_0 and $\varrho(x_0) = 0$. We may assume that $x_0 = 0$ and that $p < n + \kappa$: one can always make p smaller in order that it be in $(n, n+\kappa)$. Let us fix a smooth function η with support in the ball of radius 1 centered at the origin such that $0 \leqslant \eta \leqslant 1$, $\eta(0) = 1$, and $|\nabla\eta/\eta|^r\eta \in L^1(\mathbb{R}^n)$. Given $\varepsilon > 0$, let $\eta_\varepsilon(x) = \eta(\varepsilon^{-1}x)$. Let us consider the measure ν_ε with density $\eta_\varepsilon\varrho_\nu$. It has support in the ball U_ε of radius ε and its density with respect to $\mu_\varepsilon := \eta_\varepsilon \cdot \mu$ equals ψ. We have
$$d_{e_i}\nu_\varepsilon = g_i \cdot \mu_\varepsilon + \frac{\partial_{x_i}\eta_\varepsilon}{\eta_\varepsilon}\mu_\varepsilon.$$
Therefore, on account of (3.4.2) we obtain
$$\sup_{x \in U_\varepsilon}|\eta_\varepsilon(x)\varrho_\nu(x)| \leqslant C \sup_{x \in U_\varepsilon}|\eta_\varepsilon(x)\varrho(x)|^{1-1/p}\sum_{i=1}^n\|g_i + \partial_{x_i}\eta_\varepsilon/\eta_\varepsilon\|_{L^p(|\mu_\varepsilon|)}^{n/p},$$
where C is independent of ε. Since
$$\int_{U_\varepsilon}\left|\frac{\partial_{x_i}\eta_\varepsilon}{\eta_\varepsilon}\right|^p \eta_\varepsilon|\varrho|\,dx \leqslant \varepsilon^{n-p}\sup_{x \in U_\varepsilon}|\varrho(x)|\int_{U_1}\left|\frac{\partial_{x_i}\eta}{\eta}\right|^p \eta\,dx,$$
we obtain that $\|g_i + \partial_{x_i}\eta_\varepsilon/\eta_\varepsilon\|_{L^p(|\mu_\varepsilon|)}$ is estimated by
$$\|g_i\|_{L^p(|\mu_\varepsilon|)} + \|\partial_{x_i}\eta_\varepsilon/\eta_\varepsilon\|_{L^p(|\mu_\varepsilon|)} \leqslant C_1 + C_1\varepsilon^{n-p}\sup_{x \in U_\varepsilon}|\varrho(x)|,$$
where C_1 is independent of ε. Since $\varrho(0) = 0$ and ϱ is κ-Hölder continuous in U_ε for sufficiently small $\varepsilon > 0$, we have $\sup_{x \in U_\varepsilon}|\varrho(x)| \leqslant C_2\varepsilon^\kappa$. Therefore, the previous estimates show that
$$|\varrho_\nu(0)| \leqslant C_3\varepsilon^\alpha, \quad \alpha := \kappa(1 - 1/p) + (n + \kappa - p)n/p > 0,$$
where C_3 is independent of ε. Letting $\varepsilon \to 0$, we obtain $\varrho_\nu(0) = 0$. □

3.4.12. Corollary. *Let μ be a measure on $\mathbb{R}^n$ with a bounded density ϱ and let a measure ν on $\mathbb{R}^n$ be such that one has $\partial^{(\alpha)}\nu = g_\alpha \cdot \mu$ for every multi-index α with $0 < |\alpha| \leqslant m$, where $g_\alpha \in L^p(|\mu|)$ and $p > n$. Then ν has a density $\varrho_\nu \in C_b^{m-1}(\mathbb{R}^n)$ for any α with $0 < |\alpha| < m$ satisfying the estimates*
$$\sup_x|\partial^{(\alpha)}\varrho_\nu(x)| \leqslant C(n,p,m)\|\varrho\|_\infty^{1-1/p}\|g_\alpha\|_{L^p(|\mu|)}^{1-n/p}\sum_{j=1}^n\|g_{\alpha+\delta_j}\|_{L^p(|\mu|)}^{n/p},$$

where $C(n,p,m)$ depends only on n, p, and m, and the multi-index $\alpha + \delta_j$ is defined by $\partial^{(\alpha+\delta_j)} := \partial_{x_j} \partial^{(\alpha)}$. In addition, the functions $\partial^{(\alpha)} \varrho_\nu$ with $|\alpha| < m$ are locally Hölder continuous.

PROOF. We apply the previous corollary to $\partial^{(\alpha)} \nu$ with $|\alpha| < m$ and observe that $d_{e_j} \partial^{(\alpha)} \nu = g_{\alpha + \alpha_j} \cdot \mu$. □

3.5. Characterization by conditional measures

The examples in the previous section hint to characterize directional differential properties of measures by means of conditional measures. Let μ be a Radon measure on a locally convex space X and $h \in X$, $h \neq 0$. Then there exists a closed hyperplane Y in X for which $X = \mathbb{R}^1 h \oplus Y$. In addition, there exists a continuous linear mapping $\pi \colon X \to Y$, $th + y \mapsto y$, called the natural projection. Let ν denote the image of the measure $|\mu|$ under this projection.

Let us recall that according to §1.3 one can choose Borel measures μ^y on the straight lines $y + \mathbb{R}^1 h$, $y \in Y$, such that for every Borel set $B \subset X$ the following equality holds:

$$\mu(B) = \int_Y \mu^y(B) \, \nu(dy). \tag{3.5.1}$$

These measures are determined uniquely up to a redefinition for points y from a set of ν-measure zero.

3.5.1. Theorem. *Let μ be a Radon measure on a locally convex space X and let $h \in X$. Then*

(i) *the measure μ is continuous along h if and only if the measures μ^y on the straight lines $y + \mathbb{R}^1 h$ are continuous along h for ν-a.e. y;*

(ii) *the measure μ is Fomin differentiable along h if and only if the measures μ^y on the straight lines $y + \mathbb{R}^1 h$ are Fomin differentiable along h for ν-a.e. y and one has*

$$\int_Y \|d_h \mu^y\| \, \nu(dy) < \infty;$$

(iii) *the measure μ is Skorohod differentiable along h if and only if for ν-a.e. y the measures μ^y on the straight lines $y + \mathbb{R}^1 h$ are Skorohod differentiable along h and one has*

$$\int_Y \|d_h \mu^y\| \, \nu(dy) < \infty;$$

(iv) *the measure μ is L^p-differentiable along h if and only if the measures μ^y on the straight lines $y + \mathbb{R}^1 h$ are L^p-differentiable along h for ν-a.e. y and one has*

$$\int_Y \int_{y+\mathbb{R}^1 h} |\beta_h^{\mu^y}(x)|^p \, |\mu^y|(dx) \nu(dy) < \infty;$$

(v) *the measure μ is quasi-invariant along h if and only if the measures μ^y on the straight lines $y + \mathbb{R}^1 h$ are quasi-invariant along h for ν-a.e. y.*

3.5. Characterization by conditional measures

In the case of Fomin or Skorohod differentiability we have

$$d_h\mu(B) = \int_Y d_h\mu^y(B)\,\nu(dy) \qquad (3.5.2)$$

for all Borel sets B.

PROOF. The sufficiency of the listed requirements on conditional measures for the respective properties of μ is easily verified using formulae (1.3.1) and (1.3.3). For example, if almost all conditional measures μ^y are continuous along h, then, for every $A \in \mathcal{B}(X)$, we have $\mu^y(A + t_n h) \to \mu^y(A)$ as $t_n \to 0$, which by the Lebesgue dominated convergence theorem yields that $\mu(A + t_n h) \to \mu(A)$.

The necessity of these conditions is not obvious. Let the measure μ be Fomin differentiable along h. Since $d_h\mu \ll \mu$, according to (1.3.1) and (1.3.4) there exist measures μ^y and $\xi^y \ll \mu^y$ on $y + \mathbb{R}^1 h$ such that

$$\mu(A) = \int_Y \mu^y(A)\,\nu(dy),$$

$$d_h\mu(A) = \int_Y \xi^y(A)\,\nu(dy)$$

for all Borel sets A. Hence for all $s \in \mathbb{R}^1$ we have

$$d_h\mu(A + sh) = \int_Y \xi^y(A + sh)\,\nu(dy),$$

$$\mu(A + th) - \mu(A) = \int_Y [\mu^y_{th}(A) - \mu^y(A)]\,\nu(dy),$$

whence by (3.3.1) we find

$$\int_Y [\mu^y_{th}(A) - \mu^y(A)]\,\nu(dy) = \int_Y \int_0^t \xi^y(A + sh)\,ds\,\nu(dy).$$

Since A was arbitrary and conditional measures are determined uniquely ν-a.e., we obtain the equality

$$\mu^y_{th} - \mu^y = \int_0^t \xi^y_{sh}\,ds$$

for ν-a.e. y. For such y, taking into account the relationship $\xi^y \ll \mu^y$, we obtain Fomin differentiability of μ^y along h and the equality $d_h\mu^y = \xi^y$. Therefore, assertion (ii) is proven, and (iv) is its direct corollary. Similarly we verify (v), which gives (i) due to the fact that any measure continuous along h is absolutely continuous with respect to some measure quasi-invariant along h. Finally, (iii) is proven analogously to (ii) with the only difference being that now the role of ν is played by the projection of the measure $|\mu| + |d_h\mu|$ to the hyperplane Y. $\square$

A straightforward modification of our reasoning leads to the following more general result.

3.5.2. Theorem. *Let μ be a Radon measure on a locally convex space X that is a direct topological sum of a locally convex space Y and a Souslin locally convex space F and let S be a finite or countable subset of F. Suppose that for some n there exist derivatives $d_{h_1}\cdots d_{h_k}\mu$ (in the sense of Fomin or Skorohod) for every $h_1,\ldots,h_k \in S$, $k \leqslant n$. Then one can find conditional measures μ^y, $y \in Y$, on $y + F$ such that, for every $h_1,\ldots,h_k \in S$, $k \leqslant n$, the derivatives $d_{h_1}\cdots d_{h_k}\mu^y$ exist in the corresponding sense and we have*

$$d_{h_1}\cdots d_{h_k}\mu(B) = \int_Y d_{h_1}\cdots d_{h_k}\mu^y(B)\,\nu(dy) \quad \text{for all } B \in \mathcal{B}(X),$$

where ν is the image of $|\mu|$ under the natural projection to Y. Analogous assertions are true for the infinite differentiability, continuity, and quasi-invariance.

PROOF. In order to extend (ii) of Theorem 3.5.1 to this case, we denote by ν the projection of $|\mu|$ on F and choose conditional measures μ^y and $\xi^{h,y}$ for μ and $d_h\mu$, where $h \in S$, on the sets $y + F$, $y \in Y$. Since S is at most countable, we arrive as above to the equality

$$\mu_{th}^y - \mu^y = \int_0^t \xi_{sh}^{h,y}\,ds$$

for ν-a.e. $y \in Y$ simultaneously for all $h \in S$ and all t. As above, this shows that μ^y is Fomin differentiable along any $h \in S$ and $d_h\mu^y = \xi_h$. Other assertions from the previous theorem are extended similarly. The case of higher differentiability follows by induction. For example, if, for some $h_1, h_2 \in S$, the measure $d_{h_2}\mu$ is differentiable along h_1, the same reasoning as above shows that for ν-a.e. y one has

$$d_{h_2}\mu_{th_1}^y - d_{h_2}\mu^y = \int_0^t \eta_{sh_1}^{h_1,h_2,y}\,ds$$

for all t, where $\{\eta^{h_1,h_2,y}\}_{y\in Y}$ is the family of conditional measures on the sets $y + F$ corresponding to the measure $d_{h_1}d_{h_2}\mu$. $\square$

3.5.3. Corollary. *In the situation of the previous theorem, the measure μ is analytic along all vectors $h \in S$ precisely when one can find conditional measures μ^y on $y+F$ such that, for each $h \in S$, the measures μ^y are analytic along h and for some $c(h) > 0$ one has*

$$\sum_{n=0}^{\infty} \int_Y \frac{t^n}{n!} \|d_h^n \mu^y\|\,\nu(dy) < \infty \quad \forall t \in [0, c(h)].$$

PROOF. If this condition is fulfilled, then one readily verifies that the series $\sum_{n=0}^{\infty} z^n d_h^n \mu / n!$ converges in variation if $|z| < c(h)$. Conversely, convergence of this series yields convergence of the series $\sum_{n=0}^{\infty} t^n \|d_h^n \mu\| / n!$ whenever $t \in [0, c(h)]$. Since we have

$$\|d_h^n \mu\| = \int_Y \|d_h^n \mu^y\|\,\nu(dy),$$

the conclusion follows by the monotone convergence theorem. $\square$

3.6. Skorohod differentiability

Let us recall that conditional measures (in the form of conditional expectations) arise also in expressions for logarithmic derivatives of linear images of differentiable measures (Proposition 3.3.13).

3.6. Skorohod differentiability

Here we consider measures differentiable in the Skorohod sense. By using our characterization in terms of conditional measures we shall derive a result from [174] which guarantees the existence of a Radon (and not just Baire, as noted above) weak derivative for every Skorohod differentiable Radon measure on an arbitrary locally convex space. The proof below differs from that of [174].

3.6.1. Theorem. *Let a Radon measure μ on a locally convex space X be Skorohod differentiable along a vector $h \neq 0$. Then, there exists a Radon measure ν which is its Skorohod derivative, i.e., satisfies (3.1.4) for all bounded continuous functions f on X.*

PROOF. Let us decompose X in the sum $X = Y \oplus \mathbb{R}^1 h$, where Y is a closed hyperplane, denote by μ_0 the projection of $|\mu|$ to Y. On the straight lines $y + \mathbb{R}^1 h$ we can take conditional measures μ^y that are Skorohod differentiable along h. Now it suffices to show that the measure $d_h \mu$ is given by the equality

$$d_h\mu(A) = \int_Y d_h\mu^y(A)\,\mu_0(dy),$$

where the right side generates a Radon measure. We first observe that the right side defines a Baire measure that agrees with $d_h\mu$ on all Baire sets A. To see this, we recall that for every $f \in C_b(X)$, the integral of f against $d_h\mu^y$ is the limit of the integral of $n[(f(x + n^{-1}h) - f(x)]$ against μ^y as $n \to \infty$. In addition, the latter integral is majorized in the absolute value by $\sup_x |f(x)| \|d_h\mu^y\|$, which is a μ_0-integrable function. On the other hand, the integral of $n[(f(x + n^{-1}h) - f(x)]$ against μ tends to the integral of f against $d_h\mu$. Since μ_0 is a Radon measure and the function $\|d_h\mu^y\|$ is μ_0-integrable, we conclude that the measure $d_h\mu$ is tight. Indeed, we can take increasing compact sets $K_n \subset Y$ such that $\mu_0(Y \backslash K_n) \to 0$. Then the sets $S_n = K_n + \{th : t \in [-n,n]\}$ are compact and it is easily seen from the established identity on $\mathcal{B}a(X)$ that $|d_h\mu|^*(X \backslash S_n) \to 0$. Now we take for ν the Radon extension of $d_h\mu$ (see [193, Corollary 7.3.3] on its existence). Finally, we obtain that the above disintegration formula holds for all Borel sets A. Indeed, the Radon measure ν admits some disintegration with respect to the projection of $|\nu|$ to Y. The latter is absolutely continuous with respect to μ_0 (if $A_0 \subset Y$ is a Baire set of μ_0-measure zero, then we have $d_h\mu(A_0 \oplus \mathbb{R}^1 h) = 0$). Hence there exist measures σ^y on the straight lines $y + \mathbb{R}^1 h$ such that

$$\nu(A) = \int_Y \sigma^y(A)\,\mu_0(dy)$$

for all Borel sets A. Then we obtain $\sigma^y = d_h \mu^y$ for μ_0-a.e. y, which yields the desired equality and completes the proof. □

It is readily verified that if a measure μ on the real line is Skorohod differentiable along 1, then the measure $|\mu|$ is also. Indeed, the measure μ is given by a density p of bounded variation, and the function $|p|$ is also of bounded variation. In addition, we have $\operatorname{Var}|p| \leqslant \operatorname{Var} p$ (Exercise 3.9.2). The functions p^+ and p^- have the same property. In terms of Skorohod differentiability this means Skorohod differentiability of the measures $|\mu|$, μ^+, μ^- and the validity of the inequalities $\|d_1|\mu|\| \leqslant \|d_1\mu\|$, $\|d_1\mu^+\| \leqslant \|d_1\mu\|$, $\|d_1\mu^-\| \leqslant \|d_1\mu\|$. Note that $\|d_1|\mu|\|$ can be strictly smaller than $\|d_1\mu\|$: take μ on $\mathbb{R}^1$ with the density $I_{[0,1]} - I_{[1,2]}$; then $d_1\mu = \delta_0 - 2\delta_1 + \delta_2$, $d_1|\mu| = \delta_0 - \delta_2$. It is clear from this example that in general there is no estimate of the type $\|d_1\mu\| \leqslant C\|d_1|\mu|\|$. Note also that $d_1|\mu| \ll d_1\mu$ (see the next lemma), but these two measures need not be equivalent unlike the case of the Fomin derivative.

3.6.2. Lemma. *For any Borel set $B \subset \mathbb{R}^1$, the following inequality holds*: $|d_1|\mu||(B) \leqslant |d_1\mu|(B)$.

PROOF. Clearly, it suffices to show that the desired estimate holds for every finite collection of disjoint intervals. In turn, it is enough to show that
$$|d_1|\mu||([a,b)) \leqslant |d_1\mu|([a,b))$$
for all a, b. It remains to observe that the left side equals $||\varrho(b)| - |\varrho(a)||$ and the right side equals $|\varrho(b) - \varrho(a)|$, where ϱ is a left-continuous density of μ of bounded variation. □

By using conditional measures we easily extend this to the general case and obtain the following result.

3.6.3. Theorem. *Suppose that a Radon measure μ on a locally convex space X is Skorohod differentiable along a vector h. Then, its positive part μ^+, its negative part μ^-, and its variation $|\mu|$ are Skorohod differentiable along h as well and*
$$\|d_h|\mu|\| \leqslant \|d_h\mu\|, \quad \|d_h\mu^+\| \leqslant \|d_h\mu\|, \quad \|d_h\mu^-\| \leqslant \|d_h\mu\|.$$

PROOF. Let Y and μ_0 be the same as in the proof of the previous theorem. We take Skorohod differentiable conditional measures μ^y. The measures $d_h[(\mu^y)^+]$ and $d_h[(\mu^y)^-]$ exist and their variations do not exceed $\|d_h\mu^y\|$. Note that the measures μ^+ and μ^- are mutually singular and possess conditional measures $(\mu^+)^y$ and $(\mu^-)^y$, which are mutually singular for μ_0-a.e. y. Then it follows that $(\mu^y)^+ = (\mu^+)^y$ and $(\mu^y)^- = (\mu^-)^y$ for μ_0-a.e. y. So the only technicality is to show that $d_h[(\mu^y)^+]$ and $d_h[(\mu^y)^-]$ depend measurably on y. □

The next result was also obtained in [**174**]; we give another proof.

3.6. Skorohod differentiability

3.6.4. Theorem. *A Radon measure μ on a locally convex space X is Skorohod differentiable along a vector h if and only if for every $A \in \mathcal{B}(X)$ there exists a number $c(A)$ such that*

$$|\mu(A + th) - \mu(A)| \leqslant c(A)|t|. \tag{3.6.1}$$

In this case, the numbers $c(A)$ can be chosen uniformly bounded.

PROOF. It is clear that Skorohod differentiability gives estimate (3.6.1) with a common constant $C = \|d_h\mu\|$. Conversely, if we have (3.6.1), then by the Nikodym theorem one can take a common constant, whence we obtain $\|\mu_{th} - \mu\| \leqslant C|t|$ with some number C. Let us show that this ensures Skorohod differentiability.

First we consider the case $X = \mathbb{R}^1$. We observe that it suffices to have an estimate of the type

$$\|\mu_{t_i h} - \mu\| \leqslant C|t_i|, \quad \text{where } t_i \to 0.$$

Indeed, this estimate yields that for every $\varphi \in C_0^\infty(\mathbb{R}^1)$ one has

$$\left| \int_{\mathbb{R}} t_i^{-1} [\varphi(x - t_i) - \varphi(x)] \, \mu(dx) \right| = \left| t_i^{-1} \int_{\mathbb{R}} \varphi(x) \, [\mu_{t_i} - \mu](dx) \right| \leqslant C \sup_x |\varphi(x)|,$$

hence

$$\left| \int_{\mathbb{R}} \varphi'(x) \, \mu(dx) \right| \leqslant C \sup_x |\varphi(x)|.$$

This shows that the generalized derivative of μ is a bounded measure. To be more precise, the previous estimate ensures the existence of measures on all intervals $(-n, n)$ that are generalized derivatives of μ on $(-n, n)$, but the same estimate shows that the total variations of these derivatives on $(-n, n)$ are uniformly bounded. Hence μ is Skorohod differentiable. In addition, $\|d_1\mu\| \leqslant C$.

Let us consider the general case. Let $h \neq 0$. Let us write X in the form $X = Y \oplus \mathbb{R}^1 h$, where Y is a closed hyperplane, denote by μ_0 the projection of $|\mu|$ to Y and take conditional measures μ^y on the straight lines $y + \mathbb{R}^1 h$, $y \in Y$. We may assume that μ_0 is a probability measure. Points $x \in X$ will be written in the form $x = (y, s)$, $y \in Y$, $s \in \mathbb{R}^1$. Then the conditional measures μ^y can be regarded as measures on $\mathbb{R}^1$. Let us set $t_n = n^{-1}$. Then

$$\int_Y t_n^{-1} \|\mu^y_{t_n h} - \mu^y\| \, \mu_0(dy) \leqslant C.$$

By Fatou's theorem

$$\int_Y \liminf_n t_n^{-1} \|\mu^y_{t_n h} - \mu^y\| \, \mu_0(dy) \leqslant C.$$

Hence for μ_0-a.e. y we have

$$C(y) := \liminf_n t_n^{-1} \|\mu^y_{t_n h} - \mu^y\| < \infty.$$

By the one-dimensional case for every such y the measure μ^y is Skorohod differentiable and $\|d_h\mu^y\| \leqslant C(y)$. Now we apply Theorem 3.5.1. □

Thus, most of the properties introduced in §3.1 can be described in terms of the functions $t \mapsto \mu(A + th)$.

A direct corollary of Theorem 3.6.4 is the fact that if X is linearly and continuously embedded into a locally convex space Y, then the collection of vectors of Skorohod differentiability for μ remains the same independently of whether μ is considered as a measure on X or on Y, although the classes of continuous functions in these two cases may differ. It is clear that μ extends to $\mathcal{B}(Y)$ since $X \cap B$ is a Borel set in X for every $B \in \mathcal{B}(Y)$. For other differential properties considered above this is obvious.

The following result gives a characterization of Skorohod differentiability in the integral form.

3.6.5. Theorem. *Let μ be a Radon measure on a locally convex space X and let $\mathcal{A} = \mathcal{B}(X)$. A measure μ is Skorohod differentiable along a vector h if and only if there exists a Radon measure ν such that for every set $A \in \mathcal{B}(X)$ one has*

$$\mu(A + th) - \mu(A) = \int_0^t \nu(A + sh)\,ds. \qquad (3.6.2)$$

This is also equivalent to the equality

$$\int_X [f(x - th) - f(x)]\,\mu(dx) = \int_0^t \int_X f(x - sh)\,\nu(dx)\,ds \qquad (3.6.3)$$

for all bounded Borel functions f or for all functions $f \in C_b(X)$. In this case $\nu = d_h\mu$. An analogous assertion with Baire functions in place of Borel ones is true for the class of Baire measures. For Radon measures or measures on $\sigma(X)$ it suffices to have (3.6.3) for $f \in \mathcal{F}C_b^\infty$.

PROOF. The equivalence of (3.6.2) and (3.6.3) is obvious since the former is just the latter for $f = I_A$. Moreover, (3.6.3) for all $f \in C_b(X)$ yields (3.6.3) for all bounded Borel functions f in the case of a Radon measure and all bounded Baire functions f in the case of a Baire measure (or for $f \in \mathcal{F}C_b^\infty$ in the cases of measures on $\sigma(X)$ and Radon measures). Finally, let us show that equality (3.6.3) for all $f \in C_b(X)$ is equivalent to the fact that ν is the Skorohod derivative of the measure μ. Indeed, dividing both sides of (3.6.3) by t and letting $t \to 0$, in the limit we obtain on the left the integral of f against ν since the integrals of $t^{-1}f(x - sh)$ in s over $[0, t]$ are uniformly bounded and converge to $f(x)$. On the other hand, if ν is the Skorohod derivative of μ along h, then both sides of (3.6.3) are differentiable in t and their derivatives at any point t_0 coincide with the integral of $-f(x - t_0 h)$ against ν. Vanishing at $t = 0$, they are equal everywhere. □

By the proven theorem the Skorohod derivative of any Fomin differentiable measure coincides with its Fomin derivative. This justifies our usage of the symbol $d_h\mu$ also for weak derivatives.

Note that, by Theorem 3.6.4, under the hypotheses of Theorem 3.6.5 and its corollary in the sufficiency part, in place of the assumption that ν is a Radon measure it suffices to require that it be defined only on $\sigma(X)$

3.6. Skorohod differentiability

and satisfy the corresponding equalities only for $\sigma(X)$-measurable sets and functions. In that case, such a measure can be chosen Radon, which follows from Theorem 3.6.4 and the necessary condition from Theorem 3.6.5.

3.6.6. Corollary. *If a Radon measure μ on a locally convex space X is Skorohod differentiable along a vector h, then it is Fomin differentiable along this vector if and only if at least one of the following conditions is fulfilled (then all of them are fulfilled):*
(i) *its weak derivative ν is absolutely continuous with respect to μ;*
(ii) *its weak derivative ν is continuous along h.*
The same is true for measures on $\sigma(X)$.

3.6.7. Corollary. *Suppose that a Radon measure μ on a locally convex space X is Skorohod differentiable along a vector h. Then, for every bounded Borel function f possessing a uniformly bounded partial derivative $\partial_h f$ we have*
$$\int_X \partial_h f(x)\,\mu(dx) = -\int_X f(x)\,d_h\mu(dx). \tag{3.6.4}$$

Applying (3.6.4) to the functions $f(x) = \exp(il(x))$ with $l \in X^*$ and recalling that the Fourier transform $\widetilde{\mu}$ of the measure μ is defined by the formula
$$\widetilde{\mu}\colon X^* \to \mathbb{C}, \quad \widetilde{\mu}(l) = \int_X \exp(il(x))\,\mu(dx),$$
we obtain the identity
$$\widetilde{d_h\mu}(l) = -il(h)\widetilde{\mu}(l). \tag{3.6.5}$$

As observed in [**166**], this identity is actually a criterion of differentiability in terms of the Fourier transform (which solves a problem posed by Fomin [**443**]). Namely, we have the following necessary and sufficient condition for differentiability.

3.6.8. Theorem. *Let μ be a measure on the σ-algebra $\sigma(X)$ in a locally convex space X and let $h \in X$. Then*
(i) *the measure μ has a Skorohod derivative along the vector h if and only if there exists a measure ν on $\sigma(X)$ such that*
$$\widetilde{\nu}(l) = -il(h)\widetilde{\mu}(l), \quad \forall l \in X^*; \tag{3.6.6}$$
in this case ν coincides with the weak derivative;
(ii) *the measure μ is Fomin differentiable along h if and only if there exists a measure ν on $\sigma(X)$ satisfying equality (3.6.6) and absolutely continuous with respect to μ. In this case $\nu = d_h\mu$.*

PROOF. From (3.6.6) we obtain (3.6.3) for all functions $f = \exp(il)$, where $l \in X^*$, which gives (3.6.3) for all $f \in \mathcal{F}C_b^\infty$ since two measures on $\sigma(X)$ with equal Fourier transforms coincide. $\square$

3.6.9. Corollary. *Suppose that a Radon measure μ is Skorohod differentiable along vectors h and k. Then it is Skorohod differentiable along*

every linear combination of these vectors and for the weak derivatives we have
$$d_{sh+tk}\mu = sd_h\mu + td_k\mu.$$

A trivial corollary of Theorem 3.6.8 is the uniform boundedness of the function $\varphi\colon l \mapsto -il(h)\widetilde{\mu}(l)$ on X^*. Let us observe that if X is a Hilbert space, then this function is always continuous in the Sazonov topology (even for nondifferentiable measures μ). We recall that the Sazonov topology is generated by the family of seminorms of the form $x \mapsto \|Tx\|$, where T is a Hilbert–Schmidt operator, and the Fourier transform of any Radon measure on X is continuous in this topology (see [**193**, §7.13]). So it is natural to ask which conditions on the function φ guarantee that it is the Fourier transform of a bounded measure. Unfortunately, such a measure must be signed. For this reason the classical Minlos–Sazonov theorem (which describes the Fourier transforms of nonnegative measures as positive definite functions continuous in the Sazonov topology) is not applicable to this situation. There is an analogue of the Minlos–Sazonov theorem for signed measures (see Tarieladze [**1115**]). However, the corresponding hypotheses include the assumption that φ is the Fourier transform of a cylindrical measure ν of bounded variation. In our special case such analogues are useless since the boundedness of variation of ν is precisely what is needed for Skorohod differentiability. Indeed, if we know that there exists a cylindrical measure ν of bounded variation such that $\varphi = \widetilde{\nu}$, then its finite dimensional projections are the weak derivatives of the corresponding projections of μ. This gives the uniform Lipschitzness of the functions $t \mapsto \mu(C + th)$ for all cylinders, which by Theorem 3.6.4 implies Skorohod differentiability.

One might try to find other conditions of differentiability in terms of the Fourier transform not involving a priori assumptions of bounded variation type. Here, for example, it is not enough that the function φ be uniformly bounded.

3.6.10. Example. Let μ be the countable product of Cauchy probability measures μ_n on the real line given by densities
$$p_n(y) = \frac{2^n}{\pi} \frac{1}{1 + 2^{2n}t^2}.$$

Since $\mu(l^2) = 1$ (Exercise 3.9.6), this measure can be regarded as a measure on l^2. Let $h = (h_n)$, where $h_n = 2^{-n}$. Then the measure μ is not Skorohod differentiable along h, but the function $\varphi(y) = -i(y,h)\widetilde{\mu}(y)$ is uniformly bounded and continuous in the Sazonov topology on l^2.

PROOF. By Corollary 4.1.2 below, the set $D_C(\mu)$ of all vectors of S-differentiability of μ is the space of all sequences (x_n) such that $\sum_{n=1}^{\infty} 2^{2n}x_n^2 < \infty$. Hence $h \notin D_C(\mu)$. The Fourier transform of μ is given by the equality
$$\widetilde{\mu}(y) = \exp\Big(\sum_{n=1}^{\infty} 2^{-n}|y_n|\Big), \quad y \in l^2.$$

3.6. Skorohod differentiability

It is easily verified that the function φ defined above is uniformly bounded and continuous in the Sazonov topology. However, this function cannot be the Fourier transform of a cylindrical measure of bounded variation. □

The same example shows that even exponential decay of the function $\widetilde{\mu}$ does not guarantee the differentiability of the measure μ along h. However, as the following nice result due to A.A. Belyaev shows, the measure μ is analytic along h provided that $\widetilde{\mu}$ has support in a suitable strip. This result can be regarded as an infinite dimensional version of the classical Paley–Wiener theorem (see Rudin [976, Ch. 7]).

3.6.11. Theorem. *Let μ be a measure on $\sigma(X)$ and $h \in X$. Let $R > 0$ be such that $\widetilde{\mu}$ vanishes outside the strip $\Pi_R = \{l \in X^* \colon |l(h)| \leqslant R\}$. Then the measure μ is analytic and quasi-invariant along h. In addition, for every bounded $\sigma(X)$-measurable function f the function*

$$t \mapsto \int_X f(x - th)\, \mu(dx)$$

extends to an entire function ψ_f on the complex plane satisfying the estimate

$$|\psi_f(z)| \leqslant \|\mu\| \sup_X |f(x)| \exp(R \cdot \operatorname{Im} z). \qquad (3.6.7)$$

Conversely, if estimate (3.6.7) is valid for every bounded $\sigma(X)$-measurable function f, then $\widetilde{\mu}$ has support in Π_R.

PROOF. 1) Let $\varepsilon > 0$ and let $g \in C_0^\infty(\mathbb{R}^1)$ be a function with support in $[-R - \varepsilon, R + \varepsilon]$ and equal to 1 on $[-R, R]$. Then g is the Fourier transform of a function $p_0 \in \mathcal{S}(\mathbb{R}^1)$. Set $p := p_0/\sqrt{2\pi}$. We have

$$p^{(n)}(x) = \frac{i^n}{2\pi} \int_{-R-\varepsilon}^{R+\varepsilon} e^{itx} g(t)\, dt,$$

hence

$$x^2 p^{(n)}(x) = -\frac{i^n}{2\pi} \int_{-R-\varepsilon}^{R+\varepsilon} e^{itx} [t^n g''(t) + 2nt^{n-1} g'(t) + n(n-1) g(t)]\, dt,$$

which yields the estimate $|p^{(n)}(x)| \leqslant C_1 n^2 R^n (1 + x^2)^{-1}$ with some C_1 independent of n. Consequently, $\|p^{(n)}\|_{L^1} \leqslant C_2 n^2 R^n$. Therefore, the measure λ with density p is analytic (note that from the Paley–Wiener theorem it only follows that p is analytic). Note also that for all complex numbers z we have

$$|p(x + z)| \leqslant \frac{1}{2\pi} \int_{-R-\varepsilon}^{R+\varepsilon} |e^{itz}|\, |g(t)|\, dt \leqslant C_3 \exp\bigl[(R + \varepsilon) |\operatorname{Im} z|\bigr].$$

Let ψ be a bounded Borel function on $\mathbb{R}^1$ with bounded support. Then the function $\psi(-\cdot) * p$ is analytic and the previous estimate yields

$$\left| \int \psi(x - z) p(x)\, dx \right| = \left| \int \psi(x) p(x + z)\, dx \right| \leqslant C(\psi, \lambda) \exp\bigl[(R + \varepsilon) |\operatorname{Im} z|\bigr].$$

However, we need a more explicit bound on $C(\psi, \lambda)$. Since for real z we have
$$\left|\int \psi(x-z)\,\lambda(dx)\right| \leqslant \|\lambda\| \cdot \|\psi\|_\infty,$$
the Phragmén–Lindelöf theorem gives the estimate
$$\left|\int \psi(x) p(x+z)\,dx\right| \leqslant \|\lambda\| \cdot \|\psi\|_\infty \exp\bigl[(R+\varepsilon)|\operatorname{Im} z|\bigr]. \qquad (3.6.8)$$
Indeed, denoting the convolution on the left by φ and considering the function $\varphi_1(z) := \varphi(z)\exp[i(R+\varepsilon)z]$ in the upper half-plane, we obtain a bounded holomorphic function such that $|\varphi_1(x)| \leqslant \|\lambda\| \cdot \|\psi\|_\infty$ on the real line. By the Phragmén–Lindelöf principle (see [**778**, Ch. 7, §1]) this estimate holds in the whole upper half-plane, which yields our claim, and the same applies to the lower half-plane.

Now if ψ is an arbitrary bounded Borel function, then it also satisfies the above estimate. Indeed, letting $\psi_n(t) = \psi(t) I_{[-n,n]}(t)$, we have a common estimate (3.6.8) for the corresponding convolutions $\psi_n(-\cdot) * p$, which for real z converge to $\psi(-\cdot)*p$. Hence the entire functions $\psi_n(-\cdot)*p$ converge uniformly on compact sets in $\mathbb{C}$ to an entire function that coincides with $\psi(-\cdot)*p$ on the real line and satisfies (3.6.8).

Let ν be the image of λ under the mapping $t \mapsto th$. Then $\widetilde{\nu}(l) = 1$ if $|l(h)| \leqslant R$ since by the change of variables formula one has
$$\widetilde{\nu}(l) = \widehat{p_0}\bigl(-l(h)\bigr) = g\bigl(-l(h)\bigr).$$
Since $\widetilde{\mu}(l) = 0$ if $|l(h)| \geqslant R$, we arrive at the equality $\widetilde{\mu}(l) = \widetilde{\mu}(l)\widetilde{\nu}(l)$, which means that $\mu = \mu * \nu$. In particular, the measure μ is analytic. Moreover, one has
$$\int_X f(x-th)\,\mu(dx) = \int_X f(x-th)\,\mu*\nu(dx)$$
$$= \int_X\int_X f(x+y-th)\,\mu(dy)\,\nu(dx) = \int\int_X f(sh-th+y)\,\mu(dy)\,\lambda(ds),$$
where the function
$$F\colon t \mapsto \int_X f(th+y)\,\mu(dy)$$
is Borel measurable and its absolute value is estimated by $\|\mu\| \cdot \sup_x |f(x)|$. By what has already been proven, the function
$$\psi(t) = \int F(s-t)\,\lambda(ds) = \int_X f(x-th)\,\mu(dx)$$
extends to an entire function satisfying (3.6.8). Applying the Phragmén–Lindelöf theorem once more we arrive at the estimate
$$|\psi(z)| \leqslant \|\mu\| \cdot \sup_x |f(x)| \exp\bigl[(R+\varepsilon)|\operatorname{Re} z|\bigr].$$
Since ε is arbitrary, we obtain the desired estimate.

2) We prove the converse. Let $l \in X^*$, $l(h) = R$, $|\alpha| \geqslant 1$. We shall show that $\widetilde{\mu}(\alpha l) = 0$. Considering the measure $\lambda = \mu \circ l^{-1}$ we reduce the assertion

3.6. Skorohod differentiability

to the one-dimensional case. The measure λ has analytic density p and by the Paley–Wiener theorem (see Rudin [**976**, Ch. 7]) the function

$$\varphi(t) = \int f(x - tR)p(x)\,dx, \quad f(x) = \exp(-x^2),$$

is the Fourier transform of a generalized function $G \in \mathcal{S}'$ with support in $[-R, R]$. Since $\varphi \in L^1(\mathbb{R}^1)$ and $\widehat{f} > 0$ we see that the support of $\widehat{p}$ is contained in $[-R, R]$. $\square$

It is worth noting that the following property of a measure μ along h has recently been studied by A.V. Shaposhnikov [**992**]: the absolute continuity of all functions $t \mapsto \mu(A + th)$. He has constructed an example of a probability measure on the real line showing that this property does not imply Skorohod differentiability. Namely, such a measure is given by the density

$$\varrho(x) := \sum_{m=1}^{\infty} (m^{2/3+1} 2^m)^{-1} I_{[2\pi 4^m, 2\pi(4^m + 2^m)]}(x) \sin^2 mx.$$

On the other hand, he proved the following positive assertions.

3.6.12. Theorem. *A Radon measure μ on a locally convex space X is Skorohod differentiable along a vector h if and only if the mapping $t \mapsto \mu_{th}$ with values in the space of measures equipped with the variation norm is absolutely continuous on $[0, 1]$.*

3.6.13. Theorem. *Let X be a locally convex space, $h \in X$, and let μ be a bounded Borel measure on X. Then the following conditions are equivalent:*

(1) for every Borel set A the function $t \mapsto \mu(A + th)$ is absolutely continuous on $[0, 1]$;

(2) for every open set $U \subset X$ the function $t \mapsto \mu(U + th)$ is absolutely continuous on $[0, 1]$;

(3) for every bounded Borel function f the function

$$\varphi_f(t) = \int_X f(x)\,\mu_{th}(dx)$$

is absolutely continuous on $[0, 1]$.

The next result of A.V. Shaposhnikov shows that Skorohod differentiability follows from the "global" absolute continuity of the considered functions.

3.6.14. Theorem. *Let h be a fixed vector in a locally convex space X. Suppose that a Radon measure μ on X has the following property: for every $A \in \mathcal{B}(X)$ the function $t \mapsto \mu(A + th)$ is absolutely continuous on the whole real line in the sense that, for every $\varepsilon > 0$, there is $\delta > 0$ such that for each finite collection of pairwise disjoint closed intervals $[s_1, t_1], \ldots, [s_k, t_k]$ of total length $\sum_{i=1}^{k} |t_i - s_i| < \delta$ one has $\sum_{i=1}^{k} |\mu_{t_i h}(A) - \mu_{s_i h}(A)| < \varepsilon$. Then μ is Skorohod differentiable along h.*

3.7. Higher order differentiability

For Skorohod or Fomin differentiability, higher order derivatives $d_h^n \mu$ and mixed derivatives $d_{h_1} \cdots d_{h_n} \mu$ are defined inductively. However, there is an alternative to call a measure μ n times Fomin differentiable along $h_1, \ldots, h_n$ if the functions

$$F\colon (t_1, \ldots, t_n) \mapsto \mu(A + t_1 h_1 + \cdots + t_n h_n) \qquad (3.7.1)$$

have partial derivatives $\partial_{t_1} \cdots \partial_{t_n} F$ for all sets $A \in \mathcal{B}(X)$. Here one can take $h_1 = \cdots = h_n$. In the case of the Skorohod derivative one considers the functions

$$(t_1, \ldots, t_n) \mapsto \int_X f(x + t h_1 + \cdots + t_n x) \, \mu(dx), \quad f \in C_b(X). \qquad (3.7.2)$$

3.7.1. Proposition. *Let μ be a measure on a locally convex space X (Radon or defined on $\sigma(X)$) and let $h \in X$.*

(i) Suppose that for every $A \in \sigma(X)$ the function $t \mapsto \mu(A + th)$ is n-fold differentiable. Then the measure μ is n-fold Fomin differentiable along h.

(ii) Let μ be a Baire measure such that, for every function $f \in C_b(X)$, the function

$$t \mapsto \int_X f(x - th) \, \mu(dx)$$

is n-fold differentiable. Then the measure μ is n-fold Skorohod differentiable along h and $(n-1)$-fold Fomin differentiable along h.

(iii) Let $h_1, \ldots, h_n \in X$. The mixed Fomin derivatives $d_{h_{i_1}} \cdots d_{h_{i_k}} \mu$, $1 \leqslant i_j \leqslant n$, exist if and only if the functions (3.7.1) have partial derivatives $\partial_{t_{i_1}} \cdots \partial_{t_{i_k}} F$ for all sets $A \in \mathcal{B}(X)$. In this case, these functions on $\mathbb{R}^n$ have continuous derivatives of order n. An analogous assertion is true in the case of Skorohod differentiability and functions (3.7.2).

PROOF. (i) The assertion follows by induction since the derivative of the indicated function is $d_h \mu(A + th)$. (ii) Here induction also applies since the derivative of the indicated function equals the integral of $f(x - th)$ with respect to $d_h \mu$. The $(n-1)$-fold Fomin differentiability follows from the fact that the Skorohod derivative of $d_h^{n-2} \mu$ is continuous being Skorohod differentiable. (iii) Clearly, the existence of all mixed Fomin derivatives yields the existence and continuity of the corresponding mixed partial derivatives of F, which gives its n-fold differentiability on $\mathbb{R}^n$. Conversely, if the indicated partial derivatives of F exist for all A, then, arguing by induction, we obtain the existence of the mixed Fomin partial derivatives of μ up to order $n - 1$ and the equality

$$\partial_{t_{i_1}} \cdots \partial_{t_{i_{k-1}}} F(t_1, \ldots, t_n) = d_{h_{i_1}} \cdots d_{h_{i_{k-1}}} \mu(A + t_1 h_1 + \cdots + t_n h_n).$$

The case of Skorohod differentiability is similar. □

It will be shown in Proposition 6.3.3 in Chapter 6 that for any Radon measure $\mu \geqslant 0$ Fomin differentiable along h, a sufficient condition for the differentiability of the measure $d_h\mu$ is the existence of a version of β_h^μ which is locally absolutely continuous (or everywhere differentiable) on the straight lines $x + \mathbb{R}^1 h$ such that $\partial_h \beta_h^\mu$ is integrable with respect to μ. Then we have $\beta_h^\mu \in L^2(\mu)$ and $d_h^2\mu = [\partial_h \beta_h^\mu + (\beta_h^\mu)^2] \cdot \mu$. According to Proposition 4.3.12, for any convex measure μ this is equivalent to the inclusion $\beta_h^\mu \in L^2(\mu)$. Under some additional conditions the equality for $d_h^2\mu$ can be differentiated, which yields certain expressions for the subsequent derivatives.

3.7.2. Remark. It is readily seen that if a measure μ is n-fold differentiable, then its nth derivative is symmetric, i.e., the measure $d_{h_1} \cdots d_{h_n}\mu$ does not depend on the ordering of $h_1, \ldots, h_n$ (this follows at once from the characterization of differentiability by means of the Fourier transform). Moreover, if, for example, the derivatives $d_h\mu$, $d_k\mu$, and $d_h d_k\mu$ exist in the Fomin or Skorohod sense, then $d_k d_h\mu$ exists in the same sense and equals $d_h d_k\mu$ since the function $l \mapsto -il(k)\widetilde{d_h\mu}(l)$ coincides with the Fourier transform of the measure $d_h d_k\mu$. However, the existence of a partial derivative $d_h d_k\mu$ does not imply the existence of $d_h\mu$. One can construct the corresponding example on the plane (Exercise 3.9.7).

3.8. Convergence of differentiable measures

The situation with the differentiability of limits of sequences of differentiable measures is similar to the one for functions: the limit is differentiable provided that we are given some convergence or boundedness of the derivatives of the convergent elements. The following results illustrate this general rule.

3.8.1. Theorem. *Let $\{\mu_n\}$ be a sequence of Radon measures on a locally convex space X possessing the Skorohod derivatives $d_{h_n}\mu_n$. Suppose that there exists a Radon measure on X such that*

$$\widetilde{\mu}(l) = \lim_{n \to \infty} \widetilde{\mu_n}(l), \quad \forall l \in X^*.$$

Suppose, in addition, that the sequence $\{h_n\}$ converges to a vector h in the weak topology. Then

(i) *if $\sup_n \|d_{h_n}\mu_n\| \leq C < \infty$, then the measure μ is Skorohod differentiable along h and $\|d_h\mu\| \leq C$;*

(ii) *if $h_n \equiv h$, the measures μ_n are Fomin differentiable along the vector h and the limit $\lim_{n \to \infty} d_h\mu_n(A)$ exists for every set $A \in \mathcal{B}(X)$, then the measure μ is Fomin differentiable;*

(iii) *if $\mu_n \geq 0$ and $\sup_n \|\beta_{h_n}^{\mu_n}\|_{L^p(\mu_n)} \leq C < \infty$, where $p > 1$, then the measure μ is L^p-differentiable along h and $\beta_h^\mu \in L^p(\mu)$. In this case one has $\|\beta_h^\mu\|_{L^p(\mu)} \leq C$.*

PROOF. It suffices to prove (i) in the one-dimensional case due to Theorem 3.6.8. Then the claim follows at once from Theorem 3.5.2 and Theorem 3.6.4. Assertion (ii) follows from (i) and Corollary 3.6.6 since the measure $d_h\mu$ turns out to be absolutely continuous with respect to μ. (iii) Let

us consider a functional F on the space $\mathcal{F}C_b^\infty(X)$ defined by the formula
$$F(f) = -\int_X \partial_h f(x)\, \mu(dx).$$
By convergence of the Fourier transforms of the nonnegative measures μ_n we have weak convergence of their finite dimensional projections, whence we obtain the equalities
$$F(f) = -\lim_{n\to\infty} \int_X \partial_{h_n} f(x) \mu_n(dx) = \lim_{n\to\infty} \int_X f(x)\, d_{h_n}\mu_n(dx)$$
$$= \lim_{n\to\infty} \int_X f(x) \beta_{h_n}^{\mu_n}(x)\, \mu_n(dx).$$
Therefore, for $q = p/(p-1)$ one has
$$|F(f)| \leqslant \limsup_{n\to\infty} C \left(\int_X |f(x)|^q\, \mu_n(dx) \right)^{1/q}.$$
The right side equals $C\|f\|_{L^q(\mu)}$. Thus, $|F(f)| \leqslant C\|f\|_{L^q(\mu)}$. Therefore, there exists a function $G \in L^p(\mu)$ such that
$$F(f) = \int_X f(x) G(x)\, \mu(dx) \quad \text{and} \quad \|G\|_{L^p(\mu)} \leqslant C.$$
This yields L^p-differentiability of the measure μ along h and the equality $\beta_h^\mu = G$ (see Theorem 3.6.8). □

An analogous theorem is true for measures on $\sigma(X)$ if we define Skorohod differentiability of μ along h as the existence of a measure ν on $\sigma(X)$ such that $\widetilde{\nu}(l) = -il(h)\widetilde{\nu}(l)$. Then the above proof of (i) covers the finite dimensional case, which gives the desired measure ν of bounded variation on the algebra of cylindrical sets. Its countable additivity follows from the case $X = \mathbb{R}^\infty$, to which the claim reduces since any countable set of cylinders is determined by a countable family of functionals $l_i \in X^*$, which define a map into $\mathbb{R}^\infty$.

It is clear from the proof that this theorem extends to nets in the following formulation.

3.8.2. Theorem. *Assertions* (i)–(iii) *of the previous theorem are valid also for nets $\{\mu_\alpha\}$ and $\{h_\alpha\}$ in place of countable sequences with the difference that in* (ii) *one should require additionally that the measures $d_{h_\alpha}\mu_\alpha$ be uniformly bounded and the limit $\lim_\alpha d_{h_\alpha}\mu_\alpha(A)$ define a measure absolutely continuous with respect to μ.*

Similarly one proves the following result.

3.8.3. Proposition. *Let a Radon measure μ on X be Fomin differentiable along vectors h_α which converge weakly to a vector h. If the measures $d_{h_\alpha}\mu$ are uniformly bounded and converge on every Borel set to a measure $\nu \ll \mu$ (for example, converge in variation), then μ is Fomin differentiable along h.*

3.9. Comments and exercises

Differentiable measures were introduced more than 40 years ago by S.V. Fomin [**440**]–[**443**]. Close ideas in terms of the distributions of random processes differentiably depending on parameters had been earlier developed by Pitcher [**905**]–[**908**]. For Gaussian measures, integration by parts was applied in Daletskiĭ [**316**], Gross [**504**]. The first detailed investigation of differentiable measures on linear spaces was undertaken in Averbukh, Smoljanov, Fomin [**82**], [**83**]. In particular, the existence of logarithmic derivatives of Fomin differentiable measures and the important Theorem 3.3.1 were established in [**82**]. One should also note Skorohod's book [**1046**], in which one more important kind of differentiability of measures was introduced. The principal results of all these investigations are briefly presented in Smolyanov [**1050**] and Dalecky, Fomin [**319**]. In the 1970s-80s considerable progress in this area was achieved in the works of Smolyanov's school. A thorough survey of the main achievements in the theory of differentiable measures over the first 20 years of its existence and a detailed discussion of connections with the conceptually close Malliavin calculus was given in Bogachev, Smolyanov [**233**]. These two directions along with subsequent developments are discussed in the extensive survey Bogachev [**189**]. The book Dalecky, Fomin [**319**] includes many results of Yu.L. Daletskiĭ's students. Some aspects of the theory of differentiable measures are presented in the books Norin [**834**] and Uglanov [**1141**], concerned with applications of this theory. At present, hundreds of papers are published on the theory of differentiable measures and its applications. Many of these works are mentioned in different chapters of this book in relation with concrete results or in comments. Continuous measures were implicitly introduced in Averbukh, Smoljanov, Fomin [**82**], where the continuity of differentiable measures was shown; explicitly the term was introduced in Romanov [**965**], where a study of continuous measures was initiated (see also Romanov [**966**]–[**971**]). Theorem 3.5.1 accumulates the results of several works. Its assertion (v), which implies (i), was obtained in Skorohod [**1045**], [**1046**] (see also Gihman, Skorohod [**477**, Ch. VII, §2]). The similar assertion (ii) was proved in Yamasaki, Hora [**1198**], and (iv) is its direct corollary (later it was also proved in Albeverio, Kusuoka, Röckner [**49**]). Finally, (iii) is analogous to (ii) and was noted in Bogachev, Smolyanov [**233**], Khafizov [**602**]. Similar questions are considered in Uglanov [**1138**]. Theorem 3.5.2 was obtained in [**233**]. Among other works related to general problems of the theory of differentiable measures, we mention Bogachev [**166**], [**169**], [**179**], [**181**], Kuo [**663**]–[**670**], Shimomura [**1027**], Uglanov [**1130**], [**1131**]. The absence of Radon probability measures on infinite dimensional spaces equivalent to all its shifts was established in Girsanov, Mityagin [**481**] and Sudakov [**1087**], [**1088**] under some restrictions on measures or spaces; Proposition 3.2.8 generalizes these results. On applications of quasi-invariant measures to representations of the canonical commutation relations in quantum field theory and spectral theory, see Gelfand, Vilenkin [**473**], Samoilenko [**981**], Araki [**77**], [**78**], Hegerfeld [**527**]–[**529**], Hegerfeld, Melsheimer [**530**], and Shimomura [**1030**].

Exercises

3.9.1. Justify assertion (iv) of Proposition 3.4.1.

3.9.2. Let f be a function of bounded variation on $[a, b]$. Prove that the function $|f|$ is of bounded variation as well and $\operatorname{Var}|f| \leqslant \operatorname{Var} f$.

3.9.3. Let $p > 1$. Prove that a function f on $[a, b]$ is absolutely continuous and $f' \in L^p[a, b]$ precisely when $\sup \sum_{i=1}^n |f(b_i) - f(a_i)|^p |b_i - a_i|^{1-p} < \infty$, where sup is taken over all disjoint intervals $[a_i, b_i] \subset [a, b]$, and this supremum equals $\|f'\|_{L^p}^p$.

3.9.4. (Hora [**553**], Shimomura [**1028**]) Let ξ be an integrable random variable on a probability space with $\mathbb{E}\xi = 0 < \mathbb{E}|\xi|$. Prove that there exists a probability measure μ with an absolutely continuous density ϱ on the real line such that its logarithmic derivative ϱ'/ϱ as a random variable on the real line with the measure μ has the same distribution as ξ.

3.9.5. Let μ be a Radon probability measure on a locally convex space X that is Skorohod differentiable along two vectors h and k such that $d_h\mu = d_k\mu$. Prove that $h = k$.

HINT: Let $l \in X^*$; taking a sufficiently small $t > 0$ with $\widetilde{\mu}(tl) > 0$ we obtain $l(h) = l(k)$ since $\widetilde{d_h\mu}(tl) = \widetilde{d_k\mu}(tl)$, i.e., $l(th)\widetilde{\mu}(tl) = l(tk)\widetilde{\mu}(tl)$.

3.9.6. Let μ be the countable product of Cauchy probability measures μ_n on the real line given by densities $p_n(y) = 2^n \pi^{-1}(1 + 2^{2n}t^2)^{-1}$. Prove that $\mu(l^2) = 1$.

3.9.7. (Yamasaki, Hora [**1198**]) Construct an example of a probability measure μ on $\mathbb{R}^2$ for which the Fomin derivative $d_{e_1}d_{e_2}\mu$ exists, but $d_{e_1}\mu$ does not.

HINT: Take a probability density $\varphi \in C_0^\infty(\mathbb{R}^2)$ and consider a measure with density $\varrho(x, y) = c \sum_{n=1}^\infty n^{-2}\varphi(n^{-1}x + a_n, ny + b_n)$, where a_n, b_n are chosen in such a way that the functions $\varphi(n^{-1}x + a_n, ny + b_n)$ have disjoint supports.

3.9.8. Let a Radon probability measure μ on a locally convex space X be Skorohod differentiable along $h \in X$ and let $l \in X^*$, $l(h) = 1$. Prove the inequality

$$\|d_h\mu\| \cdot \int_X |l(x)|\, \mu(dx) \geqslant 1/8.$$

3.9.9. Let μ be a nonzero Radon measure on a locally convex space X such that, for some nonzero $h \in X$, one has $\mu \perp \mu_{th}$ for every $t \neq 0$. Show that $\mu \perp \nu$ for every Radon measure ν that is continuous along h.

HINT: Otherwise there is a Radon probability measure ν continuous along h such that $\mu = \sigma + \eta$, $\sigma \ll \nu$, $\sigma \perp \eta$, and $\|\sigma\| > 0$; then σ is continuous along h, hence $\|\sigma_{th} - \sigma\| \to 0$ as $t \to 0$, which is impossible since

$$2\|\mu\| = \|\mu_{th} - \mu\| \leqslant \|\sigma_{th} - \sigma\| + \|\eta_{th} - \eta\| \leqslant \|\sigma_{th} - \sigma\| + 2\|\eta\|, \quad \|\eta\| < \|\mu\|.$$

3.9.10. Let μ be a Radon measure on a locally convex space X such that $\mu \sim \mu_h$ for some h. Suppose that a sequence of μ-measurable functions converges in measure μ. Show that the functions $f_n(\,\cdot\, - h)$ converge in measure μ.

HINT: Use Exercise 1.6.23.

CHAPTER 4

Some classes of differentiable measures

In this chapter we discuss the differentiability of some special classes of measures on infinite dimensional spaces. We consider product measures, stable and Gaussian measures, convex measures, distributions of diffusion processes, Gibbs measures, and mixtures of measures.

4.1. Product measures

In this section we study the differentiability of the so-called product measures, i.e., products of measures. Suppose that for every n we are given a Radon probability measure μ_n on a locally convex space X_n differentiable along a vector h_n. A typical situation is where all X_n are one-dimensional. It is clear from Proposition 3.3.13 that the product measure $\mu = \bigotimes_{n=1}^{\infty} \mu_n$ on the space $X = \prod_{n=1}^{\infty} X_n$ is differentiable along all "finite" vectors of the form $h = (h_1, \ldots, h_n, 0, 0, \ldots)$. However, it is also clear that the collection of vectors of this form do not exhaust the subspace $D(\mu)$ of all vectors of differentiability since it is a Banach space (see Theorem 5.1.1 below). We shall employ the following convention: if f_n is a function defined on X_n, then the same symbol f_n denotes the function defined on X by the equality $x = (x_n) \mapsto f_n(x_n)$.

The differentiability of product measures is closely connected to convergence of sums of independent random variables (see Theorem 4.1.1), so the martingale methods are efficient in its study. Let us recall that every μ-integrable function f on X is the L^1-limit of its conditional expectations with respect to the σ-algebras generated by the first n coordinates (see §1.3). In the case of a product measure these conditional expectations are given by the simple explicit formulae (1.3.8).

4.1.1. Theorem. *The product measure μ on X is differentiable along the vector $h = (h_n)$ precisely when the series*

$$\sum_{n=1}^{\infty} \beta_{h_n}^{\mu_n}$$

converges in the norm of $L^1(\mu)$. An equivalent condition:

$$\sup_n \Big\| \sum_{i=1}^{n} \beta_{h_i}^{\mu_i} \Big\|_{L^1(\mu)} < \infty.$$

PROOF. It is verified directly that the measure μ is differentiable along all vectors $h = (h_1, \ldots, h_n, 0, 0, \ldots)$ and $\beta_h^\mu = \sum_{i=1}^n \beta_{h_i}^{\mu_i}$. So convergence of the series of $\beta_{h_i}^{\mu_i}$ in $L^1(\mu)$ yields convergence in variation of the derivatives $d_{P_n h}\mu$, where we set $P_n h := (h_1, \ldots, h_n, 0, 0, \ldots)$. This gives the differentiability of μ and the equality of the logarithmic derivative along h to the sum of the indicated series. Conversely, suppose that the measure μ is differentiable along h. Then the conditional expectation of the function β_h^μ with respect to the σ-algebra $\mathcal{B}_n$ generated by the first n variables coincides with $S_n = \sum_{i=1}^n \beta_{h_i}^{\mu_i}$. Indeed, for every function f of the form

$$f(x) = \varphi\bigl(l_1(x_1, \ldots, x_n), \ldots, l_n(x_1, \ldots, x_n)\bigr),$$

where l_i are continuous linear functionals on $X_1 \times \cdots \times X_n$ and $\varphi \in C_b^\infty(\mathbb{R}^n)$, we have

$$\int_X f(x) \beta_h^\mu(x)\, \mu(dx) = -\int_X \partial_h f(x)\, \mu(dx) = -\int_{X_1 \times \cdots \times X_n} \partial_h f(x) \bigotimes_{i=1}^n \mu_i(dx_i)$$

$$= \int_{X_1 \times \cdots \times X_n} f(x) \sum_{i=1}^n \beta_{h_i}^{\mu_i}(x_i) \bigotimes_{i=1}^n \mu_i(dx_i) = \int_X f(x) \sum_{i=1}^n \beta_{h_i}^{\mu_i}(x_i)\, \mu(dx).$$

Hence $\{S_n\}$ is a closable martingale and converges in $L^1(\mu)$. The equivalence of the condition $\sup_n \|S_n\|_{L^1} < \infty$ is a known result in probability theory; see Vakhania, Tarieladze, Chobanyan [**1167**, p. 293, Corollary 2; p. 351, Corollary]. $\square$

4.1.2. Corollary. *If for all $n \in \mathbb{N}$ we have $X_n = X_1$, $\mu_n = \mu_1$, $h_n = c_n h_1$ and $\beta_{h_1}^{\mu_1} \in L^2(\mu_1)$, then the differentiability of μ along $h = (h_n)$ is equivalent to convergence of the series $\sum_{n=1}^\infty c_n^2$. In particular, if μ_n are identical probability measures on $\mathbb{R}^1$ with finite Fisher's information I, then the collection $D(\mu)$ of all vectors of differentiability of μ coincides with l^2. In addition, for some $c > 0$, whenever $|c_n| \leq 1$ one has*

$$\|\mu_h - \mu\| \geq 2 - 2\exp\Bigl(-c \sum_{n=1}^\infty c_n^2\Bigr). \quad (4.1.1)$$

PROOF. For any vector $h = (c_1 h_1, \ldots, c_n h_n, 0, \ldots)$ we have

$$\beta_h^\mu(x) = c_1 \beta_{h_1}^{\mu_1}(x_1) + \cdots + c_n \beta_{h_1}^{\mu_1}(x_n).$$

Hence $\|\beta_h^\mu\|_{L^2(\mu)}^2 = \sum_{i=1}^n c_i^2 I$, where $I = \|\beta_{h_1}^{\mu_1}\|_{L^2(\mu_1)}^2$. It will be shown in Chapter 5 that the series $\sum_{n=1}^\infty c_n^2$ converges if $(c_n h_1) \in D(\mu)$, but we directly verify convergence of this series for $(c_n h_1) \in D(\mu)$. We know that the series $\sum_{n=1}^\infty c_n \beta_{h_1}^{\mu_1}(x_n)$ converges in $L^1(\mu)$. It follows from our assumptions that the random variables $\xi_n(x) := \beta_{h_1}^{\mu_1}(x_n)$ on the space (X, μ) are independent, have zero mean and finite second moment I. Hence the characteristic functional φ of the random variable ξ_1 is twice differentiable at zero and $|1 - \varphi(t)| \geq \delta t^2$ in a neighborhood of zero, where $\delta > 0$. Mean convergence of the series $\sum_{n=1}^\infty c_n\xi_n$ and independence of ξ_n yield convergence of the product $\prod_{n=1}^\infty \varphi(c_n t)$. By the above estimate this gives convergence of the series of c_n^2. Estimate (4.1.1) follows from Corollary 1.2.2 since for some $\kappa > 0$ we have $\|(\mu_1)_{rh_1} - \mu_1\| \geqslant \kappa |r|$ for all $r \in [-1, 1]$. □

Since in Theorem 4.1.1 we have dealt with independent centered random variables $\xi_n = \beta_{h_n}^{\mu_n}$ on (X, μ), it is worth mentioning that there are various necessary and sufficient conditions for convergence in L^1 of such series. Let us formulate the following criterion (see Shiryaev [**1033**, Ch. 4]).

4.1.3. Proposition. *Let ξ_n be independent mean zero random variables. The series $\sum_{n=1}^\infty \xi_n$ converges in mean if and only if*

$$\sum_{n=1}^\infty \mathbb{E}\big[\xi_n^2/(1+|\xi_n|)\big] < \infty.$$

An equivalent condition: for some (and then for all) $\tau > 0$ one has

$$\sum_{n=1}^\infty \mathbb{E}\big(|\xi_n| I_{|\xi_n|>\tau}\big) < \infty.$$

It is natural to ask which random variables ξ can be obtained as logarithmic derivatives of probability measures. Clearly, $0 < \mathbb{E}|\xi| < \infty$ and $\mathbb{E}\xi = 0$ are necessary conditions. According to Hora [**553**], Shimomura [**1028**], for every random variable with these two properties there exists a probability measure μ on the real line such that it is differentiable and the distribution of β_1^μ on $(\mathbb{R}^1, \mu)$ coincides with the distribution of ξ (a proof of this fact is Exercise 3.9.4).

The following Kakutani alternative [**589**] is useful in the study of product measures (its proof can be found, e.g., in [**193**, Theorem 10.3.6]).

4.1.4. Theorem. *Suppose that for every n we are given a measurable space $(X_n, \mathcal{B}_n)$ and probability measures μ_n and ν_n such that $\nu_n \ll \mu_n$ and ϱ_n is the Radon–Nikodym density of ν_n with respect to μ_n. Set $\mu := \bigotimes_{n=1}^\infty \mu_n$, $\nu := \bigotimes_{n=1}^\infty \nu_n$. Then either $\nu \ll \mu$ or $\nu \perp \mu$ and the latter is equivalent to the following condition:*

$$\prod_{n=1}^\infty \int_{X_n} \sqrt{\varrho_n}\, d\mu_n := \lim_{N\to\infty} \prod_{n=1}^N \int_{X_n} \sqrt{\varrho_n}\, d\mu_n = 0.$$

Some interesting estimates for shifts of product measures can be found in Barthe, Cordero-Erausquin, Fradelizi [**99**], Bobkov [**159**]; see also Chapter 5 and Exercise 5.5.5.

4.2. Gaussian and stable measures

A Radon probability measure μ on a locally convex space X is called *stable of order* $\alpha \in (0, 2]$ if, for every n, there exists a vector $a_n \in X$ such that the distribution of the random vector $n^{-1/\alpha}(X_1 + \cdots + X_n) - a_n$ coincides with μ, where X_i are independent and have the same distribution μ. In terms of convolutions this can be written as

$$\mu(B) = \mu_n * \cdots * \mu_n(B + a_n), \quad B \in \mathcal{B}(X), \tag{4.2.1}$$

where on the right one has the n-fold convolution of the Radon measure μ_n given by the equality $\mu_n(B) := \mu(n^{1/\alpha}B)$. If $a_n = 0$, then μ is called *strictly stable*.

It is known (see Dudley, Kanter [**373**]) that a measure μ is stable of order α precisely when all of its finite dimensional projections are stable of order α (for strictly stable measures this is obvious). If all one-dimensional projections of μ are strictly stable of order α, then so is μ; however, for arbitrary stable distributions this is false (see Marcus [**777**]).

Gaussian measures are stable of order 2. A density of a stable measure is explicitly expressed via elementary functions only in several exceptional cases. However, for a symmetric stable of order α density p the characteristic functional has the form

$$\int_{\mathbb{R}^1} \exp(iyx)p(x)\,dx = \exp(-c|y|^\alpha).$$

If a measure μ is stable of order α, then the measure

$$\mu_s(B) := \int_X \mu(2^{1/\alpha}B + x)\,\mu(dx),$$

which is a homothetic image of the convolution of the measures μ and $\mu(-\cdot\,)$, is *symmetric* (i.e., $\mu_s(B) = \mu_s(-B)$) and stable of order α. We have $\widetilde{\mu_s} = |\widetilde{\mu}|$. If the space X is quasicomplete (for example, is a Banach or Fréchet space), then the Fourier transform of a stable of order α measure μ has the form

$$\widetilde{\mu}(l) = \exp\!\left(il(a) - \int_X |l(x)|^\alpha\,\Gamma(dx) + i\beta Q(\alpha, \Gamma, l)\right),$$

where $a \in X$, Γ is a nonnegative Radon measure on X with bounded support (called *spectral*), $\beta \in [0, 1]$, and

$$Q(\alpha, \Gamma, l) = \begin{cases} \tan(\pi\alpha/2) \displaystyle\int_X l(x)|l(x)|^{\alpha-1}\,\Gamma(dx) & \text{if } \alpha \neq 1, \\ -\dfrac{2}{\pi} \displaystyle\int_X l(x) \ln|l(x)|\,\Gamma(dx) & \text{if } \alpha = 1. \end{cases}$$

It is shown in Bogachev [**171**] that there is the smallest $\beta \in [0, 1]$ with which such a representation is possible. This number $\beta = \beta(\mu)$ is called the *index of asymmetry* of the measure μ. For example, the measure μ is symmetric if $\beta(\mu) = 0$ and $a = 0$. If $\beta(\mu) = 1$, then μ is called *completely asymmetric*. If $\beta(\mu) \in (0, 1)$, then there exists a completely asymmetric stable of order α measure ν such that

$$\mu = \mu_s\!\left(\left(1 - \beta(\mu)^{-1/\alpha}\right)\!\cdot\right) * \nu.$$

4.2. Gaussian and stable measures

If $\beta(\mu) = 0$, then $\mu = (\mu_s)_{-a}$. Since homotheties preserve the sets of vectors of continuity, quasi-invariance and differentiability and formation of convolutions does not reduce these sets, we conclude that for a stable measure μ with $\beta(\mu) < 1$ the indicated sets are the same as for the measure μ_s. The following fact is proven in Sztencel [**1098**].

4.2.1. Theorem. *Let μ be a symmetric stable Radon measure on a locally convex space X. Then there exist a Radon probability measure σ on the space $T = X^\infty \times \mathbb{R}^\infty$ and a family of centered Gaussian Radon measures γ_t on X, where $t \in T$, such that for every $A \in \mathcal{B}(X)$ the function $t \mapsto \gamma_t(A)$ is measurable with respect to σ and*

$$\mu(A) = \int_T \gamma_t(A)\,\sigma(dt), \quad A \in \mathcal{B}(X).$$

One can see from the proof of this theorem in [**1098**] that if the measure μ is not concentrated on a proper closed linear subspace, then the measures γ_t for σ-a.e. t have the same property, i.e., one can choose all measures γ_t to be nondegenerate.

On stable measures on locally convex spaces, see Bogachev [**171**], Tortrat [**1121**] (the case of Banach spaces is thoroughly studied in Linde [**725**]). Analytic properties of one-dimensional stable measures are studied in Zolotarev [**1226**]. It is easily verified by means of the Fourier transform that every stable density p on $\mathbb{R}^n$ is infinitely differentiable. However, it is not obvious and it has remained unknown for some time whether the partial derivatives of p belong to $L^1(\mathbb{R}^n)$. This question was answered positively by the author in [**169**], where the following general fact was established.

4.2.2. Theorem. *Let a Radon measure μ on a locally convex space X be stable of order α. Suppose that μ is continuous along a vector h. Then*
  (i) *the measure μ is infinitely differentiable along h and*

$$\|d_h^n \mu\| \leqslant n^{n/\alpha} \|d_h \mu\|^n;$$

  (ii) *if $\alpha \geqslant 1$, then the measure μ is analytic along h;*
  (iii) *if $\alpha \geqslant 1$, then μ is quasi-invariant along h.*

PROOF. Suppose that the measure μ is differentiable along h. Formula (4.2.1) yields the n-fold differentiability of μ and the estimate

$$\|d_h^n \mu\| \leqslant \|d_h \mu_n\|^n \leqslant n^{n/\alpha} \|d_h \mu\|.$$

If $\alpha \geqslant 1$, then this estimate along with Stirling's formula gives convergence of the series of $|z|^n \|d_h^n \mu\|/n!$ whenever $|z| < (e\|d_h \mu\|)^{-1}$ (if $\alpha > 1$, then for all $|z|$), which means the analyticity along h, whence the quasi-invariance follows.

We now show that the continuity along h implies the differentiability. Let us take a measure on the real line that is symmetric stable of order α and has a density p and the characteristic functional $\exp(-|y|^\alpha)$. By using the duality formulae from [**1226**, §2.3] one readily verifies that $p' \in L^1(\mathbb{R}^1)$.

This can be also derived from the unimodality of p (see [**1226**, §2.7]). Denote by ν the image of this measure under the mapping $t \mapsto th$ with values in X. Set $\nu_n(B) := \nu(nB)$ and $\mu_n := \mu * \nu_n$. It is readily seen that the measures μ_n are stable of order α. According to what we have proved above they are infinitely differentiable. As $n \to \infty$ we have

$$\|\mu_n - \mu\| \leqslant 2 \sup_{B \in \mathcal{B}(X)} |\mu_n(B) - \mu(B)|$$

$$\leqslant 2 \sup_{B \in \mathcal{B}(X)} \int_{\mathbb{R}^1} |\mu(B - n^{-1}th) - \mu(B)|\, p(t)\, dt$$

$$\leqslant 2 \int_{\mathbb{R}^1} \|\mu - \mu_{n^{-1}th}\|\, p(t)\, dt \to 0$$

by the continuity of μ along h and the Lebesgue dominated convergence theorem. Let $r_n := \|d_h \mu_n\|$. We show that $\sup_n r_n < \infty$. Otherwise we may assume that $r_n \to \infty$. By Taylor's formula for the mapping $t \mapsto (\mu_n)_{th}$ from $\mathbb{R}^1$ to the Banach space of measures, for all $t > 0$ we obtain

$$\|(\mu_n)_{th} - \mu\| \geqslant t\|d_h \mu_n\| - t^2\|d_h^2 \mu_n\| \geqslant t\|d_h \mu_n\| - t^2 4^{1/\alpha}\|d_h \mu_n\|^2,$$

where we have used the estimate established above. We find $\xi \in (0,1)$ such that $\xi - 4^{1/\alpha} \xi^2 = \delta > 0$. Set $t_n := \xi/r_n$. Then $\|(\mu_n)_{t_n h} - \mu_n\| \geqslant \delta$. On the other hand, $t_n \to 0$, hence

$$\|(\mu_n)_{t_n h} - \mu_n\| \leqslant \|(\mu_n)_{t_n h} - \mu_{t_n h}\| + \|\mu_{t_n h} - \mu\| + \|\mu - \mu_n\|$$
$$= \|\mu_n - \mu\| + \|\mu_{t_n h} - \mu\| + \|\mu - \mu_n\| \to 0.$$

The obtained contradiction shows that $\sup_n r_n = c < \infty$. Therefore, we have $\|d_h^m \mu_n\| \leqslant m^{m/\alpha} c^m$. It follows that μ is infinitely differentiable along h. $\square$

If μ is stable and $\mu_h \sim \mu$, then $\mu_{th} \sim \mu$ for all t (see Zinn [**1225**]).

4.2.3. Corollary. *Let μ be a stable measure on $\mathbb{R}^d$ not concentrated on a proper affine subspace. Then μ has an infinitely differentiable density ϱ such that all its partial derivatives $\partial_{x_{i_1}} \ldots \partial_{x_{i_k}} \varrho$ are integrable and*

$$\left\|\frac{\partial^n \varrho}{\partial x_i^n}\right\|_{L^1(\mathbb{R}^d)} \leqslant n^{n/\alpha} \left\|\frac{\partial \varrho}{\partial x_i}\right\|_{L^1(\mathbb{R}^d)}.$$

Estimates of derivatives of a density of a stable measure in terms of its Fourier transform are obtained in Bentkus, Juozulynas, Paulauskas [**135**].

However, a stable measure on an infinite dimensional space may fail to have nonzero vectors of differentiability or continuity.

4.2.4. Example. Let a random process $\xi(t)$, $t \in [0,1]$, be stable of order $\alpha \in (0,2)$ and have independent increments with $\mathbb{E} \exp(i\lambda \xi(t)) = \exp(-t|\lambda|^\alpha)$ and $\mathbb{E} \exp(i\lambda[\xi(t+h) - \xi(t)]) = \exp(-h|\lambda|^\alpha)$. Denote by μ the distribution of this process in the space of all trajectories or in $L^2[0,1]$. Then μ is stable of order α and, for every nonzero vector h, the measures μ_h and μ are mutually singular.

4.2. Gaussian and stable measures

PROOF. We shall consider μ in the space of all trajectories. It is clear that all finite dimensional projections are stable of order α and symmetric, whence one can easily derive that the measure μ is stable of order α. Let h be a nonconstant function. According to Exercise 4.6.8 we have

$$\sup \sum_{i=1}^{n} |h(t_i) - h(t_{i-1})|^2 (t_i - t_{i-1})^{-2/\alpha} = \infty,$$

where sup is taken over all n and all $0 = t_0 < t_1 < \cdots < t_n \leqslant 1$. The projection μ_n of the measure μ under the mapping $x \mapsto (x(t_1), \ldots, x(t_n))$ from X to $\mathbb{R}^n$ has density $\prod_{i=1}^{n} (t_i - t_{i-1})^{-1/\alpha} p((x_i - x_{i-1})(t_i - t_{i-1})^{-1/\alpha})$, where p is the distribution density for $\xi(1)$. Now the equality $\|\mu_h - \mu\| = 2$ follows from estimate (4.1.1). The same is true for any nonzero constant h since $\xi(0) = 0$. □

It should be noted that, according to Corollary 4.2.3, all finite dimensional projections of the measure μ considered in this example (corresponding to $t_i > 0$) have smooth densities with respect to the corresponding Lebesgue measures.

Gaussian measures possess especially nice differentiability properties; as we have seen in Theorem 3.1.9, for any Gaussian measure γ the logarithmic derivative β_h^γ coincides with $-\hat{h}$. Differentiating the identity $d_h \gamma = -\hat{h} \cdot \gamma$ we obtain some explicit formulae for higher order derivatives of the measure γ. Their Radon–Nikodym densities with respect to γ are certain finite dimensional polynomials in measurable linear functionals (for example, the Radon–Nikodym density of $d_h^n \gamma$ is a polynomial of degree n of the functional $\hat{h}$). The analyticity of Gaussian measures was observed in Albeverio, Høegh-Krohn [41] and Bentkus [131], [132].

4.2.5. Remark. Let μ be a stable measure on a locally convex space X. We shall see in Corollary 6.2.6 below that if μ is differentiable along h, then

$$\beta_h^\mu \in L^{3-\varepsilon}(\mu) \quad \text{for every } \varepsilon \in (0, 3).$$

Hence for any stable measure μ with a density ϱ on $\mathbb{R}^n$ we obtain

$$\int_{\mathbb{R}^n} \left|\frac{\nabla \varrho}{\varrho}\right|^{3-\varepsilon} \varrho(x)\, dx < \infty.$$

The following questions naturally arise.

(i) Is it true that $\beta_h^\mu \in L^p(\mu)$ for all $p \geqslant 1$?

(ii) Is it true that μ and μ_h are mutually singular for all vectors h such that μ is not continuous along h?

(iii) For stable product measures one has the following estimate (see Bogachev, Smolyanov [233]):

$$\|\mu_h - \mu\| \geqslant 2(1 - \theta(\alpha)/\|d_h\mu\|^{\alpha/4}).$$

For a proof we consider first countable products of identical measures, employ estimates (4.1.1) and (1.2.1), and study the behavior of $\|\mu_h - \mu\|$ for measures on the real line for large h. Is there an analogous estimate in the general case? Such an estimate would give a positive answer to the previous question.

4.3. Convex measures

Convex measures (also known as logarithmically concave measures) are very interesting for many applications in stochastic analysis. For example, many measures described as Gibbs states in statistical mechanics belong to this class. In the finite dimensional case this class consists of all measures with densities of the form $\exp(-V)$, where V is a convex function. Thus, the measure with density $C\exp(-|x|^k)$ belongs to this class. A simple infinite dimensional example is any Gaussian measure. As we shall see below, certain important properties of convex measures are similar to those of Gaussian measures.

4.3.1. Definition. *A probability measure μ on the σ-algebra $\sigma(X)$ in a locally convex space X is called convex (or logarithmically concave) if, for all sets A, B from $\sigma(X)$ and all $\lambda \in [0,1]$, one has*

$$\mu(\lambda A + (1-\lambda)B) \geqslant \mu(A)^\lambda \mu(B)^{1-\lambda}.$$

A Radon probability measure μ on X is called convex if its restriction to $\sigma(X)$ is convex.

Let us remark that for a convex Radon measure the determining inequality is fulfilled for all compact sets. For arbitrary Borel sets we have only

$$\mu_*(\lambda A + (1-\lambda)B) \geqslant \mu(A)^\lambda \mu(B)^{1-\lambda}.$$

If X is a Souslin space, then the set $\lambda A + (1-\lambda)B$ is Souslin, hence in place of μ_* we can write μ. In the case of $\sigma(X)$ this reasoning yields the measurability of $\lambda A + (1-\lambda)B$ with respect to the completion of μ on $\sigma(X)$.

We recall that V is a convex function on $\mathbb{R}^n$ if it is defined and finite on some open convex set $C = \mathrm{Dom}\, V$ and

$$V(\lambda x + (1-\lambda)y) \leqslant \lambda V(x) + (1-\lambda)V(y), \quad \forall x, y \in C, \forall \lambda \in [0,1].$$

If $C \neq \mathbb{R}^n$, then it is convenient to set $V(x) = +\infty$ outside C. Redefining the function V on the boundary of C, one can make it closed, i.e., having the closed subgraph. On convex functions, see Rockafellar [**957**].

Let us mention several general facts related to convex measures.

4.3.2. Theorem. *A measure ν on a finite dimensional space X is convex if and only if there exist an affine mapping $T\colon \mathbb{R}^n \to X$ and a probability measure μ with density $p = e^{-V}$ on $\mathbb{R}^n$, where V is a convex function (outside the domain of definition of V we set $p = 0$) such that one has $\nu = \mu \circ T^{-1}$.*

4.3.3. Proposition. *A probability measure μ on the σ-algebra $\sigma(X)$ in a locally convex space X (or a Radon probability measure on X) is convex if and only if all its finite dimensional projections are convex.*

4.3.4. Corollary. (i) *The image of a convex measure under a continuous linear mapping (or a measurable linear mapping) is convex.*

(ii) *Convexity of measures is preserved by products and convolutions with convex measures.*

(iii) *Convexity of measures is preserved by weak limits, i.e., if a net of convex measures is weakly convergent, then its limit is convex as well.*

PROOF. Assertion (i) follows from the relation
$$T^{-1}(\lambda A + (1-\lambda)B) \supset \lambda T^{-1}(A) + (1-\lambda)T^{-1}(B)$$
valid for any linear mapping T. Assertion (ii) follows from (i), and (iii) is readily verified in the finite dimensional case. $\square$

4.3.5. Theorem. *Let μ be a convex Radon measure on a locally convex space X.*

(i) *Suppose that G is an additive subgroup in X. Then either $\mu_*(G) = 0$ or $\mu_*(G) = 1$.*

(ii) *Suppose that for every continuous linear functional f the support of the measure $\mu \circ f^{-1}$ is either a point or the whole real line. Then the topological support of μ is a closed affine subspace.*

Proofs of all these facts can be found in Borell [**239**] (see also [**193**, §7.14]).

4.3.6. Theorem. *Let μ be a convex Radon measure on the product $X = Y \times Z$, where Z is a separable Fréchet space, and let μ_Y be the projection of μ to Y. Then the conditional measures μ_y on the space Z are convex for μ_Y-a.e. $y \in Y$.*

PROOF. Let $A, B \in \mathcal{B}(Z)$, $\lambda \in (0,1)$. Set $\widehat{A} = Y \times A$, $\widehat{B} = Y \times B$. For every convex cylindrical set $C \in \mathcal{B}(Y)$ we put $\widehat{C} := C \times Z$. Due to the inclusion
$$\lambda(\widehat{C} \cap \widehat{A}) + (1-\lambda)(\widehat{C} \cap \widehat{B}) \subset \widehat{C} \cap (\lambda \widehat{A} + (1-\lambda)\widehat{B}),$$
convexity of μ and Hölder's inequality we have
$$\int_C \mu_y(\lambda A + (1-\lambda)B)\, \mu_Y(dy) = \mu\big(\widehat{C} \cap (\lambda \widehat{A} + (1-\lambda)\widehat{B})\big)$$
$$\geqslant \mu(\widehat{C} \cap \widehat{A})^\lambda \mu(\widehat{C} \cap \widehat{B})^{1-\lambda} = \left(\int_C \mu_y(A)\, \mu_Y(dy)\right)^\lambda \left(\int_C \mu_y(B)\, \mu_Y(dy)\right)^{1-\lambda}$$
$$\geqslant \int_C \mu_y(A)^\lambda \mu_y(B)^{1-\lambda}\, \mu_Y(dy).$$

Since a convex set C is arbitrary, according to Exercise 4.6.5 for μ_Y-a.e. y we have
$$\mu_y(\lambda A + (1-\lambda)B) \geqslant \mu_y(A)^\lambda \mu_y(B)^{1-\lambda}.$$

According to Exercise 4.6.3, there exists a countable class $\mathcal{K}$ of Borel sets such that this inequality for all $A, B \in \mathcal{K}$ yields convexity of μ_y. Countability of $\mathcal{K}$ enables us to find a set Y_0 of full μ_Y-measure for which the obtained estimate is true for all $A, B \in \mathcal{K}$ and $y \in Y_0$. Therefore, for all such y the measure μ_y is convex. □

Borell [**240**] used a similar reasoning to show convexity of conditional measures generated by measurable linear mappings.

4.3.7. Theorem. *Let μ be a convex measure on a locally convex space X and let q be a μ-measurable seminorm that is finite μ-a.e. Then we have $\exp(cq) \in L^1(\mu)$ provided that $\mu(q \leqslant c^{-1}) > e^2/(1+e^2)$.*

PROOF. Let $A \in \mathcal{B}(X)$ be an absolutely convex set such that we have $\theta := \mu(A) > 0$. We show that

$$\mu(X \backslash tA) \leqslant \left(\frac{1-\theta}{\theta}\right)^{t/2}, \quad t \geqslant 1. \tag{4.3.1}$$

To this end we observe that $2(t+1)^{-1}(X \backslash tA) + (t-1)(t+1)^{-1}A \subset X \backslash A$, which is easily verified. Hence

$$\mu(X \backslash A) \geqslant \mu(X \backslash tA)^{2/(t+1)} \mu(A)^{(t-1)/(t+1)},$$

whence we obtain $\mu(X \backslash tA)^{2/(t+1)} \leqslant \theta^{(t-1)/(t+1)}(1-\theta)$, which yields (4.3.1).

Let $\mu(q \leqslant 1) > e^2/(1+e^2)$. Let us take a compact set $K \subset \{q \leqslant 1\}$ with $\mu(K) > e^2/(1+e^2)$. Let A be the absolutely convex hull of K. Then $A \in \mathcal{B}(X)$ (see Exercise 1.6.15), $A \subset \{q \leqslant 1\}$ and $\theta := \mu(A) > e^2/(1+e^2)$. In addition, $\{q \geqslant n\} \subset X \backslash nA$. Then $\mu(\exp q \geqslant n) \leqslant \bigl((1-\theta)/\theta\bigr)^{-\ln n/2} = n^{-\alpha}$, where $\alpha > 1$ since $(1-\theta)/\theta < e^{-2}$. □

4.3.8. Proposition. *Let μ be a convex Radon measure on a locally convex space X and let V be a continuous convex function with $\exp(-V) \in L^1(\mu)$. Then the measure $\nu := c\exp(-V) \cdot \mu$ is convex, where the number c is such that ν is a probability measure. Moreover, the assumption that V is continuous can be replaced by the weaker assumption that V is a limit of a sequence of continuous convex functions convergent a.e.*

PROOF. It suffices to prove our claim for convex functions bounded from below and apply this to the functions $V_k := \max(V, -k)$, $k \in \mathbb{N}$. Hence we may assume that $V \geqslant 0$. In the finite dimensional case the assertion is true by Theorem 4.3.2. Now the general case follows by Corollary 4.3.4. Indeed, we can take a net of convex measures μ_α concentrated on finite dimensional subspaces and weakly convergent to μ (for instance, by using finite dimensional projections of μ). Then the measures $e^{-V} \cdot \mu_\alpha$ converge weakly to $\exp(-V) \cdot \mu$, whence we obtain weak convergence also for the normalized measures. The last assertion of the proposition is easily deduced from the first one. □

4.3. Convex measures

There is a simple criterion of integrability of the function $\exp(-V)$ for a convex function V on $\mathbb{R}^n$ obtained by Krugova [**646**]. The recessive cone $\widehat{C}$ of the set $C = \mathrm{Dom}\, V$ is defined as the set of all vectors $h \in \mathbb{R}^n$ such that $x + \lambda h \in C$ for all $x \in C$ and $\lambda \geqslant 0$.

4.3.9. Lemma. *Let V be a convex function on $\mathbb{R}^n$. The function $\exp(-V)$ defined by zero outside $C = \mathrm{Dom}\, V$ is integrable on $\mathbb{R}^n$ if and only if the sets $\{V \leqslant c\}$ are bounded, which is equivalent to the following: $V(x + th) \to \infty$ as $t \to +\infty$ for all $x \in C$ and all nonzero $h \in \widehat{C}$. In particular, if $C = \mathbb{R}^n$, then the function $\exp(-V)$ is integrable precisely when $V(x) \to \infty$ as $|x| \to \infty$.*

PROOF. Let $h \in \widehat{C}$ and $h \neq 0$. Suppose that for some $x \in C$ the quantities $V(x + th)$ do not tend to infinity as $t \to +\infty$. Since the function $t \mapsto V(x + th)$ is convex, these quantities are bounded from above by some number $M > 0$. We may assume that $V(x) < M$. There is a number $r > 0$ such that $V(z) < M$ whenever $|z - x| \leqslant 2r$. If $|z - x| \leqslant r$, then one has $z + th = (x + 2th)/2 + (2z - x)/2$ and $|(2z - x) - x| = 2|z - x| \leqslant 2r$, whence

$$V(z + th) \leqslant V(x + 2th)/2 + V(2z - x)/2 \leqslant M/2 + M/2 = M.$$

Therefore, $e^{-V} \geqslant e^{-M}$ on a set of infinite Lebesgue measure formed by the shifts of the ball by the vectors th. Hence e^{-V} does not belong to $L^1(\mathbb{R}^n)$. In the considered situation the set $\{V \leqslant M\}$ is unbounded. If for some c the set $\{V \leqslant c\}$ is unbounded, then, since it is closed and convex, its recessive cone contains a nonzero vector h (see Rockafellar [**957**, Theorem 8.4]). Suppose that all sets $\{V \leqslant c\}$ are bounded. We may assume that $0 \in C$. Since $\{V \leqslant V(0) + 1\}$ is bounded, there exists $M > 0$ such that $V(x) > V(0) + 1$ as $|x| \geqslant M$. Then $V(y) > V(0) + M^{-1}|y|$ whenever $|y| > M$. Indeed, let $V(y) \leqslant V(0) + M^{-1}|y|$. Then

$$V(M|y|^{-1}y) \leqslant M|y|^{-1}V(y) + (1 - M|y|^{-1})V(0) \leqslant V(0) + 1$$

contrary to the fact that $V(M|y|^{-1}y) > V(0) + 1$. The obtained estimate obviously yields the integrability of e^{-V}. $\square$

The following important result was proved by Krugova [**646**].

4.3.10. Theorem. *Suppose that μ is a convex measure on $\mathbb{R}^n$ with a density e^{-V}. Then the following assertions are true.*

(i) The measure μ is Skorohod differentiable along all vectors and, for every vector $h \in \mathbb{R}^n$, we have

$$d_h \mu = -\partial_h V e^{-V}\, dx - e^{-V}(n_x, h)|_{\partial C}\, \sigma,$$

where σ is the standard Lebesgue surface measure on the boundary ∂C of the domain $C = \mathrm{Dom}\, V$, n_x is the outer normal (defined σ-a.e.) on ∂C, and a closed version of the function V is considered.

(ii) *Suppose, in addition, that* $\lim_{y\to x} V(y) = \infty$ *for almost all points x at the boundary of* $\mathrm{Dom}\, V$. *Then the measure μ is Fomin differentiable and*

$$d_h\mu = -\partial_h V e^{-V}\, dx.$$

In particular, this is true if $\mathrm{Dom}\, V = \mathbb{R}^n$.

PROOF. We prove only the last assertion, in which the function V is everywhere finite. It suffices to verify the integrability of $|\partial_{x_1} V|\exp(-V)$. We observe that in the one-dimensional case, taking a minimum point t_0 of V, we see that the integral of $|V'|\exp(-V)$ equals $2\exp(-V(t_0))$. Hence

$$\int_{\mathbb{R}^n} |\partial_{x_1} V| \exp(-V)\, dx = 2 \int_{\mathbb{R}^{n-1}} \exp(-W)\, dy,$$

where $W(y_1,\ldots,y_{n-1}) := \inf_{x_n} V(y_1,\ldots,y_{n-1},x_n)$. The function W is convex. In addition, $W(y) \to +\infty$ as $|y| \to \infty$. By Lemma 4.3.9 we have the inclusion $\exp(-W) \in L^1(\mathbb{R}^{n-1})$. In the general case, some technical nuances appear due to the boundary of C; they are considered in detail in [**646**]. □

The following example was constructed in Krugova [**646**].

4.3.11. Example. Let $n=1$ and let $V'(x) = c_i$ on the interval Δ_i, where $c_0 = 0$, $c_k = e^{kc_{k-1}}$, $\delta_k = (c_k - c_{k-1})c_k^{-1}$, $\Delta_k := \left[\sum_{i=1}^{k-1}\delta_i, \sum_{i=1}^{k}\delta_i\right]$. For $x \geqslant 0$ we define $V(x)$ as the integral of V' over $[0,x]$; for $x < 0$ we set $V(x) = V(-x)$. Then the function V is convex and the associated measure $\mu = \exp(-V)\,dx$ is finite, but the function $|V'|^p \exp(-V)$ is not integrable for any $p > 1$, i.e., the logarithmic derivative does not belong to $L^p(\mu)$ whenever $p > 1$. This example can be modified in order to make V uniformly convex, i.e., satisfying the condition $V''(x) \geqslant \delta > 0$.

Let us give a sufficient condition for the quadratic integrability of the logarithmic derivative of a convex measure.

4.3.12. Proposition. *Suppose that μ is a convex measure on $\mathbb{R}^n$ with a density $\exp(-V)$, where $\partial_{x_i} V \in W^{1,1}_{\mathrm{loc}}(\mathbb{R}^n)$, $i=1,\ldots,n$. Let $h \in \mathbb{R}^n$. The function $\partial_h^2 V = V''(\,\cdot\,)(h,h)$ belongs to $L^1(\mu)$ if and only if the function $\beta_h^\mu = \partial_h V$ belongs to $L^2(\mu)$. In this case there exists $d_h^2\mu$.*

PROOF. If $\partial_h^2 V = \partial_h \beta_h^\mu \in L^1(\mu)$, then we have

$$\int_{\mathbb{R}^n} |\beta_h^\mu(x)|^2 \mu(dx) = -\int_{\mathbb{R}^n} \partial_h\beta_h^\mu(x)\mu(dx) = \int_{\mathbb{R}^n}\partial_h^2 V(x)\mu(dx).$$

In order to justify this calculation we take a sequence of functions f_k on the real line defined as follows: $f_k(t) = t$ if $|t| \leqslant k$, $f_k(t) = k$ if $t > k$, $f_k(t) = -k$ if $t < -k$. Then

$$\int_{\mathbb{R}^n} f_k\big(\beta_h^\mu(x)\big)\beta_h^\mu(x)\mu(dx) = -\int_{\mathbb{R}^n} f_k'\big(\beta_h^\mu(x)\big)\partial_h\beta_h^\mu(x)\mu(dx),$$

which converges to the integral of $\partial_h^2 V = \partial_h \beta_h^\mu$ with respect to the measure μ because this integral is finite. Since $f_k(t) t \geqslant 0$, this yields $\partial_h V \in L^2(\mu)$ (here no convexity of V is used). If V is convex, then $V''(x)(h,h) \geqslant 0$. Hence the μ-integrability of $|\partial_h V|^2$ in a similar manner yields the μ-integrability of $\partial_h^2 V$. In this situation Proposition 6.3.3 from Chapter 6 ensures the existence of $d_h^2 \mu$. □

4.3.13. Corollary. *If in the previous proposition we have $\beta_{e_i}^\mu \in L^2(\mu)$ for all e_i, then*

$$\int_{\mathbb{R}^n} |\nabla V(x)|^2 \, \mu(dx) = \int_{\mathbb{R}^n} \operatorname{trace} D^2 V(x) \, \mu(dx).$$

The next theorem of Krugova shows that, for convex measures, the square integrability of the logarithmic derivative β_h^μ yields the existence of $d_h^2 \mu$ in the Skorohod sense. This result can be seen from the reasoning in the previous proposition, but we give a direct proof, which employs a technical lemma.

4.3.14. Lemma. *Let μ and ν be two Borel probability measures on $\mathbb{R}^n$, let $f \in L^2(\mu)$, and let g be the density of $(f \cdot \mu) * \nu$ with respect to μ. Then*

$$\int_{\mathbb{R}^n} |g|^2 \, d(\mu * \nu) \leqslant \int_{\mathbb{R}^n} |f|^2 \, d\mu.$$

The same is true for $\mathbb{R}^d$-valued mappings.

PROOF. Let $\xi \colon \mathbb{R}^n \times \mathbb{R}^n \to \mathbb{R}^n$, $\xi(x,y) = x+y$. It is readily seen that $g \circ \xi$ coincides with the conditional expectation of f regarded on $\mathbb{R}^n \times \mathbb{R}^n$ with respect to the measure $\mu \otimes \nu$ and the σ-algebra generated by ξ. This yields our claim. □

4.3.15. Corollary. *Suppose that in the above lemma we have $\mu = \varrho \, dx$, where $\varrho \in W^{1,1}_{\mathrm{loc}}(\mathbb{R}^n)$ and $|\nabla \varrho / \varrho| \in L^2(\mu)$. Then*

$$\int_{\mathbb{R}^n} \left| \frac{\nabla(\varrho * \nu)}{\varrho * \nu} \right|^2 d\mu * \nu \leqslant \int_{\mathbb{R}^n} \left| \frac{\nabla \varrho}{\varrho} \right|^2 d\mu.$$

PROOF. We apply the lemma to the mapping $f = \nabla \varrho / \varrho$ and use the equalities $(f \cdot \mu) * \nu = (\nabla \varrho) * \nu = \nabla(\varrho * \nu) = \bigl[\nabla(\varrho * \nu)/(\varrho * \nu)\bigr] \cdot (\mu * \nu)$. □

4.3.16. Theorem. *Let μ be a convex measure on a locally convex space X differentiable along a vector h and let $\beta_h^\mu \in L^2(\mu)$. Then the measure $d_h \mu$ is Skorohod differentiable along h. In addition,*

$$\|d_h^2 \mu\| \leqslant 2 \|\beta_h^\mu\|_{L^2(\mu)}^2. \tag{4.3.2}$$

PROOF. We prove this theorem in the case $X = \mathbb{R}^1$. The general case is easily deduced by using conditional measures. Suppose first that μ has a smooth density $\varrho = e^{-V}$, where

$$|V'|^2 e^{-V} = |\varrho'|^2/\varrho \in L^2(\mathbb{R}^1), \quad \varrho'' \in L^1(\mathbb{R}^1).$$

Since $\varrho'' = [-V'' + |V'|^2]e^{-V}$, we have $V''e^{-V} \in L^1(\mathbb{R}^1)$, and the integrals of $(V')^2 e^{-V}$ and $V'' e^{-V}$ are equal (the integral of ϱ'' vanishes). Hence the integral of $|\varrho''|$ does not exceed the integral of $2(V')^2 e^{-V}$, i.e., one has (4.3.2). In the general one-dimensional case we consider the probability measures $\mu_k := \varrho_k\, dt$, where $\varrho_k = \varrho * p_k$, $p_k(t) = ke^{-t^2/k}/(2\sqrt{\pi})$. These measures are convex. We observe that by Corollary 4.3.15 we have

$$\int_{\mathbb{R}} \frac{|\varrho_k'(t)|^2}{\varrho_k(t)}\, dt \leqslant \int_{\mathbb{R}} \frac{|\varrho'(t)|^2}{\varrho(t)}\, dt.$$

In addition, $\varrho_k'' \in L^1(\mathbb{R}^1)$. Hence $\|d_1^2 \mu_k\| \leqslant 2\|\beta_1^\mu\|_{L^2(\mu)}^2$. Letting $k \to \infty$ we obtain the existence of $d_1^2 \mu$ and estimate (4.3.2). $\square$

The converse assertion is unfortunately false. Let us consider the following example: let $e^{-V(x)} = x/|\ln x|$ if $0 < x < 1/4$. It is readily verified that the corresponding function V is convex on this interval. We can extend it to the interval $(0, \infty)$ in such a way that the function V will be continuously differentiable on $(0, \infty)$ and have a piecewise continuous second derivative and the measure $\mu = e^{-V}\, dx$ will be twice Fomin differentiable. However, the logarithmic derivative of this measure is not square-integrable on the interval $(0, 1/4)$ with respect to the measure μ. In particular, in this example V'' and $|V'|^2$ are not in $L^1(\mu)$, although we have $V'' - |V'|^2 \in L^1(\mu)$.

4.3.17. Theorem. *Suppose that a convex Radon measure μ on a locally convex space X is differentiable along vectors h and k and β_h^μ, $\beta_k^\mu \in L^2(\mu)$. Then the Skorohod derivatives $d_h d_k \mu$ and $d_k d_h \mu$ exist and coincide.*

PROOF. We show that the measure $\zeta := 2^{-1}[d_{h+k}^2 \mu - d_h^2 \mu - d_k^2 \mu]$, which exists by the above theorem, serves as a Skorohod derivative of the measure $d_k \mu$ along the vector h. By Theorem 3.6.8 it suffices to prove the equality $\widetilde{\zeta}(l) = -il(h)\widetilde{d_k \mu}(l)$ for all $l \in X^*$. To this end, we observe that

$$\widetilde{d_h^2 \mu}(l) = -l(h)^2 \widetilde{\mu}(l), \quad \widetilde{d_k^2 \mu}(l) = -l(k)^2 \widetilde{\mu}(l),$$

$$\widetilde{d_{h+k}^2 \mu}(l) = -l(h+k)^2 \widetilde{\mu}(l), \quad -il(h)\widetilde{d_k \mu}(l) = -l(h)l(k)\widetilde{\mu}(l).$$

The same is true for $d_k d_h \mu$ (we can also use Remark 3.7.2). $\square$

We recall that in the general case the existence of the repeated derivative $d_h d_g \mu$ does not follow from the existence of the derivative $d_g d_h \mu$ (see Exercise 3.9.7).

One more remarkable result of Krugova [**647**] gives for convex measures an analog of the well-known Gaussian dichotomy.

4.3.18. Theorem. *Let μ be a Radon convex measure on a locally convex space X and $h \in X$. Then*

$$\|\mu_h - \mu\| \geqslant 2 - \exp(-\|d_h \mu\|/2),$$

where $\|d_h\mu\| := \infty$ if $d_h\mu$ does not exist. In particular, the measure μ is either Skorohod differentiable along the vector h or mutually singular with all its shifts μ_{th}, where $t \neq 0$.

PROOF. We first observe that our claim in the infinite dimensional case is valid if it is fulfilled for all finite dimensional projections of the measure μ since whenever $\|d_h\mu\| < \infty$ the indicated inequality obviously extends from finite dimensional projections to the measure μ (the variation of every Radon measure equals the supremum of variations of its finite dimensional projections). If $d_h\mu$ does not exist, the supremum of the quantities $\|d_{Ph}\mu \circ P^{-1}\|$ over finite dimensional projections $P\colon X \to \mathbb{R}^n$ is infinite (otherwise the Skorohod derivative would exist), whence we see that the supremum of the quantities $\|\mu \circ P_{Ph}^{-1} - \mu \circ P^{-1}\|$ equals 2. This means that $\|\mu_h - \mu\| = 2$, hence μ_h and μ are mutually singular.

Therefore, we have to justify the finite dimensional case. In this case it suffices to deal with convex measures with smooth positive densities. Indeed, the measure μ is given by a density on some affine subspace in $\mathbb{R}^n$. If $d_h\mu$ does not exist, then $\mu \perp \mu_h$. If $d_h\mu$ exists, then we may assume that μ is given by a density on $\mathbb{R}^n$ and take a sequence of centered Gaussian measures γ_k for which $\|\mu * \gamma_k - \mu\| \to 0$ and $\|d_h(\mu * \gamma_k) - d_h\mu\| \to 0$. Thus, it suffices to obtain the desired estimate for the measures $\mu * \gamma_k$, which are convex and have smooth positive densities.

Let us consider the case $n = 1$. Then $\mu = e^{-V} dx$, where V is a smooth convex function on the real line, and our construction of μ in the form of convolution yields that the function V is strictly convex, i.e., is constant on no interval. It is easily seen that $d_1\mu = -V'e^{-V} dx$. Let x_0 be the point of minimum of V (it is unique in our case). Then $V'(x) \leq 0$ if $x \leq x_0$ and $V'(x) \geq 0$ if $x \geq x_0$. Therefore,

$$\|d_1\mu\| = \int_{-\infty}^{+\infty} |V'(x)|e^{-V(x)} dx = 2e^{-V(x_0)}.$$

Let us find $\|\mu_h - \mu\|$ for $h > 0$. The function V' increases, so we have the inequality $V'(x) - V'(x - h) \geq 0$. Hence the function $x \mapsto V(x) - V(x - h)$ increases as well. Since the integrals of $e^{-V(x)}$ and $e^{-V(x-h)}$ are equal, there exists a number y_0 such that $V(y_0) = V(y_0 - h)$. Therefore,

$$\|\mu_h - \mu\| = \int_{-\infty}^{+\infty} |e^{-V(x-h)} - e^{-V(x)}| dx$$
$$= \int_{-\infty}^{y_0} \left(e^{-V(x)} - e^{-V(x-h)}\right) dx + \int_{y_0}^{+\infty} \left(e^{-V(x-h)} - e^{-V(x)}\right) dx$$
$$= 2\int_{y_0-h}^{y_0} e^{-V(x)} dx = 2\max_{\alpha \in \mathbb{R}} \mu([\alpha, \alpha + h]).$$

Thus, we have to establish the inequality

$$\max_{\alpha \in \mathbb{R}} \mu([\alpha, \alpha + h]) \geq 1 - \exp\bigl[-\exp(-V(x_0))\bigr]. \qquad (4.3.3)$$

Without loss of generality, we may assume that $x_0 = 0$. We observe that there exists an increasing convex function W on the ray $[0, +\infty)$ such that $W(0) = V(0)$ and the integral of e^{-W} over $[0, +\infty)$ equals 1. For W we take the inverse function to the function $t \mapsto \lambda(\{x \colon V(x) \leqslant t\})$, where λ is Lebesgue measure, defined on $[V(0), +\infty)$ and having the range $[0, +\infty)$, i.e., the set $\{W \leqslant t\}$ is the interval of the same length as $\{V \leqslant t\}$ but with the left end at the origin. It is easy to show that the function W is convex and increasing and $W(0) = V(0)$. Set $W(x) = +\infty$ if $x < 0$. By construction we have $\lambda(\{x \colon V(x) \leqslant t\}) = \lambda(\{x \colon W(x) \leqslant t\})$ and

$$\lambda\big(\{x \colon e^{-V(x)} \geqslant t\}\big) = \lambda\big(\{x \colon e^{-W(x)} \geqslant t\}\big).$$

Hence the integral of e^{-W} equals the integral of e^{-V}, i.e., equals 1. Let us consider the probability measure $\nu = e^{-W}\, dx$, concentrated on $[0, \infty)$. Its density is the nonincreasing rearrangement of the density of the measure μ, i.e., the inverse function to the function $t \mapsto \lambda(\{x \colon e^{-V(x)} \geqslant t\})$, defined on the half-line $[0, +\infty)$. We have

$$\lambda\big(\{x \colon a \leqslant e^{-V(x)} \leqslant b\}\big) = \lambda\big(\{x \colon a \leqslant e^{-W(x)} \leqslant b\}\big)$$

for all a, b. Let us take $b := e^{-V(0)} = e^{-W(0)}$ and $a := e^{-W(h)}$. Then the set $\{x \colon a \leqslant e^{-V(x)} \leqslant b\}$ is an interval of length h, i.e., an interval of the form $[\alpha, \alpha + h]$ since $\{x \colon a \leqslant e^{-W(x)} \leqslant b\} = [0, h]$. Therefore,

$$\max_{\alpha \in \mathbb{R}} \mu([\alpha, \alpha + h]) = \int_0^h e^{-W(x)}\, dx.$$

Hence for the proof of (4.3.3) it suffices to show that

$$\int_0^h e^{-W(x)}\, dx \geqslant 1 - \exp\big[-\exp(-W(0))\big] \qquad (4.3.4)$$

for every increasing convex function W on $[0, +\infty)$ such that the integral of e^{-W} is 1. Let us consider the linear function $V_0(x) = cx$, $c = e^{-W(0)}$, on $[0, +\infty)$. Then the integral of $e^{-V_0(x)}$ over $[0, +\infty)$ equals 1. We show that this function delivers a minimum of the left-hand side of (4.3.4) provided that $W(0)$ is fixed. Let $V_0(h) < W(h)$. Since W is increasing on $[0, +\infty)$, one has $W(h) - W(0) \leqslant W'(h)h$. Hence we obtain

$$W'(h) \geqslant \frac{W(h) - W(0)}{h} > \frac{V_0(h) - V_0(0)}{h} = V_0'(h),$$

whence

$$\int_h^\infty e^{-W(x)}\, dx \leqslant \int_h^\infty e^{-W'(h)(x-h) - W(h)}\, dx$$
$$< \int_h^\infty e^{-V_0'(h)(x-h) - V_0(h)}\, dx = \int_h^\infty e^{-V_0(x)}\, dx.$$

4.3. Convex measures

Since e^{-W} and e^{-V_0} have the unit integral over $[0, +\infty)$, the integral of e^{-W} over $[0, h]$ is not less than the integral of e^{-V_0} over $[0, h]$, which yields the desired estimate. Now let $W(h) \leqslant V_0(h)$. Then for $x = \alpha h$, where $\alpha \in (0, 1)$, we have

$$W(x) \leqslant (1-\alpha)W(0) + \alpha W(h) \leqslant (1-\alpha)V_0(0) + \alpha V_0(h) = V_0(\alpha h) = V_0(x).$$

Therefore, $e^{-W(x)} \geqslant e^{-V_0(x)}$ if $x \in [0, h]$, which gives the same inequality for the integrals.

Let us proceed to the case $n > 1$. The measure μ has a density e^{-V}, where V is a smooth convex function, $d_h \mu = -\partial_h V e^{-V} dx$. We may assume that $h = se_1$, $s > 0$. Therefore, according to what has been said above

$$\|d_h \mu\| = s \int_{\mathbb{R}^n} |\partial_{x_1} V(x)| e^{-V(x)} dx = 2s \int_{\mathbb{R}^{n-1}} \exp\bigl(-\min_{t \in \mathbb{R}} V(y + te_1)\bigr) dy,$$

$$\|\mu_h - \mu\| = \int_{\mathbb{R}^n} |e^{-V(x-h)} - e^{-V(x)}| dx$$

$$= \int_{\mathbb{R}^{n-1}} \left(\int_{\mathbb{R}^1} |e^{-V(x-h)} - e^{-V(x)}| dx_1 \right) dx_2 \cdots dx_n$$

$$= 2 \int_{\mathbb{R}^{n-1}} \max_{\alpha \in \mathbb{R}^1} \left\{ \int_\alpha^{\alpha+s} e^{-V(x)} dx_1 \right\} dx_2 \cdots dx_n.$$

Now we construct a function W on $[0, +\infty) \times \mathbb{R}^{n-1}$ similar to the one-dimensional case. Points in $\mathbb{R}^n$ will be denoted by single symbols as well as in the form $(x_1, \ldots, x_n)$. The set

$$G := \{(x, y) \in \mathbb{R}^{n+1} : y \geqslant V(x)\},$$

i.e., the super-graph of the function V, is convex and closed. Let us consider the set

$$G_n := \{(x_2, \ldots, x_n, y) \in \mathbb{R}^n : \exists x_1 \text{ with } V(x_1, \ldots, x_n) \leqslant y\},$$

which is also convex and closed (the latter can be easily derived from the fact that $V(x) \to +\infty$ as $|x| \to +\infty$ due to the integrability of e^{-V}). On G_n we define the function $F := F_1 - F_2$, where

$$F_1(x_2, \ldots, x_n, y) := \max_{(x_1, \ldots, x_n, y) \in G} x_1, \quad F_2(x_2, \ldots, x_n, y) := \min_{(x_1, \ldots, x_n, y) \in G} x_1.$$

The function F is nonnegative and concave on G_n, which is verified directly. Set

$$D := \{(x_1, x_2, \ldots, x_n, y) \in \mathbb{R}^{n+1} : 0 \leqslant x_1 \leqslant F(x_2, \ldots, x_n, y)\}.$$

Then D is a convex closed subset of $\mathbb{R}^{n+1}$. It is the super-graph of some convex function W on $\mathbb{R}^n$ (equal $+\infty$ outside the projection of the set D). Indeed, if $(x_1, \ldots, x_n, y^0) \in D$, then $(x_1, \ldots, x_n, y) \in D$ for every $y > y^0$. Let us consider the measure $\nu = e^{-W} dx$. The equality

$$\max_{\alpha \in \mathbb{R}^1} \int_\alpha^{\alpha+s} e^{-V(x_1, x_2, \ldots, x_n)} dx_1 = \max_{\alpha \in \mathbb{R}^1} \int_\alpha^{\alpha+s} e^{-W(x_1, x_2, \ldots, x_n)} dx_1$$

yields that
$$\int_{\mathbb{R}^1} e^{-V(x_1,x_2,\ldots,x_n)}\,dx_1 = \int_{\mathbb{R}^1} e^{-W(x_1,x_2,\ldots,x_n)}\,dx_1$$
for all $(x_2,\ldots,x_n) \in \mathbb{R}^{n-1}$. Since the integral of the left side over $\mathbb{R}^{n-1}$ equals 1, the same is true for the right side. Thus, letting
$$e^{-U(x_1)} = \int_{\mathbb{R}^{n-1}} e^{-W(x_1,x_2,\ldots,x_n)}\,dx_2\cdots dx_n,$$
we obtain a probability density on the half-line. It is well known (see [**193**, §3.10(vi)] or [**248**]) that the function U is convex as well. In addition, the function e^{-U} decreases on $[0,\infty)$ since the function $x_1 \mapsto e^{-W(x_1,x_2,\ldots,x_n)}$ decreases for every fixed $(x_2,\ldots,x_n)$. Therefore, we obtain a convex measure $\lambda = e^{-U}\,dx_1$ on $\mathbb{R}^1$, concentrated on $[0,+\infty)$. Taking into account the equality
$$\min_{x_1} V(x_1,x_2,\ldots,x_n) = \min_{x_1} W(x_1,x_2,\ldots,x_n) = W(0,x_2,\ldots,x_n)$$
we obtain the following expression for the norms in question:
$$\|\nu_h - \nu\| = 2\int_{\mathbb{R}^{n-1}} \left(\int_0^s e^{-W(x_1,\ldots,x_n)}\,dx_1\right) dx_2\cdots dx_n$$
$$= 2\int_0^s \left(\int_{\mathbb{R}^{n-1}} e^{-W(x_1,\ldots,x_n)}\,dx_2\cdots dx_n\right) dx_1 = 2\int_0^s e^{-U(x_1)}\,dx_1,$$
$$\|d_h\nu\| = 2s\int_{\mathbb{R}^{n-1}} e^{-W(0,x_2,\ldots,x_n)}\,dx_2\cdots dx_n = 2se^{-U(0)}.$$
Thus,
$$\|\lambda_s - \lambda\| = \|\nu_h - \nu\| = \|\mu_h - \mu\|, \quad \|d_s\lambda\| = \|d_h\nu\| = \|d_h\mu\|.$$
For the measure λ our claim is already proven, hence it is valid for the measure μ as well. $\square$

It is still unknown whether every convex measure on an infinite dimensional space has a nonzero vector of continuity.

4.4. Distributions of random processes

Here we briefly discuss differentiability of measures in functional spaces generated by such widely used random processes as diffusion processes and processes with independent increments. We do not mention Gaussian processes since their distributions are Gaussian measures, and the situation with their differentiability has been discussed above.

Let ξ_t be the diffusion process generated by the stochastic differential equation
$$d\xi_t = A(t,\xi_t)dW_t + b(t,\xi_t)dt, \quad \xi_0 = x_0, \qquad (4.4.1)$$
where W_t is a Wiener process in $\mathbb{R}^n$ and continuous mappings
$$A\colon [0,T]\times\mathbb{R}^n \to \mathcal{L}(\mathbb{R}^n), \quad b\colon [0,T]\times\mathbb{R}^n \to \mathbb{R}^n$$

4.4. Distributions of random processes

are uniformly Lipschitzian in the second variable. Under these conditions equation (4.4.1) has a unique solution (see Gihman, Skorohod [**477**, V. 3, Ch. III]).

The differentiability of the measure $\mu = \mu^\xi$ generated by the process ξ_t in the space $X = C([0,T], \mathbb{R}^n)$ was first studied by Pitcher [**905**]. By using the Girsanov theorem he proved the following assertion (a similar proof in our terminology is given in Bogachev, Smolyanov [**233**]). Let us recall (Remark 3.2.7) that the differential properties of the measure μ do not depend on whether we consider it on $C([0,T], \mathbb{R}^n)$, on $L^2([0,T], \mathbb{R}^n)$ or on the space $\mathbb{R}^{[0,T]}$ of all trajectories on $[0,T]$ equipped with the σ-algebra generated by the evaluation functionals $t \mapsto x(t)$. As we shall now see, nonzero vectors of differentiability exist in the case of the diffusion matrix dependent only on t; in the case of its dependence on x such vectors exist only in some very special cases. For example, if $n = 1$ and the diffusion coefficient is nonconstant, then there are no such vectors.

Let us consider the case where $A(t,x) \equiv A_0(t)$ and the operators $A_0(t)^{-1}$ are uniformly bounded. Set

$$z(t) := \int_0^t A_0(s)^{-1} dx(s) - \int_0^t A_0(s)^{-1} b(s, x(s)) ds.$$

We observe that $z(\,\cdot\,)$ is a function of $x(\,\cdot\,)$. The stochastic integral of the form

$$\int_0^T \psi(t, x(t)) dz(t)$$

is defined via the integral with respect to $dx(t)$ by using the stochastic differential $dz(t) = A_0(t)^{-1} dx(t) - A_0(t)^{-1} b(t, x(t)) dt$. Hence this stochastic integral is again a function of $x(\,\cdot\,)$.

The derivative of the mapping $b(t,x)$ in x (this is an operator on $\mathbb{R}^n$) is denoted by $Db(t,x)$. Let $\|Db(t,x)\| \leqslant C < \infty$.

4.4.1. Theorem. *Under the stated assumptions the measure μ is differentiable along all vectors $h \in W_0^{1,2}([0,T], \mathbb{R}^n)$ and*

$$\beta_h^\mu(x) = -\int_0^T \Big(A_0(t)^{-1} Db(t, x(t)) h(t) + A_0(t)^{-1} h'(t), dz(t)\Big).$$

PROOF. For every fixed $\lambda \in \mathbb{R}^1$ the process $\eta_t = \xi_t - \lambda h(t)$ satisfies the stochastic equation

$$d\eta_t = A_0(t) dW_t + b(t, \eta_t + \lambda h(t)) dt - \lambda h'(t) dt, \quad \eta_0 = x_0.$$

By the Girsanov Theorem 1.5.6 the measure $\mu_{\lambda h}$, which equals the distribution of the process η_t, has the following density with respect to μ:

$$\varrho_\lambda(x) = \exp\Big\{\int_0^T (u_\lambda(t), dz(t)) - \frac{1}{2}\int_0^T |u_\lambda(t)|^2 dt\Big\},$$

where

$$u_\lambda(t) = A_0(t)^{-1}\Big[b\big(t,x(t)\big) - b\big(t,x(t)+\lambda h(t)\big)\Big] - \lambda A_0(t)^{-1}h'(t).$$

As $\lambda \to 0$ the ratio $\lambda^{-1}[\varrho_\lambda - 1]$ converges in $L^1(\mu)$ to

$$-\int_0^T \Big(A_0(t)^{-1}Db\big(t,x(t)\big)h(t) + A_0(t)^{-1}h'(t), dz(t)\Big).$$

This can be verified by using the uniform boundedness of the functions u_λ, which follows from our assumptions. For $A_0 = I$ this formula can be obtained differently: the measure μ has the density

$$f(x) = \exp\bigg\{\int_0^T \Big(b\big(t,x(t)\big), dx(t)\Big) - \frac{1}{2}\int_0^T \big|b\big(t,x(t)\big)\big|^2\,dt\bigg\}$$

with respect to the Wiener measure. Now we can explicitly calculate $\partial_h f/f$ and use the fact that for the Wiener measure the logarithmic derivative is the stochastic integral of the integrand $-\big(h'(t), dx(t)\big)$. □

It is clear that assuming sufficient differentiability of b one can obtain higher differentiability of μ.

In the same paper [**905**], Pitcher conjectured that in the case of a spatially nonconstant diffusion coefficient the differentiability result may fail. The problem remained open about 30 years and only in [**173**], [**179**] the following result was proved (the multidimensional case was considered later in [**184**]).

4.4.2. Theorem. *Let $A(t,x) \equiv A(x)$, where A is Lipschitzian and continuously differentiable, A^{-1} is uniformly bounded, and for every nonzero vector $v \in \mathbb{R}^n$ there exists a basis vector e_i such that $\partial_v(Ae_i, Ae_i) \not\equiv 0$. Then μ has no nonzero vectors of continuity. In particular, this is true if $n = 1$ and A is not constant.*

PROOF. A detailed justification in the multidimensional case is given in [**184**]. So here we prove only the last assertion, which is technically simpler, but more clearly explains the nature of the phenomenon. Let us assume that $x_0 = 0$. There is a function $f \in C^1(\mathbb{R}^1)$ such that $f(0) = 0$ and $f'(x) = A\big(f(x)\big)$. By Itô's formula the process $\eta_t = f(W_t)$ satisfies the equation

$$d\eta_t = f'(W_t)dW_t + \frac{1}{2}f''(W_t)dt = A(\eta_t)dW_t + \frac{1}{2}A'(\eta_t)A(\eta_t)dt$$

with $\eta_0 = 0$. By the Girsanov theorem the measure μ is equivalent to the distribution ν of the process η_t, which is the image of the Wiener measure P under the mapping $F\colon C[0,T] \to C[0,T]$, $F(x)(t) = f\big(x(t)\big)$. We show that the measure ν has no nonzero vectors of continuity (then the same is also true for μ). Our verification is based on the law of iterated logarithm

4.4. Distributions of random processes

(see Wentzell [**1191**, §7.3], Kallenberg [**590**, p. 259]), according to which, for every fixed $t_0 \in (0, T)$, for P-a.e. x one has

$$L(x) = 1, \quad L(x) := \limsup_{t \to 0} |x(t_0 + t) - x(t_0)|(-2t \ln \ln t)^{-1/2}.$$

Suppose that the measure ν is continuous along h and $h(t_0) > 0$ for some $t_0 \in (0, T]$. If $L(x) = 1$, then

$$L(f(x)) = A(y(t_0)), \quad y(t) = f(x(t)),$$

since

$$f(x(t_0 + t)) - f(x(t_0)) = [x(t_0 + t) - x(t_0)]f'(x(t_0) + \theta),$$

where

$$|\theta| \leq |x(t_0 + t) - x(t_0)| \to 0 \quad \text{as } t \to 0,$$

which yields that

$$\limsup_{t \to 0} |f(x(t_0 + t)) - f(x(t_0))|(-2t \ln \ln t)^{-1/2} = f'(x(t_0)) = A(x(t_0)).$$

Hence the measure ν is concentrated on the set Ω of trajectories y with the property that

$$L(y) = A(y(t_0)).$$

Let $A' \geq m > 0$ (this condition is not essential, but it simplifies the proof). For sufficiently small $\lambda > 0$ the sets Ω and $\Omega + \lambda h$ must intersect (otherwise ν is not continuous along h). Let us take $y \in \Omega$ with $y + \lambda h \in \Omega$. Then

$$L(y + \lambda h) = A(y(t_0) + \lambda h(t_0)) \geq A(y(t_0)) + m\lambda h(t_0) > L(y).$$

On the other hand,

$$L(y + \lambda h) \leq L(y) + L(\lambda h).$$

If we show that $L(h) = 0$, then we obtain a contradiction. Actually, we have more:

$$\limsup_{t \to 0} |h(t_0 + t) - h(t_0)|t^{-1/2} = 0.$$

Indeed, otherwise one can find numbers $t_n \to 0$ with

$$M_n := [h(t_0 + t_n) - h(t_0)]t_n^{-1/2} \to +\infty$$

(the case of $-\infty$ is similar). Set $s_n := M_n^{-1}$. Then, as $n \to \infty$, uniformly in $l \in C[0,T]^*$ we have

$$\big|(\exp[is_n l(h)] - 1)\widetilde{\nu}(l)\big| = |\widetilde{\nu_{s_n h}}(l) - \widetilde{\nu}(l)| \leq \|\nu_{s_n h} - \nu\| \to 0.$$

Let us consider the functionals $l_n(x) = [x(t_0 + t_n) - x(t_0)]t_n^{-1/2}$. Then

$$\nu(l_n) = \int_X \exp\Big[i\big(x(t_0 + t_n) - x(t_0)\big)t_n^{-1/2}\Big]\,\nu(dx)$$

$$= \int_X \exp\Big[it_n^{-1/2}\big(f \circ x(t_0 + t_n) - f \circ x(t_0)\big)\Big]\,P(dx)$$

$$= \int_{-\infty}^{+\infty}\int_{-\infty}^{+\infty} \exp\Big[it_n^{-1/2}\big(f(u) - f(v)\big)\Big]\exp\big(-u^2/(2t_0)\big)$$
$$\times \exp\big(-|u-v|^2/(2t_n)\big)(2\pi t_0 2\pi t_n)^{-1/2}\,dv\,du$$

$$= \frac{1}{2\pi}t_0^{-1/2}\int_{-\infty}^{+\infty}\int_{-\infty}^{+\infty} \exp\Big(it_n^{-1/2}\big(f(t_n^{1/2}z + s) - f(s)\big)\Big)$$
$$\times \exp\big(-z^2/2 - s^2/(2t_0)\big)\,dz\,ds.$$

The right-hand side obviously tends to a positive limit

$$\frac{1}{2\pi}t_0^{-1/2}\int_{-\infty}^{+\infty}\int_{-\infty}^{+\infty} \exp\big(if'(s)z\big)\exp\big(-z^2/2 - s^2/(2t_0)\big)\,dz\,ds$$

$$= \frac{1}{\sqrt{2\pi t_0}}\int_{-\infty}^{+\infty} \exp\big(-|f'(s)|^2/2\big)\exp\big(-s^2/(2t_0)\big)\,ds.$$

Hence the numbers $\big(\exp[is_n l_n(h)] - 1\big)\widetilde{\nu}(l_n) = (e^i - 1)\widetilde{\nu}(l_n)$ do not tend to zero, which is a contradiction. $\square$

4.4.3. Remark. (i) Certainly, if $n > 1$, then the measure μ may have nonzero vectors of differentiability even in the case of nonconstant A. For example, we can take a function with values in the set of diagonal 2×2 matrices such that the first eigenvalue is identically 1 and the second one is not constant (this corresponds to a system of two independent diffusion equations). However, as it follows from the proof in Bogachev [**184**], the set of vectors of continuity $C(\mu)$ is not dense in X if A is not constant.

(ii) Suppose that A is Hölder continuous in t. A closer look at the proof in [**184**] shows that a vector h does not belong to $C(\mu)$ if there is no interval $U \subset [0, T]$ such that for all $t \in U$ and all $z, e \in \mathbb{R}^n$ the functions $\lambda \mapsto \|A(t, z + \lambda h(t))e\|^2$ are constant in a neighborhood of zero.

(iii) It should be noted that finite dimensional projections of the measure μ considered in Theorem 4.4.2 are absolutely continuous (in the case of smooth coefficients they even have smooth densities with respect to the corresponding Lebesgue measures). Indeed, if μ is considered on $\mathbb{R}^{[0,T]}$, then this is especially simple since every continuous finite dimensional linear projection L has the form $L = L_0 \circ \pi$, where $\pi \colon x \mapsto \big(x(t_1), \ldots, x(t_d)\big)$ and $L_0 \colon \mathbb{R}^d \to \mathbb{R}^m$, $m = \dim L(X)$, is linear. Thus, it suffices to verify the differentiability of the measure $\mu \circ \pi^{-1}$, which is done directly because this measure admits an explicit expression via transition probabilities of ξ_t. In the case where $X = C([0, T], \mathbb{R}^n)$ and the functionals are generated by

4.5. Gibbs measures and mixtures of measures

measures, the same property can be derived from the results of Chapter 9 on the regularity of induced measures (see Exercise 9.8.18).

Another interesting property of distributions of diffusion processes will be mentioned in Chapter 10 in relation with Aronszajn's exceptional measures.

A proof of the following theorem can be found in Davydov, Lifshits [**335**] and Skorohod [**1042**, §47].

4.4.4. Theorem. *Let ξ_t be a random process on $[0, T]$ with independent increments without Gaussian component. Let μ be its distribution in the space $X = L^2[0, T]$ (or in the space of all trajectories $X = \mathbb{R}^{[0,T]}$). Then, for every nonzero $h \in X$, the measures μ_h and μ are mutually singular. Hence μ has no nonzero vectors of continuity.*

Such an unfortunate situation with differentiability of the considered frequently encountered non-Gaussian processes seems to be particularly annoying for the distributions of diffusion processes since they are smooth images of the Wiener measure (in the sense discussed in Chapters 8–10). Let us consider the example already encountered in the proof of Theorem 4.4.2. Let $f \in C^\infty(\mathbb{R}^1)$ be a diffeomorphism which is not affine. Suppose that $f' \geq c > 0$. Denote by g the inverse function to f. By Itô's formula the process $\xi_t = f(w_t)$ satisfies the equation

$$d\xi_t = A(\xi_t)dw_t + b(\xi_t)dt, \quad \xi_0 = f(0),$$

where $A(x) = f'(g(x))$, $b(x) = 2^{-1}f''(g(x))$. The measure μ coincides with the image of the Wiener measure under the diffeomorphism F of the space $C[0, 1]$ defined by $F(x)(t) = f(x(t))$. Nevertheless, the measure μ has no nonzero vectors of continuity! We shall see in Chapters 9 and 11 that the differentiability along nonconstant vector fields is a much more flexible concept. In our example the measure μ is differentiable along the images of constant tangent vector fields under the mapping F.

4.5. Gibbs measures and mixtures of measures

Let V be a finite dimensional subspace of a locally convex space X such that $X = Y + V$, where Y is a closed linear subspace of X. If a Radon measure μ on X is differentiable along V, then one can choose conditional measures μ^y on the slices $V + y$, $y \in Y$, differentiable along V. If the functions β_v^μ, where $v \in V$, are exponentially integrable, then the measures μ^y admit densities $\exp(U_y)$ with respect to fixed Lebesgue measures on $V + y$. This observation relates logarithmic derivatives to Gibbs measures. The concept of a Gibbs distribution is dual in a sense to the concept of a consistent system of finite dimensional distributions. A typical situation connected with a Gibbs measure μ is this: its finite dimensional projections are unknown, but we are given its conditional measures on finite dimensional subspaces. For example, let μ be a Borel probability measure on $\mathbb{R}^\infty$. Suppose that for every fixed n we are given conditional measures $\mu^{n,y}$ on the straight lines $y + \mathbb{R}^1 e_n$. If these conditional measures possess densities $t \mapsto \varrho_n(y,t)$, then we say of conditional densities. If these densities are locally absolutely continuous, then we obtain logarithmic derivatives $\beta_n(y,t) = \partial_t \varrho_n(y,t)/\varrho_n(y,t)$. As we know, the integrability of the function $|\beta_n(y,t)|$ with respect to $\mu^{n,y}(dt)\nu_n$, where ν_n is the projection of μ to the hyperplane $\{x_n = 0\}$, is equivalent to the differentiability of μ along e_n. In the case of the local Lebesgue integrability of $t \mapsto \beta_n(y,t)$ the measure $\mu^{n,y}$ on the real line is uniquely determined by its logarithmic derivative. So specifying conditional measures $\mu^{n,y}$ for all n is equivalent to specifying the logarithmic derivatives $\beta_{e_n}^\mu$ along the vectors e_n. Thus, under rather broad assumptions the problem of existence or uniqueness of a measure with given one-dimensional conditional measures can be regarded as the problem of existence or uniqueness of a measure with given logarithmic derivatives along e_n. We shall return to these questions in Chapters 7 and 12. Note also that in the case of the strict positivity of densities of conditional measures on finite dimensional subspace it suffices to know conditional measures on one-dimensional subspaces (see [**193**, Ch. 10]).

Let us discuss mixtures of differentiable measures. Let σ be a finite measure on a measurable space (T, Σ) and let μ^t, $t \in T$, be a family of Radon measures on a locally convex space X such that for every $B \in \mathcal{B}(X)$ the function $t \mapsto \mu^t(B)$ is integrable with respect to σ. Suppose also that the function $t \mapsto \|\mu^t\|$ is integrable against σ as well. Then the measure

$$\mu(B) = \int_T \mu^t(B)\, \sigma(dt) \tag{4.5.1}$$

is well-defined and called the *mixture of the measures* μ^t.

The following examples show that even if all measures μ^t are nondegenerate Gaussian, the measure μ may fail to have nonzero vectors of continuity. Moreover, it may be mutually singular with all its nonzero shifts.

4.5.1. Example. (i) Let H be an infinite dimensional separable Hilbert space. Let A and K be nonnegative Hilbert–Schmidt operators with dense ranges such that $A(H) \cap K(H) = 0$ (see Exercise 1.6.5). Suppose that v is a vector not contained in $A(H) + K(H)$ and γ is the Gaussian measure with the Fourier transform $\exp[-(K^4 y, y)]$. Let us define a measure μ by the formula

$$\mu(B) = \int_0^1 \gamma_t(B)\, dt,$$

where γ_t is the image of γ under the mapping $F(t)\colon x \mapsto \exp(tA)x + tv$, which is affine. Then γ_t is a nondegenerate Gaussian measure for each t, but $\mu_h \perp \mu$ for every $h \neq 0$; in particular, according to Exercise 3.9.9, μ is mutually singular with every measure which has a nonzero vector of continuity.

(ii) Let μ be the distribution in $L^2[0,1]$ of the process ξ_t which has independent symmetric increments and is stable of order $\alpha < 2$. According to Theorem 4.2.1, the measure μ can be written in the form (4.5.1) with nondegenerate centered Gaussian measures μ^t, but $\mu \perp \mu_h$ for every $h \neq 0$ due to Example 4.2.4.

PROOF. (i) The set $D = \bigcup_{t \in [0,1]} F(t)(K(H))$ is Souslin, hence measurable, since it is the image of the Borel set $H \times [0,1]$ under the continuous mapping $(x,t) \mapsto F(t)(x)$. In addition, $\mu(D) = 1$. Let us show that $(D+h) \cap F(t)(K(H))$ contains at most one element for each $t \in [0,1]$. This will yield the equality

$$\mu(D+h) = \int_0^1 \gamma_t(D+h)\,dt = 0$$

since γ_t is concentrated on $F(t)(K(H))$. Suppose that $(D+h) \cap F(t)(K(H))$ contains two different points x and y. Then

$$x = e^{tA}Kx_1 + tv = e^{rA}Kx_2 + rv + h, \quad y = e^{tA}Ky_1 + tv = e^{qA}Ky_2 + rv + h,$$

which implies that

$$(q-r)v = e^{tA}K(y_1 - x_1) + e^{rA}Kx_2 - e^{qA}Ky_2.$$

Since $v \notin K(H) + A(H)$ we have $q = r$ and $e^{tA}K(y_1 - x_1) = e^{rA}K(y_2 - x_2)$. If $t \neq r$ (for example, $\delta = t - r > 0$), then

$$K(y_2 - x_2)e^{(t-r)A}K(y_1 - x_1) = K(y_1 - x_1) + A\sum_{n=1}^{\infty} \delta^n A^{n-1}K(y_1 - x_1)/n!.$$

By the choice of A and K, this is possible only if $y_2 - x_2 = y_1 - x_1 = z$; but then $e^{\delta A}Kz = Kz$, whence $Kz = 0$, and $z = 0$. Thus, $x = y$, a contradiction. If $t = r$, then $h \in e^{tA}K(H)$. In this case we have $h \notin e^{pA}K(H)$ for $p \neq t$, since $e^{pA}K(H) \cap e^{tA}K(H) = 0$, as we established above. This proves our claim. Assertion (ii) is already explained. $\square$

As follows from the results in § 3.2, the measure μ is continuous along a vector h provided that almost every measure μ^t has this property. Note the following obvious sufficient condition for the differentiability of μ.

4.5.2. Proposition. *Suppose that almost every measure μ^t is n-fold differentiable along all vectors from a linear subspace L and*

$$\int_T \|d_{h_1} \cdots d_{h_n}\mu^t\|\,|\sigma|(dt) < \infty, \quad \forall\, h_1, \ldots, h_n \in L.$$

Then μ is n-fold differentiable along all vectors from the subspace L and

$$d_{h_1}\cdots d_{h_n}\mu(B) = \int_T d_{h_1}\cdots d_{h_n}\mu^t(B)\,\sigma(dt).$$

In some situations this condition is not only sufficient, but also necessary. Let $T = (0, \infty)$ and let σ be a probability measure.

4.5.3. Proposition. (i) *Let $\mathcal{B}(X)$ be generated by a countable collection of sets and let $\pi\colon X \to T$ be a Borel function. Suppose that all probability measures μ^t are differentiable along some vector h and $\mu^t(\pi^{-1}(t)) = 1$. Then the condition*

$$\int_T \|d_h\mu^t\|\, \sigma(dt) < \infty$$

is necessary and sufficient for the differentiability of μ along h.

(ii) *Let $\mu^t(B) = \mu^1(t^{-1}B)$, where the measure μ^1 is differentiable along some vector $h \in X$. Suppose that there exists an element $l \in X^*$ with $l(h) \neq 0$ such that the measure $\mu^1 \circ l^{-1}$ on $\mathbb{R}^1$ has a density p which is symmetric and decreasing on $(0, \infty)$. Then the condition*

$$\int_T t^{-1} \sigma(dt) < \infty \qquad (4.5.2)$$

is necessary and sufficient for the differentiability of μ along h.

PROOF. (i) The necessity of the indicated condition is verified similarly to Theorem 3.5.1 (see also §11.2). (ii) Since $\|d_h\mu^t\| = t^{-1}\|d_h\mu^1\|$, condition (4.5.2) is sufficient. Suppose that the measure μ is differentiable along h. The measures $\mu^t \circ l^{-1}$ have densities $p_t(x) = t^{-1}p(x/t)$. Hence the measure $\mu \circ l^{-1}$ has the following density:

$$f(x) = \int_0^\infty t^{-1} p(x/t)\, \sigma(dt).$$

Since $l(h) \neq 0$, the measures $\mu \circ l^{-1}$ and $\mu^1 \circ l^{-1}$ are Fomin differentiable along 1. We may assume that p is an absolutely continuous version of a density of the measure $\mu^1 \circ l^{-1}$. Then f is continuous, hence coincides with an absolutely continuous version of a density of the measure $\mu \circ l^{-1}$, whence we have $f' \in L^1(\mathbb{R}^1)$. Let $x > 0$. Set $Dp(x) := \limsup_{n\to\infty} n[p(x+1/n) - p(x)]$. We observe that $Dp(x) = p'(x)$ almost everywhere. Since the density p is decreasing on $(0, \infty)$ and we have $\lim_{n\to\infty} n[f(x+1/n) - f(x)] = f'(x)$ for almost all $x > 0$, the Fatou theorem shows that the nonpositive function

$$x \mapsto \int_0^\infty t^{-2} Dp(x/t)\, \sigma(dt)$$

is estimated from below by $f'(x)$ and hence is integrable on $(0, +\infty)$. By Fubini's theorem its integral equals

$$\int_0^\infty Dp(z)\, dz \int_0^\infty t^{-1}\sigma(dt) = \int_0^\infty p'(z)\, dz \int_0^\infty t^{-1}\sigma(dt).$$

A similar reasoning applies to $(-\infty, 0)$. $\square$

We observe that every convex measure μ^1 and every symmetric stable measure μ^1 satisfy the condition mentioned in the previous example for every nonzero functional $l \in X^*$. In particular, this is true if μ^1 is a centered

Gaussian measure. In the latter case, if μ is infinitely differentiable along h, then
$$d_h\mu = -\int_0^\infty \widehat{h}(\,\cdot\,)t^{-2}\mu^t\,\sigma(dt) = -\widehat{h}(\,\cdot\,)\int_0^\infty t^{-2}\mu^t\,\sigma(dt), \qquad (4.5.3)$$
whence
$$d_h^n\mu = p_n(\widehat{h})\cdot\nu_n,$$
where ν_n is a measure equivalent to μ and p_n is a polynomial on the real line (certainly, they can be written explicitly). If μ is only once differentiable, then in place of the above formula one can write
$$d_h\mu = c(\,\cdot\,)\widehat{h}(\,\cdot\,)\cdot\mu, \qquad (4.5.4)$$
where $c(\,\cdot\,)$ is a μ-measurable function (not necessarily integrable). This readily follows from formula (4.5.3) if we take the measures $\mu_n \ll \mu$ corresponding to the measures $\sigma_n := I_{[1/n,\infty)}\cdot\sigma$ and observe that the measures $d_h\mu_n = c_n(\,\cdot\,)\widehat{h}(\,\cdot\,)\cdot\mu_n$ converge in variation to the measure $d_h\mu$ and the measures μ_n converge in variation to μ, whence we obtain convergence of the sequence $\{c_n\}$ in measure μ to some function c. If H is infinite dimensional, then this function can be written, e.g., in the following form suggested in Norin, Smolyanov [835]. Let $\{e_n\}$ be an orthonormal basis in $H(\mu^1)$. By the law of large numbers a limit
$$\tau(x) = \lim_{n\to\infty} n^{-1}\sum_{i=1}^n \widehat{e}_i(x)^2 \qquad (4.5.5)$$
exists and μ^t-a.e. equals t^2 for each t. Thus, τ is a Borel function on X and one has $\mu^t\bigl(x\colon \tau(x) = t^2\bigr) = 1$ for all t. With this notation we obtain
$$d_h\mu(B) = -\int_T\int_B t^{-2}\widehat{h}(x)\,\mu^t(dx)\,\sigma(dt) = -\int_T\int_B \tau(x)^{-1}\widehat{h}(x)\,\mu^t(dx)\,\sigma(dt).$$
Therefore, we can set $c(x) := \tau(x)^{-1}$.

If $H = H(\mu^1)$ is infinite dimensional, then the mixtures considered in the last proposition admit a different characterization (see the notion of an H-spherically invariant measure in Definition 7.7.4). We shall see in §7.7 that both in the finite and infinite dimensional cases smooth spherically invariant measures can be described as measures possessing logarithmic gradients of the form $\beta_H(x) = -c(x)x$, where $c(\,\cdot\,)$ is a real function.

4.6. Comments and exercises

An observation relating the differentiability of measures with martingale convergence was used in Bogachev [169], [175] (see also Bogachev, Smolyanov [233]) for constructing measures with non-Hilbert subspaces of differentiability. These investigations were continued in Hora [553], Khafizov [601], [605]. On convex measures, see also Bobkov [160], [161], Bogachev [193, §7.14], Borell [239], Brascamp, Lieb [248], Krugova [646], [647], Kulik [652], [653], Leindler [714], Prékopa [919], [920].

There is an extensive literature devoted to Gibbs measures, (see Georgii [**474**]). Some problems related to their differentiability are discussed in Antoniouk, Antoniouk [**72**], Bogachev, Röckner [**222**], Bogachev, Röckner, Wang [**231**], Deuschel, Föllmer [**344**], Doss, Royer [**358**], Föllmer [**437**], Fritz [**451**], Holley, Stroock [**552**], Norin, Smolyanov [**836**], Nowak [**845**], [**846**]. Quasi-invariance, continuity and differentiability of mixtures of measures were studied, in particular, in Gihman, Skorohod [**477**, v. 1, Ch. VII, §2], Skorohod [**1046**], Sudakov [**1089**], Sytaja [**1097**]. On spherically symmetric measures, see also Knutova [**617**], Savinov [**986**], Shatskih [**997**]. Differentiability of some measures arising in quantum field theory is studied in Albeverio, Liang, Zegarlinski [**50**].

Hariya [**520**] and Zambotti [**1217**]–[**1219**] investigated integration by parts for some processes connected with the Wiener process, for example, for the process $|w_t|$.

Exercises

4.6.1. Construct an example of a uniformly convex measure μ on the real line such that its logarithmic derivative is not square-integrable with respect to μ.

4.6.2. Let μ be a convex Radon measure on a locally convex space X and let E be a convex Borel set with $\mu(E) > 0$. Prove that the measure $\mu(E)^{-1}\mu|_E$ is convex.

HINT: Use that $\alpha(A \cap E) + (1-\alpha)(B \cap E) \subset \bigl(\alpha A + (1-\alpha)B\bigr) \cap E$.

4.6.3. (i) Prove that for the convexity of a probability measure μ on $\mathbb{R}^n$ it suffices to have the inequality from Definition 4.3.1 for all sets A and B that are finite unions of cubes with edges of rational length parallel to the coordinate axis. (ii) Let X be a separable Fréchet space. Prove that there is a countable class $\mathcal{K}$ of cylindrical Borel sets in X such that for the convexity of a Borel probability measure μ on X it suffices to have the inequality from Definition 4.3.1 for all $A, B \in \mathcal{K}$.

4.6.4. Prove that a stable measure is convex only when it is Gaussian.

4.6.5. Let μ and ν be two Radon measures on a locally convex space such that $\mu(C) \leqslant \nu(C)$ for all convex Borel cylindrical sets. Prove that $\mu \leqslant \nu$.

HINT: It suffices to consider $\mathbb{R}^n$; using convolutions reduce to the case of measures with continuous densities and take balls for C.

4.6.6. Suppose that convex measures μ_j on $\mathbb{R}^d$ converge weakly to an absolutely continuous measure μ. Show that $\|\mu_j - \mu\| \to 0$ and that the measures μ_j are absolutely continuous for all sufficiently large j.

HINT: See Medvedev [**798**].

4.6.7. Construct an example of a one-dimensional diffusion process with coefficients $\sigma, b \in C_b^\infty(\mathbb{R}^1)$ and $\sigma \geqslant 1$ whose distribution is mutually singular with every Gaussian measure.

HINT: Take $b = 0$ and $\sigma(x) = 1 + \sin^2 x$.

4.6.8. Prove that for every nonconstant function h on $[0,1]$ and all $p > 1$ one has $\sup \sum_{i=1}^n |h(t_i) - h(t_{i-1})|^2 (t_i - t_{i-1})^{-p} = \infty$, where sup is taken over all n and all $0 = t_0 < t_1 < \cdots < t_n \leqslant 1$.

CHAPTER 5

Subspaces of differentiability of measures

The set of all vectors of differentiability of a given measure, called its subspace of differentiability, is some analogue of the Cameron–Martin space of a Gaussian measure. However, this space need not be Hilbert. Here we investigate basic properties of subspaces of differentiability and consider some applications.

5.1. Geometry of subspaces of differentiability

Let μ be a nonzero Radon measure on a locally convex space X. Let us introduce the following sets:

$$D_C(\mu) := \{h\colon \mu \text{ is Skorohod differentiable along } h\},$$
$$D(\mu) := \{h\colon \mu \text{ is Fomin differentiable along } h\},$$
$$H(\mu) := \{h \in D(\mu)\colon \beta_h^\mu \in L^2(\mu)\},$$
$$H_p(\mu) := \{h \in D(\mu)\colon \beta_h^\mu \in L^p(\mu)\},$$
$$C(\mu) := \{h\colon \mu \text{ is continuous along } h\},$$
$$Q(\mu) := \{h\colon \mu \text{ is quasi-invariant along } h\}.$$

All of these sets, as we already know, are linear subspaces in X. In addition, they possess certain natural metrics which are determined intrinsically in the sense that they do not depend on our choice of the space carrying the measure (see Remark 3.2.7). In this section we introduce and discuss such metrics.

The spaces $D_C(\mu)$ and $D(\mu)$ are equipped with the norm

$$\|h\|_D := \|d_h\mu\|.$$

As follows from the results in Chapter 3, $\|\cdot\|_D$ is indeed a norm since

$$d_{\alpha h}\mu = \alpha d_h\mu, \quad d_{h+k}\mu = d_h\mu + d_k\mu.$$

In addition, the equality $d_h\mu = 0$ is only possible if $h = 0$; otherwise for every compact set K we have $\mu(K + th) = \mu(K)$ for all t, which gives $\mu(K) = 0$ because for sufficiently large m the sets $K + mh$ are disjoint. Alternatively, one can use that $l(h)\widetilde{\mu}(l) = 0$ for all $l \in X^*$. For the space $D(\mu)$ this norm can be written as

$$\|h\|_D = \|\beta_h^\mu\|_{L^1(\mu)}.$$

The space $H(\mu)$ is equipped with the inner product
$$(h,k)_{H(\mu)} = \bigl(\beta_h^\mu, \beta_k^\mu\bigr)_{L^2(|\mu|)},$$
and the space $H_p(\mu)$ is equipped with the norm
$$\|h\|_{H_p(\mu)} = \|\beta_h^\mu\|_{L^p(\mu)}.$$
The spaces $C(\mu)$ and $Q(\mu)$ are equipped with the metric
$$\varrho(a,b) = \sup_{t\in[-1,1]} \|\mu_{ta} - \mu_{tb}\|.$$
The triangle inequality in $C(\mu)$ and $Q(\mu)$ is readily verified.

The space $D(\mu)$ with the above norm is called the *subspace of differentiability* of the measure μ. This is an important characteristic of the measure (something analogous to the Cameron–Martin space for a Gaussian measure). The subspaces of differentiability of measures were introduced in the author's works [166], [169], [175], where it was discovered that they are always Banach spaces (with rather peculiar properties).

5.1.1. Theorem. *Let μ be a nonzero Radon measure on a locally convex space X. Then*

(i) the spaces $D(\mu)$ and $D_C(\mu)$ with the norm $\|\cdot\|_D$ are Banach spaces compactly embedded into X;

(ii) the closed balls of $D_C(\mu)$ are compact in X;

(iii) $D(\mu)$ is a closed linear subspace in $D_C(\mu)$;

(iv) the space $H(\mu)$ is Hilbert and the spaces $H_p(\mu)$ are Banach; these spaces are continuously embedded into $D(\mu)$; if $p > 1$, then the closed balls of $H_p(\mu)$ are compact in X.

PROOF. It has been noted above that the indicated sets are linear spaces and $\|\cdot\|_D$ is a norm. Let us take a compact set K with $|\mu|(K) > c + \|\mu\|/2$, where $c > 0$. Then the set $K - K$ is compact and $h \in K - K$ whenever $\|h\|_D < c$. Indeed, the estimate $|\mu(K-h) - \mu(K)| \leqslant \|d_h\mu\| < c$ yields that $|\mu(K-h)| > \|\mu\|/2$, whence we have $(K-h) \cap K \neq \varnothing$. Thus, the balls with respect to the norm $\|\cdot\|_D$ are contained in compact sets. Moreover, the closed unit ball of $D_C(\mu)$ is closed in X. Indeed, if a net $h_\alpha \in D_C(\mu)$ with $\|d_{h_\alpha}\mu\| \leqslant 1$ converges to a point h, then Theorem 3.8.1 yields that $h \in D_C(\mu)$ and $\|d_h\mu\| \leqslant 1$. We verify the completeness of $D_C(\mu)$. If a sequence $\{h_n\}$ is fundamental in $D_C(\mu)$, then, according to what has been proven above, it converges in X to some $h \in D_C(\mu)$. The sequence of measures $d_{h_n}\mu$ is Cauchy in variation, so it converges in variation to some measure ν due to the completeness of the space of measures with the variation norm. It is clear that $\nu = d_h\mu$, i.e., $\|h_n - h\|_D \to 0$.

The subspace $D(\mu)$ in $D_C(\mu)$ is distinguished by the condition $d_h\mu \ll \mu$ and hence is closed with respect to the norm $\|\cdot\|_D$. The completeness of the space $H_p(\mu)$ with respect to the norm induced by its embedding into $L^p(\mu)$ is easily deduced from the completeness of $L^p(\mu)$. Moreover, $H_p(\mu)$ is linearly isometric to a closed linear subspace in $L^p(\mu)$, which for

$p=2$ shows that it is Hilbert and for $p>1$ yields its reflexivity. Since all closed balls in reflexive spaces are compact in the weak topology, their images under compact embeddings are norm compact. $\square$

We observe that assertion (i) means only that the closed unit ball of $D(\mu)$ (considered with the norm $\|\cdot\|_D$) is contained in a compact set in X. Unlike assertions (ii) and (iv), it is still unknown whether the ball itself is compact. In other words, it is not clear whether it is closed as a subset of X. This problem is open even for measures on separable Hilbert spaces (however, apparently, the topology of X is irrelevant for this problem). Certainly, if $D(\mu)$ is reflexive, then its closed balls are compact in X.

If X is a Fréchet space, then Theorem 5.1.1 and Banach's inverse mapping theorem show that the spaces $D(\mu)$, $D_C(\mu)$ and $H(\mu)$ are not closed in X provided that they are infinite dimensional.

The equality $\|h\|_D = \|\beta_h^\mu\|_{L^1(\mu)}$ shows that the Banach space $D(\mu)$ is linearly isometric to a closed linear subspace in $L^1(\mu)$. Now from the theory of L^p-spaces (see Lindenstrauss, Zafriri [**727**]) we obtain several interesting conclusions concerning the geometry of subspaces of differentiability. In these corollaries under isomorphic copies of Banach spaces we understand linearly homeomorphic ones.

5.1.2. Corollary. *Every closed (with respect to the norm $\|\cdot\|_D$) infinite dimensional subspace in $D(\mu)$ contains either an isomorphic copy of l^1 or a copy of l^p with $p\in(1,2]$. In addition, $D(\mu)$ is either reflexive or contains an isomorphic copy of l^1.*

5.1.3. Corollary. *The space $D(\mu)$ has cotype 2 (see Lindenstrauss, Zafriri [**727**]). In particular, it contains no subspaces isomorphic to the spaces l^p or infinite dimensional spaces L^p with $p>2$.*

5.1.4. Corollary. *If $L^1(\mu)$ is separable, then $D(\mu)$ is isomorphic to a closed subspace in $L^1[0,1]$.*

5.1.5. Corollary. *The space $D_C(\mu)$ is isomorphic to the dual space Y^* of some Banach space Y. In particular, it has the Radon–Nikodym property provided that it is separable.*

PROOF. Denote by Z the space $D_C(\mu)$ with the topology from X and by Y the space Z^* with the norm $\|y\|:=\sup\{y(h)\colon \|d_h\mu\|\leqslant 1\}$. In this way Y is equipped with the topology of uniform convergence on the balls from $D_C(\mu)$ that are absolutely convex and compact in Z, and so they are in the topology $\sigma(Z,Z^*)$. By the Mackey–Arens theorem (see [**379**, p. 504]), the space dual to Y coincides with $D_C(\mu)$. Finally, it is known that all separable duals have the Radon–Nikodym property. $\square$

It is worth noting that Corollary 5.1.5 is also valid for those spaces $D(\mu)$ whose closed unit balls are compact in X. In particular, all separable spaces $D(\mu)$ of this kind have the Radon–Nikodym property.

It is unknown whether the space $D(\mu)$ always has the Radon–Nikodym property. The answer is positive if it is separable and coincides with $D_C(\mu)$. Certainly, the space $D(\mu)$ has the Radon–Nikodym property if it has no subspaces isomorphic to l^1 (since it will be then reflexive by Corollary 5.1.2). It is also open whether $D_C(\mu)$ is always separable if X is a separable Banach space.

In the next section we consider examples showing that all spaces l^p with $1 \leqslant p \leqslant 2$ are isomorphic to some subspaces of differentiability. In particular, $D(\mu)$ need not be linearly homeomorphic to a Hilbert space.

5.1.6. Theorem. *Let μ be a nonzero Radon measure on a locally convex space X. Then*

(i) the space $Q(\mu)$ is an F-space (a complete metrizable topological vector space) that is compactly embedded into X; in addition, it is isometric to a subset of $L^1(\mu)$;

(ii) the space $C(\mu)$ is also an F-space compactly embedded into X.

In particular, the algebraic operations are continuous in the spaces $Q(\mu)$ and $C(\mu)$ with the metric ϱ. It should be noted that Shimomura [**1022**] constructed examples showing that the metric

$$\varrho_0(a,b) = \|\mu_a - \mu_b\|$$

on $Q(\mu)$ may fail to be consistent with the linear structure. As shown by Khafizov [**605**], the space $C(\mu)$ is isomorphic to a linear subspace in the F-space $L^0(\Omega, P)$ of measurable functions with the topology of convergence in measure for some probability space (Ω, P); see details in Exercise 5.5.4. One should keep in mind that the spaces $C(\mu)$ and $Q(\mu)$ may fail to be locally convex (see Exercise 5.5.2).

5.1.7. Proposition. *Let μ be a Radon measure on a locally convex space X and let F be a Fréchet space (or, more generally, a barrelled space, see [**379**, p. 427]) continuously embedded into X. If $F \subset D(\mu)$, then the mapping $\Psi\colon h \mapsto \beta_h^\mu$, $F \to L^1(\mu)$, is continuous. If $F \subset H(\mu)$, then the same is true for the mapping $\Psi\colon F \to L^2(\mu)$.*

PROOF. By the closed graph theorem the natural inclusion $F \to D(\mu)$ is continuous since $D(\mu)$ is a Banach space continuously embedded into X. On the other hand, the mapping $h \mapsto \beta_h^\mu$, $D(\mu) \to L^1(\mu)$ is an isometry. This reasoning is valid in the second case as well. □

5.2. Examples

Combining Theorem 4.2.2 and Corollary 6.2.2 we obtain the following result for stable measures.

5.2.1. Theorem. *Let a Radon probability measure μ on a locally convex space X be stable of some order α. Then the subspaces of continuity $C(\mu)$ and differentiability $D(\mu)$ coincide with the Hilbert space $H(\mu)$, the measure*

5.2. Examples

μ is infinitely differentiable along $H(\mu)$ and if $\alpha \geq 1$, then it is analytic along $H(\mu)$. If $\alpha \geq 1$, then $H(\mu) = Q(\mu)$.

5.2.2. Remark. Let μ be a stable Radon measure. It is unknown whether $H(\mu)$ is always separable.

5.2.3. Example. If m is a Borel probability measure on the real line with finite Fisher information, $\mu_n(A) = m(A/a_n)$, where a_n are positive numbers, and $\mu = \bigotimes_{n=1}^{\infty} \mu_n$, then

$$D(\mu) = H(\mu) = \Big\{ x \in \mathbb{R}^{\infty} \colon \sum_{n=1}^{\infty} a_n^{-2} x_n^2 < \infty \Big\}.$$

This follows from Corollary 4.1.2.

It is worth noting that for the product μ of arbitrary probability measures μ_n with finite Fisher numbers I_n the subspace of differentiability may be larger than the set $\big\{ h = (h_n) \colon \sum_{n=1}^{\infty} h_n^2 I_n < \infty \big\}$, which coincides with $H(\mu)$ and is contained in $D(\mu)$.

5.2.4. Example. Let μ_n be given by the density p_n defined on $\mathbb{R}^1$ as follows: $p_n(t) = 2n^{-1} - |t|$ if $|t| \leq n^{-1}$, $p_n(t) = n^{-1}$ if $n^{-1} < |t| < c_n$, $p_n(t) = c_n^2 n^{-1} t^{-2}$ if $|t| \geq c_n$, where $c_n = (4n)^{-1}(n^2 - 1)$ and $n \geq 2$. Let $\mu_1 = \mu_2$ and $\mu = \bigotimes_{n=1}^{\infty} \mu_n$. Then $H(\mu) = l^2$ does not coincide with $D(\mu)$.

PROOF. It follows from Theorem 4.1.1 that if a vector h is such that the series of $I_n h_n^2$ converges, then $h \in H(\mu)$ because the series of $\beta^{\mu}_{h_n e_n}$ converges in $L^2(\mu)$, where $\{e_n\}$ is the standard basis in l^2. Conversely, for any $h \in H(\mu)$, the sums $\beta^{\mu}_{h_1 e_1} + \cdots + \beta^{\mu}_{h_n e_n}$ form a martingale convergent in $L^2(\mu)$, which yields convergence of the series of $h_n^2 I_n$. We observe that $|p'_n(t)|^2 / p_n(t) = (2n^{-1} - |t|)^{-1} \in [n/2, n]$ if $|t| \leq n^{-1}$. Hence the integral of $|p'_n(t)|^2 / p_n(t)$ over $[-1/n, 1/n]$ belongs to $[1, 2]$. In addition, the integral of $|p'_n(t)|^2 / p_n(t)$ over $[c_n, +\infty)$ equals $48(n^2 - 1)^{-1}$ and the integral over $[1/n, c_n]$ vanishes. This shows that $1 \leq I_n \leq 34$. Hence $H(\mu) = l^2$. However, the vector $h = (h_n)$ with $h_n = n^{-1/2}$ belongs to $D(\mu)$ and does not belong to l^2. Indeed, let us show that the series of $\beta^{\mu}_{h_n e_n}(x) = h_n p'_n(x_n)/p_n(x_n)$ converges in $L^1(\mu)$. We have

$$\int_X |\beta^{\mu}_{h_n e_n}(x)| \, \mu(dx) = \frac{1}{\sqrt{n}} \int_{-\infty}^{+\infty} |p'_n(t)| \, dt = n^{-1/2}(2n^{-1} + 4n^{-1} c_n^2 c_n^{-2}) = 6n^{-3/2},$$

hence we have convergence of norms. $\square$

5.2.5. Example. Let μ be the countable product of identical measures on the real line with the following density f: $f(t) = 2^{-1} p(1 - |t|)^{p-1}$ on the interval $(-1, 1)$, $f(t) = 0$ outside $(-1, 1)$ and $p \in (1, 2)$. Then $D(\mu)$ coincides as a set with l^p and its norm is equivalent to the norm in l^p.

PROOF. Let $h = (h_n) \in l^p$ and let
$$\xi_n(x) = h_n f'(x_n)/f(x_n) = -h_n(p-1)\operatorname{sign} x_n(1-|x_n|)^{-1}$$
if $|x_n| < 1$, $\xi_n(x) = 0$ if $|x_n| \geq 1$. The random variables ξ_n on $(\mathbb{R}^\infty, \mu)$ are independent. We show that the series $\sum_{n=1}^\infty \xi_n$ converges almost everywhere and in $L^1(\mu)$. We may assume that $|h_n| \leq 1$. Let us observe that
$$\mu(|\xi_n| > 1) = (p-1)^p |h_n|^p$$
since this quantity is the integral of $2f$ over the interval $[1 - (p-1)|h_n|, 1]$. Hence the series of these quantities converges. The functions $\eta_n = \xi_n I_{\{|\xi_n| \leq 1\}}$ have zero integrals. The integral of the function η_n^2 is estimated by the quantity $\mu(|\xi_n| > 1) = (p-1)^p |h_n|^p$, hence the series of these integrals converges. By the Kolmogorov three series theorem we obtain convergence of $\sum_{n=1}^\infty \xi_n$ almost everywhere. If $\varepsilon = (p-1)/2$, then the series of the integrals of $|\xi_n(x)|^{1+\varepsilon} I_{\{|\xi_n| \geq 1\}}$ converges. Indeed, the condition $|\xi_n| \geq 1$ means that $|x_n| < 1$ and $1 - |x_n| \leq (p-1)|h_n|$. Hence
$$|\xi_n|^{1+\varepsilon} f(x_n) \leq \frac{1}{2} p(p-1)^{1+\varepsilon} |h_n|^{1+\varepsilon} (1-|x_n|)^{\varepsilon - 1}.$$
The integral of the right-hand side over the set $1 - (p-1)|h_n| \leq |x_n| < 1$ is estimated by $p(p-1)^{1+2\varepsilon} \varepsilon^{-1} |h_n|^{1+2\varepsilon}$. Since $1 + 2\varepsilon = p$, we obtain convergence of the indicated series. According to Vakhania, Tarieladze, Chobanyan [**1167**, Corollary 2, p. 293], the series $\sum_{n=1}^\infty \xi_n$ converges in $L^{1+\varepsilon}(\mu)$, which gives convergence in $L^1(\mu)$. Therefore, $h \in D(\mu)$.

Conversely, let $h \in D(\mu)$. Then the series $\sum_{n=1}^\infty \xi_n$ converges almost everywhere to β_h^μ by the martingale convergence theorem since its partial sums $\sum_{i=1}^n \xi_i$ are the conditional expectations of β_h^μ with respect to the σ-algebras generated by the first n coordinate functions. Kolmogorov's theorem gives convergence of the series of $\mu(|\xi_n| > 1) = (p-1)^p |h_n|^p$. □

5.2.6. Example. Let μ be the countable product of identical probability measures with the density $p(t) = (1-|t|)I_{[-1,1]}(t)$. Then $D(\mu)$ coincides as a set with the Orlicz space of all sequences (h_n) such that
$$\sum_{n=1}^\infty |h_n|^2 (1 + |\ln|h_n||) < \infty, \text{ where } 0 \ln 0 := 0.$$

PROOF. It is easily seen that the measure with density p is differentiable and its logarithmic derivative equals $\beta(t) = -\operatorname{sign} t/p(t)$ on $[-1,1]$ and vanishes outside $[-1,1]$. Suppose that $\sum_{n=1}^\infty |h_n|^2 (1 + |\ln|h_n||) < \infty$, where $h_n \neq 0$. We prove that $h = (h_n) \in D(\mu)$ by showing that the series $\sum_{n=1}^\infty h_n x_n^{-1} I_{[-1,1]}(x_n)$ converges in $L^1(\mu)$. Clearly, the random variables $\xi_n(x) := h_n \beta(x_n)$ are independent. Setting $\eta_n := \xi_n I_{\{|\xi_n| \leq 1\}}$ we obtain mean zero random variables with
$$\int_X \eta_n^2(x) \mu(dx) = 2h_n^2 \int_0^{1-|h_n|} (1-t)^{-1} dt = -2h_n^2 \ln|h_n|$$

5.2. Examples

for any n such that $|h_n| < 1$. In addition,

$$\mu(x\colon |\xi_n| > 1) = 2\int_{1-|h_n|}^{1}(1-t)\,dt = h_n^2.$$

Therefore, on account of the Kolmogorov three series theorem our assumption yields convergence of the series of ξ_n almost everywhere. Moreover, for any $\varepsilon \in (0,1)$ the series of the integrals of $|\xi_n|^{1+\varepsilon}I_{\{|\xi_n|\geqslant 1\}}$ converges since we have

$$\int_{\{|\xi_n|\geqslant 1\}}|\xi_n(x)|^{1+\varepsilon}\,\mu(dx) = 2|h_n|^{1+\varepsilon}\int_{1-|h_n|}^{1}(1-t)^{-1-\varepsilon}(1-t)\,dt = 2h_n^2.$$

Therefore, as in the previous example, we obtain convergence of the series of ξ_n in $L^1(\mu)$, which proves that $h \in D(\mu)$. As above, the converse follows by the Kolmogorov three series theorem. □

Any Gaussian measure has a Hilbert subspace of differentiability. However, for a measure absolutely continuous with respect to a Gaussian measure this may be false.

5.2.7. Proposition. *Let γ be a nondegenerate Gaussian measure on an infinite dimensional separable Banach space X. Then there exists a nonnegative function $f \in L^1(\gamma)$ such that the measure $\mu = f\cdot\gamma$ is probabilistic and its subspace of differentiability $D(\mu)$ is isomorphic to l^1.*

PROOF. It suffices to construct a centered nondegenerate Gaussian measure γ with this property on $X = l^2$ and use the existence of a measurable linear isomorphism taking this measure to the given one. We take the measure with the Fourier transform $\exp\bigl(-\sum_{n=1}^{\infty}2^{-n}x_n^2\bigr)$. Set

$$A_n := \Bigl\{x\colon 2\leqslant x_n \leqslant 3,\ \sum_{i\neq n}x_i^2 \leqslant 1\Bigr\},\quad B_n := \Bigl\{x\in A_n\colon \sum_{i\neq n}x_i^2 \leqslant 1/2\Bigr\}.$$

The sets A_n are pairwise disjoint. Hence the sets B_n are pairwise disjoint as well. In addition, A_n and B_n are closed and have boundaries of γ-measure zero. Let $\{e_n\}$ be the standard basis in l^2. We shall construct a bounded nonnegative locally Lipschitzian function f on X with the following properties: $\partial_{e_i}f(x) = 0$ for all interior points in B_n and all $i \neq n$,

$$\int_X |\partial_{e_n}f(x)|\,\gamma(dx) \leqslant C\|d_{e_n}\gamma\|,\quad \int_{A_n}|\partial_{e_n}f(x)|\,\gamma(dx) \geqslant 2\|d_{e_n}\gamma\|$$

for all n, where C is independent of n.

Let $f = 0$ outside $\bigcup_n A_n$. On the set B_n we define f by $f(x) = \varphi_n(x_n)$, where $\varphi_n(t) = 2m(t-2)$ if $2 \leqslant t \leqslant 2 + (2m)^{-1}$, $\varphi_n(t) = 2m(2 + m^{-1} - t)$ if $2 + (2m)^{-1} < t \leqslant 2 + m^{-1}$, φ_n is extended periodically with period $1/m$, and m is a natural number such that $2\|d_{e_n}\gamma\| \leqslant 2m\gamma(B_n) \leqslant 4\|d_{e_n}\gamma\|$. For example, we can take $m = \bigl[\|d_{e_n}\gamma\|/\gamma(B_n)\bigr] + 1$; note that $\|d_{e_n}\gamma\| \to \infty$.

Clearly, $|\partial_{e_n} f(x)| = 2m$ for all $x \in B_n$ with $|x_n| \neq 1 + j/(2m)$, i.e., for γ-a.e. $x \in B_n$. Hence

$$\int_{B_n} |\partial_{e_n} f(x)| \, \gamma(dx) = 2m\gamma(B_n) \geqslant 2\|d_{e_n}\gamma\|.$$

By construction $\partial_{e_i} f(x) = 0$ in the interior of B_n for all $i \neq n$. We now extend f to A_n as follows: if $x \in A_n \backslash B_n$, then $f(x) := 2\bigl(1 - \sum_{i \neq n} x_i^2\bigr)\varphi_n(x_n)$. Now f is defined everywhere. It is readily seen that f is locally Lipschitzian being Lipschitzian on each A_n. In the interior of the set $A_n \backslash B_n$ we have the estimates $|\partial_{e_n} f(x)| \leqslant 4m$ and $|\partial_{e_i} f(x)| \leqslant 4$ if $i \neq n$.

Let us estimate $\|\partial_{e_n} f\|_{L^1(\gamma)}$ from above. First, we note that there is M independent of n such that $\gamma(A_n) \leqslant M\gamma(B_n)$. Indeed, the measure γ can be regarded as the product of the Gaussian measure with the Fourier transform $\exp(-2^{-n} x_n^2)$ and the Gaussian measure γ_n with the Fourier transform $\exp\bigl(-\sum_{i \neq n} 2^{-i} x_i^2\bigr)$. By Fubini's theorem $\gamma(A_n)/\gamma(B_n) = \gamma_n(S)/\gamma_n(S/\sqrt{2})$, where S is the unit ball in l^2. It is readily seen that the measures γ_n converge weakly to γ and

$$\gamma_n(S)/\gamma_n(S/\sqrt{2}) \to \gamma(S)/\gamma(S/\sqrt{2}).$$

Therefore, since $|\partial_{e_n} f(x)| \leqslant 4$ on A_i if $i \neq n$, we find that

$$\int_X |\partial_{e_n} f(x)| \, \gamma(dx) = \sum_{i=1}^\infty \int_{A_i} |\partial_{e_n} f(x)| \, \gamma(dx)$$

$$\leqslant 4 \sum_{i=1}^\infty \gamma(A_i) + \int_{A_n} |\partial_{e_n} f(x)| \, \gamma(dx)$$

$$\leqslant 4 + 2m\gamma(A_n) \leqslant 4 + 2mM\gamma(B_n) \leqslant 4 + 4M\|d_{e_n}\gamma\| \leqslant C\|d_{e_n}\gamma\|$$

with some $C > 0$ independent of n.

We define μ by $\mu = f \cdot \gamma$ and show that

$$D(\mu) = \Bigl\{x \colon \sum_{n=1}^\infty 2^{n/2} |x_n| < \infty\Bigr\}.$$

The space on the right equipped with the norm given by the above series is a Banach space continuously embedded into l^2 and linearly homeomorphic to l^1; since $D(\mu)$ with its norm is also continuously embedded into l^2, we conclude that $D(\mu)$ is isomorphic to l^1. To prove the above equality we denote its right-hand side by L. The measure μ is differentiable along all vectors e_n and $d_{e_n} \mu = \partial_{e_n} f \cdot \gamma + f \cdot d_{e_n} \gamma$. Hence

$$\|\partial_{e_n} f\|_{L^1(\gamma)} + \|d_{e_n}\gamma\| \leqslant \|d_{e_n}\mu\| \leqslant \|\partial_{e_n} f\|_{L^1(\gamma)} + \|d_{e_n}\gamma\|,$$

whence we obtain

$$\|d_{e_n}\gamma\| \leqslant \|d_{e_n}\mu\| \leqslant (C+1)\|d_{e_n}\gamma\|.$$

Note that
$$D(\gamma) = \left\{h = (h_n): \sum_{n=1}^{\infty} 2^n h_n^2 < \infty\right\}$$
and the norm in $D(\gamma)$ is equivalent to the Hilbert norm $\left(\sum_{n=1}^{\infty} 2^n h_n^2\right)^{1/2}$. In particular, $a2^{n/2} \leqslant \|d_{e_n}\gamma\| \leqslant b2^{n/2}$ with some $a, b > 0$ independent of n. Now let $h \in L$. By the above estimates, the series of $h_n d_{e_n}\mu$ converges in variation since $|h_n|\|d_{e_n}\mu\| \leqslant b(C+1)|h_n|2^{n/2}$. Hence $h \in D(\mu)$. Conversely, let $h \in D(\mu)$. Then $h \in D(\gamma) = H(\gamma)$ since $\mu \ll \gamma$ and otherwise we would have $\mu_h \perp \mu$. We show that $h \in L$. To this end, we observe that the functions $\beta_{h^n}^{\mu}$ with $h^n = (h_1, \ldots, h_n, 0, 0, \ldots)$ have uniformly bounded norms in $L^1(\mu)$. Let us estimate their norms from below. We have
$$\beta_{h^n}^{\mu} \cdot \mu = \sum_{i=1}^{n} h_i(\partial_{e_i} f \cdot \gamma + f \cdot d_{e_i}\gamma)$$
and $\partial_{e_i} f = 0$ a.e. on B_j if $j \neq i$. Hence for any $j \leqslant n$ we have
$$\int_{B_j} |\beta_{h^n}^{\mu}(x)|\mu(dx) \geqslant \int_{B_j} \left|\sum_{i=1}^{n} h_i \partial_{e_i} f(x)\right|\gamma(dx) - \left|\sum_{i=1}^{n} h_i d_{e_i}\gamma\right|(B_j)$$
$$\geqslant |h_j| \int_{B_j} |\partial_{e_j} f(x)|\gamma(dx) - \left|\sum_{i=1}^{n} h_i d_{e_i}\gamma\right|(B_j).$$
Therefore,
$$\|\beta_{h^n}^{\mu}\|_{L^1(\mu)} \geqslant 2\sum_{j=1}^{n} |h_j| \|d_{e_j}\gamma\| - \|d_{h^n}\gamma\| \geqslant 2a\sum_{j=1}^{n} |h_j| 2^{j/2} - \|d_h\gamma\|,$$
which shows that the sums $\sum_{j=1}^{n} |h_j| 2^{j/2}$ are uniformly bounded. $\square$

Yamasaki, Hora [**1198**] and Khafizov [**601**] constructed examples of product-measures with the subspaces of differentiability isomorphic to l^1.

5.3. Disposition of subspaces of differentiability

We have studied above the subspaces of differentiability as separate Banach spaces, but it is also very useful and interesting to study their disposition in the space carrying the measure. For example, it is interesting how large the subspace of differentiability of a measure can be in a given space. Certainly, by forming the convolution of two measures $\mu * \nu$ we obtain a measure whose subspace of differentiability contains $D(\mu)$ and $D(\nu)$ since $D(\mu) + D(\nu) \subset D(\mu * \nu)$. But are there any restrictions from above on the subspace of differentiable except that in the infinite dimensional case it does not coincide with the whole space? These questions are interesting also in relation to the fact that the convolution of a function with a measure is differentiable along the subspace of differentiability of this measure, so measures with large subspaces of differentiability enable us to construct functions differentiable along large subspaces. Say, on the space $C[0, 1]$ the class of infinitely Fréchet differentiable functions is rather narrow (for example, it does not contain nonzero functions with bounded support), but there are many functions which are smooth along dense subspaces. It turns out that in the case of a Hilbert space X any subspace of differentiability is contained in the range of a Hilbert–Schmidt operator, and every such range is the subspace of differentiability of a Gaussian measure. So from the point of view of disposition in the original space (but not from the point of view of their geometry!) the subspaces of differentiability in a Hilbert space are similar to the ranges of Hilbert–Schmidt operators. For general Banach spaces, only some fragments of this picture are preserved. For example, the space $C[0,1]$ carries probability measures whose subspaces of differentiability are so large that they are not contained in the Cameron–Martin spaces of Gaussian measures.

A measure differentiable along vectors from a sequence with a dense linear span will be called *densely differentiable*. Similarly, we define *densely continuous* measures. Since on every separable Fréchet space there is a nondegenerate Gaussian measure, we see that such spaces possess many densely differentiable measures. However, not every separable locally convex space X carries a nonzero densely differentiable measure. A necessary and sufficient condition is that X contains a densely embedded separable Banach space. For example, a countable union of finite dimensional spaces does not satisfy this condition. The space $\mathcal{D}(\mathbb{R}^1)$ has no densely embedded Banach spaces either because any such space is contained in one of the subspaces D_n of functions with support in $[-n, n]$.

Let μ be a Radon probability measure on a locally convex space X such that $X^* \subset L^2(\mu)$. Set

$$E_\mu := \{v \in X : |v|_\mu < \infty\},$$

$$|v|_\mu := \sup\Bigl\{ l(v) \in X^* : \int_X |l(x)|^2 \, \mu(dx) \leqslant 1 \Bigr\}.$$

In the case of a centered Gaussian measure μ we have $E_\mu = H(\mu)$. As shown in Vakhania, Tarieladze [**1166**], in the case of a quasicomplete space X (see Exercise 1.6.9) the set E_μ coincides with $R_\mu(X_\mu^*)$, where X_μ^* is the closure of X^* in $L^2(\mu)$ and the operator $R_\mu \colon X_\mu^* \to X$, called the *covariance operator* of the measure μ, is defined by the formula

$$R_\mu f := \int_X f(x) x \, \mu(dx),$$

where the integral is understood in the weak sense, i.e., for each $l \in X^*$ the number $l(R_\mu f)$ is the integral of $l(x) f(x)$. If the space X is not quasicomplete, then the operator R_μ takes values in the completion of X. The covariance operator R_μ can be regarded just on X^*; usually $R_\mu(X^*) \neq R_\mu(X_\mu^*)$. If ν is yet another probability measure with $X^* \subset L^2(\nu)$ such that μ has weak mean 0 (i.e., the integrals of all $f \in X^*$ vanish), then $R_{\mu*\nu} = R_\mu + R_\nu$.

5.3. Disposition of subspaces of differentiability

We observe that if X is a separable Hilbert space and the measure μ has bounded support (or the square of the norm is μ-integrable), then the operator R_μ on the space $X^* = X$ is determined from the identity

$$(R_\mu f, g) = \int_X (f, x)(g, x)\, \mu(dx), \quad f, g \in X.$$

Let $\{e_n\}$ be an orthonormal basis in X. Then

$$(R_\mu e_n, e_n) = \int_X (x, e_n)^2 \, \mu(dx).$$

The square integrability of the norm shows that

$$\sum_{n=1}^\infty (R_\mu e_n, e_n) = \int_X (x, x)^2 \, \mu(dx) < \infty,$$

i.e., the operator $R_\mu\colon X \to X$ is nuclear. We may assume that it has a trivial kernel, replacing X by the smallest closed linear subspace of full μ-measure. Then the space X_μ^* is the completion of X with respect to the Euclidean norm $(R_\mu x, x)^{1/2}$ and $|v|_\mu$ coincides with $\sup\{(v, u)\colon (R_\mu u, u) \leqslant 1\}$. This means that $E_\mu = \sqrt{R_\mu}(X)$ (Exercise 1.6.6). In this case the operator $\sqrt{R_\mu}$ in X is a Hilbert–Schmidt operator.

A measure μ is called *pre-Gaussian* (or with a *Gaussian covariance*) if there is a Gaussian measure γ with the same covariance, i.e.,

$$\int_X l(x)^2 \, \mu(dx) = \int_X l(x)^2 \, \gamma(dx), \quad l \in X^*.$$

If a_γ is the mean of the measure γ and $a_\gamma = 0$, then $E_\mu = E_\gamma = H(\gamma)$; in the general case $E_\mu = E_\gamma = H(\gamma) + \mathbb{R}^1 a_\gamma$ (Exercise 5.5.6).

5.3.1. Theorem. *Let μ be a Radon probability measure on a locally convex space X with $X^* \subset L^2(\mu)$. Then $C(\mu) \subset E_\mu$.*

PROOF. Let $h \in C(\mu)$. We show that $|v|_\mu < \infty$. Otherwise there exist functionals $l_n \in X^*$ with $\|l_n\|_{L^2(\mu)} \leqslant 1$ and $l_n(h) \geqslant n$. Set $s_n := l_n(h)^{-1}$. Then $\|\mu_{s_n h} - \mu\| \to 0$. Hence

$$\big|[\exp(-il_n(s_n h)) - 1]\widetilde{\mu}(s_n l_n)\big| = |\widetilde{\mu_{s_n h}}(l_n) - \widetilde{\mu}(l_n)| \leqslant \|\mu_{s_n h} - \mu\| \to 0.$$

Since $s_n l_n(h) = 1$, one has $\widetilde{\mu}(s_n l_n) \to 0$. However, this is false because $\widetilde{\mu}(s_n l_n) \to 1$ by the relationship $\|s_n l_n\|_{L^2(\mu)} \to 0$. □

5.3.2. Theorem. *Let μ be a nonzero Radon measure on a locally convex space X and let $\{h_i\} \subset C(\mu)$ be a countable set. Then*
  (i) there exists a sequence of Radon measures μ_n infinitely differentiable along all vectors h_i and convergent to μ in variation;
  (ii) the measure μ is absolutely continuous with respect to some Radon probability measure ν infinitely differentiable along all vectors h_i.

PROOF. (i) Since the space $C(\mu)$ is metrizable and complete with respect to the metric ϱ defined in §5.1 (which is translation invariant), there exists a continuous linear operator $T\colon l^2 \to C(\mu)$ with $\{h_i\} \subset T(l^2)$. For example, one can take $T(x_i) := \sum_{i=1}^{\infty} c_i x_i h_i$, where $c_i > 0$ are chosen in such a way that $\varrho(c_i h_i, 0) \leqslant 2^{-i}$. We take a centered Gaussian measure γ_0 on l^2 with $e_i \in H(\gamma)$. Set $\gamma := \gamma_0 \circ T^{-1}$, $\gamma_n(B) := \gamma(nB)$. The measures $\mu_n := \mu * \gamma_n$ are infinitely differentiable along all vectors h_i since the Gaussian measures γ_n are also. In addition, $\|\mu - \mu_n\| \to 0$ as $n \to 0$. Indeed, for any fixed $\varepsilon > 0$ we have $\|\mu_x - \mu\| \leqslant \varepsilon$ whenever

$$x \in U_\varepsilon := \{x \in C(\mu)\colon \varrho(x,0) \leqslant \varepsilon\}.$$

For sufficiently large n one has the inequality $\gamma_n(U_\varepsilon) \geqslant 1 - \varepsilon$. Hence

$$\int_X \|\mu_x - \mu\| \gamma_n(dx) \leqslant \varepsilon + 2\|\mu\|\varepsilon = \varepsilon(2\|\mu\| + 1).$$

This gives $\|\mu_n - \mu\| \leqslant \varepsilon(2\|\mu\| + 1)$.

(ii) We may assume that $\mu \geqslant 0$. Then the measures μ_n constructed above are nonnegative. Hence $\mu \ll \sum_{n=1}^{\infty} \varepsilon_n \mu_n$ if $\varepsilon_n > 0$ and $\sum_{n=1}^{\infty} \varepsilon_n < \infty$. Now we choose ε_n approaching zero so rapidly that the sum of the series becomes infinitely differentiable along all vectors h_i. To this end, it suffices to ensure convergence of the series of measures $\varepsilon_n \|d_{h_1}^{k_1} \cdots d_{h_m}^{k_m} \mu_n\|$ for all $k_1, \ldots, k_m$. This can be easily done. □

It should be noted that in general one cannot find measures μ_n and ν differentiable along all vectors from $C(\mu)$ (see Exercise 5.5.2).

5.3.3. Theorem. *Let μ be a nonzero Radon measure on a locally convex space X. Then there exists a Radon probability measure ν with completely bounded (and compact if X is complete or quasicomplete) support such that $D(\mu) \subset D(\nu)$ and $C(\mu) \subset C(\nu)$.*

PROOF. Since $D(|\mu|) \supset D(\mu)$ (if $d_h\mu$ exists, then $d_h|\mu|$ does also), we may assume that μ is a probability measure. Let K be a compact set with $\mu(K) > 0$ and let V be the closed absolutely convex hull of K. The set V is completely bounded and if X is quasicomplete, then V is compact. Denote by E the linear span of V and by p_V the Minkowski functional of V on E. Outside E we put $p_V = 0$. It is clear that $D(\mu) \subset E$ and the function p_V is Borel. Set $f(x) = 2 - p_V(x)$ if $p_V(x) \leqslant 2$ and $f(x) = 0$ if $p_V(x) > 2$. Since $f(x) \geqslant 1$ on V, the integral of the function f is positive. Set $\nu := cf \cdot \mu$, where $c > 0$ is such that the measure ν is a probability measure. Since $D(\mu) \subset E$, for every $h \in D(\mu)$ the functions $t \mapsto f(x+th)$ are Lipschitzian for all x. Hence the measure ν is differentiable along h (see Proposition 3.3.12). Since $\nu \ll \mu$, one has $C(\mu) \subset C(\nu)$. □

5.3.4. Corollary. *In the previous theorem one has $C(\mu) \subset E_\nu$. If the space X is Hilbert, then $C(\mu)$ is contained in the range of some Hilbert–Schmidt operator.*

PROOF. We should only observe that in the case of a Hilbert space X the set E_ν is contained in the range of a Hilbert–Schmidt operator. $\square$

5.3.5. Theorem. *Let μ be a nonzero Radon measure on a locally convex space X. Then there exists a Hilbert space H compactly embedded into X with the following properties:* (i) $D(\mu) \subset H$, (ii) *for every n one can find a probability measure ν_n on X which has completely bounded (compact in the case of a complete or quasicomplete X) support and is n-fold Fréchet differentiable along H provided the space of measures is equipped with the variation norm. If μ is separable, then so is H.*

PROOF. By the previous theorem there exists a Borel probability measure ν on X which has completely bounded (compact in the case of quasicomplete X) support and is differentiable along all vectors from $D = D(\mu)$. Passing to the measure $\nu*\nu*\nu*\nu$, we may assume that it is four times differentiable along all vectors in D. Then $D \subset H(\nu) \subset D(\nu)$. This follows from Corollary 6.2.2 below, but in our special case this can be verified directly. Denote $H(\nu)$ by H and put $\nu_n = \nu * \cdots * \nu$, where the convolution is taken $n+1$ times. Then the measures ν_n have completely bounded (compact in the case of quasicompleteness) support and are n-fold Fréchet differentiable along the space H. Indeed, for all vectors $v_1, \ldots, v_n \in H$ there exists the derivative $d_{v_1} \cdots d_{v_n} \nu_n$, equal to $d_{v_1} \nu * \cdots * d_{v_n} \nu * \nu$. Since
$$\|v\|_{D(\nu)} \leqslant C|v|_H,$$
for the n-fold Gâteaux differentiable mapping $T \colon a \mapsto (\nu_n)_a$ we have
$$\|T^{(n)}(a) - T^{(n)}(b)\|$$
$$\leqslant \sup\{\|d_{v_1}\nu * \cdots * d_{v_n}\nu * \nu_a - d_{v_1}\nu * \cdots * d_{v_n}\nu * \nu_b\|,\ |v_i|_H \leqslant 1\},$$
which is estimated by $C^n \|d_{a-b}\nu\| \leqslant C^{n+1}|a-b|_H$. This estimate implies the n-fold Fréchet differentiability of T. We observe that if μ is separable, then the measure ν constructed in the previous theorem is separable as well, which yields the separability of the subsequent convolutions and results in the separability of H. $\square$

Theorem 5.3.5 gives a positive answer to the question raised by V. Bentkus and mentioned in the survey Bogachev, Smolyanov [**233**, Problem 8-e]. It remains unknown whether one can choose ν_n infinitely differentiable or analytic along H. As we have seen in Chapter 5, the space $D(\mu)$ itself may fail to be isomorphic to a Hilbert space, unlike the case of Gaussian measures. As shown in [**170**], one cannot always find a Gaussian measure γ for which $D(\mu) \subset D(\gamma)$. Let us give this example.

5.3.6. Example. Let p be a probability density on the real line such that $p(t) = c\exp\bigl((t^2-1)^{-1}\bigr) I_{[-1,1]}(t)$, let μ_0 be the countable power of the measure with density p, and let μ be the image of the measure μ_0 under the linear mapping $T\colon (x_n) \mapsto \bigl(x_n/\ln(n+1)\bigr)$. Then the measure μ is concentrated on c_0 and has the following properties:

(i) there is no Gaussian measure γ on c_0 with $D(\mu) \subset D(\gamma)$;

(ii) the measure μ is mutually singular with every Borel probability measure on c_0 with a Gaussian covariance, in particular, with every measure σ satisfying the central limit theorem, i.e., such that if X_i are independent random vectors with distribution σ, then the distributions of the sums $(X_1 + \cdots + X_n)n^{-1/2}$ converge weakly to a Radon Gaussian measure.

PROOF. (i) Let γ be a Gaussian measure on c_0 with $D(\mu) \subset D(\gamma)$. We may assume that γ is centered since $D(\gamma_a) = D(\gamma)$. Let ν be the image of the countable power of the standard Gaussian measure under the mapping T. We observe that $D(\mu_0) = l^2$, which coincides with the Cameron–Martin space of the countable power of the standard Gaussian measure. Hence $D(\mu) = D(\nu) = T(l^2)$. Therefore, $H(\nu) = D(\nu) \subset D(\gamma) = H(\gamma)$. Since $\gamma(c_0) = 1$, this yields the equality $\nu(c_0) = 1$ (see [**191**, Theorem 3.3.2]). However, it is easily seen that $\nu(c_0) = 0$, which is a contradiction.

(ii) We observe that if σ is a Borel probability measure on c_0 and $\sigma \ll \mu$, then, according to what we have already shown, the measure σ cannot be pre-Gaussian since otherwise, by Theorem 5.3.1, we would have a Gaussian measure γ on c_0 with $D(\mu) \subset C(\mu) \subset C(\sigma) \subset D(\gamma)$. Now let σ_0 be an arbitrary Borel probability measure on c_0 whose covariance coincides with the covariance of some Gaussian measure γ on c_0. Let us show that $\sigma_0 \perp \mu$. Otherwise we see that the Lebesgue decomposition $\sigma_0 = \mu' + \mu''$, where $\mu' \ll \mu$ and $\mu'' \perp \mu$, contains a measure $\mu' \neq 0$. Let us consider the probability measure $\sigma_1 = c\mu'$. We have $\sigma_1 \ll \sigma_0$ and $\sigma_1 \ll \mu$. According to Theorem 5.3.1 the set $C(\sigma_1)$ is contained in the subspace E_{σ_1}. By the estimate
$$\int_X |l(x)|^2 \, \sigma_1(dx) \leqslant c \int_X |l(x)|^2 \, \sigma_0(dx)$$
and the equality $R_{\sigma_0} = R_\gamma$ we have
$$C(\sigma_1) \subset E_{\sigma_1} \subset E_{\sigma_0} = E_\gamma,$$
which is impossible according to what we have proven since $E_\gamma \subset H(\gamma_1)$ for some Gaussian measure γ_1 (if γ has nonzero mean a, then $\gamma_1 = \gamma_{-a} * \eta$, where η is the standard Gaussian measure on the straight line $\mathbb{R}^1 a$). □

5.3.7. Remark. A measure with the properties listed in this example also exists on $C[0,1]$ since c_0 can be embedded into $C[0,1]$ as a closed linear subspace admitting a continuous linear projection. Finite dimensional projections of the constructed measure possess smooth densities. This shows that the observed "extremal" absence of the central limit theorem property is caused not by some pathological singularity of the measure μ, but by its very nice smoothness. It is not clear whether such examples exist in spaces not containing c_0. However, on every infinite dimensional separable Banach space X there is a Borel probability measure differentiable along a dense subspace and mutually singular with all Gaussian measures on X: it suffices to embed c_0 with the constructed measure into X by means of an injective continuous linear operator (not isomorphically in general) and use the zero-one law for Gaussian measures.

The following result enables one to establish the absence of Hilbert supports for measures with broad subspaces of continuity.

5.3.8. Theorem. *Let (Ω, P) be a probability space and let μ be a Radon measure on $L^2(P)$ such that $|\mu|(H) > 0$ for some Hilbert space H continuously embedded into $L^2(P)$ and contained in $L^\infty(P)$. Then $C(\mu)$ is contained in the range of some nuclear operator in $L^2(P)$.*

PROOF. Closed balls from H are weakly compact in $X = L^2(P)$. Hence H is a Borel set in X. Replacing the measure μ by the restriction of $|\mu|$ to H we obtain a positive Radon measure on H whose subspace of continuity is larger or the same. As shown above, there is a Hilbert–Schmidt operator T on H for which $C(\mu) \subset T(H)$. The inclusion $H \subset L^\infty(P)$ yields that there is a Hilbert–Schmidt operator S on X with $H \subset S(X)$ (see, for example, Bogachev, Smolyanov [**234**, §7.10]). Then the operator TS is nuclear. □

5.3.9. Example. Let ξ_t^α be a continuous Gaussian process on $[0, 1]$ with zero mean and the covariance function $K(t, s) = (s^\alpha + t^\alpha - |s - t|^\alpha)/2$, where $\alpha \in (0, 2]$ (the fractional Brownian motion; ξ_t^1 is the Wiener process). Denote by γ_α the distribution of this process in $L^2[0,1]$ or in $C[0,1]$. Then, in the case $\alpha \in (0, 1]$, for every Hilbert space $H \subset L^\infty[0, 1]$ continuously embedded into $L^2[0,1]$ we have $\gamma_\alpha(H) = 0$ (in particular, this is true for the Wiener measure γ_1).

PROOF. One can show that $H(\gamma_1) \subset H(\gamma_\alpha)$ if $\alpha \in (0, 1)$ (see Davydov, Lifshits, Smorodina [**336**, p. 45]). The fact that $H(\gamma_1) = W_0^{2,1}[0, 1]$ is not contained in the range of a nuclear operator follows from an explicit expression for the eigenvalues of the covariance operator K of the Wiener measure: $\lambda_n = \pi^{-2}(n - 1/2)^{-2}$. The operator $\sqrt{K}$ is not nuclear, hence its range cannot belong to the range of a nuclear operator (see [**191**, Proposition A.2.12, p. 369] or [**234**, §7.10]). □

One can show that in the case $\alpha \in (1, 2)$ the measure γ_α is concentrated on a Hilbert space continuously embedded into $C[0, 1]$ (Exercise 5.5.10).

Let ω be a positive increasing function on $(0, 1]$ with $\lim_{t\to 0}\omega(t) = 0$. Denote by H_ω the set of functions f in $C[0, 1]$ having the modulus of continuity ω, i.e., satisfying the estimate $|f(t) - f(s)| \leqslant C(f)\omega(|t - s|)$. The space H_ω is Banach with the norm $|f(0)| + M(f)$, where $M(f)$ is the infimum of numbers $C(f)$ for which the indicated estimate holds. Let us study when is H_ω contained in a subspace of differentiability.

5.3.10. Theorem. (i) *Let $\omega(t) = O(\sqrt{t}/|\ln t|^\beta)$, where $\beta > 3/2$. Then there exist a Hilbert space H continuously embedded into $C[0, 1]$ and containing H_ω and a Borel probability measure μ on $C[0, 1]$ with compact support infinitely differentiable along H. In the case $\beta > 5/2$ there is a Gaussian measure μ with $H_\omega \subset H(\mu)$.*

(ii) *If ω is a modulus of continuity such that $H_\omega \subset C(\mu)$ for some nonzero measure μ on $C[0,1]$ or on $L^2[0,1]$, then $\sum_{n=1}^\infty \omega(1/n)/\sqrt{n} < \infty$.*

PROOF. (i) Let f_n be the Faber–Schauder basis in $C[0,1]$ (see Kashin, Saakyan [**592**, Ch. VI, §1, p. 185]). Let $\beta = 3/2 + 2\varepsilon$. Set $a_1 = 1$ and $a_n = (\ln n)^{-1-\varepsilon}$, $n > 1$. We define a linear operator $T\colon c_0 \to C[0,1]$ by the formula
$$Tx = \sum_{n=1}^\infty a_n x_n f_n, \quad x = (x_n).$$
This series converges absolutely and uniformly since
$$\sum_{n=2^k+1}^{2^{k+1}} |a_n x_n f_n| \leqslant \sup_n |x_n| \max(|a_n|\colon 2^k < n \leqslant 2^{k+1})$$
in view of the fact that the functions $f_{2^k+1}, \ldots, f_{2^{k+1}}$ have disjoint supports and are bounded in the absolute value by 1 (see [**592**, Ch. VI, §1, p. 185]). We can construct a Borel probability measure μ_0 with compact support in c_0 that is infinitely differentiable along the Hilbert space
$$E = \Big\{(x_n)\colon x_1^2 + \sum_{n=2}^\infty (\ln n)^\varepsilon x_n^2 < \infty\Big\}.$$
To this end we take the countable power ν of the probability measure with a smooth density p with support in $[0,1]$. This measure is infinitely differentiable along l^2 and is concentrated on l^∞. For μ_0 we take the image of ν under the linear mapping $(x_n) \mapsto (\varepsilon_n x_n)$, where $\varepsilon_1 = 1$, $\varepsilon_n = (\ln n)^{-\varepsilon/2}$. Denote by μ the image of μ_0 under the mapping T. The measure μ is infinitely differentiable along $H := T(E)$. We show that $H_\omega \subset H$. For any $f \in H_\omega$ according to [**592**, Ch. VI, §1, Corollary 1, p. 188] we have $f = \sum_{n=1}^\infty c_n f_n$, where $|c_n| \leqslant c \cdot \omega(1/n) \leqslant c \cdot n^{-1/2}(\ln n)^{-\beta}$, $n \geqslant 2$. Thus, one has $f = \sum_{n=1}^\infty a_n b_n f_n$, where $b_n = c_n/a_n$, $|b_n| \leqslant c \cdot n^{-1/2}(\ln n)^{-\varepsilon-1/2}$. Since the series of $n^{-1}(\ln n)^{-1-\varepsilon}$ converges, we have $(b_n) \in E$ and $f \in T(E)$. In the case $\beta > 5/2$ in place of μ_0 one can take a Gaussian measure with the desired properties.

(ii) Let $H_\omega \subset C(\mu)$. As we know, there exists a nonnegative Hilbert–Schmidt operator T on $L^2[0,1]$ such that $C(\mu) \subset T(L^2[0,1])$. Then we have $E_\omega \subset T(L^2[0,1])$. Let $\{t_n\}$ and $\{\varphi_n\}$ be the eigenvalues and eigenfunctions of T. For any $f \in H_\omega$ we obtain $f = \sum_{n=1}^\infty t_n x_n \varphi_n$, where $\{x_n\} \in l^2$. Hence one has $\sum_{n=1}^\infty |t_n x_n| < \infty$. Finally, according to the well-known Bochkarev theorem (see [**164**]), the series of $\omega(1/n)n^{-1/2}$ converges. □

It is unknown whether the condition $\sum_{n=1}^\infty \omega(1/n) n^{-1/2} < \infty$ is sufficient for the existence of a nonzero measure μ on $C[0,1]$ with $H_\omega \subset D(\mu)$. There is such a measure on $L^2[0,1]$. It is not clear whether convergence of the indicated series is equivalent to the absolute summability of the natural embedding of H_ω into $L^2[0,1]$. Note also that if H_ω is contained in a Hilbert space continuously embedded into $C[0,1]$, then this series converges. It is unknown whether the converse is true.

In the papers Bentkus [**134**], Bentkus, Rachkauskas [**136**], [**137**] and Paulauskas, Rachkauskas [**876**] measures with large subspaces of differentiability are applied to estimates of the rate of convergence in the central limit theorem. With the aid of the results obtained in this section and the methods of these papers one can prove the following estimate. Suppose that a separable Banach space B is continuously embedded into a separable Banach space X and we are given centered equally distributed independent random vectors $X_1, \ldots, X_n$ on a probability space (Ω, P) with values in B having the same covariance as a given centered Gaussian random element Y in B. Set $S_n := (X_1 + \cdots + X_n)/\sqrt{n}$.

5.3.11. Theorem. *Suppose that a function f on X is Hölder continuous of order α and that $P(r-\varepsilon \leqslant f(Y) \leqslant r+\varepsilon) \leqslant c(r)\varepsilon$. Let $\mathbb{E}\|X_1\|_B^3 < \infty$. If there is a nonzero measure μ on X with $B \subset D(\mu)$, then*
$$\bigl|P\bigl(f(S_n) < r\bigr) - P\bigl(f(Y) < r\bigr)\bigr| \leqslant C \cdot n^{-\alpha/(2\alpha+6)},$$
where C is independent of n.

5.3.12. Example. Let $\omega(t) = O(\sqrt{t}/|\ln t|^{-\beta})$, where $\beta > 3/2$. If the elements $X_1, \ldots, X_n$ and Y take values in H_ω and $\mathbb{E}\|X_1\|_{H_\omega}^3 < \infty$, then the estimate from the previous theorem holds (if f satisfies the indicated conditions). We observe that the space H_ω is nonseparable, but one can find a separable Banach space B continuously embedded into $C[0,1]$, containing H_ω and contained in some $H_{\omega'}$, where $\omega'(t) = O(\sqrt{t}/|\ln t|^{-\beta'})$ and $\beta' > 3/2$.

5.4. Differentiability along subspaces

Let E be a locally convex space continuously embedded into X and let $\mathcal{M}$ be some locally convex space of measures on X. We shall assume that for mappings from E to $\mathcal{M}$ some type of differentiability $\mathcal{K}$ is chosen (see Chapter 1). In this section we are mainly concerned with Fréchet differentiability.

Usually the space of measures $\mathcal{M}$ is either the space of all measures on X (with some of our standard σ-algebras) or its subspace consisting, for example, of measures absolutely continuous with respect to some fixed measure λ. In the first case the most frequently used topologies on $\mathcal{M}$ are the following:

a) the topology generated by the variation norm $\mu \mapsto \|\mu\|$, when $\mathcal{M}$ becomes a Banach space;

b) the topology of convergence on every set (from the σ-algebras $\sigma(X)$, $\mathcal{B}(X)$ and so on);

c) the topology generated by the duality with the space of all bounded Borel functions, i.e., the seminorms generating the topology are given by
$$\mu \mapsto \left|\int_X f(x)\,\mu(dx)\right|,$$

where f is a bounded Borel function;

d) the weak topology generated by the duality with $C_b(X)$;

e) the topology generated by the duality with $\mathcal{FC}_b^\infty(X)$, which is similar to the topology in the space of distributions.

In the second case our mapping takes values in the space of λ-integrable function, so one can use diverse topologies in functional spaces.

5.4.1. Definition. *We shall say that a measure μ is $\mathcal{K}$-differentiable along E if the mapping $h \mapsto \mu_h$ from E to $\mathcal{M}$ is also.*

One naturally defines higher order differentiability along a subspace.

We observe that the differentiability along a given subspace is stronger than the differentiability along all vectors from this subspace.

5.4.2. Example. For every n we take a probability measure μ_n on the real line given by its density p_n defined as follows: $p_n(t) = n^2 t^2$ if $t \in [0, n^{-1}]$, $p_n(t) = 1$ if $t \in [n^{-1}, a_n]$, $p_n(t) = 1 + a_n - t$ if $t \in [a_n, a_n + 1]$, and $p_n(t) = 0$ outside $[0, a_n + 1]$, where a_n is chosen such that the integral of p_n equals 1. Note that the integral of $|p_n'|$ equals 2. Let $\mu = \bigotimes_{n=1}^{\infty} \mu_n$. Then μ is not Fréchet differentiable along $D(\mu)$. To see this, we observe that the integral of the function $\left|n[p_n(x+n^{-1}) - p(x)] - p_n'(x)\right|$ is not less than 1 since this function equals n on $[0, n^{-1}]$. Therefore, taking $t_n = n^{-1}$ and $e_n = (0, \ldots, 0, 1, 0, \ldots)$, we see that $\|t_n^{-1}[\mu_{t_n e_n} - \mu] - d_{t_n e_n}\mu\|$ does not tend to zero as $n \to \infty$. Since $\|d_{e_n}\mu\| = 2$, our claim follows.

5.4.3. Proposition. *Let E be a Banach space (or a Fréchet space) and let $n \in \mathbb{N}$ be such that for all $h_1, \ldots, h_n \in E$ there exists $d_{h_1} \cdots d_{h_n}\mu$. Then the mapping*
$$(h_1, \ldots, h_n) \mapsto d_{h_1} \cdots d_{h_n}\mu$$
from E^n to the space of measures with the variation norm is polylinear and continuous.

PROOF. It suffices to observe that this mapping is polylinear and continuous in every variable separately. The latter follows from the fact that for any fixed $h_1, \ldots, h_{n-1}$ the space E is contained in $D(d_{h_1} \cdots d_{h_{n-1}}\mu)$. By the closed graph theorem the embedding is continuous, whence the continuity of the mapping $h \mapsto d_h(d_{h_1} \cdots d_{h_{n-1}}\mu)$ follows. Therefore, by the known result on the continuity of a separately continuous polylinear mapping our mapping is jointly continuous. □

5.4.4. Corollary. *Suppose that a measure μ is twice differentiable along all vectors from a Banach space E continuously embedded into X. Then the measure μ is Fréchet differentiable along E provided the space of measures is equipped with the variation norm.*

PROOF. It suffices to observe that by the mean value theorem for all h and k from E one has the inequality
$$\|(d_h\mu)_k - d_h\mu\| \leqslant \|d_{hk}^2\mu\|.$$

Applying Theorem 3.3.7 we obtain
$$\left\|\frac{\mu_{th} - \mu}{t} - d_h\mu\right\| \leqslant \sup_{0<\tau<t} \|(d_h\mu)_{\tau h} - d_h\mu\| \leqslant |t| \|d^2_{hh}\mu\|,$$
which tends to zero as $t \to 0$ uniformly in h from every bounded set in E, since by Proposition 5.4.3 the mapping $h \mapsto \|d^2_{hh}\mu\|$ is bounded on bounded subsets in the space E. □

5.4.5. Corollary. *Let μ be a stable (for example, Gaussian) Radon measure on a locally convex space X and let E be a Banach space continuously embedded into X such that $E \subset D(\mu)$. Then μ is infinitely Fréchet differentiable (in variation) along the subspace E. In particular, it is infinitely Fréchet differentiable along the subspace $D(\mu) = H(\mu)$.*

Sometimes the differentiability property mentioned in the previous theorem is called the *differentiability in variation*. Another natural type of differentiability corresponds to the setwise convergence topology in the space of measures. We shall call it the *setwise differentiability*.

Denote by $B_n(E)$ the Banach space of n-linear continuous functions on a Banach space E continuously embedded into a locally convex space X. By Proposition 5.4.3 any measure μ on X that is n-fold differentiable along all vectors from E determines a $B_n(E)$-valued measure $\mu^{(n)}$ by the formula
$$\mu^{(n)}(A)(h_1, \ldots, h_n) = d_{h_1} \cdots d_{h_n} \mu(A).$$
Now the question arises whether this measure is countably additive. If E is a reflexive space, then the measure $\mu^{(1)} = \mu'$ is countably additive (this is proved in Lemma 7.3.1 and follows also from Proposition 5.4.7). However, for $n > 1$ this is false even if E is a separable Hilbert space. The corresponding example is constructed in Kats [**593**]. The following results obtained in Kats [**593**]–[**595**] contain some positive assertions in this direction (see also Exercise 6.8.8).

5.4.6. Proposition. *Let μ be a Radon measure on X having partial derivatives up to order $n + 1$ along all vectors from E. Then the measure $\mu^{(n)}$ is countably additive.*

5.4.7. Proposition. *Let μ have partial derivatives up to order n along all vectors from E. Then*

(i) if $B_n(E)$ is equipped with the topology of convergence on compact sets, then the measure $\mu^{(n)}$ is countably additive;

(ii) if the measure $\mu^{(n)}$ takes values in a separable subspace of $B_n(E)$, then it is countably additive;

(iii) if $B_n(E)$ is weakly sequentially complete, then the measure $\mu^{(n)}$ is countably additive;

(iv) if either $E = c_0$ or $E = l^p$, where $p > n$, then the measure $\mu^{(n)}$ is countably additive;

(v) if $X = \mathbb{R}^\infty$ and $E = l^\infty$, then the measure $\mu^{(n)}: \mathcal{B}(X) \to B_n(l^p)$ is countably additive for every $p \geqslant 1$.

Let us prove a useful result on the differentiability of convolutions.

5.4.8. Lemma. *Let μ be a Radon measure on a locally convex space X, let $H \subset X$ be a continuously embedded Hilbert space with $H \subset H(\mu)$, and let $f\colon X \to E$ be a bounded μ-measurable mapping with values in a separable Hilbert space E. Then the mapping*

$$F(x) = \int_X f(x+y)\,\mu(dy)$$

is Gâteaux differentiable along H, its derivative $D_H F(x)$ is a Hilbert–Schmidt operator, and one has

$$\|D_H F(x)\|_{\mathcal{H}(H,E)} \leqslant c\sqrt{\|\mu\|}\sup_y |f(y)|_E,$$

where c is the norm of the embedding $H \to H(\mu)$ (which is finite by the closed graph theorem).

PROOF. Suppose for notational simplicity that $\mu \geqslant 0$. It follows from our assumptions that

$$\|T\|_{\mathcal{H}(H,E)} \leqslant c\|T\|_{\mathcal{H}(H(\mu),E)}$$

for every Hilbert–Schmidt operator $T\colon H(\mu) \to E$. Hence it suffices to prove our claim for $H = H(\mu)$. Then $c = 1$. The existence of the Gâteaux derivative follows from Example 3.3.14. Let $\{e_n\}$ be an orthonormal sequence in $H = H(\mu)$, i.e., the sequence $\{\beta^\mu_{e_n}\}$ is orthonormal in $L^2(\mu)$. We have

$$\partial_{e_n} F(x) = -\int_X f(x+y)\beta^\mu_{e_n}(y)\,\mu(dy).$$

Let $\{h_n\}$ be an orthonormal basis in the subspace of E generated by the vectors $D_H F(x)e_n$, where x is fixed. Then we have $f(z) = \sum_{j=1}^\infty f_j(z)h_j$. By Bessel's inequality, we have

$$\sum_{n=1}^\infty \left(\int_X f_j(x+y)\beta^\mu_{e_n}(y)\,\mu(dy)\right)^2 \leqslant \int_X f_j(x+y)^2\,\mu(dy).$$

Therefore,

$$\sum_{n=1}^\infty |D_H F(x)e_n|^2_E = \sum_{n=1}^\infty \sum_{j=1}^\infty \left(\int_X f_j(x+y)\beta^\mu_{e_n}(y)\,\mu(dy)\right)^2$$

$$\leqslant \sum_{j=1}^\infty \int_X f_j(x+y)^2\,\mu(dy) = \int_X |f(x+y)|^2_E\,\mu(dy),$$

which does not exceed $\sup_y |f(y)|^2_E \|\mu\|$. □

5.4.9. Remark. (i) Similar reasoning shows that if a mapping $f\colon X \to E$ is linear, measurable with respect to μ and $|f(\,\cdot\,)|_E \in L^2(\mu)$, then one has $f|_H \in \mathcal{H}(H,E)$, which is well known for Gaussian measures.

5.4. Differentiability along subspaces

(ii) More generally, if $|f(\,\cdot\,)|_E \in L^2(\mu * \mu)$, where $\mu \geq 0$, then $D_H F(x)$ exists μ-a.e. and

$$\|D_H F\|_{L^2(\mu, \mathcal{H}(H,E))} \leq c\|f\|_{L^2(\mu * \mu)}.$$

Shavgulidze [**1000**] obtained the following decomposition.

5.4.10. Theorem. *Let μ be a Radon measure on a locally convex space X and let E be a Banach space continuously embedded into X such that μ is infinitely differentiable along all vectors from E. Suppose that γ is a Radon Gaussian measure on E. Then one can find nonnegative Radon measures μ_1 and μ_2 on X infinitely differentiable along $H := H(\gamma)$ such that $\mu = \mu_1 - \mu_2$.*

PROOF. We may assume that γ is centered and that

$$\int_E \|h\|_E \, \gamma(dh) \leq 1.$$

Let γ_t be the image of ν under the mapping $x \mapsto t^{1/2}x$, $t > 0$. It is readily seen that $\gamma_{t+s} = \gamma_t * \gamma_s$. Let B be the unit ball in E. Then, for every n, there is a number c_n such that

$$\|d_{h_1} \cdots d_{h_n} \mu\| \leq c_n \quad \forall h_1, \ldots, h_n \in B.$$

Let us set

$$\nu_{0,1}^n := \mu - \mu * \gamma_{4^{-(n+1)}}, \quad \nu_{k,1}^n := \mu - \mu * \gamma_{4^{-(n+1)}(1-4^{-k})}, \quad k \geq 1.$$

By the recurrence relations

$$\nu_{i,m+1}^n := \nu_{i,m}^n - \nu_{i,m}^n * \gamma_{4^{-(m+n+1)}}, \quad i = 0, 1, \ldots,$$

we define the measures $\nu_{i,m}^n$. Let δ be Dirac's measure at the origin. We observe that

$$\nu_{k,k}^n * \gamma_{4^{-(n+k+1)}} = \mu * (\delta - \gamma_{4^{-(n+1)}(1-4^{-k})}) * (\delta - \gamma_{4^{-(n+2)}}) * \cdots$$
$$* (\delta - \gamma_{4^{-(n+k)}}) * \gamma_{4^{-(n+k+1)}}$$
$$= \mu * (\delta - \gamma_{4^{-(n+2)}}) * \cdots$$
$$* (\delta - \gamma_{4^{-(n+k)}}) * (\delta - \gamma_{4^{-(n+1)}(1-4^{-k})}) * \gamma_{4^{-(n+k+1)}}$$
$$= \mu * (\delta - \gamma_{4^{-(n+2)}}) * \cdots * (\delta - \gamma_{4^{-(n+k)}}) * (\gamma_{4^{-(n+k+1)}} - \gamma_{4^{-(n+1)}})$$
$$= \mu * (\delta - \gamma_{4^{-(n+2)}}) * \cdots * (\delta - \gamma_{4^{-(n+k)}}) * (\gamma_{4^{-(n+k+1)}} - \delta)$$
$$+ \mu * (\delta - \gamma_{4^{-(n+2)}}) * \cdots * (\delta - \gamma_{4^{-(n+k)}}) * (\delta - \gamma_{4^{-(n+1)}})$$
$$= -\nu_{0,k}^{n+1} + \mu * (\delta - \gamma_{4^{-(n+1)}}) * (\delta - \gamma_{4^{-(n+2)}}) * \cdots * (\delta - \gamma_{4^{-(n+k)}}) = -\nu_{0,k}^{n+1} + \nu_{0,k}^n,$$

whence we obtain

$$\nu_{0,k}^n = \nu_{k,k}^n * \gamma_{4^{-(n+k+1)}} + \nu_{0,k}^{n+1}.$$

Let η be a Radon measure on X that is infinitely differentiable along all vectors in E, let $A \in \mathcal{B}(X)$, and let $h_1, \ldots, h_n \in E$. Then, for any, $t > 0$, one has

$$|d_{h_1}\cdots d_{h_n}(\nu - \nu * \gamma_t)|(A)|$$
$$\leqslant \int_E |d_{h_1}\cdots d_{h_n}\nu(A+h) - d_{h_1}\cdots d_{h_n}\nu(A)|\,\gamma_t(dh)$$
$$\leqslant \sup_{h\in B}\|d_h d_{h_1}\cdots d_{h_n}\nu\| \int_E \|h\|_E\,\gamma_t(dh) \leqslant t^{1/2}\sup_{h\in B}\|d_h d_{h_1}\cdots d_{h_n}\nu\|.$$

Therefore,

$$|\nu_{i,k}^n(A)| \leqslant \prod_{m=1}^k 2^{-(n+m)} \sup_{h_1,\ldots,h_k\in B} \|d_{h_1}\cdots d_{h_k}\mu\| \leqslant c_k 2^{-k(n+(k+1)2)},$$

whence we obtain

$$\|\nu_{i,k}^n\| \leqslant 2c_k 2^{-k(n+(k+1)2)}. \tag{5.4.1}$$

Let us find an increasing sequence of integers n_k such that $c_{k+1} < 2^{n_k}$. For each m let $k(m)$ be the minimal number among the numbers l with $m \leqslant n_l$. Let us set

$$\nu_1 := \mu, \quad \nu_{n+1} := \nu_{0,k(n)}^{n-k(n)+1}, \ n \geqslant 1,$$
$$\eta_1 := \mu, \quad \eta_{m+1} := \nu_{k(m),k(m)}^{m-k(m)+1}$$

whenever $m \geqslant 1$ and $m \neq n_{k(m)}$, and $\eta_{n_k+1} := \nu_{0,k}^{n_k-k+1}$. With this notation, we have

$$\nu_n = \eta_n * \gamma_{4^{-(n+1)}} + \nu_{n+1}. \tag{5.4.2}$$

Due to (5.4.1) and the estimate $c_{k+1} < 2^{n_k}$ for all $n > n_2$ we have

$$\|\nu_n\| \leqslant 2^{-(k(n)-1)(n-k(n)/2)}, \quad \|\eta_n\| \leqslant 2^{-(k(n)-1)(n-k(n)/2)}. \tag{5.4.3}$$

Therefore, setting $\xi_n(A) := \|\eta_n\| * \gamma_{4^{-(n+1)}}$, we obtain a nonnegative Radon measure $\xi := \sum_{n=1}^\infty \xi_n$. By using (5.4.2) one can show that for every n the following identity holds: $\mu = \eta_1 * \gamma_{4^{-(n+1)}} + \cdots + \eta_n * \gamma_{4^{-(n+1)}} + \nu_{n+1}$. It follows from the first estimate in (5.4.3) that $\mu(A) \leqslant \xi(A)$ for all sets $A \in \mathcal{B}(X)$. Hence $\mu_2 := \xi - \mu \geqslant 0$. Finally, since for any measure ν, any vectors $h_1, \ldots, h_n \in H$, and any $t > 0$ we have

$$\|d_{h_1}\cdots d_{h_n}(\eta * \gamma_t)\| = \|\eta * d_{h_1}\cdots d_{h_n}\gamma_t\| \leqslant t^{-n/2}\|\eta\|\,\|d_{h_1}\cdots d_{h_n}\gamma_1\|,$$

the second estimate in (5.4.3) shows that the measure ξ is infinitely differentiable along the subspace H. $\square$

If X is a Hilbert space, then $D(\mu)$ is contained in $T(X)$ for some Hilbert–Schmidt operator, so taking another Hilbert–Schmidt operator S we can find a suitable Gaussian measure γ with $H(\gamma) = ST(X)$. It should be noted that this result is not trivial even in the case of the real line, where it says that any integrable function f with integrable derivatives of all orders is a difference of two nonnegative integrable functions with integrable derivatives of all orders. Note also that it is not clear whether in the above theorem one can find measures μ_1 and μ_2 infinitely differentiable along the whole space E.

It would be also of interest to investigate the situation where in place of a Gaussian measure γ we are given a probability measure ν on E infinitely differentiable along some continuously embedded Banach space E_0.

5.5. Comments and exercises

Subspaces of differentiability of measures were introduced by the author [166] and further investigated in [169], [175]. The structure of the spaces $Q(\mu)$ was studied in Chatterji, Mandrekar [280], Shimomura [1022]–[1025]. The metric ϱ on $C(\mu)$, defined at the beginning of §5.1, was introduced in [169]. Corollary 5.1.5, following directly from assertion (ii) of Theorem 5.1.1, was noted in Khafizov [602], Bogachev [179], and Bogachev, Smolyanov [233]. Part (i) of Theorem 5.1.6 was obtained in [1022], [1025], and part (ii) was proved in [169]. Under some additional assumptions Proposition 5.1.7 was proved in Albeverio, Høegh-Krohn [41] by more complicated reasoning. The structure of subspaces of differentiability, continuity and quasi-invariance for measures on infinite dimensional spaces is an important and interesting object of research.

Connections between differential properties of measures with their supports were considered in the author's paper [172]. The fact that the Wiener measure on the classical Wiener space $C[0,1]$ has no Hilbert support was first established in Guerquin [510] and Smolyanov, Uglanov [1054].

Proposition 5.4.3 and Corollary 5.4.4 are obtained in Averbukh, Smolyanov, Fomin [82].

Diverse applications of differentiable measures and their subspaces of differentiability to approximations in Banach spaces were considered in Bentkus [133], Bogachev [181], Goodman [494], Kuo [666], Paulauskas, Rachkauskas [876], Piech [896], Uglanov [1135]. For example, for every uniformly continuous function F and every n, by using convolutions with differentiable measures, one can approximate F by uniformly continuous functions having n bounded Fréchet derivatives along a given subspace $D(\mu)$ (see Exercise 5.5.7). This is in great contrast to approximations by functions differentiable along the whole space, which do not always exist even in the case of a Hilbert space and $n = 3$. A related application is constructing bump-functions and partitions of unity (see Exercise 5.5.8). Let us recall that many Banach spaces do not possess nontrivial Fréchet differentiable bump-functions, say, on $C[0,1]$ there are no nonzero functions with bounded support and bounded Fréchet derivative; see Deville, Godefroy, Zizler [345]. This issue is important also for the Sobolev classes introduced in Chapter 8. Some further results will be given in Chapters 8 and 9.

Exercises

5.5.1. Let μ on $\mathbb{R}^\infty$ be the countable power of Lebesgue measure on $[0,1]$. Show that $D_C(\mu) = C(\mu) = l^1$.

HINT: Using Hellinger's integral show that $\|\mu - \mu_h\| = 2$ if $h \notin l^1$.

5.5.2. (i) Let p be the symmetric probability density on the real line defined as follows: $p(t) = t^{-1/2}/6$ if $t \in (0, 1]$, $p(t) = t^{-2}/6$ if $t > 1$. Let μ be the countable power of the measure with density p. Prove that $C(\mu) = Q(\mu) = l^{1/2}$. In particular, this space is not locally convex.

(ii) Prove that there is no sequence of Radon measures μ_n with $C(\mu) \subset D(\mu_n)$ convergent to μ on every Borel set. To this end verify that otherwise we would obtain $C(\mu) = \bigcap_{n=1}^{\infty} D(\mu_n)$.

(iii) Prove that there is no Radon measure ν with $\mu \ll \nu$ and $C(\mu) \subset D(\nu)$.

5.5.3. Let λ^∞ be the countable power of Lebesgue measure on $[0, 1]$, considered on l^∞, let μ be the image of λ^∞ under the mapping $T \colon (x_n) \mapsto (x_n/\ln(n+1))$. Show that similarly to Example 5.3.6 the measure μ is concentrated on c_0 and is mutually singular with every Gaussian measure on c_0.

5.5.4. (Khafizov [**605**]) Let μ be a Radon probability measure on a locally convex space X. Show that the function $\varphi(h) := H(\mu_h, \mu)$ on the metric space $L = C(\mu)$ is positive definite. Hence on the space L' algebraically dual to L there exists a probability measure P defined on the σ-algebra generated by L such that its Fourier transform coincides with φ (see Vakhania, Tarieladze, Chobanyan [**1167**, Proposition 4.4.1]). Let $\Phi \colon C(\mu) \to L^0(P)$ be the identical embedding. Show that $\Phi(h_n) \to 0$ in measure P precisely when $h_n \to 0$ in the metric of $C(\mu)$, i.e., Φ is a linear homeomorphism between $C(\mu)$ and a closed subspace in $L^0(P)$.

5.5.5. (Khafizov [**605**]) Let μ be the countable power of a probability measure ν on the real line. Then $C(\mu) = l^2$ precisely when ν has an absolutely continuous density p with $|p'|^2/p \in L^1(\mathbb{R}^1)$. This reinforces the result of Shepp [**1006**], who considered the case $Q(\mu) = l^2$.

5.5.6. Show that in the situation described before Theorem 5.3.1 one, indeed, has the equality $E_\mu = E_\gamma = H(\gamma) + \mathbb{R}^1 a_\gamma$.

5.5.7. Let F be a uniformly continuous function on a Banach space X, let μ be a Radon probability measure on X, and let $n \in \mathbb{N}$. Show that there exist uniformly convergent to F uniformly continuous functions F_j having n bounded Fréchet derivatives along $D(\mu)$.

5.5.8. Let μ be a Radon probability measure on a locally convex space X, let E be a Banach space continuously embedded into $D(\mu)$, let $K \subset X$ be compact, and let $U \supset K$ be open. Prove that for every n there exists a Borel function $f \colon X \to [0, 1]$ such that $f(x) = 1$ if $x \in K$, $f(x) = 0$ if $x \notin U$, and there exist bounded Fréchet derivatives $D_E^k f$, $k = 1, \ldots, n$.

HINT: Use the reasoning from the proof of Theorem 5.3.5.

5.5.9. (Kats [**596**]) Let μ be a Radon measure on a locally convex space X and let $E \subset X$ be a continuously embedded bornological separable locally convex space (e.g. a normed space) such that $E \subset D(\mu)$. Then there is a nonnegative measure ν that is differentiable and quasi-invariant along vectors in E such that $\mu \ll \nu$.

HINT: Take a dense sequence $\{h_n\} \subset E$ and the measure $\nu = \sum_{n=1}^{\infty} 2^{-n} |\mu| * \delta_{h_n}$; use that by the closed graph theorem, if $h_{n_i} \to h$ in E, then $h_{n_i} \to h$ in $D(\mu)$, hence $\nu(A + th_{n_i}) \to \nu(A + th)$ for any $A \in \mathcal{B}(X)$ and $t \in \mathbb{R}^1$.

5.5.10. Show that in the situation of Example 5.3.9 in case $\alpha \in (1, 2)$ the measure γ_α is concentrated on a Hilbert space continuously embedded into $C[0, 1]$.

CHAPTER 6

Integration by parts and logarithmic derivatives

In this chapter we consider integration by parts formulae more delicate than in Chapter 3 and investigate logarithmic derivatives of measures from the point of view of their integrability and differentiability.

6.1. Integration by parts formulae

It has been noted in Chapter 3 (see formula (3.3.3)) that, for any measure μ differentiable along h and certain functions f possessing the partial derivative $\partial_h f$ defined by the formula $\partial_h f(x) := \lim\limits_{t \to 0} t^{-1}[f(x+th) - f(x)]$, under certain conditions one has the equality

$$\int_X \partial_h f(x)\,\mu(dx) = -\int_X f(x)\,d_h\mu(dx). \qquad (6.1.1)$$

This equality is called an integration by parts formula. The classical integration by parts formulas for real functions assert that, under certain conditions on two functions f and g with primitives F and G (in some sense) one has the relationship

$$\int_a^b F(t)g(t)\,dt = FG\big|_a^b - \int_a^b f(t)G(t)\,dt. \qquad (6.1.2)$$

Certainly, the question arises of making (6.1.2) meaningful. In what sense should one understand the integrals? In what sense are the primitives defined? These issues turn out to be very subtle and we are not going to discuss them in detail. We only give several possible solutions to our problem, which cover all examples arising in applications. A more thorough discussion of the one-dimensional case is given in Saks [**980**].

If the functions F and G are absolutely continuous and $f = F'$, $G' = g$, then equality (6.1.2), where the integrals are Lebesgue, are well known. But what can be said in the case where the function F is *everywhere differentiable*, but is not necessarily absolutely continuous (i.e., $f = F'$ is not assumed to be Lebesgue integrable)? Then (6.1.2) remains valid for every absolutely continuous function G such that the function Gf is integrable, but the proof of this fact is less elementary. Below we employ the Denjoy integrals $\mathcal{D}$ and $\mathcal{D}^*$ considered in [**980**], however, we do not recall the corresponding (rather involved) definitions. We only note that every function that is locally absolutely continuous or everywhere differentiable is integrable in either sense. We also recall that every such function f has Lusin's property (N), i.e., $f(E)$ has measure zero for every set E of measure zero. First we establish a formula involving only the Lebesgue integral.

6.1.1. Theorem. *Let ϱ be a locally absolutely continuous function on the real line, let $\varrho \in L^1(\mathbb{R}^1)$, and let f be a continuous almost everywhere differentiable function possessing Lusin's (N)-property (for example, everywhere differentiable). Suppose that the functions $f'\varrho$ and $f\varrho'$ belong to $L^1(\mathbb{R}^1)$. Then one has*

$$\int_{-\infty}^{+\infty} f'(t)\varrho(t)\,dt = -\int_{-\infty}^{+\infty} f(t)\varrho'(t)\,dt. \qquad (6.1.3)$$

Moreover, the same is true if f is continuous, almost everywhere differentiable and has (N)-property on every interval containing no zeros of ϱ.

PROOF. Suppose that f is bounded. Let ϱ have no zeros on some closed interval $[a,b]$. Then $f' \in L^1[a,b]$. This gives the absolute continuity of f on $[a,b]$ (see [**980**, Theorem 7.7, p. 285] or [**193**, Exercise 5.8.58]). Hence the usual integration by parts formula works on $[a,b]$. Clearly, this formula remains valid also in the case where $\varrho(a) = \varrho(b) = 0$ and ϱ has no zeros in (a,b) (it suffices to consider the intervals $[a+\varepsilon, b-\varepsilon]$ and let ε go to zero). Therefore, $f'\varrho$ and $-f\varrho'$ have equal integrals over such intervals. If $\varrho(a) = 0$ and on $(a, +\infty)$ there are no zeros of ϱ, then everything remains true for $[a, +\infty)$. This is seen from the integration by parts formula for the intervals $[a, b_n]$, where $b_n \to +\infty$ are chosen such that $\varrho(b_n) \to 0$ (such b_n exist by the integrability of ϱ). The case of negative values is similar. If ϱ does not vanish at all, then we also take points $a_n \to -\infty$ and $b_n \to +\infty$ with $\varrho(a_n) \to 0$ and $\varrho(b_n) \to 0$ and write the integration by parts formula for $[a_n, b_n]$. Since $\varrho' = 0$ a.e. on the set $\{\varrho = 0\}$, the desired formula is established for bounded f. In the general case we put $f_n := \theta_n \circ f$, where $\theta_n \in C_b^1(\mathbb{R}^1)$, $\theta_n(t) = t$ if $t \in [-n, n]$, $0 \leqslant \theta_n' \leqslant 1$. It is clear that the function f_n has the same properties as f, but is bounded. Hence (6.1.3) is valid for f_n. Since $|\theta_n(t)| \leqslant |t|$, one has $|f_n| \leqslant |f|$. Hence the Lebesgue dominated convergence theorem applies and yields (6.1.3) for f. $\square$

By using differentiable conditional measures μ^x on the straight lines $x + \mathbb{R}^1 h$ and applying the one-dimensional result to them, we easily extend it to the infinite dimensional case.

6.1.2. Theorem. *Suppose that a Radon measure μ on a locally convex space X is Fomin differentiable along a vector h and that a μ-measurable function f satisfies the following conditions:*

(i) *for μ-almost all $x \in X$ the function $t \mapsto f(x+th)$ is continuous, almost everywhere differentiable and has Lusin's (N)-property (for example, it is absolutely continuous or everywhere differentiable);*

(ii) $f \in L^1(d_h\mu)$ and $\partial_h f \in L^1(\mu)$. Then one has

$$\int_X \partial_h f(x)\,\mu(dx) = -\int_X f(x)\,d_h\mu(dx). \tag{6.1.4}$$

For an extension of formula (6.1.4) to the case of Skorohod differentiability we need a more subtle one-dimensional integration by parts formula. Suppose that a function F on $[a,b]$ has bounded variation and a function g on $[a,b]$ is $\mathcal{D}$- or $\mathcal{D}^*$-integrable in the Denjoy sense (see Saks [**980**, Ch. VIII]; however, we only need to know the fact that this class contains absolutely continuous functions as well as everywhere differentiable functions). The following result holds (see [**980**, p. 245, Theorem 2.5]). Under the stated conditions, the function Fg is integrable in the same Denjoy sense and one has

$$(\mathcal{D})\int_a^b Fg\,dx = GF\,|_a^b - (S)\int_a^b G(x)\,dF(dx),$$

where G is the indefinite integral of g, the integral on the left is taken in the Denjoy sense, and the integral on the right is the Stieltjes integral. In particular, this formula covers the case where the function G is either locally absolutely continuous or everywhere differentiable and one has $G \in L^1(dF)$, $g \in L^1(F dx)$.

6.1.3. Proposition. *Suppose that a measure μ is Skorohod differentiable along a vector h and a function $f \in L^1(d_h\mu)$ is such that for μ-a.e. x the function $t \mapsto f(x+th)$ is everywhere differentiable or locally absolutely continuous. Suppose also that $\partial_h f \in L^1(\mu)$. Then (6.1.4) is true.*

PROOF. For measures on $\mathbb{R}^1$ this follows from the formula given above since μ has a density F of bounded variation and $d_1\mu = dF$. For G we take f and set $g = f'$. It is important here that the function f' is Denjoy integrable both in the case of an absolutely continuous function f (then it will be Lebesgue integrable) and in the case where f is everywhere differentiable (here the Lebesgue integrability may fail). As above, for obtaining the integration by parts formula on the whole real line we first assume that the function f is bounded and take points $a_n \to -\infty$ and $b_n \to +\infty$ with $F(a_n) \to 0$, $F(b_n) \to 0$. A passage to the case of an unbounded function f is completely analogous to our reasoning in the proof of Theorem 6.1.1. We observe that at this stage it suffices to have boundedness of variation of F on intervals and not on the whole real line (but the integrability of F on the whole real line is necessary). In the general case, assuming that $h \neq 0$, we take a hyperplane Y complementing $\mathbb{R}^1 h$ and Skorohod differentiable conditional measures μ^y on the straight lines $y + \mathbb{R}^1 h$, $y \in Y$. Then we apply the established fact to every measure μ^y. □

6.1.4. Corollary. (i) *Suppose that a measure μ is Fomin differentiable along a vector h and a μ-integrable function f satisfies the hypotheses of Theorem 6.1.2. Then the measure $\nu = f \cdot \mu$ is also differentiable along h and*

$$d_h \nu = f \cdot d_h \mu + \partial_h f \cdot \mu.$$

(ii) *Suppose that a measure μ has the Skorohod derivative $d_h\mu$ and a function $f \in \mathcal{L}^1(\mu)$ satisfies the hypotheses of Proposition 6.1.3. Then the measure $\nu = f \cdot \mu$ has the Skorohod derivative*

$$d_h\nu = f \cdot d_h\mu + \partial_h f \cdot \mu.$$

(iii) *Suppose that a measure μ is Fomin differentiable along a vector h and functions f and g satisfy condition* (i) *in Theorem 6.1.2. Suppose also that the functions $\partial_h f g$, $f \partial_h g$, and $f g \beta_h^\mu$ are μ-integrable. Then*

$$\int_X \partial_h f(x) g(x) \, \mu(dx) = -\int_X f(x) \partial_h g(x) \, \mu(dx) - \int_X f(x) g(x) \beta_h^\mu(x) \, \mu(dx). \tag{6.1.5}$$

6.1.5. Example. Suppose that γ is a centered Radon Gaussian measure on a locally convex space X, functions f and g satisfy condition (i) in Theorem 6.1.2, and $f\partial_h g, g\partial_h f, fg\widehat{h} \in L^1(\mu)$. Then

$$\int_X \partial_h f(x) g(x) \, \gamma(dx) = -\int_X f(x) \partial_h g(x) \, \gamma(dx) + \int_X f(x) g(x) \widehat{h}(x) \, \gamma(dx). \tag{6.1.6}$$

6.1.6. Proposition. *Let a measure μ on a locally convex space X be Fomin differentiable along a vector $h \neq 0$. Then, for every compact set K and every open set U containing K, there exists a measure $\lambda \ll \mu$ that is Fomin differentiable along h, has the same restriction to K as μ, and has compact support in U.*

PROOF. As in Theorem 3.5.1, take a hyperplane Y complementing $\mathbb{R}^1 h$. By compactness of K there exists a number $\delta > 0$ such that $x + th \in U$ for every $x \in K$ and every t with $|t| \leq \delta$. The set $Q = K + \delta I$, where I is the closed interval joining the points $-h$ and h, is compact and contained in U. Set $V = X \backslash Q$ and consider the function

$$f(x) = r_V(x)[r_K(x) + r_V(x)]^{-1},$$

where $r_V(x) = \inf\{|t|\colon x + th \in V\}$ and $r_K = \inf\{|t|\colon x + th \in K\}$ (in the case where there is no such t, we set $f(x) = 0$). It is readily seen that f is a Borel function, $f(x) = 1$ on K and $f(x) = 0$ outside Q. In addition, the function f is Lipschitzian along h and $|\partial_h f| \leq \delta^{-1}$. To see this, it suffices to consider the case of $\mathbb{R}^1$ and $h = 1$, by taking the intersections of $x + \mathbb{R}^1 h$ with the sets K and V. It remains to set $\lambda := f \cdot \mu$. $\square$

We observe that in the proof we have actually constructed a Borel function $f\colon X \to [0,1]$ with compact support equal to 1 on K and satisfying the Lipschitz condition along h. A minor modification of the construction gives the following result.

6.1.7. Proposition. *Let μ be a nonzero Radon measure on a locally convex space X. Then, for every compact set K and every open set U containing K, there exists a Borel function $f\colon X \to [0,1]$ vanishing outside* *some compact set in U, equal to 1 on K and satisfying the Lipschitz condition along $D_C(\mu)$.*

PROOF. We find a convex neighborhood of the origin W such that one has $K + W \subset U$. Then $B_r := \{h \in D_C(\mu) \colon \|d_h\mu\| \leqslant r\} \subset W$ for some $r > 0$. As we know, B_r is compact. Hence the set $Q := K + B_r \subset U$ is compact. Set $V := X \backslash Q$ and

$$r_V(x) := \inf\{\|d_h\mu\| \colon x + h \in V\}, \quad r_K(x) := \inf\{\|d_h\mu\| \colon x + h \in K\}$$

if $x \in K + D_C(\mu)$ and $r_V(x) = r_K(x) = 0$ if $x \notin K + D_C(\mu)$. Then the set $L := K + D_C(\mu)$ is Borel by compactness of $K + B_n$. It is readily verified that r_V and r_K are Borel. For example, $\{x \in L \colon r_K(x) \leqslant c\} = K + B_c$. Indeed, if $\|d_h\mu\| \leqslant c$ and $x \in K$, then $r_K(x + h) \leqslant c$ since $x + h - h \in K$. Conversely, if $r_K(x) \leqslant c$ and $x \in L$, then there exist $h_n \in D_C(\mu)$ with $x + h_n \in K$ and $\|d_{h_n}\mu\| \leqslant c + n^{-1}$. By compactness of $K - x$ the sequence $\{h_n\}$ has a limit point h. Then $h \in D_C(\mu)$ and $\|d_h\mu\| \leqslant c$. In addition, $x + h \in K$. For r_V we have

$$\{x \in L \colon r_V(x) < c\} = L \cap (V + B_c^\circ),$$

where $B_c^\circ := \{h \in D_C(\mu) \colon \|d_h\mu\| < c\}$. The set $V + B_c$ is open since V is open. Let $f(x) := r_V(x)(r_V(x) + r_K(x))^{-1}$ if $x \in L$ and $f(x) = 0$ if $x \notin L$. Then $0 \leqslant f \leqslant 1$, $f(x) = 1$ if $x \in K$, $f(x) = 0$ if $x \notin Q$. It is easily verified that the function f is Lipschitzian along $D_C(\mu)$. □

6.2. Integrability of logarithmic derivatives

In this section we discuss some integrability properties of logarithmic derivatives.

The following rather elementary but not obvious lemma was stated in Uglanov [1137].

6.2.1. Lemma. *There exists a number C such that for every nonnegative twice differentiable function $\varphi \colon \mathbb{R}^1 \to \mathbb{R}^1$ with an absolutely continuous second derivative φ'' the following inequality holds, where we set $0/0 = 0$:*

$$J(\varphi) = \int_{-\infty}^{+\infty} \frac{|\varphi'(t)|^2}{\varphi(t)}\, dt \leqslant C \int_{-\infty}^{+\infty} \bigl[|\varphi'(t)| + |\varphi''(t)| + |\varphi'''(t)|\bigr]\, dt.$$

Below we prove a more general fact.

6.2.2. Corollary. *If a nonnegative measure μ on a locally convex space X is three times differentiable along a vector h, then $\beta_h^\mu \in L^2(\mu)$.*

Taking $\varphi(t) = |t|\,\bigl|\ln|t|\bigr|^{-1}$ in $(-\delta, \delta)$, we see that $J(\varphi)$ cannot be estimated via $\|\varphi'\|_{L^1}$ and $\|\varphi''\|_{L^1}$. Lemma 6.2.1 was reinforced in Bogachev [185], [186] in the following way.

6.2.3. Lemma. (i) *If $\varphi\colon \mathbb{R}^1 \to [0,+\infty)$ is twice differentiable and φ'' has bounded variation, then*

$$J(\varphi) = \int_{-\infty}^{+\infty} \frac{|\varphi'(t)|^2}{\varphi(t)}\, dt \leqslant 8\|\varphi'\|_{L^1} + 6\|\varphi''\|_{L^1} + 2\operatorname{Var}\varphi''.$$

(ii) *If a nonnegative measure μ is three times Skorohod differentiable along h, then*

$$\|\beta_h^\mu\|_{L^2(\mu)} \leqslant 8\|d_h\mu\| + 6\|d_h^2\mu\| + 2\|d_h^3\mu\|.$$

PROOF. (i) By the aid of convolution the general case reduces to the case where the function φ'' is continuous. The integral of the function $(\varphi')^2/\varphi$ over the set $E := \{(\varphi')^2 \leqslant 2\varphi\varphi''\}$ is not greater than the integral of $2|\varphi''|$ over E. Let us consider the open set $\mathbb{R}^1\setminus E$ and take an interval (a,b) on which $(\varphi')^2/\varphi > 2\varphi''$ and $\varphi' > 0$, but at least at one of the endpoints of the interval this is false. Then $g = \varphi'/\varphi^{1/2}$ decreases on (a,b) since $g' < 0$. Hence $|\varphi'(a)|^2/\varphi(a) = 2\varphi''(a)$ and for $J := \|(\varphi')^2/\varphi\|_{L^1[a,b]}$ we have

$$J \leqslant \sqrt{2\varphi''(a)} \int_a^b \varphi'(t)\varphi(t)^{-1/2}\, dt = 2\sqrt{2\varphi''(a)}[\varphi(b)^{1/2} - \varphi(a)^{1/2}].$$

Hence

$$J^2 \leqslant 8\varphi''(a)[\varphi(b) - \varphi(a)] \leqslant 8\varphi''(a)\int_a^b \varphi'(t)\, dt.$$

Then either $J \leqslant \varphi''(a)$ or $J \leqslant 8\int_a^b \varphi'(t)\, dt$. In the first case one also has two possibilities:

1) $\varphi''(t) \geqslant \varphi''(a)/2$ on (a,b) and then $(\varphi')^2/\varphi \leqslant 2\varphi''(a) \leqslant 4\varphi''$ on (a,b), which gives the estimate

$$\int_a^b |\varphi'(t)|^2/\varphi(t)\, dt \leqslant 4\int_a^b \varphi''(t)\, dt;$$

2) there exists $c \in (a,b)$ with $\varphi''(c) \leqslant \varphi''(a)/2$, and in this case we obtain

$$J \leqslant \varphi''(a) \leqslant 2[\varphi''(a) - \varphi''(c)] = 2\int_a^c \varphi'''(t)\, dt \leqslant 2\int_a^b |\varphi'''(t)|\, dt.$$

Hence in either case we have

$$\int_a^b |\varphi'(t)|^2/\varphi(t)\, dt \leqslant 8\int_a^b |\varphi'(t)|\, dt + 4\int_a^b |\varphi''(t)|\, dt + 2\int_a^b |\varphi'''(t)|\, dt.$$

The same is true for any interval with $(\varphi')^2/\varphi > 2\varphi''$ and $\varphi' < 0$. Finally, we have $\|\varphi'''\|_{L^1} = \operatorname{Var}\varphi''$. Assertion (ii) follows from (i) due to the existence of three times Skorohod differentiable conditional measures μ^x on the straight lines $x + \mathbb{R}^1 h$ and the estimates obtained for these measures. □

These lemmas do not give a full picture of interrelations between the smoothness of a measure and the integrability of its logarithmic derivative.

6.2. Integrability of logarithmic derivatives

The problem was solved completely by E.P. Krugova [645], who proved the following nice result.

6.2.4. Theorem. *If μ is a Radon measure on a locally convex space X (or a measure on $\sigma(X)$), then the following assertions are true.*

(i) If μ is twice differentiable along h, then one has $\beta_h^\mu \in L^{2-\varepsilon}(\mu)$ for all $\varepsilon \in (0, 2)$. In addition,

$$\|\beta_h^\mu\|_{L^{2-\varepsilon}(\mu)} \leqslant (1+\varepsilon^{-1})\varepsilon^{-1}\|d_h\mu\| + (1-\varepsilon)\varepsilon^{-1}\|d_h^2\mu\|. \tag{6.2.1}$$

(ii) If μ is nonnegative and three times differentiable along h, then one has $\beta_h^\mu \in L^{3-\varepsilon}(\mu)$ for all $\varepsilon \in (0, 3)$. In addition, there exists a number $C(\varepsilon)$, dependent only on ε, such that

$$\|\beta_h^\mu\|_{L^{3-\varepsilon}(\mu)} \leqslant C(\varepsilon)\bigl(\|d_h\mu\| + \|d_h^2\mu\| + \|d_h^3\mu\|\bigr). \tag{6.2.2}$$

The main difficulty in the proof is to obtain the corresponding estimates for functions on the real line (the general case then follows from the results on differentiable conditional measures). Let us state this important assertion as a separate result. As we shall now see, its proof is elementary, but requires a much more thorough investigation of diverse cases than in the proof of Lemma 6.2.3.

6.2.5. Theorem. *Let φ be a differentiable function on the real line and let φ' be absolutely continuous. Then, for every $\varepsilon \in (0, 1]$, we have Krugova's inequalities:*

$$\int_{-\infty}^{+\infty} \left(\frac{|\varphi'(t)|}{\varphi(t)}\right)^{2-\varepsilon} \varphi(t)\, dt \leqslant \frac{1+\varepsilon}{\varepsilon}\int_{-\infty}^{+\infty} |\varphi'(t)|\, dt + \frac{1-\varepsilon}{\varepsilon}\int_{-\infty}^{+\infty} |\varphi''(t)|\, dt.$$

Let $\varphi \geqslant 0$ be a twice differentiable function on the real line and let φ'' be absolutely continuous. Then, for every $\varepsilon \in (0, 3)$, we have

$$\int_{-\infty}^{+\infty} \left(\frac{|\varphi'(t)|}{\varphi(t)}\right)^{3-\varepsilon} \varphi(t)\, dt$$

$$\leqslant C(\varepsilon)\left(\int_{-\infty}^{+\infty} |\varphi'(t)|\, dt + \int_{-\infty}^{+\infty} |\varphi''(t)|\, dt + \int_{-\infty}^{+\infty} |\varphi'''(t)|\, dt\right).$$

PROOF. (i) Let $\dot\varphi := \varphi'$, $\ddot\varphi := \varphi''$. Let us estimate the integral of the function $|\dot\varphi/\varphi|^{2-\varepsilon}|\varphi|$ over $\mathbb{R}^1$. We set $0/0 = 0$, so we do not consider the set $\{\varphi = \dot\varphi = 0\}$. Observe that the set

$$M = \{t\colon \varphi(t) = 0,\ \dot\varphi(t) \neq 0\}$$

is at most countable since it contains no limit points. Indeed, let a be a limit point of M. Then there exists a sequence $\{t_n\} \subset M$ such that $t_n \to a$ as $n \to \infty$. We can choose a monotone sequence $\{t_n\}$ with this property. Say, let $t_n < t_{n+1}$. We have $\varphi(t_n) = 0$ and $\varphi \in C^1(\mathbb{R}^1)$ since the function $\dot\varphi$ is absolutely continuous. Hence there exist points t_n' for which $t_n \leqslant t_n' \leqslant t_{n-1}$

and $\dot\varphi(t'_n) = 0$. Therefore, $\varphi(a) = \dot\varphi(a) = 0$, which shows that $a \notin M$. So we shall consider the integral over the set $\{t\colon \varphi(t) \neq 0\} = A \cup B$, where

$$A = \{t\colon \varphi(t) \neq 0, |\dot\varphi(t)| \leq |\varphi(t)|\}, \quad B = \{t\colon \varphi(t) \neq 0, |\dot\varphi(t)| > |\varphi(t)|\}.$$

The integral over A does not exceed the variation of the function φ:

$$\int_A \left|\frac{\dot\varphi}{\varphi}\right|^{2-\varepsilon} |\varphi|\, dt = \int_A \left|\frac{\dot\varphi}{\varphi}\right|^{1-\varepsilon} |\dot\varphi|\, dt \leq \int_A |\dot\varphi|\, dt \leq \int_{\mathbb{R}^1} |\dot\varphi|\, dt.$$

It is somewhat more involved to estimate the integral over B. The set B is open and consists of at most countably many disjoint open intervals (possibly unbounded): $B = \bigcup_{i=1}^\infty (a_i, b_i)$. In each of these intervals the functions φ and $\dot\varphi$ preserve their signs by the continuity. Let us consider an interval $(a, b) = (a_i, b_i)$ on which one has $0 < \varphi < \dot\varphi$. In this case

$$\int_a^b \left|\frac{\dot\varphi}{\varphi}\right|^{2-\varepsilon} |\varphi|\, dt = \int_a^b \left|\frac{\dot\varphi}{\varphi}\right|^{2-\varepsilon} \varphi\, dt = \int_a^b \left|\frac{\dot\varphi}{\varphi}\right|^{1-\varepsilon} \dot\varphi\, dt$$

$$= \varphi \left|\frac{\dot\varphi}{\varphi}\right|^{1-\varepsilon} \bigg|_a^b + \int_a^b \varphi(1-\varepsilon) \left|\frac{\dot\varphi}{\varphi}\right|^{-\varepsilon} \frac{\dot\varphi^2 - \varphi\ddot\varphi}{\varphi^2}\, dt.$$

Therefore,

$$\varepsilon \int_a^b \left|\frac{\dot\varphi}{\varphi}\right|^{2-\varepsilon} \varphi\, dt = \varphi^\varepsilon \dot\varphi^{1-\varepsilon}\bigg|_a^b - (1-\varepsilon)\int_a^b \left|\frac{\dot\varphi}{\varphi}\right|^{-\varepsilon} \ddot\varphi\, dt.$$

Let us estimate both terms on the right-hand side. We have

$$\left|\int_a^b \left|\frac{\dot\varphi}{\varphi}\right|^{-\varepsilon} \ddot\varphi\right| \leq \int_a^b |\ddot\varphi|\, dt.$$

At the boundary of the interval (a, b) the following cases are possible:
 1) $\varphi = \dot\varphi \neq 0$, then $\varphi^\varepsilon \dot\varphi^{1-\varepsilon} = \varphi$;
 2) $\varphi = 0$, then $\varphi^\varepsilon \dot\varphi^{1-\varepsilon} = 0 = \varphi$.
In both cases

$$\varphi^\varepsilon \dot\varphi^{1-\varepsilon}\bigg|_a^b = \varphi\bigg|_a^b = \int_a^b \dot\varphi\, dt.$$

Therefore,

$$\int_a^b \left|\frac{\dot\varphi}{\varphi}\right|^{2-\varepsilon} |\varphi|\, dt \leq \frac{1}{\varepsilon}\int_a^b |\dot\varphi|\, dt + \frac{1-\varepsilon}{\varepsilon}\int_a^b |\ddot\varphi|\, dt.$$

Certainly, our intervals may be unbounded. In this case it is easily seen that $\varphi(\infty) = \dot\varphi(\infty) = 0$. This follows by the fact that if φ and $\dot\varphi$ are integrable, then $\varphi \to 0$ as $t \to \pm\infty$. Finally, we obtain

$$\int_B \left|\frac{\dot\varphi}{\varphi}\right|^{2-\varepsilon} |\varphi|\, dt \leq \frac{1}{\varepsilon}\int_B |\dot\varphi|\, dt + \frac{1-\varepsilon}{\varepsilon}\int_B |\ddot\varphi|\, dt \leq \frac{1}{\varepsilon}\int_{\mathbb{R}} |\dot\varphi|\, dt + \frac{1-\varepsilon}{\varepsilon}\int_{\mathbb{R}} |\ddot\varphi|\, dt,$$

$$\int_{\mathbb{R}} \left|\frac{\dot\varphi}{\varphi}\right|^{2-\varepsilon} |\varphi|\, dt \leq \frac{1+\varepsilon}{\varepsilon}\int_{\mathbb{R}} |\dot\varphi|\, dt + \frac{1-\varepsilon}{\varepsilon}\int_{\mathbb{R}} |\ddot\varphi|\, dt.$$

6.2. Integrability of logarithmic derivatives

(ii) Let $\varphi \geq 0$, $\varphi, \dot\varphi, \ddot\varphi, \varphi^{(3)} \in L^1(\mathbb{R})$, $0 < \varepsilon < 1$. Since $\varphi \geq 0$, the function $\dot\varphi$ vanishes at the zero points of φ. So we only consider the set $\{t\colon \varphi(t) > 0\}$. It is the union of the following three sets:

$$\{\varphi > 0\} = \{\varphi > 0, \dot\varphi > 0\} \cup \{\varphi > 0, \dot\varphi = 0\} \cup \{\varphi > 0, \dot\varphi < 0\}.$$

The integral of $|\dot\varphi/\varphi|^{3-\varepsilon}$ over the second set vanishes. Let us consider the integral over the set $\{\varphi > 0, \dot\varphi > 0\}$ (the remaining case is similar). This set is open and is a union of at most countably many disjoint intervals, at the endpoints of which one has either $\varphi = \dot\varphi = 0$ or $\dot\varphi = 0, \varphi \neq 0$. Let (a,b) be some of such intervals. Integrating by parts we obtain the relationship

$$\varepsilon \int_a^b \left|\frac{\dot\varphi}{\varphi}\right|^{3-\varepsilon} \varphi\, dt = -\varphi^{\varepsilon-1}\dot\varphi^{2-\varepsilon}\Big|_a^b + (2-\varepsilon) \int_a^b \left|\frac{\dot\varphi}{\varphi}\right|^{1-\varepsilon} \ddot\varphi\, dt.$$

We observe that the boundary term vanishes. Indeed, if at some of the endpoints we have $\dot\varphi = 0, \varphi \neq 0$, then $\varphi^{\varepsilon-1}\dot\varphi^{2-\varepsilon} = 0$. Suppose that $\dot\varphi = \varphi = 0$ at $t = b$ (where $b = \infty$ is possible). Let $\dot\varphi(t_n) \leq \varphi(t_n)$ for some sequence $t_n \to b$, where $t < b$. Then we obtain the relationship

$$\lim_{n\to\infty} \dot\varphi(t_n)^{2-\varepsilon}/\varphi(t_n)^{1-\varepsilon} = \lim_{n\to\infty} \bigl(\dot\varphi(t_n)/\varphi(t_n)\bigr)^{1-\varepsilon} \dot\varphi(t_n) = 0.$$

Now suppose that $\dot\varphi(t) \geq \varphi(t)$ as $t \to b-$. If $\ddot\varphi(b) \neq 0$, then by l'Hôpital's rule we find

$$\lim_{t\to b-} \frac{\dot\varphi(t)^2}{\varphi(t)} = \lim_{t\to b-} \frac{2\dot\varphi(t)\ddot\varphi(t)}{\dot\varphi(t)} = 2\ddot\varphi(b), \quad \lim_{t\to b-} \frac{\varphi(t)}{\dot\varphi(t)} = \lim_{t\to b-} \frac{\dot\varphi(t)}{\ddot\varphi(t)} = 0,$$

whence

$$\lim_{t\to b-} \frac{\dot\varphi(t)^{2-\varepsilon}}{\varphi(t)^{1-\varepsilon}} = \lim_{t\to b-} \frac{\dot\varphi(t)^2}{\varphi(t)} \left(\frac{\varphi(t)}{\dot\varphi(t)}\right)^\varepsilon = 0.$$

The same is true in the case $\ddot\varphi(b) = 0$, which is seen from the estimate $\varphi(t)/\dot\varphi(t) \leq 1$ and the first equality from the two inequalities obtained above by the l'Hôpital's rule. Let us proceed to estimating the integral of $|\dot\varphi/\varphi|^{1-\varepsilon}|\ddot\varphi|$. To this end we partition the real line into several sets (possibly overlapping) and show that for each of them one has the desired estimate.

1. Let us consider the set $\{|\ddot\varphi| \leq |\dot\varphi|\}$. The interior of this set contains all zeros of $\ddot\varphi$ which are not zeros of $\dot\varphi$. As shown above, we have

$$\int_{\{|\ddot\varphi|\leq|\dot\varphi|\}} \left|\frac{\dot\varphi}{\varphi}\right|^{1-\varepsilon} |\ddot\varphi|\, dt \leq \int_{\{|\ddot\varphi|\leq|\dot\varphi|\}} \left|\frac{\dot\varphi}{\varphi}\right|^{1-\varepsilon} |\dot\varphi|\, dt$$

$$\leq \frac{1+\varepsilon}{\varepsilon} \int_{\mathbb{R}} |\dot\varphi|\, dt + \frac{1-\varepsilon}{\varepsilon} \int_{\mathbb{R}} |\ddot\varphi|\, dt.$$

Note that here we do not use the nonnegativity of φ.

2. Let us consider the set $\{|\dot\varphi| \leq \varphi\}$. The interior of this set contains all zeros of $\dot\varphi$ which are not zeros of φ (including all points for which we have $\dot\varphi(t) = \ddot\varphi(t) = 0$). We have

$$\int_{\{|\dot\varphi|\leq\varphi\}} \left|\frac{\dot\varphi}{\varphi}\right|^{1-\varepsilon} |\ddot\varphi|\, dt \leq \int_{\{|\dot\varphi|\leq\varphi\}} |\ddot\varphi|\, dt \leq \int_{\mathbb{R}} |\ddot\varphi|\, dt.$$

The considered sets contain all zeros of $\ddot{\varphi}$.

3. For the set $\{\dot\varphi^2 \leqslant 4\varphi|\ddot\varphi|\}$ we obtain

$$\int_{\{\dot\varphi^2\leqslant 4\varphi|\ddot\varphi|\}} \left|\frac{\dot\varphi}{\varphi}\right|^{1-\varepsilon} |\ddot\varphi|\, dt \leqslant 4^{1-\varepsilon} \int_{\{\dot\varphi^2\leqslant 4\varphi|\ddot\varphi|\}} \left|\frac{\ddot\varphi}{\dot\varphi}\right|^{1-\varepsilon} |\ddot\varphi|\, dt$$

$$\leqslant 4^{1-\varepsilon}\left(\frac{1+\varepsilon}{\varepsilon}\int_{\mathbb{R}}|\ddot\varphi|\, dt + \frac{1-\varepsilon}{\varepsilon}\int_{\mathbb{R}}|\varphi^{(3)}|\, dt\right).$$

Here we apply Case 1 to $\dot\varphi$ in place of φ, which is possible since in that case no positivity of φ was used. The interior of the set $\{\dot\varphi^2 \leqslant 4\varphi|\ddot\varphi|\}$ contains all points for which $\varphi = \dot\varphi = 0$, $\ddot\varphi \neq 0$. Indeed, let $\varphi(c) = \dot\varphi(c) = 0$, $\ddot\varphi(c) \neq 0$. Then by l'Hôpital's rule, we have

$$\lim_{t\to c}\frac{\dot\varphi(t)^2}{\varphi(t)\ddot\varphi(t)} = \lim_{t\to c}\frac{\dot\varphi(t)^2}{\varphi(t)}\cdot\lim_{t\to c}\frac{1}{\ddot\varphi(t)} = \frac{2\ddot\varphi(c)}{\ddot\varphi(c)} = 2 < 4.$$

4. It remains to consider the following open sets:

$$\{0 < \varphi < |\dot\varphi| < |\ddot\varphi|,\ 4\varphi|\ddot\varphi| < \dot\varphi^2\} = \bigcup_{i=1}^{\infty}(a_i, b_i), \quad \text{where } (a_i, b_i) \text{ are disjoint}.$$

In each of these intervals $\dot\varphi$ and $\ddot\varphi$ preserve their signs. At the endpoints of the intervals we can have $0 < \delta \leqslant \varphi \leqslant |\dot\varphi| \leqslant |\ddot\varphi|$ or $\varphi = \dot\varphi = \ddot\varphi = 0$. Suppose that $\varphi(a) = \dot\varphi(a) = \ddot\varphi(a) = 0$ (possibly, $a = -\infty$), where (a, b) is one of the intervals (a_i, b_i). Since $\varphi(a) = 0$ and $\varphi > 0$ in (a, b), we have $\dot\varphi(t) > 0$ on (a, b). Similarly, $\ddot\varphi > 0$ on (a, b). Let us consider the function $\dot\varphi^2/\varphi$ on (a, b). We have

$$\lim_{t\to a+}\frac{\dot\varphi(t)^2}{\varphi(t)} = 2\ddot\varphi(a) = 0, \quad \frac{\dot\varphi(t)^2}{\varphi(t)} > 0,\quad t \in (a, b).$$

However, $d(\dot\varphi^2/\varphi)/dt = \dot\varphi(2\ddot\varphi\varphi - \dot\varphi^2)/\varphi^2 < 0$, which leads to a contradiction. Therefore, it is only possible that on the whole interval $[a, b]$ we have the inequalities $\delta \leqslant \varphi \leqslant |\dot\varphi| \leqslant |\ddot\varphi|$. We consider the case where $0 < \varphi < \dot\varphi < \ddot\varphi$ on (a, b). Other cases are considered similarly. Integrating by parts we obtain the following relation:

$$\varepsilon\int_a^b\left(\frac{\dot\varphi}{\varphi}\right)^{1-\varepsilon}\ddot\varphi\, dt = \left(\frac{\varphi}{\dot\varphi}\right)^{\varepsilon}\ddot\varphi\bigg|_a^b + \varepsilon\int_a^b\left(\frac{\varphi}{\dot\varphi}\right)^{\varepsilon}\frac{\ddot\varphi^2}{\dot\varphi}\, dt - \int_a^b\left(\frac{\varphi}{\dot\varphi}\right)^{\varepsilon}\varphi^{(3)}\, dt.$$

Let us estimate every term on the right. Taking into account the inequalities fulfilled on (a, b) and the facts established above we find

$$\left|\int_a^b\left|\frac{\varphi}{\dot\varphi}\right|^{\varepsilon}\varphi^{(3)}\, dt\right| \leqslant \int_a^b|\varphi^{(3)}|\, dt,$$

$$\left|\int_a^b\left|\frac{\varphi}{\dot\varphi}\right|^{\varepsilon}\frac{\ddot\varphi^2}{\dot\varphi}\, dt\right| \leqslant \frac{1}{4^{\varepsilon}}\int_a^b\left|\frac{\dot\varphi}{\ddot\varphi}\right|^{\varepsilon}\frac{\ddot\varphi^2}{\dot\varphi}\, dt \leqslant \frac{1}{4^{\varepsilon}}\left(\frac{1+\varepsilon}{\varepsilon}\int_a^b\ddot\varphi\, dt + \frac{1-\varepsilon}{\varepsilon}\int_a^b|\varphi^{(3)}|\, dt\right).$$

It remains to estimate the boundary term. Let us consider the open set

$$\{0 < \varphi < \dot\varphi < \ddot\varphi,\ 4\varphi\ddot\varphi < \dot\varphi^2\}.$$

6.2. Integrability of logarithmic derivatives

On the boundary of this set the equalities $\dot\varphi = \ddot\varphi$ or $4\varphi\ddot\varphi = \dot\varphi^2$ are possible, but the equality $\varphi = \dot\varphi$ combined with $\ddot\varphi \ne \dot\varphi$ cannot occur since then we would have $4\ddot\varphi \le \dot\varphi$. Again we investigate all possible cases.

1. Suppose that $\dot\varphi = \ddot\varphi$ at both endpoints. Then the integration by parts formula yields

$$\left(\frac{\varphi}{\dot\varphi}\right)^\varepsilon \ddot\varphi \Big|_a^b = \left(\frac{\varphi}{\dot\varphi}\right)^\varepsilon \dot\varphi \Big|_a^b = \varphi^\varepsilon \dot\varphi^{1-\varepsilon}\Big|_a^b = \varepsilon \int_a^b \frac{\dot\varphi^{2-\varepsilon}}{\varphi^{1-\varepsilon}}\,dt + (1-\varepsilon)\int_a^b \left(\frac{\varphi}{\dot\varphi}\right)^\varepsilon \ddot\varphi\,dt$$

$$\le (1+\varepsilon)\int_a^b \dot\varphi\,dt + 2(1-\varepsilon)\int_a^b \ddot\varphi\,dt.$$

2. If we have $4\varphi\ddot\varphi = \dot\varphi^2$ at both endpoints, then

$$\left(\frac{\varphi}{\dot\varphi}\right)^\varepsilon \ddot\varphi\Big|_a^b = \frac{1}{4^\varepsilon}\left(\frac{\dot\varphi}{\ddot\varphi}\right)^\varepsilon \ddot\varphi\Big|_a^b = \frac{1}{4^\varepsilon}\dot\varphi^\varepsilon \ddot\varphi^{1-\varepsilon}\Big|_a^b.$$

Due to our hypotheses this case is similar to the previous one with $\dot\varphi$ in place of φ.

3. Let us consider the case where different equalities take place at different endpoints of the interval. Suppose, say, that $\ddot\varphi(a) = 4^{-1}\dot\varphi^2(a)/\varphi(a)$, $\ddot\varphi(b) = \dot\varphi(b)$. Then

$$\left(\frac{\varphi}{\dot\varphi}\right)^\varepsilon \ddot\varphi\Big|_a^b = (\ddot\varphi(b) - \ddot\varphi(a))\left(\frac{\varphi(a)}{\dot\varphi(a)}\right)^\varepsilon + \ddot\varphi(b)\left[\left(\frac{\varphi(b)}{\dot\varphi(b)}\right)^\varepsilon - \left(\frac{\varphi(a)}{\dot\varphi(a)}\right)^\varepsilon\right]$$

$$= (\ddot\varphi(b) - \ddot\varphi(a))\left(\frac{\varphi(a)}{\dot\varphi(a)}\right)^\varepsilon + \dot\varphi(b)\left(\frac{\varphi(b)}{\dot\varphi(b)}\right)^\varepsilon - \dot\varphi(a)\left(\frac{\varphi(a)}{\dot\varphi(a)}\right)^\varepsilon$$

$$- (\dot\varphi(b) - \dot\varphi(a))\left(\frac{\varphi(a)}{\dot\varphi(a)}\right)^\varepsilon.$$

Therefore, taking into account the estimate $0 \le \varphi/\dot\varphi \le 1$ and the facts established above we obtain

$$\left|\ddot\varphi\left(\frac{\varphi}{\dot\varphi}\right)^\varepsilon \Big|_a^b\right| \le (1+\varepsilon)\int_a^b |\dot\varphi|\,dt + (3-2\varepsilon)\int_a^b |\ddot\varphi|\,dt + \int_a^b |\varphi^{(3)}|\,dt.$$

Thus, we have considered all possible cases and obtained the desired estimates. $\square$

What happens if μ is four times differentiable? Can one obtain similar estimates with $4 - \varepsilon$? The answer is negative! Even analyticity is not enough, as the following simple example (suggested by A.Yu. Popov) shows: $\varphi(t) = t^2 \exp(-t^2)$, the function $|\varphi'(t)|^p \varphi(t)^{1-p}$ is not integrable at the origin for all $p \ge 3$. It would be interesting to find conditions ensuring the inclusion $\beta_h^\mu \in L^p(\mu)$ for all $p > 1$.

The next result gives a positive answer to the question raised in Bogachev, Smolyanov [**233**, Problem 6].

6.2.6. Corollary. *Let μ be a stable measure continuous along h. Then it is infinitely differentiable along h and $\beta_h^\mu \in L^{3-\varepsilon}(\mu)$ for every $\varepsilon \in (0,3)$.*

6.2.7. Corollary. *Let μ be a probability measure on $\mathbb{R}^n$ with a density ϱ from the class $W^{1,1}(\mathbb{R}^n)$. Then, for all $\varepsilon \in (0,1]$, we have*

$$\int_{\mathbb{R}^n} \left|\frac{\nabla \varrho(x)}{\varrho(x)}\right|^{2-\varepsilon} \varrho(x)\, dx \leqslant 2n^{1-\varepsilon/2}\varepsilon^{-1} \sum_{i=1}^n [\|d_{e_i}\mu\| + \|d_{e_i}^2\mu\|],$$

provided the right-hand side is finite, and for all $\varepsilon \in [1,3)$ we have

$$\int_{\mathbb{R}^n} \left|\frac{\nabla \varrho(x)}{\varrho(x)}\right|^{3-\varepsilon} \varrho(x)\, dx \leqslant C(\varepsilon) n^{\frac{1-\varepsilon}{3-\varepsilon}} \sum_{i=1}^n [\|d_{e_i}\mu\| + \|d_{e_i}^2\mu\| + \|d_{e_i}^3\mu\|],$$

provided the right-hand side is finite.

PROOF. For $p = 2 - \varepsilon$ we use the inequality $|v|^p \leqslant n^{p/2}\sum_{i=1}^n |v_i|^p$, $v \in \mathbb{R}^n$, and estimate the integral of $|\partial_{x_i}\varrho|^{2-\varepsilon}$ with the aid of (6.2.1). The second case is similar. □

If $\beta_h^\mu \in L^p(\mu)$, then $|d_h\mu|(B) \leqslant |\mu|(B)^{1/q}\|\beta_h^\mu\|_{L^p(\mu)}$ for all $B \in \mathcal{B}(X)$, where $q = p/(p-1)$. Hence we obtain the following estimate.

6.2.8. Corollary. *One has $|d_h\mu|(B) \leqslant |\mu|(B)^{1/3}\bigl(6\|d_h\mu\| + \|d_h^2\mu\|\bigr)$ for all $B \in \mathcal{B}(X)$ if $d_h^2\mu$ exists.*

6.3. Differentiability of logarithmic derivatives

In this section, the differentiability of measures will always mean Fomin's differentiability.

6.3.1. Proposition. *Suppose that two Radon measures μ and λ on a locally convex space X are n-fold differentiable along all vectors from a finite dimensional linear space L and $\mu \ll \lambda$. Then, there exists a version F of the Radon–Nikodym density of the measure μ with respect to λ which is λ-a.e. n-fold differentiable along all vectors in L. If λ admits strictly positive conditional measures on the straight line parallel to vectors in L, then F can be chosen with absolutely continuous derivatives $\partial_h^{n-1} F$ for such vectors h.*

PROOF. For notational simplicity we consider the case $n = 1$, where one has $L = \mathbb{R}^1 h$. Let us choose a hyperplane Y complementing L and differentiable conditional measures μ^y and λ^y with absolutely continuous densities f^y, g^y on the straight lines $y + \mathbb{R}^1 h$. It is readily seen that $\mu \ll \nu \otimes \ell$, $\lambda \ll \nu \otimes \ell$, where ν is the projection of $|\lambda|$ to Y and ℓ is the standard Lebesgue measure on $\mathbb{R}^1$. Thus, $\mu = f \cdot (\nu \otimes \ell)$, $\lambda = g \cdot (\nu \otimes \ell)$, $d_h\mu = \varphi \cdot (\nu \otimes \ell)$, $d_h\lambda = \psi \cdot (\nu \otimes \ell)$. Then ν-a.e. we have

$$f(y + th) = f^y(t) = \int_{-\infty}^t \varphi(y + sh)\, ds, \quad t \in \mathbb{R}^1,$$

$$g(y + th) = g^y(t) = \int_{-\infty}^t \psi(y + sh)\, ds, \quad t \in \mathbb{R}^1.$$

It remains to observe that $F = f/g$. □

6.3.2. Corollary. *If a measure μ is twice differentiable along h, then β_h^μ has a version for which μ-a.e. there exists the partial derivative $\partial_h \beta_h^\mu$.*

Certainly, it is not always true that F has an everywhere differentiable version even if μ and λ are smooth: consider the standard Gaussian measure μ on the real line and take $\lambda = t \cdot \mu$ or $\lambda = t^2 \cdot \mu$. In particular, the logarithmic derivative need not be everywhere continuous.

6.3.3. Proposition. *Suppose that a measure $\mu \geq 0$ is differentiable along h and for μ-a.e. x the function $t \mapsto \beta_h^\mu(x+th)$ is locally absolutely continuous or everywhere differentiable and one has $\partial_h \beta_h^\mu \in L^1(\mu)$. Then μ is twice Fomin differentiable along h, $\beta_h^\mu \in L^2(\mu)$, $d_h^2 \mu = [\partial_h \beta_h^\mu + (\beta_h^\mu)^2] \cdot \mu$, and one has*

$$\int_X |\beta_h^\mu(x)|^2 \, \mu(dx) = -\int_X \partial_h \beta_h^\mu(x) \, \mu(dx). \tag{6.3.1}$$

In addition, $\|d_h^2 \mu\| \leq 2 \|\partial_h \beta_h^\mu\|_{L^1(\mu)}$.

PROOF. With the aid of conditional measures μ^x on the straight lines $x + \mathbb{R}^1 h$ the assertion reduces to the one-dimensional case, for the justification of which we consider an interval $[a, b]$ such that $\varrho(a) = \varrho(b) = 0$ and $\varrho(t) > 0$ for all $t \in (a, b)$, where ϱ is an absolutely continuous density of μ. Then $\beta_1^\mu = \beta = \varrho'/\varrho$. We need an even weaker condition: the function ϱ'/ϱ is absolutely continuous on all inner intervals in $[a, b]$. This gives the absolute continuity of ϱ' on such intervals and the equality $\beta'\varrho = \varrho'' - (\varrho')^2/\varrho$. We show that

$$-\int_a^b \beta'(t)\varrho(t) \, dt = \int_a^b \frac{|\varrho'(t)|^2}{\varrho(t)} \, dt. \tag{6.3.2}$$

For all $[x, y] \subset (a, b)$ we have

$$-\int_x^y \beta'(t)\varrho(t) \, dt = \varrho'(x) - \varrho'(y) + \int_x^y \frac{|\varrho'(t)|^2}{\varrho(t)} \, dt. \tag{6.3.3}$$

If there exist points $a_n \to a+$ and $b_n \to b-$ with $\varrho'(a_n) \to 0$ and $\varrho'(b_n) \to 0$, then (6.3.2) follows in the limit from (6.3.3) for $x = a_n$, $y = b_n$. Suppose that there are no such sequences $\{a_n\}$ or $\{b_n\}$. For example, suppose that $\varrho'(t) \geq \varepsilon > 0$ in (a, c), where $c \in (a, b)$ (due to the continuity of ϱ' on (a, b) and the equality $\varrho(a) = 0$ the case $\varrho'(t) < 0$ is impossible). Then by the integrability of $\beta'\varrho$ and equality (6.3.3) the integral of $|\varrho'|^2/\varrho$ over (a, c) is finite. Hence $\varrho'(t) \leq M$ on (a, c). This gives the relationship $\varepsilon(t-a) \leq \varrho(t) \leq M(t-a)$ on $[a, c]$, which shows that the function $\varrho'/\varrho \geq \varepsilon/\varrho$ cannot be integrable on (a, c). Similarly we consider the point b, which proves equality (6.3.2). This equality is true for $a = -\infty$ and $b = +\infty$ if ϱ has no zeros at all, since by the integrability of ϱ' there exist points $y_n \to +\infty$, $x_n \to -\infty$ with $\varrho'(y_n) \to 0$ and $\varrho'(x_n) \to 0$. The same reasoning applies in the case where only one of the points a or b is finite. On the set $\{\varrho = 0\}$ the functions $\beta'\varrho$ and $|\varrho'|^2/\varrho$ vanish. Therefore, we arrive at (6.3.1). It can be seen from what we have said above that the function ϱ' is locally absolutely continuous (certainly, if we assume the local absolute continuity of β at once, then this is trivial). On the set $\{\varrho = 0\}$ the functions ϱ' and ϱ'' vanish almost everywhere. Hence on account of the obtained estimates the integral of $|\varrho''|$ is majorized by the integral of $|\beta'|\varrho + |\varrho'|^2/\varrho$, which yields existence of $d_1^2\mu$ and the required estimate for $\|d_1^2\mu\|$. □

6.3.4. Remark. As one can see from the proof, in place of absolute continuity or differentiability of β_h^μ along h we can require that for μ-a.e. x the functions $t \mapsto \beta_h^\mu(x+th)$ are absolutely continuous on the intervals inside the sets $\{t\colon \varrho^x > 0\}$, where ϱ^x is a continuous version of a density of the conditional measure μ^x on the straight line $x + \mathrm{I\!R}^1 h$.

6.3.5. Corollary. *Let a measure $\mu \geqslant 0$ be three times differentiable along h. Then β_h^μ admits a version β that satisfies the equality*

$$d_h^2\mu = \beta \cdot d_h\mu + (\partial_h\beta) \cdot \mu = \beta^2 \cdot \mu + (\partial_h\beta) \cdot \mu$$

in the sense that $\beta \in L^1(d_h\mu)$, $\partial_h\beta \in L^1(\mu)$, $\beta \in L^2(\mu)$.

PROOF. We have $\beta_h^\mu \in L^2(\mu)$, so the desired equality follows from the proof of the previous proposition due to the remark above. □

However, if μ is only twice differentiable or is signed, then this corollary may fail to be true (Exercise 6.8.5).

6.4. Quasi-invariance and differentiability

There exists an interesting connection between the differentiability and quasi-invariance of measures discovered by Skorohod [**1045**], [**1046**] (see also Gihman, Skorohod [**477**, Ch. VII, §2]). The following reinforcement of Skorohod's result was obtained in Bogachev, Mayer-Wolf [**211**].

6.4.1. Proposition. (i) *Let μ be a nonzero Borel measure on $\mathrm{I\!R}^n$ Fomin differentiable along all vectors. Suppose that $\exp(\varepsilon|\beta_{e_i}^\mu|) \in L^1(\mu)$ for some $\varepsilon > 0$, where $e_1, \ldots, e_n$ is a basis in $\mathrm{I\!R}^n$. Then the measure $|\mu|$ admits a continuous density p that is strictly positive. In particular, the measure μ itself has a continuous density which differs from zero everywhere. In addition, for every $h \in \mathrm{I\!R}^n$, the measure μ_h has a Radon–Nikodym density r_h with respect to μ expressed by the equality*

$$r_h(x) = \exp\left(\int_0^1 \beta_h^\mu(x - sh)\,ds\right). \qquad (6.4.1)$$

More generally, the measure μ admits a density that is continuous and differs from zero in any ball in which the functions $\exp(\varepsilon|\beta_{e_i}^\mu|)$ are integrable with respect to μ with some $\varepsilon > 0$.

(ii) *If μ is a Radon measure on a locally convex space X which is Fomin differentiable along a vector h and $\exp(\varepsilon|\beta_h^\mu|) \in L^1(\mu)$ for some $\varepsilon > 0$, then*

6.4. Quasi-invariance and differentiability

μ is quasi-invariant along h and the Radon–Nikodym density r_h of μ_h with respect to μ satisfies the equality

$$r_h(x) = \exp\left(\int_0^1 \beta_h^\mu(x - sh)\,ds\right). \qquad (6.4.2)$$

PROOF. Assertion (ii) follows from (i) and Theorem 3.5.1. Let us prove assertion (i). By the differentiability along basis vectors the measure μ has a density $\varrho \in W^{1,1}(\mathbb{R}^n)$ and $\beta_i := \beta_{e_i}^\mu = \partial_{x_i}\varrho/\varrho$. It is clear that our condition (global or local) is fulfilled also for the measures μ^+ and μ^- according to (3.3.4). Hence our claim reduces to the case $\mu \geqslant 0$. By the inequality $x^k e^{-x} \leqslant k^k e^{-k}$, for every ball B on which our condition is fulfilled we obtain

$$\varepsilon^k \|\beta_i\|_{L^k(\mu_B)}^k \leqslant k^k e^{-k} \int_B \exp\bigl(\varepsilon|\beta_i(x)|\bigr)\,\mu(dx) \leqslant k^k e^{-k} M,$$

where μ_B is the restriction of μ to B. Set $f_k = \varrho^{1/k}$. Then one can easily verify that $f_k \in W^{1,k}(B)$ and the Sobolev derivatives $\partial_{x_i} f_k$ of f_k satisfy the equalities $\partial_{x_i} f_k = k^{-1} p^{1/k - 1} \partial_{x_i}\varrho$. This follows from the fact that the right-hand side of this equality belongs to $L^k(B)$, so it suffices to consider the approximations $(\varrho + \delta)^{1/k}$ with $\delta \to 0$. Therefore, we have

$$\|\partial_{x_i} f_k\|_{L^k(B)}^k = k^{-k} \|\partial_{x_i}\varrho/\varrho\|_{L^k(\mu_B)}^k = k^{-k} \|\beta_i\|_{L^k(\mu_B)}^k \leqslant \varepsilon^{-k} e^{-k} M.$$

Thus, $\|\partial_{x_i} f_k\|_{L^k(B)} \leqslant \varepsilon^{-1} e^{-1} M^{1/k} \leqslant C$ with C independent of k. Let us recall that for every ball B the embedding of the Sobolev space $W^{1,q}(B)$ into $C(B)$ is compact whenever $q > n$ (see Chapter 2). Hence the functions f_k have continuous modifications on B; consequently, ϱ has the same property. In addition, the sequence f_k contains a subsequence that converges uniformly on B. Since the limit is continuous and can assume only the values 1 and 0, it is identically 1. Hence ϱ is strictly positive on B. □

6.4.2. Corollary. *Suppose that a nonzero Radon measure $\mu \geqslant 0$ on a locally convex space X is differentiable along vectors $h_1, \ldots, h_n \in X$ and for some $\varepsilon > 0$ one has $\exp(\varepsilon|\beta_{h_i}^\mu|) \in L^1(\mu)$. If $T\colon X \to \mathbb{R}^n$ is a continuous linear operator and $Th_1, \ldots, Th_n$ are linearly independent, then the measure $\mu \circ T^{-1}$ has a continuous density in $W^{1,1}(\mathbb{R}^n)$ that has no zeros.*

PROOF. We can apply Proposition 3.3.13 and the fact that the exponential integrability is inherited by conditional expectations. □

6.4.3. Remark. The measure μ is h-quasi-invariant and equality (6.4.2) is true under the following weaker condition: μ has a logarithmic derivative β_h^μ such that the function $s \mapsto \beta_h^\mu(x - sh)$ is locally integrable for μ-a.e. x. This condition follows from the exponential integrability of β_h^μ with respect to μ since, according to what has been proven above, on the straight lines $x + \mathbb{R}^1 h$ the measure μ has conditional measures μ^x with positive continuous densities f^x (with respect to the natural Lebesgue measures on these lines) satisfying the equalities $\beta_h^\mu(x + th) = \partial_t f^x(x + th)/f^x(x + th)$.

PROOF. It suffices to consider the one-dimensional case and $h = 1$. In this case μ has an absolutely continuous density f. Then we obtain
$$f(x+t) = \exp\left(\int_0^t f'(x+s)/f(x+s)\,ds\right) f(x).$$
Indeed, we choose some x with $f(x) \neq 0$. We may assume that $f(x) > 0$. Then, by the integrability of f'/f, the above formula is valid for small t since $(\ln f)' = f'/f$ near x. Both sides of this equality are continuous and the right-hand side is strictly positive by the integrability of f'/f. If there is t_0 such that $f(x + t_0) = 0$, then the equality remains true also at $t = t_0$ since it is true for all $t \in (0, t_0)$. Obviously, this is impossible. This shows that f has no zeros. □

Nevertheless, it should be noted that the (less general) Skorohod condition with the exponential integrability of β_h^μ can be easier for verification. In Albeverio, Röckner [53] formula (6.4.2) was proved under the following conditions: $\beta_h^\mu \in L^2(\mu)$ and μ admits conditional measures on the straight lines $y + \mathbb{R}^1 h$ with densities f^y such that the functions $1/f^y$ are locally Lebesgue integrable; if $\beta_h^\mu \in L^{1+\varepsilon}(\mu)$, then it suffices to have the local integrability of $(1/f^y)^{1/\varepsilon}$. This assertion follows at once from the previous remark since the integral of the function $|f' f^{-1}| = |f' f^{-\varepsilon/(1+\varepsilon)}| |f|^{-1/(1+\varepsilon)}$ is estimated by Hölder's inequality with $p = 1 + \varepsilon$ and $p' = (1 + \varepsilon)\varepsilon^{-1}$. In particular, the aforementioned is true if the function β_h^μ is continuous along h (which was established in Bell [105] under some additional conditions). There are interesting generalizations of formula (6.4.2) for shifts along vectors fields. We discuss this question in §10.4.

Finally, note that the condition of the exponential integrability of $|\beta_h^\mu|$ cannot be substantially weakened if we seek a sufficient condition in terms of the integrability of $\Psi(|\beta_h^\mu|)$ with a convex function Ψ (see Exercise 6.8.3). However, under the global exponential integrability of $|\beta_h^\mu|$ we have the following lower bounds.

6.4.4. Theorem. *Let $\mu = \varrho\,dx$ be a probability measure, $\varrho \in W^{1,1}_{\mathrm{loc}}(\mathbb{R}^d)$.*
(i) *Suppose that $\exp(\kappa|\nabla\varrho/\varrho|) \in L^1(\mu)$ with some $\kappa > 0$. Then, the continuous version of ϱ satisfies the inequality*
$$\varrho(x) \geqslant \exp\bigl(-c_1\exp(c_2|x|)\bigr), \quad \text{where } c_1, c_2 > 0.$$
(ii) *Suppose that $\exp(\kappa|\nabla\varrho/\varrho|^r) \in L^1(\mu)$ with some $\kappa > 0$. Then*
$$\varrho(x) \geqslant c_1 \exp\bigl(-c_2|x|^{r/(r-1)}\bigr), \quad \text{where } c_1, c_2 > 0.$$

This result is a special case of a more general theorem on solutions to elliptic equations in Bogachev, Röckner, Shaposhnikov [228] (for related results, see Bogachev, Röckner, Shaposhnikov [227], Shaposhnikov [994]); assertion (ii) was proved also in Malliavin, Nualart [773]. The one-dimensional case had been earlier considered in Nualart [861], where the formulation and proof in case (ii) need some corrections.

6.5. Convex functions

Here we briefly discuss the integration by parts formula for convex functions on infinite dimensional spaces with measures. This question is also not trivial in the finite dimensional case (see A.D. Alexandroff's Theorem 2.3.2). Let X be a locally convex space. A function F on X will be called convex along a linear subspace E (or E-convex) if the function $h \mapsto F(x+h)$ is convex on E for every $x \in X$. Suppose that the space E is equipped with some norm $|\cdot|_E$. Let $\partial_E F(x)$ denote the subdifferential of F at x along E, i.e., the set of all functionals $l \in E^*$ such that $f(x+e) - f(x) \geqslant l(e)$ for all e from some neighborhood of zero in E. We shall say that an E-convex function F has the second order derivative along E at a point x if there exist $l_x \in \partial_E F(x)$ and a bounded linear operator $T_x \colon E \to E^*$ such that for every $h \in E$ we have

$$F(x+th) - F(x) = t l_x(h) + t^2 T_x(h)(h)/2 + o(t^2), \quad t \to 0.$$

If the function F is Gâteaux differentiable along E at the point x, then $\partial F(x)$ consists of a single element $D_E F(x) \in E^*$. Set $D^2 F(x) := T_x$.

Let μ be a nonnegative Radon measure on X (which is assumed throughout this section), let L be the linear span of two vectors h and k along which the measure μ is continuous, and let F be a μ-measurable function on X such that the restriction of F to $x+L$ is convex for μ-a.e. $x \in X$. By Alexandroff's theorem, this restriction has the second order derivative at almost every point $x \in L$, where L is equipped with Lebesgue measure induced by an isomorphism between L and $\mathbb{R}^d$, $d = \dim L$ (here $d=1$ or $d=2$). This yields that the limit

$$\partial_h \partial_k F(x) := \lim_{t \to 0} 2^{-1} t^{-2} \big[F(x+th+tk) + F(x-th-tk) - F(x+th)$$
$$- F(x-th) - F(x+tk) - F(x-tk) + 2F(x) \big]$$

exists almost everywhere on L. Hence $\partial_h \partial_k F$ in the indicated sense exists μ-a.e. For the same reason

$$\partial_h^2 F(x) := \lim_{t \to 0} t^{-2} [F(x+th) + F(x-th) - 2F(x)]$$

exists μ-a.e. In addition, $\partial_h^2 F(x) \geqslant 0$. Indeed, by convexity

$$F(x+th) + F(x-th) - 2F(x) \geqslant 0.$$

We observe that $\partial_h^2 F$ may differ from the second derivative of F in the sense of distributions. For example, if f is the classical Cantor function on $[0,1]$ and F is its indefinite integral, then $F'' = 0$ a.e., but F'' in the sense of distributions is not zero. If F is a finite convex function on $\mathbb{R}^n$, then there exist locally bounded Borel measures F_{ij} on $\mathbb{R}^n$ such that the generalized derivative $\partial_{x_i} \partial_{x_j} F$ is the measure F_{ij}, and the matrix with the entries $F_{ij}(B)$ is nonnegative for every set $B \in \mathcal{B}(\mathbb{R}^n)$ (see §2.3). We take the decomposition

$$F_{ij} = F_{ij}^{\mathrm{ac}}\, dx + F_{ij}^{\mathrm{sing}}, \quad F_{ij}^{\mathrm{ac}} \in L^1_{\mathrm{loc}}(\mathbb{R}^n),$$

into absolutely continuous and singular parts. Then almost everywhere one has

$$F_{ij}^{\mathrm{ac}}(x) = \lim_{t\to 0} t^{-2}\bigl[F(x+te_i+te_j) + F(x-te_i-te_j)$$
$$- F(x+te_i) - F(x+te_j) - F(x-te_i) - F(x-te_j) + 2F(x)\bigr],$$

where $e_1, \ldots, e_n$ is the standard basis in $\mathbb{R}^n$. In order to obtain a similar decomposition in the infinite dimensional case we need an analogue of the second order generalized derivative. If in $\mathbb{R}^n$ we take a probability measure μ with a nice density ϱ, then the second order derivative of F is expressed via ϱ. Namely, for the measure $F \cdot \mu$ we consider its generalized second order derivatives $\partial_{x_i}^2 (F \cdot \mu)$, which are locally finite measures. If F is twice differentiable in the usual sense, then one can reconstruct $\partial_{e_i}^2 F$ from the expression

$$(\partial_{e_i}^2 F) \cdot \mu = \partial_{x_i}^2 (F \cdot \mu) - 2\partial_{x_i} F \cdot \partial_{x_i} \mu - F \cdot \partial_{x_i}^2 \mu,$$

where the right-hand side exists as a locally bounded measure for every convex function F. In the general case, the right-hand side is not absolutely continuous with respect to μ, and $\partial_{e_i}^2 F$ must be reconstructed from its absolutely continuous part.

Although pointwise second order derivatives do not completely characterize a function, it is interesting to know when the function $\partial_h^2 F$ is integrable. The following two lemmas give sufficient conditions for the integrability of $\partial_h F$ and $\partial_h^2 F$ along with a somewhat stronger version of the aforementioned finite dimensional differentiability.

6.5.1. Lemma. *Suppose that the measure μ is continuous along a vector h and a function $F \in L^p(\mu)$ is convex on μ-almost all straight lines $x + \mathbb{R}^1 h$. Then $\partial_h F$ exists μ-a.e. If the measure μ is quasi-invariant along h and the Radon–Nikodym densities ϱ_h and ϱ_{-h} of the measures μ_h and μ_{-h} with respect to μ belong to $L^s(\mu)$, where $s = p(p-r)^{-1}$, $r \in [1, p)$, then $\partial_h F$ belongs to $L^r(\mu)$. For the inclusion $\partial_h F \in L^1(\mu)$ it suffices to have either the inclusions $F \ln |F|$, $\varrho_h \exp \varrho_h$, $\varrho_{-h} \exp \varrho_{-h} \in L^1(\mu)$ or*

$$|F| \exp \sqrt{|\ln|F||/\varepsilon},\ \varrho_h \exp(\varepsilon |\ln \varrho_h|^2),\ \varrho_{-h} \exp(\varepsilon |\ln \varrho_{-h}|^2) \in L^1(\mu)$$

with some $\varepsilon > 0$.

If μ is a centered Gaussian measure and $h \in H(\mu)$, then $\partial_h F \in L^{p-\varepsilon}(\mu)$ for all $\varepsilon > 0$. If $F \exp \sqrt{|c \ln |F||} \in L^1(\mu)$, where $c > 2|h|_H^2$, then one has $\partial_h F \in L^1(\mu)$.

PROOF. In the one-dimensional case the function F is locally Lipschitzian, and by the convexity of F almost everywhere one has the estimate

$$|F'(x)| \leqslant |F(x-1)| + |F(x)| + |F(x+1)|.$$

Since the conditional measures on the straight lines parallel to h are absolutely continuous, the partial derivative $\partial_h F(x)$ exists μ-a.e. and

$$|\partial_h F(x)| \leqslant |F(x-h)| + |F(x)| + |F(x+h)|$$

for μ-a.e. x. Let us estimate the L^r-norm of the function on the right-hand side by using Hölder's inequality and the fact that the integral of $|F(x+h)|^r$ equals the integral of $|F(x)|^r \varrho_h(x)$. If $F \ln |F| \in L^1(\mu)$, then we use the estimate $|F|\varrho_h \leqslant |F \ln |F|| + \varrho_h \exp \varrho_h$ and a similar estimate with ϱ_{-h}. The second sufficient condition of integrability of $\partial_h F$, mentioned in the lemma, is considered similarly. In the Gaussian case $\ln \varrho_h = \widehat{h} - |h|_H^2/2$, so $\varrho_h \exp(c|\ln \varrho_h|^2)$ belongs to $L^1(\mu)$ if one has $c < (2|h|_H^2)^{-1}$. □

6.5.2. Lemma. (i) Let $\{h_n\} \subset C(\mu)$ and let L be the linear span of the sequence $\{h_n\}$. Suppose that a measurable function F is convex along L. Then μ-a.e. F has the second order derivative in the aforementioned sense along every finite dimensional subspace in L.

(ii) Let $F \in L^p(\mu)$ for some $p > 1$, $h \in X$, let F be convex on μ-almost all straight lines $x + \mathbb{R}^1 h$, and let μ be quasi-invariant along h such that the Radon–Nikodym density ϱ_{th} of μ_{th} with respect to μ satisfies the condition

$$\left|t^{-2}[\varrho_{th}(x) + \varrho_{-th}(x) - 2]\right| \leqslant G(x),$$

where $G \in L^{p'}(\mu)$, $p' = p/(p-1)$. Then the limit

$$\partial_h^2 F(x) = \lim_{t \to 0} \frac{F(x + th) + F(x - th) - 2F(x)}{t^2},$$

which exists μ-a.e., defines a nonnegative function from $L^1(\mu)$. In particular, this is true if μ is a centered Gaussian measure with the Cameron–Martin space H and $h \in H$.

PROOF. (i) In the proof of Alexandroff's theorem in Evans, Gariepy [**400**, Ch. 6] it is verified that the second order derivative of a convex function on $\mathbb{R}^n$ exists at a point x provided that x is a Lebesgue point for the first order derivative of F and for the absolutely continuous part of the second order derivative and, in addition, the singular component ζ of the second order derivative satisfies the condition $\lim_{r \to 0} |\zeta|(B(x,r))r^{-n} = 0$, where $B(x,r)$ is the closed ball of radius r centered at x. By using the absolute continuity of the conditional measures, we obtain all three conditions a.e. on every finite dimensional subspace which is a translate of the linear span of $h_1, \ldots, h_n$. (ii) The fact that $\partial_h^2 F(x) \geqslant 0$ as soon as $\partial_h^2 F(x)$ exists follows by convexity. In order to show that the function $\partial_h^2 F$ is integrable, it suffices (due to Fatou's theorem) to obtain an estimate from above for the integrals of the nonnegative functions

$$g_n(x) := n^2[F(x + n^{-1}h) + F(x - n^{-1}h) - 2F(x)].$$

We have

$$\int_X g_n(x)\,\mu(dx) = \int_X F(y)n^2[\varrho_h(y - n^{-1}h) + \varrho_h(y + n^{-1}h) - 2]\,\mu(dy)$$

$$\leqslant \int_X |F(y)|G(y)\,\mu(dy).$$

In the Gaussian case $\varrho_{th}(x) = \exp(t\widehat{h}(x) - t^2|h|_H^2/2)$, where $\widehat{h}$ is the measurable linear functional generated by h. Now it suffices to observe that $\exp|\widehat{h}| \in L^s(\mu)$ for all $s < \infty$. □

6.5.3. Remark. It is clear from the proof of Lemma 6.5.1 that for every $n \geqslant 1$, $p > 1$ and $r < p$ there exists a number $C(n, p, r)$ such that, for every convex function F from $L^p(\mu)$ with respect to the standard Gaussian measure μ on $\mathbb{R}^n$, one has the inequality

$$\left(\int_{\mathbb{R}^n} |\nabla F(x)|^r \, \mu(dx)\right)^{1/r} \leqslant C(n, p, q)\left(\int_{\mathbb{R}^n} |F(x)|^p \, \mu(dx)\right)^{1/p}.$$

Similar inequalities are valid for many other measures.

We shall now see that if a function F is convex along h and integrable to a suitable power with respect to a sufficiently smooth measure μ, then the measure $F \cdot \mu$ is twice differentiable along h. This fact enables us to introduce the second order generalized derivative of F along h.

6.5.4. Theorem. (i) *Suppose that the measure μ is twice Skorohod differentiable along a vector $h \in X$ and that a function F is convex on μ-almost all straight lines $x + \mathbb{R}^1 h$. Suppose also that the function F is integrable with respect to the measures μ and $d_h^2 \mu$ and that the function $\partial_h F$ is integrable with respect to the measure $d_h \mu$. Then the measure $F \cdot \mu$ is twice Skorohod differentiable along h. Moreover, one has the inequality*

$$\|d_h^2(F \cdot \mu)\| \leqslant 2\|F\|_{L^1(d_h^2 \mu)} + 2\|\partial_h F\|_{L^1(d_h \mu)}. \qquad (6.5.1)$$

(ii) *If, in addition, $F \geqslant 0$ and $F^p, F|\beta_h^\mu|^p \in L^1(\mu)$ for some $p > 1$, then*

$$\int_X |\partial_h F|^r \, d\mu < \infty \qquad (6.5.2)$$

for some $r > 1$. Finally, if $F \in L^\alpha(\mu)$ for all $\alpha \in [1, \infty)$ and $\beta_h^\mu \in L^2(\mu)$, then $\partial_h F \in L^r(\mu)$ for every $r < 2$.

PROOF. Let $X = \mathbb{R}^1$ and let the support of μ belong to $[a, b]$. The function F is Lipschitzian on $[a, b]$. The measure μ has an absolutely continuous density ϱ such that ϱ' has bounded variation. Hence the measure $F \cdot \mu$ is differentiable and satisfies the equality $d_1(F \cdot \mu) = F' \cdot \mu + F \cdot d_1 \mu$. Suppose additionally that F and ϱ are smooth. Then, certainly, the measure $d_1(F \cdot \mu)$ is Skorohod differentiable, but we need an estimate of the variation of its derivative. By convexity $F'' \geqslant 0$. Therefore,

$$0 \leqslant \int F''(x) \varrho(x) \, dx = \int F(x) \varrho''(x) \, dx \leqslant \int |F(x)| \, |d_1^2 \mu|(dx).$$

Since $d_1^2(F \cdot \mu) = F'' \cdot \mu + 2F' \cdot d_1 \mu + F \cdot d_1^2 \mu$, we arrive at the estimate

$$\|d_1^2(F \cdot \mu)\| \leqslant \|F'' \cdot \mu\| + 2\|F' \cdot d_1 \mu\| + \|F \cdot d_1^2 \mu\| \leqslant 2\|F \cdot d_1^2 \mu\| + 2\|F' \cdot d_1 \mu\|.$$

Therefore, estimate (6.5.1) is established in the present special case. Now, assuming that ϱ is only absolutely continuous with ϱ' of bounded variation,

6.5. Convex functions

but still assuming that F is smooth and μ has bounded support, we can find a sequence of smooth probability densities ϱ_j with support in a fixed interval such that
$$\lim_{j\to\infty} \int \Big[|\varrho_j(x) - \varrho(x)| + |\varrho_j'(x) - \varrho'(x)|\Big]\, dx = 0$$
and $\sup_j \|d_1^2 \mu_j\| < \infty$. According to (6.5.1) we have
$$\limsup_{j\to\infty} \|d_1^2(F \cdot \mu_j)\| \leqslant 2\|F \cdot d_1^2 \mu\| + 2\|F' \cdot d_1 \mu\|.$$
Hence $d_1^2(F \cdot \mu)$ exists and one has
$$\|d_1^2(F \cdot \mu)\| \leqslant \limsup_{j\to\infty} \|d_1^2(F \cdot \mu_j)\| \leqslant 2\|F \cdot d_1^2 \mu\| + 2\|F' \cdot d_1 \mu\|.$$

The next step is to weaken the assumption of smoothness of F, still assuming that μ has bounded support in $[a,b]$. To this end it suffices to observe that there exists a sequence of smooth convex functions F_j which converges uniformly to F on $[a,b]$ such that the functions F_j' converge to F' in $L^1[a,b]$. As above, it is verified that (6.5.1) is true. Now we drop the assumption of boundedness of support of μ. Let $\zeta_j \in C_0^\infty(\mathbb{R}^1)$, $0 \leqslant \zeta_j \leqslant 1$, $\zeta_j(x) = 1$, if $|x| \leqslant j$, $\zeta_j(x) = 0$, if $|x| \geqslant j+1$, $\sup_j \sup_x \big[|\zeta_j'(x)| + |\zeta_j''(x)|\big] < \infty$. Let $\mu_j := \zeta_j \cdot \mu$. Then the measures $F \cdot \mu_j$ converge to $F \cdot \mu$ in variation. In addition, $d_1 \mu_j = \zeta_j' \cdot \mu + \zeta_j \cdot d_1 \mu$, $d_1^2 \mu_j = \zeta_j'' \cdot \mu + 2\zeta_j' \cdot d_1 \mu + \zeta_j \cdot d_1^2 \mu$. It is readily seen from this expression that $F' \cdot d_1 \mu_j \to F' \cdot d_1 \mu$ and $F \cdot d_1^2 \mu_j \to F \cdot d_1^2 \mu$ in variation. Thus, we arrive at (6.5.1) in the general one-dimensional case.

Let us write X as the topological sum $X = \mathbb{R}^1 h + Y$ for some closed hyperplane Y in X. Let ν be the projection of μ on Y. There exist conditional measures μ^y on the straight lines $y + \mathbb{R}^1 h$, $y \in Y$, which are twice Skorohod differentiable along h. For ν-a.e. $y \in Y$ the restriction of the function F to $y + \mathbb{R}^1 h$ is integrable with respect to $d_h^2 \mu^y$ and the restriction of $\partial_h F$ is integrable with respect to $d_h \mu^y$. By using the one-dimensional case we obtain
$$\|d_h^2(F \cdot \mu)\| \leqslant 2\|F \cdot d_h^2 \mu\| + 2\|\partial_h F \cdot d_h \mu\|,$$
which is (6.5.1). Now suppose that $F \geqslant 0$ and $|F|^p, |\beta_h^\mu|^p F \in L^1(\mu)$ for some $p > 1$. According to Krugova's inequality (see §6.3), we have the estimate
$$\left(\int_X |\beta_h^{F\cdot\mu}|^{2-\varepsilon} F\, \mu(dx)\right)^{1/(2-\varepsilon)} \leqslant 2\varepsilon^{-1}\big[\|d_h(F \cdot \mu)\| + \|d_h^2(F \cdot \mu)\|\big]$$
for every $\varepsilon \in (0,1)$. Let us observe that $\beta_h^{F\cdot\mu} = \beta_h^\mu + \partial_h F/F$ a.e. with respect to the measure $F \cdot \mu$. Since $\beta_h^\mu \in L^p(F \cdot \mu)$ with some $p \in (1,2)$, one has $\partial_h F/F \in L^p(F \cdot \mu)$. Let
$$r \in (1,p), \quad s = pr - 1, \quad \alpha = (p-1)s^{-1} = r(p-1)p^{-1}, \quad t = s(s-1)^{-1}.$$
Then $\alpha t = r(p-1)(p-r)^{-1}$. Since $p(p-r)(p-1)^{-1} \to p$ as $r \to 1$, there exists $r > 1$ such that $r \leqslant p(p-r)(p-1)^{-1}$. For this r we have $\alpha t \leqslant p$,

hence $|F|^{\alpha t} \in L^1(\mu)$. By Hölder's inequality, due to the equalities $rs = p$ and $\alpha s = p - 1$, we obtain

$$\int_X |\partial_h F|^r \, \mu(dx) \leqslant \left(\int_X \frac{|\partial_h F|^{rs}}{F^{\alpha s}} \mu(dx) \right)^{1/s} \left(\int_X F^{\alpha t} \, \mu(dx) \right)^{1/t} < \infty.$$

This proves the first assertion in (ii). The second assertion in (ii) is seen from the previous reasoning. $\square$

6.5.5. Corollary. *Let a probability measure μ be quasi-invariant and twice Skorohod differentiable along h and let a function $F \in \mathcal{L}^p(\mu) \cap \mathcal{L}^1(d_h^2\mu)$ be convex on μ-almost all straight lines $x + \mathbb{R}^1 h$. If one has $\beta_h^\mu \in L^{p'}(\mu)$, $\varrho_h, \varrho_{-h} \in L^s(\mu)$, where $s = p(p-r)^{-1}$, then the measure $F \cdot \mu$ is twice Skorohod differentiable along h and inequalities (6.5.1) and (6.5.2) hold.*

If μ is twice Fomin differentiable along h, then the inclusion $F \in L^1(d_h^2\mu)$ follows from the inclusions $F \in L^p(\mu)$ and $\beta_{h,h}^\mu \in L^{p'}(\mu)$, where $\beta_{h,h}^\mu$ is the density of $d_h^2\mu$ with respect to μ, in particular, it is fulfilled in the Gaussian case.

6.5.6. Corollary. *Let μ be a centered Gaussian measure, $h \in H(\mu)$, and let a μ-measurable function F be convex on μ-almost all straight lines $x + \mathbb{R}^1 h$.*

(i) If $F(1+|\widehat{h}|^2), \partial_h F \widehat{h} \in L^1(\mu)$, then the measure $F \cdot \mu$ is twice Skorohod differentiable along h and inequality (6.5.1) is fulfilled.

(ii) If $F \in L^p(\mu)$ for some $p > 1$, then (i) is true and inequality (6.5.2) holds.

PROOF. We have $\beta_h^\mu = -\widehat{h}$ and $\beta_{h,h}^\mu = |\widehat{h}|^2 - |h|_H^2$. $\square$

As already mentioned, it may occur that the pointwise second order derivative vanishes almost everywhere, but the corresponding part of the measure $d_h^2(F \cdot \mu)$ is nontrivial. Let us explain how $\partial_h^2 F$ can be interpreted in the generalized sense. Set

$$F_{hh} := d_h^2(F \cdot \mu) - 2\partial_h F \cdot d_h\mu - F \cdot d_h^2\mu$$

if each of the three measures on the right exists separately.

6.5.7. Theorem. *Suppose that the hypotheses of Theorem 6.5.4(i) are fulfilled and $\partial_h F \in L^1(\mu)$ (which is the case under the assumptions of Lemma 6.5.1 or Theorem 6.5.4(ii)). Then the measure F_{hh} is finite and nonnegative. In addition,*

$$F_{hh} = d_h(\partial_h F \cdot \mu) - \partial_h F \cdot d_h\mu. \qquad (6.5.3)$$

If the measure μ is twice Fomin differentiable along h, then $\partial_h^2 F$ is the Radon–Nikodym density of the absolutely continuous part of the measure

$$F_{hh} = d_h^2(F \cdot \mu) - 2\partial_h F \cdot d_h\mu - F \cdot d_h^2\mu$$

with respect to μ.

6.5. Convex functions

PROOF. The integral of a smooth cylindrical function $\zeta \geqslant 0$ against the measure F_{hh} can be written as follows by the integration by parts formula for $d_h^2(F \cdot \mu)$ and $d_h^2 \mu$:

$$\int_X \zeta \, d_h^2(F \cdot \mu)(dx) - 2\int_X \zeta \partial_h F \, d_h \mu(dx) - \int_X \zeta F \, d_h^2 \mu(dx)$$
$$= -\int_X \partial_h \zeta \partial_h F \, \mu(dx) - \int_X \zeta \partial_h F \, d_h \mu(dx) = -\int_X \partial_h F \, d_h(\zeta \cdot \mu)(dx).$$

The right-hand side is nonnegative. This follows from the existence of differentiable conditional measures for μ on the straight lines $x + \mathbb{R}^1 h$ and the observation (Exercise 6.8.6) that if ϱ is an absolutely continuous probability density on $\mathbb{R}^1$ and G is a convex function such that $G'\varrho'$ and $G'\varrho$ are integrable, then

$$\int_{-\infty}^{+\infty} G'(t)\varrho'(t)\, dt \leqslant 0.$$

Equality (6.5.3) is seen from the chain of equalities above. Let us prove the last assertion in the one-dimensional case. Then $\mu = \varrho \, dx$, where the functions ϱ and ϱ' are absolutely continuous on bounded intervals. Calculating the first and second derivatives in the sense of distributions we obtain the equalities $(F\varrho)' = F'\varrho + F\varrho'$ and

$$(F\varrho)'' = (F'\varrho)' + F'\varrho' + F\varrho'' = \varrho F'' + 2F'\varrho' + F\varrho'' = \varrho F''_{\text{ac}} + \varrho F''_{\text{sing}} + 2F'\varrho' + F\varrho''.$$

It remains to recall that F''_{ac} coincides almost everywhere with the limit

$$\lim_{t \to 0}[F(x+t) + F(x-t) - 2F(x)]t^{-2}.$$

In the general case we take a closed hyperplane Y complementing $\mathbb{R}^1 h$ to X. One can find a Radon measure $\sigma \geqslant 0$ on Y, measures μ^y and twice Skorohod differentiable measures $(F \cdot \mu)^y$ on $\mathbb{R}^1 h$ for which

$$\mu = \int_Y \mu^y \sigma(dy), \quad d_h^2(F \cdot \mu) = \int_Y \partial_h^2(F \cdot \mu)^y \sigma(dy).$$

By Proposition 1.3.3 we see that the absolutely continuous part $[d_h^2(F \cdot \mu)]_{\text{ac}}$ of the measure $d_h^2(F \cdot \mu)$ with respect to μ admits the representation

$$[d_h^2(F \cdot \mu)]_{\text{ac}} = \int_Y \partial_h^2 F(y + \cdot h) \cdot \mu^y \sigma(dy) = \partial_h^2 F \cdot \mu,$$

which completes the proof. □

6.5.8. Proposition. *Suppose that a measure μ is twice Fomin differentiable along h, $\Psi \geqslant 0$ is a function in $\mathcal{L}^1(\mu)$ such that the sets $\{\Psi \leqslant c\}$ have compact closures, and a sequence $\{F_n\}$ is bounded in $L^p(\mu)$ for some $p > 1$. Suppose also that the functions F_n are convex and twice differentiable along h on almost all straight lines parallel to h, there exist $\partial_h \Psi$, $\partial_h^2 \Psi$, and the functions*

$$F_n[|\partial_h^2 \Psi| + |\partial_h \Psi \beta_h^\mu| + |\Psi \beta_{h,h}^\mu|], \; \partial_h F_n[|\partial_h \Psi| + |\Psi \beta_h^\mu|], \; \partial_h^2 F_n \Psi \qquad (6.5.4)$$

belong to $L^1(\mu)$. If $d_h^2(\Psi \cdot \mu)$ has a density $g \in L^{p'}(\mu)$ with respect to μ, then the sequence of measures $\partial_h^2 F_n \cdot \mu$ is uniformly tight.

PROOF. Under our assumptions one has

$$\int_X \partial_h^2 F_n \, \Psi \, \mu(dx) = \int_X F_n \, d_h^2(\Psi \cdot \mu)(dx) = \int_X F_n g \, \mu(dx),$$

which is estimated by $\sup_n \|F_n\|_{L^p(\mu)} \|g\|_{L^{p'}(\mu)}$. Since $\partial_h^2 F_n \geqslant 0$ a.e., the integrals on the left are uniformly bounded, hence by Chebyshev's inequality the sequence of measures $(\partial_h^2 F_n) \cdot \mu$ is uniformly tight. □

If $\Psi \geqslant 1$, then the sequence of measures $(\partial_h^2 F_n) \cdot \mu$ is uniformly bounded. It is also clear that (6.5.4) is fulfilled if the measure μ is quasi-invariant along h and $\beta_h^\mu, \beta_{h,h}^\mu, \varrho_h, \varrho_{-h}, \Psi, \partial_h \Psi, \partial_h^2 \Psi \in L^s(\mu)$ for sufficiently large s.

6.5.9. Remark. One can also consider similarly mixed second order partial derivatives F_{hk}, setting

$$F_{hk} := \frac{F_{h+k\,h+k} - F_{hh} - F_{kk}}{2}$$

for suitable h and k. This notation is consistent from the point of view of generalized derivatives.

If μ is a centered Gaussian Radon measure and H is its Cameron–Martin space, then H-convex μ-measurable functions admit the following description obtained in Feyel, Üstünel [**429**], where functions with values in the set $\mathbb{R}^1 \cup \{+\infty\}$ were considered.

6.5.10. Theorem. *A function $f \in \mathcal{L}^0(\mu)$ has a modification that is convex along the space H precisely when for every $h \in H$ the inequality*

$$f(x) \leqslant f(x+h)/2 + f(x-h)/2$$

is fulfilled μ-a.e. (where a set of measure zero may depend on h). Moreover, there exists a Borel modification with values in $(-\infty, +\infty]$ that is convex on X.

A useful concept of a c-convex function, $c \in \mathbb{R}^1$, is introduced in [**429**]. This is a μ-measurable function f with the following property: for μ-a.e. x the function $f(x+h) + c|h|_H^2/2$ is convex on H.

A number of corollaries of the results obtained here will be given in §8.5 for Gaussian measures.

6.6. Derivatives along vector fields

So far we have considered integration by parts for constant directions. However, already elementary calculus provides useful integration by parts formulae for vector fields. For example, if a measure μ on $\mathbb{R}^n$ is given by a smooth density ϱ, then, for every continuously differentiable vector field

6.6. Derivatives along vector fields

$v = (v^1, \ldots, v^n)\colon \mathbb{R}^n \to \mathbb{R}^n$ and every function $\varphi \in C_0^\infty(\mathbb{R}^n)$, one has the equality

$$\int_{\mathbb{R}^n} (\nabla\varphi, v)\, \mu(dx) = -\int_{\mathbb{R}^n} \varphi(\operatorname{div} v)\varrho\, dx - \int_{\mathbb{R}^n} \varphi(v, \nabla\varrho)\, dx$$

$$= -\int_{\mathbb{R}^n} \varphi[\operatorname{div} v + (v, \nabla\varrho/\varrho)]\, \mu(dx),$$

where $\operatorname{div} v := \sum_{i=1}^n \partial_{x_i} v^i$, $\nabla\varrho(x)/\varrho(x) := 0$ if $\varrho(x) = 0$. Infinite dimensional analogues of this formula play a very important role in the Malliavin calculus. At first sight, it could seem that we used essentially the existence of Lebesgue measure and density ϱ; indeed, the right-hand side involves $\nabla\varrho/\varrho$. However, we shall see in the next chapter that this object, the logarithmic gradient of μ, has a natural infinite dimensional analogue in spite of the fact that ϱ and $\nabla\varrho$ are not meaningful separately. Here we discuss only some immediate corollaries of the already obtained integration by parts formulae.

Let X be a locally convex space and let a vector field be given, i.e., a mapping $v\colon X \to X$. The derivative $\partial_v\varphi(x)$ of the function φ on X along v is defined by

$$\partial_v\varphi(x) := \lim_{t\to 0} \frac{\varphi(x + tv(x)) - \varphi(x)}{t}$$

if a finite limit exists. For example, if $\varphi(x) = \varphi_0(l_1(x), \ldots, l_n(x))$, where $l_i \in X^*$ and $\varphi_0 \in C^1(\mathbb{R}^n)$, then

$$\partial_v\varphi(x) = \sum_{i=1}^n l_i(v(x))\partial_{x_i}\varphi_0(l_1(x), \ldots, l_n(x)).$$

6.6.1. Definition. *Let μ be a Radon measure on X. We shall say that a Radon measure ν is the Skorohod derivative of the measure μ along the field v if, for all $\varphi \in \mathcal{F}C_b^\infty(X)$, we have $\partial_v\varphi \in \mathcal{L}^1(\nu)$ and*

$$\int_X \partial_v\varphi(x)\, \mu(dx) = -\int_X \varphi(x)\, \nu(dx). \tag{6.6.1}$$

Set $d_v\mu := \nu$. If $d_v\mu \ll \mu$, then we shall call μ *Fomin differentiable along v* and the density of the measure $d_v\mu$ with respect to μ will be called the *logarithmic derivative of μ along v* or the *divergence of the field v with respect to μ* and denoted by the symbols β_v^μ or $\delta_\mu v$. Let $\mathfrak{D}(v, \mu)$ be the class of all functions $\varphi \in \mathcal{L}^1(\nu)$ for which $\partial_v\varphi \in \mathcal{L}^1(\mu)$ and (6.6.1) holds.

6.6.2. Example. Suppose that a measure μ is Skorohod or Fomin differentiable along a vector h and a Borel function $\psi \in L^1(d_h\mu)$ has a version for which the functions $t \mapsto \psi(x + th)$ are absolutely continuous on all intervals and $\partial_h\psi \in \mathcal{L}^1(\mu)$. Then the measure μ is differentiable in the respective sense along the field $v(x) = \psi(x)h$ and $d_v\mu = \psi \cdot d_h\mu + \partial_h\psi \cdot \mu$. In the case of Fomin differentiability one has $\beta_v^\mu = \psi\beta_h^\mu + \partial_h\psi$.

PROOF. This is clear from the equalities
$$\int_X \partial_v \varphi \, \mu(dx) = \int_X \psi \partial_h \varphi \, \mu(dx) = \int_X \partial_h(\psi\varphi) \, \mu(dx) - \int_X \varphi \partial_h \psi \, \mu(dx)$$
$$= -\int_X \varphi\psi \, d_h\mu(dx) - \int_X \varphi \partial_h \psi \, \mu(dx),$$
which hold for $\varphi \in \mathcal{FC}_b^\infty(X)$ according to the results obtained in §6.1. □

Suppose we are given functions $u^1, \ldots, u^n, v^1, \ldots, v^n \in \mathcal{L}^1(\mu)$ and vectors $e_1, \ldots, e_n$ such that there exist $\beta_{e_i}^\mu$ and, for fixed i and j, the functions u^i, v^j, and $\beta_{e_j}^\mu$ have versions for which the functions $t \mapsto u^i(x + te_j)$, $t \mapsto v^i(x + te_j)$, $t \mapsto \beta_{e_j}^\mu(x + te_j)$ are absolutely continuous on all intervals. Let the functions $\partial_{e_i} u^j$ possess the same property. Set $\Phi_{ij} := \partial_{e_j} \beta_{e_i}^\mu$. We shall regard
$$D\beta(x) := \bigl(\Phi_{ij}(x)\bigr)_{i,j \leqslant n}, \quad Du(x) := \bigl(\partial_{e_j} u^i(x)\bigr)_{i,j \leqslant n}, \quad Dv(x) := \bigl(\partial_{e_j} v^i(x)\bigr)_{i,j \leqslant n}$$
as operators on $\mathbb{R}^n$ (where i is a line number). Let $\|\cdot\|_{\mathcal{L}(\mathbb{R}^n)}$ and $\|\cdot\|_{\mathcal{H}(\mathbb{R}^n)}$ denote the operator norm and the Hilbert–Schmidt norm respectively.

Let us consider vector fields $u := \sum_{i=1}^n u^i e_i$ and $v := \sum_{j=1}^n v^j e_j$. Set
$$|u(x)|^2 := \sum_{i=1}^n |u^i(x)|^2.$$

6.6.3. Theorem. *Suppose that*
$$\partial_{e_i} u^j \partial_{e_l} v^m, \; v^j \beta_{e_j}^\mu \partial_{e_i} u^i, \; u^i \beta_{e_i}^\mu \partial_{e_j} v^j, \; v^j \partial_{e_i} \partial_{e_j} u^i, \; u^i v^j \Phi_{ij} \in L^1(\mu).$$
Then one has
$$\int_X \beta_u^\mu \beta_v^\mu \, \mu(dx) = \int_X \Bigl(\mathrm{trace}\,[Du \cdot Dv] - \sum_{i,j \leqslant n} \Phi_{ij} u^i v^j\Bigr) \mu(dx)$$
$$= \int_X \sum_{i,j \leqslant n} \bigl[\partial_{e_j} u^i \partial_{e_i} v^j - \Phi_{ij} u^i v^j\bigr] \mu(dx). \tag{6.6.2}$$
In addition, if $\mu \geqslant 0$, then one has
$$\int_X |\beta_u^\mu(x)|^2 \, \mu(dx) \leqslant \int_X \Bigl[\|D\beta(x)\|_{\mathcal{L}(\mathbb{R}^n)} |u(x)|^2 + \|Du(x)\|_{\mathcal{H}(\mathbb{R}^n)}^2\Bigr] \mu(dx). \tag{6.6.3}$$

PROOF. For notational simplicity we shall denote the integral with respect to the measure μ by $\mathbb{E}$ and omit the summation sign when summing over repeated upper and lower indices. Then
$$u = u^i e_i, \; v = v^j e_j, \; \beta_u^\mu = \partial_{e_i} u^i + u^i \beta_{e_i}^\mu, \; \beta_v^\mu = \partial_{e_j} v^j + v^j \beta_{e_j}^\mu,$$
$$\mathbb{E}\beta_u^\mu \beta_v^\mu = \mathbb{E}\partial_{e_i} u^i \partial_{e_j} v^j + \mathbb{E}v^j \beta_{e_j}^\mu \partial_{e_i} u^i + \mathbb{E}u^i \beta_{e_i}^\mu \partial_{e_j} v^j + \mathbb{E}u^i v^j \beta_{e_i}^\mu \beta_{e_j}^\mu.$$
The last term equals
$$-\mathbb{E}v^j \beta_{e_i}^\mu \partial_{e_j} u^i - \mathbb{E}u^i \beta_{e_i}^\mu \partial_{e_j} v^j - \mathbb{E}u^i v^j \partial_{e_j} \beta_{e_i}^\mu$$

by the integration by parts formula. Similarly,
$$\mathbb{E} v^j \beta^\mu_{e_j} \partial_{e_i} u^i = -\mathbb{E} \partial_{e_j} \partial_{e_i} u^i - \mathbb{E} \partial_{e_i} u^i \partial_{e_j} v^j,$$
$$-\mathbb{E} v^j \beta^\mu_{e_i} \partial_{e_j} u^i = \mathbb{E} v^j \partial_{e_i} \partial_{e_j} u^i + \mathbb{E} \partial_{e_j} u^i \partial_{e_i} v^j.$$

Taking into account that $\partial_{e_j} \partial_{e_i} u^i = \partial_{e_i} \partial_{e_j} u^i$ a.e., we arrive at (6.6.2). Inequality (6.6.3) follows from the estimate $|\text{trace}\,(T^2)| \leq \|T\|^2_{\mathcal{H}(\mathbb{R}^n)}$ for every operator T on $\mathbb{R}^n$. □

Let γ be a centered Radon Gaussian measure on X and let $\{e_i\}$ be an orthonormal basis in its Cameron–Martin space $H := H(\gamma)$. Suppose that a function $f \in \mathcal{L}^2(\gamma)$ is such that the functions $t \mapsto f(x+te_i)$ are locally absolutely continuous and the function $\sum_{i=1}^\infty (\partial_{e_i} f)^2$ belong to $\mathcal{L}^1(\gamma)$. Set
$$D_H f(x) := \sum_{i=1}^\infty \partial_{e_i} f(x) e_i, \quad \text{trace}_H A := \sum_{i=1}^\infty (Ae_i, e_i)_H, \quad A \in \mathcal{L}_{(1)}(H).$$

For every vector field v of the form $v(x) = \sum_{i=1}^n v_i(x) e_i$, where
$$v_i(x) = \varphi_i(\widehat{e}_1(x), \ldots, \widehat{e}_n(x)), \quad \varphi_i \in C_b^\infty(\mathbb{R}^n),$$
by the formula proved above we obtain
$$\int_X \left(D_H f(x), v(x)\right)_H \gamma(dx) = -\int_X f(x) \delta v(x) \, \gamma(dx), \qquad (6.6.4)$$
where $\delta v(x) = \text{trace}_H D_H v(x) - \sum_{i=1}^n \widehat{e}_i(x)(e_i, v(x))_H$. For any two such vector fields v and u we have
$$\int_X \delta v(x) \delta u(x) \, \gamma(dx)$$
$$= \int_X (v(x), u(x))_H + \int_X \text{trace}_H (D_H v(x) D_H u(x)) \, \gamma(dx), \quad (6.6.5)$$
$$\int_X |\delta v(x)|^2 \, \gamma(dx) \leq \int_X |v(x)|^2_H \, \gamma(dx) + \int_X \|D_H v(x)\|^2_{\mathcal{H}(H)} \, \gamma(dx).$$

In Chapter 8 these formulae will be extended to all H-valued vector fields v belonging to the completion of the class of smooth vector fields considered above with respect to the Sobolev norm $\|\cdot\|_{2,1}$.

6.7. Local logarithmic derivatives

Since we consider only bounded measures, the logarithmic derivatives (if they exist) are integrable over the whole space. However, already the finite dimensional case shows that it is useful to extend this notion. For example, a probability measure on $\mathbb{R}^n$ with a smooth density ϱ may fail to be differentiable because of the global nonintegrability of $\nabla \varrho$, but locally the mapping $\nabla \varrho / \varrho$ is defined. Such a straightforward generalization is possible also in the infinite dimensional case, but here the following notion is more fruitful. Let X be a locally convex space and let $\mathcal{K}$ be some class of Borel functions on X.

6.7.1. Definition. Let μ be a Radon measure on X, $h \in X$, and $\partial_h \varphi \in L^1(\mu)$ for all $\varphi \in \mathcal{K}$. If $\beta_h^{\mu,\mathrm{loc}}$ is a μ-measurable function such that $\beta_h \varphi \in L^1(\mu)$ for all $\varphi \in \mathcal{K}$ and one has

$$\int_X \partial_h \varphi(x)\, \mu(dx) = -\int_X \varphi(x)\beta_h(x)\, \mu(dx),$$

then we shall call μ differentiable along h with respect to the class $\mathcal{K}$ and call $\beta_h^{\mu,\mathrm{loc}}$ a logarithmic derivative of μ along h with respect to $\mathcal{K}$.

Let us emphasize that $\beta_h^{\mu,\mathrm{loc}}$ may not belong to $L^1(\mu)$. Certainly, the real contents of this concept depends on $\mathcal{K}$. In particular, if $\mathcal{K}$ is very narrow, it may occur that a logarithmic derivative exists but is not unique.

6.7.2. Example. (i) Let γ be a Gaussian measure on a separable Hilbert space X and let $f \in \mathcal{L}^1(\gamma)$ be a continuously Fréchet differentiable function. Then the measure $\mu = f \cdot \gamma$ is differentiable along every vector $h \in H(\gamma)$ with respect to the class $\mathcal{K}$ of continuously differentiable functions with bounded support and $\beta_h^{\mu,\mathrm{loc}} = \beta_h^\gamma + \partial_h f/f$.

(ii) Let X_0 be a separable Hilbert space of full γ-measure compactly embedded into X and let $\mathcal{K}$ be the class of Lipschitzian functions on X_0 with bounded support in X_0 extended by zero outside X_0. If f is a Lipschitzian function on X_0 and $f \in L^1(\gamma)$, then the assertion in (i) remains valid. In addition, the function $\beta_h^{\mu,\mathrm{loc}}$ need not be integrable over all compact sets in X.

6.7.3. Theorem. Let $X = \mathbb{R}^n \times Y$, where Y is a locally convex space, let μ be a Radon measure on X with conditional measures μ^y on $\mathbb{R}^n \times \{y\}$, and let ν be the projection of $|\mu|$ on Y. Suppose that $\mathcal{K}$ is a class of bounded Borel functions such that

(i) for all $\psi \in \mathcal{K}$ and all $y \in Y$ the function $x \mapsto \psi(x,y)$ has a bounded derivative $\nabla_x \psi$,

(ii) $(x,y) \mapsto \psi(x+v,y) \in \mathcal{K}$ and $\varphi \circ \psi \in \mathcal{K}$ for all $\psi \in \mathcal{K}$, $v \in \mathbb{R}^n$, $\varphi \in C_0^\infty(\mathbb{R}^1)$, $\varphi(0) = 0$, moreover, $\psi_1 \psi_2 \in \mathcal{K}$ if $\psi_1, \psi_2 \in \mathcal{K}$,

(iii) the class $\mathcal{K}$ separates Radon measures on X.

Let $\beta \colon X \to \mathbb{R}^n$ be a μ-measurable mapping such that for all $\psi \in \mathcal{K}$ and $v \in \mathbb{R}^n$ we have $\psi|\beta| \in L^1(\mu)$ and

$$\int_X (\nabla_x \psi, v)\, \mu(dx) = -\int_X \psi(\beta, v)\, \mu(dx). \tag{6.7.1}$$

Then, for ν-a.e. y, the measure μ^y has a density f^y on $\mathbb{R}^n \times \{y\}$ such that

$$f^y \in W^{1,1}_{\mathrm{loc}}(\mathbb{R}^n) \quad \text{and} \quad \beta(x,y) = \nabla_x f^y(x)/f^y(x) \quad \mu^y\text{-a.e.} \tag{6.7.2}$$

PROOF. We can find $A_j \in \mathcal{B}(X)$ such that $\bigcup_{j=1}^\infty A_j$ has full measure and there exist functions $\varphi_j \in \mathcal{K}$ with $\varphi_j > 0$ on A_j. To this end we put

$$\mathcal{K}_0 := \{\psi \in \mathcal{K} \colon 0 \leqslant \psi \leqslant 1\}.$$

6.7. Local logarithmic derivatives

According to [**193**, Theorem 4.7.1], there exists a sequence $\varphi_j \in \mathcal{K}_0$ such that, for every function $\psi \in \mathcal{K}_0$, we have $\psi \leq \sup_j \varphi_j$ μ-a.e. The union of the sets $A_j = \{\varphi_j > 0\}$ has full measure. Indeed, if $\sup_j \varphi_j = 0$ on a set A of positive $|\mu|$-measure, then, for any $\varphi \in \mathcal{K}_0$, we have $\varphi = 0$ μ-a.e. on A. Hence the same is true for all $\varphi \in \mathcal{K}$, which is easily verified by considering compositions with functions from $C_0^\infty(\mathbb{R}^1)$ vanishing at the origin. However, by our assumption this is possible only if $\mu|_A$ is the zero measure. We may even assume that $\varphi_j = 1$ on A_j. For this purpose one should replace every function φ_j by a sequence of functions $\theta_k \circ \varphi_j$, where $\theta_k \in C_0^\infty(\mathbb{R}^1)$, $0 \leq \theta_k \leq 1$, $\theta_k(t) = 0$ if $t \leq 0$ or $t \geq k+1$ and $\theta_k(t) = 1$ if $k^{-1} \leq t \leq k$. Then the sets $\{\theta_k \circ \varphi_j = 1\}$ cover $\{\varphi_j > 0\}$. Let us consider the measures $\mu_j := \varphi_j \cdot \mu$. Let $\beta_j = \beta + \nabla_x \varphi_j / \varphi_j$. We observe that (6.7.1) is true for $\psi = \psi_1 \psi_2$ whenever $\psi_1, \psi_2 \in \mathcal{K}$. Hence

$$\int_X (\nabla_x \psi, v)\, \mu_j(dx) = -\int_X \psi(\beta_j, v)\, \mu_j(dx), \quad \psi \in \mathcal{K}, v \in \mathbb{R}^n.$$

In addition, $|\beta_j| \in L^1(\mu_j)$ since $|\beta|\varphi_j \in L^1(\mu)$. Let $v \in \mathbb{R}^n$, $t \in \mathbb{R}^1$. Then one has

$$\int_X \bigl[\psi(x+tv, y) - \psi(x,y)\bigr]\, \mu_j(dx)$$
$$= -\int_0^t \int_X \psi(x+sv, y)\bigl(\beta_j(x,y), v(x)\bigr)\, \mu_j(dx)\, ds \quad (6.7.3)$$

for all $\psi \in \mathcal{K}$, which is verified in the following way. Both sides of (6.7.3) are continuously differentiable in t and vanish at $t = 0$. It follows from (6.7.1) that their derivatives coincide since $(x,y) \mapsto \psi(x+tv, y) \in \mathcal{K}$ by our assumption. The left-hand side of (6.7.3) equals the integral of ψ against the measure $(\mu_j)_t - \mu_j$, where $(\mu_j)_t$ is the image of the measure μ_j under the shift $(x,y) \mapsto (x+tv, y)$. The right-hand side of (6.7.3) is the integral of ψ against the measure

$$\sigma_j^t := \int_0^t \bigl((\beta_j, v)\mu_j\bigr)_s ds.$$

By our assumption on $\mathcal{K}$ we have

$$(\mu_j)_t - \mu_j = \sigma_j^t. \qquad (6.7.4)$$

Therefore, (6.7.3) is fulfilled for all bounded Borel functions ψ. Set

$$\mu_j^y = \varphi_j(\cdot, y)\mu^y, \quad \mu_j(B) = \int_Y \mu_j^y(B)\, \nu(dy).$$

Equality (6.7.4) yields the absolute continuity of the measures μ_j^y for ν-a.e. y. Indeed, let p be a probability density on $\mathbb{R}^n$ with support in the unit ball U, $p_\varepsilon(t) = \varepsilon^{-n} p(t/\varepsilon)$, $\gamma_\varepsilon = p_\varepsilon\, dx$, $\varepsilon \in (0,1)$,

$$\pi_\varepsilon(B) = \int_Y \mu_j^y * \gamma_\varepsilon(B)\, \nu(dy).$$

Then, for every bounded Borel function g, we have

$$\int_X g(x,y)\, d\pi_\varepsilon = \int_Y \int_{\mathbb{R}^n \times \{y\}} \int_{\mathbb{R}^n} g(x+\varepsilon z, y) p(z)\, dz\, \mu_j^y(dx)\, \nu(dy)$$

$$= \int_{\mathbb{R}^n} \int_X g(x+\varepsilon z, y) p(z)\, d\mu_j\, dz. \quad (6.7.5)$$

Now (6.7.4) and (6.7.5) yield

$$\left| \int_X g\, d\mu_j - \int_X g\, d\pi_\varepsilon \right| = \left| \int_U \int_X g\, [d(\mu_j) - d(\mu_j)_{\varepsilon z}]\, p(z)\, dz \right|$$

$$= \left| \int_U \int_0^\varepsilon \int_X g(x+sz, y)(\beta_j, z)\, d\mu_j\, ds\, p(z)\, dz \right| \leqslant \varepsilon \|g\|_{L^\infty(\mu)} \|\beta_j\|_{L^1(\mu_j, \mathbb{R}^n)}$$

since $|(\beta_j, z)| \leqslant |\beta_j|$ on the support of p. Therefore,

$$\|\mu_j - \pi_\varepsilon\| \leqslant 2\varepsilon \|\beta_j\|_{L^1(\mu_j, \mathbb{R}^n)}.$$

It is clear that every measure π_ε has absolutely continuous conditional measures on $\mathbb{R}^n \times \{y\}$. Hence, for ν-a.e. y, the conditional measure μ_j^y admits a density $q_j^y(x)$ with respect to Lebesgue measure. Thus, on account of (6.7.4) we obtain that there exists a measurable set Y_0 of full ν-measure such that, for every $i = 1, \ldots, n$ and every rational t, for all $y \in Y_0$ we have

$$q_j^y(x+te_i) - q_j^y(x) = \int_0^t \left[(\beta_j, e_i)\, q_j^y \right](x+se_i)\, ds \quad \text{for a.e. } x.$$

Hence $q_j^y \in W_{\text{loc}}^{1,1}(\mathbb{R}^n)$ and $\nabla_x q_j^y(x)/q_j^y(x) = \beta_j(x,y)$ for every $y \in Y_0$. Let us recall that $\varphi_j \cdot \mu^y = \mu_j^y$ for ν-a.e. $y \in Y$. Since the union of the sets A_j has full μ-measure, the set $(\mathbb{R}^n \times \{y\}) \cap (\bigcup_{j=1}^\infty A_j)$ has full μ^y-measure for ν-a.e. $y \in Y$. Therefore, for ν-a.e. y the measure μ^y possesses a density f^y such that $\varphi_j(x,y) f^y(x) = q_j^y(x)$ for all j and a.e. x. In addition, μ-a.e. on A_j we have $\nabla_x \varphi_j = 0$ since the derivative of any differentiable function F on the space $\mathbb{R}^n$ vanishes almost everywhere on the set $\{F = 1\}$. Hence we obtain a set Y_1 of full ν-measure such that, for every $i = 1, \ldots, n$, every rational t and every $y \in Y_1$, we have

$$f^y(x+te_i) - f^y(x) = \int_0^t \left[(\beta, e_i)\, f^y \right](x+se_i)\, ds \quad \text{for a.e. } x,$$

which gives $f^y \in W_{\text{loc}}^{1,1}(\mathbb{R}^n)$ and $\nabla_x f^y(x)/f^y(x) = \beta(x,y)$ μ^y-a.e. $\square$

Local logarithmic derivatives enable us to obtain integration by parts formulae for considerably broader classes of functions than the initial class $\mathcal{K}$.

6.7.4. Corollary. *Let μ satisfy the conditions in the previous theorem and let f be a μ-measurable function such that, for every $y \in Y$, the function $x \mapsto f(x,y)$ belongs to $W_{\text{loc}}^{1,1}(\mathbb{R}^n)$ and $|\nabla_x f|, f|\beta| \in L^1(\mu)$. Then one has*

$$\int_X (\nabla_x f, v)\, \mu(dx) = -\int_X f(v, \beta)\, \mu(dx), \quad \forall v \in \mathbb{R}^n. \quad (6.7.6)$$

The same is true if in place of the inclusion $x \mapsto f(x,y) \in W^{1,1}_{\mathrm{loc}}(\mathbb{R}^n)$ the usual partial derivatives $\partial_{x_i} f(x,y)$ exist for all x.

The proof is delegated to Exercise 6.8.7.

6.7.5. Remark. (i) It is easily seen that the separability condition (iii) can be weakened; for example, it suffices to replace it by the following condition:

(iii′) there exists a measurable set $\Omega \subset X$ of full measure with respect to all shifts $(\mu)_t$ generated by the vectors tv such that $\mathcal{K}$ separates measures on Ω.

In particular, this is fulfilled if Ω has full μ-measure and is mapped into itself by the shifts $(x,y) \mapsto (x+v,y)$.

(ii) The condition that $\varphi \circ \psi \in \mathcal{K}$ for all $\psi \in \mathcal{K}$ and $\varphi \in C_0^\infty(\mathbb{R}^1)$ with $\varphi(0) = 0$ in (ii) can be replaced by the following assumption:

there exist functions $\psi_j \in \mathcal{K}$ such that the sets $\{\psi_j = 1\}$ cover $\mathbb{R}^n \times Y$ up to a set of μ-measure zero.

Finally, we observe that if $\mathcal{K}$ is a linear space closed with respect to compositions with C_0^∞-functions vanishing at zero, then $\psi_1 \psi_2 \in \mathcal{K}$ for all $\psi_1, \psi_2 \in \mathcal{K}$.

It is useful to remember that Lipschitzian functions with bounded support on a Banach space X separate Radon measures. If X^* is separable (say, X is separable and reflexive), then Radon measures are separated by continuously Fréchet differentiable Lipschitzian functions with bounded support (Exercise 1.6.18).

6.8. Comments and exercises

The first integration by parts formulas in the infinite dimensional case for Fomin differentiable measures on linear spaces were proved in Averbukh, Smolyanov, Fomin [82] under the condition of existence of both sides in (6.1.4) and some additional technical conditions of "stable integrability". In this special case, during some time the formula from [82] was the most general, although in the author's paper [167] it was shown that the condition of differentiability of f can be replaced by the Lipschitz condition. Finally, it was observed by Khafizov [602] that in the situation of Theorem 6.1.2 the existence of differentiable conditional measures reduces the problem to the one-dimensional case. The extension of the integration by parts formula to the case of Skorohod differentiability, accomplished in Bogachev [185], [186] by means of conditional measures, required somewhat more subtle results in the one-dimensional case. The differentiability of the Radon–Nikodym derivatives of smooth measures was studied in Uglanov [1137], where among other things Proposition 6.3.1 and Corollary 6.3.5 were obtained. Our exposition in §6.5 follows the paper Bogachev, Goldys [196]. On second order derivatives of convex functions see also Matoušková, Zajíček [786]. Some comments on derivatives of measures along vectors fields are given in Chapter 9, where we consider a more general situation.

Exercises

6.8.1. Construct an example of an everywhere differentiable function on an interval that is not absolutely continuous.

6.8.2. Let ϱ be an absolutely continuous probability density on the real line, let $\varrho' \in \mathcal{L}^1(\mathbb{R}^1)$, and let f be a bounded measurable function that is absolutely continuous on every closed interval on which ϱ has no zeros. Suppose also that the function $f'\varrho$ is integrable on the real line with Lebesgue measure (on the set of zeros of ϱ we set $f'\varrho = 0$). Prove that for every bounded Lipschitzian function g the following equality holds:
$$\int_{-\infty}^{+\infty} g(x)f'(x)\varrho(x)\,dx = -\int_{-\infty}^{+\infty} g'(x)f(x)\varrho(x)\,dx - \int_{-\infty}^{+\infty} g(x)f(x)\varrho'(x)\,dx.$$

6.8.3. (Scheutzow, Weizsäcker [**987**]) Let a function $\psi > 0$ be convex on the real line. Suppose that μ is a probability measure on the real line with an absolutely continuous density ϱ. Prove that if $\psi(|\varrho'/\varrho|) \in L^1(\mu)$ and the function $x^{-2}\ln\psi(x)$ is not integrable at zero, then $\varrho > 0$ a.e. If the function $x^{-2}\ln\psi(x)$ is integrable at zero, then there exists an absolutely continuous probability density ϱ vanishing on a set of positive Lebesgue measure such that the function $\psi(|\varrho'/\varrho|)$ is integrable with respect to $\mu = \varrho\,dx$.

6.8.4. Let μ be a Radon measure on a locally convex space X and let $h \in X$ be such that there exist numbers C_n for which
$$\left|\int_X \partial_h^n \varphi(x)\,\mu(dx)\right| \leq C_n |\sup_x \varphi(x)|, \quad \forall \varphi \in \mathcal{FC}_b^\infty.$$
Prove that μ is infinitely differentiable along h.

6.8.5. (Uglanov [**1137**]) Construct an example of a twice differentiable measure $\lambda \geq 0$ on $\mathbb{R}^1$ with a logarithmic derivative β such that $\beta \notin L^1(d_1\lambda)$ and $\beta' \notin L^1(\lambda)$. Construct an example of a signed three times differentiable measure λ with the same properties.

6.8.6. Let ϱ be an absolutely continuous probability density on $\mathbb{R}^1$ and let G be a convex function such that $G'\varrho'$ and $G'\varrho$ are integrable on $\mathbb{R}^1$. Prove that
$$\int_{-\infty}^{+\infty} G'(t)\varrho'(t)\,dt \leq 0.$$

6.8.7. Prove Corollary 6.7.4.

6.8.8. Prove Proposition 5.4.6.

HINT: Let $n = 2$ (the general case is similar); if $\mu^{(2)}$ is not countably additive, then one can find Borel sets $B_n \downarrow \varnothing$ and $u_n, v_n \in E$ such that $\|u_n\|_E \leq 1$, $\|v_n\|_E \leq 1$, $|d_{u_n}d_{v_n}\mu(B_n)| \geq c > 0$; since $d_u d_v \mu = (d_{u+v}d_{u+v}\mu - d_u^2\mu - d_v^2\mu)/2$, there are $h_n \in E$ with $\|h_n\|_E \leq 1$ and $|d_{h_n}^2\mu(B_n)| \geq c/3$; set $\nu_n := d_{h_n}\mu$; observe that $\|d_{h_n}\nu_n\| \leq C$ and $\|d_{h_n}^2\nu_n\| \leq C$ for some $C > 0$ since μ is three times differentiable along E; finally, use Corollary 6.2.8 to show that $|\nu_n|(B_n) \to 0$ and obtain a contradiction.

CHAPTER 7

Logarithmic gradients

This chapter is devoted to one of the key objects of the theory of differentiable measures: vector logarithmic derivatives or logarithmic gradients. Their importance for applications can be seen, for example, from the fact that logarithmic gradients are precisely drift coefficients of symmetrizable diffusions (see Chapter 12). For a measure μ on $\mathbb{R}^n$ with a density ϱ from the Sobolev class $W^{1,1}(\mathbb{R}^n)$ the logarithmic gradient is the vector field $\nabla\varrho/\varrho$. In the infinite dimensional case, where there are no analogues of ϱ and $\nabla\varrho$, in a remarkable way it turns out to be possible to define an analogue of $\nabla\varrho/\varrho$ for many measures important for applications.

7.1. Rigged Hilbert spaces

Let X be a locally convex space. We shall say that $H \subset X$ is a continuously embedded Hilbert space if H is a linear subspace equipped with a Hilbert norm such that the natural embedding $H \to X$ is continuous. If H is dense in X, one can define an embedding $j_H \colon X^* \to H$ in the following way. For every $k \in X^*$ the restriction of k to H is continuous in the Hilbert norm of H. Therefore, by the Riesz representation theorem, there is an element $j_H(k) \in H$ such that

$$_{X^*}\langle k, h\rangle_X = \bigl(j(k), h\bigr)_H, \quad \forall h \in H.$$

Since H is dense, such an element is unique. If H is fixed, the mapping j_H is denoted by j. Identifying X^* with $j_H(X^*)$, we obtain a triple $X^* \subset H \subset X$.

In the described situation H is sometimes called a rigged Hilbert space and (X^*, H, X) is called a standard triple or rigged triple (or Gelfand triple).

If $\{e_\alpha\}$ is an orthonormal basis in H, we obtain

$$j_H(k) = \sum_\alpha k(e_\alpha) e_\alpha, \quad k \in X^*.$$

7.1.1. Example. (i) Let $X = \mathcal{S}'(\mathbb{R}^n), H = L^2(\mathbb{R}^n)$. We obtain a triple $\mathcal{S}(\mathbb{R}^n) \subset H \subset \mathcal{S}'(\mathbb{R}^n)$ with the natural embeddings.

(ii) Let $X = \mathbb{R}^\infty, H = l^2$. Then $X^* = \mathbb{R}_0^\infty \subset H$ is the set of all finite sequences. We observe that in both examples we have regarded the dual spaces already as subspaces in H; each element $j(k)$ has been represented by k itself.

(iii) Let X be a separable Hilbert space, let T be a continuous injective nonnegative operator in X, and let the space $H = T(X)$ be equipped with its natural Hilbert norm $|h|_H = |T^{-1}h|_X$. Identifying X^* with X by the Riesz theorem, we obtain the equalities $j(X^*) = T^2(X)$ and $j(k) = T^2k$ since

$$(k,h)_X = \bigl(j(k), h\bigr)_H = \bigl(T^{-1}j(k), T^{-1}h\bigr)_X = \bigl(T^{-2}j(k), h\bigr)_X, \quad h \in H.$$

In particular, the element $j(k) = T^2k$ can differ from k if $T \neq I$.

(iv) Let $H = l^2$, $\lambda_n > 0$, $\lambda_n \to 0$, and let X be the Hilbert space of sequences (x_n) with $\sum_{n=1}^{\infty} \lambda_n^2 x_n^2 < \infty$. This time we identify the space X^* not with X but with the space Y of sequences (y_n) for which $\sum_{n=1}^{\infty} \lambda_n^{-2} y_n^2 < \infty$, and the action of $y = (y_n)$ on X will be defined by $x \mapsto \sum_{n=1}^{\infty} y_n x_n$. Then the space Y is already naturally embedded into X and j_H is the identical embedding $Y \to X$. In fact, this case is analogous to (ii), just in place of the space $\mathbb{R}^\infty$ a smaller Hilbert space is taken.

(v) If $X = \mathbb{R}^\infty$, $H = \{(h_n) \colon \sum_{n=1}^{\infty} \alpha_n^2 h_n^2 < \infty\}$, then X^* is naturally identified with $\mathbb{R}_0^\infty$, but $j(k)$ is represented by the element $(\alpha_n^{-2} k_n)$ and not by k itself (except for the case $\alpha_n \equiv 1$).

7.2. Definition of logarithmic gradient

Let us now describe a construction associating to a given differentiable measure μ on a linear space some vector-valued mapping: its logarithmic gradient. In the space $\mathbb{R}^n$ equipped with a smooth measure $\mu = \varrho\, dx$ this mapping is exactly $\nabla \varrho/\varrho$.

Let X be a locally convex space, let $H \subset X$ be a continuously and densely embedded Hilbert space, and let $j_H \colon X^* \to H$ be the above defined embedding, which we shall denote also by j if the space H is fixed.

7.2.1. Definition. *Let a Radon measure μ on X be differentiable along all vectors from $j_H(X^*)$. If there exists a Borel mapping $\beta_H^\mu \colon X \to X$ such that*

$$_{X^*}\langle k, \beta_H^\mu\rangle_X = \beta_{j(k)}^\mu, \quad \forall k \in X^*, \qquad (7.2.1)$$

then this mapping β_H^μ is called a logarithmic gradient of the measure μ along H (or a vector logarithmic derivative along H).

It is clear that β_H^μ is uniquely determined by the measure μ.

We observe that the definition of β_H^μ can be reformulated as the integration by parts formula

$$\int_X \partial_{j(k)} f(x)\, \mu(dx) = -\int_X f(x)\,_{X^*}\langle k, \beta_H^\mu(x)\rangle_X\, \mu(dx), \quad f \in \mathcal{FC}_b^\infty(X). \qquad (7.2.2)$$

In this way logarithmic gradients of measures were introduced in Albeverio, Høegh-Krohn [41], [42].

7.2.2. Example. Let $X = H = \mathbb{R}^n$ and let μ be a measure on $\mathbb{R}^n$ differentiable along all vectors. Denoting by p the density of μ, we obtain

$$\beta_H^\mu(x) = \nabla p(x)/p(x).$$

PROOF. Identifying $(\mathbb{R}^n)^*$ with $\mathbb{R}^n$ we obtain the identity $j(k) = k$. In addition, we have $\partial_k p(x) = (k, \nabla p(x))$. □

7.2.3. Example. (i) Let γ be a centered Radon Gaussian measure on a locally convex space X and let its Cameron–Martin space H be dense. Then $\beta_H^\gamma(x) = -x$.

(ii) Let $X = \mathbb{R}^\infty$, $H = l^2$, and let μ be the countable product of probability measures μ_n given by smooth densities p_n on the real line. Then $\beta_H^\mu \colon \mathbb{R}^\infty \to \mathbb{R}^\infty$ is given by the equality

$$\bigl(\beta_H^\mu(x)\bigr)_n = \frac{p_n'(x_n)}{p_n(x_n)} = \frac{d}{dx_n}\ln p_n(x_n), \text{ where } \frac{p_n'(s)}{p_n(s)} := 0 \text{ if } p_n(s) = 0.$$

PROOF. (i) We know that for any $h \in H$ the measure γ is differentiable along h and $\beta_h^\gamma(x) = -\widehat{h}(x)$, where $\widehat{h}$ is the measurable linear functional generated by h. For every functional $k \in X^*$ we have $j(k) \in H$ and $\widehat{j(k)} = k$. Hence $\beta_{j(k)}^\gamma = -k(x)$. Therefore, the operator $-I \colon x \mapsto -x$ serves as β_H^γ.

(ii) In this case X^* is the space $\mathbb{R}_0^\infty$ of all finite sequences and $j(k) = k$ for every $k \in \mathbb{R}_0^\infty$. The logarithmic derivative of the measure μ along a vector $k = (k_1, \ldots, k_n, 0, 0, \ldots)$ is the function

$$x \mapsto \sum_{i=1}^n k_i p_i'(x_i)/p_i(x_i).$$

Therefore, the mapping indicated above can be taken for β_H^μ. □

One should keep in mind that a logarithmic gradient depends essentially on the geometry of the Hilbert space H and not only on H as a subspace in X. If we change the inner product in such a way that the topology of $E = H$ remains the same, then we obtain, as a rule, another mapping for a logarithmic gradient. Indeed, the vector $j_E(k)$ may not coincide with $j_H(k)$. This is seen even in the one-dimensional case. In the general case let μ be a Radon probability measure on a locally convex space X having a logarithmic gradient β_H^μ along some Hilbert space H. Let $A \in \mathcal{L}(H)$ be an invertible operator. We define a new Hilbert space E as follows: $E = H$, $(a,b)_E := (Aa, Ab)_H$. Then we have the equality $j_H(k) = A^* A j_E(k)$. To see this, we observe that for every $k \in X^*$ we have

$$\langle k, h \rangle = \bigl(j_E(k), h\bigr)_E = \bigl(A j_E(k), Ah\bigr)_H.$$

The same quantity equals $\bigl(j_H(k), h\bigr)_H$, which proves the aforementioned equality. Thus, if the operator A is not unitary, then the vectors $j_H(k)$ and $j_E(k)$ do not coincide for some $k \in X^*$. This means that the functions $\beta_{j_H(k)}^\mu$

and $\beta^\mu_{j_E(k)}$ do not coincide (see Exercise 3.9.5). Therefore, if there exists also a logarithmic gradient β^μ_E, then it differs from β^μ_H.

It would be interesting to investigate connections between logarithmic gradients along H and E. In particular, it would be interesting to know whether the existence of one of them always implies the existence of the other. In the case of a Gaussian measure this connection can be easily studied.

7.2.4. Example. Let μ be a centered Radon Gaussian measure on a locally convex space X and let its Cameron–Martin space H be dense. Suppose that we are given an invertible operator $A \in \mathcal{L}(H)$. As in the previous example, we equip $E = H$ with the inner product $(a,b)_E := (Aa, Ab)_H$. Then the logarithmic gradient β^γ_E along E exists and equals

$$\beta^\gamma_E = \widehat{T}\beta^\gamma_H,$$

where $\widehat{T}$ is the measurable linear operator in X generated by the bounded operator $T := (AA^*)^{-1} \in \mathcal{L}(H)$ (see §1.4).

PROOF. Let us verify that the mapping $\Lambda := \widehat{T}\beta^\gamma_H$ serves as a logarithmic gradient along E. Let $k \in X^*$. Then, by the calculations from the previous example, we have the equality $j_E(k) = A^{-1}(A^*)^{-1}j_H(k) = Tj_H(k)$. In addition, we have $j_H(k \circ \widehat{T}) = j_E(k)$ since for every $v \in H$ one has the equality

$$\bigl(j_H(k\circ\widehat{T}),v\bigr)_H = (k\circ\widehat{T})(v) = k(\widehat{T}v) = k(Tv)$$
$$= \bigl(j_H(k), Tv\bigr)_H = \bigl(Tj_H(k),v\bigr)_H = \bigl(j_E(k),v\bigr)_H.$$

Here we have used the fact that the mapping j_H extends to measurable linear functionals and for every proper linear version f of such a functional, one has $\bigl(j_H(f),v\bigr)_H = f(v)$, $v \in H$ (see §1.4). Therefore,

$$k\bigl(\widehat{T}\beta^\gamma_H(x)\bigr) = -k(\widehat{T}x) = \beta^\gamma_{j_H(k\circ\widehat{T})} = \beta^\gamma_{j_E(k)},$$

which proves our assertion. $\square$

Let us emphasize that the space H in Definition 7.2.1 need not coincide with $D(\mu)$ or $H(\mu)$. The only restriction is the inclusion $j_H(X^*) \subset D(\mu)$. This is, certainly, a restriction on H. In particular, the set $H \cap D(\mu)$ is dense in H since it contains the set $j_H(X^*)$ that is dense in H. Let us consider an example where H differs indeed from $D(\mu)$.

7.2.5. Example. Let γ be a centered Gaussian measure on a separable Hilbert space X with the Cameron–Martin space $D(\gamma) = T(X)$ and let $H = Q(X) \subset T(X)$, where T and Q are injective nonnegative operators with $Q^2(X) \subset T^2(X)$. We set $(a,b)_H := (Q^{-1}a, Q^{-1}b)_X$. Then

$$\beta^\gamma_H = -R^*,$$

where R is the operator satisfying the equation $Q^2 = T^2R$.

7.2. Definition of logarithmic gradient

PROOF. First we observe that by our assumptions the operator R exists algebraically. Therefore, by the closed graph theorem it is bounded. We have

$$(k, -R^*x)_X = (Rk, -x)_X = (T^{-2}T^2Rk, -x)_X = \beta^\gamma_{T^2Rk}(x) = \beta^\gamma_{Q^2k}(x),$$

which is $\beta^\gamma_{j(k)}(x)$ since $j(k) = Q^2k$ according to Example 7.1.1(iii). □

Let us consider a more general example with a linear mapping.

7.2.6. Example. Let H be the Cameron–Martin space of a nondegenerate Radon centered Gaussian measure on a locally convex space X and let A be a selfadjoint operator on a dense domain $D(A) \subset H$. Suppose that there is a bounded inverse operator A^{-1}. Set $T := A^{-1/2}$. By Theorem 1.4.5 there exists a Radon centered Gaussian measure γ^A on X with the Cameron–Martin space $T(H) = D(A^{1/2})$ and the Fourier transform

$$\int_X \exp(il(x))\,\gamma^A(dx) = \exp\left(-\frac{1}{2}|Tj_H(l)|^2_H\right), \quad l \in X^*.$$

The operator A is an isomorphism between Hilbert spaces $H(\gamma^A)$ and H. According to Corollary 1.4.6, the operator A extends to a γ^A-measurable linear mapping from X to X, which we denote here by $\Gamma(A)\colon X \to X$ and which takes γ^A to γ. This extension is called the second quantization of the operator A. In some cases it is continuous. For example, if the operator A in l^2 takes (x_n) to $(\lambda_n x_n)$, where $\lambda_n > 0$ and $\lambda_n \to +\infty$, then the operator $\Gamma(A)$ in $\mathbb{R}^\infty$ is given by the same formula.

For every $h \in H$ let ξ_h be the γ^A-measurable functional generated by the vector $A^{-1}h \in H(\gamma^A)$. In other words, this is the extension of the continuous linear functional $v \mapsto (h, v)_H$ on $H(\gamma^A)$. Let $l \in X^*$ be such that $j_H(l) \in H(\gamma^A) = T(H)$. Then

$$\beta^{\gamma^A}_{j_H(l)} = -\xi_{Aj_H(l)}$$

since for every $h \in H(\gamma^A)$ we have $\bigl(Aj_H(l), h\bigr)_H = \bigl(j_H(l), h\bigr)_{H(\gamma^A)}$. Hence if $j_H(X^*) \subset T(H)$, then

$$\beta^{\gamma^A}_H = -\Gamma(A).$$

Suppose additionally that

$$j_H(X^*) \subset D(A), \quad A\bigl(j_H(X^*)\bigr) \subset j_H(X^*).$$

Then we have

$$\beta^{\gamma^A}_{j_H(l)} = -\bigl(j_H^{-1}Aj_H\bigr)(l) \in X^*,$$

i.e., we obtain a linear mapping

$$\Lambda\colon l \mapsto \bigl(j_H^{-1}Aj_H\bigr)(l), \quad X^* \to X^*.$$

If X is a reflexive Banach space, then by the closed graph theorem the operator Λ is continuous and its adjoint (acting in $X^{**} = X$) coincides with $\Gamma(A)$. This gives broad sufficient conditions for the continuity of $\Gamma(A)$.

Say, if X is Hilbert and $H = S(X)$, where S is a Hilbert–Schmidt operator in X, then the inclusion $S^2(X) \subset T(H)$ is a sufficient condition.

Let us now discuss the behaviour of logarithmic gradients under linear mappings.

7.2.7. Proposition. *Let μ be a Radon measure on a locally convex space X having a logarithmic gradient β_H^μ along some continuously embedded Hilbert space $H \subset X$. Let $A\colon X \to Y$ be an injective continuous linear operator with dense range in a locally convex space Y (or a μ-measurable linear mapping such that $A(H)$ is dense in Y). We define a new Hilbert space E by setting $E = A(H)$, $(a,b)_E = (A^{-1}a, A^{-1}b)_H$. Let $\nu = \mu \circ A^{-1}$. Then*
$$\beta_E^\nu(y) = A\beta_H^\mu(A^{-1}y).$$

PROOF. For every $l \in Y^*$ we have
$$j_E(l) = A j_H(l \circ A).$$
Indeed, for every $h \in H$ one has the equality
$$\bigl(j_E(l), Ah\bigr)_E = {}_{Y^*}\langle l, Ah\rangle_Y = {}_{X^*}\langle l \circ A, h\rangle_X = \bigl(j_H(l \circ A), h\bigr)_H = \bigl(A j_H(l \circ A), Ah\bigr)_E.$$
According to Proposition 3.3.13, we have $\beta_{Av}^\nu(y) = \beta_v^\mu(A^{-1}y)$. Therefore, letting $v = j_H(l \circ A)$, we arrive at the equality
$${}_{Y^*}\langle l, A\beta_H^\mu(A^{-1}y)\rangle_Y = {}_{X^*}\langle l \circ A, \beta_H^\mu(A^{-1}y)\rangle_X = \beta_v^\mu(A^{-1}y) = \beta_{Av}^\nu(y) = \beta_{j_E(l)}^\nu(y),$$
which completes the proof. $\square$

In the infinite dimensional case, the logarithmic gradient of a nonzero measure usually cannot be represented as the H-gradient of a function.

7.2.8. Example. Let $X = \mathbb{R}^\infty$, $H = l^2$, and let μ be the countable product of the standard Gaussian measures on the real line. Then again $\beta_E^\mu(x) = -x$, but there exists no function f with $\nabla_H f(x) = -x$. Indeed, otherwise $f(x) = -\sum_{i=1}^\infty x_i^2/2 + C$, but this series obviously diverges μ-a.e.

Perhaps, the following assertion is true: if μ is a probability measure on $\mathbb{R}^\infty$ with $H(\mu) = l^2$, then there exists no Borel function W such that for μ-a.e. x the functions $t \mapsto W(x + te_n)$ are locally absolutely continuous and $\partial_{e_n} W = \beta_{e_n}^\mu$ μ-a.e. for all n (however, see Exercise 7.9.6).

7.3. Connections with vector measures

There is another construction of the vector logarithmic derivative of a measure, which determines it as the Radon–Nikodym derivative of the vector-valued derivative of this measure.

Suppose that on a σ-algebra $\mathcal{A}$ of subsets of a space X we are given an additive set function η with values in a Banach space Y; we shall call η

7.3. Connections with vector measures

a *vector measure*. Its *semivariation* $V(\eta)$ is defined as

$$V(\eta) = \sup \Big\| \sum_{i=1}^n \alpha_i \eta(X_i) \Big\|_Y,$$

where sup is taken over all finite partitions of X into disjoint parts $X_i \in \mathcal{A}$ and all finite collections of real numbers α_i with $|\alpha_i| \leqslant 1$. If $\Gamma \subset Y^*$ is a set such that $\|v\|_Y = \sup_{l \in \Gamma} \langle l, v \rangle$ for all $v \in Y$, then

$$V(\eta) = \sup_{l \in \Gamma} \|\langle l, \eta \rangle\|,$$

where $\|\langle l, \eta \rangle\|$ is the variation of the real measure $\langle l, \eta \rangle$. A semivariation may be finite in the case of infinite *variation* of η, which is defined by the formula

$$\mathrm{Var}\,(\eta) := \sup \sum_{i=1}^n \|\eta(X_i)\|_Y,$$

where sup is taken over all finite partitions of X into disjoint parts $X_i \in \mathcal{A}$. According to the classical Pettis theorem (see Dunford, Schwartz [**376**, Theorem 4.10.1]), if for every $l \in Y^*$ the real set function $\langle l, \eta \rangle$ is countably additive, then the measure η is countably additive in the norm of Y and has finite semivariation.

Let μ be a Radon measure on a locally convex space X differentiable along all vectors from a locally convex space E continuously embedded into $D(\mu)$. We observe that the embedding is automatically continuous if E is a Fréchet space and $E \subset D(\mu)$ (for example, this is the case if E is a Banach space, see Proposition 5.1.7).

7.3.1. Lemma. *The formula*

$$\langle D\mu, v \rangle = d_v \mu, \quad v \in E, \tag{7.3.1}$$

defines a vector measure $D\mu$ with values in E^* such that for every $v \in E$ the real measure $\langle D\mu, v \rangle$ is countably additive.

If E is a Banach space, then the additive measure $D\mu$ has bounded semivariation. If, in addition, E is a reflexive Banach space, then $D\mu$ is countably additive with respect to the norm.

PROOF. The continuity of the embedding $E \to D(\mu)$ yields that the functionals $v \mapsto d_v \mu(A)$, where $A \in \mathcal{B}(X)$, are continuous on E. Hence formula (7.3.1) defines an E^*-valued measure. If, in addition, E is a Banach space, then this measure has finite semivariation. Indeed, the norm of any element $G \in E^*$ equals $\sup\{G(v) \colon \|v\|_E \leqslant 1\}$, whence

$$V(D\mu) = \sup_{\|v\|_E \leqslant 1} \|\langle D\mu, v \rangle\| = \sup_{\|v\|_E \leqslant 1} \|d_v \mu\| \leqslant C < \infty$$

since the identical embedding of E into the Banach space $D(\mu)$ is a bounded operator. If E is a reflexive Banach space, then the dual to E^* is E. Hence the measure $D\mu$ is weakly countably additive since we $\langle D\mu, v \rangle = d_v \mu$. By the Pettis theorem it is countably additive in the norm. □

It would be nice to define the vector logarithmic derivative as a vector-valued density of $D\mu$ with respect to μ, since this is precisely what we have in the finite dimensional case. However, in infinite dimensions one cannot do the same without special tricks. Indeed, let μ be a centered Gaussian measure with the infinite dimensional Cameron–Martin space H. Letting $E = H$, we obtain the following explicit expression: $\langle D\mu, v \rangle = -\widehat{v} \cdot \mu$, $v \in H$. However, there is no mapping $\Lambda \colon X \to H$ such that $\bigl(\Lambda(x), v\bigr)_H = -\widehat{v}(x)$ μ-a.e. for every $v \in H$. Indeed, suppose that such a mapping exists. Taking for v the elements of an orthonormal basis $\{e_i\}$ in H, we obtain on the right-hand side of this hypothetic equality independent standard Gaussian variables. Hence $\sum_{i=1}^{\infty} \bigl(\Lambda(x), e_i\bigr)_H^2 = \infty$ μ-a.e. The following result, proved in Bogachev [**169**] for Hilbert subspaces of differentiability and extended in Khafizov [**604**] to Banach subspaces of differentiability, shows that the general situation is similar if $E = D(\mu)$.

7.3.2. Proposition. *Suppose that μ is a probability measure and that the space $E = D(\mu)$ is infinite dimensional. Then the $D(\mu)^*$-valued measure $D\mu$ has unbounded variation.*

PROOF. We consider only the case where E has an equivalent Hilbert norm defined by an inner product $(\cdot, \cdot)_E$, which makes our reasoning very simple but clearly shows the essence of the phenomenon. If $D\mu$ has bounded variation, then, identifying E^* with E, we obtain an E-valued measure of bounded variation, which has a vector Radon–Nikodym density F with respect to μ since $\mathrm{Var}\,(D\mu) \ll \mu$ and the space E has the Radon–Nikodym property. Hence for every $h \in E$ we have $\bigl(F(x), h\bigr)_E = \beta_h^\mu(x)$ μ-a.e. Let $\{e_n\}$ be an orthonormal sequence in E. Then $\lim_{n \to \infty} \bigl(F(x), e_n\bigr)_E = 0$ and $\bigl|\bigl(F(x), h\bigr)_E\bigr| \leqslant |F(x)|_E$, whence $\lim_{n \to \infty} \|(F, e_n)_E\|_{L^1(\mu)} = 0$ by the Lebesgue dominated convergence theorem. However, by the equivalence of norms $\|\cdot\|_{D(\mu)}$ and $\|\cdot\|_E$ we have $(F, e_n)_E = \beta_{e_n}^\mu$ and $\|\beta_{e_n}^\mu\|_{L^1(\mu)} = \|e_n\|_{D(\mu)} \geqslant c$. Therefore, we obtain a contradiction. □

Thus, if we want to define a vector density of $D\mu$, then we should either choose E smaller than $D(\mu)$ or consider this vector measure as taking values in some other space. The definition below combines both possibilities. We assume further that $E \subset D(\mu)$ is a Hilbert space. Then by the Riesz representation theorem the measure $D\mu$ turns out to be an E-valued measure. Therefore, it becomes an X-valued measure.

7.3.3. Proposition. *Let X be a Banach space, let $E \subset D(\mu)$ be a Hilbert space, and let an E-valued measure $D\mu$ be defined by*

$$\bigl(D\mu(A), v\bigr)_E = d_v\mu(A), \ A \in \mathcal{B}(X).$$

Then, regarded as an X-valued measure, $D\mu$ has bounded variation provided that the embedding $E \to X$ is an absolutely summing operator. In particular, this is true if the space X is Hilbert.

7.3. Connections with vector measures

PROOF. We already know that the E-valued measure $D\mu$ is countably additive and has bounded semivariation. Let us employ the following simple fact from the theory of vector measures: if Z is a Banach space, then, for every Z-valued measure φ of bounded semivariation and every absolutely summing mapping $T\colon Z \to X$, the measure $T(\varphi)$ has bounded variation (Exercise 7.9.7). Hence in our case the measure $D\mu$ regarded as an X-valued measure has bounded variation. $\square$

7.3.4. Definition. *Suppose that the measure $D\mu$ has a vector density Λ_E^μ with respect to μ in the following weak sense: $\Lambda_E^\mu\colon X \to X$ is a μ-measurable mapping such that Λ_E^μ is Pettis integrable and for every measurable set A one has*

$$\int_A \Lambda_E^\mu(x)\,\mu(dx) \in E \quad \text{and} \quad \left(\int_A \Lambda_E^\mu(x)\,\mu(dx), v\right)_E = d_v\mu(A),\ \forall v \in E.$$

Then Λ_E^μ is called the weak vector density of $D\mu$ with respect to μ.

7.3.5. Proposition. *Let $E \subset D(\mu)$ be a Hilbert space dense in X. The mapping Λ_E^μ exists precisely when β_E^μ exists. In this case $\Lambda_E^\mu = \beta_E^\mu$.*

PROOF. If β_E^μ exists, then, for every measurable set A and every $k \in X^*$, we have

$$\int_A \langle k, \beta_E^\mu(x)\rangle\,\mu(dx) = \int_A \beta_{j(k)}^\mu(x)\,\mu(dx) = d_{j(k)}\mu(A),$$

where $j := j_E$. Hence β_E^μ is Pettis integrable. The functional $v \mapsto d_v\mu(A)$ is continuous on E since the embedding $E \to D(\mu)$ is continuous. Therefore, there exists an element $b \in E$ such that $(v,b)_E = d_v\mu(A)$ for all $v \in E$. Letting $v = j(k)$, we obtain $d_{j(k)}\mu(A) = \langle k,b\rangle$. Hence the integral

$$\int_A \beta_E^\mu(x)\,\mu(dx),$$

which is an element of X, coincides with b and belongs to E. The same reasoning shows that we have the second necessary property. Thus, β_E^μ can be taken for Λ_E^μ. Conversely, if Λ_E^μ exists, then for any $v = j(l)$, where $l \in X^*$, by the definition of the Pettis integral we find

$$d_{j(l)}\mu(A) = \left(\int_A \Lambda_E^\mu(x)\,\mu(dx), j(l)\right)_E = l\left(\int_A \Lambda_E^\mu(x)\,\mu(dx)\right) = \int_A l(\Lambda_E^\mu(x))\,\mu(dx).$$

Hence $\langle l, \Lambda_E^\mu\rangle$ serves as the Radon–Nikodym density of $d_{j(l)}\mu$ with respect to μ, i.e., coincides with $\beta_{j(l)}^\mu$. $\square$

The difference between β_E^μ and Λ_E^μ is that in the definition of β_E^μ we do not require that $E \subset D(\mu)$.

We observe that the vector measure $D\mu$ is absolutely continuous with respect to μ since $d_v\mu \ll \mu$ for each v. Hence it is natural to ask whether it has a vector density with respect to μ. It is seen from what has been said above that if X is a Banach space and the mapping β_E^μ exists and its norm is integrable, then the countably additive X-valued measure $D\mu$ has a Bochner integrable density with respect to μ (but according to Proposition 7.3.2 this is false for $D\mu$ regarded as an E-valued measure).

7.4. Existence of logarithmic gradients

We shall start with an example showing that in infinite dimensions even a Gaussian measure may fail to have a logarithmic gradient along a densely embedded Hilbert space contained in its Cameron–Martin space. Certainly, it is easy to find a large space H not generating logarithmic derivatives. For example, if the space X is Hilbert, then for $H = X$ there is no logarithmic gradient for any nonzero measure on X since a nonzero measure cannot be differentiable along all vectors in X.

7.4.1. Example. Let $X = \{(x_n)\colon \sum_{n=1}^\infty n^{-2} x_n^2 < \infty\}$, $H = l^2$. Then

$$j_H(X^*) = \left\{(x_n)\colon \sum_{n=1}^\infty n^2 x_n^2 < \infty\right\}.$$

We take for μ the countable product of Gaussian measures on $\mathbb{R}^1$ with densities

$$p_n(t) = n(2\pi)^{-1/2} \exp(-n^2 t^2/2).$$

Here $D(\mu) = j_H(X^*)$, but there is no logarithmic gradient β_H^μ.

PROOF. Otherwise $\bigl(\beta_H^\mu(x)\bigr)_n = -n^2 x_n$ for each n since for the nth coordinate functional $l_n(x) = x_n$ we have $j_H(l_n) = e_n$, $e_n = (0,\ldots,0,1,0,\ldots)$, and $\beta_{e_n}^\mu(x) = p_n'(x_n)/p_n(x_n) = -n^2 x_n$. Thus, we have

$$\sum_{n=1}^\infty n^{-2} \bigl|\bigl(\beta_H^\mu(x)\bigr)_n\bigr|^2 = \sum_{n=1}^\infty n^2 x_n^2.$$

The series on the right diverges μ-a.e., which contradicts to $\beta_H^\mu(x) \in X$. □

Let us give some sufficient conditions for the existence of logarithmic gradients.

7.4.2. Theorem. *Let μ be a Borel measure on a separable Banach space X with the Radon–Nikodym property and let $H \subset X$ be a densely embedded Hilbert space such that its natural embedding into X is an absolutely summing mapping (for example, let X be a Hilbert space whose embedding $H \to X$ is a Hilbert–Schmidt operator).*
(i) *If $H \subset D(\mu)$, then β_H^μ exists and $\|\beta_H^\mu(\cdot)\|_X \in L^1(\mu)$.*
(ii) *Let X be a Hilbert space and let $j_H(X^*) \subset H(\mu)$. Assume that this embedding is a Hilbert–Schmidt operator. Then the logarithmic gradient β_H^μ exists and $\|\beta_H^\mu(\cdot)\|_X \in L^2(\mu)$.*

7.4. Existence of logarithmic gradients

PROOF. (i) It was shown above that there exists an H-valued measure $D\mu$ of bounded semivariation which has bounded variation as an X-valued measure and

$$\bigl(D\mu(B), h\bigr)_H = d_h\mu(B) = \int_B \beta_h^\mu(x)\,\mu(dx), \quad B \in \mathcal{B}(X),\ h \in H.$$

Since X has the Radon–Nikodym property and $D\mu \ll \mu$, the measure $D\mu$ as an X-valued measure has a vector density Λ with respect to μ. This means that there exists a Bochner μ-integrable mapping $\Lambda\colon X \to X$ such that, for every Borel set B, the vector integral

$$\int_B \Lambda(x)\,\mu(dx)$$

is an element of H and, for every $h \in H$, one has the equality

$$\left(\int_B \Lambda(x)\,\mu(dx), h\right)_H = \bigl(D\mu(B), h\bigr)_H = d_h\mu(B).$$

Now we can set $\beta_H^\mu = \Lambda$. Let us recall that Hilbert spaces have the Radon–Nikodym property and that absolutely summing operators between Hilbert space are precisely Hilbert–Schmidt operators. This proves assertion (i).

(ii) We take an orthonormal basis $\{e_n\}$ in X and observe that the series $\sum_{n=1}^\infty \beta_{j(e_n)}^\mu(x)^2$ converges μ-a.e., where $j := j_H$. Indeed, since the embedding $j := j_H\colon X^* \to H(\mu)$ is a Hilbert–Schmidt operator, we have

$$\sum_{n=1}^\infty |j(e_n)|_{H(\mu)}^2 < \infty.$$

Hence $\sum_{n=1}^\infty \|\beta_{j(e_n)}^\mu\|_{L^2(\mu)}^2 < \infty$. Therefore, $\sum_{n=1}^\infty \beta_{j(e_n)}^\mu(x)^2 < \infty$ a.e. Hence the mapping

$$\beta_H^\mu(x) = \sum_{n=1}^\infty \beta_{j(e_n)}^\mu(x) e_n$$

is well defined and belongs to $L^2(\mu, X)$. It is clear that we have obtained the desired logarithmic gradient. $\square$

It is worth noting that, according to Example 7.4.1, it is not enough that in Theorem 7.4.2(ii) just the embedding $H \to X$ is a Hilbert–Schmidt operator (without the condition that the embedding $j_H(X^*) \subset H(\mu)$ is Hilbert–Schmidt).

Let us discuss in more detail logarithmic gradients of Gaussian measures. As we shall see in Chapter 12, this is useful for the study of linear stochastic differential equations. For Gaussian measures, there are simple sufficient conditions for the existence of logarithmic gradients.

7.4.3. Proposition. *Let γ be a centered Radon Gaussian measure on a locally convex space X and let H be a separable Hilbert space continuously* and densely embedded into X such that $H \subset H(\gamma)$. Then β_H^γ exists. Moreover, there exists an operator $T \in \mathcal{L}(H(\gamma))$ such that $H = T(H(\gamma))$ and β_H^γ coincides with the measurable linear extension of $-TT^*$.

PROOF. By the closed graph theorem the natural embedding $H \to H(\gamma)$ is continuous. Denote by T the composition of an arbitrary unitary isometry $H(\gamma) \to H$ with the embedding of H into $H(\gamma)$. Then one has $T \in \mathcal{L}(H(\gamma))$, $H = T(H(\gamma))$, $(a,b)_{H(\gamma)} = (Ta, Tb)_H$. As explained in §1.4, every continuous linear operator A on $H(\gamma)$ admits a unique extension to a measurable linear mapping $\widehat{A} \colon X \to X$. Let us apply this result to the operator $A = -TT^*$. For every $k \in X^*$ and every $h \in H(\gamma)$ we have

$$-{}_{X^*}\langle k, TT^*h\rangle_X = -\big(j_H(k), TT^*h\big)_H = -\big(T^{-1}j_H(k), T^*h\big)_{H(\gamma)} = -\big(j_H(k), h\big)_{H(\gamma)}$$

since $j_H(k) \in H = T(H(\gamma))$. According to Example 7.2.3(i), the functional $h \mapsto -\big(j_H(k), h\big)_{H(\gamma)}$ is the restriction to $H(\gamma)$ of the measurable linear functional β_v^γ, where $v = j_H(k)$. This means that the measurable linear functionals $x \mapsto -\langle k, \widehat{TT^*}(x)\rangle$ and β_v^γ coincide γ-a.e. since they coincide on the space $H(\gamma)$. Now we can set $\beta_H^\gamma(x) := -\widehat{TT^*}(x)$. □

7.4.4. Proposition. *Let γ be a centered Radon Gaussian measure on a locally convex space X with the dense Cameron–Martin space $H := H(\gamma)$ and let $A \colon X \to X$ be a γ-measurable everywhere defined linear mapping. We define a linear operator $A^* \colon X^* \to H$ by the formula*

$$(A^*k, h)_H := \langle k, Ah\rangle := {}_{X^*}\langle k, Ah\rangle_X.$$

Suppose that X is sequentially complete. Then the following conditions are equivalent:

(i) there exists a Hilbert space E densely embedded into X such that one has $j_E(X^*) \subset H(\gamma)$ and $A = \beta_E^\gamma$;

(ii) the function

$$(f, g) \mapsto -\langle f, A^*g\rangle$$

defines an inner product on X^*.

Under either of these conditions one has $A^* = A \circ j_H$ and

$$\big(Aj_H(f), j_H(g)\big)_H = \big(j_H(f), Aj_H(g)\big)_H, \quad f, g \in X^*.$$

PROOF. We observe that the mapping A^* is well defined since the function $h \mapsto \langle k, Ah\rangle$ is continuous on H being the restriction of the measurable linear function $\langle k, Ax\rangle$ (see §1.4). Let condition (i) be fulfilled. By definition, for every $k \in X^*$ we have γ-a.e.

$$\langle k, Ax\rangle = \beta_{j_E(k)}^\gamma(x).$$

Therefore,

$$\langle k, Ah\rangle = -\big(j_E(k), h\big)_H$$

7.4. Existence of logarithmic gradients

for all $h \in H$ because two everywhere defined measurable linear functions that are almost everywhere equal coincide on H and every proper linear version of β_v^γ has the form $h \mapsto -(v, h)_H$ on H. Therefore,
$$(A^*k, h)_H = \langle k, Ah \rangle = -\bigl(j_E(k), h\bigr)_H, \quad \forall h \in H.$$
Thus, $A^*k = -j_E(k)$ and
$$-\langle k, A^*k \rangle = \langle k, j_E(k) \rangle = \bigl(j_E(k), j_E(k)\bigr)_E = |j_E(k)|_E^2.$$
Conversely, let condition (ii) be fulfilled. Set
$$E_0 := A^*(X^*), \quad (A^*f, A^*g)_E := -\langle f, A^*g \rangle.$$
By our assumption, E_0 is a Euclidean space with the inner product $(\cdot, \cdot)_E$. We observe that the unit ball in the space E_0 is bounded in X. It suffices to verify that it is weakly bounded. Let $l \in X^*$. Then, for every $u \in E_0$ with $|u|_E \leqslant 1$, we have
$$|l(u)| = |\langle l, u \rangle| \leqslant |A^*l|_E |u|_E \leqslant |A^*l|_H.$$
Thus, the embedding $(E_0, |\cdot|_E) \to X$ is continuous. Since X is sequentially complete, this embedding extends to an embedding of the completion of E_0. Denote by E the image of this completion in X. In this case the unit ball U_E from E is the closure of the unit ball from E_0 in the space X. We observe that U_E is an absolutely convex weakly compact set in X since the embedding $E \to X$ is continuous. By the identity
$$\bigl(j_E(k), A^*f\bigr)_E = \langle k, A^*f \rangle = -(A^*k, A^*f)_E, \quad f \in X^*,$$
we have $j_E(k) = -A^*k$, $k \in X^*$. Hence, for every $h \in H$, we obtain
$$\langle k, Ah \rangle = (A^*k, h)_H = -\bigl(j_E(k), h\bigr)_H = \beta_{j_E(k)}^\gamma(h),$$
where we choose a version of $\beta_{j_E(k)}^\gamma$ that is linear on all of X. This means that γ-a.e. we have $\langle k, Ax \rangle = \beta_{j_E(k)}^\gamma(x)$. Thus, $A = \beta_E^\gamma$.

However, we have to verify that E is dense in X. It suffices to show that $A^*(X^*)$ is dense, which follows from the fact that if $f \in X^*$ and $\langle f, A^*g \rangle = 0$ for all $g \in X^*$, then $f = 0$ since $\langle f, A^*g \rangle$ is an inner product by assumption.

Under the equivalent conditions (i) and (ii) one has the equality
$$Aj_H(f) = A^*f, \quad f \in X^*.$$
In particular, $A\bigl(j_H(X^*)\bigr) \subset H$. Indeed, for all $g \in X^*$ we have
$$\langle g, Aj_H(f) \rangle = \bigl(A^*g, j_H(f)\bigr)_H = \langle f, A^*g \rangle = \langle g, A^*f \rangle,$$
where the last equality is fulfilled by the symmetry of the inner product mentioned in (ii). In addition,
$$\bigl(Aj_H(f), j_H(g)\bigr)_H = \bigl(j_H(f), Aj_H(g)\bigr)_H, \quad f, g \in X^*.$$
This also follows from the symmetry of the inner product and the equality
$$\bigl(Aj_H(f), j_H(g)\bigr)_H = \bigl(A^*f, j_H(g)\bigr)_H = \langle g, A^*f \rangle.$$

In many cases E is automatically separable. For example, this is true if X is a separable Fréchet space or, more generally, if the measure $\gamma \circ A^{-1}$ is Radon. It is also sufficient to have the condition $A(H) \subset H$. It is not clear whether E is always separable. A simple sufficient condition for the separability of E is the existence of a countable collection of functionals $f_n \in X^*$ separating the points of X. □

7.4.5. Corollary. *In the situation of the proposition, condition* (ii) *is equivalent to the condition that the function* $(f,g) \mapsto -\langle f, Aj_H g\rangle$ *is an inner product on* X^*.

PROOF. This condition follows from condition (ii) since $A^* = A \circ j_H$. If it is fulfilled, then $\langle f, Aj_H g\rangle = \langle g, Aj_H f\rangle$ on X^*, whence
$$(Aj_H g, j_H f)_H = \langle f, Aj_H g\rangle = \langle g, Aj_H f\rangle.$$
This yields the equality $(Aj_H g, h)_H = \langle g, Ah\rangle$ for all $h \in H$ by the continuity of both sides on H and the fact that $j_H(X^*)$ is dense in H. Hence $A \circ j_H$ is the operator A^* from the proposition. Hence (ii) is fulfilled. □

This proposition describes also the subspaces E generating logarithmic gradients.

7.4.6. Example. Let X, H, and γ be the same as in the proposition above. Suppose that all Borel measures on X are Radon. Let $E \subset X$ be a continuously embedded separable Hilbert space such that $j_E(X^*) \subset H$. The existence of β_E^γ is equivalent to the existence of an operator $A \in \mathcal{L}(H, X)$ satisfying the equality $Aj_H = -j_E$ and extendible to a γ-measurable operator $\widehat{A}$ with values in X. In this case $\beta_E^\gamma = \widehat{A}$. Indeed, $\langle f, j_E g\rangle = (j_E f, j_E g)_E$ is an inner product on X^*.

7.4.7. Example. Let X be a Hilbert space, $H = R(X)$, $E = T(X)$, where $R \in \mathcal{H}(X)$, $T \in \mathcal{L}(X)$, R and T are positive, $T^2(X) \subset R(X)$. We have $j_E(g) = T^2 g$ for $g \in X^* = X$. By Corollary 1.4.7, the characterization from the previous example takes the form $AR^2 = -T^2$ and $AR(X) \subset S(X)$ with some $S \in \mathcal{H}(X)$. Since $T^2 = RB$, where $B \in \mathcal{L}(X)$, we have the inequalities $T^2 = B^* R$ and $Ax = -B^* R^{-1} x$ if $x \in R(X)$. The condition $A \in \mathcal{H}(H, X)$ is equivalent to the following condition: $B \in \mathcal{H}(X)$. Therefore, the condition for the existence of β_E in terms of the operators R and T is this: $T^2(X) \subset R(X)$ and $R^{-1} T^2 \in \mathcal{H}(X)$.

The following result shows that logarithmic gradients in $\mathbb{R}^n$ are generalized gradients.

7.4.8. Theorem. *Let* $\mu = \varrho\,dx$ *be a probability measure on* $\mathbb{R}^n$ *such that* $\varrho \in W^{1,1}(\mathbb{R}^n)$ *and* $|\beta^\mu| = |\nabla\varrho/\varrho| \in L^2(\mu)$. *Then the mapping* β^μ *belongs to the closure of the set* $\{\nabla\psi\colon \psi \in C_0^\infty(\mathbb{R}^n)\}$ *in the space* $L^2(\mu, \mathbb{R}^n)$ *of mappings with values in* $\mathbb{R}^n$ *that are square-integrable with respect to* μ.

7.4. Existence of logarithmic gradients

PROOF. Let $f_k = \ln[\min(\varrho, k) + k^{-1}]$. Clearly, $f_k \in W^{1,1}_{\text{loc}}(\mathbb{R}^n) \cap L^\infty(\mathbb{R}^n)$ and $|\nabla f_k| \leqslant |\nabla \varrho|/\varrho$ a.e. Then $f_k \in W^{2,1}(\mu)$ (see §2.6). It remains to observe that $|\nabla f_k - \nabla \varrho/\varrho| \to 0$ in $L^2(\mu)$ by the dominated convergence theorem. $\square$

Some analogue of Theorem 7.4.8 holds for infinite dimensional spaces. Let X be a separable Hilbert space and let $H \subset X$ be a densely embedded Hilbert space such that the natural embedding is a compact operator. Denote by $G_2(H, \mu, X)$ the closure of the class of vector mappings $\{\nabla_H f \colon f \in \mathcal{FC}_b^\infty\}$ in $L^2(\mu, X)$.

7.4.9. Corollary. *Let μ be a Borel probability measure on the space X. Suppose that the logarithmic gradient β_H^μ exists and $\beta_H^\mu \in L^2(\mu, X)$. Then we have $\beta_H^\mu \in G_2(H, \mu, X)$.*

PROOF. The space H has the form $H = T(X)$, $(Tx, Ty)_H = (x, y)_X$, where T is an injective symmetric compact operator on X. Let $\{\varphi_n\}$ be the eigenbasis of T, $T\varphi_n = \lambda_n \varphi_n$. The vectors $e_n := \lambda_n \varphi_n$ form an orthonormal basis in H; they are mutually orthogonal in X. Let us identify H with l^2 setting in correspondence to each e_n the nth vector of the standard basis in l^2. The space X identifies with the weighted Hilbert space of sequences (x_n) such that $\sum_{n=1}^\infty \lambda_n^2 x_n^2 < \infty$, and X^* identifies with the space Y of sequences (y_n) such that $\sum_{n=1}^\infty \lambda_n^{-2} y_n^2 < \infty$; see Example 7.1.1(iv). Then j_H is the identical embedding of Y into X. The mapping β_H^μ has the components β_i. In addition, $\beta_i = \beta_{e_i}^\mu$, which follows from the definition of a logarithmic gradient since for the coordinate functionals $l_i \colon x \mapsto x_i$ we have $l_i(\beta_H^\mu) = \beta_i$ and $j(l_i) = e_i$. Denote by π_n the orthogonal projection from X to the subspace X_n generated by $e_1, \ldots, e_n$. The space X_n can be identified with $\mathbb{R}^n$, however, it carries two different inner products: one, induced from H, is the usual inner product of $\mathbb{R}^n$, and the other one is induced from X and given by the formula $\sum_{i=1}^n \lambda_i^2 x_i y_i$. Naturally, when considering orthogonal projections in X we use the inner product in X, but below we also employ the inner product in H. For every $y \in X$ we have $|\pi_n y - y|_X \to 0$. Hence by the Lebesgue dominated convergence theorem we obtain convergence $|\pi_n \beta_H^\mu - \beta_H^\mu|_X \to 0$ in $L^2(\mu)$. Let B^n denote the conditional expectation of the mapping β_H^μ with respect to the σ-algebra σ_n generated by the projection π_n. In the coordinate form we have $B^n = (B_i^n)$, where the component B_i^n is the conditional expectation of β_i with respect to σ_n. It is known that the mappings B^n form an X-valued square-integrable martingale convergent to β_H^μ in $L^2(\mu, X)$ (see Vakhania, Tarieladze, Chobanyan [**1167**, Ch. II, Theorem 4.1]). Therefore, the mappings $\pi_n B^n$ also converge to β_H^μ in $L^2(\mu, X)$ since

$$\|\pi_n B^n - \beta_H^\mu\|_{L^2(\mu, X)} \leqslant \|\pi_n B^n - \pi_n \beta_H^\mu\|_{L^2(\mu, X)} + \|\pi_n \beta_H^\mu - \beta_H^\mu\|_{L^2(\mu, X)}$$
$$\leqslant \|B^n - \beta_H^\mu\|_{L^2(\mu, X)} + \|\pi_n \beta_H^\mu - \beta_H^\mu\|_{L^2(\mu, X)} \to 0.$$

On the other hand, we have the equality $\pi_n B^n(x) = \beta^n(\pi_n x)$, where the Borel mapping $\beta^n = (b_i^n) \colon X_n \to X_n$ is the logarithmic gradient of the projection μ_n of the measure μ to the n-dimensional space X_n regarded now with the inner product from H (i.e., from l^2 and not from X). Indeed, for every $C \in \mathcal{B}(X_n)$ and every $i = 1, \ldots, n$ we have

$$d_{e_i}\mu_n(C) = d_{e_i}\mu(\pi_n^{-1}C) = \int_X I_C(\pi_n x)\beta_{e_i}^\mu(x)\,\mu(dx)$$
$$= \int_X I_C(\pi_n x) B_i^n(x)\,\mu(dx) = \int_{X_n} I_C(z)\beta_i^n(z)\,\mu_n(dz).$$

Now we apply the finite dimensional result and find a function $\psi_n \in C_b^\infty(X_n)$ with $\|\beta^n - \nabla\psi\|_{L^2(\mu_n, H)} \leqslant 2^{-n}$, where the gradient of φ_n is taken with respect to the standard inner product on $X_n = \mathbb{R}^n$, i.e., with respect to the inner product in H. Since $|h|_X \leqslant M|h|_H$, we have $\|\beta^n - \nabla\psi\|_{L^2(\mu_n, X)} \leqslant 2^{-n}M$, whence our assertion follows. □

7.5. Measures with given logarithmic gradients

Here we discuss the problem of existence of measures with a given logarithmic gradient. This problem has different nonequivalent settings. The most important for applications and discussed below assumes that we have a fixed space H determining the embedding $j_H \colon X^* \to X$. Then the question of existence of a measure μ with a given logarithmic gradient β along H reduces to the problem of existence of a measure with given partial logarithmic derivatives β_h, $h = j_H(k)$, $k \in X^*$. It is clear that the problem will change if one can also vary H. Finally, we need not involve any H at all and consider the problem of reconstructing a measure from its partial logarithmic derivatives β_h along vectors h from some subspace $\mathcal{D}$ in X (the first option corresponds to the subspace $\mathcal{D} = j_H(X^*)$).

We first discuss the following important question. When is a continuous mapping $\beta = (\beta^i)\colon \mathbb{R}^n \to \mathbb{R}^n$ a local logarithmic gradient of a probability measure, i.e., $\beta = \nabla\varrho/\varrho$, where $\varrho \in C^1(\mathbb{R}^n)$ is a probability density? It is clear that the mapping β must be of the form $\beta = \nabla V$, where $V \in C^1(\mathbb{R}^n)$ and $\exp V \in L^1(\mathbb{R}^n)$. The first condition is sufficiently constructive; for example, if β is continuously differentiable, then this condition reduces to the equalities $\partial_{x_j}\beta^i = \partial_{x_i}\beta^j$. But how can we verify the integrability of $\exp V$ in terms of β? Since we have the equality

$$V(x) - V(0) = \int_0^1 \bigl(\nabla V(tx), x\bigr)\,dt,$$

we need the exponential integrability of the right-hand side. If it is given, then for the desired measure μ we can take $c \cdot \exp V\,dx$. This reasoning leads to the following somewhat more general statement.

7.5.1. Proposition. *A Borel mapping β on $\mathbb{R}^n$ is the local logarithmic gradient of a probability measure with a continuous positive density precisely*

7.5. Measures with given logarithmic gradients

when $\beta = \nabla V$, $V \in W^{1,1}_{\text{loc}}(\mathbb{R}^n)$, and the function

$$\varrho(x) := \exp \int_0^1 \bigl(\beta(tx), x\bigr)\, dt$$

belongs to $L^1(\mathbb{R}^n)$. This probability measure is unique and has density const$\cdot \varrho$. If one has $|\beta|\varrho \in L^1(\mathbb{R}^n)$, then β is the logarithmic gradient of μ.

In many situations easily verified sufficient conditions are more useful than necessary and sufficient. Such sufficient conditions are, for example, the following: $\beta \in C^1(\mathbb{R}^n, \mathbb{R}^n)$, $\partial_{x_j}\beta^i = \partial_{x_i}\beta^j$ and

$$\bigl(\beta(x), x\bigr) \leqslant -C \quad \text{if } |x| \geqslant 1,\ \text{where } C > n.$$

Let us give a less obvious sufficient condition, from which the previous one follows with $V(x) = k(x,x)$ for some $k > 0$.

7.5.2. Proposition. *Suppose that $\beta = \nabla G$, where $G \in W^{1,1}_{\text{loc}}(\mathbb{R}^n)$ and $|\nabla G| \exp G \in L^1_{\text{loc}}(\mathbb{R}^n)$, and that we are given a function $V \in C^2(\mathbb{R}^n)$ such that the sets $\{V \leqslant c\}$ are compact, there exist numbers $c_k \to +\infty$ for which the sets $V^{-1}(c_k)$ are Lipschitzian surfaces, and $LV := \Delta V + (\beta, \nabla V) \leqslant -1$ outside some compact set. Then $\exp G \in L^1(\mathbb{R}^n)$, i.e., β is the local logarithmic gradient of the probability measure with a density const $\cdot \exp G$.*

PROOF. It easily follows from our assumptions that $\exp G \in L^1_{\text{loc}}(\mathbb{R}^n)$ (see Exercise 2.8.15). Let $U_k := \{V \leqslant c_k\}$. Since $\beta = \nabla G$ and the outer unit normal η on $\partial U_k = V^{-1}(c_k)$ is collinear with ∇V, one has

$$\int_{U_k} [\Delta V + (\beta, \nabla V)] \exp G \, dx = \int_{\partial U_k} (\eta, \nabla V) \exp G \, dS \geqslant 0.$$

Outside some compact set K we have the estimate $LV \leqslant -1$. Since the open sets $\{V < c_k\}$ cover K, we obtain $LV \leqslant -1$ outside U_{k_1} for some k_1. By assumption we have $LV \exp G \in L^1_{\text{loc}}(\mathbb{R}^n)$. Therefore,

$$\int_{U_k \setminus U_{k_1}} \exp G \, dx \leqslant \int_{U_{k_1}} |LV| \exp G \, dx, \quad k \geqslant k_1.$$

Hence $\exp G \in L^1(\mathbb{R}^n)$. Of course, it suffices to have $\sup_{x \notin K} LV(x) < 0$. □

A substantial advantage of this sufficient condition in terms of Lyapunov functions (suggested by A.I. Kirillov [**607**]) as compared to the necessary and sufficient conditions from Proposition 7.5.1 is that this condition is local. This becomes particularly important for applications in the infinite dimensional case.

For example, letting $V(x) = (x,x)$, we obtain $LV(x) = 2n + 2\bigl(x, \beta(x)\bigr)$. So it suffices to have $\bigl(\beta(x), x\bigr) \leqslant -C$ outside some ball with some $C > n$. In particular, it suffices that $\bigl(\beta(x), x\bigr) \to -\infty$ as $|x| \to \infty$.

The described method enables one to estimate some integrals with respect to the measure $\mu = \exp G \, dx$.

7.5.3. Proposition. *Let Φ_1 and Φ_2 be two Borel functions such that $\Phi_1 \in L^1(\mu)$, $\Phi_2 \geq 0$ and $LV \leq \Phi_1 - \Phi_2$ a.e., where $V \in C^2(\mathbb{R}^n)$. Suppose that the sets $\{V \leq c\}$ are compact and there exist numbers $c_k \to +\infty$ for which the sets $V^{-1}(c_k)$ are Lipschitzian surfaces. Then one has*

$$\int_{\mathbb{R}^n} \Phi_2(x)\,\mu(dx) \leq \int_{\mathbb{R}^n} \Phi_1(x)\,\mu(dx). \tag{7.5.1}$$

PROOF. First we obtain (7.5.1) for the functions $\min(\Phi_2, k)$. This is done similarly to the previous proposition by integrating the estimate $LV \leq \Phi_1 - \min(\Phi_2, k)$ over U_k against the measure μ. Then we let $k \to \infty$. □

If $V \in C^\infty(\mathbb{R}^n)$, then by Sard's theorem there exist $c_k \to +\infty$ such that the sets $V^{-1}(c_k)$ do not contain critical points and are smooth surfaces.

Let us consider the infinite dimensional case. Suppose that X is a locally convex space, a sequence $\{l_n\} \subset X^*$ separates the points in X, $\{e_n\} \subset X$, and one has $l_n(e_k) = \delta_{nk}$. Suppose, in addition, that we are given a sequence $q = \{q_n\} \in l^1$, where $q_n > 0$, such that the set

$$X_0 := \left\{ x\colon |x|_0^2 = \sum_{n=1}^{\infty} q_n l_n(x)^2 < \infty \right\}$$

is a separable Hilbert space compactly embedded into X. Let E_n denote the linear span of $e_1, \ldots, e_n$. As a model situation one should have in mind the case where $X = \mathbb{R}^\infty$, $l_n(x) = x_n$, the vector e_n has zero components except for 1 at the nth place.

7.5.4. Theorem. *Suppose we are given functions $\beta^n\colon X_0 \to \mathbb{R}^1$ that are continuous on balls in X_0 in the topology of X and there exist continuously differentiable functions G_n on E_n for which $\beta^i = \partial_{e_i} G_n$ on E_n whenever $i \leq n$. Assume also that*

$$\sum_{n=1}^{\infty} q_n l_n(x) \beta^n(x) \leq C - \Theta(|x|_0), \quad x \in \bigcup_{n=1}^{\infty} E_n, \tag{7.5.2}$$

where Θ is a nonnegative locally bounded Borel function on $[0, +\infty)$ and we have $\lim_{t \to \infty} \Theta(t) = +\infty$. In addition, let

$$|\beta^n(x)| \leq C_n + K_n |x|_0^{d_n}, \quad \forall\, x \in X_0. \tag{7.5.3}$$

Then, there exists a Radon probability measure μ on X_0 such that for each n we have $\beta_{e_n}^\mu = \beta^n \in L^1(\mu)$.

PROOF. The theorem reduces to the case $X = \mathbb{R}^\infty$ since X is injectively embedded into $\mathbb{R}^\infty$ by means of the mapping $x \mapsto \{l_n(x)\}$. The balls from X_0 are compact in X, hence the topology from X coincides on them with the topology from $\mathbb{R}^\infty$ (i.e., the topology generated by the functionals l_n). The space E_n can be identified with $\mathbb{R}^n$ by declaring the vectors $e_1, \ldots, e_n$

7.5. Measures with given logarithmic gradients

an orthonormal basis. Set $V(x) := |x|_0^2 = \sum_{i=1}^n q_i x_i^2$. Then one has

$$LV(x) := \sum_{i=1}^n [\partial_{e_i} V(x) + \partial_{e_i} V(x) \partial_{e_i} G_n(x)]$$

$$= 2\sum_{i=1}^n q_i + 2\sum_{i=1}^n q_i x_i \partial_{e_i} G_n(x) \leqslant 2C + 2\sum_{i=1}^n q_i - 2\Theta(|x|_0).$$

According to two previous propositions the function $\exp G_n$ is integrable on E_n and letting $M := C + \sum_{i=1}^\infty q_i$ and taking z_n such that the measure $\mu_n := z_n \exp G_n \, dx$ is of total unit mass we have

$$\int_{E_n} \Theta \circ V(x) z_n \exp G_n(x) \, dx \leqslant M.$$

The measures μ_n are uniformly tight on the space X since the indicated estimate and the condition $\lim_{R \to \infty} \Theta(R) = +\infty$ along with Chebyshev's inequality yield $\lim_{R \to \infty} \sup_n \mu_n(x \colon |x|_0 \geqslant R) = 0$, where the sets $K_R := \{x \colon |x|_0 \leqslant R\}$ are compact in X. Passing to a subsequence, we may assume that the measures μ_n converge weakly to some probability measure μ on X and $\mu(X_0) = 1$. If in place of V we take the function V^m, then the same reasoning gives the uniform estimate

$$\sup_n \int_{E_n} V^{m-1}(x) \Theta \circ V(x) \, \mu_n(dx) \leqslant M_m < \infty.$$

Hence $\sup_n \|V^{m-1}\|_{L^1(\mu_n)} < \infty$ and $V^{m-1} \in L^1(\mu)$. By estimate (7.5.3) we obtain $\lim_{R \to \infty} \sup_n \|\beta^i I_{\{|\beta^i| \geqslant R\}}\|_{L^1(\mu_n)} = 0$ and $\beta^i \in L^1(\mu)$ for all i. Along with the continuity of β^i on the compact sets $\{x \colon |x|_0 \leqslant R\}$ this ensures convergence of the integrals of $\psi \beta^i$ against the measure μ_n to the integral of $\psi \beta^i$ against the measure μ for every function $\psi \in \mathcal{FC}_b^\infty(X)$ (see [**193**, Lemma 8.4.3]). Since

$$\int_X \partial_{e_i} \psi(x) \, \mu_n(dx) = -\int_X \psi(x) \beta^i(x) \, \mu_n(dx),$$

the same is true for μ in place of μ_n. □

7.5.5. Remark. If X_0 is compactly embedded into a Banach space X_1 continuously embedded into X and the functions β^i extend to bounded continuous functions on balls in X_1, then without condition (7.5.3) we obtain the equality $\beta^i = \beta_{e_i}^{\mu, \text{loc}}$ with respect to the class $\text{Lip}_0(X_0)$ of Lipschitzian functions on X_0 with bounded support. To this end, we first obtain this equality with respect to the class $\text{Lip}_0(X_1)$ by the same reasoning and then apply Corollary 6.7.4 to $X = \mathbb{R}^1 e_n \times Y$.

Condition (7.5.2) looks like $(x, \beta(x))_{X_0} \leqslant C - \Theta(|x|_0)$. However, here we do not assume that β^i corresponds to a mapping with values in X_0; in addition, the indicated condition is fulfilled only on the union of E_n. In place of this condition one can require the following: $|\beta^n(x)| \leqslant C_n + C_n\Theta(|x|_0)$ on X_0 and there exists a function V on $\bigcup_{n=1}^{\infty} E_n$ for which one has the estimate $LV(x) \leqslant C - \Theta(|x|_0)$ on each E_n and the restrictions of V to E_n satisfy the hypotheses of Proposition 7.5.2. This is proved by the same reasoning.

The continuity of β^n on balls in X_0 in the topology from X is equivalent to the continuity on balls in the weak topology from X_0. It is not clear whether it is sufficient in Theorem 7.5.4 to require only the continuity with respect to the norm of X_0 (which is weaker than the stated condition). If we reinforce the dissipativity condition, this becomes possible.

7.5.6. Theorem. *Suppose that in the previous theorem we require the continuity of β^n only with respect to the norm of X_0, but condition (7.5.2) is replaced by the following one: for every $M > 0$ there is a compact set K_M in X_0 such that*

$$\sum_{n=1}^{\infty} q_n l_n(x)\beta^n(x) \leqslant -M \quad \text{whenever } x \in \bigcup_{n=1}^{\infty} E_n \backslash K_M$$

and $\sum_{n=1}^{\infty} q_n l_n(x)\beta^n(x) \leqslant M_0 < +\infty$ on $\bigcup_{n=1}^{\infty} E_n$. Then, there exists a Radon probability measure μ on X_0 for which $\beta_{e_n}^\mu = \beta^n \in L^1(\mu)$ for all n.

PROOF. The only difference with the proof of the previous theorem concerns justification of the uniform tightness of $\{\mu_n\}$. This is done by Proposition 7.5.3, where one should take the function $\Phi_1(x) = M_0 + 2\sum_{n=1}^{\infty} q_n$ and define Φ_2 as follows: let $Q_n := \bigcup_{j=1}^n K_j$, let $\Phi_2(x) = M$ if $x \in Q_{M+1} \backslash Q_M$, and let $\Phi_2(x) = 0$ if $x \notin \bigcup_{j=1}^{\infty} K_j$. Then the integrals of Φ_2 against the measures μ_n turn out to be uniformly bounded, which gives the uniform tightness of $\{\mu_n\}$. $\square$

7.5.7. Example. Let $X = W^{2,-1}[0,1]$ and $X_0 = W_0^{2,1}[0,1]$. We take the orthonormal basis $\varphi_n(t) = \sqrt{2}\sin\pi n t$ in $L^2[0,1]$. The elements of X have the form $x = \sum_{n=1}^{\infty} x_n \psi_n$, where $\psi_n(t) = n\varphi_n(t)$ and the series converges in $W^{2,-1}[0,1]$. The functions ψ_n form an orthonormal basis in X. Let $l_n(x) = n^4 x_n$, $q_n = n^{-2}$, $e_n = n^{-3}\varphi_n$. Then $l_n(e_n) = 1$. We wish to find a measure μ on X whose logarithmic gradient along $H = X_0$ has a heuristic expression $\beta(x) = x'' - x^3$. This expression is not defined on all of X, but is meaningful on X_0, where its components β^n in the basis $\{\psi_n\}$ have the form $\beta^n(x) = -\pi^2 n^2 (x, \varphi_n)_2 - (x^3, \varphi_n)$, $x \in X_0$, since the operator $x \mapsto x''$ is diagonal in the basis $\{\psi_n\}$ and has the eigenvalues $\{-\pi^2 n^2\}$. Due to the equality $(x'', x)_2 = -(x', x')_2$ we have

$$\sum_{n=1}^{\infty} q_n l_n(x) \beta^n(x) = (x'', x)_2 - (x^3, x)_2 \leqslant -(x', x')_2 = -|x|_0^2$$

for all x in the linear span $e_1, \ldots, e_n$. In addition, the functions β^n are continuous on balls in X_0 in the weak topology by the compactness of the embedding of X_0 into $C[0,1]$. Hence we obtain a probability measure whose logarithmic derivatives along the vectors e_n are the functions β^n.

7.6. Uniqueness problems

The uniqueness problem for logarithmic gradients is the problem of the unique determinateness of a measure by its logarithmic gradient. The uniqueness problem, as well as the existence problem, admits several different nonequivalent formulations. Suppose that we are going to say that "two measures μ and ν have one and the same logarithmic derivative β". What does it mean? Should it mean that there exist β_H^μ and β_H^ν such that $\beta_H^\mu = \beta$ μ-a.e. and $\beta_H^\nu = \beta$ ν-a.e.? This formulation is not always reasonable. For example, suppose that μ and ν are mutually singular (say, one of them has a smooth density with support in $[0,1]$ and a smooth density of the other has support in $[2,3]$). Then we must conclude that both measures have equal logarithmic derivatives. Indeed, we can define β as β^μ on $[0,1]$ and as β^ν on $[2,3]$. Certainly, for measures on the real line there are many possibilities to exclude such trivial examples (e.g., considering only measures with positive continuous densities). In infinite dimensions the situation is worse, and we have to accept a convention in the sense that we should understand the equality of logarithmic derivatives. Unfortunately, there is no unique natural way of doing that.

We shall discuss three possibilities corresponding to fixing either a measure, or a version of β or a class of admissible measures (and mappings β). The first one is ensured by the following definition.

Let X be a locally convex space, let $H \subset X$ be a separable Hilbert space continuously and densely embedded into X, and let $j_H \colon X^* \to H$ be the embedding defined in §7.1. This embedding will be further denoted by j. Let us fix a linear subspace K in X^* separating the points in X and a Radon probability measure μ on X which has logarithmic derivatives $\beta_{j(l)}^\mu$ for all $l \in K$. For every $l \in K$ we choose and fix a Borel version of $\beta_{j(l)}^\mu$, denoted by the same symbol. We warn the reader that the various objects introduced below may depend on this choice. Let us define $\mathcal{G}_K^{\beta^\mu}$ as the set of all Radon probability measures ν on X such that, for every $l \in K$, the logarithmic derivative $\beta_{j(l)}^\nu$ of the measure ν exists and satisfies the equality $\beta_{j(l)}^\nu = \beta_{j(l)}^\mu$ ν-a.e. Here we *do not assume* the existence of a mapping β^μ for which $\beta_{j(l)}^\mu = l(\beta^\mu)$. The question of the validity of the equality

$$\sharp \mathcal{G}_K^{\beta^\mu} = 1, \qquad (7.6.1)$$

where $\sharp$ denotes cardinality, is of interest, in particular, in relation to the uniqueness problem for Gibbs states in statistical physics. In this section we also investigate the question of whether the equality

$$\sharp \mathcal{G}_{K,ac}^{\beta^\mu} = 1 \qquad (7.6.2)$$

is fulfilled, where
$$\mathcal{G}^{\beta^\mu}_{K,ac} := \{\nu \in \mathcal{G}^{\beta^\mu}_K : \nu \ll \mu\}.$$
We observe that $\mathcal{G}^{\beta^\mu}_K$ and $\mathcal{G}^{\beta^\mu}_{K,ac}$ are convex sets.

One can consider a yet weaker type of uniqueness.

7.6.1. Definition. *We shall say that a Radon probability measure μ on X is uniquely determined by its logarithmic gradient β^μ_H if β^μ_H has a Borel version β for which $\sharp\mathcal{G}^\beta_K = 1$, where $\beta^\mu_{j(k)} := k(\beta)$, $k \in K$.*

We shall say that a Borel mapping $\beta\colon X \to X$ satisfies the condition of weak uniqueness if there exists a Radon probability measure μ on X such that, for some μ-version β_0 of the mapping β, we have $\sharp\mathcal{G}^{\beta_0}_K = 1$, i.e., $\beta^\mu_H = \beta$ μ-a.e. and μ is uniquely determined by its logarithmic gradient.

Generally speaking, (7.6.1) and (7.6.2) are not fulfilled even if $\dim X = 1$, as the following trivial example shows.

7.6.2. Example. Suppose that $K = H = X = \mathbb{R}^1$. Let a function φ from $\mathcal{S}(\mathbb{R}^1)$ be such that $\varphi(0) = 0$, $\varphi > 0$ on $R \setminus \{0\}$ and
$$\int \varphi \, dx = 1.$$
Let us set
$$\psi := \frac{1}{2}\varphi \quad \text{on } [0, \infty) \quad \text{and} \quad \psi := c\varphi \quad \text{on } (-\infty, 0],$$
where $c > 0$ is such that the integral of ψ equals 1. Let $\mu := \varphi \, dx$, $\nu := \psi \, dx$. Then $\mu \neq \nu$ and one has the equality
$$\beta^\mu_H = \frac{\varphi'}{\varphi} = \beta^\nu_H,$$
where we set $\beta^\mu_H(0) = \beta^\nu_H(0) = 0$. Hence $\sharp\mathcal{G}^{\beta^\mu} > 1$. Say, we can take the function $\varphi(x) := x^2 g(x)$, $x \in \mathbb{R}^1$, where g is the standard Gaussian density on $\mathbb{R}^1$. In this case $\beta^\mu_H(x) = -x + 2x^{-1}$.

In this example the measures μ and ν are not uniquely determined by their logarithmic derivative. However, the latter satisfies the condition of weak uniqueness! Indeed, let us take the measure μ_0 with density $2x^2 g(x) I_{[0,+\infty)}(x)$. Its logarithmic derivative β coincides with β^μ_H on $(0, \infty)$ and vanishes on $(-\infty, 0]$. Since the measure μ_0 is zero on $(-\infty, 0]$, we see that β is a μ_0-version of β^μ_H. There are no other probability measures with a logarithmic derivative β: any such measure must be zero on $(-\infty, 0]$ and equal $\mathrm{const}\, x^2 g(x) \, dx$ on $(0, +\infty)$, whence we obtain $\mathrm{const} = 2$. On the real line, the logarithmic derivative of every differentiable probability measure satisfies the condition of weak uniqueness (Exercise 7.9.2). It would be interesting to investigate the condition of weak uniqueness in $\mathbb{R}^n$.

On the other hand, there is the following positive result in the finite dimensional case.

7.6.3. Theorem. *Suppose that $K = H = X = \mathbb{R}^n$ is equipped with the standard inner product, $\varrho \in W^{1,1}(\mathbb{R}^n)$ and $\mu := \varrho\, dx$. If $|\nabla \varrho|/\varrho \in L^1_{\mathrm{loc}}(\mathbb{R}^n)$, where we set $\nabla \varrho(x)/\varrho(x) := 0$ if $\varrho(x) = 0$, then $\mathcal{G}_K^{\beta\mu} = \{\mu\}$.*

PROOF. By Lemma 2.2.6 we have $\varrho(x) > 0$ a.e. and $\ln \varrho \in W^{1,1}_{\mathrm{loc}}(\mathbb{R}^n)$. Suppose we have one more probability measure ν with density f possessing a logarithmic gradient $\nabla f/f$ which ν-a.e. equals $\nabla \varrho/\varrho$. In particular, the equality holds almost everywhere on the set $\{f > 0\}$, and on the set $\{f = 0\}$ we have $\nabla f/f = 0$ a.e. Hence the function $|\nabla f/f|$ is locally integrable. Therefore, $f(x) > 0$ almost everywhere and $\ln f \in W^{1,1}_{\mathrm{loc}}(\mathbb{R}^n)$. Since we have $\nabla \ln f = \nabla \ln \varrho$ a.e. and both logarithms are locally Sobolev functions, we obtain $\ln f = \ln \varrho + C$ a.e., hence $f = e^C \varrho$ a.e. Since μ and ν are probability measures, we have $C = 1$. □

Let us consider some infinite dimensional Gaussian cases in which uniqueness holds or not.

7.6.4. Theorem. *Let us consider the situation from Example 7.2.6. Let A and γ^A be the objects defined there. Suppose that $K := j_H^{-1}(D(A))$ separates the points in the space X and $Aj_H(K) \subset j_H(X^*)$. Set*

$$M := \bigcap_{l \in K} \left\{ x \in X : \ _{X^*}\langle j_H^{-1} A j_H(l), x\rangle_X = 0 \right\}.$$

(i) Let σ be a Radon probability measure on X such that $\sigma(M) = 1$. Then $\gamma^A * \sigma \in \mathcal{G}_K^{\gamma^A}$.

(ii) One has $\mathcal{G}_K^{\gamma^A} = \{\gamma^A\}$ if and only if $M = \{0\}$.

PROOF. As explained in Example 7.2.6, for any $l \in K$ we have

$$\beta := \beta_{j_H(l)}^{\gamma^A} = -\bigl(j_H^{-1} A j_H\bigr)(l) \in X^*.$$

Set $h := j_H(l)$ and $\mu := \gamma^A$. Let σ be a Radon probability measure on M. Then we obtain the equality $d_h(\mu * \sigma) = d_h\mu * \sigma = (\beta \cdot \mu) * \sigma$. We show that $(\beta \cdot \mu) * \sigma = \beta \cdot (\mu * \sigma)$. Since $\beta(x+y) = \beta(x)$ whenever $y \in M$ and $\sigma(M) = 1$, for every $B \in \mathcal{B}(X)$ one has

$$[\beta \cdot (\mu * \sigma)](B) = \int_X \int_X I_A(x+y) \beta(x+y)\, \mu(dx)\sigma(dy)$$
$$= \int_X \int_X I_A(x+y) \beta(x)\, \mu(dx)\sigma(dy) = (\beta \cdot \mu) * \sigma(B),$$

as required. Thus, the equality $\beta_h^{\mu*\sigma} = \beta$ is fulfilled. Then we show that one has $\mathcal{G}_K^{\gamma^A} = \{\gamma^A\}$ if $M = \{0\}$. Let ν be a probability measure such that $\beta_h^\nu = \beta \in X^*$ in the above notation. Then one has $\partial_h \exp[it\beta(x)] = it\beta(h)\exp[it\beta(x)]$. The integration by parts formula gives

$$\frac{d}{dt}\widetilde{\nu}(t\beta) = i\int_X \exp[it\beta(x)]\beta(x)\,\nu(dx) = t\beta(h)\int_X \exp[it\beta(x)]\,\nu(dx),$$

i.e., $d[\widetilde{\nu}(t\beta)]/dt = t\beta(h)\widetilde{\nu}(t\beta)$. For μ the same relationship is true, hence one has $\widetilde{\nu}(t\beta) = \widetilde{\mu}(t\beta)$. Since the functionals of the form β separate the points in X, we obtain the equality $\mu = \nu$. □

7.6.5. Remark. We have been unable to prove in Theorem 7.6.4 that
$$\mathcal{G}_K^{\gamma^A} = \{\gamma^A * \sigma\colon \sigma \in \mathcal{P}_r(X),\ \sigma(M) = 1\}. \qquad (7.6.3)$$
However, this is true if X is a separable Hilbert space and the semigroup $\exp(-tA)$ generated by $-A$ on H extends to a strongly continuous semigroup on X (see Bogachev, Röckner [**219**]).

Although the operator A in Theorem 7.6.4 is strictly positive, it may occur that $M \neq \{0\}$, as the following example shows.

7.6.6. Example. Let us set $X := \mathcal{D}'(\mathbb{R}^d)$, $X^* = K := C_0^\infty(\mathbb{R}^d)$, $H := L^2(\mathbb{R}^d)$, $A := -\Delta + 1$, $D(A) := W^{2,2}(\mathbb{R}^d)$. Then γ^A is the so-called *free Euclidean field of quantum field theory*, which has been intensively studied (see [**948**]), and M is the set of $(-\Delta + 1)$-harmonic functions in the space $\mathcal{D}'(\mathbb{R}^d)$, i.e., the set of elements $F \in \mathcal{D}'(\mathbb{R}^d)$ with $-\Delta F + F = 0$, which is, certainly, nonempty. We observe, however, that if we take $X = \mathcal{S}'(\mathbb{R}^d)$, then $\gamma^A\bigl(\mathcal{S}'(\mathbb{R}^d)\bigr) = 1$, but $M \cap \mathcal{S}'(\mathbb{R}^d) = \{0\}$. Taking for σ Dirac's measure at a nonzero point in the case $M \neq \{0\}$, by Theorem 7.6.4 we see that $\mathcal{G}_K^{\gamma^A}$ contains not only centered Gaussian measures.

It is surprising that even if X is a separable Hilbert space, there exist two *centered* Gaussian measures on X having the same logarithmic gradient which is a continuous linear operator on X. This is in contrast to the finite dimensional case. The fact that this is possible is based on the observation that the range of R in Example 7.2.5 need not be dense (as is the case in the following theorem).

7.6.7. Theorem. *Let X be an infinite dimensional separable Hilbert space. Then one can find a Hilbert space H that is densely embedded into X by means of a Hilbert–Schmidt operator and two distinct nondegenerate centered Gaussian measures μ_1 and μ_2 on X such that there exists a bounded linear operator A on X which is a common Borel version of $\beta_H^{\mu_1}$ and $\beta_H^{\mu_2}$.*

PROOF. We shall use Example 7.2.5. Let us take $X = L^2[0,1]$ and let $H = W_0^{4,2}[0,1]$ be the Sobolev space of real functions f with absolutely continuous third derivatives such that
$$f^{(4)} \in L^2[0,1],\ f^{(j)}(0) = f^{(j)}(1) = 0\ ,\ 1 \leqslant j \leqslant 3\ .$$
We define two operators $(A_1, D(A_1))$ and $(A_2, D(A_2))$ on X by
$$D(A_1) := \{u \in W^{2,2}[0,1]\colon u(0) = u(1) = 0\},$$
$$D(A_2) := \{u \in W^{2,2}[0,1]\colon u(1) - u'(1) = 0 = u(0) + u'(0)\},$$
where both operators on their domains act as $-d^2/dt^2$. It is well known that $(A_i, D(A_i))$, $i = 1, 2$, are injective nonnegative selfadjoint operators

7.6. Uniqueness problems

on X and that $T_i := A_i^{-1}$, $i = 1, 2$, are injective nonnegative selfadjoint Hilbert–Schmidt operators on X. To see that they are injective, we observe that for any $u \in D(A_1)$ we have

$$-\int_0^1 u''(t)u(t)\, dt = \int_0^1 u'(t)^2\, dt + u'u|_0^1 = \int_0^1 u'(t)^2\, dt \geq \int_0^1 u(t)^2\, dt,$$

and for any $u \in D(A_2)$ we have

$$-\int_0^1 u''(t)u(t)\, dt = \int_0^1 u'(t)^2\, dt + u'u|_0^1$$

$$= \int_0^1 u'(t)^2\, dt + u(1)^2 + u(0)^2 \geq \frac{1}{2}\int_0^1 u(t)^2\, dt.$$

Since the embedding $H \subset X$ is a Hilbert–Schmidt operator, there exists an injective nonnegative selfadjoint Hilbert–Schmidt operator Q on X such that $Q(X) = H$ and $|\cdot|_H = |Q^{-1} \cdot|_X$. It is readily seen that we have

$$Q(X) = H \subset T_1^2(X) \cap T_2^2(X) \quad \text{and} \quad T_1^{-2}h = T_2^{-2}h = h^{(4)} \quad \forall h \in H.$$

Since $Q^2(X) \subset Q(X)$ and $T_i^2(X) \subset T_i(X)$, $i = 1, 2$, our claim follows from Example 7.2.5 if we choose for μ_1 and μ_2 the centered Gaussian measures on the space X with covariance operators T_1^2 and T_2^2 and we take the operator $A = R^*$ defined by the relation $R = T_1^{-2}Q^2 = T_2^{-2}Q^2$. $\square$

7.6.8. Corollary. *Let $\mu = (\mu_1 + \mu_2)/2$, where μ_1 and μ_2 are centered Gaussian measures from the previous theorem. Then μ is a non-Gaussian probability measure and one has $\beta_H^\mu = \beta_H^{\mu_1} = A$, where A is a bounded linear operator.*

The situation from Theorem 7.6.7 is impossible if H is the Cameron–Martin space of these two measures (see Proposition 7.6.10 below). However, if A need not be a continuous linear operator, then one has the following trivial example.

7.6.9. Example. Let γ be a nondegenerate symmetric Gaussian measure on a separable Hilbert space X and let Y be a separable Hilbert space compactly embedded into X and having full γ-measure. We take an arbitrary vector v not belonging to Y and consider the non-Gaussian probability measure $\mu = (\gamma_v + \gamma)/2$. Then both measures μ and γ admit equal versions of the logarithmic gradient along $H = H(\gamma)$. Indeed, put

$$\beta_H^\mu(x) = -x \quad \text{on } Y \quad \text{and} \quad \beta_H^\mu(x) = -x + v \quad \text{outside } Y.$$

For every $l \in X^*$ and every function $\varphi \in \mathcal{F}C_b^\infty(X)$ we have

$$\int_X \partial_{j_H(l)}\varphi(x)\, \mu(dx) = -\int_X \varphi(x)\, d_{j_H(l)}\mu(dx)$$

$$= \frac{1}{2}\int_X \varphi(x)l(x)\, \gamma(dx) + \frac{1}{2}\int_X \varphi(x)l(x-v)\, \gamma_v(dx),$$

which coincides with the integral of $-\varphi l(\beta_H^\mu)$ against the measure μ since γ and γ_v are concentrated on disjoint sets Y and $Y - v$ (we observe that $\beta_h^{\gamma_v}(x) = \beta_h^\gamma(x - v)$).

The following result is easily deduced from the proof of Theorem 7.6.4, but we give a direct justification employing the same idea in this simple case.

7.6.10. Proposition. *Let γ be a centered Radon Gaussian measure on a locally convex space X and let μ be a Radon probability measure on X differentiable along some linear space $D \subset H(\gamma)$ such that for all $h \in D$ the functions β_h^γ and β_h^μ admit equal modifications which are continuous linear functionals on X. Suppose additionally that these functionals separate the points in X. Then $\gamma = \mu$.*

PROOF. Denote by β_h the continuous linear functional serving as a common version for β_h^γ and β_h^μ, where $h \in D \subset H(\gamma)$. Then $h = -j_H(\beta_h)$ and hence we have $\beta_h(h) = -|h|_H^2$. If $h \neq 0$, then we may assume that $\beta_h(h) = -1$. Proposition 3.3.13 yields that the one-dimensional projection of the measure μ defined by the functional $l = -\beta_h$ has the logarithmic derivative $\beta \colon t \mapsto -t$ along the vector $v = -\beta_h(h)$ since $\beta(l) = \beta(-\beta_h) = \beta_h$ coincides with the conditional expectation β_h with respect to the σ-algebra generated by the functional $l = -\beta_h$. Hence $\mu \circ l^{-1}$ is the standard Gaussian measure on the real line. Certainly, this can also be seen directly from the relations

$$\frac{d}{dt}\widetilde{\mu}(tl) = i\int_X l(x)\exp[itl(x)]\,\mu(dx) = -i\int_X \exp[itl(x)]\,d_h\mu(dx)$$
$$= i\int_X \partial_h \exp[itl(x)]\,\mu(dx) = -tl(h)\widetilde{\mu}(tl),$$

whence we find $\widetilde{\mu}(tl) = \exp\bigl(-l(h)t^2/2\bigr)$. The same is true for $\gamma \circ l^{-1}$. This means that both measures assign equal integrals to every function of the form $f = \exp(i\beta_h)$. Lemma 1.2.10 yields the equality $\mu = \gamma$. □

The same reasoning also applies in the following more general situation.

7.6.11. Theorem. *Let μ and ν be Radon probability measures on X. Suppose that there exists a set $K \subset X^*$ separating the points in X and possessing the following property: for every $l \in K$ there are functionals $l_1, \ldots, l_n \in K$ and vectors $h_1, \ldots, h_n$ such that l is a linear combination of $l_1, \ldots, l_n$, $l_i(h_j) = \delta_{ij}$ and the logarithmic derivatives $\beta_{h_i}^\mu$ and $\beta_{h_i}^\nu$ exist and admit equal versions of the form $f_i(l_1, \ldots, l_n)$, where Borel functions f_i are locally integrable on $\mathbb{R}^n$. Then $\mu = \nu$.*

PROOF. Set $Px = \bigl(l_1(x), \ldots, l_n(x)\bigr)$. Similarly to the previous example, the function f_i can be taken for a logarithmic derivative along $P(h_i)$ for both measures $\mu \circ P^{-1}$ and $\nu \circ P^{-1}$. By Theorem 7.6.3 we obtain $\mu \circ P^{-1} = \nu \circ P^{-1}$. Due to our assumption concerning K this gives the equality $\mu = \nu$. □

7.6. Uniqueness problems

We observe that in this theorem and the previous proposition we have an equality of two measures with equal logarithmic derivatives along some collection of vectors, but logarithmic gradients are not involved.

As we have seen in Corollary 7.6.8, it may occur that a centered Gaussian measure γ and a non-Gaussian probability measure μ have the same logarithmic gradient β_H which, in addition, is a continuous linear operator. Hence it is reasonable to ask under what condition on a linear operator A a probability measure μ is Gaussian if its logarithmic gradient β_H^μ coincides with A μ-a.e. The next result is a corollary of Theorem 7.6.4, but we give a simple direct proof suggested by our justification of Proposition 7.6.10.

7.6.12. Proposition. *Let μ be a Radon probability measure on X and let A be a continuous linear operator on X such that $\beta_H^\mu = A$ μ-a.e. If the functionals of the form $k \circ A$, where $k \in X^*$, separate the points in X, then μ is a centered Gaussian measure. In addition, it is a unique Radon probability measure whose logarithmic gradient along H is equal to A.*

PROOF. Let $k \in X^*$ and $l = k \circ A$. Our proof of Proposition 7.6.10 shows that $\mu \circ l^{-1}$ is a centered Gaussian measure on the real line. Let us equip X with the weakest locally convex topology τ with respect to which all functionals of the form $k \circ A$, where $k \in X^*$, are continuous. It follows from our assumption that this topology is Hausdorff. The measure μ remains Radonian also in this weaker topology τ. In addition, it remains Gaussian on (X, τ) since the dual to (X, τ) is the linear span of the indicated functionals. The obtained centered Radon Gaussian measure on (X, τ) will be denoted by μ^τ. Now we can conclude that for every $f \in X^*$ the measure $\mu \circ f^{-1}$ is centered Gaussian since the functional f is measurable with respect to μ^τ. This is clear from the fact that the measure μ^τ is concentrated on a countable union of compact sets in the original topology, but on such compact sets the original topology coincides with the topology τ, hence the functional f on these compact sets is continuous in the topology τ. The assertion about uniqueness is clear from the justification of Proposition 7.6.10. □

We observe that actually in place of existence of β_H^μ we could use the following condition: every functional $k \circ A$ is a logarithmic derivative along some $h \in X$. In such a formulation no H is required. Hence we arrive at the following result.

7.6.13. Corollary. *Let μ be a Radon probability measure on X and let A be a continuous linear operator on X such that the functionals $k \circ A$, where $k \in X^*$, are logarithmic derivatives of μ along some vectors. Suppose that the operator A is injective on a linear subspace of full μ-measure. Then μ is a centered Gaussian measure.*

7.6.14. Corollary. *Suppose that in the previous proposition A is a bijection. Then μ is a centered Gaussian measure. In addition, it is a unique Radon probability measure whose logarithmic gradient along H equals A.*

Thus, one can discuss the uniqueness problem for logarithmic derivatives when we fix a measure μ with such logarithmic derivatives and fix their Borel versions (say, a Borel version of β_H^μ), and then a measure ν is considered as a measure with the same logarithmic gradient provided that β_H^ν coincides ν-a.e. with *this fixed version* of β_H^μ. It is clear that the situation may be different for other versions of β_H^μ. One more possibility is that, given a mapping β, we do not fix a measure μ, but seek for any measure ν for which $\beta_H^\nu = \beta$ ν-a.e., where the mapping β may have another ν-version with the uniqueness property (the condition of weak uniqueness from Definition 7.6.1). The mapping β may be fixed due to some specific features of the problem under consideration, and measures with a logarithmic derivative β may be sought in some prescribed class. However, the mapping β does not always have a prescribed expression. For example, in quantum field theory, it is often part of the problem to give a rigorous definition of the mapping β which was originally given only heuristically on some small domain of definition. It is clear from the very beginning that the choice of H is important for most problems of this kind. Finally, the third possibility is to consider, say, only continuous logarithmic gradients or logarithmic gradients with some other special properties, or admit only measures from certain special classes. Some considerations in this direction are presented in the next section.

7.6.15. Remark. The problem of uniqueness of a measure with a given logarithmic derivative is close to the problem of uniqueness of a measure with given multipliers of quasi-invariance, i.e., a measure μ for which $\mu_h \sim \mu$ for all h from some set $K \subset X$, where we are given the Radon–Nikodym densities r_h of the measures μ_h with respect to μ (these densities are called multipliers of quasi-invariance). If logarithmic derivatives β_h^μ exist and $\exp[c(h)|\beta_h^\mu|] \in L^1(\mu)$ for some $c(h) > 0$, then from §6.5 we know an explicit expression for r_h (in §6.5 some other conditions for reconstructing r_h from β_h are indicated). On the other hand, if the functions $t \mapsto r_t(x+th)$ are locally absolutely continuous, then by the functions r_h we can find the logarithmic derivatives $\beta_h(x) = [r_t(x+th)]'_{t=0}$ (assuming their integrability). This shows the equivalence of both problems under broad conditions (certainly, the density r_h is not always differentiable).

Now we have the following trivial condition for the equality of measures in terms of multipliers of quasi-invariance.

7.6.16. Example. Let μ and ν be two Radon probability measures quasi-invariant along vectors from some set $S \subset X$ such that both measures possess equal Borel versions of multipliers of quasi-invariance r_h for all $h \in S$ and the functions r_h separate the measures on X with respect to which all r_h are integrable. Then one has $\mu = \nu$.

PROOF. The integral of $r_h(\mu - \nu)$ over X equals $\mu_h(X) - \nu_h(X) = 0$, whence $\mu = \nu$ by our assumption. □

Some of the results established above can be obtained by applying this condition to functions r_h of the type $\exp g$, where $g \in X^*$. Such functions arise for linear logarithmic derivatives. However, excepting such simple cases, considerably more subtle considerations are required for obtaining the uniqueness results.

7.7. Symmetries of measures and logarithmic gradients

As already mentioned above, logarithmic gradients are precisely the drift coefficients of symmetrizable diffusion processes (see Chapter 12 for a discussion of this subject). In applications (for example, in quantum field theory), diffusion processes often possess some symmetries. It would be interesting to understand which properties of the corresponding drifts are responsible for such symmetries. In particular, it is worth discussing which symmetries of invariant measures of diffusion processes can be expressed by means of the properties of logarithmic gradients.

One of the simplest symmetries is the central one. Let us recall that a Radon probability measure μ on a locally convex space X is called centrally symmetric (or just symmetric) if $\mu(B) = \mu(-B)$ for all Borel sets B. It is clear that for the logarithmic gradient of a symmetric measure μ one has the equality

$$\beta_H^\mu(-x) = -\beta_H^\mu(x) \quad \mu\text{-a.e.}$$

Is the converse true? In general, not. For example, let us take a probability density $f \in C_0^\infty(\mathbb{R}^1)$ with support in $[0,1]$ and put $\varrho(x) = \frac{1}{3}f(x)$ on $[0,\infty)$, $\varrho(x) = \frac{2}{3}f(-x)$ on $(-\infty, 0]$. Then ϱ is an asymmetric probability density on the real line such that $\varrho'(-x)/\varrho(-x) = -\varrho'(x)/\varrho(x)$ on the set $\{\varrho > 0\}$, i.e., $\varrho\, dx$-a.e. On the other hand, the answer is positive if μ is a unique probability with the logarithmic gradient β_H^μ. Indeed, in that case we obtain that the measure $\nu(B) = \mu(-B)$ coincides with μ since it has the same logarithmic gradient: $\beta_H^\nu(x) = -\beta_H^\mu(-x) = \beta_H^\mu(x)$.

7.7.1. Example. Let μ be a probability measure on $\mathbb{R}^n$ such that its logarithmic gradient β^μ exists, is locally integrable with respect to Lebesgue measure and $\beta^\mu(-x) = -\beta^\mu(x)$ μ-a.e. Then the measure μ is centrally symmetric.

In the infinite dimensional case one could require the continuity of β_H^μ. However, this does not help much. Example 7.6.6 gives such a situation. We have found there a centered Radon Gaussian measure γ on the space $X = \mathcal{D}'(\mathbb{R}^n)$ (one can realize this example also on some Hilbert space embedded into $\mathcal{D}'(\mathbb{R}^n)$), a densely embedded Hilbert space $H \subset X$ and a vector v such that the measures γ and γ_v have the same logarithmic gradient β along H, which is a continuous linear operator on X. The probability measure $\mu = (\gamma + \gamma_v)/2$ has the same logarithmic gradient β along H, but is not Gaussian. Existence of such examples is probably related to the smallness of the group of central symmetries. One could hope to get more for larger groups of symmetries. Let X be a locally convex space and let $T\colon G \to \mathcal{L}(X)$ be a measurable representation of a measurable group G. In place of continuous linear operators we can, of course, consider more general measurable nonlinear transformations; then we obtain a measurable action of G.

A Radon measure μ on X is called G-invariant if
$$\mu(T(g)B) = \mu(B) \quad \text{whenever } g \in G, B \in \mathcal{B}(X).$$
How are the properties of β_H^μ connected with the G-symmetry of μ? If G possesses a finite left-invariant measure Γ (for example, if this is a compact topological group), then the measure
$$\mu^G(B) = \int_\Gamma \mu(T(g)B)\,\Gamma(dg),$$
called the G-symmetrization of the measure μ, is G-symmetric. Suppose, in addition, that $T(g)\colon D(\mu) \to D(\mu)$ and
$$\int_\Gamma \|d_{T(g)h}\mu\|\,\Gamma(dg) < \infty, \quad \forall h \in j(X^*).$$
For example, let $\|T(g)\|_{\mathcal{L}(D(\mu))} \leqslant C$. Then the measure μ^G is differentiable along $j_H(X^*)$. It is natural to call its logarithmic gradient the G-symmetrization of β_H^μ.

To give a concrete example, we describe spherically invariant differentiable measures on $\mathbb{R}^n$ in terms of their logarithmic gradients. Certainly, any measure μ of this type has a density of the form $p(x) = g((x,x))$, where g is a function on the real line. This follows from the fact that its Fourier transform $\widetilde{p}$ is a continuous spherically invariant function, i.e., $\widetilde{p}$ is a function of the norm, and then p has a version that is a function of the norm. Therefore, $\nabla p(x) = 2g'((x,x))x$, whence we obtain
$$\nabla p(x)/p(x) = c(x)x,$$
where c is a real function. It turns out that this property is a complete characterization of spherically invariant differentiable measures.

7.7.2. Proposition. *A measure μ on $\mathbb{R}^n$ with the logarithmic gradient β^μ (with respect to $\mathbb{R}^n$) is spherically invariant if and only if there exists a Borel real function $c(\,\cdot\,)$ on $\mathbb{R}^n$ such that $\beta^\mu(x) = c(x)x$ μ-a.e. Such a function $c(\cdot)$ has a spherically invariant modification.*

PROOF. It only remains to prove the implication "if". Denote by p a density of μ. In order to show that p has a spherically invariant version, it is enough to prove that its Fourier transform is spherically invariant. To this end, it suffices to show that the function p is spherically invariant in the sense that the equality $p(Sx) = p(x)$ holds a.e. for every fixed orthogonal operator S that is a rotation in a two-dimensional plane. Thus, let T_t be a family of orthogonal transformations acting as follows: T_t is the rotation by

7.7. Symmetries of measures and logarithmic gradients

angle t in a two-dimensional plane L and T_t is the identity mapping on the orthogonal complement to L. According to Proposition 3.4.6, we can choose a version of p which is absolutely continuous on the circles $\{T_t x, t \in [0, 2\pi]\}$ for almost all x. Then, for such x, for almost all t we have

$$\partial p(T_t x)/\partial t = \bigl(\nabla p(T_t x), \partial T_t x/\partial t\bigr) = c(T_t x)(T_t x, \partial T_t x/\partial t) = 0$$

since $(T_t x, T_t x) \equiv (x, x)$ and so $\partial(T_t x, T_t x)/\partial t = 0$ for almost all t. By the absolute continuity of $p(T_t x)$ we arrive at the desired identity. $\square$

Similarly one can investigate other symmetries of measures μ.

7.7.3. Proposition. *Suppose that T is a representation of a connected Lie group G in $\mathcal{L}(\mathrm{I\!R}^n)$ and μ is a measure on $\mathrm{I\!R}^n$ with a logarithmic gradient β. Suppose that for every x the orbit $\{T(g)x\colon g \in G\}$ has a unique normal at the point $T(e)x$, given by a vector $A(x)$. Then μ is G-symmetric if and only if*

$$\beta(x) = c(x) A(x),$$

where $c(\,\cdot\,)$ is a real function on $\mathrm{I\!R}^n$.

PROOF. We employ reasoning completely analogous to that given above. Let $\beta(x) = c(x) Ax$ be the logarithmic gradient of μ. We take an element $\mathfrak{g}$ in the Lie algebra $\mathfrak{G}$ of the group G and put $g(t) = \exp(t\mathfrak{g})$, $S_t = T\bigl(g(t)\bigr)$. As in Remark 2.1.11, one can show that μ admits a density p such that the functions $F\colon t \mapsto p(S_t x)$ are absolutely continuous on $[-1, 1]$ for almost all x. Therefore,

$$\frac{\partial F}{\partial t} = \left(\nabla p(S_t x), \frac{\partial S_t x}{\partial t}\right) = c(S_t x) p(S_t x)\left(A(S_t x), \frac{\partial S_t x}{\partial t}\right) = 0.$$

This implies that μ is G-invariant. The converse is analogous. $\square$

In infinite dimensions there is no translation invariant Lebesgue measure and it is more difficult to detect nontrivial symmetries of measures. It is readily seen that Dirac's measure at the origin is a unique spherically invariant Radon probability measure on l^2. However, one can define H-spherically invariant measures on a locally convex space X as measures invariant with respect to the action of the group of orthogonal operators on H, where $H \subset X$ is a continuously and densely embedded Hilbert space generating the corresponding embedding $j\colon X^* \to H$.

7.7.4. Definition. *A Radon probability measure μ on the space X is called H-spherically invariant if its Fourier transform has the form*

$$\widetilde{\mu}(l) = \varphi\bigl(|j_{\scriptscriptstyle H}(l)|_{\scriptscriptstyle H}\bigr), \quad l \in X^*,$$

where φ is a continuous function on $\mathrm{I\!R}^1$.

An equivalent definition: if $l_1, \ldots, l_n \in X^*$ and $\bigl(j_{\scriptscriptstyle H}(l_i), j_{\scriptscriptstyle H}(l_j)\bigr)_{\scriptscriptstyle H} = \delta_{ij}$, then the image of the measure μ under the mapping $(l_1, \ldots, l_n)\colon X \to \mathrm{I\!R}^n$ is spherically symmetric (Exercise 7.9.9).

If H is infinite dimensional, then a probability measure μ on X without an atom at the origin is H-spherically invariant if and only if it is a mixture of Gaussian measures μ^t defined by $\mu^t(B) = \mu^1(B/t) = \gamma(B/t)$, where γ is the centered Gaussian measure with the Cameron–Martin space H, i.e.,

$$\mu(B) = \int_0^\infty \mu^t(B)\,\sigma(dt), \quad B \in \mathcal{B}(X),$$

where σ is a Borel probability measure on $(0, +\infty)$. For a proof, see Vakhania, Tarieladze, Chobanyan [**1167**, Theorem 4.2, p. 239]. To be more precise, in order to apply the cited theorem, we have to show that the function $\exp\bigl(-|j_H(l)|_H^2/2\bigr)$ on X^* is the Fourier transform of a centered Radon Gaussian measure γ because the theorem only gives the representation

$$\widetilde{\mu}(l) = \int_0^\infty \exp\bigl(-t^2|j_H(l)|_H^2/2\bigr)\sigma(dt).$$

Let us consider the embedding $X \to \mathbb{R}^T$, $T = X^*$, $x \mapsto \bigl(l(x)\bigr)_{l\in X^*}$. By Kolmogorov's theorem, on the space $\mathbb{R}^T$ we obtain a centered Gaussian measure γ with the required Fourier transform, so that the desired representation of μ holds on $\sigma(\mathbb{R}^T)$. We show that γ has a Radon extension on $\mathbb{R}^T$. It suffices to show that this is true at least for one of the measures μ_t with the Fourier transforms $\exp\bigl(-t^2|j_H(l)|_H^2/2\bigr)$. Let K be a compact set with $\mu(K) > 0$. Replacing K by its absolutely convex closed hull in $\mathbb{R}^T$ we may assume that K is absolutely convex and compact. By Theorem 1.4.11 it suffices to show that there is $t > 0$ with $\mu_t^*(K) > 0$. Suppose this is false. For every rational t we can find an absolutely convex set $C_t \in \sigma(\mathbb{R}^T)$ such that $K \subset C_t$ and $\mu_t(C_t) = 0$. The intersection C of this countable collection of sets contains K, belongs to $\sigma(\mathbb{R}^T)$, and is absolutely convex. In addition, it is readily seen that $\mu_t(C) = 0$ for all t (in fact, one has $\mu_t(C) \leqslant \mu_s(C)$ if $t \geqslant s$, see [**191**, §3.3]). Then $\mu(K) = 0$, which is a contradiction. Therefore, the measure γ extends to a Radon Gaussian measure γ_1 on $\mathbb{R}^T$. Then μ on $\mathbb{R}^T$ will coincide with the mixture of the measures $\gamma_t \colon B \mapsto \gamma_1(B/t)$, which follows from the equality of the corresponding Fourier transforms. Finally, we see that γ_1 is concentrated on the original space. Indeed, take a compact set $K \subset X$ with $\mu(K) > 0$. Then for its linear span L we obtain $\gamma_1(L) = 1$ since otherwise $\gamma_1(L) = 0$, hence $\gamma_t(L) = 0$ for all t.

There is no similar characterization in $\mathbb{R}^n$ since such a mixture must have a positive density. However, Proposition 7.7.2 enables us to describe differentiable H-spherically invariant measures both in the finite dimensional and infinite dimensional cases as measures having logarithmic gradients along H of the form $\beta_H(x) = c(x)x$, where $c(\,\cdot\,)$ is a real function.

7.7.5. Proposition. *Suppose that a Radon probability measure μ on the space X is differentiable along all vectors from $j_H(X^*)$ and that*

$$\beta_H(x) = c(x)x,$$

where c is a real Borel function on X. Then there exists a Borel probability measure σ on $(0, \infty)$ for which

$$\mu(B) = \int_0^\infty \gamma(B/t)\, \sigma(dt).$$

Conversely, if the measure μ defined by this formula has a logarithmic gradient β_H^μ along H, then β_H^μ has the form indicated above.

PROOF. It is clear from what has been said after Definition 7.7.4 that it suffices to show the following: if functionals $l_1, l_2 \in X^*$ are such that the vectors $e_i = j_H(l_i)$ are orthogonal in H and have unit norm, then $\widetilde{\mu}(t_1 l_1 + t_2 l_2)$ is a function of $t_1^2 + t_2^2$. Set $Px = l_1(x)e_1 + l_2(x)e_2$. Let β_P be the conditional expectation of $P\beta_H^\mu$ with respect to the σ-algebra generated by l_1, l_2. Then $\beta_P(x) = \beta(Px)$, where β is the logarithmic gradient of the measure $\mu \circ P^{-1}$ on the space $P(H)$ with the inner product from H. It is clear that β_P has the form $x \mapsto c_2(Px)Px$, where c_2 is the corresponding conditional expectation of the function c (it suffices to observe that $P\beta_H^\mu(x) = c(x)Px$). Thus, $\beta(z) = c_2(z)z$. It remains to apply Proposition 7.7.2.

Suppose that a measure μ is the mixture of measures $B \mapsto \gamma(B/t)$ and its logarithmic gradient β_H^μ exists. We show that $\beta_H^\mu(x) = c(x)x$. This follows at once from equality (4.5.4), but can be easily seen also from the finite dimensional case. Indeed, let us take a sequence $\{l_n\} \subset X^*$ separating the points in some linear subspace of full γ-measure. We may assume that the elements $e_n = j_H(l_n)$ form an orthonormal basis in the space H. Set $P_n(x) := \sum_{i=1}^n l_i(x)e_i$. Let B_n be the conditional expectation of the mapping $P_n \beta_H^\mu$ with respect to the σ-algebra generated by $l_1, \ldots, l_n$. Then $B_n(x) = \beta_n(P_n x)$, where β_n is the logarithmic gradient of the measure $\mu \circ P_n^{-1}$ on the space $H_n := P_n(X)$ with the inner product from H. By the spherical symmetry of $\mu \circ P_n^{-1}$ we have $\beta(z) = c_n(z)z$, where c_n is some Borel function on H_n. For every fixed i, the sequence $\langle l_i, \beta_n \circ P_n \rangle = \beta_{e_i}^{\mu \circ P_n^{-1}} \circ P_n$ converges in $L^1(\mu)$ to $\beta_{e_i}^\mu$ by the martingale convergence theorem. In addition, $\beta_n(P_n x) = c_n(P_n x) P_n x$. Since the functions $l_i \circ P_n$ converge in measure μ to the function l_i, the functions $c_n \circ P_n$ converge in measure μ to some limit $c(\,\cdot\,)$. Then μ-a.e. we have $\beta_{e_i}^\mu(x) = c(x)l_i(x)$, whence $\beta_H(x) = c(x)x$. $\square$

Formula (4.5.5) in Proposition 4.5.3 gives an explicit expression for the function c: $c = \tau^{-1}$, where $\tau(x) = \lim_{n \to \infty} n^{-1} \sum_{i=1}^\infty \widehat{e}_i(x)^2$ (this limit exists μ-a.e.). It is seen from this that c does not depend on σ, hence different mixtures lead to the same mapping β_H^μ. Though, such mappings are discontinuous, so they do not give nontrivial examples of nonuniqueness of a measure with a given logarithmic gradient (we recall that without requirement of continuity such measures exist even on the real line, which produces trivial examples in all dimensions by considering products). In general, there is no explicit expression for the logarithmic gradient of a mixture of measures.

7.8. Mappings and equations connected with logarithmic gradients

Let ν be a Radon probability measure on a locally convex space X having a logarithmic gradient β_H^ν along a densely embedded Hilbert space $H \subset X$. We shall say that ν is *H-ergodic* if, for every function $f \in L^1(\nu)$, the condition $\partial_h f = 0$ for all $h \in H$ implies that $f = \text{const}$, where $\partial_h f$ is defined in the sense of the integration by parts formula

$$\int_X \varphi(x) \partial_h f(x)\, \nu(dx) = -\int_X f(x) [\partial_h \varphi(x) + \varphi(x) \beta_h^\nu(x)]\, \nu(dx)$$

for all $\varphi \in \mathcal{FC}_b^\infty$, provided that $f \beta_h^\nu \in L^1(\nu)$.

7.8.1. Example. If μ and ν are two Radon probability measures with $\mu \ll \nu$ and $\beta_H^\mu = \beta_H^\nu$ such that ν is H-ergodic, then $\mu = \nu$.

PROOF. Let $\mu = f \cdot \nu$, $\varphi \in \mathcal{FC}_b^\infty$, $h \in H$. Then

$$\int_X f(x) \partial_h \varphi(x)\, \nu(dx) = \int_X \partial_h \varphi(x)\, \mu(dx) = -\int_X \varphi(x) \beta_h^\mu(x)\, \mu(dx)$$
$$= -\int_X f(x) \varphi(x) \beta_h^\nu(x)\, \nu(dx),$$

whence

$$\int_X \varphi(x) \partial_h f(x)\, \nu(dx) = -\int_X f(x) [\partial_h \varphi(x) + \varphi(x) \beta_h^\nu(x)]\, \nu(dx) = 0.$$

Thus, $\partial_h f = 0$ and so $f = 1$ by the ergodicity of ν. $\square$

If ν is not H-ergodic, then for a suitable function f the measure $f \cdot \nu$ has the same logarithmic gradient along H.

Sometimes the following trivial observation helps in finding a measure ν with a given logarithmic gradient β. Let μ be a probability measure with a logarithmic gradient β_H^μ and let a function $f \in L^1(\mu)$ be such that $f \in L^1(d_h \mu)$, $\partial_h f \in L^1(\mu)$ for all $h \in H$. We define a measure ν by $\nu := f \cdot \mu$ and observe that the equality $d_h \nu = (\partial_h f / f + \beta_h^\mu) \cdot \nu$ holds. Therefore,

$$\beta_H^\nu = \nabla_H f / f + \beta_H^\mu.$$

Hence the initial problem for β reduces to a linear differential equation

$$\nabla_H f = f(\beta - \beta_H^\mu).$$

In terms of the vector measure $D\mu$ the problem of finding a measure with the logarithmic gradient β consists in finding a measure satisfying the equation

$$D\mu = \beta \cdot \mu.$$

Also of interest is the inverse problem of finding a vector field v for which

$$d_v \mu = \beta \cdot \mu$$

for a given measure μ and a given function $\beta \in L^1(\mu)$, where the differentiability along a vector field is defined in a natural way (see §§6.6, 8.10, 11.1).

One can define vector logarithmic derivatives of higher order. Denote by $SL_n(X^*)$ the space of all n-linear symmetric mappings $Q\colon (X^*)^n \to \mathbb{R}^1$, equipped with the weak topology. Let H be the same as in Definition 7.2.1. Set $j := j_H$. Suppose that a measure μ is n-fold differentiable along all vectors in $j(X^*)$. The Radon–Nikodym density of the measure $d_{h_1}\cdots d_{h_n}\mu$ with respect to μ will be denoted by $\beta^{(n)}_{h_1,\ldots,h_n}$.

7.8.2. Definition. *A measurable mapping $\Lambda_n\colon X \to SL_n(X^*)$ will be called a vector logarithmic derivative of the measure μ of order n along H if, for all $k_i,\ldots,k_n \in X^*$, one has the equality*

$$\Lambda_n(x)(k_1,\ldots,k_n) = \beta^{(n)}_{j(k_1),\ldots,j(k_n)}(x) \quad \mu\text{-a.e.}$$

A somewhat different definition is obtained if we require that the mapping Λ be with values in the space of symmetric operators from $SL_{n-1}(X^*)$ to X. In the most interesting cases this gives the same definition. For example, if X is a Hilbert space and $n=2$, then we can consider Λ_2 as a mapping with values in $\mathcal{L}(X^*,X)$. Under some additional technical conditions one has

$$\Lambda_2(x)(k,l) = \partial_{j(k)}\beta_{j(l)}(x) + \beta_{j(k)}(x)\beta_{j(l)}(x).$$

Therefore,

$$\Lambda_2(x) = D_x\beta_H(x) \circ j + \beta_H(x) \otimes \beta_H(x).$$

Similar, but more complicated identities relate vector logarithmic derivatives of higher order with derivatives of the mapping β_H. It should be noted that expressions of this kind may be meaningful only as a whole without making sense for every separate ingredient. It would be interesting to find sufficient conditions for a given mapping to be a vector logarithmic derivative of order n (this would be a variant of an infinite dimensional Frobenius-type theorem). A related problem is uniqueness of a probability measure with a given logarithmic derivative of order n. As we already know, such uniqueness may fail even for centered Gaussian measures (see Theorem 7.6.7).

7.9. Comments and exercises

Logarithmic gradients of measures were introduced in Albeverio, Høegh-Krohn [41], [42] and later studied by many authors. The first survey on this was given in Bogachev [188]. Example 7.2.6 is borrowed from Bogachev, Röckner [219]. Proposition 7.4.4, characterizing logarithmic gradients of Gaussian measures, was obtained in Bogachev [189, §6.2] (see also Bogachev [191, Proposition 7.6.3]); its partial case for Hilbert spaces noted in Example 7.4.7 was later given in Weizsäcker, Smolyanov [1190]. Theorem 7.4.8 was derived in Bogachev, Röckner [219] from a result of Röckner, Zhang [963]. Corollary 7.4.9 in the case of continuous β_H^μ was noted in Kirillov [607], [611], however, the proof employed the continuity of the logarithmic gradients of the finite dimensional projections of the measure. It remains open whether the latter follows from the continuity of β_H^μ. Sufficient conditions for the continuity of the logarithmic gradients of finite dimensional projections are given in Exercise 8.14.3; it would be interesting to find weaker conditions.

The method of Lyapunov functions is a standard tool for constructing stationary distributions of diffusion processes; when applied to a symmetric process with a constant diffusion coefficient it gives exactly a measure with a given logarithmic gradient. However, directly for constructing measures with given logarithmic gradients, without constructing any diffusions, the method of Lyapunov functions was first applied by A.I. Kirillov (see Kirillov [**607**]–[**615**]). This enabled us to considerably reinforce the efficiency of the method and gave the first constructive conditions for existence of measures with given logarithmic gradients. Lyapunov functions were employed in Bogachev, Röckner [**217**], [**219**], [**220**], where elliptic equations for measures were introduced and studied, the partial case of which (corresponding to symmetric operators) is the problem of finding a measure with a given logarithmic gradient; in these works, various a priori estimates like Proposition 7.5.3 are obtained. For further developments of these ideas along with applications for constructing Gibbs distributions, see Albeverio et al. [**44**], [**47**] and Bogachev, Röckner [**222**]. The latter work, which we follow in our presentation, contains some additional results on the existence problem. Closely related problems are considered in Cattiaux, Roelly, Zessin [**273**].

Theorem 7.6.3 was proved in Bogachev, Röckner [**219**]. The assertion that in Theorem 7.6.4 one has $\mathcal{G}_K^{\gamma^A} = \{\gamma^A\}$ if $M = \{0\}$, which follows directly from the condition for equality of two measures with equal multipliers of quasi-invariance, was noted in the papers Bogachev [**185**], [**186**] under a redundant assumption of continuity of $\beta_{j(k)}^{\gamma^A}$, and in the special case $A = I$ this was noted in Norin, Smolyanov [**835**], Roelly, Zessin [**964**]. Trivial examples of distinct probability measures on the real line with equal logarithmic derivatives give, of course, examples also for infinite dimensional spaces by forming products. The first nontrivial examples in the infinite dimensional case (i.e., with continuous logarithmic gradients) were constructed in Bogachev [**187**], Bogachev, Röckner [**217**], [**219**]. An explicit form of the function c in the expression $c(x)x$ for the logarithmic gradient of a mixture of Gaussian measures $\gamma(t\,\cdot\,)$ was indicated in Norin, Smolyanov [**835**]. Logarithmic derivatives of measures are considered also in Mayer-Wolf [**792**].

Exercises

7.9.1. Let μ and ν be two probability measures on $\mathbb{R}^n$ and let β be a Borel vector field serving as their common logarithmic gradient. Suppose that we have $\exp(c|\beta|) \in L^1_{\mathrm{loc}}(\mu)$ for some $c > 0$. Prove that $\mu = \nu$.

HINT: The measure μ has a continuous strictly positive density, whence we obtain the local integrability of $|\beta|$ with respect to Lebesgue measure, which enables us to apply Theorem 7.6.3.

7.9. Comments and exercises

7.9.2. Show that on the real line the logarithmic derivative of every differentiable probability measure satisfies the condition of weak uniqueness from Definition 7.6.1.

HINT: In the case where a continuous version ϱ of a density of the given measure is strictly positive, the claim is obvious. If $\varrho(a) = 0$ and $\varrho(x) > 0$ for $x > a$ sufficiently close to a, then one can take the measure with a density ϱ_0, where $\varrho_0(x) = c\varrho(x)$ if $x \in [a,b]$, where b is the nearest zero of ϱ in $(a, +\infty)$, and $\varrho_0 = 0$ outside $[a, b]$.

7.9.3. Give an example of a countably additive measure with values in l^2 having unbounded variation.

7.9.4. Let μ be a Borel probability measure μ on $\mathbb{R}^\infty$ and let h be such that $\mu_h \sim \mu$. Let $f \in L^1(\mu)$ and let $\mathbb{E}_n$ denote the conditional expectation with respect to the σ-algebra generated by the first n coordinates. Show that the functions $(\mathbb{E}_n f)(x - h)$ converge in measure to the function $f(x - h)$.
HINT: Use Exercise 3.9.10.

7.9.5. Suppose that a convex probability measure μ on $\mathbb{R}^\infty$ has a logarithmic gradient $\beta = (\beta^i)$ with respect to l^2 and is quasi-invariant along all vectors with finitely many nonzero coordinates. Prove that the mapping β is monotone in the following sense:
$$\sum_{i=1}^{k} \bigl(\beta^i(x+h) - \beta^i(x)\bigr) h_i \leqslant 0 \quad \text{for any } k \text{ and any } h = (h_1, \ldots, h_k, 0, 0, \ldots).$$
HINT: Consider first a convex measure on $\mathbb{R}^n$ with a density $\exp(-V)$, where V is a finite convex function, and show that $\bigl(\nabla V(x+h) - \nabla V(x), h\bigr) \leqslant 0$ using that the function $t \mapsto V(x + th)$ is convex; then apply the previous exercise and use that the conditional expectations of β^i, $i \leqslant n$, with respect to the σ-algebra generated by the first n coordinates coincide with the logarithmic derivatives along the standard basis vectors of the projection of μ on $\mathbb{R}^n$.

7.9.6. Let p_n be a probability density on $\mathbb{R}^1$ vanishing outside of the closed interval $[-2^{-n}, 1]$, equal to 1 on $[0, 1 - 2^{-n}]$, and defined as an affine function on the remaining two intervals so that it is globally continuous. Let $\mu = \bigotimes_{n=1}^\infty \mu_n$.
(i) Prove that $D(\mu) = l^1$, the series $W(x) = \sum_{n=1}^\infty \ln p_n(x_n)$ converges μ-a.e. and in $L^2(\mu)$ and that $\partial_{e_n} W = \beta^\mu_{e_n}$ μ-a.e., where $\{e_n\}$ is the standard basis in l^1.
(ii) Let H_0 be the weighted Hilbert space of sequences $x = (x_n)$ such that $\|x\|_0^2 = \sum_{n=1}^\infty t_n^{-2} x_n^2 < \infty$, where $t_n > 0$ and $\sum_{n=1}^\infty t_n^2 < \infty$. Define a continuous linear operator T on $\mathbb{R}^\infty$ by $Tx = (t_n^{-1} x_n)$. Set $W_0(x) := W(T^{-1}x)$. Show that the measure $\nu := \mu \circ T^{-1}$ is differentiable along $H = l^2 = T(H_0)$ and $\partial_{e_n} W_0 = \beta^\nu_{e_n}$ ν-a.e. for all n.
HINT: (i) Observe that $\sup_n \|p_n'\|_{L^1} < \infty$, hence for any vector $h = (h_n) \in l^1$ the series of $h_n p_n'(x_n)/p_n(x_n)$ converges in $L^1(\mu)$; conversely, show that convergence of this series in $L^1(\mu)$ yields convergence of the series of $|h_n|$; show that the series of $\|\ln p_n(x_n)\|_{L^2(\mu)} = \|\ln p_n(x_n)\|_{L^2(\mu_n)}$ is convergent and that one has the equality $\beta^\mu_{e_n}(x) = p_n'(x_n)/p_n(x_n)$. (ii) Observe that $\beta^\nu_{Th}(x) = \beta^\mu_h(T^{-1}x)$.

7.9.7. Let Z be a Banach space, let φ be a Z-valued measure of bounded semivariation, and let $T : Z \to X$ be an absolutely summing mapping. Prove that the measure $T(\varphi)$ has bounded variation.

HINT: Let φ be defined on a measurable space $(\Omega, \mathcal{A})$; observe that for any countable family of disjoint sets $A_i \in \mathcal{A}$, the series of $\varphi(A_i)$ converges unconditionally in Z, hence the series of $\|T\varphi(A_i)\|_X$ converges; assuming that $T\varphi$ has unbounded variation, obtain a contradiction.

7.9.8. Let μ be a Radon probability measure on a locally convex space X and let E be a finite dimensional subspace in X such that the logarithmic derivatives β_h^μ for $h \in E$ are bounded on compact sets. Prove that conditional measures μ^y on the slices $y + E$ have continuous positive densities. In particular, this is true if $E \subset j_H(X^*)$, where H is a separable Hilbert space densely embedded into X and μ has a logarithmic gradient β_H^μ bounded on compact sets.

7.9.9. Show that a measure μ is H-spherically invariant (see Definition 7.7.4) if and only for all $l_1, \ldots, l_n \in X^*$ with $\bigl(j_H(l_i), j_H(l_j)\bigr)_H = \delta_{ij}$, the image of the measure μ under the mapping $(l_1, \ldots, l_n)\colon X \to \mathbb{R}^n$ is spherically symmetric.

HINT: Use that a measure on $\mathbb{R}^n$ is spherically invariant precisely when its Fourier transform is spherically invariant, i.e., is a function of norm.

7.9.10. Let H be a separable Hilbert space densely embedded into a locally convex space X, let $j\colon X^* \to H$ be the associated embedding, and let μ be a Radon measure on X differentiable along all vectors in $j(X^*)$ and possessing a logarithmic gradient β along H. Suppose that $T\colon X \to X$ is a continuous invertible linear operator such that $\mu \circ T^{-1} = \mu$. Assume additionally that $T(H) \subset H$ and that there is a continuous linear operator $S\colon X \to X$ such that $S(H) \subset H$ and $S|_H = (T|_H)^*$. Prove that $S\beta(Tx) = \beta(x)$ a.e.

HINT: Observe that $T\colon H \to H$ is continuous by the closed graph theorem; use the invariance of μ with respect to T to show that for every $l \in X^*$ one has $\beta_{Tj(l)}^\mu(Tx) = \beta_{j(l}^\mu(x)$ a.e.; then show that $Tj(l) = j(l \circ S)$ using that
$$\bigl(j(l\circ S), v\bigr)_H = l(Sv) = l\bigl((T|_H)^* v\bigr) = \bigl(Tj(l), v\bigr), \quad v \in H.$$

7.9.11. Show that if a probability measure μ on $\mathbb{R}^n$ has an infinitely differentiable logarithmic gradient, then it has an infinitely differentiable density.

7.9.12. Suppose that a probability measure μ on $\mathbb{R}^n$ has a logarithmic gradient β and that F is a diffeomorphism of $\mathbb{R}^n$. Find the local logarithmic gradient of the measure $\mu \circ F^{-1}$.

7.9.13. Let ϱ be a continuous probability density on $\mathbb{R}^1$ of the class $W^{1,1}(\mathbb{R}^1)$. Find all extreme points in the set of probability measures on $\mathbb{R}^1$ having the logarithmic derivative $\beta = \varrho'/\varrho$, where we set $\beta = 0$ on the set of zeros of ϱ.

7.9.14. Let H be a separable Hilbert space densely embedded into a locally convex space X, let $j\colon X^* \to H$ be the associated embedding, and let μ be a Radon measure on X differentiable along all vectors in $j(X^*)$ and possessing a logarithmic gradient β_H^μ along H. Let T be an invertible operator on H and let E be the Hilbert space obtained by introducing the inner product $(u,v)_E = (Tu, Tv)_H$ on H. Study the existence of the logarithmic gradient β_E^μ along E.

CHAPTER 8

Sobolev classes on infinite dimensional spaces

This chapter is devoted to development of the principal ideas of the theory of Sobolev spaces in the case of infinite dimensional spaces with measures. The key constructions are similar, but the infinite dimensionality brings considerable specifics. The Sobolev classes $W^{p,k}$ still consist of integrable functions with finite integral norms of derivatives, but the derivatives, as well as their norms, can be understood in diverse ways. In addition, here we have no canonical Lebesgue measure; hence in place of measures with smooth densities differentiable measures come into play. Another interesting Sobolev class arises in connection with the integration by parts formula and generalized derivatives. Finally, there are schemes leading to continuous scales of spaces and using semigroups or interpolation spaces.

8.1. The classes $W^{p,r}$

Let $\mathcal{FC}_b^\infty = \mathcal{FC}_b^\infty(X)$ denote the class of all functions f on a given locally convex space X of the form

$$f(x) = \varphi\big(l_1(x),\ldots,l_n(x)\big), \ l_i \in X^*, \ \varphi \in C_b^\infty(\mathbb{R}^n).$$

Let $H \subset X$ be a densely embedded separable Hilbert space and let $j\colon X^* \to H$ be the corresponding embedding (see §7.1). Suppose that we are given a nonnegative Radon measure μ on X. We shall discuss several definitions of Sobolev classes with respect to μ, which turn out to be equivalent in the Gaussian case. As in the finite dimensional case, there are at least three possibilities to introduce Sobolev classes. The first one is to take completions of some class of nice functions (for example, $\mathcal{FC}_b^\infty$) with respect to some Sobolev norms. Another definition arises if we consider functions with sufficiently integrable partial derivatives (defined as directional derivatives or by means of integration by parts formulas). Finally, one can define Sobolev classes by means of a certain semigroup associated to the logarithmic gradient β_H^μ (provided it exists). We discuss all these approaches consequently. As above, let

$$\partial_h f(x) := \lim_{t \to 0} t^{-1}[f(x+th) - f(x)].$$

Suppose that the measure μ satisfies the following condition (which is automatically fulfilled if μ has full support, that is, is positive on all nonempty open sets):

$$\text{if } f, g \in \mathcal{F}C_b^\infty, \ f = g \ \mu\text{-a.e., then } D_H f = D_H g \ \mu\text{-a.e.} \tag{8.1.1}$$

Let $\{e_n\}$ be an orthonormal basis in H. For $p \geqslant 1$ and $r \in \mathbb{N}$ the Sobolev norm $\|\cdot\|_{p,r}$ is defined by the formula

$$\|f\|_{p,r} := \sum_{k=0}^{r} \left(\int_X \left[\sum_{i_1,\ldots,i_k=1}^{\infty} (\partial_{e_{i_1}} \cdots \partial_{e_{i_k}} f)^2 \right]^{p/2} \mu(dx) \right)^{1/p}. \tag{8.1.2}$$

The same norm can be written as

$$\|f\|_{p,r} = \sum_{k=0}^{r} \|D_H^k f\|_{L^p(\mu, \mathcal{H}_k)}.$$

For example, for $r = 1$ we obtain

$$\|f\|_{p,1} = \|f\|_{L^p(\mu)} + \|D_H f\|_{L^p(\mu, \mathcal{H})}.$$

If $f \in \mathcal{F}C_b^\infty$, then $\|f\|_{p,r} < \infty$ since $\|D_H^k f\|_{\mathcal{H}_k} \in C_b(X)$. Denote by $W^{p,r}(\mu)$ or by $W_H^{p,r}(\mu)$ the completion of $\mathcal{F}C_b^\infty$ with respect to the norm $\|\cdot\|_{p,r}$. Set

$$W^\infty(\mu) := W_H^\infty(\mu) := \bigcap_{p>1, r>1} W^{p,r}(\mu), \quad W^{\infty,r}(\mu) := \bigcap_{p>1} W^{p,r}(\mu),$$

$$W^{p,\infty}(\mu) := \bigcap_{r>1} W^{p,r}(\mu).$$

The classes $W^{p,r}$ are obtained as abstract completions of $\mathcal{F}C_b^\infty$ with respect to some norms majorizing the norm of L^p. So we would like to consider these classes as embedded into $L^p(\mu)$ in such a way that to every element $f \in W^{p,r}(\mu)$ we associate generalized derivatives $D_H^k f \in L^p(\mu, \mathcal{H}_k)$, $k \leqslant r$. For instance, for every function $f \in W^{p,1}(\mu)$ we obtain at least one mapping

$$D_H f \colon X \to H$$

since there exists a sequence of smooth cylindrical functions φ_j convergent to f in $L^p(\mu)$ and fundamental in the norm $\|\cdot\|_{p,1}$, therefore, we can set $D_H f = \lim_{j\to\infty} D_H \varphi_j$. However, the following unpleasant thing can occur: another sequence of smooth cylindrical functions ψ_j may exist that is also convergent to f in $L^p(\mu)$ and fundamental in the Sobolev norm, but having some other limit for the mappings $D_H \psi_j$ in $L^p(\mu, H)$. This may occur even for measures with full support on the real line: recall Example 2.6.1. In order that a mapping $D_H f$ be unique in $L^p(\mu, H)$, certain additional restrictions on the measure are needed.

Similarly one defines Sobolev spaces $W^{p,r}(\mu, E)$ of mappings with values in a separable Hilbert space E. The corresponding norms are denoted by $\|\cdot\|_{p,r,E}$. In place of $D_H^n F$ we also write $\nabla_H^n F$.

8.1. The classes $W^{p,r}$

8.1.1. Definition. *We shall say that the norm $\|\cdot\|_{p,r}$ is closable if, for every sequence $\{\varphi_j\} \subset \mathcal{F}C_b^\infty$ that is Cauchy in this norm and converges to 0 in $L^p(\mu)$, we have $\|\varphi_j\|_{p,r} \to 0$.*

For a closable norm $\|\cdot\|_{p,r}$, the elements of $W^{p,r}(\mu)$ can be regarded as functions in $L^p(\mu)$, and for every such element f, there are uniquely defined mappings $D_H^k f \in L^p(\mu, \mathcal{H}_k)$, $k = 1, \ldots, r$. As in the finite dimensional case (see Chapter 2), the closability of $\|\cdot\|_{p,1}$ implies that of $\|\cdot\|_{p,r}$ for all finite $r > 1$. If $f \in W^{p,r}(\mu)$ and $h_1, \ldots, h_r \in H$, then we put $\partial_{h_1} \cdots \partial_{h_r} f(x) := D_H f^r(x)(h_1, \ldots, h_r)$.

8.1.2. Proposition. *Let $p \in [1, +\infty)$, $q = p(p-1)^{-1}$. Let $\beta_{e_i}^\mu \in L^q(\mu)$ for some orthonormal basis $\{e_i\}$ in H. Then the norm $\|\cdot\|_{p,r}$ is closable. If for all $k \leqslant r$ the measure μ has derivatives $d_{v_1} \cdots d_{v_k} \mu$ along all vectors $v_1, \ldots, v_k$ from the linear span of $\{e_i\}$ and these derivatives possess densities with respect to μ of the class $L^q(\mu)$, then for any function $f \in W^{p,r}(\mu)$ the mappings $D_H^k f$, $k = 1, \ldots, r$, are Sobolev derivatives, that is, for all $\psi \in \mathcal{F}C_b^\infty$ we have*

$$\int_X D_H^k f(x)[e_{i_1}, \ldots, e_{i_k}] \psi(x)\, \mu(dx) = (-1)^k \int_X f(x)\, d_{e_{i_1}} \cdots d_{e_{i_k}}(\psi \cdot \mu)(dx). \tag{8.1.3}$$

PROOF. Let $\varphi_j \in \mathcal{F}C_b^\infty$ be such that $\varphi_j \to 0$ in $L^p(\mu)$ and $\partial_{e_i} \varphi_j \to F_i$ in $L^p(\mu)$ for each i. We show that $F_i = 0$. For every function $\psi \in \mathcal{F}C_b^\infty$ we have

$$\int_X F_i(x)\psi(x)\,\mu(dx) = \lim_{j \to \infty} \int_X \partial_{e_i}\varphi_j(x)\psi(x)\,\mu(dx)$$
$$= -\lim_{j \to \infty} \int_X \varphi_j(x)[\partial_{e_i}\psi(x) + \psi(x)\beta_{e_i}^\mu(x)]\,\mu(dx) = 0,$$

whence $F_i = 0$ a.e. This yields a similar assertion for mappings with values in any Banach space, which shows the closability of the norms $\|\cdot\|_{p,r}$. Equality (8.1.3) is verified similarly since the measures $d_{e_{i_1}} \cdots d_{e_{i_k}}(\psi \cdot \mu)$ have densities with respect to μ belonging to $L^q(\mu)$. □

As in the one-dimensional case, the differentiability along e_i is not necessary for closability. Theorem 2.6.4 gives a weaker condition of closability.

8.1.3. Theorem. *Suppose that for every fixed i the conditional measures μ^x on the straight lines $x + \mathbb{R}^1 e_i$ satisfy the condition of Theorem 2.6.4 with some $p > 1$ for μ-a.e. x. Then the norms $\|\cdot\|_{p,r}$ are closable.*

For example, it suffices that for every fixed i the conditional measures μ^x on the straight lines $x + \mathbb{R}^1 e_i$ have densities that are locally separated from zero or continuous. Our reasoning in Example 2.6.3 shows that if for the measure μ the norm $\|\cdot\|_{p,r}$ is closable, then for the measure $\nu = \varrho \cdot \mu$, where $\varrho^{-q/p} \in L^1(\mu)$, the norm $\|\cdot\|_{p,r}$ corresponding to the measure ν is closable as well. It was shown in Albeverio, Röckner [53] that for $p = 2$ and fixed i the Hamza condition from Theorem 2.6.4 for almost all conditional measures on the straight lines $x + \mathbb{R}^1 e_i$ is necessary for the closability of the partial norm $f \mapsto \|f\|_{L^2(\mu)} + \|\partial_{e_i} f\|_{L^2(\mu)}$, i.e., the closability of the indicated partial norm is equivalent to the Hamza condition for conditional measures on the straight lines parallel to e_i. It follows that the same is true also for $p \geqslant 2$ (apparently, for all $p > 1$, but this is not proved yet). However, we recall that by virtue of Pugachev's example the closability of the entire Sobolev norm does not imply the closability of partial norms even on the two-dimensional plane, as we have already noted in Chapter 2.

It is natural to consider also the class $W_*^{p,r}(\mu)$ obtained as the completion of the class of all bounded Borel functions f on X which are infinitely differentiable along all vectors in H and have finite norms $\|f\|_{p,r}$. It is not known whether the equality $W_*^{p,r}(\mu) = W^{p,r}(\mu)$ is true for differentiable measures in the general case. In the Gaussian case, this equality holds.

8.2. The classes $D^{p,r}$

As in the finite dimensional case, one can introduce Sobolev classes by means of directional differential properties. Let μ be a nonnegative Radon measure on a locally convex space X and let $H \subset X$ be a continuously and densely embedded separable Hilbert space. Let E be one more separable Hilbert space.

8.2.1. Definition. *Let $F\colon X \to E$ be μ-measurable. The mapping F is called absolutely ray continuous if, for every $h \in H$, the mapping F has a modification F_h such that for every $x \in X$ the mapping $t \mapsto F_h(x + th)$ is absolutely continuous on bounded intervals.*

8.2.2. Definition. *A μ-measurable mapping $F\colon X \to E$ is called stochastically Gâteaux differentiable if there exists a measurable mapping $D_H F\colon X \to \mathcal{H}(H, E)$ such that for every $h \in H$ we have*

$$\frac{F(x+th) - F(x)}{t} - D_H F(x)(h) \xrightarrow[t \to 0]{} 0 \quad \text{in measure } \mu.$$

The derivative of the nth order $D_H^n F$ is defined inductively as $D_H(D_H^{n-1} F)$. An alternative notation is $\nabla_H^n F$.

8.2.3. Definition. *Let $1 \leqslant p < \infty$. The space $D^{p,1}(\mu, E)$ is defined as the class of all mappings $f \in L^p(\mu, E)$ such that f is ray absolutely continuous, stochastically Gâteaux differentiable and $D_H f \in L^p(\mu, \mathcal{H}(H, E))$. The space $D^{p,1}(\mu, E)$ (another notation is $D_H^{p,1}(\mu, E)$) is equipped with the norm*

$$\|f\|_{p,1,E}^0 := \|f\|_{L^p(\mu)} + \|D_H f\|_{L^p(\mu, \mathcal{H}(H,E))}.$$

For $r = 2, 3, \ldots$ *we define the classes* $D^{p,r}(\mu, E) = D_H^{p,r}(\mu, E)$ *inductively:*

$$D^{p,r}(\mu, E) := \left\{ f \in D^{p,r-1}(\mu, E) \colon D_H f \in D^{p,r-1}(\mu, \mathcal{H}(H, E)) \right\}.$$

8.2. The classes $D^{p,r}$

The corresponding norms are defined by the equalities

$$\|f\|^0_{p,r,E} := \|f\|_{L^p(\mu)} + \|D_H f\|_{L^p(\mu,\mathcal{H}(H,E))} + \cdots + \|D_H^r f\|_{L^p(\mu,\mathcal{H}_r(H,E))}.$$

We put $D^{p,r}(\mu) := D_H^{p,r}(\mu) := D^{p,r}(\mu, \mathbb{R}^1)$. Set

$$D^\infty(\mu) := D_H^\infty(\mu) := \bigcap_{r\geqslant 1, p\in[1,+\infty)} D^{p,r}(\mu), \quad D^{\infty,r}(\mu) := \bigcap_{p\in[1,+\infty)} D^{p,r}(\mu),$$

$$D^{p,\infty}(\mu) := \bigcap_{r\geqslant 1} D^{p,r}(\mu).$$

These spaces can be naturally equipped with countable families of norms from the respective classes $D^{p,r}(\mu)$ making them Fréchet spaces.

If the measure μ is continuous along all vectors in H, then for a version of $F \in D^{p,1}(\mu, E)$ that is absolutely continuous along $h \in H$ the partial derivative $\partial_h F(x)$ exists μ-a.e. (since on the straight lines $x + \mathbb{R}^1 h$ conditional measures are absolutely continuous and E has the Radon–Nikodym property, see §1.1), in addition, one has $\partial_h F(x) = D_H f(x)[h]$ μ-a.e.

We observe that the gradient $D_H \colon D^{p,r}(\mu, E) \to D^{p,r-1}(\mu, \mathcal{H}(H, E))$ is a continuous operator.

8.2.4. Theorem. *Let $p \geqslant 1$, $r \in \mathbb{N}$, and let the measure μ satisfy the following condition: for every $h \in H$, conditional measures μ^x on the straight lines $x + \mathbb{R}^1 h$ possess densities ϱ^x such that the functions $t \mapsto \varrho^x(t)$ are locally separated from zero (or in the case $p > 1$ the functions $t \mapsto |\varrho^x(t)|^{-q/p}$ are locally Lebesgue integrable). Then the spaces $D^{p,r}(\mu, E)$ are complete.*

PROOF. In the case $X = E = \mathbb{R}^1$ this is clear from our reasoning in Example 2.6.3. This reasoning applies also to any separable Hilbert space E. In the case of a general space X and $r = 1$, any sequence $\{f_n\}$ that is fundamental in $D^{p,1}(\mu, E)$ has a limit f in $L^p(\mu, E)$ and $\{D_H f_n\}$ has some limit G in $L^p(\mu, \mathcal{H}(H, E))$. Let $h \in H$ and let Y be a hyperplane topologically complementing the line $\mathbb{R}^1 h$. Denote by ν the projection of μ on Y. Passing to a subsequence we may assume that for ν-a.e. $y \in Y$ the mappings $t \mapsto f_n(y + th)$ converge to $t \mapsto f(y + th)$ in $L^p(\mu^y, E)$, and the mappings $t \mapsto D_H f_n(y + th)(h)$ converge to $t \mapsto G(y + th)(h)$ in $L^p(\mu^y, \mathcal{H}(H, E))$ (see Proposition 1.3.1). For such y we obtain convergence of the mappings $t \mapsto f_n(y + th)$ and their derivatives in $L^1[-R, R]$ for all R. We take the version f defined by the equality $f(y + th) = \lim_{n\to\infty} f_n(y + th)$ at the points where this limit exists and by the equality $f(y + th) = 0$ at all other points. For ν-a.e. $y \in Y$, there is a point of the first kind on the straight line $y + \mathbb{R}^1 h$. Therefore, on such lines we have pointwise convergence of f_n to f, the absolute continuity of $t \mapsto f(y + th)$ on all bounded intervals, and the equality $\partial_t f(y + th) = G(y + th)$ a.e. Thus, $f \in D^{p,1}(\mu, E)$. The case $r > 1$ follows by induction. □

8.2.5. Example. (i) Let f be a bounded Borel function on X which has bounded (with respect to the Hilbert–Schmidt norm) Fréchet derivatives of all orders along H. Then $f \in D^\infty(\mu)$.

(ii) Let μ be continuous along all vectors in H and let $f \in L^p(\mu)$ be such that $|f(x+h) - f(x)| \leqslant C|h|_H$, whenever $x \in X$, $h \in H$. Then $f \in D^{p,1}(\mu)$.

(iii) Suppose that the measure μ is r-fold differentiable along H. For any bounded Borel function f we put

$$F(x) = \int_X f(x+y)\, \mu(dy).$$

Then $F \in D^{p,r}(\mu)$.

PROOF. Assertions (i) and (ii) are obvious and (iii) follows by induction from the fact that for any bounded mapping $g\colon X \to E$, where E is a Hilbert space, the derivative $D_H G(x)$ of the mapping

$$G(x) = \int_X g(x+y)\, \mu(dy)$$

is a Hilbert–Schmidt operator and we have $\|D_H G(x)\|_{\mathcal{H}(H,E)} \leqslant c \sup_y |g(y)|_E$ (this is seen from Lemma 5.4.8). □

Another interesting example arises in the theory of stochastic differential equations. Let γ be the classical Wiener measure on the space $L^2([0,1], \mathbb{R}^d)$ of mappings with values in $\mathbb{R}^d$. Let $\sigma\colon \mathbb{R}^d \to \mathcal{L}(\mathbb{R}^d)$ and $b\colon \mathbb{R}^d \to \mathbb{R}^d$ be C_b^∞-mappings and let ξ_t be the solution of the stochastic differential equation

$$d\xi_t = \sigma(\xi_t) dw_t + b(\xi_t) dt,\ \xi_0 = x_0,$$

with respect to the standard Wiener process w_t. Denote by μ^ξ the measure induced by the process ξ_t on the path space $X = C([0,1], \mathbb{R}^d)$ (or on the Hilbert space $X = L^2([0,1], \mathbb{R}^d)$). It is known (see §1.5) that there exists a Borel mapping $F\colon X \to X$ for which $\mu^\xi = \gamma \circ F^{-1}$. We shall assume that the process w_t is defined as $w_t(\omega) = \omega(t)$ on the probability space (X, γ).

8.2.6. Theorem. *For every fixed t_0 the mapping $\omega \mapsto \xi_{t_0}(\omega)$ belongs to the class $D^\infty(\gamma, \mathbb{R}^d)$.*

PROOF. One can show that under our assumptions the $\mathbb{R}^d$-valued random processes $\eta_t := \partial_h \xi_t$, where $h \in W_0^{2,1}([0,1], \mathbb{R}^d)$, satisfy the equation

$$\eta_t = \int_0^t D\sigma(\xi_s)\eta_s\, dw_s + \int_0^t \sigma(\xi_s) h'(s)\, ds + \int_0^t Db(\xi_s)\eta_s\, ds.$$

By induction one can obtain formulae for all subsequent derivatives and then estimate their $L^p(\gamma)$-norms. Details can be found in the books Ikeda, Watanabe [**571**, Ch. V, §8], Nualart [**848**, §2.2.2], and Shigekawa [**1017**, Ch. 6]. □

8.3. Generalized derivatives and the classes $G^{p,r}$

Yet another natural way of introducing Sobolev classes is based on generalized derivatives. As above, let μ be a nonnegative Radon measure on a locally convex space X, let $H \subset X$ be a continuously and densely embedded separable Hilbert space, and let $j\colon X^* \to H$ be the corresponding embedding.

8.3.1. Definition. *We shall say that a function $f \in \mathcal{L}^1(\mu)$ has a generalized partial derivative $g = \partial_h f \in L^1(\mu)$ along a vector $h \in H$ if there exists β_h^μ such that $f\beta_h^\mu \in \mathcal{L}^1(\mu)$ and for every $\varphi \in \mathcal{F}C_b^\infty$ one has*

$$\int_X \partial_h \varphi(x) f(x)\, \mu(dx) = -\int_X g(x)\varphi(x)\, \mu(dx) - \int_X f(x)\varphi(x)\beta_h^\mu(x)\, \mu(dx).$$

It is clear that a generalized partial derivative is uniquely determined as an element of $L^1(\mu)$.

8.3.2. Definition. *Let $G^{p,1}(\mu) = G_H^{p,1}(\mu)$ be the class of all real functions $f \in L^p(\mu)$ possessing generalized partial derivatives along all vectors in $j(X^*)$ and having finite norms*

$$\|f\|_{p,1} := \|f\|_{L^p(\mu)} + \|D_H f\|_{L^p(\mu,H)} < \infty,$$

where $D_H f$ is defined as follows: there is a mapping $T\colon X \to H$ such that for every $l \in X^*$ we have $\langle l, T(x)\rangle = \partial_{j(l)} f(x)$ a.e. Then we set $D_H f := \nabla_H f := T$. The space $G^{p,1}(\mu)$ is equipped with the norm $\|\cdot\|_{p,1}$. Similarly we define the class $G^{p,1}(\mu, E) = G_H^{p,1}(\mu, E)$ of mappings with values in a Hilbert space E. Hence one can inductively introduce the classes $G^{p,r}(\mu) = G_H^{p,r}(\mu)$ of functions $f \in L^p(\mu)$ with $D_H^k f \in L^p(\mu, \mathcal{H}_k)$ whenever $k \leqslant r$ equipped with the norms

$$\|f\|_{p,r} := \|f\|_{L^p(\mu)} + \sum_{k=1}^r \|D_H^k f\|_{L^p(\mu,\mathcal{H}_k)}.$$

Finally, one can define (see Bogachev, Mayer-Wolf [**211**]) larger Sobolev classes $G_{\mathrm{oper}}^{p,1}(\mu, E)$ if the derivative is allowed to take values in $\mathcal{L}(H, E)$ with the operator norm. Then we set

$$\|f\|_{p,1,0} := \|f\|_{L^p(\mu)} + \|D_H f\|_{L^p(\mu,\mathcal{L}(H,E))}.$$

By using the completeness of the L^p-spaces of vector mappings one can readily show the following result.

8.3.3. Proposition. *Let $p \geqslant 1$ and $\beta_h^\mu \in L^{p/(p-1)}(\mu)$ for all $h \in j(X^*)$. Then the spaces $G^{p,1}(\mu, E)$ and $G_{\mathrm{oper}}^{p,1}(\mu, E)$ with the respective norms are complete.*

Let us establish a useful technical result on differentiability of conditional expectations.

8.3.4. Proposition. *Suppose that a measure $\mu \geq 0$ is differentiable along a vector h and that a bounded measurable function f has a generalized partial derivative $\partial_h f$. Denote by $\mathbb{E}^{\mathcal{A}}$ the conditional expectation with respect to the σ-algebra $\mathcal{A}$ generated by n functionals $l_1, \ldots, l_n \in X^*$ and the measure μ. Then, the function $\mathbb{E}^{\mathcal{A}} f$ has the generalized partial derivative*

$$\partial_h \mathbb{E}^{\mathcal{A}} f = \mathbb{E}^{\mathcal{A}} \partial_h f + \mathbb{E}^{\mathcal{A}}(f \beta_h^{\mu}) - \mathbb{E}^{\mathcal{A}} f \mathbb{E}^{\mathcal{A}} \beta_h^{\mu}. \tag{8.3.1}$$

PROOF. First we consider the case where h belongs to the linear subspace in L generated by $j(l_1), \ldots, j(l_n)$. We may assume that $f \geq 1$ by adding a constant. Our condition implies that the measure $\nu = f \cdot \mu$ is differentiable along h. Hence its projection ν_1 under the linear mapping $T = (l_1, \ldots, l_n) \colon X \to \mathbb{R}^n$ is differentiable along Th. The projection μ_1 of the measure μ is differentiable along Th as well. Let E be the orthogonal complement to Th in $\mathbb{R}^n$. Both measures μ_1 and ν_1 have absolutely continuous conditional densities $\psi(y, \cdot)$ and $\varphi(y, \cdot)$ on the straight lines $y + tTh$, $y \in E$, with respect to the natural Lebesgue measures. Hence

$$\mathbb{E}^{\mathcal{A}} f(x) = \varphi(y, t)/\psi(y, t), \quad Tx = y + tTh, \ y \in E.$$

Thus, for every x the function

$$s \mapsto \mathbb{E}^{\mathcal{A}} f(x + sh) = \varphi(y, t+s)/\psi(y, t+s)$$

is absolutely continuous on every closed interval not containing zeros of the function $\psi(y, t+s)$. In addition,

$$\mathbb{E} \beta_h^{\mu} = \partial_t \psi(y, t)/\psi(y, t), \quad \mathbb{E} \beta_h^{\nu} = \partial_t \varphi(y, t)/\varphi(y, t).$$

Hence $\partial_h \mathbb{E}^{\mathcal{A}} f(x)$ coincides μ-a.e. with

$$\frac{\partial_t \varphi(y, t)}{\psi(y, t)} - \frac{\varphi(y, t)}{\psi(y, t)} \frac{\partial_t \psi(y, t)}{\psi(y, t)} = \mathbb{E}^{\mathcal{A}} f(x) \mathbb{E}^{\mathcal{A}} \beta_h^{\nu}(x) - \mathbb{E}^{\mathcal{A}} f(x) \mathbb{E}^{\mathcal{A}} \beta_h^{\mu}(x).$$

Since the function $\mathbb{E}^{\mathcal{A}} f$ is bounded, $\beta_h^{\nu} = \beta_h^{\mu} + \partial_h f/f$ and $f \geq 1$, we see that β_h^{ν} is μ-integrable. Hence $\partial_h \mathbb{E}^{\mathcal{A}} f$ is μ-integrable as well. Let $g = G \circ T$, where $G \in C_b^{\infty}(\mathbb{R}^n)$. Applying the integration by parts formula to $\mathbb{E}^{\mathcal{A}} f$ (which is justified by Exercise 6.8.2), and taking into account the equality

$$\int_X \partial_h g(x) f(x) \mu(dx) = \int_X \partial_h g(x) \mathbb{E}^{\mathcal{A}} f(x) \mu(dx)$$

we obtain

$$\int_X g \partial_h \mathbb{E}^{\mathcal{A}} f \, \mu(dx) = -\int_X \partial_h g \, f \, \mu(dx) - \int_X g [\mathbb{E}^{\mathcal{A}} f] \beta_h^{\mu} \, \mu(dx)$$

$$= \int_X g \partial_h f \, \mu(dx) + \int_X g f \beta_h^{\mu} \, \mu(dx) - \int_X g [\mathbb{E}^{\mathcal{A}} f] \beta_h^{\mu} \, \mu(dx)$$

$$= \int_X g \big(\mathbb{E}^{\mathcal{A}} \partial_h f + \mathbb{E}^{\mathcal{A}}(f \beta_h^{\mu}) - \mathbb{E}^{\mathcal{A}} f \mathbb{E}^{\mathcal{A}} \beta_h^{\mu} \big) \mu(dx).$$

Since the function $\partial_h \mathbb{E}^{\mathcal{A}} f$ is $\mathcal{A}$-measurable by construction, we arrive at the desired identity in the case $h \in L$. It remains to observe that if h is orthogonal to L in H, then $l_i(h) = 0$ for $i = 1, \ldots, n$. Hence $\partial_h g = 0$ for every function $g = G \circ T$. Then $\partial_h \mathbb{E}^{\mathcal{A}} f = 0$. Therefore, we have

$$\int_X g\bigl(\mathbb{E}^{\mathcal{A}}\partial_h f + \mathbb{E}^{\mathcal{A}}[f\beta_h^\mu] - \mathbb{E}^{\mathcal{A}} f \mathbb{E}^{\mathcal{A}} \beta_h^\mu\bigr) \mu(dx)$$
$$= \int_X \bigl(g\partial_h f + gf\beta_h^\mu - g\mathbb{E}^{\mathcal{A}} f \beta_h^\mu\bigr) \mu(dx) = 0$$

by the integration by parts formula since $\partial_h(g\mathbb{E}^{\mathcal{A}} f) = 0$ and

$$\int_X gf\beta_h^\mu \, \mu(dx) = -\int_X g\partial_h f \, \mu(dx)$$

due to the equality $\partial_h g = 0$. □

8.3.5. Corollary. *Let $h_1, \ldots, h_n \in X$ be such that $l_i(h_j) = \delta_{ij}$ and $\exp(c|\beta_{h_i}^\mu|) \in L^1(\mu)$ for some $c > 0$. If a bounded function f has generalized partial derivatives $\partial_{h_i} f$, then $\mathbb{E}^{\mathcal{A}} f = f_0(l_1, \ldots, l_n)$, where $f_0 \in W^{1,1}_{\mathrm{loc}}(\mathbb{R}^n)$.*

PROOF. We have the equality $\mathbb{E}^{\mathcal{A}} f = f_0(l_1, \ldots, l_n)$, where $f_0 \in L^1(\mu_n)$ and μ_n is the image of μ under the mapping $(l_1, \ldots, l_n)$. The measure μ_n possesses a continuous strictly positive density from the class $W^{1,1}(\mathbb{R}^n)$ (Corollary 6.4.2). As shown above, there exist generalized partial derivatives $\partial_{h_i} \mathbb{E}^{\mathcal{A}} f \in L^1(\mu)$. Along with Proposition 3.3.13 this shows that f_0 has generalized partial derivatives with respect to μ_n, so $f_0 \in W^{1,1}_{\mathrm{loc}}(\mathbb{R}^n)$. □

8.4. The semigroup approach

By using a continuous operator semigroup on $L^2(\mu)$ or on $L^1(\mu)$ one can define a continuous scale of spaces. An advantage of this approach is its applicability to abstract measures defined on abstract spaces without any analytic or topological structure.

Let $(X, \mathcal{A}, \mu)$ be a probability space and let $\{T_t\}_{t \geq 0}$ be a strongly continuous contracting semigroup in $L^1(\mu)$ (this means that $\lim_{t \to 0} \|T_t f - f\|_{L^1(\mu)} = 0$ and $\|T_t\| \leq 1$ for all $f \in L^1(\mu)$) which is *sub-Markovian*, that is, $0 \leq T_t f \leq 1$ if $0 \leq f \leq 1$ (if additionally $T_t 1 = 1$, then $\{T_t\}_{t \geq 0}$ is called *Markovian*). It is known (see [**234**, §6.10(viii)] or [**376**, Theorem VI.10.11]) that $T_t(L^p(\mu)) \subset L^p(\mu)$ for all $p \in [1, +\infty)$. Then $\{T_t\}_{t \geq 0}$ is automatically a strongly continuous semigroup on $L^p(\mu)$. There is a standard construction of Sobolev classes associated with the semigroup $\{T_t\}$. Set

$$V_r f := \Gamma(r/2)^{-1} \int_0^\infty t^{r/2 - 1} e^{-t} T_t f \, dt.$$

One can verify that for every $p \geq 1$ the mapping V_r is a bounded linear operator on $L^p(\mu)$. Hence the space

$$H^{p,r}(\mu) := V_r\bigl(L^p(\mu)\bigr), \quad \|f\|_{p,r}^* := \|V_r^{-1} f\|_{L^p(\mu)},$$

is complete. Set
$$H^\infty(\mu) := \bigcap_{p\in[1,+\infty), r\geqslant 1} H^{p,r}(\mu).$$

Let L be the generator of $\{T_t\}_{t\geqslant 0}$ in $L^p(\mu)$. This means that the domain $D_p(L)$ of the operator L consists of all $f \in L^p(\mu)$ for which a limit
$$Lf := \lim_{t\to 0} t^{-1}(T_t f - f)$$
exists in $L^p(\mu)$. It is known that $D_p(L)$ is a dense linear subspace in $L^p(\mu)$. In the case where L is a nonpositive seldadjoint operator in $L^2(\mu)$ (this is equivalent to that the operators T_t are selfadjoint contractions), one can define selfadjoint operators $(I - L)^{-r/2}$, $r > 0$. On $L^2(\mu)$ we have the equality
$$(I - L)^{-r/2} = V_r.$$
By using this equality $(I - L)^{-r/2}$ can be defined on $L^p(\mu)$. Then
$$\|f\|^*_{p,r} = \|(I - L)^{-r/2} f\|_{L^p(\mu)}.$$

Connections with the constructions from the previous sections steam from the following important result of Albeverio, Röckner [**54**].

Let μ be a Radon probability measure on a separable Banach space X, let H be a Hilbert space continuously and densely embedded into X, let $j \colon X^* \to H$ be the canonical embedding, and let $\{e_n\}$ be an orthonormal basis in H, where $e_n = j(l_n)$, $l_n \in X^*$, and the functionals l_n separate the points in X. Set
$$\Delta_H f(x) := \sum_{n=1}^\infty \partial^2_{e_n} f(x)$$
for all functions f of the class $\mathcal{F}_{\{l_n\}}$ consisting of functions of the form
$$f(x) = f_0(l_1(x), \ldots, l_k(x)), \quad f_0 \in C_b^\infty(\mathbb{R}^k).$$

8.4.1. Theorem. *Suppose that β^μ_H exists and the function $x \mapsto \|\beta^\mu_H(x)\|^2_X$ is integrable with respect to μ. Then there exists a strongly continuous sub-Markovian semigroup $\{T_t\}_{t\geqslant 0}$ in $L^2(\mu)$ consisting of selfadjoint operators, the domain of generator L of this semigroup contains the class $\mathcal{F}_{\{l_n\}}$ and on this class*
$$Lf(x) = \left[\Delta_H f(x) + {}_{X^*}\langle f'(x), \beta^\mu_H(x)\rangle_X\right] = \sum_{n=1}^\infty \left[\partial^2_{e_n} f(x) + \partial_{e_n} f(x) \beta^\mu_{e_n}(x)\right].$$

Moreover, this semigroup extends to a strongly continuous semigroup on all $L^p(\mu)$, where $p \in [1, +\infty)$.

Under the hypotheses of this theorem, the question arises about relations between the spaces $H^{p,r}(\mu)$ and $W^{p,r}(\mu)$ for $r \in \mathbb{N}$ (for example, for $r = 1$). Few known results will be discussed below. Among those very rare cases in

8.4. The semigroup approach

which the semigroup T_t is determined explicitly, we single out the particularly important *Ornstein–Uhlenbeck semigroup*, which, for a centered Radon Gaussian measure γ on a locally convex space X, is given by the formula

$$T_t f(x) = \int_X f\bigl(e^{-t}x - \sqrt{1 - e^{-2t}}\, y\bigr)\, \gamma(dy), \quad f \in \mathcal{L}^p(\gamma).$$

A simple verification of the fact that $\{T_t\}_{t\geqslant 0}$ is a strongly continuous semigroup on all $L^p(\gamma)$, $1 \leqslant p < \infty$, can be found in book [**191**]. An important feature of this semigroup is that the measure γ is invariant for it, that is,

$$\int_X T_t f(x)\, \gamma(dx) = \int_X f(x)\, \gamma(dx).$$

It is verified directly that, as $t \to \infty$, the functions $T_t f$ converge in $L^p(\gamma)$ to the integral of f for all $f \in L^p(\gamma)$.

The generator L of the Ornstein–Uhlenbeck semigroup is called the *Ornstein–Uhlenbeck operator* (more precisely, for every $p \in [1, +\infty)$, there is such a generator on the corresponding domain in $L^p(\gamma)$; if p is not explicitly indicated, then usually $p = 2$ is meant).

For all functions $f \in \mathcal{FC}_b^\infty(X)$ one has the equality

$$Lf = \Delta_H f - \langle f'(x), x \rangle = \sum_{n=1}^\infty [\partial_{e_n}^2 f - \widehat{e}_n \partial_{e_n} f],$$

where $\{e_n\}$ is an orthonormal basis in $H = H(\gamma)$.

Let us mention a useful result of Stein, which helps in the study of the pointwise behaviour of rather general semigroups. Let μ be a probability measure on a measurable space $(X, \mathcal{A})$ and let $\{T_t\}_{t\geqslant 0}$ be a strongly continuous semigroup of bounded selfadjoint operators on $L^2(\mu)$ with $\|T_t\| \leqslant 1$. Suppose that T_t extends to an operator in $\mathcal{L}(L^p(\mu))$ of norm at most 1 for all $t > 0$ and $p \in [1, +\infty]$. We observe that the latter is fulfilled by the interpolation theorem if $T_t 1 = 1$ and $T_t f \geqslant 0$ as $f \geqslant 0$; such semigroups are called *symmetric diffusion* semigroups. It is shown in Stein [**1078**, p. 70] that if $f \in L^p(\mu)$, then, for every $t > 0$, one can take a version $\widetilde{T_t f}$ of the function $T_t f(\,\cdot\,)$ such that, for every fixed x, the function $t \mapsto \widetilde{T_t f}(x)$ will be real-analytic on $(0, +\infty)$. The following important fact is proved there.

8.4.2. Theorem. *Let $Mf(x) := \sup_{t>0} |\widetilde{T_t f}(x)|$ and $p > 1$. Then, for some $C_p > 0$, for all $f \in \mathcal{L}^p(\mu)$ the inequality $\|Mf\|_p \leqslant C_p \|f\|_p$ holds. In addition, $\lim_{t\to 0} \widetilde{T_t f}(x) = f(x)$ a.e.*

8.4.3. Example. Let γ be a centered Gaussian measure on a locally convex space X, let $\{T_t\}_{t\geqslant 0}$ be the Ornstein–Uhlenbeck semigroup, and let $f \in \mathcal{L}^p(\gamma)$ be Borel measurable, where $p > 1$. Then, for γ-a.e. x, we have

$$\lim_{t\to 0} T_t f(x) = f(x), \quad \lim_{t\to +\infty} T_t f(x) = \int_X f(x)\, \gamma(dx).$$

PROOF. It suffices to consider the case $p < \infty$. The first equality follows from Theorem 8.4.2 and the fact (verified by the monotone class argument) that one can take $\widetilde{T_t f}(x) = T_t f(x)$ for Borel f. The second equality does not follow from that theorem even in the case where the measure μ is invariant with respect to T_t. Here we need the obvious fact that this equality is true for $f \in C_b(X)$. In the general case we denote the integral of f by Jf and put $\psi(x) := \limsup_{t \to \infty} |T_t f(x) - Jf|$. Then, for any $g \in C_b(X)$, we have

$$\psi(x) \leqslant \limsup_{t \to \infty} |T_t f(x) - T_t g(x)| + \limsup_{t \to \infty} |T_t g(x) - Jg| + |Jg - Jf|,$$

whence $\|\psi\|_p \leqslant (C_p + 1)\|f - g\|_p$ since the second term on the right vanishes. Taking $g_n \in C_b(X)$ such that $\|f - g_n\|_p \to 0$, we obtain $\|\psi\|_p = 0$, hence we have $\psi(x) = 0$ a.e. □

If the space X is finite dimensional, then the proven assertions remain valid also for $p = 1$; here the maximal operator in $L^1(\gamma)$ has weak order $(1,1)$. In the infinite dimensional case the problem is open. Various interesting results connected with this operator can be found in the survey Sjögren [**1040**] and in Aimar, Forzani, Scotto [**14**], Mauceri, Meda, Sjögren [**790**], Fabes, Gutierrez, Scotto [**405**], Forzani et al. [**445**], and Gutiérrez, Segovia, Torrea [**512**].

Let us mention an approach based on Dirichlet forms. This approach also enables one to deal with measures that are not necessarily differentiable. It turns out to be particularly natural in the case $r = 1, p = 2$. For simplicity we consider only the case $r = 1$. In a similar manner one can consider higher order derivatives, although this is more complicated and less natural in this approach. Let H be a separable Hilbert space continuously and densely embedded into a locally convex space X and let a Radon probability measure μ on X satisfy the following condition on the class $\mathcal{F}C_b^\infty$:

$$\varphi = \psi \ \mu\text{-a.e. implies that } \nabla_H \varphi = \nabla_H \psi \ \mu\text{-a.e.}$$

This condition is fulfilled, for instance, if μ has full support.

Suppose that we are given a mapping $A\colon X \to \mathcal{L}(H)$ belonging to the space $L^p(\mu, \mathcal{L}(H))$. Then the operator

$$D\colon \mathcal{F}C_b^\infty \to L^p(\mu, H), \quad D\varphi(x) = A(x)\nabla_H \varphi(x),$$

is well defined. Suppose also that this operator is closable in $L^p(\mu)$. We shall use the same symbol for its closure with the domain of definition $F^{p,1}(\mu)$. Thus, the space $F^{p,1}(\mu)$ is complete with respect to the norm

$$\|f\|_{A,p} = \left\{ \int_X |f(x)|^p \, \mu(dx) + \int_X |Df(x)|_H^p \, \mu(dx) \right\}^{1/p}.$$

Any element of $F^{p,1}(\mu)$ can be identified with a limit of a sequence $\{\varphi_n\}$ from $\mathcal{F}C_b^\infty$ convergent in $L^p(\mu)$ such that the sequence $\{D\varphi_n\}$ converges in the space $L^p(\mu, H)$. Set $Df := \lim\limits_{n\to\infty} D\varphi_n$. Sometimes also the notation $A(x)\nabla_H f(x)$ is used, however, $\nabla_H f$ may not exist separately.

For sufficient conditions for the closability of D, see Röckner [**958**]. For example, if the measure μ is differentiable along all vectors in H, then D is closable. We observe that in the case $A(x) \equiv I$ the compactness of the embedding $H \to X$ is not necessary for the closability of D (an obvious example is a nondegenerate centered Gaussian measure μ on an infinite dimensional Hilbert space X and $H = X$).

The connection with the semigroup approach stems from the following observation. Suppose that the operator D is closable in $L^2(\mu)$. Let us define a nonnegative quadratic form $\mathcal{E}_A$ on $\mathcal{F}C_b^\infty$ by the equality
$$\mathcal{E}_A(f, f) = \int_X \bigl(Df(x), Dg(x)\bigr)_H \mu(dx).$$
By the closability of this form and a well-known result in the theory of operators, there exists a nonnegative selfadjoint operator L on $L^2(\mu)$ such that the equality $(Lf, f) = \mathcal{E}_A(f, f)$ holds on $\mathcal{F}C_b^\infty$. The domain of definition $D(\sqrt{L})$ of the operator $\sqrt{L}$ contains the class $F^{2,1}(\mu)$. Let $T_t = \exp(-tL)$. Suppose that $\{T_t\}$ is a sub-Markovian semigroup. Then, by the interpolation theorem, the semigroup $\{T_t\}$ extends to a strongly continuous contracting semigroup on each $L^p(\mu)$, $p > 2$. This enables us to introduce the classes $H^{p,r}(\mu)$.

8.5. The Gaussian case

Let γ be a centered Radon Gaussian measure on a locally convex space X and let $H = H(\gamma)$. In this special case Sobolev classes have additional interesting properties. The proofs of the results given below can be read in the author's book [**191**, Ch. 5].

8.5.1. Theorem. *For $p > 1$ and $r \in \mathbb{N}$ the Sobolev classes $W^{p,r}(\gamma)$, $D^{p,r}(\gamma)$, $G^{p,r}(\gamma)$, and $H^{p,r}(\gamma)$, defined above, coincide. The same is true for analogous classes of mappings with values in a separable Hilbert space.*

The most difficult part of the proof of this theorem is to establish the equivalence of norms in $W^{p,r}(\gamma)$ and $H^{p,r}(\gamma)$, the so-called Meyer equivalence, which can be formulated as follows.

8.5.2. Theorem. *Let $p > 1$ and $r > 0$. There exist positive numbers $m_{p,r}$ and $M_{p,r}$ such that, for every smooth cylindrical function f, we have*
$$m_{p,r} \|D_H^r f\|_{L^p(\gamma, \mathcal{H}_r)} \leqslant \|(I - L)^{r/2} f\|_{L^p(\gamma)}$$
$$\leqslant M_{p,r} \Bigl[\|D_H^r f\|_{L^p(\gamma, \mathcal{H}_r)} + \|f\|_{L^p(\gamma)}\Bigr]. \quad (8.5.1)$$

We recall (see §1.4) that if $\{e_n\}$ is an orthonormal basis in H, then the functions
$$H_{k_1,\ldots,k_n, j_1,\ldots,j_n} = H_{k_1}(\widehat{e}_{j_1}) \cdots H_{k_n}(\widehat{e}_{j_n}),$$

where H_k is the kth Hermite polynomial, form an orthonormal basis in the space $L^2(\gamma)$. The order of such a function equals $k_1 + \cdots + k_n$. For example, if γ is the countable power of the standard Gaussian measure, then
$$H_{k_1,\ldots,k_n,j_1,\ldots,j_n}(x) = H_{k_1}(x_{j_1}) \cdots H_{k_n}(x_{j_n}).$$
Denoting by $\mathcal{P}_n(\gamma)$ the closed subspace in $L^2(\gamma)$ generated by all Hermite functions of order at most n, and by $\mathcal{X}_n$ the orthogonal complement of $\mathcal{P}_{n-1}(\gamma)$ in $\mathcal{P}_n(\gamma)$, $\mathcal{X}_0 = \mathcal{P}_0(\gamma)$, we obtain the orthogonal decomposition
$$L^2(\gamma) = \bigoplus_{n=0}^{\infty} \mathcal{X}_n$$
and the corresponding decompositions
$$F = \sum_{n=0}^{\infty} F_n, \quad F_n := I_n(F), \quad F \in L^2(\gamma), \tag{8.5.2}$$
where I_n is the projector on $\mathcal{X}_n$. Elements of $\mathcal{P}_n(\gamma) := \bigoplus_{k=0}^n \mathcal{X}_k$ are called γ-measurable polynomials of degree at most n (see [**191**, §5.10] for their equivalent description as γ-measurable functions f which, for each $h \in H$, have versions such that $t \mapsto f(x+th)$ is a polynomial of degree n). Many (but not all) objects of the theory of Sobolev classes over Gaussian measures can be described by means of decomposition (8.5.2). For example, the Ornstein–Uhlenbeck semigroup is given by
$$T_t F = \sum_{n=0}^{\infty} e^{-nt} F_n.$$
Its generator — the Ornstein–Uhlenbeck operator — is
$$LF = -\sum_{n=0}^{\infty} n F_n$$
on the domain in $L^2(\gamma)$ consisting of all F with $\sum_{n=0}^{\infty} n^2 \|F_n\|_2^2 < \infty$, which coincides with the Sobolev class $W^{2,2}(\gamma)$. The domain of definition of $\sqrt{-L}$ in $L^2(\gamma)$ equals $W^{2,1}(\gamma)$. The only property of the elements $\widehat{e}_n$ employed here is that they are independent standard Gaussian random variables. So one can start with any separable Hilbert space H consisting of centered Gaussian random variables (defined on some fixed probability space) and then introduce Hermite polynomials in these variables and the subsequent objects. This kind of Gaussian analysis is developed in many works (see Dorogovtsev [**354**], [**355**], [**356**], Janson [**585**], Nualart [**848**] and the references therein). In this approach, there are no topological concepts imported by the space X on which the Gaussian measure γ is defined (since there is no necessity at all to introduce this measure). For some problems this may be an advantage, but for some problems this may be a loss.

It is worth noting that $\|D_H^r f\|_{L^p(\gamma, \mathcal{H}_r)} \leqslant c_p \|f\|_p + c_p \|D_H^{r+1} f\|_{L^p(\gamma, \mathcal{H}_{r+1})}$ for all $f \in W^{p,1}(\gamma)$, see Shigekawa [**1017**, p. 85].

It is also useful to introduce local Sobolev classes.

8.5.3. Definition. *Let $W^{p,r}_{\text{loc}}(\gamma)$ be the class of all functions f for which there exist an increasing sequence of sets $X_n \in \mathcal{B}(X)$ and a sequence of functions $\chi_n \in W^{p,r}(\gamma)$ such that one has $\chi_n|_{X_n} = 1$, $\gamma(X\backslash \bigcup_n X_n) = 0$, and $\chi_n f \in W^{p,r}(\gamma)$.*

In the same manner one defines the classes $W^{p,r}_{\text{loc}}(\gamma, E)$ of mappings with values in a separable Hilbert space E.

8.5.4. Lemma. *Let $f \in W^{p,r}_{\text{loc}}(\gamma)$. For any $m \geqslant n$ the mappings $\nabla^k_H(\chi_m f)$ and $\nabla^k_H(\chi_n f)$, where $k \leqslant r$, coincide γ-a.e. on X_n. This makes $\nabla^k_H f$ meaningful independently of our choice of $\{\chi_n\}$.*

8.5.5. Theorem. *Let $f \in W^{p,r}_{\text{loc}}(\gamma, E)$ and $\nabla^r_H f \in L^p(\gamma, \mathcal{H}_r)$. Then we have $f \in W^{p,r}(\gamma, E)$.*

We prove also several results on convex functions on a space with a centered Radon Gaussian measure γ (see also §6.5). Let $H = H(\gamma)$.

8.5.6. Lemma. *If the space X is quasicomplete, then there exist a linear subspace $E \subset X$ with $\gamma(E) = 1$ and a convex function $\Psi \geqslant 0$ on E such that the sets $\{\Psi \leqslant c\}$ have compact closures and the functions $\Psi, \partial_h \Psi, \partial^2_h \Psi$ for any $h \in H$ belong to all $L^p(\gamma)$ with $p < \infty$ and Ψ is Lipschitz along H.*

PROOF. We take a metrizable compact set of positive measure. Its closed absolutely convex hull K is compact since X is quasicomplete. Let E be the linear span of the set K and let q_K be the Minkowski functional of K (see §1.1). By Theorem 1.4.10 we have $\exp(\varepsilon q_K) \in L^1(\gamma)$ for some $\varepsilon > 0$. The function $\Psi := T_1 q_K$ is convex on E, $\Psi \leqslant q_K + M$, where M is the integral of q_K, is infinitely differentiable along H, and all its partial derivatives along vectors in H belong to all $L^p(\gamma)$. In addition, the set $E \cap \{\Psi \leqslant c\}$ has compact closure since it is contained in $(3c+3M)K$, which follows by integrating in y the estimate $q_K(e^{-1}x - (1-e^{-2})^{1/2}y) \geqslant e^{-1}q_K(x) - (1-e^{-2})^{1/2}q_K(y)$. The function q_K is H-Lipschitzian. Hence Ψ is H-Lipschitzian as well. □

8.5.7. Proposition. *Let $F \in L^p(\gamma)$, where $p > 1$, be an H-convex function and let $h \in H$. Then the measures $(\partial^2_h T_\varepsilon F) \cdot \gamma$, $\varepsilon \in (0,1)$, are uniformly tight and converge weakly to F_{hh} as $\varepsilon \to 0$.*

PROOF. Let Ψ be the function from the previous lemma. Integrating by parts twice we find that

$$\int_X \Psi \partial^2_h T_\varepsilon F \, \gamma(dx) = \int_X [\partial^2_h \Psi + 2\partial_h \Psi \widehat{h} + \Psi \widehat{h}^2 - \Psi] T_\varepsilon F \, \gamma(dx),$$

which is uniformly bounded in ε due to Hölder's inequality since we have the estimate $\|T_\varepsilon F\|_{L^p(\gamma)} \leqslant \|F\|_{L^p(\gamma)}$. Hence $(\partial^2_h T_\varepsilon F \cdot \gamma)(\Psi > R) \leqslant CR^{-1}$, where the set $\{\Psi \leqslant R\}$ up to a measure zero set is contained in some compact set K_R. Thus, the measures $(\partial^2_h T_\varepsilon F) \cdot \gamma$ are uniformly tight. Their uniform boundedness is clear from a similar estimate with 1 in place of Ψ. Hence, as $\varepsilon_n \to 0$, the sequence of measures $\partial_h^2 T_{\varepsilon_n} F \cdot \gamma$ has a limit point ν in the weak topology. This limit point is the measure F_{hh} since for every smooth cylindrical function f we have

$$\int_X f\partial_h^2(T_\varepsilon F)\,\gamma(dx) = \int_X T_\varepsilon F[\partial_h^2 f + 2\partial_h f\widehat{h} + f(\widehat{h}^2 - 1)]\,\gamma(dx).$$

As $\varepsilon \to 0$, the right-hand side tends to

$$\int_X F[\partial_h^2 f + 2\partial_h f\widehat{h} + f(\widehat{h}^2 - 1)]\,\gamma(dx)$$
$$= \int_X f\,d_h^2(F\cdot\gamma)(dx) - 2\int_X f\partial_h F\,d_h\gamma(dx) - \int_X Ff\,d_h^2\gamma(dx),$$

which coincides with the integral of f with respect to the measure F_{hh}. □

8.5.8. Corollary. *If a function $F \in L^2(\gamma)$ is H-convex, then the formula $(T_B h, h)_H := F_{hh}(B)$, $B \in \mathcal{B}(X)$, defines a countably additive measure $B \mapsto T_B$ with values in the space $\mathcal{H}$ of Hilbert–Schmidt operators on H equipped with the Hilbert–Schmidt norm $\|\cdot\|_\mathcal{H}$. In addition, $V(T_B) \leqslant \sqrt{2}\|F\|_{L^2(\gamma)}$ and this measure has bounded semivariation estimated by $2\sqrt{2}\|F\|_{L^2(\gamma)}$.*

PROOF. Suppose first that $F \in W^{2,2}(\gamma)$. Then its second derivative $D_H^2 F(x)$ is a nonnegative Hilbert–Schmidt operator on H. Let $B \in \mathcal{B}(X)$. The formula

$$(T_B h, h)_H = \int_B D_H^2 F(x)(h, h)\,\gamma(dx)$$

defines a nonnegative Hilbert–Schmidt operator such that $\|T_B\|_\mathcal{H}^2$ does not exceed the integral of $\|D_H^2 F(x)\|_\mathcal{H}^2$. Let $\{e_j\}$ be an orthonormal basis in H such that $T_B e_j = t_j e_j$. The functions $\xi_j := -1 + \widehat{e_j}^2$ are mutually orthogonal in $L^2(\gamma)$ and have the same norm $\sqrt{2}$ in $L^2(\gamma)$. Since $\partial_{e_j}^2 F \geqslant 0$, the Hilbert–Schmidt norm of the operator T_B can be estimated as follows:

$$\sum_{j=1}^\infty t_j^2 = \sum_{j=1}^\infty \left|\int_B \partial_{e_j}^2 F\,\gamma(dx)\right|^2 \leqslant \sum_{j=1}^\infty \left|\int_X \partial_{e_j}^2 F\,\gamma(dx)\right|^2$$
$$= \sum_{j=1}^\infty \left|\int_X F\xi_j\,\gamma(dx)\right|^2 \leqslant 2\int_X F^2\,\gamma(dx).$$

Therefore, our assertion is true for all functions $T_{1/k}F$. Letting $k \to \infty$ and observing that $T_{1/k}F \to F$ in $L^2(\gamma)$, we obtain our assertion by the above proposition. □

8.6. The interpolation approach

Now we turn to the interpolation approach. Let $\mathcal{E}^{p,1}(\mu)$ be a Banach space continuously embedded into $L^p(\mu)$, where $p \in (1, +\infty)$. The norm in

8.6. The interpolation approach

$\mathcal{E}^{p,1}(\mu)$ is denoted by $\|\cdot\|_{p,1}$, and the norm in L^p is denoted by $\|\cdot\|_p$. We shall assume that $\|f\|_p \leqslant \|f\|_{p,1}$. For every function $f \in L^p(\mu)$ we put
$$K_t(f) := \inf\{\|f_1\|_p + t\|f_2\|_{p,1} : f = f_1 + f_2, f_1 \in L^p(\mu), f_2 \in \mathcal{E}^{p,1}\}.$$
Given $\alpha \in (0,1)$, let $\mathcal{E}^{p,\alpha}(\mu)$ denote the space of all elements $f \in L^p(\mu)$ of finite norm
$$\|f\|_{p,\alpha} := \left(\int_0^\infty |t^{-\alpha} K_t(f)|^p \frac{dt}{t}\right)^{1/p} < \infty.$$
It is known that $\|\cdot\|_{p,\alpha}$ is indeed a norm and $\mathcal{E}^{p,\alpha}(\mu)$ is complete with respect to it and is an interpolation space between $L^p(\mu)$ and $\mathcal{E}^{p,1}(\mu)$ (see Triebel [1123, §1.3]). Let us consider an important example, where X is a locally convex space equipped with a centered Radon Gaussian measure μ with the Cameron–Martin space H. We take the scale of fractional spaces $H^{p,r}(\mu)$, $0 < r < \infty$, defined above by means of the Ornstein–Uhlenbeck semigroup $(T_t)_{t \geqslant 0}$ on $L^p(\mu)$. Then we take for $\mathcal{E}^{p,1}(\mu)$ the space $W^{p,1}(\mu)$. The following theorem is proved in Watanabe [1184].

8.6.1. Theorem. *For all $p \in (1, +\infty)$, $r \in \mathbb{R}^1$, and $\varepsilon > 0$ we have continuous embeddings $H^{p,r+\varepsilon}(\mu) \subset \mathcal{E}^{p,r}(\mu) \subset H^{p,r-\varepsilon}(\mu)$.*

Suppose now that our probability measure μ is defined on the product $(X \times Y, \mathcal{B}_X \otimes \mathcal{B}_Y)$, where $(X, \mathcal{B}_X)$ and $(Y, \mathcal{B}_Y)$ are two measurable spaces. Denote by μ_Y the projection of μ on Y. Suppose that the measure μ is separable and possesses regular conditional measures μ_y on $(X, \mathcal{B}_X)$, $y \in Y$ (or conditional measures μ^y on $X \times \{y\}$). Let us recall that regular conditional probability measures exist if, for example, X and Y are Souslin topological spaces with their Borel σ-algebras; in addition, in this case μ is separable. For every function f on $X \times Y$ and every $y \in Y$ put $f_y : x \mapsto f(x,y)$. The functions f_y are called sections or restrictions of the function f.

We shall say that the space $\mathcal{E}^{p,1}(\mu)$ satisfies *condition* (D) *of decomposition* if, for every $y \in Y$, there exists a continuously embedded Banach space $(\mathcal{E}_y^{p,1}, \|\cdot\|_{p,1,y}) \subset L^p(\mu_y)$ with the following property: a function f from $L^p(\mu)$ belongs to $\mathcal{E}^{p,1}(\mu)$ if and only if for μ_Y-a.e. $y \in Y$ we have $f_y \in \mathcal{E}_y^{p,1}$, the function $y \mapsto \|f_y\|_{p,1,y}$ belongs to the class $L^p(\mu_Y)$, and the norm in $\mathcal{E}^{p,1}(\mu)$ is equivalent to the norm
$$f \mapsto \left(\int_Y \|f_y\|_{p,1,y}^p \mu_Y(dy)\right)^{1/p}.$$

We observe that the latter condition of equivalence of norms is automatically fulfilled by the closed graph theorem if one has the estimate $\|f\|_{L^p(\mu_y)} \leqslant \|f_y\|_{p,1,y}$.

Condition (D) is fulfilled if μ is a centered Radon Gaussian measure on $X \times Y$ and Y is a finite dimensional subspace in $H(\mu)$ (Exercise 8.14.6).

For every $y \in Y$, replacing μ and $\mathcal{E}^{p,1}$ by μ_y and $\mathcal{E}_y^{p,1}$, respectively, we define the corresponding function $K_t^y(f)$ and the fractional Sobolev space $(\mathcal{E}^{p,\alpha}(\mu_y), \|\cdot\|_{p,\alpha,y})$ in the same manner as $K_t(f)$ and $\mathcal{E}^{p,\alpha}(\mu)$.

8.6.2. Theorem. *Suppose that the space $\mathcal{E}^{p,1}(\mu)$ satisfies condition* (D) *and $f \in L^p(\mu)$. Then $f \in \mathcal{E}^{p,\alpha}(\mu)$ if and only if for μ_Y-a.e. y the function f_y belongs to $\mathcal{E}^{p,\alpha}(\mu_y)$ and $y \mapsto \|f_y\|_{p,\alpha,y} \in L^p(\mu_Y)$.*

PROOF. Let $f \in \mathcal{E}^{p,\alpha}(\mu)$. It is easily verified that there exist measurable mappings $t \mapsto \varphi_t$ and $t \mapsto \psi_t$ on $(0, +\infty)$ with values in $L^p(\mu)$ and $\mathcal{E}^{p,1}(\mu)$, respectively, such that the relationships

$$f = \varphi_t + \psi_t \quad \text{and} \quad \|\varphi_t\|_{L^p(\mu)} + t\|\psi_t\|_{p,1} \leqslant K_t(f) + \min(t,1)$$

are fulfilled for all $t \in (0, +\infty)$. In fact, if $\mathcal{E}^{p,1}(\mu)$ is reflexive (as is the case in many applications), then one can even find φ_t and ψ_t such that

$$\|\varphi_t\|_{L^p(\mu)} + t\|\psi_t\|_{p,1} = K_t(f).$$

It is clear that the functions $\|\varphi_t\|_p^p$ and $t^p \|\psi_t\|_{p,1}^p$ are integrable with weight $t^{-\alpha p - 1}$ if and only if the function $K_t(f)$ is also. We observe that the integral of the function g against the measure μ equals

$$\int_Y \int_X g(x,y) \, \mu_y(dx) \, \mu_Y(dy).$$

Therefore, the function

$$(y,t) \mapsto t^{-\alpha p - 1} \int_X |\varphi_t(x,y)|^p \, \mu_y(dx)$$

is integrable with respect to $\mu_Y \otimes dt$. In addition, by condition (D) the function

$$(y,t) \mapsto t^{-\alpha p - 1 + p} \|\psi_t(x,y)\|_{p,1,y}^p$$

is $\mu_Y \otimes dt$-integrable as well. Since

$$f(x,y) = \varphi_t(x,y) + \psi_t(x,y) \quad \mu_y\text{-a.e. for } \mu_Y\text{-a.e. } y \in Y,$$

we conclude that $f_y \in \mathcal{E}^{\alpha,p}(\mu_y)$ for μ_Y-a.e. $y \in Y$.

Conversely, let a function $f \in L^p(\mu)$ be such that for μ_Y-a.e. y the function $f_y \colon x \mapsto f(x,y)$ belongs to $\mathcal{E}^{p,\alpha}(\mu_y)$ and the function $y \mapsto \|f_y\|_{p,\alpha,y}$ belongs to $L^p(\mu_Y)$. It is readily shown that there exist two measurable functions $(y,t,x) \mapsto \varphi_{y,t}(x)$ and $(y,t,x) \mapsto \psi_{y,t}(x)$ on $Y \times (0,+\infty) \times X$ such that for all t and y the function $x \mapsto \varphi_{y,t}(x)$ belongs to $L^p(\mu_y)$, the function $x \mapsto \psi_{y,t}(x)$ belongs to $\mathcal{E}^{p,1}(\mu_y)$, $f_y = \varphi_{y,t} + \psi_{y,t}$ and

$$\|\varphi_{y,t}\|_{L^p(\mu_y)} + t\|\psi_{y,t}\|_{p,1,y} \leqslant K_t^y(f_y) + \min(t,1).$$

By our assumption the function

$$y \mapsto \|f_y\|_{p,\alpha,y}^p = \int_0^\infty t^{-\alpha p - 1} K_t^y(f_y)^p \, dt$$

is μ_Y-integrable. Therefore, the functions

$$y \mapsto \int_0^\infty t^{-\alpha p - 1} \|\varphi_{y,t}\|_{L^p(\mu_y)}^p \, dt, \quad y \mapsto \int_0^\infty t^{-\alpha p - 1 + p} \|\psi_{y,t}\|_{p,1,y}^p \, dt$$

are μ_Y-integrable as well. It follows that the functions
$$t \mapsto \int_Y \|\varphi_{y,t}\|_{L^p(\mu_y)}^p \mu_Y(dy), \quad t \mapsto \int_Y \|\psi_{y,t}\|_{p,1,y}^p \mu_Y(dy)$$
are integrable on $(0, +\infty)$ with weights $t^{-\alpha p - 1}$ and $t^{-\alpha p - 1 + p}$, respectively. Set
$$\varphi_t(x, y) := \varphi_{t,y}(x), \quad \psi_t(x, y) := \psi_{t,y}(x)$$
for all $t \in (0, +\infty)$ such that the integrals above are finite. It is clear that for every such t we obtain that $f = \varphi_t + \psi_t$ μ-a.e. and $\varphi_t \in L^p(\mu)$, $\psi_t \in \mathcal{E}^{p,1}(\mu)$ due to condition (D). Then the inequality
$$K_t(f) \leqslant \|\varphi_t\|_p + t\|\psi_t\|_{p,1}$$
is true, which completes the proof. $\square$

Now we consider some examples where condition (D) is fulfilled. The Gaussian case has already been mentioned.

8.6.3. Corollary. *Let μ be a centered Radon Gaussian measure on a locally convex space X and let H be its Cameron–Martin space. Suppose that $E \subset H$ is a finite dimensional subspace and Y is a closed linear subspace in X such that $X = E \oplus Y$. Let $f \in \mathcal{E}^{p,\alpha}(\mu)$. Then for μ_Y-a.e. $y \in Y$ the function f_y belongs to $\mathcal{E}^{p,\alpha}(\mu_y)$. Hence f_y belongs to $H^{p,\alpha-\varepsilon}(\mu_y)$ for every $\varepsilon > 0$ and in the case $\alpha p > \dim E$ admits a version which is locally $(\alpha - \dim E/p)$-Hölder.*

If $f \in H^{p,\alpha}(\mu)$, then for every $\varepsilon > 0$ for μ_Y-a.e. $y \in Y$ the function f_y belongs to $H^{p,\alpha-\varepsilon}(\mu_y)$ and, in the case $(\alpha - \varepsilon)p > \dim E$, has a version which is locally $(\alpha - \varepsilon - \dim E/p)$-Hölder.

Suppose now that μ is a Radon probability measure on a locally convex space X and $E \subset X$ is a finite dimensional linear subspace. Then there exists a closed linear subspace $Y \subset X$ such that $X = E \oplus Y$. We equip E with a fixed inner product $(\,\cdot\,,\,\cdot\,)_E$; the corresponding norm will be denoted by $|\cdot|_E$. Again we denote by μ_Y the projection of μ on Y and by μ_y, $y \in Y$, conditional measures on E (which exist since μ is Radon and E is finite dimensional). Assume that these conditional measures possess continuous positive densities (with respect to some Lebesgue measure on E) majorized by a common constant M. Let us define the Sobolev class $W_E^{p,1}(\mu)$ as the completion of the space of smooth cylindrical functions with respect to the norm
$$\|f\|_{p,1} := \|f\|_p + \big\| |D_E f|_E \big\|_p,$$
where $D_E f$ is defined by the relation $(D_E f, h)_E = \partial_h f$, $h \in E$. We shall assume that the class $W_E^{p,1}(\mu)$ is well defined (i.e., the operator $D_E f$ is closable). Now we can set
$$\mathcal{E}^{p,1}(\mu) := W^{p,1}(\mu), \quad \mathcal{E}^{p,1}(\mu_y) := W^{p,1}(\mu_y)$$
and construct the fractional classes $\mathcal{E}^{p,\alpha}(\mu)$ and $\mathcal{E}_y^{p,\alpha}(\mu_y)$, $0 < \alpha < 1$. Verification of condition (D) is delegated to Exercise 8.14.5.

8.6.4. Corollary. *Suppose that under the stated assumptions we have $f \in \mathcal{E}^{p,\alpha}(\mu)$. Then, for μ_Y-a.e. $y \in Y$, the function f_y belongs to $\mathcal{E}^{p,\alpha}(\mu_y)$.*

We observe that the Sobolev classes $W^{p,1}(\mu)$ and $W^{p,1}(\mu_y)$ are well defined provided that μ is quasi-invariant along E, i.e., the shifts $\mu(\,\cdot\, - h)$ are equivalent to μ for all $h \in E$, and, in addition, the densities ϱ^y of conditional measures μ_y with respect to the natural Lebesgue measures on $E+y$ (existing by the quasi-invariance) are locally strictly positive (or, more generally, the functions $(\varrho^y)^{1/(1-p)}$ are locally Lebesgue integrable).

It should be emphasized that the results established here differ from the statements on restrictions to a fixed surface known in the finite dimensional case. The point is that, for a fixed element y, no estimate of $\|f_y\|_{p,\alpha,y}$ via the norm of f is given. It would be interesting to also study restrictions of Sobolev functions on infinite dimensional spaces to nonlinear surfaces (cf. Airault, Van Biesen [**31**]). Properties of restrictions of functions from fractional Sobolev classes can be important also in other problems (see Lescot, Malliavin [**720**], Malliavin [**767**]). See also Exercise 8.14.18.

8.7. Connections between different definitions

We have already seen above that in the Gaussian case certain Sobolev classes coincide. Here we consider connections between Sobolev classes over non-Gaussian measures.

Let us note a common useful property of the classes $W^{p,r}$, $G^{p,r}$ and $H^{p,r}$.

8.7.1. Proposition. *If $\{f_n\}$ is bounded in one of the spaces $W^{p,r}(\mu, H)$, $G^{p,r}(\mu, H)$ or $H^{p,r}(\mu)$, where $p > 1$, then there exists a weakly convergent subsequence $\{f_{n_i}\}$, for which the arithmetic means*

$$k^{-1}(f_{n_1} + \cdots + f_{n_k})$$

converge in the norm of the corresponding space.

PROOF. This can be easily deduced from the uniform convexity of the spaces $L^p(\mu, \mathcal{H}_k)$ (see, e.g., [**193**, Theorem 4.7.15]). □

The following theorem generalizes part of the results given above for Gaussian measures. Let μ be a Radon probability measure on a locally convex space X.

8.7.2. Theorem. (i) *If $\beta_h^\mu \in L^{p/(p-1)}(\mu)$ for all $h \in j(X^*)$, then we have $W^{p,r}(\mu) \subset G^{p,r}(\mu)$ and $D^{p,r}(\mu) \subset G^{p,r}(\mu)$ for all $p \in [1, \infty), r \in \mathbb{N}$.*

(ii) *Suppose that for every $h \in j(X^*)$ the conditional measures μ^y on the straight lines $x + \mathbb{R}^1 h$ have continuous positive densities (for example, there exist $c_h > 0$ such that $\exp(c_h \beta_h^\mu) \in L^1(\mu)$). Then*

$$D^{p,1}(\mu) = G^{p,1}(\mu).$$

8.7. Connections between different definitions

(iii) *Suppose that condition* (ii) *is fulfilled and there exists a sequence* $\{k_n\} \subset X^*$ *generating* $\sigma(X)$ *such that the vectors* $e_n = j(k_n)$ *form an orthonormal basis in* H *and*

$$\sup_n \left\| P_n \beta_H^\mu - \mathbb{E}_n(P_n \beta_H^\mu) \right\|_{L^p(\mu, H)} < \infty, \qquad (8.7.1)$$

where $P_n \beta_H^\mu = \sum_{i=1}^n \beta_{e_i}^\mu e_i$ *and* $\mathbb{E}_n$ *is the conditional expectation with respect to the* σ*-algebra generated by* $k_1, \ldots, k_n$. *Then*

$$W^{p,1}(\mu) = D^{p,1}(\mu) = G^{p,1}(\mu) \quad \text{for all } p > 1.$$

PROOF. (i) The inclusion $W^{p,r}(\mu) \subset G^{p,r}(\mu)$ has been verified in Proposition 8.1.2. The inclusion $D^{p,r}(\mu) \subset G^{p,r}(\mu)$ follows from the integration by parts formula.

(ii) Let $f \in G^{p,1}(\mu)$. Let us consider the measure $\nu = f \cdot \mu$. By our hypothesis, it is differentiable along every $h \in j(X^*)$ and $d_h \nu = g \cdot \mu + f \cdot d_h \mu$, where g is the generalized partial derivative of f along h. According to Proposition 6.3.1 (respectively, Proposition 6.4.1), the measure ν admits a density ϱ with respect to the measure μ such that for every $x \in X$ the function $t \mapsto \varrho(x + th)$ is absolutely continuous. We have $f = \varrho$ μ-a.e. In addition, $\partial_h \varrho$ coincides a.e. with the generalized partial derivative $\partial_h f$ of the function f, hence belongs to $L^p(\mu)$.

(iii) Suppose that condition (8.7.1) is fulfilled and $f \in G^{p,1}(\mu) = D^{p,1}(\mu)$. It suffices to prove our claim for bounded functions f, by taking compositions $\varphi \circ f$ with $\varphi \in C_b^\infty(\mathbb{R}^1)$. Without loss of generality, we may assume that $\sup_x |f(x)| \leqslant 1$. Let $\{k_n\}$ be a sequence in X^* generating the σ-algebra $\sigma(X)$ such that the vectors $e_n = j(k_n)$ form an orthonormal basis in H. Denote by σ_n the σ-algebra generated by $\{k_1, \ldots, k_n\}$ and by $\mathbb{E}_n$ the conditional expectation with respect to σ_n. By the martingale convergence we have $f_n := \mathbb{E}_n f \to f$ in $L^p(\mu)$. Let us show that $f_n \in W^{p,1}(\mu)$ and $\nabla_H f_n \to \nabla_H f$ in $L^p(\mu, H)$. Proposition 8.3.4 yields that one has $f_n \in G^{p,r}(\mu)$ and

$$\nabla_H f_n = \mathbb{E}_n(\nabla_H f) + \mathbb{E}_n(f P_n \beta_H^\mu) - f_n \mathbb{E}_n(P_n \beta_H^\mu),$$

where $P_n \beta_H^\mu = \sum_{i=1}^n \beta_{e_i}^\mu e_i$. Hence the inclusion $f_n \in W^{p,r}(\mu)$ is a result for measures on $\mathbb{R}^n$ (see Chapter 2). Now we shall find a subsequence of functions in $\{f_n\}$ whose arithmetic means converge in $W^{p,1}(\mu)$. By the martingale convergence one has $\partial_h f_n \to \partial_h f$ in $L^p(\mu)$ for each h. By convergence of vector martingales we have $\mathbb{E}_n(\nabla_H f) \to \nabla_H f$ in $L^p(\mu, H)$. We observe that

$$\sup_n \left\| \mathbb{E}_n(f P_n \beta_H^\mu) - f_n \mathbb{E}_n(P_n \beta_H^\mu) \right\|_{L^p(\mu, H)} < \infty.$$

Indeed, since $\mathbb{E}_n f \mathbb{E}_n(P_n \beta_H^\mu) = \mathbb{E}_n[f \mathbb{E}_n(P_n \beta_H^\mu)]$, $|f| \leqslant 1$, and $\mathbb{E}_n$ is a contraction on $L^p(\mu, H)$, we have

$$\left\| \mathbb{E}_n(f P_n \beta_H^\mu) - f_n \mathbb{E}_n(P_n \beta_H^\mu) \right\|_{L^p(\mu, H)} \leqslant \left\| P_n \beta_H^\mu - \mathbb{E}_n(P_n \beta_H^\mu) \right\|_{L^p(\mu, H)}.$$

Now we can apply Proposition 8.7.1. □

8.7.3. Remark. Condition (8.7.1) is fulfilled if $P_n \beta_H^\mu$ is a function of the functionals $k_1, \ldots, k_n$. This is the case if μ is a Gaussian measure with $H(\gamma) = H$.

The situation with the class $H^{p,r}(\mu)$ is less studied. For example, the inclusion $H^{p,r}(\mu) \subset G^{p,r}(\mu)$ has been only proven so far under very restrictive conditions. One can verify that in the case of a Gaussian measure μ the classes $W^{1,1}(\mu)$ and $D^{1,1}(\mu)$ still coincide, whereas the class $H^{1,1}(\mu)$ differs from them (even on the real line: see Exercise 8.14.7).

8.7.4. Proposition. *Let $f \in W^{p,r}(\mu)$, $g \in W^{s,r}(\mu)$. Then one has $fg \in W^{t,r}(\mu)$, where $t = sp/(p+s)$. A similar result holds for $D^{p,r}(\mu)$.*

PROOF. Hölder's inequality with exponents p/t and $p/(p-t)$ along with the equality $tp/(p-t) = s$ yields that $fg \in L^t(\mu)$. The same is true for the product $\|D_H^{r-l} f\|_{\mathcal{H}_{r-l}} \|D_H^l g\|_{\mathcal{H}_l}$. In the case of $W^{p,r}(\mu)$ these estimates can be applied to approximating functions from $\mathcal{FC}_b^\infty(X)$, and in the case of $D^{p,r}(\mu)$ directly to f and g. □

8.7.5. Proposition. *Let f be an element of one of the Sobolev classes $W^{p,r}(\mu)$ or $D^{p,r}(\mu)$ and let $\varphi \in C_b^r(\mathbb{R}^1)$. Then $\varphi(f)$ belongs to the same Sobolev class and $D_H \varphi(f) = \varphi'(f) D_H f$. In the case $r = 1$ this is true also for Lipschitzian φ.*

PROOF. Let us consider the case $r = 1$. For $W^{p,1}(\mu)$ the assertion follows from the finite dimensional case considered in Chapter 2. Let $f \in D^{p,1}(\mu)$ and $h \in H$. We may assume that the functions $t \mapsto f(x + th)$ are absolutely continuous on all bounded intervals. Then the functions $t \mapsto \varphi\big(f(x+th)\big)$ are absolutely continuous as well and

$$\partial_t\big[\varphi\big(f(x+th)\big)\big] = \varphi'\big(f(x+th)\big)\partial_t[f(x+th)]$$

a.e. in t for every fixed x. Set $D_H(\varphi \circ f) := \varphi'(f) D_H f$. Then

$$t^{-1}\big[\varphi\big(f(x+th)\big) - \varphi\big(f(x)\big)\big] - D_H(\varphi \circ f)(x)(h) \xrightarrow[t \to 0]{} 0 \quad \text{in measure } \mu.$$

The case $r > 1$ is similar. □

8.7.6. Corollary. *Let f and g be elements of one of the Sobolev classes $W^{p,1}(\mu)$ or $D^{p,1}(\mu)$. Then the functions $\max(f, g)$ and $\min(f, g)$ belong to the same Sobolev class.*

PROOF. We have

$$\max(f, g) = (|f - g| + f + g)/2, \quad \min(f, g) = (|f - g| - f - g)/2,$$

which yields our claim. □

8.7.7. Corollary. *Let $f \in D^{p,r}(\mu)$, $f \neq 0$ a.e. and let $1 \leqslant m \leqslant r$, $q > 1$, $s > 2q(m+1)$ be such that $p \geqslant 2mqs/(s - 2qm - 2q)$ and $1/f \in L^s(\mu)$. Then one has $1/f \in G^{q,m}(\mu)$. If $f \geqslant 0$, then it suffices to have $s > q(m+1)$ and $p \geqslant mqs/(s - qm - q)$.*

8.7. Connections between different definitions

PROOF. Set $g_\varepsilon := f(\varepsilon + f^2)^{-1}$, $\varepsilon > 0$. By the above proposition one has $g_\varepsilon \in D^{p,r}(\mu)$. As $\varepsilon \to 0$ we have $g_\varepsilon \to 1/f$ in $L^q(\mu)$ since $s > q$. Now it suffices to verify that the functions g_ε with $\varepsilon \in (0,1)$ have uniformly bounded norms in $G^{q,m}(\mu)$. To this end, we observe, repeatedly differentiating g_ε, that the quantity $\|D_H^m g_\varepsilon(x)\|_{\mathcal{H}_m}$ is estimated by a sum of products of the form $(f^2(x) + \varepsilon)^{-m-1} \psi_1(x) \cdots \psi_{2m}(x)$, where the functions ψ_i are the norms of derivatives of f of orders from 0 to m. It remains to estimate the $L^q(\mu)$-norm of such a product by Hölder's inequality with exponents $s/(2qm + 2q)$ and $s/(s - 2qm - 2q)$ and use the completeness of $G^{p,r}(\mu)$. If $f \geq 0$, then a similar reasoning applies to the functions $g_\varepsilon = (f + \varepsilon)^{-1}$. □

8.7.8. Proposition. *Let $u, v \in W^{2p,r}(\mu, H)$. Then $(u,v)_H \in W^{p,r}(\mu)$. This is true also for the pair $D^{2p,r}(\mu, H)$ and $D^{p,r}(\mu)$.*

PROOF. Let $r = 1$. Since $|(u,v)_H| \leq |u|_H |v|_H$, one has $(u,v)_H \in L^p(\mu)$. In addition, $|D_H(u,v)_H|_H \in L^p(\mu)$ since
$$|D_H(u,v)_H|_H \leq \|D_H u\|_{\mathcal{H}} |v|_H + \|D_H v\|_{\mathcal{H}} |u|_H.$$
Indeed, for any orthonormal basis $\{e_i\}$ in H, letting $u^j = (u, e_j)_H$ and $v^j = (v, e_j)_H$, we have
$$|D_H(u,v)_H|_H^2 = \sum_{i=1}^\infty \Big| \sum_{j=1}^\infty (v^j \partial_{e_i} u^j + u^j \partial_{e_i} v^j) \Big|^2.$$
This yields the assertion for $W^{p,r}$ by finite dimensional approximations. In the case $D^{p,r}$ we take suitable versions of u and v and also apply these estimates. In the case $r > 1$ the norm of the mapping $D_H^r(u,v)_H$ in $\mathcal{H}_r$ is estimated by a sum of products $\|D_H^{r-k} u\|_{\mathcal{H}_{r-k}} \|D_H^k v\|_{\mathcal{H}_k}$ with some coefficients, which gives a function from $L^p(\mu)$. □

8.7.9. Example. Let γ be a Radon Gaussian measure on a locally convex space X with the Cameron–Martin space H and let $u, v \in W^\infty(\gamma, H)$. Then one has $(u,v)_H \in W^\infty(\gamma)$.

8.7.10. Proposition. *Let a measure μ be continuous along vectors from a subspace dense in H and $f \in D^{1,1}(\mu)$. Then $D_H f = 0$ a.e. on the set $\{f = 0\}$.*

PROOF. It suffices to recall that if φ is absolutely continuous on an interval, then $\varphi' = 0$ almost everywhere on the set $\{\varphi = 0\}$. □

A measure $\mu \geq 0$ is called *H-ergodic* with respect to one of the Sobolev classes introduced above if the equality $D_H f = 0$ in the sense of this class implies that f is a constant.

8.7.11. Example. Suppose that a measure μ is H-ergodic with respect to $D^{1,1}(\mu)$ (or $W^{1,1}(\mu)$). If the indicator function I_A of a set A is contained in $D^{1,1}(\mu)$, then either $I_A = 0$ a.e. or $I_A = 1$ a.e. Indeed,
$$D_H I_A = D_H(I_A^2) = 2 I_A D_H I_A = 2 D_H I_A \quad \text{a.e.}$$

since $D_H I_A = 0$ a.e. on the set $\{I_A = 0\}$. Hence we have $D_H I_A = 0$ a.e., whence $I_A = \mathrm{const}$ by the ergodicity of μ.

The classes $W^{p,r}$ and $D^{p,r}$ can be defined also for restrictions of the measure μ to certain sets Ω of positive measure passing to the measure $\mu|_\Omega$. It would be interesting to study such classes. Another problem which has not yet been studied concerns extensions of functions from classes on domains in infinite dimensional spaces. In some simple cases such an extension can be explicitly constructed. For example, if γ is a centered Radon Gaussian measure and $\Omega = \{\xi > 0\}$, where $\xi = \widehat{h}$, $h \in H = H(\gamma)$, $|h|_H = 1$, then any function $f \in W^{p,1}(\gamma|_\Omega)$ can be extended to a function $Ef \in W^{p,1}(\gamma)$ by the formula $Ef(x) := f(x - 2\xi(x)h)$ if $\xi(x) < 0$. It readily follows from Exercise 1.6.19 that $\|Ef\|_{W^{p,1}(\gamma)} = 2\|f\|_{W^{p,1}(\gamma|_\Omega)}$. Hence $f \mapsto Ef$ is a bounded linear operator. It would be interesting to study extensions from noncylindrical domains. Shaposhnikov [**993**] obtained a Lusin type theorem for vector fields on the Wiener space, which generalizes Alberti [**34**].

Some results on extensions of H-Lipschitzian functions were obtained in Kobanenko [**618**] and Bogachev [**192**]. For scalar functions, the following result was obtained in [**618**]. Suppose we are given a Souslin locally convex space X, a linear subspace E in X equipped with a norm $\|\cdot\|_E$ such that the closed unit ball is a Souslin set, a Souslin set $\Omega \subset X$, and a function $f \colon \Omega \to \mathbb{R}^1$ such that one has $|f(x+h) - f(x)| \leqslant C\|h\|_E$ whenever $x \in \Omega$, $h \in E$, $x + h \in \Omega$, and the sets $\{f > c\}$ are Souslin. Then f can be extended to a function $\widetilde{f}$ on X such that $|\widetilde{f}(x+h) - \widetilde{f}(x)| \leqslant C\|h\|_E$ for all $x \in X$, $h \in E$ and the sets $\{\widetilde{f} > c\}$ are Souslin. If γ is a centered Radon Gaussian measure on a locally convex space X, $H = H(\gamma)$, Ω is a γ-measurable set and $F \colon \Omega \to H$ is a γ-measurable mapping such that $|F(x) - F(y)|_H \leqslant C|x - y|_H$ whenever $x, y \in \Omega$, $x - y \in H$, then, according to [**192**], there exists a γ-measurable mapping $\widetilde{F} \colon X \to H$ a.e. equal to F on Ω such that $|\widetilde{F}(x+h) - \widetilde{F}(x)|_H \leqslant C|h|_H$ for all $x \in X$, $h \in H$.

8.8. The logarithmic Sobolev inequality

Sobolev inequalities and embedding theorems do not extend to the infinite dimensional case. This is related to the fact that in the finite dimensional case the corresponding estimates depend on dimension. However, there are very useful inequalities independent of dimension that remain valid also in the infinite dimensional case. Here we consider such inequalities for Gaussian measures.

Let γ be a centered Radon Gaussian measure on a locally convex space X, $H = H(\gamma)$, and let I_n be the projector to $\mathcal{X}_n$ (see §8.5). The following important result of Gross [**505**] is called the *logarithmic Sobolev inequality*.

8.8.1. Theorem. *For all $f \in W^{2,1}(\gamma)$ we have*

$$\int_X f^2 \ln |f| \, d\gamma \leqslant \int_X |D_H f|_H^2 \, d\gamma + \frac{1}{2}\left(\int_X f^2 \, d\gamma\right) \ln\left(\int_X f^2 \, d\gamma\right).$$

8.8.2. Corollary. *The Ornstein–Uhlenbeck semigroup $\{T_t\}$ is hypercontractive, i.e., one has*

$$\|T_tf\|_q \leqslant \|f\|_p \text{ if } t > 0,\, p > 1,\, q > 1,\, e^{2t} \geqslant (q-1)/(p-1).$$

The same is true for T_t on $L^p(\gamma, E)$, where E is a separable Hilbert space.

8.8.3. Corollary. *For any $p \in [1, \infty)$ the operator $I_n \colon f \mapsto I_n(f)$ from $L^2(\gamma)$ to $L^p(\gamma)$ is continuous and for any $p \geqslant 2$ one has the inequality*

$$\|I_n(f)\|_p \leqslant (p-1)^{n/2}\|f\|_2. \tag{8.8.1}$$

In addition, for every $p \in (1, +\infty)$ the operator I_n is continuous on $L^p(\gamma)$ and

$$\|I_n\|_{\mathcal{L}(L^p(\gamma))} \leqslant (M-1)^{n/2}, \quad M := \max\bigl(p, p(p-1)^{-1}\bigr).$$

8.8.4. Corollary. *Let $f \in \mathcal{X}_n$ and $p \geqslant 2$, $r \in \mathbb{N}$. Then*

$$\|f\|_{H^{2,r}} \leqslant \|f\|_{H^{p,r}} \leqslant (p-1)^{n/2}\|f\|_{H^{2,r}} = (p-1)^{n/2}(n+1)^{r/2}\|f\|_{L^2(\gamma)}.$$

Hence the norms $\|\cdot\|_{H^{p,r}}$ are equivalent to the $L^2(\gamma)$-norm on $\mathcal{X}_n$.

8.8.5. Corollary. *Suppose that $f \in \mathcal{X}_n$ and $\alpha \in \bigl(0, n/(2e)\bigr)$. Let us set $c(\alpha, n) = 2\exp\alpha + n/(n - 2e\alpha)$. Then*

$$\gamma\Bigl(x\colon |f(x)| \geqslant t\|f\|_2\Bigr) \leqslant c(\alpha, n)\exp(-\alpha t^{2/n}).$$

8.8.6. Corollary. *The subspace $\mathcal{X}_n$ is closed in the space of measurable functions with respect to convergence in measure. Moreover, convergence in measure of a sequence from the class $\mathcal{P}_n(\gamma) := \bigoplus_{k=0}^n \mathcal{X}_k$ is equivalent to its convergence in every space $L^p(\gamma)$ with $p \in [1, +\infty)$. In addition,*

$$\|f\|_{L^p(\gamma)} \leqslant \|f\|_{L^q(\gamma)} \leqslant \Bigl(\frac{q-1}{p-1}\Bigr)^{n/2}\|f\|_{L^p(\gamma)}, \quad f \in \mathcal{X}_n,\, 1 < p < q.$$

8.8.7. Corollary. *Let E be a separable Hilbert space. A sequence F_j in $\bigoplus_{k=0}^n \mathcal{X}_k(E)$ converges in measure (then also in all $L^p(\gamma, E)$) precisely when for every $k = 0, \ldots, n$ the integrals of $D_H^k F_j$ over X converge in $\mathcal{H}_k(H, E)$.*

8.8.8. Theorem. *Suppose that γ is a Radon Gaussian measure on a locally convex space X and let $\gamma(A) > 0$. Then, for every $d \geqslant 0$, on the space $\mathcal{P}_d(\gamma)$ of γ-measurable polynomials of degree at most d every $L^p(\gamma)$-norm with $1 \leqslant p < \infty$ is equivalent to every $L^r(\gamma|_A)$-norm with $1 \leqslant r < \infty$.*

PROOF. It suffices to verify that convergence of $f_j \in \mathcal{P}_d(\gamma)$ in measure on A yields convergence in measure on X. Passing to a subsequence, we may assume that $\{f_j\}$ converges a.e. on A. Then, according to [**191**, Proposition 5.10.9], one has convergence a.e. on X. □

This result of Berezhnoy [**141**] (proved in Dorogovtsev [**357**] for balls in a Hilbert space) was generalized in Berezhnoy [**142**] to countable products of convex measures on finite dimensional spaces. It is not clear whether the same is true for arbitrary convex measures.

8.8.9. Theorem. *For all $f \in W^{2,1}(\gamma)$ the following Poincaré inequality is fulfilled:*
$$\int_X \left(f(x) - \int_X f(y)\gamma(dy) \right)^2 \gamma(dx) \leq \int_X |D_H f(x)|_H^2 \gamma(dx).$$
More generally, let a function $f \in W^{1,1}_{\text{loc}}(\gamma)$ be such that
$$\int_X |D_H f(x)|_H^p \gamma(dx) = N_p$$
for some $p \in \mathbb{N}$, $p > 1$. Then $f \in W^{p,1}(\gamma)$ and
$$\int_X \left| f(x) - \int_X f \, d\gamma \right|^p \gamma(dx) \leq (\pi/2)^p \sqrt{p!} N_p.$$

Let us mention several results on the exponential integrability of Sobolev functions.

8.8.10. Theorem. *If $\exp(\alpha |D_H f|_H^2) \in L^1(\gamma)$ for some $\alpha > 1/2$, then we have $\exp(|f|) \in L^1(\gamma)$ and*
$$\int_X \exp\left(f - \int f \, d\gamma \right) d\gamma \leq \left(\int_X \exp(\alpha |D_H f|_H^2) \, d\gamma \right)^{1/(2\alpha-1)}.$$
If $|D_H f|_H \leq C$, where C is a constant, then $\exp(\alpha f^2) \in L^1(\gamma)$ for all $\alpha < (2C^2)^{-1}$. In particular,
$$\int_X \exp\left(f - \int_X f \, d\gamma \right) d\gamma \leq \int_X \exp(|D_H f|_H^2) \, d\gamma.$$

Proofs of all these results can be found in [**191**, Ch. 5]).

There are other classes of measures μ for which one has logarithmic Sobolev inequalities of the form
$$\int_X f^2 \ln |f| \, d\mu \leq C \int_X |D_H f|_H^2 \, d\mu + C \left(\int_X f^2 \, d\mu \right) \ln \left(\int_X f^2 \, d\mu \right)$$
for all functions $f \in \mathcal{F}C_b^\infty$. For example, if $X = \mathbb{R}^\infty$ and $H = l^2$, then this is true for the measure μ that is a countable product of measures μ_n on the real line satisfying the same inequality.

Interesting results on the exponential integrability of functions from the Sobolev classes $F^{2,1}(\mu)$ associated with Dirichlet forms are obtained in Aida, Masuda, Shigekawa [**11**], Aida, Shigekawa [**12**], Aida, Stroock [**13**]. For example, if the measure μ satisfies the logarithmic Sobolev inequality, a function $f\colon X \to \mathbb{R}^1$ is Lipschitzian along H and $\varrho \in F^{2,1}(\mu)$, then for sufficiently small $a > 0$ we have
$$\int_X e^{a|f(x)|^2} \varrho^2(x) \, \mu(dx) < \infty.$$

In relation to logarithmic Sobolev inequalities one should also mention the so-called isoperimetric inequalities. For example, if γ is a centered Radon Gaussian measure on a locally convex space X with the Cameron–Martin

subspace H, and U_H is the closed unit ball in H, then, for any γ-measurable set A one has
$$\gamma(A+tU_H) \geq \Phi(a+t) \quad \forall t \geq 0, \qquad (8.8.2)$$
where Φ is the standard Gaussian distribution function and a is chosen in such a way that $\Phi(a) = \gamma(A)$. If $\gamma(A) \geq 1/2$, then
$$\gamma(A+rU_H) \geq \Phi(r) \geq 1 - \frac{1}{2}\exp\left(-\frac{1}{2}r^2\right). \qquad (8.8.3)$$
Note that inequality (8.8.2) can be written as
$$\gamma(A+rU_H) \geq \Phi(a+r),$$
where $a = \Phi^{-1}\bigl(\gamma(A)\bigr)$. Therefore, $\Phi(a+r)$ is the measure of the set $\Pi + rU_H$, where Π is a half-space having the same measure as A. If we define the surface measure of A as the limit of the ratio $r^{-1}\bigl(\gamma(A+rU) - \gamma(A)\bigr)$ as $r \to 0$ (see §9.5), then (8.8.2) shows that the half-spaces possess the minimal surface measures among the sets of given positive measure. See also Remark 9.5.8.

8.9. Compactness in Sobolev classes

Embedding theorems for Sobolev classes on $\mathbb{R}^n$ discussed in Chapter 2 do not extend to the Sobolev classes introduced above on infinite dimensional spaces. For example, if a measure γ on $X = \mathbb{R}^\infty$ is the countable product of the standard Gaussian measures on $\mathbb{R}^1$, then, for every nonzero smooth function φ with bounded support on the real line, the sequence of functions $f_n(x) = \varphi(x_n)$ is bounded in $W^{2,1}(\gamma)$, but is not precompact in $L^2(\gamma)$ since the mutual distances between these elements are equal. Diverse logarithmic Sobolev type inequalities play a role of a certain substitute for embedding theorems since they ensure uniform integrability of functions from sets that are bounded in Sobolev norms. However, it is also of interest to have sufficient conditions for compactness of sets from Sobolev classes in L^p or in other Sobolev classes. Such conditions in the case of a Gaussian measure on an infinite dimensional space were obtained by N.N. Frolov (see [**452**], [**453**], [**457**]) and Da Prato, Malliavin, and Nualart (see [**311**]).

Let γ be a centered Radon Gaussian measure on a locally convex space X with the Cameron–Martin space $H = H(\gamma)$.

8.9.1. Theorem. *Let C be a compact injective linear operator on H. Then, for every $R > 0$, the set*
$$F = \{f \in W^{2,1}(\gamma) \colon D_H f(x) \in C(H) \text{ } \gamma\text{-a.e.}, \|f\|_{L^2(\gamma)} + \|C^{-1}D_H f\|_{L^2(\gamma, H)} \leq R\}$$
is compact in $L^2(\gamma)$.

8.9.2. Theorem. *A set $F \subset W^{2,\infty}(\gamma)$ has compact closure in the space $W^{2,\infty}(\gamma)$ equipped with its natural topology of a Fréchet space if and only if it is bounded in $L^2(\gamma)$ and, for every $n \geq 1$, there exists a selfadjoint compact operator C_n on $\mathcal{H}_n$ such that for all $f \in F$ we have*
$$D_H^n f(x) \in C_n(\mathcal{H}_n) \text{ } \gamma\text{-a.e. and } \sup_{f \in F} \|C_n^{-1} D_H^n f\|_{L^2(\gamma, \mathcal{H}_n)} < \infty.$$

8.9.3. Theorem. *A set $F \subset W^\infty(\gamma)$ has compact closure in the space $W^\infty(\gamma)$ equipped with its natural topology of a Fréchet space if and only if the following two conditions are fulfilled:*

(i) $\sup_{f \in F} \|f\|_{p,r} < \infty$ *for all $p, r \geqslant 1$;*

(ii) *for every $n \geqslant 1$, there exists a selfadjoint compact operator C_n on $\mathcal{H}_n$ such that for all $f \in F$ we have*
$$D_H^n f \in C_n(\mathcal{H}_n) \ \gamma\text{-a.e. and } \sup_n \|C^{-1} D_H^n f\|_{L^2(\gamma, \mathcal{H}_n)} < \infty.$$

8.9.4. Corollary. *Let X be a Hilbert space. The set of continuously Fréchet differentiable functions f on X with $\|f\|_{L^2(\gamma)} + \|D_X f\|_{L^2(\gamma, X)} \leqslant C$ has compact closure in $L^2(\gamma)$.*

Compactness in Gaussian Sobolev classes is investigated also in Bally, Saussereau [97], Peszat [880]. Conditions for compactness in Sobolev classes over more general smooth measures is an important open problem. For some results, see Peszat [881]. In Da Prato, Debussche, Goldys [309], one can find conditions for compactness of embeddings to $L^p(\nu)$ for differently defined Sobolev classes over a measure ν which is absolutely continuous with respect to a Gaussian measure on a Hilbert space.

8.10. Divergence

A useful property of the Gaussian Sobolev classes is the existence of divergence for vector fields from $W^{2,1}(\gamma, H)$, where γ is a centered Radon Gaussian measure on a locally convex space X having the Cameron–Martin space H. The next important fact follows from the results in §6.6.

8.10.1. Theorem. *Suppose that $v \in W^{2,1}(\gamma, H)$. There exists a function $\delta v = \beta_v^\gamma \in L^2(\gamma)$, called the divergence of v, such that*
$$\int_X \bigl(D_H f(x), v(x)\bigr)_H \gamma(dx) = - \int_X f(x) \delta v(x) \, \gamma(dx), \quad f \in W^{2,1}(\gamma).$$
In addition, $\|\delta v\|_{L^2(\gamma)}^2 \leqslant \|v\|_{L^2(\gamma, H)}^2 + \|D_H v\|_{L^2(\gamma, \mathcal{H})}^2$.

The reader is warned that many authors define divergence without a minus in this formula in order to be able to identify the divergence with the Skorohod integral (see §8.11). Here the divergence coincides with the logarithmic derivative β_v^γ of the field v. If $\{e_n\}$ is an orthonormal basis in the space H and $v = \sum_{n=1}^\infty v_n e_n$, then
$$\delta v(x) = \sum_{n=1}^\infty \bigl(\partial_{e_n} v_n(x) - v_n(x) \widehat{e_n}(x)\bigr). \qquad (8.10.1)$$
However, the two parts of this series may fail to be summable separately. The determining formula for δv can be written as
$$\int_X \partial_v f(x) \, \gamma(dx) = - \int_X \delta v(x) f(x) \, \gamma(dx).$$

8.10. Divergence

An advantage of this representation is that it does not employ any auxiliary Hilbert space. It would be interesting to describe broader classes of nonconstant vector fields of differentiability for Gaussian measures.

8.10.2. Example. The measure γ is not differentiable along the vector field $v(x) = x$ if H is infinite dimensional.

PROOF. Suppose that γ is differentiable along v. Let $\{e_n\}$ be an orthonormal basis in the space $H(\gamma)$, $P_n x = \sum_{i=1}^{n} \widehat{e}_i(x) e_i$, $v_n(x) = P_n v(x)$. Let σ_n denote the σ-algebra generated by P_n. For every smooth function φ of $P_n x$ the function $\partial_v \varphi = \partial_{v_n} \varphi$ is σ_n-measurable. Hence the measure $\gamma \circ P_n^{-1}$ is differentiable along v_n and the corresponding logarithmic derivative β_n coincides with the conditional expectation of β_v^γ with respect to σ_n. Therefore, $\{\beta_n\}$ is a martingale convergent in $L^1(\gamma)$. On the other hand, direct calculations show that

$$\beta_n = \sum_{i=1}^{n} (1 - \widehat{e}_i^{\,2}).$$

The random variables $1 - \widehat{e}_i^{\,2}$ are independent and have a common distribution. Hence the series $\sum_i (1 - \widehat{e}_i^{\,2})$ does not converge. This contradiction proves our claim. $\square$

It is easily verified that for every function $f \in W^{2,2}(\gamma)$ one has the equality

$$\delta D_H f = Lf.$$

Proofs of the following two theorems can be found in [191, §5.8].

The operator $\delta \colon W^{2,1}(\gamma, \mathcal{H}_k) \to W^{2,1}(\gamma, \mathcal{H}_{k-1})$ is defined by the formula

$$\int_X (D_H f, v)_{\mathcal{H}_k} \, d\gamma = -\int_X (f, \delta v)_{\mathcal{H}_{k-1}} \, d\gamma, \quad f \in W^{2,1}(\gamma, \mathcal{H}_{k-1}).$$

8.10.3. Theorem. *If $p \in (1, \infty)$, $r, k \in \mathbb{N}$, then δ extends to a continuous operator $\delta \colon W^{p,r}(\gamma, \mathcal{H}_k) \to W^{p,r-1}(\gamma, \mathcal{H}_{k-1})$. For any $v \in W^\infty(\gamma, H)$ series (8.10.1) converges in all $W^{p,r}(\gamma, H)$ and $\delta v \in W^\infty(\gamma)$.*

Convergence of series (8.10.1) in $W^{p,r}(\gamma, H)$ follows by the continuity of δ and convergence $\sum_{i=1}^{n} v_i e_i \to v$ in $W^{p,r+1}(\gamma, H)$ verified by the Lebesgue dominated convergence theorem.

8.10.4. Theorem. *For every $p \in (1, \infty)$, there exists a number N_p such that*

$$\|(I - L)^{-1/2} \delta v\|_{L^p(\gamma)} \leqslant N_p \|v\|_{L^p(\gamma, H)}, \quad v \in W^{p,1}(\gamma, H).$$

It is clear from §6.6 that the divergence exists also for some non-Gaussian measures. Suppose that H is a separable Hilbert space continuously and densely embedded into a locally convex space X, a Radon probability measure μ on X has the logarithmic gradient β_H^μ, $\{e_i\}$ is an orthonormal basis in the space H, $v = \sum_{i=1}^{\infty} v^i e_i \in W^{2,1}(\mu, H)$, the conditions of Theorem 6.6.3

are fulfilled for v^i and e_j, and the function $\|D\beta(x)\|_{\mathcal{L}(H)}|v(x)|_H^2$ belongs to the class $L^1(\mu)$. The results of §6.6 yield the following assertion.

8.10.5. Theorem. *Under the stated assumptions, there exists the divergence $\delta_\mu v \in L^2(\mu)$ and one has*

$$\|\delta_\mu v\|_{L^2(\mu)}^2 \leqslant \|D_H v\|_{L^2(\mu,\mathcal{H})} + \big\|\|D\beta(\,\cdot\,)\|_{\mathcal{L}(H)}^{1/2} v\big\|_{L^2(\mu,H)}.$$

In addition, $\delta_\mu v = \lim\limits_{n\to\infty} \delta_\mu P_n v$ in $L^2(\mu)$, where P_n is the projector in H on the linear span of $e_1, \ldots, e_n$.

8.10.6. Example. Let γ be a centered Radon Gaussian measure on a locally convex space X and let $\xi \in L^2(\gamma)$ have zero integral over X. Then there exists $v \in W^{2,1}(\gamma, H)$ with $\delta v = \xi$ and $\|\xi\|_2 \leqslant \|v\|_{W^{2,1}} \leqslant C\|\xi\|_2$, where C is a constant.

PROOF. Let $\xi = \sum_{n=1}^\infty \xi_n$ be decomposition (8.5.2). Then

$$f := -\sum_{n=1}^\infty n^{-2}\xi_n \in W^{2,2}(\gamma) \quad \text{and} \quad v := -\nabla_H f \in W^{2,1}(\gamma, H),$$

where

$$\xi = Lf = -\delta\nabla_H f.$$

It remains to apply the estimate from Theorem 8.10.1 and Meyer's estimate (8.5.1) with $r = p = 2$. □

8.10.7. Proposition. *Let X be a separable Hilbert space, let H be a Hilbert space densely embedded by a Hilbert–Schmidt operator T, and let a measure μ on X have the logarithmic gradient β_H^μ. Suppose that $u\colon X \to X^*$ is Gâteaux differentiable along X and the functions $\langle u, \beta_H^\mu\rangle$ and $\|D_X u\|_{\mathcal{L}(X,X^*)}$ are μ-integrable. Set $v(x) := j_H u(x)$. Then the measure μ is differentiable along the field v and one has the equality*

$$\beta_v^\mu = \langle u, \beta_H^\mu\rangle + \mathrm{trace}_H(D_H v). \tag{8.10.2}$$

PROOF. We may assume that $H = T(X)$, where the operator $T \in \mathcal{L}(X)$ is symmetric and has an orthonormal eigenbasis $\{\psi_i\}$ with $T\psi_i = t_i\psi_i$ and that $(a,b)_X = (Ta, Tb)_H$. Then the vectors $e_i := T\psi_i$ form an orthonormal basis in H. If we identify X^* with X by the Riesz theorem, then $j_H = T^2$. Hence

$$(D_H v e_i, e_i)_H = (T^2 \partial_{T\psi_i} u, T\psi_i)_H = (T\partial_{T\psi_i}u, \psi_i)_X$$
$$= (\partial_{T\psi_i} u, T\psi_i)_X = (D_X u T\psi_i, T\psi_i)_X,$$

which does not exceed $\|D_X u\|_{\mathcal{L}(X)}|T\psi_i|_X^2$ in the absolute value. Therefore,

$$|\mathrm{trace}_H(D_H v)| \leqslant c\|D_X u\|_{\mathcal{L}(X,X^*)}.$$

Hence the right-hand side of (8.10.2) is integrable. We have

$$\langle u, \beta_H^\mu\rangle = \sum_{n=1}^\infty \langle u, e_n\rangle \beta_{e_n}^\mu = \sum_{n=1}^\infty (v, e_n)_H \beta_{e_n}^\mu.$$

For any $\varphi \in \mathcal{F}C_b^\infty(X)$ the integration by parts formula gives
$$\int_X \partial_{e_n}\varphi(v, e_n)_H\, \mu(dx) = -\int_X \varphi\bigl[\partial_{e_n}(v, e_n)_H + (v, e_n)_H \beta_{e_n}^\mu\bigr] \mu(dx),$$
whence after summation in n we obtain
$$\int_X (D_H\varphi, v)_H\, \mu(dx) = -\int_X \varphi\bigl[\mathrm{trace}_H(D_H v) + \langle u, \beta_H^\mu\rangle\bigr] \mu(dx).$$
This gives equality (8.10.2). □

Let us note the following property of locality of divergence: $\delta_\mu v = 0$ a.e. on the set $v^{-1}(0)$ under some additional conditions. For example, this is true for any $v \in G^{2,1}(\mu, H)$ (Exercise 8.14.2); without any assumption of differentiability of $v \in L^2(\mu, H)$ this is true on sufficiently regular sets Ω on which $v = 0$. For example, according to Gomilko, Dorogovtsev [492], in the case of a Gaussian measure μ on $\mathrm{I\!R}^n$ this is true for Ω such that for a.e. $x \in \Omega$ one has
$$\lambda_n\bigl(\Omega \cap B(x, r)\bigr)/\lambda_n\bigl(B(x, r)\bigr) = 1 + o(r^2) \quad \text{as } r \to 0.$$
In the infinite dimensional case, it suffices that $t^{-p} I_\Omega(1 - T_t I_\Omega) \xrightarrow[t \to 0]{} 0$ in measure μ for some $p > 1/2$.

In general, divergence has no property of locality. Different examples are known; the one presented below was suggested by L. Ambrosio; another one, more explicit, and more technical, was constructed by A.V. Shaposhnikov.

8.10.8. Example. According to Alberti [34], there exists a function f in $C_0^1(\mathrm{I\!R}^2)$ such that the continuous vector field $u = \nabla f$ coincides with the field $w(x, y) = (y, -x)$ on a positive measure set E. Let
$$v(x, y) := (x, y) - Su(x, y),$$
where S is the rotation in the plane that takes $(1, 0)$ to $(0, 1)$. Then v vanishes on E, but $\mathrm{div}\, v = 2$ because $Su = (-\partial_y f, \partial_x f)$, hence for the distributional derivative we have $\mathrm{div}\, Su = -\partial_x \partial_y f + \partial_y \partial_x f = 0$.

8.11. An approach via stochastic integrals

In the case of the classical Wiener space the principal objects of the Malliavin calculus, including gradient and divergence, can be introduced in terms of random processes and stochastic integrals. Suppose that, as in Definition 1.5.2 in §1.5, we are given a Gaussian process on a probability space $(\Omega, P, \mathcal{F})$, indexed by sets $B \in \mathcal{B}$, where $(T, \mathcal{B})$ is a measurable space with a finite positive atomless measure μ. Set $H := L^2(\mu)$. Let us recall that in §1.5 we defined the stochastic integrals $I_n(f_n)$ and $w(h)$, $h \in H$. The *Ornstein–Uhlenbeck semigroup* $\{T_t\}$ on $L^2(P)$ is given by the formula
$$T_t F = \mathrm{I\!E} F + \sum_{n=1}^\infty e^{-nt} I_n(f_n), \quad F = \mathrm{I\!E} F + \sum_{n=1}^\infty I_n(f_n), \qquad (8.11.1)$$

where $f_n \in L^2(\mu^n)$ are symmetric and $\|I_n(f_n)\|_2^2 = n!\|f_n\|_2^2$. We have already encountered this semigroup in a different representation.

8.11.1. Definition. *The Malliavin operator L is defined on the domain*

$$D(L) = \Big\{F \in L^2(P)\colon \sum_{n=1}^{\infty} n\|I_n(f_n)\|_2^2 < \infty\Big\}$$

by the formula

$$LF = -\sum_{n=1}^{\infty} nI_n(f_n). \tag{8.11.2}$$

Clearly, the series in (8.11.2) converges in $L^2(P)$. Below we shall consider other representations of the Malliavin operator. The following properties of the operator L are easily verified.

(i) L is a densely defined nonpositive selfadjoint operator;
(ii) $\mathbb{E}(LF) = 0$ for all $F \in D(L)$;
(iii) L coincides with the generator of the semigroup T_t;
(iv) $-L$ has the square root $\sqrt{-L}$ defined by the formula

$$\sqrt{-L}F = \sum_{n=1}^{\infty} \sqrt{n}I_n(f_n)$$

on the domain

$$H^{2,1} := D(\sqrt{-L}) = \Big\{F\colon \sum_{n=1}^{\infty} n\|I_n(f_n)\|_2^2 < \infty\Big\}.$$

Now we define the Malliavin derivative. Let a real function F be defined by equality (1.5.2). For every $h \in H = L^2(\mu)$ put

$$D_h F := \sum_{n=1}^{\infty} nI_{n-1}\Big[\big(h, f_n(t_1, \ldots, t_{n-1}, \cdot)\big)_H\Big],$$

provided that the series converges in $L^2(P)$. Similarly, for $F \in H^{2,1}$ we put

$$D_t F := \sum_{n=1}^{\infty} nI_{n-1}\big(f_n(\cdot, t)\big), \quad t \in T.$$

We shall say that F has the *Malliavin derivative DF* if there exists a mapping $DF \in L^2(P, H)$ such that $(DF, h)_H = D_h F$ for all $h \in H$ in the sense of equality in $L^2(P)$. In this case $DF(\omega)(t) = D_t F(\omega)$ if DF is considered as an element of $L^2(P \otimes \mu)$. Indeed, it suffices to verify this for $F = I_n(f_n)$, where f_n has the form $f_n(t_1, \ldots, t_n) = I_{B_1}(t_1) \cdots I_{B_n}(t_n)$, $B_i \in \mathcal{B}$. Then

$$D_h F = nw(B_1) \cdots w(B_{n-1})(h, I_{B_n})_{L^2(\mu)},$$

which equals the integral of $h(t)D_t F$ with respect to the measure μ since

$$D_t F = nI_{B_n}(t)w(B_1) \cdots w(B_{n-1}).$$

8.11. An approach via stochastic integrals

8.11.2. Proposition. *A function $F \in L^2(P)$ has the Malliavin derivative DF if and only if $F \in D(\sqrt{-L})$. In this case*

$$\mathbb{E}(|DF|_H^2) = \mathbb{E}\big[(\sqrt{-L}F)^2\big] = \sum_{n=1}^{\infty} n\|I_n(f_n)\|_2^2 < \infty.$$

PROOF. Let $\{e_i\}$ be an orthonormal basis in the space H. Then the equality $\mathbb{E}\big[|DF|_H^2\big] = \sum_{i=1}^{\infty} \mathbb{E}\big[(D_{e_i}F)^2\big]$ can be written as

$$\sum_{i=1}^{\infty}\sum_{n=1}^{\infty} n^2 \mathbb{E}\bigg[\bigg(\int_T e_i(t) I_{n-1}\big(f_n(t_1,\ldots,t_{n-1},t)\big)\,\mu(dt)\bigg)^2\bigg]$$

$$= \sum_{n=1}^{\infty} n^2(n-1)!\|f_n\|_2^2 = \mathbb{E}\big[(\sqrt{-L}F)^2\big],$$

which proves our assertion. □

The following result shows that the Malliavin derivative can be interpreted as a directional derivative. To this end we suppose that $(\Omega, \mathcal{F}, P)$ is the canonical probability space associated with our Gaussian process $\{w(B),\ B \in \mathcal{B}\}$, i.e., one has $\Omega = \mathbb{R}^{\mathcal{B}}$ and $\mathcal{F}$ is the product-σ-algebra completed with respect to the distribution of the process w. In this case the transformation

$$w(B) \mapsto w(B) + \int_B h(t)\,\mu(dt)$$

is well defined for every $h \in H$; moreover, it transforms P into an equivalent measure with the Radon–Nikodym density

$$\exp\bigg(w(h) - \frac{1}{2}\int_T h(t)^2\,\mu(dt)\bigg).$$

8.11.3. Proposition. *Under the above assumptions, let $F \in L^2(P)$. Suppose that the ratio*

$$\frac{1}{\varepsilon}\bigg[F\bigg(w + \varepsilon \int_T h(t)\,\mu(dt)\bigg) - F(w)\bigg]$$

converges in $L^2(P)$ as $\varepsilon \to 0$. Then $D_h F$ exists and coincides with the limit of this ratio.

8.11.4. Remark. As indicated in Zakai [**1213**], in the case where Ω is the classical Wiener space with the Wiener measure P, LF can be interpreted as the derivative of F with respect to the scale parameter λ:

$$LF(\omega) = \partial F(\lambda\omega)/\partial\lambda\big|_{\lambda=1}.$$

Since homothetic transformations do not preserve measurability (see examples in [**191**, §2.5]), this equality is interpreted in [**1213**] in the following sense. For any $\lambda \in (-1,1)$ put

$$F_\lambda = \mathbb{E}F + \sum_{n=1}^{\infty} \lambda^n I_n(f_n).$$

Then LF exists if and only if a limit of the expression $\varepsilon^{-1}(F_{1-\varepsilon} - F)$ exists in $L^2(P)$ as $\varepsilon \to 0$; then LF coincides with the indicated limit.

8.11.5. Lemma. *The operators D and L are related by the following commutation relation:*
$$D_h LF - LD_h F = -D_h F.$$

This identity is verified directly. The following proposition contains basic rules of the calculus associated with the operators D and L.

8.11.6. Proposition. (i) *Let $\varphi \in C^2(\mathbb{R}^n)$ be a function with derivatives of an at most polynomial growth. Set*
$$F := \varphi \circ \eta, \quad \eta := \bigl(w(h_1), \ldots, w(h_n)\bigr),$$
where $h_1, \ldots, h_n \in H$. Then
$$D_h F = \sum_{i=1}^n \partial_{x_i} \varphi(\eta)(h, h_i)_H,$$
$$LF = \sum_{i,j=1}^n \partial_{x_i} \partial_{x_j} \varphi(\eta)(h_i, h_j)_H - \sum_{i=1}^n \partial_{x_i} \varphi(\eta) w(h_i).$$

(ii) *Let $\varphi \in C_b^2(\mathbb{R}^n)$ and $F = (F_1, \ldots, F_n)$, where $F_i \in H^{2,1}$. Then we have $\varphi(F) \in H^{2,1}$ and*
$$D\varphi(F) = \sum_{i=1}^n \partial_{x_i} \varphi(F) DF_i.$$

If, in addition, $F_i \in D(L)$, then $\varphi(F)$ belongs to $D(L)$ and
$$L[\varphi(F)] = \sum_{i,j=1}^n \partial_{x_i} \partial_{x_j} \varphi(F)(DF_i, DF_j)_H + \sum_{i=1}^n \partial_{x_i} \varphi(F) LF_i.$$

A proof is delegated to Exercise 8.14.9.

Let $u = \{u_t(w), (t,w) \in T \times \Omega\}$ be a $\mathcal{B} \otimes \mathcal{F}$-measurable process such that one has $\mathbb{E}[u_t^2] < \infty$ for all t. One can show (see Nualart [**848**, §1.1.2]) that, for every t, the random variable u_t admits a representation in the form of a series of multiple Wiener–Itô integrals
$$u_t = \mathbb{E} u_t + \sum_{n=1}^\infty I_n\bigl(f_n(t|t_1, \ldots, t_n)\bigr),$$
where kernels f_n can be chosen with the following properties:
(i) each f_n is a measurable function in all variables,
(ii) for every $t \in T$, the function $f(t|\cdot)$ is symmetric in the variables $t_1, \ldots, t_n$ and belongs to $L^2(T^n, \mu^n)$.

8.11. An approach via stochastic integrals

Denote by $\widetilde{f}_n$ the symmetrization of f_n as a function of $n+1$ variables. Then

$$\widetilde{f}_n(t_1,\ldots,t_n,t) = \frac{1}{n+1}\Big[f_n(t|t_1,\ldots,t_n) + \sum_{i=1}^{n} f_n(t_i|t_1,\ldots,t_{i-1},t,t_{i+1},\ldots,t_n)\Big].$$

8.11.7. Definition. *The Skorohod integral of $\{u_t\}$ is defined by*

$$\int_T u\delta w := \sum_{n=0}^{\infty} I_{n+1}(\widetilde{f}_n), \quad \widetilde{f}_0(t) := \mathbb{E}u_t,$$

provided that the multiple Itô integrals exist and the series converges in L^2. The Skorohod integral of $\{u_t\}$ will be also denoted by $-\delta u$.

We observe that the Skorohod integral of $\{u_t\}$ is defined if

$$\|u\| := \Big(\int_T (\mathbb{E}u_t)^2\,\mu(dt) + \sum_{n=1}^{\infty}(n+1)!\|\widetilde{f}_n\|_2^2\Big)^{1/2} < \infty.$$

In this case

$$\mathbb{E}\int_T u\delta w = 0 \quad \text{and} \quad \mathbb{E}\Big(\int_T u\delta w\Big)^2 = \|u\|^2.$$

8.11.8. Proposition. *Let $u \in L^2(\mu \otimes P)$ and*

$$\int_T \mathbb{E}\big[|Du_t|_H^2\big]\,\mu(dt) < \infty.$$

Then the Skorohod integral of $\{u_t\}$ exists and

$$\mathbb{E}(\delta u)^2 = \int_T \mathbb{E}(u_t^2)\,\mu(dt) + \int_T\int_T \mathbb{E}[Du_t(s)Du_s(t)]\,\mu(dt)\,\mu(ds)$$

$$\leqslant \int_T \mathbb{E}(u_t^2)\,\mu(dt) + \int_T \mathbb{E}\big[|Du_t|_H^2\big]\,\mu(dt)$$

$$= \int_T \mathbb{E}(u_t^2)\,\mu(dt) + \int_T \mathbb{E}\big[(\sqrt{-L}u_t)^2\big]\,\mu(dt).$$

8.11.9. Theorem. (i) *Let $F \in L^2(P)$. Then $LF = \delta DF$ in the sense that F belongs to the domain of definition of L if and only if $F \in H^{2,1}$ and the mapping DF is Skorohod integrable (then the indicated equality holds).*

(ii) *A process $\{u_t\}$ is Skorohod integrable if and only if the mapping*

$$F \mapsto \mathbb{E}(DF, u)_H$$

on $H^{2,1}$ is continuous in the L^2-norm. In this case one has

$$\mathbb{E}(DF, u)_H = -\mathbb{E}(F\delta u).$$

Proofs can be found in Nualart [**848**]; part (ii) of the previous theorem (obtained in Gaveau, Trauber [**471**]) shows that the Skorohod integral is adjoint to the gradient.

8.11.10. Example. (i) Let us consider the case where $T = [0,1]$, μ is Lebesgue measure and $w([0,t]) = w_t$, where w_t is the usual Wiener process; then
$$w(B) = \int_B I_B(t)\, dw_t.$$
Let $\{u_t\}$ be an adapted square-integrable process. Then the Skorohod integral of $\{u_t\}$ exists and coincides with the Itô integral.

(ii) The process $u_t(w) = w_1$ is not adapted, hence is not Itô integrable, but it is Skorohod integrable and
$$\int_0^1 w_1 \delta w = w_1^2 - 1.$$

PROOF. (i) We observe that $(DF, u)_H$ can be written as an L^2-limit
$$(DF, u)_H = \lim_{\varepsilon \to 0} \frac{1}{\varepsilon}\left[F\left(w + \varepsilon \int_0^1 u_t\, dw_t\right) - F(w)\right].$$
By the Girsanov theorem, for every $F \in H^{2,1}$ we obtain
$$\mathbb{E} F\left(w + \varepsilon \int_0^1 u_t\, dw_t\right) = \mathbb{E}\left[F \exp\left(\varepsilon \int_0^1 u_t\, dw_t - \frac{1}{2}\varepsilon^2 \int_0^1 u_t^2\, dt\right)\right].$$
Therefore,
$$\mathbb{E}(DF, u)_H = \mathbb{E}\left(F \int_0^1 u_t\, dw_t\right).$$
By Theorem 8.11.9(ii) δu exists and equals $-\int_0^1 u_t\, dw_t$.

(ii) The function $F\colon w \mapsto w_1 = \int_0^1 dw_t$ belongs to $D(L)$. According to Proposition 8.11.6, we have $DF = 1$, $D(F^2) = 2FDF = 2F$, $L(F^2) = 2 - 2w(1)^2$. Hence $\delta F = \delta D(F^2)/2 = L(F^2)/2 = 1 - w_1^2$. Note that we could also find δF directly from the definition: $f_1 = 1$, $f_n = 0$ if $n \neq 1$. □

Let $\{u_t\}$ be a measurable process such that
$$P\left(\int_T u_t^2\, \mu(dt) < \infty\right) = 1$$
and let $\{e_n\}$ be an orthonormal basis in $H = L^2(\mu)$. Yet another stochastic integral was introduced by Ogawa [**867**] (see also Nualart [**848**]).

8.11.11. Definition. *The Ogawa integral $\delta^0 u$ with respect to $\{e_n\}$ is defined as the sum of the series*
$$\sum_{n=1}^\infty (u, e_n)_{L^2(\mu)} w(e_n)$$
if this series converges in probability.

In general, the Ogawa integral (and its existence) may depend on our choice of an orthonormal basis.

8.11.12. Proposition. *Let $u \in L^2(\mu \otimes P)$ and*
$$\int_T \mathbb{E}\big[|Du_t|_H^2\big]\, \mu(dt) < \infty.$$
Suppose, in addition, that the random kernel $K(t,s) = Du_t(s)$ has finite trace P-almost surely, i.e., defines a trace class operator. Then the Ogawa integral $\delta^0 u$ exists for every orthonormal basis in H and
$$\delta^0 u = -\delta u + \operatorname{trace} K.$$

8.11.13. Example. Let us consider the situation of Corollary 1.5.5, where w_t is the Wiener process. Let $F \in H^{2,1}$. Then
$$F = \mathbb{E}F + \int_0^1 u_t\, dw_t$$
with some adapted process u_t. However, in this particular case the following Clark representation describes u_t as the so-called optional projection of the derivative of F:
$$F = \mathbb{E}F + \int_0^1 \mathbb{E}(D_t F | \mathcal{F}_t)\, dw_t, \qquad (8.11.3)$$
where $\mathcal{F}_t$ is the σ-algebra generated by the variables w_s, $s \leqslant t$. For justification of formula (8.11.3) it suffices to consider $F = I_n(f_n)$, where
$$f_n(t_1, \ldots, t_n) = I_{[a_1, b_1]}(t_1) \cdots I_{[a_n, b_n]}(t_n).$$

8.11.14. Example. Any square-integrable random variable ξ with zero mean on the classical Wiener space can be written as the stochastic integral of an adapted square-integrable process u_t. Then ξ coincides with β_v for the vector field v defined by the formula
$$v(\omega)(t) = -\int_0^t u_s(\omega)\, ds.$$
This follows from Example 8.11.10 and Theorem 8.11.9(ii).

8.12. Some identities of the Malliavin calculus

In the previous section, some differential operations such as D and L have been introduced in the framework of Gaussian orthogonal measures. We have already encountered similar objects in the theory of Sobolev classes over Gaussian measures. Now it is the right time to observe that these are indeed isomorphic realizations of the same objects. Namely, there is a natural isomorphism between the Sobolev classes $W^{2,1}(\gamma) = H^{2,1}(\gamma) = D^{2,1}(\gamma)$ and the classes $H^{2,1}$ defined in §8.11 (this isomorphism extends to the classes with higher derivatives). This isomorphism can be defined by associating to each element F with decomposition (1.5.2) the element f represented in form (8.5.2). It is readily verified that the operators D, L, T_t (and some other ones) correspond to each other in both pictures. However, each of the two representations possess some specific features which have no natural analog in the other one. For example, in the Gaussian analysis on linear spaces there is no intrinsic time scale, but there are topological properties such as continuity of certain Sobolev functions, which is missing in the approach of §8.11. Therefore, our choice of the approach should be motivated by the type of problems to which it is oriented. Following Nualart, Zakai [860], we give a short presentation of certain useful relationships between the operators $D := D_H$, δ, L, and T_t on Sobolev classes over Gaussian measures or on the classes of functions considered in §8.11 (as already noted, such relationships are the same in both cases). Justifications are straightforward.

Let α be an arbitrary real number and let k be an arbitrary natural number.

8.12.1. Theorem. *The following identities hold*:
$$\delta D = L, \quad \delta^k D^k = P^k(L), \quad P^k(L) = -L(-L-1)\cdots(-L-(k-1)I),$$
$$(I-L)^\alpha D = D(-L)^\alpha, \quad (kI-L)^\alpha D^k = D^k(-L)^\alpha,$$
$$\delta(I-L)^\alpha = (-L)^\alpha \delta, \quad \delta^k(kI-L)^\alpha = (-L)^\alpha \delta^k,$$
$$\delta^k(kI-L)^\alpha D^k = P^k(L)(-L)^\alpha = (-L)^\alpha P^k(L),$$
$$DT_t = e^{-t}T_t D, \quad D^k T_t = e^{-kt}T_t D^k.$$

8.12.2. Remark. In addition to the relationships above, the following commutation properties hold:
(i) $D\delta$ commutes with $\delta D = L$ and, more generally, with all L^α;
(ii) $D^k \delta^k$ and $(D\delta)^k$ commute with L^α;
(iii) if $\delta^k u = 0$, then $\delta^k L^\alpha u = 0$. Conversely, if $\delta^k L^\alpha u = 0$ and $\mathbb{E}u = 0$, then one has $\delta^k u = 0$.

8.12.3. Remark. Let $\Gamma_{2,1}(\gamma)$ be the subspace in $L^2(\gamma, H)$ generated by the gradients $D_H \varphi$, where $\varphi \in W^{2,1}(\gamma)$. For every element $u \in L^2(\gamma, H)$ let u^1 denote its orthogonal projection on $\Gamma_{2,1}(\gamma)$. Let $u^0 = u - u^1$. It is clear that u^0 may not belong to $W^{2,1}(\gamma, H)$, however, its divergence δu^0 vanishes by definition:
$$\int_X \bigl(D_H\varphi(x), u^0(x)\bigr)_H \gamma(dx) = 0, \quad \forall \varphi \in W^{2,1}(\gamma).$$
If $u \in W^{2,1}(\gamma, H)$, then
$$u^1 = D_H L^{-1} \delta u \quad \text{and} \quad u^1 \in W^{2,1}(\gamma, H).$$
Indeed, since $\delta u \in L^2(\gamma)$, there exists a function $f \in W^{2,2}(\gamma)$ such that $Lf = \delta u$. For any $\varphi \in W^{2,1}(\gamma)$ we have
$$\int_X \bigl(D_H\varphi(x), u(x) - u^1(x)\bigr)_H \gamma(dx)$$
$$= -\int_X \varphi(x)\delta u(x)\,\gamma(dx) + \int_X \varphi(x)Lf(x)\,\gamma(dx) = 0.$$
Similarly, for every mapping $u \in W^{2,1}(\gamma, \mathcal{H}_k)$ one can obtain an orthogonal decomposition $u = u^1 + u^0$, in which $\delta^k u^0 = 0$ and u^1 has the form $u^1 = D_H^k v$.

8.13. Sobolev capacities

Let μ be a nonnegative Radon measure on a topological space X. Then μ has the topological support, i.e., the minimal closed set of full measure.

Let $\mathcal{F}$ be some linear space of μ-measurable functions equipped with some norm $f \mapsto \|f\|_0$. There is a standard procedure that associates to $\mathcal{F}$ a set function $C_{\mathcal{F}}$ on X, called the capacity generated by $\mathcal{F}$. We shall assume that functions from $\mathcal{F}$ which coincide μ-a.e. have equal norms.

8.13.1. Definition. *For every open set $U \subset X$ put*
$$C_{\mathcal{F}}(U) := \inf\{\|f\|_0 \colon f \in \mathcal{F}, f \geqslant 0 \quad and \quad f \geqslant 1 \quad \mu\text{-a.e. on } U\}.$$
For every subset A in X put
$$C_{\mathcal{F}}(A) := \inf\{C_{\mathcal{F}}(U) \colon A \subset U, U \text{ open}\}.$$

The inequality $\|f+g\|_0 \leqslant \|f\|_0 + \|g\|_0$ yields that the capacity $C_{\mathcal{F}}$ is *subadditive*, i.e., for every two sets A and B we have
$$C_{\mathcal{F}}(A \cup B) \leqslant C_{\mathcal{F}}(A) + C_{\mathcal{F}}(B).$$

Proofs of the assertions given below can be found in [**191**, Ch. 5].

8.13.2. Example. (i) Typical examples of spaces $\mathcal{F}$ for which one constructs capacities are $W^{p,r}(\mu)$, $D^{p,r}(\mu)$, $H^{p,r}(\mu)$, $G^{p,r}(\mu)$, and $F^{p,1}(\mu)$, considered above.

(ii) An important class of examples (which includes $H^{p,r}$ and $F^{2,1}$) can be obtained in the following way. Let $T \in \mathcal{L}(L^p(\mu))$ be an injective operator which takes nonnegative functions to nonnegative functions. Set
$$\mathcal{F} = T(L^p(\mu)), \quad \|Tf\|_0 = \|f\|_p.$$

If $\mathcal{F}$ is continuously embedded into $L^p(\mu)$ (which is the case if $\mathcal{F}$ is complete, $\mathcal{F} \subset L^p(\mu)$ and the embedding $\mathcal{F} \subset L^0(\mu)$ is continuous), then
$$C_{\mathcal{F}}(B) \geqslant k^{-1}\mu(B)^{1/p}$$
for all μ-measurable sets B, where k is the norm of the continuous embedding $\mathcal{F} \to L^p(\mu)$. Indeed, let U be open and let $f \in \mathcal{F}$ be such that $f|_U \geqslant 1$ a.e. Then one has $\|f\|_p \geqslant \mu(U)^{1/p}$. By the continuity of the embedding one has $\|f\|_p \leqslant k\|f\|_0$, whence $\|f\|_0 \geqslant k^{-1}\mu(U)^{1/p}$. This yields the estimate $C_{\mathcal{F}}(U) \geqslant k^{-1}\mu(U)^{1/p}$. This estimate is true for every measurable set B since the values of the functions μ and $C_{\mathcal{F}}$ on B coincide with the infimums over open sets.

If $\mathcal{F} = L^1(\mu)$, then $C_{\mathcal{F}}(B) = \mu(B)$, but typically the function $C_{\mathcal{F}}$ is not countably additive even on the Borel σ-algebra. In applications, it is often important to know whether the capacity $C_{\mathcal{F}}$ has the following properties:

(i) the *countable subadditivity*
$$C_{\mathcal{F}}\Big(\bigcup_{n=1}^{\infty} A_n\Big) \leqslant \sum_{n=1}^{\infty} C_{\mathcal{F}}(A_n);$$

(ii) the *continuity from below*, i.e.,

$$\lim_{n\to\infty} C_{\mathcal{F}}(A_n) = C_{\mathcal{F}}\Big(\bigcup_{n=1}^{\infty} A_n\Big)$$

for every increasing sequence of sets A_n;

(iii) *tightness*, i.e., for every $\varepsilon > 0$ there is a compact set K with $C_{\mathcal{F}}(X\backslash K) \leqslant \varepsilon$.

It is worth noting that every capacity $C_{\mathcal{F}}$ automatically has the property

$$\lim_{n\to\infty} C_{\mathcal{F}}(K_n) = C_{\mathcal{F}}\Big(\bigcap_{n=1}^{\infty} K_n\Big)$$

for every decreasing sequence of compact sets K_n.

Indeed, for any open set U containing the compact set $\bigcap_n K_n$ one can find n such that $K_n \subset U$.

Note also that property (ii) implies property (i) for every capacity $C_{\mathcal{F}}$.

The capacity $C_{\mathcal{F}}$ is called a *Choquet capacity* if it has property (ii) (note that somewhat more general Choquet capacities are considered in Meyer [**804**, Ch. III, §2]).

It is known (see [**804**]) that if $C_{\mathcal{F}}$ is a Choquet capacity and X is a Souslin space, then

$$C_{\mathcal{F}}(B) = \sup\{C_{\mathcal{F}}(K)\colon K \subset B,\ K \text{ is compact}\}$$

for all Souslin sets B, in particular, for all Borel sets B. However, unlike the case of measures, this does not imply tightness of a capacity.

Recall that a normed space E is called uniformly convex if for every $\varepsilon > 0$ there exists $\delta > 0$ such that the conditions $\|u\| \leqslant 1$, $\|v\| \leqslant 1$ and $\|u - v\| > \varepsilon$ imply that $\|u + v\| \leqslant 2 - \delta$. All spaces L^p with finite $p > 1$ are uniformly convex (see [**193**, Theorem 4.7.15]).

This property enables us to obtain the following result.

8.13.3. Theorem. *Suppose that the space $\mathcal{F}$ is uniformly convex and convergence in $\mathcal{F}$ implies convergence in measure μ. Then, for every open set U, there exists a unique element $\pi_U \in \mathcal{F}$ for which $\pi_U \geqslant 1$ μ-a.e. on U and*

$$\|\pi_U\|_0 = C_{\mathcal{F}}(U).$$

If, in addition, $\mathcal{F} = T(L^p(\mu))$, where the operator $T \in \mathcal{L}(L^p(\mu))$ preserves the positivity and $\|Tf\|_0 = \|f\|_p$, then π_U admits a representation $\pi_U = Tf$ with some nonnegative function $f \in L^p(\mu)$.

PROOF. There is a sequence of nonnegative functions $f_n \in \mathcal{F}$ such that $f \geqslant 1$ a.e. on U and $\|f_n\| \to C_{\mathcal{F}}(U)$. Since the space $\mathcal{F}$ is uniformly convex, we can find a subsequence $\{g_n\}$ in $\{f_n\}$ whose arithmetic means $w_n := (g_1 + \cdots + g_n)/n$ converge in $\mathcal{F}$ to some function w (see, e.g., Diestel [**350**, Ch. 3, §7]). By our assumption, $\{w_n\}$ converges in measure. Clearly, $w \geqslant 0$, $\|w\|_0 \leqslant \limsup_n \|f_n\|_0 = C_{\mathcal{F}}(U)$, and $w \geqslant 1$ a.e. on U. Hence we have

8.13. Sobolev capacities

$\|w\|_0 \geqslant C_{\mathcal{F}}(U)$, so $\|w\|_0 = C_{\mathcal{F}}(U)$. Set $\pi_U := w$. Suppose that there is another element $v \in \mathcal{F}$ with the stated properties. Let $h = (v+w)/2$. The uniform convexity yields $\|h\|_0 < C_{\mathcal{F}}(U)$, which is impossible since $h \geqslant 0$ and $h \geqslant 1$ a.e. on U. The last claim follows from the fact that once we have $w = Tf$, then $\|T|f|\|_0 = \|Tf\|_0$ and $T|f| \geqslant Tf$, whence $Tf = T|f|$. □

We have not clarified so far the role of the topology of the space X, although open sets are explicitly involved in the definition. Surprisingly enough, in many interesting cases capacities are invariant with respect to weakening the original topology.

8.13.4. Lemma. *Let C be a nonnegative increasing subadditive set function on the family of all open subsets in a Hausdorff topological space X. For every set A put*

$$C(A) := \inf\{C(U)\colon A \subset U, U \text{ is open}\}.$$

Suppose that there exists a sequence of compact sets K_n with

$$\lim_{n\to\infty} C(X\backslash K_n) = 0.$$

Let τ be a weaker Hausdorff topology on X and let C_τ be the extension of C from the class of τ-open sets to all sets by the above formula. Then $C(A) = C_\tau(A)$ for every set A.

Usually the proof of tightness is not that trivial. We shall consider this problem below in relation to Sobolev classes.

Another problem also related to the topology of X is the property of quasicontinuity.

A function f on X is called *quasicontinuous* (with respect to $C_{\mathcal{F}}$) if, for every $\varepsilon > 0$, there exists an open set U with $C_{\mathcal{F}}(U) < \varepsilon$ such that f is continuous on the set $X\backslash U$.

We shall say that a certain property is fulfilled *quasieverywhere* if it is fulfilled outside of a set of capacity zero.

Any *open* set U of μ-measure zero has capacity zero (since the function $f \equiv 0$ satisfies the condition $f \geqslant 1$ a.e. on U). However, for closed sets this is not true (see [**191**, Exercise 5.12.42]).

8.13.5. Lemma. *If a function f is quasicontinuous and $f \geqslant 0$ μ-a.e. on an open set U, then $f \geqslant 0$ quasieverywhere on U.*

8.13.6. Theorem. *Suppose that $\mathcal{F}$ is a uniformly convex Banach space such that $C(X) \cap \mathcal{F}$ is dense in $\mathcal{F}$ and convergence in $\mathcal{F}$ implies convergence in measure μ. Then the following assertions are valid.*

(i) Every function $f \in \mathcal{F}$ has a quasicontinuous modification f^* such that

$$C_{\mathcal{F}}(x\colon f^*(x) > R) \leqslant R^{-1}\|f\|_0. \tag{8.13.1}$$

(ii) For every sequence $\{f_n\}$ convergent in $\mathcal{F}$, there exists a subsequence $\{f_n^*\}$ which converges quasieverywhere to f^*.

(iii) If f is quasicontinuous and $f \geq 0$ μ-a.e. on an open set U, then $f \geq 0$ quasieverywhere on U.

(iv) If $\mathcal{F}$ has the form indicated in Example 8.13.2(ii), then, for every set B, there exists a unique element
$$u_B \in F_B = \{u \in \mathcal{F} \colon u^* \geq 1 \text{ quasieverywhere on } B\}$$
of minimal norm. In addition, $u_B \geq 0$ and $C_{\mathcal{F}}(B) = \|u_B\|_0$. For an open set B we have $u_B = \pi_B$ (see Theorem 8.13.3). The function u_B is called the equilibrium potential of B.

PROOF. Let $\{f_n\} \subset C(X) \cap \mathcal{F}$ be a sequence convergent to f in $\mathcal{F}$. We observe that, for every $r > 0$ and every function $\varphi \in C(X) \cap \mathcal{F}$, we have
$$C_{\mathcal{F}}(x \colon \varphi(x) > r) \leq r^{-1} \|\varphi\|_0$$
since $\varphi/r \geq 1$ on the open set $\{x \colon \varphi(x) > r\}$. Hence, for every $r > 0$, we have
$$C_{\mathcal{F}}(|f_n - f_k| > r) \leq 2r^{-1} \|f_n - f_k\|_0 \to 0 \quad \text{as } n, k \to \infty.$$
Therefore, for every n, there exists k_n such that
$$C_{\mathcal{F}}(|f_j - f_i| > 2^{-n}) \leq 2^{-n}, \quad \forall i, j \geq k_n.$$
We verify that the sequence $\{g_n\} = \{f_{k_n}\}$ converges to f quasieverywhere. It suffices to show that this sequence is fundamental quasieverywhere. Let $\varepsilon > 0$. We choose n_1 such that $1/2^{n_1} < \varepsilon$. Set
$$X_\varepsilon = \bigcap_{j > n_1} \{x \colon |g_j - g_{j+1}| \leq 2^{-j}\}.$$
By the subadditivity of $C_{\mathcal{F}}$ and our choice of g_n we have
$$C_{\mathcal{F}}(X \setminus X_\varepsilon) \leq \sum_{j > n_1} 2^{-j} \leq 2^{-n_1} < \varepsilon.$$

Obviously, on X_ε we have $|g_{j+1} - g_j| \leq 2^{-j}$ for all $j > n_1$. Thus, the series $\sum_j (g_{j+1} - g_j)$, $g_0 = 0$, converges on X_ε. It is clear that $\sum_{j=0}^{n} (g_{j+1} - g_j) = g_n$. This shows convergence of $\{g_n\}$. Since ε is arbitrary, we obtain convergence quasieverywhere. The same reasoning shows that for every $\varepsilon > 0$ there exists a closed set X_ε with $C_{\mathcal{F}}(X \setminus X_\varepsilon) \leq \varepsilon$ on which convergence is uniform. Hence the limit f^* of the sequence $\{g_n\}$ is quasicontinuous. Moreover, it follows from our reasoning that, for every fixed $r > 0$ and $\varepsilon \in (0, r)$, there exist a closed set X_ε with $C_{\mathcal{F}}(X \setminus X_\varepsilon) < \varepsilon$ on which f^* is continuous and a function $g \in C(X) \cap \mathcal{F}$ such that $\|f^* - g\|_0 \leq \varepsilon$ and $|f^* - g| \leq \varepsilon$ on X_ε. Then
$$C_{\mathcal{F}}(x \colon f^*(x) > r) \leq C_{\mathcal{F}}(X_\varepsilon \cap \{x \colon f^*(x) > r\}) + \varepsilon$$
$$\leq C_{\mathcal{F}}(X_\varepsilon \cap \{x \colon g(x) > r - \varepsilon\}) + \varepsilon \leq C_{\mathcal{F}}(x \colon g(x) > r - \varepsilon) + \varepsilon$$
$$\leq (r - \varepsilon)^{-1} \|g\|_0 + \varepsilon \leq (r - \varepsilon)^{-1} (\|f\|_0 + \varepsilon) + \varepsilon.$$
Since ε is arbitrary, we arrive at (8.13.1).

8.13. Sobolev capacities

Suppose now that $\{f_n\}$ is an arbitrary sequence convergent in $\mathcal{F}$. For every n there exists k_n with $\|f_{k_n}^* - f^*\|_0 \leqslant 4^{-n}$. By using the estimate

$$C_{\mathcal{F}}\big(x\colon |f_{k_n}^*(x) - f^*(x)| > 2^{-n}\big) \leqslant 2^{-n}$$

we obtain as above quasieverywhere convergence of $\{f_{k_n}^*\}$ to f^*. Thus, assertions (i) and (ii) are proven.

For the proof of assertion (iii) we denote by S the topological support of μ. As already noted above, every open set of measure zero has capacity zero. Hence one has the equality $C_{\mathcal{F}}(X\setminus S) = 0$. Let $\varepsilon > 0$ and let Z be a closed set on which f is continuous such that $C_{\mathcal{F}}(X\setminus Z) < \varepsilon$. We observe that if $z \in Z \cap S$, then $f(z) \geqslant 0$. Indeed, otherwise by the continuity of f on Z we obtain $f(x) < 0$ on $V \cap Z$, where $V \subset U$ is a neighborhood of the point z. Hence $\mu(V) = 0$, whence $V \subset X\setminus S$, which is a contradiction. Therefore, $C_{\mathcal{F}}(U \cap \{f < 0\}) \leqslant \varepsilon$, which gives the equality $C_{\mathcal{F}}(U \cap \{f < 0\}) = 0$.

It remains to prove assertion (iv). Let $\{U_n\}$ be a sequence of open sets containing B such that $C_{\mathcal{F}}(U_n) \geqslant C_{\mathcal{F}}(B) + 2^{-n}$. Set $u_n = \pi_{U_n}$, where π_{U_n} is a nonnegative quasicontinuous function associated to U_n by Theorem 8.13.3. Since $\mathcal{F}$ is uniformly convex, there is a subsequence $\{u_{i_k}\}$ in $\{u_i\}$ for which the sequence of the arithmetic means $S_n = (u_{i_1} + \cdots + u_{i_n})/n$ converges in $\mathcal{F}$ to a nonnegative function u_B. Passing to a subsequence once again we may assume that convergence takes place quasieverywhere. By assertion (iii) quasieverywhere on U_n we have $u_n \geqslant 1$. Hence $u_B \geqslant 1$ quasieverywhere on B. It is clear that

$$\|u_B\|_0 \leqslant \limsup_n \|u_n\|_0 = C_{\mathcal{F}}(B).$$

Uniqueness of a quasicontinuous function u for which $u \geqslant 1$ on B quasieverywhere and $\|u\|_0 = C_{\mathcal{F}}(B)$ follows by the uniform convexity of $\mathcal{F}$ exactly as in Theorem 8.13.3.

In order to prove the minimality of the norm of u_B it suffices to show that for every quasicontinuous function $h \in \mathcal{F}$ such that $h \geqslant 1$ quasieverywhere on B one has the estimate $\|h\|_0 \geqslant C_{\mathcal{F}}(B)$. Suppose first that $h \geqslant 0$ a.e. Let $\varepsilon > 0$. One can find a closed set Z on which h is continuous such that $C_{\mathcal{F}}(X\setminus Z) < \varepsilon$ and $h \geqslant 1$ on $B \cap Z$. The set $G = (Z \cap \{h > 1-\varepsilon\}) \cup (X\setminus Z)$ is open. By Theorem 8.13.3 there exists a nonnegative function $h_0 = \pi_{X\setminus Z} \in \mathcal{F}$ such that $h_0 \geqslant 1$ a.e. on the set $X\setminus Z$ and $\|h_0\|_0 = C_{\mathcal{F}}(X\setminus Z) < \varepsilon$. Since $h \geqslant 0$ and $h_0 \geqslant 0$ a.e., we obtain $h + h_0 > 1 - \varepsilon$ a.e. on G. Therefore,

$$C_{\mathcal{F}}(B) \leqslant C_{\mathcal{F}}(G) \leqslant (1-\varepsilon)^{-1}\|h + h_0\|_0 \leqslant (1-\varepsilon)^{-1}\big(\|h\|_0 + \varepsilon\big).$$

Since ε is arbitrary, we obtain $C_{\mathcal{F}}(B) \leqslant \|h\|_0$. If h is not assumed to be nonnegative a.e., we can take the function $h_1 = (T|g|)^*$, where $g \in L^p(\mu)$ is such that $Tg = h$. It is clear that $h_1 \geqslant h$ a.e. since $|g| \geqslant g$. Hence $h_1 \geqslant h$ quasieverywhere by assertion (iii). Thus, $h_1|_B \geqslant 1$ quasieverywhere and $\|h_1\|_0 = \|h\|_0 = C_{\mathcal{F}}(B)$. The uniqueness result yields that $h_1 = h$. Hence $h \geqslant 0$ quasieverywhere. $\square$

8.13.7. Corollary. *Suppose that the hypotheses of Theorem 8.13.6 are fulfilled. Then, for every increasing sequence of sets B_n we have*

$$\lim_{n\to\infty} C_{\mathcal{F}}(B_n) = C_{\mathcal{F}}\Big(\bigcup_{n=1}^{\infty} B_n\Big).$$

PROOF. Let u_n be the equilibrium potential of B_n. The sequence $\{S_n\}$ of the arithmetic means of a subsequence in $\{u_n\}$ converges in $\mathcal{F}$ to a function u. It is clear that the union D of the sets $B_n \cap \{u_n < 1\}$ has capacity zero since each of these sets has capacity zero. On $B\backslash D$ we have $u \geqslant 1$. Therefore, $u \geqslant 1$ quasieverywhere on B, whence $C_{\mathcal{F}}(B) \leqslant \|u\|_0$. On the other hand,

$$\|u\|_0 \leqslant \limsup_n \|u_n\|_0 = \limsup_n C_{\mathcal{F}}(B_n) \leqslant C_{\mathcal{F}}(B),$$

which completes the proof. □

Since the property to have capacity zero is stronger than the property to have measure zero, the problem arises of describing measures vanishing on all sets of capacity zero. This problem was solved in Kazumi, Shigekawa [**599**] and Shimomura [**1029**] (see also a partial result in Sugita [**1092**]). Roughly speaking, measures ν vanishing on all sets of $C_{\mathcal{F}}$-capacity zero can be described as measures for which the functionals

$$\varphi \mapsto \int \varphi \, d\nu$$

are continuous on $\mathcal{F}$. Let us give a more precise formulation. Suppose that $\mathcal{F}$ has the form $\mathcal{F} = T(L^p(\mu))$, where the operator $T \in \mathcal{L}(L^p(\mu))$ preserves the positivity and $\|Tf\|_0 = \|f\|_p$.

8.13.8. Theorem. *Let ν be a Radon probability measure on X such that $\nu(A) = 0$ for every $A \in \mathcal{B}(X)$ with $C_{\mathcal{F}}(A) = 0$. Then, there exists a strictly positive bounded Borel function ϱ on X such that the functional*

$$f \mapsto \int_X f(x)\varrho(x)\,\nu(dx) \quad \text{is continuous on } \mathcal{F}.$$

PROOF. Let

$$h(t) = \sup\{\nu(A) \colon C_{\mathcal{F}}(A) \leqslant t\}.$$

Note that $\lim_{t\to 0} h(t) = 0$. Indeed, otherwise for some $c > 0$ there exists a sequence of Borel sets A_n such that $C_{\mathcal{F}}(A_n) \leqslant 2^{-n}$ and $\nu(A_n) \geqslant c$. This leads to a contradiction since for the set $B = \limsup A_n = \bigcap_{n\geqslant 1}\bigcup_{k\geqslant n} A_k$ we have

$$\nu(B) \geqslant \limsup \nu(A_k) \geqslant c \quad \text{and} \quad C_{\mathcal{F}}(B) \leqslant \sum_{k\geqslant n} C_{\mathcal{F}}(A_k) \leqslant 2^{-n+1}$$

for every n, whence we obtain $C_{\mathcal{F}}(B) = 0$. Let us apply the following result due to Maurey (see Vakhania, Tarieladze, Chobanyan [**1167**, Ch. VI, Lemma 5.5]). Let C be a convex set of ν-measurable nonnegative functions

which is bounded in $L^0(\nu)$. Then there exists a strictly positive ν-measurable function ϱ such that
$$\sup_{f \in C} \int_X f(x)\varrho(x)\,\nu(dx) \leqslant 1.$$
Let us take
$$C := \{f \in \mathcal{F} \colon \|f\|_0 \leqslant 1, f \geqslant 0\}.$$
Since every function $f \in \mathcal{F}$ has a quasicontinuous modification and μ-equivalent functions coincide as elements of $\mathcal{F}$, we obtain that all functions from $\mathcal{F}$ are ν-measurable. Indeed, for every such function f and every $\varepsilon > 0$ there exists $\delta < \varepsilon$ such that $\nu(A) \leqslant \varepsilon$ provided that $C_\mathcal{F}(A) < \delta$. Since there exists a closed set Z with $C_\mathcal{F}(X \backslash Z) < \delta$ on which f is continuous, we obtain $\nu(X \backslash Z) \leqslant \varepsilon$, whence the measurability of f with respect to ν follows.

Let us observe that $\nu(f > r) \leqslant h(r)$ if $\|f\|_0 \leqslant r^2$ since we have the inequality $C_\mathcal{F}(f > r) \leqslant r^{-1}\|f\|_0 \leqslant r$. Therefore, we arrive at the relationship
$$\int_X \frac{f(x)}{1+f(x)}\,\nu(dx) \leqslant h(r) + r \quad \text{for all } f \in r^2 C,$$
which along with the continuity of h at zero means that C is bounded in $L^0(\nu)$. Finally, we observe that for every function f with $\|f\|_0 \leqslant 1$ the function $g = T(|f|)$ is nonnegative, $\|g\|_0 = \|f\|_0 \leqslant 1$ and $|f| \leqslant g$ quasieverywhere. Hence by Maurey's lemma we obtain a strictly positive ν-measurable function ϱ such that
$$\int_X f(x)\varrho(x)\,\nu(dx) \leqslant \int_X g(x)\varrho(x)\,\nu(dx) \leqslant 1$$
for all $f \in C$. Therefore, the functional
$$f \mapsto \int_X f\varrho\,d\nu$$
is continuous on $\mathcal{F}$. $\square$

Measures ν satisfying the condition of Theorem 8.13.8 are called *measures of finite energy*. One might ask of the converse to this theorem. However, this question should be more precise since for integration of functions from $\mathcal{F}$ (which are μ-measurable) against the measure ν it is necessary to choose in some way ν-measurable modifications of the elements of $\mathcal{F}$. Even if all such functions are Borel, they may have different ν-versions. It would not be reasonable to assume the absolute continuity of ν with respect to μ since then ν trivially vanishes on all sets of capacity zero. The following concept looks more reasonable.

Suppose that $C_b(X) \cap \mathcal{F}$ is dense in $\mathcal{F}$. We shall say that a Radon measure ν on X generates a distribution $\Phi \in \mathcal{F}^*$ if the functional
$$\varphi \mapsto \int_X \varphi\,d\nu$$
on $C_b(X) \cap \mathcal{F}$ is continuous in the $\|\cdot\|_0$-norm.

The following result from Kazumi, Shigekawa [**599**], Shimomura [**1029**] describes measures vanishing on all sets of capacity zero as positive generalized functions. We shall say that an element $\Phi \in \mathcal{F}^*$ is nonnegative ($\Phi \geqslant 0$) if $\langle \Phi, \varphi \rangle \geqslant 0$ for every function $\varphi \in \mathcal{F}$ such that $\varphi \geqslant 0$ quasieverywhere.

8.13.9. Theorem. *Suppose that $\mathcal{F} = H^{p,r}(\mu)$, the class $C_b(X) \cap \mathcal{F}$ is dense in $\mathcal{F}$, $1 \in \mathcal{F}$, and the capacity $C_\mathcal{F}$ is tight. Then, for every nonnegative $\Phi \in \mathcal{F}^*$, exists a nonnegative Radon measure ν_Φ such that*

$$\langle \Phi, \varphi \rangle = \int_X \varphi(x)\, \nu_\Phi(dx) \quad \text{for all } \varphi \in C_b(X) \cap \mathcal{F}.$$

8.13.10. Theorem. *The capacity $C_\mathcal{F}$ is tight in either of the following cases.* (i) *X is a complete locally convex space and there is a Radon probability measure ν on X with $H \subset H(\nu)$ and $\mathcal{F} = D^{p,r}(\mu)$. For example, this is true if $H \subset H(\mu)$.*

(ii) *X is a Fréchet space, there exists a Radon probability measure ν on X with $H \subset H(\nu)$ and $\mathcal{F} = W^{p,r}(\mu)$, and the class $W^{p,r}(\mu)$ is well defined. For example, this is true if $H \subset H(\mu)$.*

(iii) *X is a Fréchet space and $\mathcal{F} = W^{p,1}(\mu)$, where the class $W^{p,1}(\mu)$ is well defined.*

Assertion (iii) is proved in Röckner, Schmuland [**962**] for Banach spaces and extended to Fréchet spaces in Pugachev [**931**]. Proofs of assertions (i) and (ii) are given in [**931**].

8.13.11. Remark. One can define capacities also on functions. Namely, for every lower semicontinuous function $f\colon X \to [0, \infty]$ we put

$$C_\mathcal{F}(f) = \inf\{\|\varphi\|_0 \colon \varphi \in \mathcal{F}, \varphi \geqslant f\ \mu\text{-a.e.}\}.$$

For an arbitrary function g we put

$$C_\mathcal{F}(g) = \inf\{C_\mathcal{F}(f) \colon f \text{ is lower semicontinuous and } |g(x)| \leqslant f(x)\}.$$

For the indicator function I_B of a set B this gives $C_\mathcal{F}(I_B) = C_\mathcal{F}(B)$. It is clear that $C_\mathcal{F}$ is a seminorm; one can take the corresponding factor space. The capacity on functions has properties similar to its properties on sets; see Feyel, La Pradelle [**428**], Kazumi, Shigekawa [**599**].

In the remainder of this section we briefly discuss basic properties of Gaussian capacities on locally convex spaces, such as invariance with respect to embeddings and existence of Souslin supports. For proofs, see Bogachev [**191**, Ch. 5]. Gaussian capacities are defined by means of the classes $W^{p,r}(\gamma)$, $H^{p,r}(\gamma)$, $D^{p,r}(\gamma)$ with $p > 1$ and $r \in \mathbb{N}$ (which coincide for fixed r and p) and are denoted by $C_{p,r}$. All conditions mentioned in the formulations of the results above are fulfilled for these capacities. It is worth noting that in many works on Gaussian capacities the definition of capacity slightly differs from our definition in that in place of the norm $\|f\|_{p,r}$ one takes $\|f\|_{p,r}^p$. Certainly, one obtains another set function, but its basic properties described in the above theorems (for example, its countable subadditivity, right continuity, tightness, etc.) will be the same which follows by the convexity of the function $t \mapsto |t|^p$.

8.13.12. Definition. *A set D is called slim if $C_{p,r}(D) = 0$ for all $p > 1$ and all $r \in \mathbb{N}$.*

The following result obtained in Bogachev, Röckner [**216**] gives a complete solution to a problem raised by K. Itô and P. Malliavin (see Fukushima [**462**]). References to previous less general results can be found in [**191**], [**216**].

8.13.13. Theorem. *For every $\varepsilon > 0$, there exists a metrizable compact set K_ε such that $C_{p,r}(X \backslash K_\varepsilon) < \varepsilon$. In addition,*

$$C_{p,r}(B) = \sup\{C_{p,r}(K)\colon K \subset B \text{ is a metrizable compact set}\}$$

for every Borel set B (this is true also for every Souslin set B).

8.13.14. Corollary. *Let E be a γ-measurable linear subspace in X such that $\gamma(E) > 0$. Then $C_{p,r}(X \backslash E) = 0$.*

8.13.15. Corollary. *Suppose that Y is a locally convex space and that $T \in \mathcal{L}(X, Y)$ is an injective operator. Denote by $C_{p,r}^T$ the capacity associated with the measure $\nu = \gamma \circ T^{-1}$. Then we have $C_{p,r}^T(B) = C_{p,r}(T^{-1}(B))$ for every set $B \in \mathcal{B}(Y)$.*

8.13.16. Corollary. *Let Y be a locally convex space and let $T\colon X \to Y$ be a γ-measurable proper linear mapping such that the measure $\nu = \gamma \circ T^{-1}$ is Radon on Y. Then T is quasieverywhere continuous and the conclusion of the previous corollary is true.*

It would be interesting to extend the previous results to more general measures.

If a γ-measurable set A is such that $A + H(\gamma) = A$, then it is readily verified that $C_{p,r}(A)$ is either 1 or 0 (Exercise 8.14.12). However, it is unknown whether the equality $C_{p,r}(X \backslash A) = 0$ holds if $C_{p,r}(A) = 1$. In general, the equality $C_{p,r}(B) = 1$ does not imply that $C_{p,r}(X \backslash B) = 0$.

Many classical results on "almost everywhere properties" are true in a more precise form as "quasieverywhere properties". This is the subject of the "quasisure analysis" (see Bouleau, Hirsch [**247**], Malliavin [**764**], Malliavin, Nualart [**771**], [**772**], Ren [**941**]–[**944**]). There are two main directions in which such refinements have been obtained:
(a) existence of mappings with certain properties that hold quasieverywhere (for example, modifications of functions from Sobolev classes);
(b) assertions of "quasieverywhere" type related to limit theorems.

We give only one typical result belonging to direction (a) and partly to direction (b), which can be easily deduced from the above results.

8.13.17. Theorem. *Suppose that γ is a centered Radon Gaussian measure on a locally convex space X, $\{e_n\}$ is an orthonormal basis in $H = H(\gamma)$, and $f \in W^\infty(\gamma)$. Let f_n be the conditional expectation of f with respect to the σ-algebra generated by $\widehat{e}_1, \ldots, \widehat{e}_n$. Then, there exist functions $\varphi_n \in C^\infty(\mathbb{R}^n)$ such that $f_n = \varphi_n(\widehat{e}_1, \ldots, \widehat{e}_n)$ and there exists a subsequence $\{n_i\}$ such that for all $p \geqslant 1$, $r \geqslant 1$, and every positive ε there is an open set U with the following properties: $C_{p,r}(U) \leqslant \varepsilon$ and for every k the sequences f_{n_i}, $D_H^k f_{n_i}$ and $L^k f_{n_i}$ converge uniformly on $X \backslash U$ to continuous limits.*

8.14. Comments and exercises

Sobolev classes over infinite dimensional spaces were first introduced and investigated by N.N. Frolov [**452**], [**453**] at the beginning of the 1970s (see also his works [**454**]–[**457**]). Somewhat later such classes and related objects were studied also by other authors (see Daletskiĭ, Paramonova [**321**]–[**323**], Krée [**637**], [**638**], Gross [**505**], Lascar [**698**]), and after appearance of the Malliavin calculus in the celebrated work Malliavin [**751**], where they play an important role, Sobolev classes became a subject of intensive research of many authors. The equivalence of different definitions of Sobolev classes over Gaussian measures was established by Meyer and Sugita (see [**806**]–[**808**] and [**1090**], [**1091**]). Pisier [**904**] gave a shorter proof of the Meyer inequalities. Let us also mention the works Decreusefond, Hu, Üstünel [**338**], Feyel, La Pradelle [**424**]–[**427**], Hirsch [**546**], Ikeda, Watanabe [**571**], Inahama [**575**], Krée, Krée [**639**], Krée [**640**], Larsson-Cohn [**697**], Paclet [**871**], [**872**], Peszat [**880**], [**881**], Potthoff [**915**], Shigekawa [**1010**], [**1012**], [**1016**], [**1017**], Shigekawa, Yoshida [**1021**], Üstünel [**1149**], Üstünel, Zakai [**1164**], Watanabe [**1178**], [**1184**]. Sobolev classes of other types on metric spaces are studied in Ambrosio, Tilli [**69**], Cheeger [**281**], Hajłasz, Koskela [**514**], Heinonen [**531**], Keith [**600**], Reshetnyak [**954**]–[**956**], Vodop'janov [**1177**].

Our exposition in §8.6 follows the paper Airault, Bogachev, Lescot [**19**]. Hölder continuity of Sobolev functions along finite dimensional subspaces is shown in Ren, Röckner [**945**], Hu, Ren [**561**]). Recently E.V. Nikitin has studied Besov classes in the infinite dimensional case.

The hypercontractivity of the Ornstein–Uhlenbeck semigroup was discovered by Nelson [**822**], and Gross [**505**] established the logarithmic Sobolev inequality. There is extensive literature devoted to these issues as well as to the Poincaré inequality and exponential integrability; see Aida, Masuda, Shigekawa [**11**], Aida, Shigekawa [**12**], Aida, Stroock [**13**], Albeverio, Kondratiev, Röckner [**45**], Bakry [**87**], [**88**], Bakry, Emery [**89**], Blanchere et al. [**158**], Bobkov, Götze [**162**], Bobkov, Ledoux [**163**], Feissner [**422**], Gross [**506**], Kulik [**653**], Ledoux [**711**], [**712**], Royer [**973**], Üstünel [**1148**], [**1150**]. See Shigekawa [**1017**, p. 90] for multiplicative inequalities.

Various properties of the Ornstein–Uhlenbeck semigroup and Ornstein–Uhlenbeck operator (spectral, asymptotic and smoothing properties, associated Sobolev spaces) are studied in many works; this direction obviously deserves a separate survey, so here we only mention the following works, in which further references can be found: Accardi, Bogachev [**1**], Bogachev, Röckner [**218**], Bogachev, Röckner, Schmuland [**225**], Chojnowska-Michalik, Goldys [**284**]–[**286**], Lunardi [**738**], Maas, Neerven [**746**], Metafune [**800**], Metafune, Pallara, Priola [**801**], Metafune et al. [**803**], Neerven [**820**].

Riesz transforms related to the Ornstein–Uhlenbeck operator and various estimates for the corresponding analogs of singular integrals can be found in Dragičević, Volberg [**359**], Fabes, Gutierrez, Scotto [**405**], García-Cuerva et al. [**467**], Gutiérrez [**511**], Gutiérrez, Segovia, Torrea [**512**], Mauceri, Meda [**789**], Mauceri, Meda, Sjögren [**790**], Sjögren [**1040**]. Analogous questions for Banach space valued mappings are considered in Maas [**744**], [**745**], Maas, Neerven [**748**]. Estimates of norms of $D_H^k T_t f$ for the Ornstein–Uhlenbeck semigroup are obtained in Lee [**713**].

Functions of bounded variation on the Wiener space are studied in Fukushima [**463**], Fukushima, Hino [**464**], Hino [**541**] and in the recent works Ambrosio, Miranda, Maniglia, Pallara [**67**], [**68**], Hino [**542**] (see Exercise 8.14.15, where more general measures are considered).

On partitions of unity formed by functions from Sobolev classes, see Albeverio, Ma, Röckner [**51**]. A useful concept of convexity on the Wiener space was introduced and investigated in Feyel, Üstünel [**429**]. On the closability of gradient, see Goldys, Gozzi, Neerven [**491**]. Problems related to Sobolev classes on infinite dimensional spaces are discussed also in Chow, Menaldi [**287**].

Complex analysis on the Wiener space is developed in Fang [**409**], Fang, Ren [**417**], [**418**], Kusuoka, Taniguchi [**693**], Nishimura [**830**], Sugita [**1093**].

It is of interest to study to what extent a smooth Wiener functional is determined by its restriction to the Cameron–Martin space. For example, linear functions have versions uniquely defined by such restrictions. This is discussed in Carmona, Nualart [**265**], Sugita [**1093**], Zakai [**1214**].

In Malliavin [**766**], Üstünel, Zakai [**1161**], one discusses a construction of a filtration on a given space with a Gaussian measure which would be analogous to the natural filtration $\mathcal{F}_t$ on the classical Wiener space generated by the random variables w_s, $s \leqslant t$.

Clark [**290**] considered representation (8.11.3) in which the functional F on the classical Wiener space was Fréchet differentiable (and satisfied some additional technical conditions). To functionals from $H^{2,1}$ Clark's result was extended by Ocone [**863**]. For related results, see Davis [**331**], Elliott, Kohlmann [**383**], Haussmann [**522**], [**523**], Karatzas, Ocone, Li [**591**], Maas, Neerven [**747**], Nualart, Zakai [**858**], Ocone, Karatzas [**866**], Wu [**1195**].

Capacities on metric spaces connected with classes $H^{p,r}$ are studied in Fukushima, Kaneko [**465**], Kazumi, Shigekawa [**599**]. Tightness of capacities associated with Dirichlet forms is related to the existence of the corresponding Markov processes. A detailed discussion of this important connection can be found in Albeverio, Röckner [**52**]–[**54**], Bouleau, Hirsch [**247**],

Fukushima [**461**], Fukushima, Oshima, Takeda [**466**], Lyons, Röckner [**742**], Ma, Röckner [**743**], Röckner [**958**]. More general Gaussian capacities associated with operator semigroups $\{T_t\}_{t\geqslant 0}$ are discussed in Bogachev, Röckner [**218**], Shigekawa [**1013**]. The topological invariance of Gaussian capacities was established in Albeverio et al. [**40**] for Banach spaces and in Bogachev, Röckner [**216**], Feyel, La Pradelle [**425**] for general locally convex spaces. Capacities are studied also in Caraman [**261**], [**262**], Denis [**340**], Hirsch [**545**], Pugachev [**935**], Ren, Röckner [**946**], Schmuland [**989**], Yoshida [**1207**].

Generalized functions on the Wiener space were considered by many authors, see Hida et al. [**538**], Huang, Yan [**564**], Körezlioğlu, Üstünel [**633**], Meyer, Yan [**809**], [**810**], Obata [**862**], Üstünel [**1144**], [**1149**]. Different approaches to generalized functions on infinite dimensional spaces (in particular, based on differentiable or other measures) are developed in Albeverio et al. [**39**], Berezanskiĭ [**139**], Berezansky, Kondratiev [**140**], Dalecky, Fomin [**319**], Kuo [**671**], Smolyanov, Shavgulidze [**1053**], Uglanov [**1130**], [**1134**], [**1141**]. In Daletskiĭ [**318**] and Albeverio et al. [**39**], there is a discussion of analogs of the Chebyshev–Hermite system for non-Gaussian measures.

Exercises

8.14.1. Let H be a Hilbert space densely embedded into a locally convex space X, let $j\colon X^* \to H$ be the associated embedding, and let μ a Radon probability measure on X quasi-invariant along all vectors from $j(X^*)$. Let $h \in j(X^*)$ and $d\mu_h/d\mu \in L^q(\mu)$. Given $f \in W^{p,1}_H(\mu)$, $p \geqslant q/(q-1)$, prove that μ-a.e. one has

$$f(x+h) - f(x) = \int_0^1 \bigl(D_H f(x+sh), h\bigr)_H \, ds.$$

HINT: Use the quasi-invariance of μ and Hölder's inequality to extend this identity from smooth cylindrical functions.

8.14.2. If $v \in W^{2,1}(\mu, H) \subset D^{2,1}(\mu, H)$, then in the situation described in Theorem 8.10.5 the equality $\delta_\mu v = 0$ is fulfilled a.e. on $v^{-1}(0)$.

8.14.3. Suppose that in Corollary 8.3.5 the functions $\beta^\mu_{h_i}$ have generalized partial derivatives along the vectors h_j belonging to $L^p(\mu)$, where $p > n$. Prove that the image of μ under the mapping $(l_1, \ldots, l_n)$ has a continuous logarithmic gradient on $\mathbb{R}^n$.

8.14.4. Give a sufficient condition for compactness in $L^2(\mu)$ of a set from $W^{2,1}(\mu)$ for a differentiable measure μ.

8.14.5. Let μ be a Radon probability measure on a locally convex space X, let $E \subset X$ be a finite dimensional linear subspace, and let Y be a closed linear subspace such that $X = E \oplus Y$. We equip E with a fixed inner product $(\,\cdot\,,\,\cdot\,)_E$; the corresponding norm will be denoted by $|\cdot|_E$. Let μ_Y be the projection of μ on Y and let μ_y, $y \in Y$, be conditional measures on E with positive continuous densities ϱ_y with respect to a fixed Lebesgue measure on E such that $\varrho_y \leqslant M$. We define the Sobolev class $W^{p,1}_E(\mu)$ as the completion of $\mathcal{FC}^\infty_b$ with respect to the norm

$$\|f\|_{p,1} := \|f\|_p + \|\,|D_E f|\,\|_p,$$

where $D_E f$ is determined by the relationship $(D_E f, h)_E = \partial_h f$, $h \in E$. Suppose that the class $W_E^{p,1}(\mu)$ is well defined (i.e., the operator $D_E f$ is closable). For the measures μ_y the analogous classes $W_E^{p,1}(\mu_y)$ are equipped with the norms $\|\cdot\|_{p,1,y}$. Prove that $f \in W_E^{p,1}(\mu)$ precisely when $f_y \in W_E^{p,1}(\mu_y)$ for μ_Y-a.e. y and the function $y \mapsto \|f_y\|_{p,1,y}$ belongs to $L^p(\mu_Y)$.

HINT: Let the latter condition be fulfilled; the general case can be easily reduced to the case where f has compact support. Let us find $f_j \in \mathcal{F}C_b^\infty$ such that $f_j \to f$ in $L^p(\mu)$ and $(f_j)_y \to f_y$ in $L^p(\mu_y)$ for μ_Y-a.e. y. Let θ be a probability density from $C_0^\infty(E)$, $\theta_j(x) = j^d \theta(jx)$, $d = \dim E$. Then there exist $\{j_n\}$ and $\{k_n\}$ such that letting $\varphi_n := f_n * \theta_{j_n} * \theta_{k_n}$, where the convolution is taken with respect to the variable from E, we have $\varphi_n \to f$ in $W_E^{p,1}(\mu)$.

8.14.6. Let X be a locally convex space, let Y be a finite dimensional space, and let μ be a centered Radon Gaussian measure on $X \times Y$. Prove that condition (D) introduced after Theorem 8.6.1 is fulfilled if $Y \subset H(\mu)$.

8.14.7. Let γ be the standard Gaussian measure on $\mathbb{R}^1$. Show that the classes $H^{1,1}(\gamma)$ and $W^{1,1}(\gamma)$ do not coincide.

8.14.8. (Rudenko [975]) Let γ be a Radon Gaussian measure on a locally convex space X, let $\{T_t\}_{t \geqslant 0}$ be the Ornstein–Uhlenbeck semigroup, and let ν be a Radon measure on X. Let $(T_t\nu)^{\mathrm{ac}}$ and ν^{ac} be the absolutely continuous components of the measures $T_t\nu$ and ν with respect to γ, where

$$(T_t\nu)(B) := \int_X T_t I_B(x)\,\nu(dx), \quad B \in \mathcal{B}(X).$$

(i) Prove that if X finite dimensional, then, as $t \to 0+$, the Radon–Nikodym densities $d(T_t\nu)^{\mathrm{ac}}/d\gamma$ converge to $d\nu^{\mathrm{ac}}/d\gamma$ a.e. with respect to γ. (ii) Prove that in the general case one has convergence in measure γ.

8.14.9. Prove Proposition 8.11.6.

8.14.10. Prove invariance of the Sobolev capacities with respect to embedding into a larger space for a differentiable measure μ.

8.14.11. Prove that in the situation of Theorem 8.13.9 a compact set K has capacity zero if and only if every measure of finite energy vanishes on K.

8.14.12. Let γ be a Radon Gaussian measure and let a γ-measurable set A be such that $A + H(\gamma) = A$. Show that $C_{p,r}(A)$ is either 1 or 0.

8.14.13. Let a Radon probability measure μ on a locally convex space X be continuous along a continuously embedded separable Hilbert space H and satisfy the Poincaré inequality

$$\int_X \left(f(x) - \int_X f\,d\mu\right)^2 \mu(dx) \leqslant C \int_X |D_H f(x)|^2\,\mu(dx), \quad f \in D^{2,1}(\mu).$$

Let $A \in \mathcal{B}(X)$ and $\mathrm{dist}_H(x, A) := \inf\{|x - y|_H : y \in A, x - y \in H\}$, where $\mathrm{dist}_H(x, A) = 0$ if $x \notin A + H$. Show that $\mathrm{dist}_H(x, A) \in D^{p,1}(\mu)$, $p \geqslant 1$.

8.14.14. Prove an analog of Theorem 8.5.5 for non-Gaussian measures satisfying the Poincaré inequality.

8.14.15. Let μ be a Radon probability measure on a locally convex space X Fomin differentiable along vectors from a densely embedded Hilbert space H. Denote by $BV_H(\mu)$ the class of all functions $f \in L^1(\mu)$ with the following properties: $f\beta_h^\mu \in L^1(\mu)$ for all $h \in H$, there exists an H-valued measure $\Lambda(f)$ of bounded semivariation $V(\Lambda(f))$ for which $d_h(f \cdot \mu) = (\Lambda(f), h)_H + f \cdot d_h\mu$ for all $h \in H$, where $d_h(f \cdot \mu)$ is the Skorohod derivative, and the following norm is finite:
$$\|f\|_{BV} := \|f\|_{L^1(\mu)} + V(\Lambda(f)) + \sup_{|h|_H \leq 1} \|f \cdot d_h\mu\|.$$
Prove that $BV_H(\mu)$ is complete with respect to this norm.

8.14.16. (E.V. Nikitin) Let γ be a centered Radon Gaussian measure on a locally convex space X, $H = H(\gamma)$, and let ν be a centered Gaussian measure on H. Let $s \in \mathbb{R}^1, p, q \in [1, +\infty)$. Set $f_h(x) := f(x+h)$,
$$\|f\|_{s,p,q,\nu} := \|f\|_{L^p(\gamma)} + \left(\int_{|h|_H \leq 1} |h|_H^s \|f - f_h\|_{L^p(\gamma)}^q \nu(dh)\right)^{1/q}.$$
Let $B(s, p, q, \nu)$ be the class of all functions $f \in L^p(\gamma)$ such that $\|f\|_{s,p,q,\nu} < \infty$. Prove that $B(s, p, q, \nu)$ is complete with respect to the norm $\|\cdot\|_{s,p,q,\nu}$ and that $W^{r,1}(\gamma) \subset B(s, p, q, \nu)$ if $r > p$.

8.14.17. (Agafontsev, Bogachev [4]) Let γ on $\mathbb{R}^\infty$ be the countable power of the standard Gaussian measure on the real line.

(i) Let a sequence of functions $f_n + g_n$, where $f_n \in \mathcal{P}_d(\gamma)$ and a γ-measurable function g_n depends only on the variables x_i with $i > n$, have the characteristic functionals $\varphi_{f_n+g_n}(t)$ equicontinuous at the origin. Let $\mathbb{E}(f_n|\sigma_n)$ denote the conditional expectation with respect to the σ-algebra generated by the variables x_i with $i > n$. Prove that the sequence of functions $\psi_n := f_n - \mathbb{E}(f_n|\sigma_n)$ is bounded in every $L^p(\gamma)$.

(ii) Let a sequence of functions $\eta_n = f_n + g_n$, where $f_n \in \mathcal{P}_d(\gamma)$ and a γ-measurable function g_n depends only on the variables x_i with $i > n$, have a finite limit η almost everywhere. Show that $\eta \in \mathcal{P}_d(\gamma)$.

(iii) Let a γ-measurable function f possess the following property: for some $d \geq 0$, for every $n \in \mathbb{N}$ and γ-a.e. x the function
$$(t_1, \ldots, t_n) \mapsto f(x_1 + t_1, x_2 + t_2, \ldots, x_n + t_n, x_{n+1}, x_{n+2}, \ldots)$$
is a polynomial of degree d. Prove that $f \in \mathcal{P}_d(\gamma)$. Moreover, in place of the latter condition it suffices that, for every finite collection of integer numbers $b_1, \ldots, b_n$, for γ-a.e. x the function
$$t \mapsto f(x_1 + tb_1, x_2 + tb_2, \ldots, x_n + tb_n, x_{n+1}, x_{n+2}, \ldots)$$
is a polynomial of degree d. This generalizes the result from [**191**, Remark 5.10.5].

8.14.18. Let μ be a Radon probability measure on a locally convex space X differentiable along a continuously embedded separable Hilbert space H and let $F \in W^{s,1}(\mu)$. Suppose that $F \geq 0$ and that $\mu \circ F^{-1}$ has a bounded density g. Set $A := \{F < 1\}$. Prove that I_A belongs to the class $\mathcal{E}^{p,\alpha}(\mu)$ defined in §8.6 if $\alpha p < 1/q$, where $s = pq/(q-1)$.

HINT: Consider $F_\varepsilon = \varphi_\varepsilon(F)$, $\varphi_\varepsilon(t) = 0$ if $t \leq 0$, $\varphi_\varepsilon(t) = t/\varepsilon$ if $t \in [0, \varepsilon]$, $\varphi_\varepsilon(t) = 1$ if $t \geq \varepsilon$; show that $\|F_\varepsilon\|_{p,1} \leq C_1 \varepsilon^{-1+1/(pq)}$ and $\|I_A - F_\varepsilon\|_p \leq C_2 \varepsilon^{1/p}$, whence $K_t(I_A) \leq C_3 \varepsilon^{1/(pq)}$.

CHAPTER 9

The Malliavin calculus

The Malliavin calculus has two components: derivative on spaces with measures and analysis of smoothness of nonlinear images of measures. The first one has already been considered in our discussion of differentiable measures and Sobolev classes. This chapter is concerned with the second component.

9.1. General scheme

The main idea of Malliavin's method has already been explained in the preface. In this section, this method is discussed in greater detail. The central problem of the Malliavin calculus is the study of smoothness of measures $\nu = \mu \circ f^{-1}$ induced by measurable functions f or mappings $f = (f_1, \ldots, f_d)$ on a measurable space (X, μ) equipped with certain differential structure, where the measure μ and the function f are smooth in a certain sense. In the case $d = 1$ the relationship $\nu = \varrho \, dx$, where ϱ is a smooth density and $\varrho^{(n)} \in L^1(\mathbb{R}^1)$ for all n, is equivalent to the existence of numbers C_n such that
$$\int_{\mathbb{R}^1} \varphi^{(n)}(t)\, \nu(dt) \leqslant C_n \sup_t |\varphi(t)|, \quad \forall \varphi \in C_0^\infty(\mathbb{R}^1).$$
The left-hand side is the integral of $\varphi^{(n)}(f(x))$ with respect to μ. Suppose that there exists a vector field v on X such that the measure $(\partial_v f)^{-1} \cdot \mu$ is differentiable along v. Then the integral
$$\int_X \varphi^{(n)}(f(x))\, \mu(dx) = \int_X \varphi^{(n)}(f(x)) \partial_v f(x) \frac{1}{\partial_v f(x)}\, \mu(dx)$$
$$= \int_X \partial_v [\varphi^{(n-1)}(f(x))] \frac{1}{\partial_v f(x)}\, \mu(dx)$$
can be written in the form
$$-\int_X \varphi^{(n-1)}(f(x))\, d_v\Big(\frac{1}{\partial_v f}\mu\Big)(dx).$$
Suppose that this procedure can be repeated n times. Then we arrive at the representation
$$\int_X \varphi^{(n)}(f(x))\, \mu(dx) = \int_X \varphi(f(x))\, \nu_n(dx)$$

with some bounded measure ν_n. Therefore, we obtain the desired estimates with $C_n = \|\nu_n\|$. A more thorough analysis shows that the measure ν_n has the form
$$\sum_{k=0}^{n} R_k d_v^k \mu,$$
where R_k is a polynomial in functions $\partial_v f, \ldots, \partial_v^{n+1} f$ divided by $(\partial_v f)^{2n-k}$. Therefore, the main problem is to control the integrability of powers of the function $1/\partial_v f$.

Let us consider the following instructive example. Let μ be a centered nondegenerate Gaussian measure on a separable infinite dimensional Hilbert space X and let T be the covariance operator of μ. We consider the function $f(x) = (x, x)$. Let us try to find a suitable vector field v. Since we are interested in finding a field along which the function f is less degenerate, it would be nice to take $v = \mathrm{grad} f$. Unfortunately, no Gaussian measure is differentiable along the field $v(x) = cx$ with $c \neq 0$ (see Example 8.10.2). Hence it is natural to choose some vector field with values in the Cameron–Martin space $H = \sqrt{T}(X)$. For example, let us take $v(x) = Tx$. Then $\partial_v f(x) = 2(x, Tx)$. It is clear that all functions $\partial_v^k f$ are polynomials on X, hence they are integrable to every power. Thus, in order to show that $\mu \circ f^{-1}$ admits a smooth density, it remains to verify that the function Q^{-1}, where $Q(x) = (Tx, x)$, belongs to all $L^p(\mu)$. This is equivalent to the following:
$$\mu(x\colon Q(x) < \varepsilon) = o(\varepsilon^n) \quad \text{as } \varepsilon \to 0 \quad \text{for every } n.$$
As we shall see below, this can be done indeed. Certainly, in this particular example one can verify the smoothness of the distribution of f directly since it admits an explicit formula (in terms of infinite convolutions). However, it should be noted that the presented method applies to much more general situations.

In the case where $f = (f_1, \ldots, f_d)\colon X \to \mathbb{R}^d$, our strategy remains the same with the only difference being that in place of a single field v one should construct d vector fields $v_1, \ldots, v_d$ on X such that the *Malliavin matrix*
$$\sigma(x) = \left(\sigma^{ij}\right)_{i,j=1}^{d} = \left(\partial_{v_i} f_j\right)_{i,j=1}^{d}$$
has the determinant $\Delta(x)$ with $\Delta^{-1} \in \bigcap_{p<\infty} L^p(\mu)$ (see §9.3).

The classical Malliavin calculus deals with the case where μ is the Wiener measure on the space $X = C_0[0,T]$ or on $L^2[0,T]$, $C_0([0,T], \mathbb{R}^d)$, $L^2([0,T], \mathbb{R}^d)$, etc. The functions f_j are assumed to belong to some Sobolev class. A standard choice of v_i is $v_i = D_H f_i$. Here
$$\sigma^{ij}(x) = \bigl(D_H f_i(x), D_H f_j(x)\bigr)_H.$$
Thus, we see that the integration by parts formula plays an important role in the described approach.

In practical realizations, major problems concern the verification of the inclusion $\det \sigma^{-1} \in L^p(\mu)$. The verification of the inclusion $f_i \in W^{p,r}(\mu)$

is usually less difficult. Typical examples are functionals of the Wiener process given by ordinary or stochastic integrals. We shall consider such examples in the subsequent sections. The situation may be different for non-Gaussian measures, especially for those which have no constant vectors of differentiability. This is the case if we consider random processes with independent increments without Gaussian component.

There are two essentially different ways of constructing vector fields v_i involved in the definition of the Malliavin matrix. The first one, developed in Shigekawa [1008], [1009], Stroock [1082], Nualart, Zakai [858], Zakai [1212], and other works, corresponds to the choice $v_i = D_H f_i$. With minor variations the same construction arises if we consider the Ornstein–Uhlenbeck operator L (or its analogs), as described in [1082].

Another approach is due to Bismut [149]. It should be noted that in the 1960s the situation considered below was already studied by Pitcher [905]–[908] in relation to estimating parameters of stochastic processes. Suppose we are given a family of transformations $T_\varepsilon \colon X \to X$ such that $\mu_\varepsilon := \mu \circ T_\varepsilon^{-1} = g_\varepsilon \cdot \mu_0$ and the mappings $\varepsilon \mapsto \varphi \circ T_\varepsilon$ with values in $L^1(\mu)$ are differentiable for sufficiently large algebra $\mathcal{E}$ of functions φ. Set

$$\partial_v(\varphi \circ T_0) := \frac{d}{d\varepsilon}\varphi \circ T_\varepsilon^{-1}\Big|_{\varepsilon=0}.$$

Suppose also that the mapping $\varepsilon \mapsto g_\varepsilon$ is differentiable at the origin as a mapping to $L^1(\mu_0)$. Then the measure μ is differentiable along v and

$$\beta_v^\mu = -\frac{\partial g_\varepsilon}{\partial \varepsilon}\Big|_{\varepsilon=0} \circ T_0.$$

Bismut considered the case

$$d\xi_t^\varepsilon = X_0(\xi_t^\varepsilon)dt + \sum_{i=1}^m X_i(\xi_t^\varepsilon) \circ dw_t^i + \varepsilon \sum_{i=1}^m X_i(\xi_t^\varepsilon)u_t^i dt, \quad \xi_0^\varepsilon = x,$$

where each X_i is a smooth vector field on $\mathbb{R}^d$, $w_t^1, \ldots, w_t^d$ are independent Wiener processes, u is an adapted process with values in $\mathbb{R}^m$ and $\circ dw_t$ denotes the Stratonovich integral (see Ch. 1). The transformations G_ε are given by $G_\varepsilon(\omega)(t) = \xi_t^\varepsilon$, where the space $C([0,T], \mathbb{R}^d)$ with the Wiener measure is chosen for (X, μ). Bismut's approach is more general in the sense that f and μ may be differentiable along a field v not belonging to $W^{2,1}(\mu)$. Thus, in principle, this approach applies to a broader class of examples.

9.1.1. Example. Let w_t be a Wiener process on $[0,1]$, let u be an adapted process with $|u| \in \bigcap_{1 \leqslant p < \infty} L^p(\Omega \times [0,1])$. Let us define ξ^ε by

$$d\xi_t^\varepsilon = \sigma(t, \xi_t^\varepsilon)dw_t + \big[b(t, \xi_t^\varepsilon) + \varepsilon \sigma^2(t, \xi_t^\varepsilon)u_t\big]dt, \quad \xi_0^\varepsilon = x,$$

where σ and b satisfy standard assumptions (for example, they are Borel measurable, bounded and Lipschitzian in the second variable). The transformations T_ε are defined by $T_\varepsilon(\omega)(t) = \xi_t^\varepsilon(\omega)$. By the Girsanov theorem the measures $Q_\varepsilon = P \circ T_\varepsilon^{-1}$ are equivalent and their densities g_ε with respect to Q_0 are given by

$$g_\varepsilon = \exp\left[\varepsilon \int_0^1 u_t \, d\xi_t^0 - \frac{1}{2}\varepsilon \int_0^1 u_t\bigl(2b(t, \xi_t^0) + \varepsilon \sigma^2(t, \xi_t^0) u_t\bigr) \, dt\right].$$

Hence $\partial g_\varepsilon / \partial \varepsilon\big|_{\varepsilon=0}$ exists and coincides with

$$\int_0^1 u_t \, d\xi_t^0 - \int_0^1 u_t b(t, \xi_t^0) \, dt.$$

Therefore,

$$\beta_v^P = -\int_0^1 \sigma(t, \xi_t^0) u_t \, dw_t.$$

In the case $\sigma = 1$, $b = 0$ we have the following expression for v:

$$v(\omega)(t) = \int_0^t u_s \, ds.$$

As we know, in the infinite dimensional case a smooth measure may not be differentiable along all smooth vector fields, where the differentiability of a measure μ along a field v is defined (see §6.6) as the integration by parts formula

$$\int_X \partial_v f(x)\,\mu(dx) = -\int_X f(x)\, d_v\mu(dx)$$

for functions $f \in \mathcal{F}C_b^\infty(X)$. For example, if μ is a Gaussian measure, then it is not differentiable along constant fields $v(x) \equiv h$ not belonging to the Cameron–Martin space $H(\mu)$ and also along the field $v(x) = x$ (as shown in Example 8.10.2). On the other hand, according to Theorem 8.10.1, such a measure μ is differentiable along every field $v \in D^{2,1}(\mu, H(\mu))$.

There are essentially two ways of constructing vector fields of differentiability. The first one is to form vector fields of the type $v(x) = \sum_n f_n(x) a_n$ on a space equipped with some measure μ differentiable along vectors a_n, where functions f_n are differentiable in a suitable sense. The second possibility is to consider fields generated by families of transformations. One can also take nonlinear images of fields of both types. Further development of these ideas is discussed in Chapter 11. In this chapter, we concentrate on measures having constant vectors of differentiability; typical examples are Gaussian measures, product measures, and Gibbs measures.

9.2. Absolute continuity of images of measures

One of the classical problems of stochastic analysis is the study of measures induced by functionals of random processes from the viewpoint of the existence of densities with respect to Lebesgue measure and differential properties of these densities. In diverse applications (e.g., in mathematical statistics, limit theorems of probability theory, and mathematical physics), the following classes of (real or vector) functionals most frequently arise:

1) solutions to stochastic differential equations at a fixed moment;

9.2. Absolute continuity of images of measures

2) polynomial functionals, in particular, quadratic forms and multiple stochastic integrals;

3) functionals of supremum type and norms;

4) integral functionals given by ordinary or stochastic integrals.

Below we consider some typical examples; in this section we discuss the existence of densities. The general formulation of the problem is this. Let μ be a differentiable measure on a locally convex space X (for example, a Gaussian measure) and let $F\colon X \to \mathbb{R}^n$ be a sufficiently regular mapping. What can we say about the induced measure $\nu = \mu\circ F^{-1}$? Is it absolutely continuous with respect to Lebesgue measure on $\mathbb{R}^n$? Does it have a bounded density? Can one choose a smooth version of this density? Of course, we need certain conditions of "nondegeneracy" of the mapping F since otherwise ν may have atoms or a singular component.

9.2.1. Example. Suppose that $X = \mathbb{R}^1$, $n = 1$, a probability measure μ is given by a smooth density ϱ and F is a smooth function. Then a necessary and sufficient condition for the absolute continuity of ν is the equality
$$\mu\bigl(x\colon F'(x) = 0\bigr) = 0.$$

PROOF. If the indicated equality is fulfilled, then every interval U in the complement of the set $\{F' = 0\}$ is diffeomorphically mapped onto $F(U)$. Hence all sets of positive Lebesgue measure in U are taken to sets of positive measure. Hence the preimage of any set E of zero Lebesgue measure has μ-measure zero. This means the absolute continuity of ν. If the set $\{x\colon F'(x) = 0\}$ has positive μ-measure, then ν is not absolutely continuous since the image of this set has Lebesgue measure zero, which is easily verified (Exercise 9.8.2 contains an even stronger assertion). □

9.2.2. Theorem. *Let $F\colon \mathbb{R}^n \to \mathbb{R}^n$ be a Borel mapping, let $E \subset \mathbb{R}^n$ be a Borel set of finite Lebesgue measure, and let $e_1, \ldots, e_n$ be a basis in $\mathbb{R}^n$ such that, for every $x \in E$, there exist*
$$\partial_{e_i} F(x) = \lim_{t\to 0} \frac{F(x + te_i) - F(x)}{t}, \quad i = 1, \ldots, n.$$
Suppose that the vectors $\partial_{e_1} F(x), \ldots, \partial_{e_n} F(x)$ are linearly independent for a.e. $x \in E$. Let $\lambda|_E$ be the restriction of Lebesgue measure to E. Then the measure $\lambda|_E \circ F^{-1}$ is absolutely continuous.

Conversely, if this measure is absolutely continuous, then, for a.e. $x \in E$, the vectors $\partial_{e_i} F(x)$ are linearly independent. Both assertions are true in a more general case where in place of usual partial derivatives one takes approximate partial derivatives.

PROOF. If the mapping F is continuously differentiable, then this assertion is proved similarly to the previous example. Indeed, for the proof of the direct assertion it suffices to observe that each point x at which the derivative of F is not degenerate has a neighborhood in which F is a diffeomorphism. The converse follows from Theorem 2.4.3 which shows that the set where the derivative of F is degenerate is mapped by F onto a set of measure zero.

In the general case, according to Theorem 2.4.1, we can find a sequence of Borel sets $E_k \subset E$ and a sequence of continuously differentiable mappings $F_k\colon \mathbb{R}^n \to \mathbb{R}^n$ with the following properties: $E\backslash \bigcup_{k=1}^\infty E_k$ has measure zero and $F_k|_{E_k} = F|_{E_k}$. Each set E_k contains a Borel subset B_k of the same measure on which $\partial_{e_i} F_k = \partial_{e_i} F$. This follows by the fact that by Fubini's theorem almost every point $x \in E_k$ is not an isolated point of the intersection of E_k with the straight line $x + \mathbb{R}^1 e_i$ for every $i = 1,\ldots,n$. Hence, for a.e. $x \in B_k$, the partial derivatives $\partial_{e_i} F_k(x)$, $i = 1,\ldots,n$, are linearly independent, provided that the vectors $\partial_{e_i} F(x)$ are linearly independent for a.e. $x \in E$, whence we obtain the absolute continuity of the measure $\lambda|_{B_k} \circ F^{-1} = \lambda|_{B_k} \circ F_k^{-1}$. Since $E\backslash \bigcup_{k=1}^\infty B_k$ has measure zero, the direct assertion is proven. It follows from our reasoning that the converse assertion also reduces to the case of a continuously differentiable mapping. $\square$

It is seen from our proof that if the set of points where the Jacobian of F vanishes is not a measure zero set, then it contains a compact K of positive measure on which F is continuous and $F(K)$ has measure zero.

A similar result holds for mappings from $\mathbb{R}^n$ to $\mathbb{R}^d$ with $d < n$.

9.2.3. Corollary. *Let $F\colon \mathbb{R}^n \to \mathbb{R}^d$ be a Borel mapping. Suppose that $E \subset \mathbb{R}^n$ is a Borel set of finite Lebesgue measure such that for almost every $x \in E$ the partial derivatives $\partial_{e_i} F(x)$ exist and span $\mathbb{R}^d$. Then the measure $\lambda|_E \circ F^{-1}$ on $\mathbb{R}^d$ is absolutely continuous. The same is true in the case of approximate partial derivatives.*

Ponomarev [**910**] investigated the absolute continuity of the images of Lebesgue measure under mappings $f \in C^r(\mathbb{R}^n, \mathbb{R}^d)$ with $d < n$ and discovered that in this case one can find a mapping $f \in C^{n-d}(\mathbb{R}^n, \mathbb{R}^d)$ which transforms Lebesgue measure on $\mathbb{R}^n$ to an absolutely continuous measure on $\mathbb{R}^d$, but the rank of $Df(x)$ is less than d on a set of positive measure. In particular, the condition of the previous corollary is not necessary (unlike the theorem). If $r \geqslant n - d + 1$, then the absolute continuity of the image of Lebesgue measure is equivalent to the equality $\operatorname{rank} Df(x) = d$ almost everywhere.

Now we proceed to the infinite dimensional case.

9.2.4. Theorem. *Suppose that μ is a Radon measure on a locally convex space X. Let $\{a_i\}$ be a sequence in $C(\mu)$ and let $F\colon X \to \mathbb{R}^m$ be a Borel mapping. Suppose that, for μ-a.e. x, there exist vectors $v_1(x),\ldots,v_m(x)$ in $\{a_i\}$ such that the vectors*

$$\partial_{v_j} F(x) = \lim_{t \to 0} \frac{F(x + tv_j(x)) - F(x)}{t}, \quad j = 1,\ldots,m,$$

exist and are linearly independent. Then the measure $\mu \circ F^{-1}$ is absolutely continuous with respect to Lebesgue measure on $\mathbb{R}^m$. In particular, this is true if μ is a Gaussian measure, the mapping F belongs to the Sobolev class $W_{\text{loc}}^{1,1}(\mu, \mathbb{R}^m)$ and $D_H F(x)$ is surjective for almost all x.

PROOF. The set of all collections consisting of m vectors in $\{a_i\}$ is countable. By our hypothesis, up to a set of μ-measure zero the space X is the union of the sets
$$E_{i_1, \ldots, i_m} := \{x \colon \partial_{a_{i_j}} F(x) \text{ exist and are linearly independent}\}.$$
These sets are measurable with respect to μ. Indeed, for any fixed a_i the set B_i of all points x at which $\Lambda_i(x) := \lim_{n \to \infty} n[F(x + n^{-1} a_i) - F(x)]$ exists is Borel, and the mapping Λ_i on B_i is Borel measurable. Outside B_i we define Λ_i by zero. In addition, the mapping $(t, x) \mapsto F(x + t a_i)$ from $\mathbb{R}^1 \times X$ to $\mathbb{R}^m$ is Borel measurable. Therefore, the function
$$g_n(x) := \sup_{|t| \in (0, 1/n)} \psi(t, x), \quad \psi(t, x) := \bigl| t^{-1}[F(x + t a_i) - F(x)] - \Lambda_i(x) \bigr|,$$
with values in $[0, +\infty]$ is measurable on X with respect to μ (see Example 1.2.9). Hence the set of points x where $\lim_{n \to \infty} g_n(x) = 0$ is measurable with respect to μ. This yields the μ-measurability of $E_{i_1, \ldots, i_m}$. Now the theorem reduces to the case where the measure μ is concentrated on $E = E_{1, \ldots, m}$. This case follows from the finite dimensional assertion due to the existence of conditional measures μ^y on the sets $y + L$ continuous along $a_1, \ldots, a_m$, where L is the linear span of $a_1, \ldots, a_m$ (the continuity along m linearly independent vectors gives the absolute continuity with respect to the natural Lebesgue measure on $y + L$). $\square$

9.2.5. Corollary. *Let a Radon measure μ on a Banach space X be continuous along vectors from a dense set and let $F \colon X \to \mathbb{R}^m$ be a locally Lipschitzian mapping such that the set of points x where the Gâteaux derivative $F'(x)$ exists but is not surjective has μ-measure zero. Then the measure $\mu \circ F^{-1}$ is absolutely continuous. In particular, this is true if $m = 1$ and F is the norm on X.*

PROOF. The first assertion is clear from the theorem, the second one follows from the fact that the norm has the Gâteaux derivative μ-a.e. (see Theorem 10.6.4 below) and, as it is readily seen, this derivative does not vanish at the points where it exists. $\square$

9.2.6. Example. Let $X = C([0, T], \mathbb{R}^d)$ and let μ be the measure generated by the solution of the stochastic differential equation
$$d\xi_t = \sigma(\xi_t) dw_t + b(\xi_t) dt, \quad \xi_0 = x,$$
with bounded Lipschitzian coefficients σ and b. Suppose that the operators $\sigma(x)$ are invertible and uniformly bounded and that $F \colon X \to \mathbb{R}^m$ satisfies the conditions of the previous corollary. Then the measure $\mu \circ F^{-1}$ is absolutely continuous.

PROOF. A direct application of the previous corollary is possible only in the case of constant σ, where the measure μ is equivalent to the Wiener measure by Girsanov's theorem. In the general case, as we know from Chapter 4, the measure μ may fail to have nonzero vectors of continuity. So here one should employ the fact that the measure μ has the form $\mu = P \circ \Phi^{-1}$, where P is the Wiener measure on X and Φ is a Borel mapping such that the components of $F \circ \Phi$ belong to the class $W^{2,1}(P)$ (see Theorem 8.2.6). Details are left as Exercise 9.8.7. □

Our next result is a version of Theorem 9.2.4 for weak derivatives.

9.2.7. Theorem. *Let μ be a Radon measure on a locally convex space X and $\{a_i\} \subset D(\mu)$. Suppose that the function $F\colon X \to \mathbb{R}^1$ satisfies the following conditions*:

(i) *the function F is a limit of an almost everywhere convergent sequence of measurable functions F_n differentiable along the vectors a_i and possessing the derivatives $\partial_{a_i} F_n$ integrable on every compact set;*

(ii) *for every i there exists a measurable function g_i integrable on every compact set $K \subset X$ such that*

$$\lim_{n \to \infty} \int_K |\partial_{a_i} F_n(x) - g_i(x)| \, |\mu|(dx) = 0;$$

(iii) *one has $|\mu|\bigl(x\colon g_i(x) = 0 \text{ for all } i\bigr) = 0$.*
Then the measure $\mu \circ F^{-1}$ is absolutely continuous.

PROOF. Let us fix a compact set $K \subset X$ and $i \in \mathbb{N}$. According to Proposition 6.1.7, there exists a Borel function $\varphi\colon X \to [0,1]$ equal to 1 on K, equal to 0 outside some compact set Q and satisfying the Lipschitz condition along a_i. The measure $\nu := \varphi \cdot \mu$ is differentiable along a_i and has compact support Q. It is clear that $\|\partial_{a_i} F_n - g_i\|_{L^1(\nu)} \to 0$ as $n \to \infty$. Hence for every function $\psi \in C_0^\infty(\mathbb{R}^1)$ we have

$$\int_X \psi'(F(x)) g_i(x) \, \nu(dx) = \lim_{n \to \infty} \int_X \psi'(F_n(x)) \partial_{a_i} F_n(x) \, \nu(dx)$$

$$= \lim_{n \to \infty} \int_X \partial_{a_i}(\psi \circ F_n)(x) \, \nu(dx) = -\lim_{n \to \infty} \int_X \psi \circ F_n(x) \, d_{a_i}\nu(dx)$$

$$= -\int_X \psi \circ F(x) \, d_{a_i}\nu(dx)$$

since $d_{a_i}\nu \ll |\mu|$. Thus, we obtain

$$\left| \int_X \psi'(F(x)) g_i(x) \, \nu(dx) \right| \leqslant \sup_t |\psi(t)| \cdot \|d_{a_i}\nu\|.$$

Hence the measure $(g_i \cdot \nu) \circ F^{-1}$ is absolutely continuous. Since K is arbitrary, our condition on g_i yields the desired conclusion. □

9.2.8. Theorem. *Let γ be a Radon Gaussian measure on a locally convex space X and let $Q = \sum_{n=0}^{\infty} Q_n$, where each Q_n is a γ-measurable polynomial of degree at most n, the series converges in $L^2(\gamma)$, and*

$$\sum_{n=0}^{\infty} \lambda^n \|Q_n\|_{L^2}^2 < \infty$$

for some $\lambda > 1$. Then either the distribution of Q is absolutely continuous or $Q = \mathrm{const}$ γ-almost everywhere.

PROOF. We can take an almost everywhere convergent subsequence S_{N_k} of partial sums of the indicated series. We shall deal with the version of Q which equals the limit of S_{N_k} where it exists and is finite and equals 0 at all other points. Let $h \in H(\gamma)$, $h \neq 0$. Denote by γ^y Gaussian conditional measures on the straight lines $y + \mathrm{I\!R}^1 h$, $y \in Y$, where Y is a hyperplane in X complementing the straight line $\mathrm{I\!R}^1 h$. Let γ_Y be the projection of γ on Y. It follows from our assumptions that for γ_Y-a.e. y the series

$$\sum_{n=1}^{\infty} \lambda^n \int Q_n(y + th)^2 \, \gamma^y(dt)$$

converges. Hence, on every interval $[a, b]$, the function $f \colon t \mapsto Q(y+th)$ has polynomial approximations $S_n(t) = \sum_{k=0}^{n} Q_k(y + th)$ of degree n with the following mean-square estimate of error: $\|f - S_n\|_{L^2[a,b]} \leq C\lambda^{-n/2}$, so one has the estimate $\limsup_{n \to \infty} \|f - S_n\|_{L^2[a,b]}^{1/n} \leq \lambda^{-1/2} < 1$. This is seen from the inequality

$$\sum_{k=n+1}^{\infty} \|Q_k(y+th)\|_{L^2[a,b]} \leq \Big(\sum_{k=n+1}^{\infty} \lambda^{-k} \Big)^{1/2} \Big(\sum_{k=n+1}^{\infty} \lambda^k \|Q_k(y+th)\|_{L^2[a,b]}^2 \Big)^{1/2}.$$

By the well-known Bernstein theorem (see Timan [**1117**, pp. 383, 413]) the function f is analytic. We take an orthonormal basis $\{h_n\}$ in $H(\gamma)$. It follows from what we have proved above that for γ-a.e. x the functions $t \mapsto Q(x + th_n)$ are analytic. Let

$$Z := \{x \in X \colon \text{the functions } t \mapsto Q(x + th_n) \text{ are constant}\}.$$

The set Z is measurable and invariant with respect to shifts by linear combinations of the vectors h_n. By the $0 - 1$ law (see Theorem 1.4.8) we have either $\gamma(Z) = 0$ or $\gamma(Z) = 1$. In the first case by the analyticity along h_n we obtain the equality $\gamma\big(x \colon \partial_{h_n} Q(x) = 0 \ \forall n\big) = 0$, which yields the absolute continuity of $\gamma \circ Q^{-1}$. In the second case we obtain that Q coincides almost everywhere with some constant. $\square$

The following elegant algebraic criterion of the absolute continuity of the joint distribution of a system of polynomials is due to Kusuoka [**676**].

9.2.9. Theorem. *Let γ be a centered Radon Gaussian measure on a locally convex space X and let $F = (f_1, \ldots, f_n)$, where $f_1, \ldots, f_n$ are*

γ-measurable polynomials. Denote by $\mathcal{I}$ the class of all polynomials Q on $\mathbb{R}^n$ such that $Q(F) = 0$ γ-a.e. Let
$$V = \{z \in \mathbb{R}^n \colon\ Q(z) = 0,\ \forall Q \in \mathcal{I}\}.$$
Then the measure $\gamma \circ F^{-1}$ is absolutely continuous with respect to the standard surface measure on the algebraic manifold V. In particular, the measure $\gamma \circ F^{-1}$ is absolutely continuous on $\mathbb{R}^n$ if and only if one has $\mathcal{I} = \{0\}$.

9.3. Smoothness of induced measures

Now we proceed to the results on smoothness of induced measures. Suppose we are given a Radon measure μ on a locally convex space X and a Borel mapping
$$F = (F_1, \ldots, F_d) \colon X \to \mathbb{R}^d.$$
Suppose also that there are Borel vector fields $v_1, \ldots, v_d$ on X along which the functions F_i are differentiable. Denote by $\sigma_{ij}(x)$ the matrix with the entries $\partial_{v_i} F_j(x)$, $i, j \leqslant d$, and put $\Delta(x) := \det(\sigma_{ij})$. If $\Delta(x) \neq 0$, then $\gamma^{ij}(x)$ will denote the entries of the matrix $\sigma(x)^{-1}$. If $d = 1$, $v_1 = v$, $F_1 = F$, we have $\Delta = \partial_v F$. The matrix $\Delta(x)$ is called the Malliavin matrix.

9.3.1. Theorem. *Suppose that $\Delta^{-1} \in L^1(\mu)$, the measure $\gamma^{ij} \cdot \mu$ is Skorohod or Fomin differentiable along the fields v_i and $\varphi \circ F \in \mathfrak{D}(v_j, \gamma^{ij} \cdot \mu)$ for all $\varphi \in C_0^d(\mathbb{R}^d)$ (see §6.6). Then $\nu := \mu \circ F^{-1}$ has a density $\varrho \in BV(\mathbb{R}^d)$, and in the case of Fomin differentiability we have $\varrho \in W^{1,1}(\mathbb{R}^d)$.*

Suppose additionally that there exist Fomin derivatives which are obtained by applying to μ the differentiations $\eta \mapsto d_{v_i}(\gamma^{ij} \cdot \eta)$ up to order m. Then we have $\varrho \in W^{1,m}(\mathbb{R}^d)$.

PROOF. Letting $\partial_i \varphi := \partial_{x_i} \varphi$ for $\varphi \in C_0^\infty(\mathbb{R}^d)$ and integrating by parts we obtain
$$\int_{\mathbb{R}^d} \partial_i \varphi(y)\, \nu(dy) = \int_X (\partial_i \varphi) \circ F(x)\, \mu(dx) = \int_X \sum_{j,k \leqslant d} \gamma^{ij} \sigma_{jk}[(\partial_k \varphi) \circ F]\, \mu(dx)$$
$$= \int_X \sum_{j=1}^d \gamma^{ij} \partial_{v_j}(\varphi \circ F)\, \mu(dx) = -\sum_{j=1}^d \int_X \varphi \circ F\, d_{v_j}(\gamma^{ij} \cdot \mu)(dx)$$
due to the equality $\partial_{v_j}(\varphi \circ F) = \sum_{k=1}^d \sigma_{jk}[(\partial_k \varphi) \circ F]$. Therefore, the generalized partial derivatives of ν are the measures $\sum_{j=1}^d d_{v_j}(\gamma^{ij} \cdot \mu) \circ F^{-1}$. In the case where $d_{v_j}(\gamma^{ij} \cdot \mu) \ll \mu$ they are absolutely continuous with respect to ν, which yields their absolute continuity with respect to Lebesgue measure (see Chapter 3). In the case $m > 1$, in place of $\partial_i \varphi$ we start with a partial derivative of order m, which gives a partial derivative of order $m-1$ on the right, but in place of the measure μ a sum of the measures $d_{v_j}(\gamma^{ij} \cdot \mu)$ will appear. The hypotheses of the theorem are chosen so that we could repeat the described procedure. After m iterations we obtain the integral of $\varphi \circ F$ with respect to a bounded measure absolutely continuous with respect

9.3. Smoothness of induced measures

to μ. This means that the generalized partial derivatives of ν up to order m belong to $L^1(\mathbb{R}^d)$. □

Let us consider in more detail the structure of measures obtained by differentiating μ in this proof. The functions γ^{ij} have the form $\gamma^{ij} = M^{ij}/\Delta$, where M^{ij} is a minor of the matrix σ. Under broad assumptions (fulfilled in the concrete situations considered below) one has the equality

$$d_{v_j}(\gamma^{ij} \cdot \mu) = \partial_{v_j}(M^{ij}\Delta^{-1}) \cdot \mu + (M^{ij}\Delta^{-1})\beta^{\mu}_{v_j} \cdot \mu = G\Delta^{-2} \cdot \mu,$$

where G is a polynomial in the functions σ_{ij}, their derivatives along v_j and the functions $\beta^{\mu}_{v_j}$. Differentiating further these measures multiplied by γ^{ij}, at the mth step we obtain measures of the form $\Delta^{-2m}G_m \cdot \mu$, where G_m is a polynomial in the functions σ_{ij}, $\beta^{\mu}_{v_j}$ and their multiple derivatives along v_k (one can indicate some more explicit dependence, but this is not needed). Usually in applications the functions G_m are integrable to any power (or to a sufficiently high power), so the main problem is to verify the integrability of $\Delta^{-2m-\varepsilon}$ with some $\varepsilon > 0$.

9.3.2. Theorem. *Suppose that μ is a Radon measure on a locally convex space X, $n, m \in \mathbb{N}$, $h_1, \ldots, h_n \in X$, there exist all Fomin derivatives $d_{h_{i_1}} \cdots d_{h_{i_m}} \mu$ and their densities with respect to μ belong to all $L^p(\mu)$ with $p < \infty$. Let a Borel function $F \colon X \to \mathbb{R}^1$ be such that there exist all partial derivatives*

$$\partial_{h_{i_1}} \cdots \partial_{h_{i_{m+1}}} F \in \bigcap_{p<\infty} L^p(\mu)$$

and $\left(\sum_{i=1}^n |\partial_{h_i} F|^2\right)^{-1} \in L^r(\mu)$ for some $r > 2m$. Then the measure $\mu \circ F^{-1}$ has a density from the class $W^{q,m}(\mathbb{R}^1)$, where $q < r/(2m)$.

PROOF. Let us consider the field $v = \sum_{i=1}^n \partial_{h_i} F \cdot h_i$. Our hypotheses imply that the operator $\eta \mapsto d_v(\eta/\partial_v F)$ can be applied to the measure μ m times. This produces a measure on X whose density with respect to μ has the form $g(\partial_v F)^{-2m}$, $\partial_v F = \sum_{i=1}^n |\partial_{h_i} F|^2$. After the kth step we obtain a sum of measures of the form $\psi \cdot \zeta$, where ζ is a partial derivative of μ of order at most k and the function ψ has the form $\psi = \psi_0 \Delta^{-2k}$, $\Delta = \partial_v F$, the function ψ_0 is a polynomial in the partial derivatives of F along the vectors h_i. Then the measure $\Delta^{-2k-1}\psi_0 \cdot \zeta$ is differentiable along v, and its derivative is a sum of measures of the form

$$\partial^2_{h_i} F \Delta^{-2k-1}\psi_0 \cdot \zeta \quad \text{and} \quad \partial_{h_i} F \Delta^{-2k-2}\psi_i \cdot \zeta + \partial_{h_i} F \Delta^{-2k-1}\psi_0 \cdot d_{h_i}\zeta,$$

where ψ_i is a polynomial in the partial derivatives of F along the vectors h_j, i.e., this derivative has the same structure as at the previous step, but Δ^{-1} enters already in power $2k + 2$. For example,

$$d_v\left(\frac{1}{\partial_v F} \cdot \mu\right) = -\frac{\partial^2_v F}{(\partial_v F)^2}\mu + \frac{1}{\partial_v F}d_v\mu.$$

Hence the generalized derivative of the measure $\mu \circ F^{-1}$ of order m is the measure $\nu \circ F^{-1}$, where ν has the form $\nu = g(\partial_v F)^{-2m} \cdot \mu$ with $g \in L^p(\mu)$ for all $p < \infty$. Therefore, the measure $\mu \circ F^{-1}$ is given by a density $\varrho \in W^{1,m}(\mathbb{R}^1)$. For $m = 1$ this is true also for $|\mu|$ in place of μ. In particular, the measure $|\mu| \circ F^{-1}$ has a density $\varrho_0 \in L^\infty(\mathbb{R}^1)$. Hölder's inequality yields the inclusion $g(\partial_v F)^{-2m} \in L^q(|\mu|)$ whenever $q < r/(2m)$. Hence the measure $\nu \circ F^{-1}$ has a density $\xi \in L^q(|\mu| \circ F^{-1})$ with $q < r/(2m)$ with respect to $|\mu| \circ F^{-1}$. Therefore, the density $\xi \varrho_0$ of the measure $\nu \circ F^{-1}$ with respect to Lebesgue measure belongs to $L^q(\mathbb{R}^1)$, which gives the inclusion $\varrho \in W^{q,m}(\mathbb{R}^1)$. □

From the proof we can obtain certain estimates of derivatives of densities of the induced measures. For example, for $d = 1$ we have

$$\|\varrho\|_{L^\infty} \leqslant \mathrm{Var}\, \varrho \leqslant \|\partial_v^2 F(\partial_v F)^{-2}\|_{L^1(\mu)} + \|(\partial_v F)^{-1}\|_{L^1(d_v\mu)}. \quad (9.3.1)$$

Similarly one can consider multidimensional mappings.

9.3.3. Theorem. *Let $F = (F_1, \ldots, F_d)$, where each F_i satisfies the hypotheses of the previous theorem and $\Delta^{-1} \in L^r(\mu)$, where $r > 2m$. Then $\mu \circ F^{-1}$ has a density ϱ from $W^{1,m}(\mathbb{R}^d)$. If $m \geqslant d$, then $\varrho \in W^{q,m}(\mathbb{R}^d)$ whenever $q < r/(2m)$.*

9.3.4. Remark. (i) If in Theorem 9.3.1 the indicated Fomin derivatives up to order $m \geqslant d$ have densities from $L^q(\mu)$ with respect to $\mu \geqslant 0$, then $\varrho \in W^{q,m}(\mathbb{R}^d)$. If this condition is fulfilled for all m and the same is true for all measures $F_i^k \cdot \mu$ in place of μ, then $\varrho \in \mathcal{S}(\mathbb{R}^d)$. Indeed, it is seen from the proof that $\varrho^{(m)}$ is the density of the measure $(g \cdot \mu) \circ F^{-1}$, where $g \in L^q(\mu)$, whence $\varrho^{(m)} = \varrho_1 \varrho$, where $|\varrho_1|^q \varrho \in L^1(\mathbb{R}^d)$. Since $\varrho \in L^\infty(\mathbb{R}^d)$ due to the condition $m \geqslant d$, one has $\varrho^{(m)} \in L^q(\mathbb{R}^d)$. The second assertion follows from Exercise 2.8.16 since the density of the measure $(F_i^k \cdot \mu) \circ F^{-1}$ equals $x_i^k \varrho$.

(ii) If in Theorem 9.3.2 and Theorem 9.3.3 we have

$$F \in \bigcap_{p<\infty} L^p(\mu) \quad \text{and} \quad F_i \in \bigcap_{p<\infty} L^p(\mu),$$

respectively, then $\varrho \in \mathcal{S}(\mathbb{R}^1)$ ($\varrho \in \mathcal{S}(\mathbb{R}^d)$). The justification is similar to (i).

(iii) In the situations considered above when the distribution of a mapping F has a density in $\mathcal{S}(\mathbb{R}^d)$, it is possible to define compositions $\Psi(F)$ with generalized functions $\Psi \in \mathcal{S}'(\mathbb{R}^d)$ by setting

$$\langle \Psi(F), \varphi \rangle := \langle \Psi, (\varphi \cdot \mu) \circ F^{-1} \rangle, \quad \varphi \in W^\infty(\mu).$$

This defines $\Psi(F)$ as an element of the dual space to $W^\infty(\mu)$, i.e., as a distribution on (X, μ), because the measure $(\varphi \cdot \mu) \circ F^{-1}$ can be identified with its density from $\mathcal{S}(\mathbb{R}^d)$, moreover, the mapping

$$\varphi \mapsto (\varphi \cdot \mu) \circ F^{-1}$$

from $W^\infty(\mu)$ to $\mathcal{S}(\mathbb{R}^d)$ is continuous provided that $W^\infty(\mu)$ is equipped with its natural topology of a Fréchet space.

9.3. Smoothness of induced measures

9.3.5. Theorem. *Let γ be a centered Radon Gaussian measure on a locally convex space X with the Cameron–Martin space $H = H(\gamma)$ and let a mapping*
$$F = (F_1, \ldots, F_d) \colon X \to \mathbb{R}^d$$
be such that
$$F_1, \ldots, F_d \in W^\infty(\gamma), \quad \Delta^{-1} \in \bigcap_{p \in (1, +\infty)} L^p(\gamma), \quad \Delta = \det\bigl((D_H F_i, D_H F_j)_H\bigr).$$
Then the measure $\gamma \circ F^{-1}$ has a density from the Schwartz class $\mathcal{S}(\mathbb{R}^d)$.

PROOF. We take $v_i := D_H F_i$. As shown in §8.8,
$$\partial_{v_i} F_j = (D_H F_i, D_H F_j)_H \in W^\infty(\gamma), \quad \delta v_i \in W^\infty(\gamma), \quad \Delta^{-1} \in W^\infty(\gamma).$$
Hence Theorem 9.3.1 and the previous remark apply. □

An analog of this theorem for nondegenerate mappings on general measurable manifolds will be presented in Theorem 11.1.11. Let us give concrete examples to which one can apply general theorems proven above.

9.3.6. Theorem. *Let $d \in \mathbb{N}$ and let a Radon measure μ and a μ-measurable function Q on a locally convex space X be such that, for every n, the space X can be represented as a direct topological sum of its closed linear subspaces $X_0, \ldots, X_n$ and*
$$Q(x) = Q_0(x_0) + Q_1(x_0, x_1) + \cdots + Q_n(x_0, x_n),$$
where x_i is the projection of x on X_i, each Q_i is a measurable function such that, for each $i \geqslant 1$, the function $x_i \mapsto Q_i(x_0, x_i)$ is a polynomial of degree d. Suppose that there exist vectors $h_i \in X_i$, $i = 1, \ldots, n$, such that $Q_i(x_0 + t h_i) = c_{i,d}(x_0) t^d + \cdots$, where $|c_{i,d}(x_0)| \geqslant c > 0$. Assume also that the measure μ is infinitely differentiable along all vectors from the linear span of the vectors h_i and all functions $\partial_{h_1}^{k_1} \cdots \partial_{h_n}^{k_n} Q$ are integrable to every power with respect to the partial derivatives of μ. Then the measure $\mu \circ Q^{-1}$ has a density in the class $\mathcal{S}(\mathbb{R}^1)$.

PROOF. Let us fix $n > 1$. For each $i = 1, \ldots, n$, we take a hyperplane Y_i in X_i complementing $\mathbb{R}^1 h_i$. The direct sum of $X_0, Y_1, \ldots, Y_n$ will be denoted by Z. Then, for all $x \in X$, we have a decomposition $x = z + x_1 h_1 + \cdots + x_n h_n$, where $z \in Z$, $x_i \in \mathbb{R}^1$, i.e., $Q(x) = F_0(z) + F_1(z, t_1) + \cdots + F_n(z, t_n)$, where $F_i(z, t_i)$ for any $i \geqslant 1$ is a polynomial in t_i of degree at most d. We take the field $v = \sum_{i=1}^n \partial_{h_i} Q \cdot h_i$. Then the function $G := \partial_v Q = \sum_{i=1}^n |\partial_{h_i} Q|^2$ has the form $G(x) = \sum_{i=1}^n G_i(z, t_i)$, where $G_i(z, t_i) = a_{i,d}(z) t_i^{2d-2} + \cdots$ is a polynomial of degree $2d-2$ in t_i and $a_{i,d}(z) \geqslant c^2 d^2$. Let $d > 1$. We prove that
$$|\mu|\bigl(x \colon G(x) \leqslant \varepsilon\bigr) \leqslant C(\mu, n, d) c^{n/(1-d)} \varepsilon^{n/(2d-2)},$$
where $C(\mu, n, d)$ is independent of ε. We take conditional measures μ^z on the n-dimensional planes $L_z := z + \mathbb{R}^1 h_1 + \cdots + \mathbb{R}^1 h_n$, $z \in Z$, having the derivatives $d_{h_1} \cdots d_{h_n} \mu^z$. The measure μ^z has a density ϱ^z with respect

to Lebesgue measure on L_z generated by the isomorphism with $\mathbb{R}^n$ under which the vectors $h_1,\ldots,h_n$ correspond to the standard basis vectors. We have $|\varrho^z| \leqslant \|d_{h_1}\cdots d_{h_n}\mu^z\|$. Exercise 9.8.12 and Fubini's theorem yield the estimate
$$|\mu^z|(G\leqslant\varepsilon) \leqslant (2d-2)^n(\varepsilon c^{-2}d^{-2})^{n/(2d-2)}\|d_{h_1}\cdots d_{h_n}\mu^z\|\varepsilon^{n/(2d-2)}.$$
Integrating this estimate with respect to the projection of $|\mu|$ on Z, we find
$$|\mu|(G\leqslant\varepsilon) \leqslant C(\mu,n,d)c^{n/(1-d)}\|d_{h_1}\cdots d_{h_n}\mu\|\varepsilon^{n/(2d-2)}.$$
If $d=1$, then $G\geqslant\varepsilon_0>0$. Thus, $G^{-1}\in L^p(\mu)$ if $n>(2d-2)p$. Theorem 9.3.2 yields that $\mu\circ Q^{-1}$ has a density $\varrho\in W^{q,m}(\mathbb{R}^1)$ for all $q,m\in\mathbb{N}$. Remark 9.3.4 gives $\varrho\in\mathcal{S}(\mathbb{R}^1)$. □

9.3.7. Example. A typical situation where the previous theorem applies is a polynomial of the integral type
$$Q(x) = \int_a^b f(t,x(t))\,\sigma(dt)$$
on the space $X = L^\infty([a,b],\sigma)$, where σ is a Borel measure and a function f has the form $f(t,z) = \sum_{j=0}^d a_j(t)z^j$, where $a_j\in L^1(\sigma)$. Let a measure μ on X be infinitely differentiable along all vectors from some subspace D and along with all its partial derivatives have all moments. Assume also that, for every n, there exist n elements $h_1,\ldots,h_n$ in D with disjoint supports and
$$\int_a^b a_d h_i^d\,d\sigma \neq 0$$
(or D is dense in $L^\infty([a,b],\sigma)$ in the $*$-weak topology). Then $\mu\circ Q^{-1}$ has a density from the class $\mathcal{S}(\mathbb{R}^1)$. For $X_1,\ldots,X_n$ we take the one-dimensional spaces $\mathbb{R}^1 h_i$, where $h_1,\ldots,h_n\in D$ have disjoint supports and for X_0 we take a closed linear subspace topologically complementing the sum of $X_1,\ldots,X_n$ to X. Then
$$x = x_0 + x_1 h_1 + \cdots + x_n h_n, \quad \text{where } x_0\in X_0,\ x_i\in\mathbb{R}^1 \text{ if } i>0.$$
The function $f(t,x(t))$ can be written in the form
$$b_0(t)x_0(t)^d + \sum_{i=1}^n g_i(t,x_0(t),x_i),$$
where each function g_i has the form
$$g_i(t,u,x_i) = \sum_{k=0}^d [g_{i,k,0}(t)+\cdots+g_{i,k,d}(t)u^d]x_i^k, \quad g_{i,k,j}\in L^1(\sigma).$$
In addition, $f(t,x_i h_i(t)) = a_d(t)h_i(t)^d x_i^d + \cdots$ is a polynomial in x_i. Hence the function $Q(x_0 + sh_i) = c_i s^d + \cdots$ is a polynomial in s and $c_i \neq 0$. More general results can be found in Bogachev [**184**].

9.3. Smoothness of induced measures

Let $X = C([0,T], \mathbb{R}^d)$ and let μ be the measure generated by the solution of the stochastic equation
$$d\xi_t = A(t, \xi_t) dw_t + b(t, \xi_t) dt, \quad \xi_0 = x_0,$$
where A and b belong to the class C_b^∞ and A^{-1} is uniformly bounded. First we consider the quadratic form of the process ξ_t defined by
$$Q(\xi) = \int_0^T \int_0^T \bigl(K(s,t)\xi_s, \xi_t\bigr) \, ds \, dt,$$
where an operator-valued mapping $K\colon [0,T]\times[0,T] \to \mathcal{L}(\mathbb{R}^d)$ satisfies Hölder's condition of order α, $K(s,t) = K(t,s) = K(s,t)^*$. On the space X this form has the representation
$$Q(x) = \int_0^T \int_0^T \bigl(K(s,t)x(s), x(t)\bigr) \, ds \, dt.$$
If the diffusion coefficient is not constant, then, as we know from §4.4, the measure μ may have no nonzero vectors of differentiability. Hence we have to consider Q as a functional of the Wiener process, but this functional will not be a quadratic form, which considerably complicates a proof, especially in the part related to the study of the Malliavin matrix. For the proof of the following theorem, see Bogachev [**184**, §4].

9.3.8. Theorem. *Suppose that for every n there exist points $t_1, \ldots, t_n$ in $[0,T]$ such that the mappings $s \mapsto K(s, t_i)$ from $[0,T]$ to $\mathcal{L}(\mathbb{R}^d)$ are linearly independent. Then the measure $\mu \circ Q^{-1}$ has a density from $\mathcal{S}(\mathbb{R}^1)$.*

Now we consider the following functional of the process ξ_t assuming that $x_0 = 0$:
$$F(\xi_t) = \int_0^T f(t, \xi_t) \, dt,$$
where a measurable function f has all derivatives $\partial_x^r f$ such that
$$\|\partial_x^r f(t, x)\| \leqslant c_r \exp(|x|^{\alpha_r}), \quad r = 0, 1, \ldots, \quad \alpha_r < 2.$$
We shall call the function f exponentially nondegenerate with exponent $\beta > 0$ if there exist a neighborhood of the origin $V \subset \mathbb{R}^{d+1}$ and a Borel set Z of finite Hausdorff measure $H_\kappa(Z)$ with some $\kappa < d+1$ such that
$$\{z = (t, x) \in V \colon \partial_x f(t, x) = 0\} \subset Z$$
and for some $c > 0$ one has
$$|\partial_x f(z)| \geqslant c \cdot \exp\bigl(-\operatorname{dist}(z, Z)^{-\beta}\bigr), \quad z = (t, x) \in V.$$
This condition is fulfilled if, for example, the function f is real analytic in V and $\partial_x f \not\equiv 0$ in V. This follows from the Lojasiewicz theorem (see Malgrange [**749**, Ch. IV, §4]), which states that for any real analytic function f on an open set $\Omega \subset \mathbb{R}^n$ and every compact set $K \subset \Omega$ there exist numbers $C > 0$ and $\alpha > 0$ such that
$$|f(x)| \geqslant C \operatorname{dist}\bigl(x, f^{-1}(0)\bigr)^\alpha, \quad x \in K.$$

We shall use the following estimates:
$$P\Big(\sup_{t\in[0,1]} |w_t| \leqslant \varepsilon\Big) \leqslant k_1 \exp(-k_2\varepsilon^{-2}), \quad \text{where } k_1, k_2 > 0;$$

in addition, for some numbers $\kappa_1, \kappa_2 > 0$, for every neighborhood of the origin $D \subset \mathbb{R}^{d+1}$ of volume $|D|$ we have
$$P\big((t, \xi_t) \in D \,\forall t \in [0, \tau]\big) \leqslant \kappa_1 \exp\big(-\kappa_2 \tau^{1+2/d} |D|^{-2/d}\big).$$

The proof of the second (more general) estimate is given in [**184**, §5]. For every ε-neighborhood Z^ε of the set $Z \subset \mathbb{R}^{d+1}$ of finite Hausdorff measure $H_\kappa(Z)$ we have $|Z^\varepsilon| \leqslant \kappa_0 \varepsilon^{d+1-\kappa}$, which gives
$$P\Big(\sup_{t\in[0,\tau]} \text{dist}\,\big((t,\xi_t), Z\big) \leqslant \varepsilon\Big) \leqslant \kappa_3 \exp\big(-\kappa_4 \tau^{1+2/d} \varepsilon^{-2(d+1-\kappa)/d}\big).$$

If f does not depend on t, then we can take for Z the intersection of a ball with $[0,\tau]\times Z_0$, $Z_0 = \{x\colon \nabla f(x) = 0\}$. If Z_0 is contained in a hypersurface, then $\kappa = d$ and $2(d+1-\kappa)/d = 2/d$.

9.3.9. Theorem. *Suppose that the function f is nondegenerate with exponent $\beta < (d+1-\kappa)/(d+1)$. Then the measure $\mu \circ F^{-1}$ has a density from $\mathcal{S}(\mathbb{R}^1)$.*

PROOF. The technical details considerably simplify if $d = 1$, ξ_t is the Wiener process, i.e., $A = 1$ and $b = 0$, f does not depend on t, $Z = \{0\}$, and one has $|f'(z)| \geqslant c\exp(-|z|^{-\beta})$ whenever $z \in [-a, a]$, where $\beta < 2$. It is instructive to first consider this case. Then F is infinitely Fréchet differentiable and
$$\partial_h F(x) = \int_0^T f'(x(t))h(t)\,dt,$$
whence
$$|D_H F(x)| = \sup\Big\{\int_0^T f'(x(t))h(t)\,dt\colon |h|_H \leqslant 1\Big\},$$

where $|h|_H = \|h'\|_{L^2[0,T]}$. The formula for $|D_H F(x)|$ explains the meaning of the term "stochastic variational calculus" which P. Malliavin gave to his calculus. If $f'(0) \neq 0$, then the function $|D_H F(x)|$ is separated from zero. The problem arises only if $f'(0) = 0$. Here the character of degeneracy of f' at the origin comes into play. Let $c = 1$ and $\varepsilon \in (0, a/4)$. With probability at least $1 - M_1 \exp(-M_2\varepsilon^{-2})$ the process $t \mapsto |x(t)|$ achieves the value 2ε. Since x has a modulus of continuity $C(x)t^{1/3}$, where $\mu(C(x) \geqslant R) \leqslant M_3 \exp(-M_4 R^2)$, we see that with probability at least $1 - M_3 \exp(-M_4\varepsilon^{-2})$ there exists an interval $J \subset [0,1]$ of length ε^6 on which $\varepsilon \leqslant |x(t)| \leqslant 3\varepsilon < a$, i.e.,
$$|f'(x(t))| \geqslant \exp(-\varepsilon^{-\beta}).$$

Assuming that $f'(x(t)) \geqslant 0$ for $t \in J$, we take the function h_ε vanishing outside J, equal to 1 on the middle third of J, and linearly extended to the two

9.3. Smoothness of induced measures

remaining intervals. Then $|h_\varepsilon|_H = \sqrt{6}\varepsilon^{-3}$ and $|\partial_h F(x)| \geq 3^{-1}\varepsilon^6 \exp(-\varepsilon^{-\beta})$, whence we obtain

$$|D_H F(x)| \geq 8^{-1}\varepsilon^9 \exp(-\varepsilon^{-\beta}).$$

Since $\beta < 2$, for every n there exists M_n such that

$$\mu(x\colon |D_H F(x)| \geq \delta) \geq 1 - M_n \delta^n,$$

i.e., $|D_H F|^{-1} \in L^p(\mu)$ for all $p < \infty$.

In the general case the idea is similar, but the technical details turn out to be more involved; we shall omit some of them referring to [**184**]. Let $T = c = 1$. Now the function F may fail to be even continuous as a function of w_t, but it belongs to $W^\infty(\gamma)$ (which follows by Theorem 8.2.6) and

$$\partial_h F(w) = \int_0^1 \bigl(\partial_x f(t, \xi_t), y_t\bigr)\, dt,$$

where the process y_t satisfies the equation

$$dy_t = \bigl(A'_x(t, \xi_t)dw_t\bigr)y_t + [b'_x(t, \xi_t)y_t + A(t, \xi_t)h'(t)]dt, \quad y_0 = 0.$$

Let us introduce an operator-valued process Z_t satisfying the equation

$$dZ_t = b'_x(t, \xi_t)Z_t dt + \bigl(A'_x(t, \xi_t)dw_t\bigr)Z_t, \quad Z_0 = I.$$

One can verify that the operators Z_t are invertible, Z_t^{-1} also satisfies some linear stochastic equation, the processes Z_t and Z_t^{-1} have finite Hölder norms of any order $\alpha < 1/2$ belonging to all $L^p(\mathrm{P})$ and that y_t has the form

$$y_t = Z_t \int_0^t Z_s^{-1} A(s, \xi_s) h'(s)\, ds.$$

Thus, similarly to the previous case, we need estimates $\mathrm{P}(\zeta \leq \varepsilon) \leq C_n \varepsilon^n$ for

$$\zeta := \sup_{\|u\|_{L^2[0,1]} \leq 1} \int_0^1 \Bigl(\partial_x f(t, \xi_t), Z_t \int_0^t Z_s^{-1} A(s, \xi_s) u(s)\, ds\Bigr) dt.$$

Let $\eta_t := Z_t^* \partial_x f(t, \xi_t)$, let ω be such that $|\eta_\tau(\omega)| = \varepsilon$ for some τ, and let

$$\sup_t \|A^{-1}(t, \xi_t) Z_t(\omega)\| \leq M, \quad |\eta_t(\omega) - \eta_s(\omega)| \leq M|t-s|^{1/3}.$$

Set $k = 42$ and

$$\varphi(t) = \int_0^1 \eta_{t+s\varepsilon^k}\sigma(s)\, ds,$$

where σ is a smooth probability density with support in $[0, 1]$ and the function η_t is extended to $[1, 2]$ by a constant. We take $u(t) = A^{-1}(t, \xi_t)Z_t\varphi'(t)$. Direct calculations show that

$$|\eta_t - \varphi(t)| \leq M\varepsilon^{14}, \quad |\varphi'(t)| \leq M\|\sigma'\|_{L^1}\varepsilon^{-k}, \quad \|u\|_{L^2} \leq M^2\|\sigma'\|_{L^1}\varepsilon^{-k}.$$

Whenever $t \in [\tau, \tau + \varepsilon^9]$ (if $\tau + \varepsilon^9 > 1$, then we consider $[\tau - \varepsilon^9, \tau]$), by the Hölder continuity we have $|\eta_t| \geqslant \varepsilon - M\varepsilon^3$. Here $\eta_0 = 0$. Hence

$$\int_0^1 \left(\eta_t, \int_0^t Z_s^{-1} A(s,\xi_s) u(s)\, ds\right) dt = \int_0^1 \left(\eta_t, \varphi(t) - \varphi(0)\right) dt$$
$$\geqslant \int_0^1 |\eta_t|^2\, dt - \int_0^1 |\eta_t|\,|\eta_t - \varphi(t) + \varphi(0)|\, dt \geqslant \varepsilon^9(\varepsilon - M\varepsilon^3)^2 - 2M^2 \varepsilon^{14},$$

which is not less than $\varepsilon^{11}/2$ for $M = \varepsilon^{-1}$ and sufficiently small ε. Therefore,

$$\zeta(\omega) \geqslant (2\|\sigma'\|_{L^1})^{-1} \varepsilon^{55}.$$

In addition, $\mathrm{P}\bigl(\sup_t \|Z_t\| \geqslant \varepsilon^{-1}\bigr)$ and $\mathrm{P}\bigl(\sup_{t,s} |t-s|^{-1/2} |\eta_t - \eta_s| \geqslant \varepsilon^{-1}\bigr)$ are $o(\varepsilon^n)$ for all n. Now it remains to verify that

$$\mathrm{P}\bigl(\sup_t |\eta_t| < \varepsilon\bigr) \leqslant M_n \varepsilon^n \quad \text{for all } n.$$

Since $\mathrm{P}(\sup_t \|Z_t\| \geqslant R) \leqslant K_n R^{-n}$ for all n, it remains to use the inequality $|\partial_x f(t, \xi_t)| \geqslant \varepsilon$, which holds if $\mathrm{dist}\bigl((t, \xi_t), Z\bigr) \geqslant |\ln \varepsilon|^{-1/\beta}$, and apply the estimate

$$\mathrm{P}\Bigl(\sup_t \mathrm{dist}\bigl((t, \xi_t), Z\bigr) \leqslant |\ln \varepsilon|^{-1/\beta}\Bigr) \leqslant \kappa_3 \exp\bigl(-\kappa_4 |\ln \varepsilon|^q\bigr),$$

where $q := 2(d+1-\kappa)/(d\beta) > 1$, which shows that the right-hand side is $o(\varepsilon^n)$ for all n. □

Smoothness of the induced measures for functionals of the Wiener process w_t in $\mathbb{R}^n$ of the form

$$f(w) = \int_0^1 \int_0^1 \Psi(w_t - w_s)\, dt\, ds$$

was established in Gaveau, Moulinier [**469**] under the assumption that Ψ is infinitely differentiable with derivatives of at most exponential growth and $\Psi'(0) \neq 0$.

In Gaveau, Moulinier [**469**] and Moulinier [**818**], functionals of the form

$$f(w) = \int_0^T A(w_t) \cdot dw_t$$

were studied, where A is an infinitely differentiable vector field on $\mathbb{R}^3$ with derivatives of at most exponential growth and w_t is the Wiener process in $\mathbb{R}^3$. Assuming that $\mathrm{div}\, A = 0$ and $\mathrm{curl}\, A$ is sufficiently nondegenerate, it was shown that the measure induced by f has a smooth density. Let us observe that the usual nondegeneracy of A is not enough here. For example, if $Ax = x$, then $f(w) = |w_T|^2/2 + c$, so the corresponding density is not infinitely differentiable.

Concerning the distributions of norms of Gaussian vectors, see Bogachev [**191**, §6.11], Davydov, Lifshits, Smorodina [**336**, §12], Paulauskas, Račkauskas [**876**]. The distributions of norms are considered also in Theorem 9.5.7

below. Some additional information about the distributions of quadratic forms can be found in the next section and Exercises 9.8.11–9.8.10.

We close this section with the following observation which shows that nondegenerate mappings are everywhere dense.

9.3.10. Proposition. *Let μ be a Radon probability measure on a locally convex space X infinitely differentiable along all vectors from a densely embedded infinite dimensional Hilbert space H. Suppose that every element of X^* belongs to all L^p with $p < \infty$ with respect to μ and all measures $d_{h_1}\cdots d_{h_m}\mu$, where $h_i \in H$. Then, given $p < \infty$ and $r \in \mathbb{N}$, for any $F = (F_1, \ldots, F_d) \in W^{p,r}(\mu, \mathbb{R}^d)$ and any $\varepsilon > 0$, there is a mapping $G = (G_1, \ldots, G_d) \in W^{p,r}(\mu, \mathbb{R}^d)$ such that*

$$\|F - G\|_{p,r} \leqslant \varepsilon, \quad \inf_{x \in X} \det\Big(\big(D_H G_i(x), D_H G_j(x)\big)\Big)_{i,j \leqslant d} > 0.$$

PROOF. It suffices to prove our assertions for mappings F that depend on finitely many functionals. Let $F = \varphi(l_1, \ldots, l_n)$, $\varphi \in C_b^\infty(\mathbb{R}^n)$, $l_i \in X^*$, $l_i = j_H(h_i)$, where $h_i \in H$, $(h_i, h_j)_H = \delta_{ij}$. We can find functionals $l_{n+1}, \ldots, l_{n+d} \in X^*$ such that $l_k = j_H(h_k)$, $h_k \in H$, $(h_k, h_j) = \delta_{kj}$. Let $G_i := F_i + \varepsilon_0 l_{n+i}$, where $\varepsilon_0 > 0$. Then $D_H G_i = D_H F_i + \varepsilon_0 h_{n+i}$, where $D_H F_i(x) \perp h_k$ whenever $i \leqslant n$ and $k > n$. If $\varepsilon_0 > 0$ is sufficiently small, we obtain

$$\|F - G\|_{p,r} \leqslant \varepsilon, \quad (D_H G_i, D_H G_j) = (D_H F_i, D_H F_j) + \varepsilon_0^2 \delta_{ij}$$

and the matrix with the entries $(D_H G_i, D_H G_j)$ majorizes $\varepsilon_0^2 \cdot \text{Id}$. □

9.4. Infinite dimensional oscillatory integrals

An efficient method of the study of induced measures is based on infinite dimensional oscillatory integrals. The idea of the method is this. Suppose we are given a smooth mapping $F = (f_1, \ldots, f_n) \colon X \to \mathbb{R}^n$. The regularity of the measure on $\mathbb{R}^n$ induced by F will follow from the estimates

$$|J(t_1, \ldots, t_n)| \leqslant c_k |t|^{-k}, \quad |t|^2 = \sum_{j=1}^n t_j^2,$$

at infinity, where

$$J(t_1, \ldots, t_n) := \int_X \exp\Big(i \sum_{j=1}^n t_j f_j(x)\Big) \mu(dx).$$

It is clear that $J(t_1, \ldots, t_n) = I_\alpha(t)$, where α is the unit vector proportional to $\tau = (t_1, \ldots, t_n)$, $t = |\tau|$, and I_α is the Fourier transform of the measure on $\mathbb{R}^1$ induced by the real function $F_\alpha = (F, \alpha)$. Therefore, our problem reduces to the study of a family of scalar functions dependent on a parameter α that is an element of the unit sphere. A practical realization of this scheme can be found in Bogachev [184]. Here we discuss only the joint distributions of quadratic forms and polynomials.

We shall say that a quadratic form Q on a locally convex space E is infinite dimensional if, for every n, there exists a finite dimensional subspace in E on which one has the inequality $\operatorname{Rank}(Q) \geqslant n$.

For example, a continuous quadratic form Q on a Hilbert space H is infinite dimensional if and only if $Q(x) = (Ax, x)$, where $A \in \mathcal{L}(H)$ is a symmetric operator such that $\dim A(H) = \infty$.

As explained above, the main step in the proof of smoothness of distributions in the Malliavin method is the verification of integrability of the inverse Malliavin matrix. In the case of quadratic forms, the next lemma is useful.

9.4.1. Lemma. *Let ν be a Radon measure on a locally convex space X and let G be a ν-measurable nonnegative quadratic form on X that is positive definite on an n-dimensional subspace $L = \operatorname{span}(a_1, \ldots, a_n)$. If there exists the Skorohod derivative $d_{a_1} \cdots d_{a_n} \nu$, then*
$$|\nu|\big(x\colon\ G(x) \leqslant \varepsilon\big) \leqslant M_G^{-n/2} \varepsilon^{n/2} \|d_{a_1} \cdots d_{a_n} \nu\|,$$
where $M_G = \min\{G(a)\colon a = \sum_{i=1}^n t_i a_i,\ \sum_{i=1}^n t_i^2 = 1\}$.

PROOF. We can take conditional measures ν^y on $y + L$, $y \in Y$, where Y is a closed linear subspace topologically complementing L to X, which have the Skorohod derivatives $d_{a_1} \cdots d_{a_n} \nu^y$. It suffices to obtain the estimate
$$|\nu^y|\big(x\colon\ G(x) \leqslant \varepsilon\big) \leqslant M_G^{-n/2} \varepsilon^{n/2} \|d_{a_1} \cdots d_{a_n} \nu^y\|.$$
The density of the measure ν^y on $y + L$ with respect to Lebesgue measure λ induced by the isomorphism $y + x_1 a_1 + \cdots + x_n a_n \mapsto (x_1, \ldots, x_n)$ between $y + L$ and $\mathbb{R}^n$ is majorized by $\|d_{a_1} \cdots d_{a_n} \nu^y\|$. In addition, the quantity $\lambda\big(x \in y + L\colon G(x) \leqslant \varepsilon\big)$ does not exceed the Lebesgue volume of the ellipsoid $\{x \in L\colon G(x) \leqslant \varepsilon\}$, which is estimated by $M_G^{-n/2} \varepsilon^{n/2}$. □

9.4.2. Theorem. *Suppose that a Radon measure μ on a locally convex space X is infinitely differentiable along all vectors in a linear subspace D, $Q_1, \ldots, Q_d$ are μ-measurable quadratic forms, and for all $(\alpha_1, \ldots, \alpha_d)$ from the unit sphere in $\mathbb{R}^d$ the form $\alpha_1 Q_1 + \cdots + \alpha_d Q_d$ is infinite dimensional on D. If $\partial_{a_1} \cdots \partial_{a_k} Q$ belongs to $L^p(d_{a_1} \cdots d_{a_m} \mu)$ for all $k, m, p \in \mathbb{N}$, $a_i \in D$, then the measure $\mu \circ Q^{-1}$, where $Q = (Q_1, \ldots, Q_d)$, has a density in $\mathcal{S}(\mathbb{R}^d)$. In particular, this is true if μ is a Gaussian measure and all nontrivial linear combinations of the forms Q_i are infinite dimensional on $H(\mu)$.*

PROOF. It is clear from what has been said above that it suffices to show that, for every $p > 1$, every point $\alpha = (\alpha_1, \ldots, \alpha_d)$ in the unit sphere has a neighborhood U_α such that the functions $|D_H f_\beta|^{-2}$ with $\beta \in U_\alpha$, where $f_\beta := \beta_1 F_1 + \cdots + \beta_d F_d$, are uniformly bounded in $L^p(\mu)$. This is done by means of the previous lemma. Indeed, for every finite dimensional linear subspace $L \subset H$ we have
$$|D_H f_\beta(x)|^2 \geqslant \sup_{h \in U \cap L} |\partial_h f_\beta(x)|^2,$$

where U is the unit ball in H. On the right we have a quadratic form which is positive definite on L, provided that the form f_β is positive or negative definite on L. The latter is seen from the following observation. If Q_0 is a positive definite quadratic form on $\mathbb{R}^n$, then $|\nabla Q_0(x)|^2 \geqslant 4\alpha^2 Q_0(x)$, where $\alpha = \min_{|x|=1} Q_0(x)$. For justification it suffices to write Q_0 in its eigenbasis. The local uniformity in β follows from the fact that if the form f_β is positive definite on L, then the same is true for β' close to β, and the minimal eigenvalue is separated from zero locally uniformly in β from the unit sphere in $\mathbb{R}^d$. □

A more precise result specifying the order of differentiability of the induced density in terms of the rank of linear combinations of these forms and the order of differentiability of μ is given in Bogachev, Smolyanov [233, §4].

9.4.3. Theorem. *Suppose that $X = C[a,b]$ or $X = C_0[a,b]$ and that $F = (F_1, \ldots, F_n)$, where*

$$F_i(x) = \int_a^b f_i(t, x(t))\, dt, \quad f_i(t, z) = \sum_{j=0}^d a_{i,j}(t) z^j, \quad a_{i,j} \in L^1[a,b].$$

Suppose that γ is a Gaussian measure on X, its Cameron–Martin space $H(\gamma)$ is dense in X and whenever $(\alpha_1, \ldots, \alpha_n) \neq 0$ at least one of the coefficients $\alpha_1 a_{1,j} + \cdots + \alpha_n a_{n,j}$ with $j \geqslant 1$ is not zero. Then $\gamma \circ F^{-1}$ has a density from $\mathcal{S}(\mathbb{R}^n)$.

The proof is given in Bogachev [184]. For the functionals considered in this theorem, exponential estimates of the type $J(t) \leqslant C \exp(-k|t|^\kappa)$, $t \to \infty$, are obtained in [184]. Such estimates yield the membership in certain Gevrey classes. However, in such examples, one cannot expect to have the real analyticity of distributions even for real analytic functionals f since, for example, for bounded f the induced measure has bounded support. In [184] similar results are obtained for joint distributions of functionals of other types mentioned in §9.3, for example, for functionals on $L^2[0,T]$ of the type

$$F(x) = \int_0^T f(t, x(t))\, \sigma(dt),$$

where σ is a measure on $[0,T]$, provided that $L^2[0,T]$ is equipped with a measure μ which is either differentiable or corresponds to a nondegenerate diffusion as in Theorem 9.3.8.

9.5. Surface measures

In the infinite dimensional case, there are several different natural possibilities to define surface measures. Here we briefly describe an approach combining the Malliavin calculus with a scheme suggested in Bogachev [181] and developed in Pugachev [928]–[930], [931], [935], where one can find the proofs.

Suppose first that a probability measure μ on X has compact support S_μ. Let $F \in G^{1,1}(\mu)$ and let v be a Borel vector field on X along which μ is differentiable and $\partial_v F > 0$ on S_μ. For every function $\varphi \in C_0^\infty(\mathbb{R}^1)$ we have

$$\int_X \varphi'(F)\,\partial_v F\,\mu(dx) = \int_X \partial_v(\varphi \circ F)\,\mu(dx)$$
$$= -\int_X \varphi(F)\,d_v\mu = -\int_X \varphi(F)\,\beta_v^\mu\,\mu(dx).$$

Therefore, the derivative of the measure $(\partial_v F \cdot \mu) \circ F^{-1}$ in the sense of generalized functions is the measure $(\beta_v^\mu \cdot \mu) \circ F^{-1}$. Hence $(\partial_v F \cdot \mu) \circ F^{-1}$ has an absolutely continuous density k whose variation does not exceed the norm $\|\beta_v^\mu\|_{L^1(\mu)}$. An analogous fact is true also for the measures $g \cdot \mu$, where $g \in \mathcal{FC}_b^\infty(X)$. The density of $(\partial_v F g \cdot \mu) \circ F^{-1}$ will be denoted by k_g. We have $|k_g(t)| \leqslant k(t) \sup_x |g(x)|$. Now we define a surface measure $\mu^{0,v}$ by the formula

$$\int g(x)\,\mu^{0,v}(dx) := k_g(0), \quad g \in \mathcal{FC}_b^\infty(X).$$

The obtained linear functional on $\mathcal{FC}_b^\infty(X)$ is generated by the Radon measure on S_μ which is a weak limit of the sequence of nonnegative measures $2n I_{|F| \leqslant 1/n} \partial_v F \cdot \mu$. This sequence is concentrated on the compact set S_μ, is uniformly bounded (since its values on 1 tend to $k_1(0)$), and converges on the class $\mathcal{FC}_b^\infty$, which is dense in $C(S_\mu)$. One can omit the assumption of compactness of support of μ under the following condition (fulfilled under very broad assumptions, in particular, for Radon measures on Fréchet spaces): there exist functions $\theta_j \colon X \to [0, 1]$ increasing to 1 such that the sets $S_j = \{\theta_j > 0\}$ are compact, $\mu(X \backslash S_j) \to 0$, $d_v(\theta_j \cdot \mu) = \partial_v \theta_j \cdot \mu + \theta_j \beta_v^\mu \cdot \mu$, $\|\partial_v \theta_j\|_{L^1(\mu)} \to 0$. In the case of a Fréchet space there is a separable reflexive Banach space E with $\mu(E) = 1$ compactly embedded into X (see §1.2), and for θ_j one can take uniformly Lipschitzian functions on E equal to 1 on the ball of radius j and equal to 0 outside the ball of radius $j+1$. The sequence of increasing measures $(\theta_j \cdot \mu)^{0,v}$ converges in variation since

$$\|(\theta_j \cdot \mu)^{0,v} - (\theta_k \cdot \mu)^{0,v}\| \leqslant \|\partial_v \theta_j - \partial_v \theta_k\|_{L^1(\mu)} + \|(\theta_j - \theta_k)\beta_v^\mu\|_{L^1(\mu)}.$$

Under reasonable assumptions, the measure $\mu^{v,0}$ is concentrated on $F^{-1}(0)$. This is readily seen from our construction if F is continuous. In a more general case this can be shown for a quasicontinuous version of F. Analogous constructions lead to surface measures $\mu^{y,v}$ on the sets $F^{-1}(y)$ (if F is continuous or quasicontinuous). One has the equality

$$\int_X u(F(x))g(x)\partial_v F(x)\,\mu(dx) = \int_{\mathbb{R}^1} u(y) \int_X g(x)\,\mu^{y,v}(dx)\,dy$$

for all bounded Borel functions u on $\mathbb{R}^1$ and $g \in C_b(X)$. This equality shows that the measures $k_1(y)^{-1}\mu^{y,v}$ serve as conditional measures for $\partial_v F \cdot \mu$ if $k_1(y) > 0$.

9.5. Surface measures

Certainly, the described construction depends on our choice of the vector field v. With its aid one can obtain diverse known constructions of surface measures. The field $v = D_H F/|D_H F|_H$ leads to the standard construction in the theory of Gaussian measures (see [**191**, §6.7]). One can also take $v = D_H F$, but then one should replace $\mu^{0,v}$ by the measure $\psi^* \cdot \mu^{0,v}$, where $\psi = |D_H F|_H^{-1}$ and ψ^* is a quasicontinuous version. Such surface measures for a Gaussian measure γ will be denoted by γ^{S_y}. For these measures one has the Stokes formula (see [**191**, §6.7]):

$$\int_{F \leq y} \delta u(x)\, \gamma(dx) = -\int_{F^{-1}(y)} (u(x), \mathrm{n}(x))\, \gamma^{S_y}(dx),$$

where $\mathrm{n} := D_H F/|D_H F|_H$.

Let us describe another approach suggested in [**181**]. Let X be a Banach space, let μ be a Radon measure on X, and let F be a measurable function on X having almost everywhere the Gâteaux derivative F' which is locally integrable. Set $S := \{x \colon F(x) = 0\}$.

9.5.1. Definition. *A locally finite measure σ on S is called a local surface measure for μ if every point $s \in S$ has a neighborhood V such that the measures*

$$\eta_t(B) := (2t)^{-1} \int_{B \cap V \cap \{-t < F < t\}} \|F'(x)\|\, \mu(dx)$$

converge weakly to $\sigma(\cdot \cap V)$ as $t \to 0$. We shall call σ the surface measure for μ and denote it by μ^S if it has bounded variation. If $\{S_n\}$ is a countable union of disjoint surfaces of the considered type, then we define the surface measure μ^S on $S = \bigcup_n S_n$ as the sum of the series $\sum_n \mu^{S_n}$ if the latter converges weakly.

It is not difficult to verify that for any continuous μ-integrable function f one has $(f \cdot \mu)^S = f \cdot \mu^S$.

9.5.2. Proposition. *Let $D(\mu)$ be dense in X and let the function F have a locally bounded Gâteaux derivative that does not vanish on the surface S. Assume also that the partial derivatives $\partial_h F$, where $h \in D(\mu)$, are continuous. Then the local surface measure on S exists.*

The next theorem relates surface integrals with integrals over the space.

9.5.3. Theorem. *Let V be an open set with boundary S that is locally of the form $F^{-1}(0)$, where the function F has a continuous derivative which does not vanish on S. Let $\mathrm{n} := \mathrm{n}_S := F'/\|F\|$ be a normal to S. Suppose that $D(\mu)$ is dense in X, the measure μ^S is finite, $\{e_n\} \subset D(\mu)$ is a finite or countable sequence, φ is a measurable function twice differentiable along the vectors e_i, $\partial_{e_i}^2 \varphi \in L^1(\mu)$ for all i, the series $G := \sum_{i=1}^\infty [\partial_{e_i}^2 \varphi + \partial_{e_i} \varphi \beta_{e_i}]$ converges in $L^1(\mu)$, and the series $\partial_\mathrm{n} \varphi := \sum_{i=1}^\infty \partial_{e_i}\varphi \langle \mathrm{n}, e_i \rangle$ converges in $L^1(\mu^S)$. Then*

$$\int_S \partial_\mathrm{n} \varphi(s)\, \mu^S(ds) = \int_V G(x)\, \mu(dx).$$

PROOF. It is enough to prove the equalities
$$\int_S \partial_{e_i}\varphi(s)\langle \mathrm{n}(s), e_i\rangle\, \mu^S(ds) = \int_V [\partial^2_{e_i}\varphi(x) + \partial_{e_i}\varphi(x)\beta^\mu_{e_i}(x)]\, \mu(dx).$$
Letting $\nu := \partial_{e_i}\varphi \cdot \mu$, we rewrite this as
$$\int_S \langle \mathrm{n}(s), e_i\rangle\, \nu^S(ds) = d_{e_i}\nu(V).$$
We can assume that ν has compact support since it can be approximated in variation by measures $\nu_j = \varphi_j \cdot \nu$, where φ_j is a Borel function with compact support differentiable along all vectors in $D(\nu)$. Moreover, it suffices to consider the case where $\langle \mathrm{n}, e_i\rangle \neq 0$ on the support of ν (if $\langle \mathrm{n}(s), e_i\rangle = 0$ at some s, then one can find a sequence of vectors a_j convergent to e_i in $D(\nu)$ such that $\langle \mathrm{n}(s), a_j\rangle \neq 0$). By using a suitable partition of unity, this reduces everything to the case where in a neighborhood W of the support of ν the set V has the form $V = \{x = z + se_i : z \in B, s < f(z)\}$, where B is an open ball in the closed hyperplane Z complementing $\mathbb{R}^1 e_i$, $\|e_i\| = 1$, f is locally Lipschitzian on Z with continuous $\partial_{e_i}f$, $f|_B > 0$, and $S \cap W$ is the graph of f. Here we have $S \cap W = \{F = 0\} \cap W$, where $F(z, s) = s - f(z)$, so $\langle \mathrm{n}, e_i\rangle = \partial_{e_i}F/\|F'\| = 1/\|F'\|$. Let
$$V_t := \{x = z + se_i : z \in B, f(z) < s < f(z) + t\}.$$
Then we obtain
$$\int_S \langle \mathrm{n}(s), e_i\rangle\, \nu^S(ds) = \lim_{t \to 0+} \frac{\nu(V_t)}{t} = \lim_{t \to 0+} \frac{\nu(V + te_i) - \nu(V)}{t} = d_{e_i}\nu(V)$$
due to our assumptions on the support of ν. □

9.5.4. Corollary. *Suppose that the condition of the previous theorem is fulfilled for the sets $V + ra$, $0 < r < t$, where $a \in X$ is fixed. Then, for such t we have*
$$\mu(V + ta) - \mu(V) = \int_0^t \int_{S+ra} \langle \mathrm{n}_{S+ra}(x), a\rangle\, \mu^{S+ra}(dx)\, dr.$$

9.5.5. Remark. Replacing $\|F'\|$ by $|D_H F|_H$, we obtain surface measures defined at the beginning of this section. A geometric difference between these two types of surface measures is that in the case of $\|F'\|$ (at least, for sufficiently regular surfaces) the surface measure is obtained as a limit of the measure of the ε-neighborhood of S generated by the normal n divided by ε. In the case of $|D_H F|_H$, which is standard for Gaussian measures, the ε-neighborhood is constructed as $S + \varepsilon U_H$, where U_H is the unit ball in $H(\gamma)$. Obviously, one approach corresponds to the geometry of the Banach space X, whereas the other one is related only to the geometry of $H(\gamma)$ and does not depend on the geometry of X at all (in particular, it applies to locally convex spaces). This modification is particularly useful for defining surface measures on level sets of functions from Sobolev spaces over infinite dimensional Gaussian (or other smooth) measures since such functions need not be even continuous. Similarly one can define surface measures on surfaces of codimension $n > 1$ (see Airault, Malliavin [20]). The construction is this. Let $F\colon X \to \mathbb{R}^n$ satisfy the conditions of Theorem 9.3.5. Then, by this theorem, the measure $\gamma \circ F^{-1}$ has an infinitely differentiable density k. In addition, on the sets $S_y = F^{-1}(y)$ there exist conditional measures $\gamma(\cdot|y)$. On S_y we have surface measures $a^{F,y}$, for which

$$\int_S f(x)\, a^{F,y}(dx) = k(y) \int_{S_y} f(x)\sqrt{\Delta(x)}\,\gamma(dx|y).$$

Then the following equality holds:

$$\int_S f(x) a^{F,0}(dx) = \lim_{\varepsilon \to 0} \frac{1}{\lambda_n(B(0,\varepsilon))} \int_X f(x)\sqrt{\Delta(x)} I_{|F| \leqslant \varepsilon}(x)\, \gamma(dx),$$

where $\lambda_n(B(0,\varepsilon))$ is the volume of the ball of radius ε in $\mathbb{R}^n$.

We now apply these results to the distribution of norm. Let us recall that the modulus of convexity of a norm q on a Banach space X is defined by the formula

$$\delta_q(\varepsilon) = \inf\bigl(1 - q(x+y)/2\colon q(x), q(y) \leqslant 1,\ q(x-y) \geqslant \varepsilon\bigr).$$

The norm q is said to be uniformly convex of order α if $\delta_q(\varepsilon) \geqslant c\varepsilon^\alpha$ for some $c > 0$ (in that case $\alpha \geqslant 2$ if $\dim X > 1$). Each super-reflexive Banach space admits an equivalent norm with the modulus of convexity $\delta_q(\varepsilon) \sim \varepsilon^\alpha$, where $\alpha \geqslant 2$ (see Deville, Godefroy, Zizler [345, Theorem 4.8, p. 154]).

9.5.6. Lemma. *The following assertions are equivalent:*
(i) *a norm q is uniformly convex of the order α;*
(ii) *there exists $C > 0$ such that $q'_-(b+te) \geqslant Ct^{\alpha-1}$ for all $t \in (0,1)$ and all vectors b, e such that $q(b) = 1$, $q(e) = 1$, $p'_-(b)(e) \geqslant 0$, where q'_- denotes the lower derivative of the function $t \mapsto q(b+te)$.*

PROOF. Let (i) be fulfilled. Note that the function $t \mapsto q(x+ta)$ has a continuous derivative at all points excepting some at most countable set. Assume that for some b, e we have $q'_-(b)(e) \geqslant 0$ but $q'_-(b+te)(e) \leqslant dt_1^{\alpha-1}$, $d = C\alpha/3$ for some $t_1 \in (0, 1/2)$. By the above argument we can assume that $q'_-(b+te) \leqslant 2dt^{\alpha-1}$ for all t from some interval $(t_2 - r, t_2 + r)$, where $r < 1/4$. We can also assume that $q'_-(b+te) \geqslant 0$ for all $t \in [t_2, t_2+r]$ since the function $t \mapsto q'_-(b+te)$ has only one zero. Replacing b by $b + te$ we arrive at the case $t_2 = 0$. Then $q(b+te) \geqslant 1$ if $t \in [0, r]$. Hence there is $s \in (0, 1]$ with $q(sb + re) = 1$. It is clear that $s > 3/4$. Taking into account that $q'_-(sb+te) = q'_-(b+te/s)$ due to the equality

$$h^{-1}[q(sb+te+he) - q(sb+te)] = (h/s)^{-1}[q(b+te/s+he/s) - q(b+te/s)],$$

we find

$$1 - q(sb + re/2) = \int_0^{r/2} q'_-(sb+te)\,dt = \int_0^{r/2} q'_-(b+te/s)\,dt$$

$$\leqslant 2d \int_0^{r/2} s^{1-\alpha} t^{\alpha-1}\,dt = 2ds^{1-\alpha}\alpha^{-1}(r/2)^\alpha \leqslant 2d\alpha^{-1}r^\alpha < cr^\alpha.$$

On the other hand, $1 - q(sb+re/2) \geqslant Cr^\alpha$ since $sb+re/2 = (sb+sb+re)/2$, $q(sb+re) = 1$, $q(sb) \leqslant 1$. This contradiction proves that (ii) is fulfilled.

Now assume that (ii) is true. Suppose that $q(x) = q(y) = 1$, $q(x-y) = \varepsilon < 1/4$, $e = (x-y)/q(x-y)$. Denote by b the point of minimal q-norm in the interval connecting the points x and y. Then we obtain $q'_-(b) = 0$. We observe that $q'_-(b/q(b)) = q'_-(b) = 0$ and $q(b) > 1 - \varepsilon > 3/4$. Hence, if $|t| \leqslant 1$, then

$$q'_-(b+te) = q(b)q'_-(b/q(b) + te/q(b)) \geqslant q(b)C|t/q(b)|^{\alpha-1} = Cq(b)^{2-\alpha}|t|^{\alpha-1}.$$

Hence, assuming that $b \in [(x+y)/2, y]$ and letting $\delta = q(b - (x+y)/2)$, we obtain

$$1 - q((x+y)/2) \geqslant \int_\delta^{\delta+\varepsilon/2} q'_-(b-te)\,dt \geqslant Cq(b)^{2-\alpha}\alpha^{-1}(\varepsilon/2)^\alpha,$$

which completes the proof. $\square$

9.5.7. Theorem. *Let X be a Banach space such that its norm q has k Lipschitzian Fréchet derivatives on the unit sphere and satisfies the condition*

$$\delta_q(\varepsilon) \geqslant C\varepsilon^\alpha,$$

where $C, \alpha > 0$. If a measure μ is infinitely differentiable along all vectors from some infinite dimensional linear subspace D and all its partial derivatives along vectors in D have finite moments of all orders, then the function $M: t \mapsto \mu(x: q(x) < t)$ is k-fold differentiable, the function $M^{(k)}$ is absolutely continuous on $\mathbb{R}^1$, and the function $Q: x \mapsto \mu(U+x)$ has k continuous Fréchet derivatives, where U is a ball in X. Moreover, the mapping $(t,x) \mapsto \mu(tU+x)$ has k continuous Fréchet derivatives.

PROOF. For the proof of the first assertion it is sufficient to construct a vector field $v: X \to X$ along which the function q and the measure μ are k times differentiable with $|\partial_v q|^{-p} \in L^1(\nu)$ for any $\nu = \partial_v^l q \cdot d_v^r \mu$, $p \in \mathbb{N}$, $l, r \leqslant k$. Let

$$v(x) = \sum_{i=1}^n \partial_{e_i} q(x) e_i,$$

where vectors $e_i \in D$ are linearly independent, $q(e_i) = 1$, and n will be chosen later. All differentiability conditions are fulfilled, so we only need to verify the inclusion $|\partial_v q|^{-p} \in L^1(\nu)$. To this end, it suffices to obtain

9.5. Surface measures

estimates

$$|\nu|(x\colon G(x)\leqslant\varepsilon)\leqslant\beta_r\varepsilon^r, \quad G:=\partial_v q=\sum_{i=1}^n(\partial_{e_i}q)^2.$$

Without loss of generality, we may assume that $\alpha\geqslant 1$. Denote by Y the topological complement to $L=\mathrm{span}(e_1,\ldots,e_n)$ in X and take smooth conditional measures ν^y on the subspaces $y+L$, $y\in Y$. Let μ_Y be the projection of $|\mu|$ on Y. The affine spaces $y+L$ are equipped with Lebesgue measures transported from L, where L is identified with $\mathbb{R}^n$ by identifying $e_1,\ldots,e_n$ with the standard basis in $\mathbb{R}^n$. Denote by $m(y)$ the point in $y+L$ where q attains its minimum. Note that there is $\beta>0$ such that, whenever $y\in Y\setminus\{0\}$, for all $z\in L$ we have

$$G(z+m(y))\geqslant\beta\bigl(|z|/q(m(y))\bigr)^{2\alpha-2},$$

where $z\mapsto|z|$ is the standard norm on $L=\mathbb{R}^n$. Indeed, denoting by ∇ the gradient along L with its Euclidean norm, we have $(\nabla q(m(y)+e),e)\geqslant 0$, hence the inequality $\bigl(\nabla q\bigl(m(y)/q(m(y))+e\bigr),e\bigr)\geqslant 0$ holds. If $|t|\leqslant 1$, this gives the estimate $\bigl(q\bigl(m(y)/q(m(y))+te\bigr),e\bigr)\geqslant C|t|^{\alpha-1}$. Therefore,

$$\bigl(\nabla q\bigl(m(y)+tq(m(y))e\bigr),e\bigr)\geqslant C|t|^{\alpha-1}, \quad |t|\leqslant 1.$$

which yields the estimate

$$\bigl(\nabla q(m(y)+se),e\bigr)\geqslant Cq(m(y))^{1-\alpha}|s|^{\alpha-1}, \quad |s|\leqslant q(m(y)).$$

Since the norm q on L is equivalent to the Euclidean norm, we obtain

$$G(m(y)+se)=\bigl|\nabla q(m(y)+se)\bigr|^2\geqslant C_1 q(m(y))^{2-2\alpha}|s|^{2\alpha-2}, \quad |s|\leqslant q(m(y)).$$

If $|s|\geqslant q(m(y))$ we have $\bigl|\nabla q(m(y)+se)\bigr|^2\geqslant C_1$. By the closed graph theorem there is $C_2>0$ such that $q(z+y)\geqslant C_2 q(z)+C_2 q(y)$ for all $z\in L$, $y\in Y$. Since $m(y)=y+z$ for some $z\in L$, we obtain $q(m(y))\geqslant C_2 q(y)$. In addition, we have $q(m(y))\leqslant q(y)$. Since $\alpha>1$, we arrive at the estimate

$$G(m(y)+z)\geqslant C_3 q(y)^{2-2\alpha}|z|^{2\alpha-2}, \quad |z|\leqslant C_2^{-1}.$$

Therefore, whenever $\varepsilon<C_2^{-1}$, we have

$$\{z\in L\colon G(y+z)\leqslant\varepsilon\}\subset\Bigl\{z\in L\colon q(z-m(y))\leqslant C_3^{1/(2-2\alpha)}\varepsilon^{1/(2\alpha-2)}q(y)\Bigr\}.$$

We know that the density ϱ^y of the measure ν^y admits the following estimate (see Theorem 3.4.7): $|\varrho^y|\leqslant\|d_{e_1}\cdots d_{e_n}\nu^y\|$. Therefore,

$$|\nu^y|\bigl(\{x\in L+y\colon G(x)\leqslant\varepsilon\}\bigr)\leqslant C_4\varepsilon^{n/(2\alpha-2)}q(y)^n\|d_{e_1}\cdots d_{e_n}\nu^y\|,$$

whence integrating in y with respect to μ_Y we find

$$|\mu|(\{x\colon G(x)\leqslant\varepsilon\})\leqslant C_4\varepsilon^{n/(2\alpha-2)}\int_Y q(y)^n\|d_{e_1}\cdots d_{e_n}\nu^y\|\,\mu_Y(dy)$$
$$\leqslant C_5\varepsilon^{n/(2\alpha-2)},$$

where C_5 may depend on n, but is independent of ε. This estimate ensures the desired integrability of G^{-1}. For example, to obtain the inclusion $G^{-1}\in L^p(\mu)$ it suffices to take $n>p(2\alpha-2)$. Similarly, if $m\leqslant k$ and $a_1,\ldots,a_m\in\{e_1,\ldots,e_n\}$, one has $G^{-1}\in L^p(d_{a_1}\cdots d_{a_m}\mu)$. In addition, our hypotheses yield that we have $\partial_v^l q\in L^p(d_{a_1}\cdots d_{a_m}\mu)$ for all $p<\infty$.

The last assertion of the theorem follows from the first one combined with Theorem 9.5.3 and Corollary 9.5.4. Indeed, letting $U:=\{q\leqslant 1\}$ and $S:=\{q=1\}$, applying Theorem 9.5.3 and its corollary to $\varphi=q$ and the finite collection $e_1,\ldots,e_n$ and taking into account that

$$\mathrm{n}=q'/\|q'\|,\quad \langle\mathrm{n},a\rangle=\partial_a q/\|q'\|,\quad \partial_\mathrm{n} q=\sum_{i=1}^n \partial_{e_i}q\langle\mathrm{n},e_i\rangle=\sum_{i=1}^n |\partial_{e_i}q|^2/\|q'\|,$$

we see that the first derivative of the function $t\mapsto\mu(U+ta)$ equals

$$\int_{S+ta}\langle\mathrm{n}_{S+ta},a\rangle\,\mu^{S+ta}(dx)=\int_{S+ta}\partial_\mathrm{n} q(x)\frac{\partial_a q(x)}{\sum_{i=1}^n|\partial_{e_i}q(x)|^2}\mu^{S+ta}(dx)$$
$$=\int_{S+ta}\partial_\mathrm{n} q(x)\,(\psi\cdot\mu)^{S+ta}(dx)=\int_{U+ta}\sum_{i=1}^n[\partial_{e_i}^2 q+\partial_{e_i}q\beta_{e_i}^{\psi\cdot\mu}]\,\psi\,\mu(dx),$$

where $\psi:=\partial_a q\bigl(\sum_{i=1}^n|\partial_{e_i}q|^2\bigr)^{-1}$. It is clear from the reasoning above that for sufficiently large n the measure $\psi\cdot\mu$ is well defined and differentiable. By induction we obtain differentiability in t of order k. Moreover, it is not difficult to obtain the continuity of the partial derivatives of the considered function in both variables, which completes the proof. □

We observe that the indicated condition on the norm is fulfilled for the spaces $L^{2n}(\sigma)$ with $n\in\mathbb{N}$ (in this case the above theorem was proved in Uglanov [**1135**]).

9.5.8. Remark. Let γ be a centered Radon Gaussian measure on a locally convex space X with the Cameron–Martin subspace H. As we observed at the end of §8.8, inequality (8.8.2) can be related to surface measures. Fang [**406**] used this relation to prove the following estimate. If $f\in W^\infty(\gamma)$ is nondegenerate, ψ is the density of the measure $(|D_H f|\cdot\gamma)\circ f^{-1}$, and σ is the surface measure generated by γ, then

$$\sigma(\{f=t\})=\psi(t)\geqslant\frac{1}{\sqrt{2\pi}}\exp\Bigl[-\frac{1}{2}\Phi^{-1}\bigl(\mu(\{f\geqslant t\})\bigr)\Bigr],\quad t\in\mathbb{R}^1.$$

9.6. Convergence of nonlinear images of measures

Suppose we are given a measure μ and a sequence of transformations F_j of this measure convergent in some sense. Then the question arises about convergence of the measures $\mu \circ F_j^{-1}$. Here we consider convergence of such measures in variation. First we consider mappings that are nondegenerate in the Malliavin sense.

Suppose that in Theorem 9.3.1 we are given a sequence of mappings
$$F_n = (F_1^n, \ldots, F_d^n),$$
for which the corresponding σ_{ij}, γ^{ij}, and Δ will be denoted by σ_{ij}^n, γ_n^{ij}, and Δ_n. Our justification of the cited theorem and Remark 9.3.4 yield the following assertion.

9.6.1. Theorem. *Let $\mu \geqslant 0$ and let the measures $d_{v_j}(\gamma_n^{ij} \cdot \mu)$ have densities ξ_n^{ij} with respect to μ that are uniformly bounded in the space $L^p(\mu)$ for some $p > d$. Then the measures $\mu \circ F_n^{-1}$ have densities ϱ_n that are uniformly bounded in $W^{p,1}(\mathbb{R}^d)$, and $\{\varrho_n\}$ contains a subsequence which converges uniformly on compact sets.*

If the measures $\mu \circ F_n^{-1}$ converge weakly (for example, if the mappings F_n converge in measure), then they converge in variation.

Under very broad assumptions $\xi_n^{ij} = \partial_{v_j}\gamma_n^{ij} + \gamma_n^{ij}\beta_{v_j}$, which equals a sum of functions of the form $\Delta_n^{-2}\partial_{v_j}\partial_{v_i}F_k^n G_1$, $\Delta_n^{-1}\partial_{v_j}\partial_{v_i}F_k^n G_2$, $\Delta_n^{-1}\beta_{v_j}^\mu G_3$, where G_1 and G_2 are products of $2d - 2$ elements σ_{ij}^n, G_3 is a product of $d - 1$ elements σ_{ij}^n (the coefficients at such products do not depend on n). For practical verification of the condition of the theorem, it suffices that for some $p > d$,
$$\sup_{n,i,j,k,s}\left[\|\Delta_n^{-2}\|_{L^{2p}(\mu)} + \||\sigma_{ij}^n|^{d-1}\beta_{v_j}^\mu\|_{L^{2p}(\mu)} + \||\partial_{v_i}\sigma_{jk}^n|\sigma_{sl}^n|^{2d-2}\|_{L^{2p}(\mu)}\right] < \infty.$$

9.6.2. Example. Let γ be a Gaussian Radon measure on a locally convex space X, let $H = H(\gamma)$, and let mappings $F_n\colon X \to \mathbb{R}^d$ converge in measure and be uniformly bounded in the norm of $W^{2p(4d-1),2}(\gamma, \mathbb{R}^d)$. Suppose that the functions Δ_n^{-1} are uniformly bounded in $L^{4p}(\gamma)$ for some $p > d$. Then the measures $\gamma \circ F_n^{-1}$ converge in variation.

Now we turn to a more general case.

9.6.3. Theorem. *Let $F\colon \mathbb{R}^n \to \mathbb{R}^n$ and $F_j\colon \mathbb{R}^n \to \mathbb{R}^n$ be measurable mappings, let λ be Lebesgue measure on $\mathbb{R}^n$, and let E be a measurable set of finite Lebesgue measure in $\mathbb{R}^n$. Suppose that at every point $x \in E$ there exist approximate partial derivatives $\mathrm{ap}D_iF(x)$ and $\mathrm{ap}D_iF_j(x)$, $i = 1, \ldots, n$, and the mappings F_j converge to F in measure on the set E and their approximate partial derivatives $\mathrm{ap}D_iF_j$, $i = 1, \ldots, n$, converge in measure on the set E to the approximate partial derivatives $\mathrm{ap}D_iF$. Suppose also that the approximate Jacobian $J(F)$ of the mapping F does not vanish on E. Then the following conditions are equivalent:*

(i) *for every measurable set $A \subset E$, the measures $\lambda|_A \circ F_j^{-1}$ converge in variation to the measure $\lambda|_A \circ F^{-1}$;*

(ii) *for every measurable set $A \subset E$ and every $\delta > 0$, there exists a compact set $K_\delta \subset A$ such that $\lambda(A \backslash K_\delta) \leqslant \delta$ and $\lim\limits_{j\to\infty} \lambda\bigl(F_j(K_\delta)\bigr) = \lambda\bigl(F(K_\delta)\bigr)$.*

PROOF. Let condition (ii) be fulfilled. It suffices to show that every subsequence $\{G_j\}$ in $\{F_j\}$ contains a subsequence $\{G_{j_k}\}$, for which our assertion is true. Hence it suffices to find such a subsequence in $\{F_j\}$. In particular, by the Riesz theorem we may assume that the sequence $\{F_j\}$ converges along with approximate partial derivatives almost everywhere on E. In addition, in the course of our proof we shall apply several times the following simple observation: it suffices to show that, for every $\varepsilon > 0$, there exists a compact set $A_\varepsilon \subset A$ such that $\lambda(A\backslash A_\varepsilon) < \varepsilon$ and for A_ε the assertion is true.

Then we note that the theorem reduces to the case where the mappings F_j and F are the restrictions to E of some continuously differentiable mappings. Indeed, it is known (see Federer [**421**, Theorem 3.1.16]) that if a mapping $f\colon A \to \mathbb{R}^n$ satisfies condition (2.4.1) for almost all $a \in E$, then, for every number $\varepsilon > 0$, there exists a continuously differentiable mapping $g\colon \mathbb{R}^n \to \mathbb{R}^n$ such that

$$\lambda\bigl(x \in E\colon f(x) \neq g(x)\bigr) < \varepsilon.$$

Given $\varepsilon > 0$, we apply this theorem to each mapping F_j and the number $\varepsilon 4^{-j}$ as well as to F and the number $\varepsilon/2$. Moreover, due to Egoroff's theorem and the observation made above, the assertion reduces to the case where E is compact, the mappings F_j converge to F uniformly on E, and their derivatives converge uniformly on E to the derivative of F. By the inverse function theorem, every point of E has a neighborhood at which F is a diffeomorphism. Hence E can be partitioned in finitely many disjoint measurable subsets possessing neighborhoods on which F is a diffeomorphism. Since the measure $\lambda|_E$ equals the sum of the restrictions of Lebesgue measure to these subsets, everything reduces to the case where F is a diffeomorphism in some neighborhood U of the compact set E. Moreover, we may assume that F is the identical mapping since our assertion is invariant with respect to diffeomorphisms. By using the uniform convergence of $\det DF_j$ to 1 on E and passing once again to a subsequence, we may assume that for all $x \in E$ and all j the inequality $|\det DF_j(x) - 1| < 2^{-j}$ is fulfilled. It is known (see Federer [**421**, Corollary 3.2.4]) that for every j there exists a compact set $Q_j \subset E$ such that F_j is injective on Q_j and one has

$$\lambda\bigl(F_j(E)\backslash F_j(Q_j)\bigr) < 2^{-j}. \qquad (9.6.1)$$

By using condition (ii) and deleting from E a subset of an arbitrarily small measure we arrive at the situation where one has the equality

$$\lim_{j\to\infty} \lambda\bigl(F_j(E)\bigr) = \lambda\bigl(F(E)\bigr) = \lambda(E). \qquad (9.6.2)$$

9.6. Convergence of nonlinear images of measures

Let us take a subsequence $\{F_{j_k}\}$ such that
$$\lambda\big(F_{j_k}(E) \triangle E\big) < \frac{1}{2^k}. \qquad (9.6.3)$$
Such a subsequence exists by (9.6.2) and the uniform convergence of F_j to F on E, due to which for every open set $U \supset E$ with $\lambda(U\backslash E) < 4^{-k}$ there exists a number n_k with $F_j(E) \subset U$ for all $j \geqslant n_k$. We may assume further that $j_k = k$. According to [**421**, Theorem 3.2.3], for every measurable set A on which F_j is injective and $\det DF_j > 0$, one has the equality
$$\lambda\big(F_j(A)\big) = \int_A \det DF_j(x)\, dx. \qquad (9.6.4)$$
Then (9.6.4) with $A = Q_j$, (9.6.1) and (9.6.3) yield
$$\lambda(Q_j) \geqslant \lambda\big(F_j(Q_j)\big) - 2^{-j}\lambda(E) \geqslant \lambda\big(F_j(E)\big) - 2^{-j} - 2^{-j}\lambda(E)$$
$$\geqslant \lambda(E) - 2^{1-j} - 2^{-j}\lambda(E) = \lambda(E) - 2^{-j}(2 + \lambda(E)).$$
The sets $E_N = \bigcap_{j=N}^{\infty} Q_j$ are compact, for all $j \geqslant N$ the mappings F_j are injective on E_N, and by the previous estimate we have $\lambda(E_N) \to \lambda(E)$ since
$$\lambda(E\backslash E_N) \leqslant \sum_{j=N}^{\infty} \lambda(E\backslash Q_j) \leqslant (2 + \lambda(E)) \sum_{j=N}^{\infty} 2^{-j}.$$
Hence it suffices to consider every E_N separately. So we shall assume that E coincides with one of the sets E_N. Passing to a new subsequence and replacing E_N by its compact part of close measure, we arrive as above to relationships (9.6.3) with $j_k = k$ for our new set E. Denote by S the set of all $y \in E$ for which there exists a number $m = m(y)$ with
$$y \in \bigcap_{j \geqslant m} F_j(E).$$
By estimate (9.6.3) we have $\lambda(S) = \lambda(E)$ since
$$\lambda\bigg(E \triangle \bigcap_{j=m}^{\infty} F_j(E)\bigg) \leqslant \sum_{j=m}^{\infty} 2^{-j} = 2^{-m+1}.$$
For completing the proof it remains to show that the densities ϱ_j of the induced measures $\lambda|_E \circ F_j^{-1}$ converge almost everywhere on E to the density $\varrho = I_E$ of the measure $\lambda|_E \circ F^{-1}$ (see [**193**, §2.8]). We observe that
$$\varrho_j(y) = \frac{I_{F_j(E)}(y)}{\det DF_j\big(F_j^{-1}(y)\big)}.$$
Indeed, since F_j is injective on E, the change of variables formula (see (2.4.2)) shows that for every bounded measurable function φ we have
$$\int_{\mathbb{R}^n} \varphi(y)\, \lambda|_E \circ F_j^{-1}(dy) = \int_E \varphi\big(F_j(x)\big)\, dx = \int_{F_j(E)} \varphi(y) \frac{1}{\det DF_j\big(F_j^{-1}(y)\big)}\, dy.$$

Let $y \in S$. Then $F_j^{-1}(y) \to F^{-1}(y) = y$. Indeed, otherwise some subsequence $F_{j_k}^{-1}(y)$ in the sequence $F_j^{-1}(y)$ in the compact set E converges to a point $x \neq y$, whence by the uniform convergence of F_j to F we obtain the relationship
$$y = F_{j_k}\bigl(F_{j_k}^{-1}(y)\bigr) \to F(x) = x$$
contrary to the fact that $x \neq y$. Furthermore, by the uniform convergence of derivatives we obtain
$$\det DF_j\bigl(F_j^{-1}(y)\bigr) \to \det DF\bigl(F^{-1}(y)\bigr) = 1.$$
If $y \notin E$, then for all sufficiently large j we have the equality $I_{F_j(E)}(y) = 0$, whence we obtain $\varrho_j(y) \to 0$.

Now let condition (i) be fulfilled. It is clear from the reasoning above that the general case reduces to the case where the set $A \subset E$ is compact, F is a continuously differentiable mapping in a neighborhood of A, the mappings F_j are continuous on A and $F_j \to F$ uniformly on A. Let $\varepsilon > 0$ and let U be a neighborhood of $F(A)$ such that $\lambda\bigl(U \backslash F(A)\bigr) < \varepsilon$. Then, for all sufficiently large j, one has $F_j(A) \subset U$, whence $\lambda\bigl(F_j(A) \backslash F(A)\bigr) < \varepsilon$. On the other hand, $\lambda\bigl(F(A) \backslash F_j(A)\bigr) \to 0$ as $j \to \infty$. Indeed, the measurable set $X_j = F(A) \backslash F_j(A)$ has measure zero with respect to $\lambda|_A \circ F_j^{-1}$, and its measure with respect to $\lambda|_A \circ F^{-1}$ is estimated from below by $c\lambda(X_j)$ since the proof of the first part of the theorem shows that the density of the measure $\lambda|_A \circ F^{-1}$ on $F(A)$ is uniformly separated from zero. $\square$

9.6.4. Remark. (i) Under the assumptions of the previous theorem the measure $\lambda|_E \circ F^{-1}$ is absolutely continuous (this is seen from the proof), so the singular components of the measures $\lambda|_E \circ F_j^{-1}$ converge to zero in variation.

(ii) The nondegeneracy of apDF on E is also necessary for the absolute continuity of the induced measure $\lambda|_E \circ F^{-1}$ (see §9.2 and also Theorem 3 and the remark after Theorem 4 in Ponomarev [**910**]).

(iii) Our proof shows that if the mappings F_j are injective on E, then condition (i) of Theorem 9.6.3 (hence also condition (ii)) follows from other assumptions of this theorem. Indeed, in this case we obtain at once the indicated expression for ϱ_j. However, in the general case, two equivalent conditions (i) and (ii) do not follow automatically from other assumptions of this theorem. Let us consider the following simple example. Let $E \subset [0,1]$ be a Cantor type set (nowhere dense compact) of positive Lebesgue measure and $F(x) = x$. For every natural number k we divide $[0,1]$ into intervals $I_{k,j}$ by the points of the form j/k. Inside every interval $I_{k,j}$ intersecting E in a set of positive measure we take disjoint intervals $\Delta_{k,j}^i = [a_{k,j}^i, b_{k,j}^i]$, $i = 1,2$, such that
$$\lambda(E \cap \Delta_{k,j}^1) = \lambda(E \cap \Delta_{k,j}^2) \geqslant \frac{1}{3}\lambda(E \cap I_{k,j}).$$
We may assume that some intervals $L_{k,j}$ and $R_{k,j}$ from the complement to the set E are adjacent to $\Delta_{k,j}^2$ from the left and right. Now we define a

function F_k as follows: $F_k(x) = x$ if $x \notin L_{k,j} \cup \Delta_{k,j}^2 \cup R_{k,j}$, $F_k(x) = x - c_{k,j}$ if $x \in \Delta_{k,j}^2$, where $c_{k,j} = b_{k,j}^2 - b_{k,j}^1$, and on $L_{k,j}$ and $R_{k,j}$ we define F_k to make it a smooth function on $[0,1]$. We have $F_k' = 1$ on E and $F_k \to F$ uniformly on E (as one can see from our construction, we can achieve the uniform convergence on the whole interval $[0,1]$), but the images of Lebesgue measure on E under F_k do not converge to the image of Lebesgue measure on E under F. This is seen from the fact that $\lambda\big(F_k(E) \triangle E\big) \geqslant \lambda(E)/3$.

Note also that, in general, condition (ii) does not imply convergence $\lambda\big(F_j(E)\big) \to \lambda\big(F(E)\big)$. For example, let E be the classical Cantor set of measure zero, let $F(x) = 0$, and let $F_j = G$, $j \geqslant 1$, where G is an arbitrary function for which $G(E) = [0,1]$. Then $\lambda|_E \circ F^{-1} = \lambda|_E \circ F_j^{-1} = 0$ for all j, and all assumptions of the theorem are fulfilled, although $\lambda(E) = 0$.

If one does not wish to deal with approximate derivatives, then in the formulation of the established theorem they can be replaced by more common usual partial derivatives (which will give a somewhat weaker assertion). For example, the theorem applies to mappings from the class $W_{\text{loc}}^{1,1}(\mathbb{R}^n, \mathbb{R}^n)$ satisfying the corresponding conditions of convergence on E.

9.6.5. Corollary. *Suppose that in the situation of Theorem 9.6.3 one of the equivalent conditions* (i) *or* (ii) *is fulfilled, where E may even have infinite Lebesgue measure. Let μ be an absolutely continuous probability measure on $\mathbb{R}^n$. Then the measures $\mu|_E \circ F_j^{-1}$ converge in variation to the measure $\mu|_E \circ F^{-1}$.*

PROOF. This follows from the next lemma. □

9.6.6. Lemma. *Let $(\Omega, \mathcal{F}, \sigma)$ be a measurable space with a finite measure σ and let F_j, where $j = 0, 1, \ldots$, be mappings with values in a measurable space $(Y, \mathcal{E})$ measurable with respect to σ such that, for every set $E \in \mathcal{F}$, the measures $\sigma|_E \circ F_j^{-1}$ converge in variation to the measure $\sigma|_E \circ F_0^{-1}$. Then, for every finite measure μ absolutely continuous with respect to σ and every set $E \in \mathcal{F}$, the induced measures $\mu|_E \circ F_j^{-1}$ converge in variation to the measure $\mu|_E \circ F_0^{-1}$.*

PROOF. Let $\mu = \varrho \cdot \sigma$. The hypotheses of the lemma imply that our assertion is true if ϱ assumes finitely many values. In the general case there exists a sequence of measures μ_i whose densities ϱ_i with respect to σ assume finitely many values such that $\|\mu - \mu_i\| \to 0$. Note that

$$\big\|\mu_i|_E \circ F_j^{-1} - \mu|_E \circ F_j^{-1}\big\| = \big\|(\mu_i - \mu)|_E \circ F_j^{-1}\big\| \leqslant \|\mu_i - \mu\|.$$

Hence

$$\lim_{i \to \infty} \big\|\mu_i|_E \circ F_j^{-1} - \mu|_E \circ F_j^{-1}\big\| = 0$$

uniformly in j. It remains to use the fact that for μ_i the assertion is true. □

Now we give conditions sufficient for (ii). These conditions, although not being necessary, may be more useful for practical verification. Their obvious advantage is that there is no need to consider images of sets.

9.6.7. Corollary. *Suppose that continuous mappings $F_j\colon \mathbb{R}^n \to \mathbb{R}^n$ converge uniformly on compact sets to a continuous mapping $F\colon \mathbb{R}^n \to \mathbb{R}^n$ and F_j and F possess Lusin's (N)-property and almost everywhere have regular approximate derivatives (for example, usual derivatives) DF_j and DF such that DF_j converge to DF in measure on some set E of finite Lebesgue measure. Suppose that $\det DF \neq 0$ on E and that on every compact set the sequence $\{|\det DF_j|\}$ is uniformly integrable. Then the measures $\lambda|_E \circ F_j^{-1}$ converge in variation to the measure $\lambda|_E \circ F^{-1}$. In addition, if μ is an absolutely continuous probability measure on $\mathbb{R}^n$, then the measures $\mu|_E \circ F_j^{-1}$ converge in variation to the measure $\mu|_E \circ F^{-1}$ (this is true also for E of infinite Lebesgue measure).*

PROOF. The uniform integrability of the functions $\det DF_j$ on compact sets and the estimate

$$\lambda\big(F_j(A)\big) \leqslant \int_A |\det DF_j(x)|\,dx \qquad (9.6.5)$$

(see Theorem 2.4.3) yield that, for every ball $U \subset \mathbb{R}^n$, one has

$$\lim_{\varepsilon \to 0}\ \sup_{A\subset U\colon\ \lambda(A)\leqslant\varepsilon}\ \sup_j \lambda\big(F_j(A)\big) = 0. \qquad (9.6.6)$$

We verify that $\lim_{j\to\infty} \lambda\big(F_j(K)\big) = \lambda\big(F(K)\big)$ for every compact set $K \subset E$. It is seen from the proof of the previous theorem that it suffices to verify our claim for compact sets K on which F is the restriction of a diffeomorphism in some neighborhood of K. Let $\varepsilon > 0$ and $4\varepsilon < \lambda(K)$. It is clear that for all sufficiently large j we have the estimate $\lambda\big(F_j(K)\big) \leqslant \lambda\big(F(K)\big) + \varepsilon$. Suppose that $\lambda\big(F_j(K)\big) \leqslant \lambda\big(F(K)\big) - \varepsilon$ for all j. According to (9.6.6) there exists a connected set Q that is a finite union of closed cubes such that $K \subset Q$ and $\lambda\big(F_j(Q)\big) \leqslant \lambda\big(F_j(K)\big) + \varepsilon/8$ for all F_j, and the same is true for F in place of F_j. Let ∂Q be the boundary of Q. Then $F(\partial Q)$ has measure zero. We take a sufficiently small closed neighborhood G of the compact set $F(\partial Q)$ such that $\lambda(G) < \varepsilon/8$ and $F_j(\partial Q) \subset G$ for all sufficiently large j. We can take G in the form of a finite union of closed cubes. Then $\mathbb{R}^n\backslash G$ has finitely many connected components $W_1, \ldots, W_m$ in which the restrictions of F_j and F to Q have constant degrees of mapping $\deg_Q F_j$ and $\deg_Q F$ (see Gol'dshteĭn, Reshetnyak [**490**, Ch. 5, §1.5]), and, by the uniform convergence of F_j to F, for all sufficiently large numbers j we have

$$\deg_Q F_j(y) = \deg_Q F(y) = d_i \qquad \text{for all } y \in W_i.$$

If $y \in Y := F(K)\backslash\big(G \cup F(Q\backslash K)\big)$, then $\deg_Q F(y) = 1$. This follows from the fact that for every such point y there exists a ball $K(y,r)$ with $\deg_{K(y,r)} F(y) = 1$ (see Radó, Reichelderfer [**938**, V.2.2, p. 329]). Therefore,

$\deg_Q F_j(y) = 1$ for all $y \in Y$ and all sufficiently large j. Since $|\deg_Q F_j(y)|$ does not exceed the cardinality of $F_j^{-1}(y) \cap Q$ (see [490, Ch. 5, §1.5, Remark]), we obtain $Y \subset F_j(Q)$, whence

$$\lambda\bigl(F_j(Q)\bigr) \geqslant \lambda\bigl(F(K)\bigr) - \lambda(G) - \lambda\bigl(F(Q\backslash K)\bigr) \geqslant \lambda\bigl(F(K)\bigr) - \varepsilon/4,$$

which gives a contradiction. The assertion for μ in place of λ follows by the same reasoning as in Corollary 9.6.5. □

9.6.8. Corollary. *In the previous corollary we have*

$$\lim_{j\to\infty} \lambda\bigl(F_j(A) \triangle F(A)\bigr) = 0 \quad \text{and} \quad \lim_{j\to\infty} \lambda\bigl(F_j(A)\bigr) = \lambda\bigl(F(A)\bigr)$$

for every measurable set $A \subset E$. The same is true for μ in place of λ.

PROOF. Suppose first that A is compact. Then the second equality has been proved in Corollary 9.6.7. Let $\varepsilon > 0$ and let V be an open set such that $F(A) \subset V$ and $\lambda\bigl(V\backslash F(A)\bigr) < \varepsilon$. Then for all sufficiently large j we have $F_j(A) \subset V$ and

$$\lambda\bigl(F_j(A)\bigr) > \lambda\bigl(F(A)\bigr) - \varepsilon \geqslant \lambda(V) - 2\varepsilon.$$

Therefore, $\lambda\bigl(F_j(A) \triangle F(A)\bigr) < 3\varepsilon$, whence we obtain the first equality to be proven. Both equalities extend to arbitrary measurable sets $A \subset E$ due to (9.6.5) and the uniform integrability of $\{|\det DF_j|\}$. The claim for μ follows by Lemma 9.6.6. □

As already noted, a sufficient condition for the existence of regular approximate derivatives almost everywhere is the membership in the Sobolev class $W^{p,1}_{\mathrm{loc}}(\mathbb{R}^n, \mathbb{R}^n)$ with some $p > n - 1$. However, this condition does not guarantee the existence of a continuous modification, which requires the estimate $p > n$ (this estimate ensures also Lusin's (N)-property, see §2.4). In the latter case, as we shall now see, the formulation of the main result simplifies. Let us consider in more detail mappings of the class $W^{p,1}_{\mathrm{loc}}(\mathbb{R}^n, \mathbb{R}^n)$.

9.6.9. Corollary. *Let $F_j, F \in W^{p,1}_{\mathrm{loc}}(\mathbb{R}^n, \mathbb{R}^n)$, where $p \geqslant n$, and let mappings F_j converge to F in the Sobolev norm $\|\cdot\|_{p,1}$ on every ball in $\mathbb{R}^n$. Suppose that $E \subset \{\det F \neq 0\}$ is a measurable set of finite Lebesgue measure. Then the measures $\lambda|_E \circ F_j^{-1}$ converge in variation to the measure $\lambda|_E \circ F^{-1}$. In the case $p > n$ the same is true if in place of convergence in the norm $\|\cdot\|_{p,1}$ on every ball we require the boundedness of $\{F_j\}$ in $W^{p,1}(U, \mathbb{R}^n)$ for every ball U and convergence of F_j to F almost everywhere.*

PROOF. Let us recall that in the case $p > n$ the mappings F_j and F possess continuous modifications with Lusin's (N)-property. By the Sobolev embedding theorem the continuous modifications of the mappings F_j converge uniformly on compact sets to the continuous modification of the mapping F. Hence we can apply the previous corollary. As will be seen from the reasoning below for the case $p = n$, we can prove our assertion without using Lusin's property (N) for F_j and F.

In the case of the local boundedness of $\{F_j\}$ in the norm $\|\cdot\|_{p,1}$ for $p > n$ and convergence to F almost everywhere along with derivatives, the same reasoning applies. Indeed, the compactness of the embedding of $W^{p,1}(U, \mathbb{R}^n)$ to $C(U, \mathbb{R}^n)$ for every ball U and convergence almost everywhere yield that the sequence of the continuous modifications of the mappings F_j converges uniformly on U to the continuous modification of F. In addition, the boundedness of $\{\|DF_j\|^p\}$ in $L^1(U)$ and the condition $p > n$ yield that the sequence $\{\det DF_j\}$ is uniformly integrable on U.

Let us proceed to the case $p = n$ (to which our previous reasoning does not apply) and consider a sequence of mappings $F_j \in W^{n,1}(U, \mathbb{R}^n)$ convergent to a mapping $F_0 = F$ in $W^{n,1}(U, \mathbb{R}^n)$, where U is a ball in $\mathbb{R}^n$. Suppose that $F_j \to F_0$ and $DF_j \to DF_0$ a.e. on U. Suppose also that we are given a measurable set

$$\Omega \subset \{x \in U|\ \det DF_0(x) \neq 0\}$$

and let $\varepsilon > 0$ be fixed. According to Theorem 2.5.5, there exist Lipschitzian mappings $\psi_j \colon U \to \mathbb{R}^n$, $j \geq 0$, such that

$$\lambda\bigl(x \in U|\ F_j(x) \neq \psi_j(x)\bigr) + \int_U \|Df(x) - D\psi(x)\|^n\, dx \leq \varepsilon 2^{-j}.$$

By Egoroff's theorem, there is a compact subset $K \subset \bigcap_{n=0}^{\infty}\{F_j = \psi_j\} \subset \Omega$ such that

$$\lim_{j\to\infty} \sup_{x\in K}\bigl(|F_j(x) - F_0(x)| + \|DF_j(x) - DF_0(x)\|\bigr) = 0, \quad \lambda(\Omega \setminus K) < 3\varepsilon.$$

We take a smooth mapping β on $\mathbb{R}^n$ such that $0 \leq |\beta(x)| \leq 2$, $\beta(x) = x$ if $|x| \leq 1$, $\beta(x) = 0$ if $|x| \geq 2$. Let $M = \sup_x \|D\beta(x)\|$. For $\alpha \in (0,1)$ we define a mapping $\beta_\alpha \colon \mathbb{R}^n \to \mathbb{R}^n$ as follows:

$$\beta_\alpha(x) = \alpha\beta(x/\alpha).$$

Let $c_j = \sup_{x\in K} |F_j(x) - F_0(x)|$. The mappings

$$G_j = \beta_{c_j}(\psi_j - \psi_0) + \psi_0$$

are Lipschitzian and converge uniformly on U to $G_0 = \psi_0$ as $j \to \infty$ since $c_j \to 0$ and $|\beta_\alpha(x)| \leq 2\alpha$. We observe that

$$\|DG_j(x)\| \leq \sup_y \|D\beta_{c_j}(y)\|\, \|D\psi_j(x) - D\psi_0(x)\| + \|D\psi_0(x)\|$$
$$\leq M\|D\psi_j(x) - D\psi_0(x)\| + \|D\psi_0(x)\|.$$

By convergence of $\{DF_j\}$ in $L^n(U, \mathbb{R}^n)$ the sequence $\{D\psi_j\}$ converges in $L^n(U, \mathbb{R}^n)$. Hence the sequence $\{\|DG_j\|^n\}$ is uniformly integrable on U. Due to the estimate $|\det DF_j(x)| \leq \|DF_j(x)\|^n$, the sequence $\{\det DG_j\}$ is uniformly integrable as well. We have $G_j = \psi_j = F_j$ on K as $j \geq 0$. Since $D\beta_c(y) = I$ if $|y| < c$ and $D\psi_j(x) = DF_j(x)$ for almost all $x \in K$ by the equality $\psi_j = F_j$ on K, one has $DG_j(x) = D\psi_j(x) = DF_j(x)$ for $x \in K$. Hence the mappings DG_j converge in measure on K to $DF_0|_K = DG_0|_K$.

9.6. Convergence of nonlinear images of measures

Since the Lipschitzian mappings G_j possess (N)-property, the previous corollary shows that the measures $\lambda|_K \circ F_j^{-1} = \lambda|_K \circ G_j^{-1}$ converge in variation to the measure $\lambda|_K \circ G_0^{-1} = \lambda|_K \circ F_0^{-1}$. Since $\varepsilon > 0$ is arbitrary, the proof is complete. $\square$

The assertion of Corollary 9.6.9 is true if convergence in $W^{p,1}(U, \mathbb{R}^n)$ is replaced by the following condition: the maps F_j belong to $W^{p,r}_{\text{loc}}(\mathbb{R}^n, \mathbb{R}^n)$ with $rp > n$ and converge to F in the norm of $W^{p,r}(U, \mathbb{R}^n)$ for every ball U and, in addition, the sequence $\{|\det DF_j|\}$ is uniformly integrable on U.

The assumption of nondegeneracy of DF on the set E is essential; for example, the sequence of functions $F_j(x) = j^{-1}$ on $[0,1]$ converges uniformly to zero along with their derivatives, but the induced measures are the Dirac measures at the points j^{-1}, which do not converge in variation to the Dirac measure at zero. Below we consider mappings between spaces of different dimensions; then the nondegeneracy of DF is replaced by the surjectivity of DF. It is clear that all results in this section with obvious changes in their formulations remain valid for mappings defined on arbitrary domains in $\mathbb{R}^n$.

We shall say that a measurable function ϱ on $\mathbb{R}^n$ is locally strictly positive if $\operatorname{essinf}_U \varrho > 0$ for every ball U.

9.6.10. Theorem. *Suppose that X is a locally convex space, Z is a finite dimensional linear subspace in X generated by vectors $e_1, \ldots, e_n$, Y is a closed linear subspace complementing Z to X, μ is Radon probability measure on X for which conditional measures μ^y on $Z + y$ for $y \in Y$ are absolutely continuous with respect to Lebesgue measures on $Z + y$ generated by some linear isomorphism between Z and $\mathbb{R}^n$ and the corresponding densities are either locally strictly positive or continuous. Let μ-measurable mappings F_j, $F \colon X \to \mathbb{R}^n$ be absolutely continuous on bounded intervals of the straight lines $x + \mathbb{R}^1 e_i$, $i = 1, \ldots, n$, for μ-almost all x and let $F_j \to F$ and $\partial_{e_i} F_j \to \partial_{e_i} F$ in $L^p(\mu, \mathbb{R}^n)$, where $p \geqslant n$. If E is a μ-measurable set such that $\det\bigl((\partial_{e_i} F, \partial_{e_k} F)\bigr)_{i,k=1}^n \neq 0$ on E, then the measures $\mu|_E \circ F_j^{-1}$ converge in variation to the measure $\mu|_E \circ F^{-1}$.*

PROOF. As in the proof of Theorem 9.6.3, it suffices to verify our claim for a subsequence in $\{F_j\}$. Hence we may assume that

$$\|F_j - F\|_{L^p(\mu, \mathbb{R}^n)} + \|\partial_{e_i} F_j - \partial_{e_i} F\|_{L^p(\mu, \mathbb{R}^n)} \leqslant 2^{-j}$$

for all $j \geqslant 1$ and $i = 1, \ldots, n$. Let ν be the image of μ under the natural projection of X onto Y. By Fubini's theorem and the integrability of the series

$$S_i(x) = \sum_{j=1}^\infty \Bigl[|\partial_{e_i} F_j(x) - \partial_{e_i} F(x)|^p + |F_j(x) - F(x)|^p\Bigr], \quad i = 1, \ldots, n,$$

we have that for $i = 1, \ldots, n$ for ν-a.e. y the series

$$Z_i(y, u) = \sum_{j=1}^{\infty} \Big[|\partial_{e_i} F_j(y+u) - \partial_{e_i} F(y+u)|^p + |F_j(y+u) - F(y+u)|^p\Big]$$

are integrable in u with respect to the measure μ^y. For every such y and every $i = 1, \ldots, n$ we obtain convergence of the sequence $\{F_j(y + \cdot)\}$ to $F(y + \cdot)$ and convergence of the sequence $\{\partial_{e_i} F_j(y + \cdot)\}$ to $\partial_{e_i} F(y + \cdot)$ in $L^p(\mu^y, \mathbb{R}^n)$. In the case of the strict local positivity of the density ϱ^y of the measure μ^y this gives convergence in $L^p(U, \mathbb{R}^n)$ for every ball $U \subset Z$, and in the case of continuity of ϱ^y the same holds for every ball whose closure is contained in the set $\{\varrho^y \neq 0\}$. Hence Corollary 9.6.9 applies. For every Borel set $B \subset X$, we have

$$\mu(E \cap B) = \int_Y \mu^y(E \cap B)\, \nu(dy).$$

Hence, for every Borel mapping $G\colon X \to \mathbb{R}^n$ and every Borel set $B \subset \mathbb{R}^n$, we have

$$\mu|_E \circ G^{-1}(B) = \int_Y \mu^y|_E \circ G^{-1}(B)\, \nu(dy). \qquad (9.6.7)$$

It remains to observe that

$$\big\|\mu|_E \circ F_j^{-1} - \mu|_E \circ F^{-1}\big\| \leqslant \int_Y \big\|\mu^y|_E \circ F_j^{-1} - \mu^y|_E \circ F^{-1}\big\|\, \nu(dy), \qquad (9.6.8)$$

which converges to zero as $j \to \infty$ by the Lebesgue dominated convergence theorem. The theorem is proven. $\square$

9.6.11. Corollary. *Let the hypotheses of the previous theorem be fulfilled and let a measure μ_0 be absolutely continuous with respect to μ. Then the measures $\mu_0 \circ F_j^{-1}$ converge to $\mu_0 \circ F^{-1}$ in variation.*

PROOF. This follows by Lemma 9.6.6. $\square$

We observe that in the situation of the previous theorem the measure $\mu|_E \circ F^{-1}$ is absolutely continuous (this is clear from the proof).

9.6.12. Corollary. *Let γ be a Radon Gaussian measure on a locally convex space X with the Cameron–Martin space H, $F_j, F \in W^{p,1}(\gamma, \mathbb{R}^n)$, where $p \geqslant n$. Assume that $F_j \to F$ in the norm of $W^{p,1}(\gamma, \mathbb{R}^n)$. Then, for any γ-measurable set $E \subset \{x\colon D_H F(x)(H) = \mathbb{R}^n\}$, the measures $\gamma|_E \circ F_j^{-1}$ converge in variation to the measure $\gamma|_E \circ F^{-1}$.*

PROOF. The set E is contained in the union of the sets

$$E_{k_1, \ldots, k_n} = \big\{x \in E\colon \det(\partial_{e_{k_i}} F, \partial_{e_{k_j}} F) \neq 0\big\},$$

where $\{e_i\}$ is a fixed orthonormal basis in H. Hence it suffices to prove our claim for every $E_{k_1, \ldots, k_n}$. This is easily done by using the previous theorem due to the existence of modifications of the mappings F_j and F with the required properties along the finite dimensional subspace generated by $e_{k_1}, \ldots, e_{k_n}$. $\square$

We observe that Theorem 9.6.10 applies to Gibbs measures possessing absolutely continuous conditional distributions with locally strictly positive densities on finite dimensional subspaces. It is clear that this theorem applies to the spaces $X = \mathbb{R}^d$ with $d \geqslant n$ and suitable measures μ (for example, measures with locally strictly positive densities). This enables us to extend the obtained results to the case of manifolds. Sobolev classes $W^{p,r}(U, M)$ of mappings between Riemann manifolds U and M are defined similarly to the case of $\mathbb{R}^n$; the corresponding local classes $W^{p,r}_{\text{loc}}(U, M)$ can be defined merely in local coordinates.

9.6.13. Corollary. *Let M_1 and M_2 be finite dimensional Riemann manifolds, $p \geqslant \dim M_2$, and let $F_j \in W^{p,1}_{\text{loc}}(M_1, M_2)$ be mappings convergent to a mapping $F \in W^{p,1}_{\text{loc}}(M_1, M_2)$ in the following sense: for every relatively compact local chart $U \subset M_1$ we have $F_j|_U \to F|_U$ in the Sobolev space $W^{p,1}(U, M_2)$. Suppose that μ is a Radon measure on M_1 that is absolutely continuous with respect to the Riemann volume on M_1 and E is a μ-measurable set. If $DF(x)\colon T_x M_1 \to T_{F(x)} M_2$ is a surjection for almost all $x \in E$, then the measures $\mu|_E \circ F_j^{-1}$ converge to the measure $\mu|_E \circ F^{-1}$ in variation.*

PROOF. As above, it suffices to prove our assertion for compact sets E. Then everything reduces to a finite collection of charts with compact closure. Hence we may assume that $M_1 = \mathbb{R}^d$ and $M_2 = \mathbb{R}^n$, where $d \geqslant n$. Let $e_1, \ldots, e_d$ be a basis in $\mathbb{R}^d$. The set E is covered by the subsets $E_{k_1, \ldots, k_n}$ of points x such that the vectors $DF(x)(e_{k_i})$, $i = 1, \ldots, n$, generate $\mathbb{R}^n$. It suffices to prove our assertion for $E \subset E_{k_1, \ldots, k_n}$. Now we apply Corollary 9.6.11. □

Let us prove an analog of Corollary 9.6.7 for mappings between spaces of different dimensions.

9.6.14. Proposition. *Let $F, F_j \colon \mathbb{R}^d \to \mathbb{R}^n$ be continuous mappings with Lusin's (N)-property possessing almost everywhere regular approximate derivatives (for example, usual derivatives) DF and DF_j and let $E \subset \mathbb{R}^d$ be a set of finite Lebesgue measure. Suppose that the mappings F_j converge to F uniformly on compact sets, the mappings DF_j converge to DF in measure on E, and that the minors of order n of the matrices DF_j either converge in $L^1(K)$ for every compact set K or are majorized in the absolute value by a locally integrable function. If the operator $DF(x)$ is surjective for almost all $x \in E$, then the measures $\lambda|_E \circ F_j^{-1}$ converge to $\lambda|_E \circ F^{-1}$ in variation. If μ is an absolutely continuous finite measure, then the measures $\mu|_E \circ F_j^{-1}$ converge to the measure $\mu|_E \circ F^{-1}$ in variation; this is true also for sets E of infinite Lebesgue measure.*

PROOF. We can apply a reasoning similar to the one used in the proof of Theorem 9.6.10. It suffices to prove our claim for $E \subset E_{k_1, \ldots, k_n}$, where $E_{k_1, \ldots, k_n}$ is the set of all points x in a fixed ball U such that the vectors

$DF(x)(e_{k_i})$, $i = 1, \ldots, n$, generate $\mathbb{R}^n$. Passing to a subsequence, we may assume that the mappings DF_j converge almost everywhere on E and that $\|\Lambda_{F_j} - \Lambda_F\|_{L^1(U)} \leqslant 2^{-j}$, where

$$\Lambda_G(x) = \det\bigl(DG(x)(e_{k_i}), DG(x)(e_{k_l})\bigr)_{i,l=1}^n.$$

Then, as in the proof of Theorem 9.6.10, we obtain that for almost all $y \in \mathbb{R}^{d-n}$ the mappings $z \mapsto F_j(y, z)$ and $z \mapsto F(y, z)$ on $\mathbb{R}^n$ satisfy the hypotheses of Corollary 9.6.7 with the sets $E_y = \{z \in \mathbb{R}^n : (y, z) \in E\}$ in place of E. Clearly, all these hypotheses, with the exception of the uniform integrability of the Jacobians, are fulfilled. The estimate for Λ_{F_j} indicated above gives L^1-convergence for almost all y, whence we obtain the uniform integrability of the Jacobians. In the case of a locally integrable majorant we also obtain the uniform integrability. Applying estimate (9.6.8) and Corollary 9.6.7 we complete the proof. $\square$

Clearly, this proposition extends immediately to mappings between Riemannian manifolds provided that the corresponding assumptions are fulfilled in local charts.

9.6.15. Remark. Equality (9.6.7) and Corollary 9.6.8 applied to the conditional measures μ^y yield that in Theorem 9.6.10, Corollary 9.6.12, Corollary 9.6.13, and Proposition 9.6.14 one has

$$\lim_{j\to\infty} \mu\bigl(F_j(A) \triangle F(A)\bigr) = 0 \quad \text{and} \quad \lim_{j\to\infty} \mu\bigl(F_j(A)\bigr) = \mu\bigl(F(A)\bigr)$$

for every measurable set $A \subset E$. The same is true for every μ-absolutely continuous measure μ_0 in place of μ.

9.6.16. Example. (i) Suppose we are given a sequence of locally Lipschitzian mappings $f_j \colon \mathbb{R}^n \to \mathbb{R}^n$ that, on every compact set, converges in measure to a locally Lipschitzian mapping f such that the partial derivatives $\partial_{x_i} f_j$ converge in measure on compact sets to $\partial_{x_i} f$ and are uniformly bounded on every compact set. Then, for every absolutely continuous probability measure μ on $\mathbb{R}^n$, the measures $\mu \circ f_j^{-1}$ converge in variation to the measure $\mu \circ f^{-1}$ provided that this measure is absolutely continuous. The latter is fulfilled if f is real analytic and the derivative of f is surjective at some point (which implies that the derivative is surjective at almost every point).

(ii) Suppose that X is a locally convex space, Z is a finite dimensional linear subspace in X generated by vectors $e_1, \ldots, e_n$, Y is a closed linear subspace complementing Z to X, a Radon measure μ on X is continuous along the vectors e_i, Borel mappings $F_j \colon X \to \mathbb{R}^n$ converge in measure $|\mu|$ to a Borel mapping F, for $|\mu|$-a.e. x the mappings $z \mapsto F_j(x + z)$ and $z \mapsto F(x + z)$ are locally uniformly Lipschitzian, and the mappings $\partial_{e_i} F_j$ converge in measure μ to $\partial_{e_i} F$. If the vectors $\partial_{e_1} F(x), \ldots, \partial_{e_n} F(x)$ are linearly independent $|\mu|$-a.e., then the measure $\mu \circ F^{-1}$ is absolutely continuous and the measures $\mu \circ F_j^{-1}$ converge to it in variation. For example, this is true if the mappings $z \mapsto F(x+z)$ are real analytic and the sets $F(x+Z)$ have inner points.

PROOF. (i) The uniform Lipschitzness on compact sets enables us to find a locally uniformly convergent subsequence. As noted in §9.1, the absolute continuity of the measure $\mu \circ F^{-1}$ is equivalent to the condition that μ-a.e. the operator $Df(x)$ is surjective. For an analytic mapping, the existence of a point of surjectivity yields the surjectivity almost everywhere. (ii) As in Theorem 9.6.10, it suffices to establish convergence in variation of the images of the conditional measures μ^y, which follows from (i). □

9.7. Supports of induced measures

In addition to the differentiability of the image of a measure under a transformation it is useful to know when the corresponding density is strictly positive or at least does not vanish on open sets. We shall mention a number of results in this direction. Our first observation is a trivial corollary of Example 8.7.11.

9.7.1. Proposition. *Let a measure $\mu \geq 0$ be H-ergodic with respect to the class $D^{1,1}(\mu)$ (or $W^{1,1}(\mu)$) and let f be in the respective class. Then the topological support of the induced measure $\mu \circ f^{-1}$ is connected, i.e., is a bounded or unbounded closed interval.*

PROOF. If the support of the induced measure is not connected, one can find an interval (a,b) of $\mu \circ f^{-1}$-measure zero such that both sets $(-\infty, a)$ and $(b, +\infty)$ are of positive $\mu \circ f^{-1}$-measure. Then there is a function $\varphi \in C_b^1(\mathbb{R}^1)$ such that $\varphi(t) = 0$ if $t \leq a$ and $\varphi(t) = 1$ if $t \geq b$. Therefore, $\varphi \circ f$ assumes only the values 0 and 1, but is not a constant a.e., which is impossible by Example 8.7.11. □

9.7.2. Proposition. *Let μ be a nonnegative H-ergodic measure (for example, let μ be a Gaussian measure with $H = H(\mu)$). Suppose that a function $f \in D^{p,1}(\mu)$, $p > 2$, is such that the measure $\mu \circ f^{-1}$ admits a locally Lipschitzian density ϱ. Then the function ϱ is strictly positive in the interior of its support.*

PROOF. Let a be a point in the support of ϱ such that $\varrho(a) = 0$, but for every n the sets $A_n = f^{-1}([a-1/n, a))$ and $B = f^{-1}((a, \infty))$ have positive μ-measures. Set $A := f^{-1}((-\infty, a])$. Since the density ϱ is Lipschitzian on $[a-1, a+1]$ with some constant C, one has $\varrho(t) \leq Cn^{-1}$ as $t \in [a - n^{-1}, a]$, whence
$$\mu(A_n) = \mu \circ f^{-1}([a - n^{-1}, a]) \leq Cn^{-2}.$$
Let us define a function f_n by $f_n = g_n \circ f$, where $g_n(t) = 1$ on $(-\infty, a - 1/n]$, $g_n(t) = 0$ on $[a, \infty)$ and $g_n(t) = -nt + na$ on $[a-1/n, a]$. Then $f_n \in D^{p,1}(\mu)$. We observe that $D_H f_n = g_n'(f) D_H f$. Letting $r = 2p/(p+2)$, $q = p/r$,

$s = p/(p-r)$, we have $r > 1$, $q > 1$ and $1/q + 1/s = 1$. In addition, $2/s = r$. Therefore, we arrive at the estimate

$$\int_X |D_H f_n(x)|_H^r \, \mu(dx) \leqslant n^r \int_{A_n} |D_H f(x)|_H^r \, \mu(dx)$$

$$\leqslant n^r \mu(A_n)^{1/s} \left(\int_X |D_H f(x)|_H^{rq} \, \mu(dx) \right)^{1/q} \leqslant C^{1/s} \left(\int_X |D_H f(x)|_H^p \, \mu(dx) \right)^{1/q}.$$

Therefore, the sequence $\{f_n\}$ is bounded in $D^{r,1}(\mu)$. Since $f_n \to I_A$ a.e., by Proposition 8.7.1 we obtain $I_A \in D^{r,1}(\mu)$, whence $I_A = 1$ a.e. according to Example 8.7.11. This contradicts our assumption that $\mu(B) > 0$. See also Exercise 9.8.17. □

The following result was obtained in Fang [**407**].

9.7.3. Proposition. *Let γ be a Radon Gaussian measure, $H = H(\gamma)$ and $F \in W^\infty(\gamma, \mathbb{R}^d)$. Then the support of the measure $\gamma \circ F^{-1}$ on $\mathbb{R}^d$ is connected.*

PROOF. We only consider the case of bounded F; for reduction of the general case to this one, see [**407**]. If the support K of $\gamma \circ F^{-1}$ is not connected, we can write $K = K_1 \cup K_2$, where K_1 and K_2 are disjoint nonempty compact sets. Then there is a function $\varphi \in C_0^\infty(\mathbb{R}^d)$ such that $\varphi|_{K_1} = 1$ and $\varphi|_{K_2} = 0$. As we know, the image of μ under the real function $f = \varphi(F)$ has connected support in $\varphi(K) = \varphi(K_1) \cup \varphi(K_2) = \{0, 1\}$. This is impossible since both values 0 and 1 must belong to the support of $\mu \circ f^{-1}$. □

9.7.4. Proposition. *In Theorem 9.3.5, for all $p \in [1, +\infty)$ and all multi-indices $\alpha = (\alpha_1, \ldots, \alpha_d)$ we have $\varrho^{-1} \partial^{(\alpha)} \varrho \in L^p(\gamma \circ F^{-1})$.*

PROOF. We have seen that for any $\varphi \in C_0^\infty(\mathbb{R}^d)$ the integral of $\partial^{(\alpha)} \varphi \varrho$ with respect to the measure γ equals the integral of $(\varphi \circ F) g$, where $g \in L^p(\gamma)$ for all $p < \infty$. Let $g_0 \circ F$ be the conditional expectation of g with respect to the σ-algebra generated by F. Then $g_0 \in L^p(\gamma \circ F^{-1})$. Hence we obtain $\partial^{(\alpha)} \varrho = (-1)^{|\alpha|} g_0 \varrho$, as required. □

9.7.5. Lemma. *If the hypotheses of Theorem 9.3.5 are fulfilled, then*

$$\lim_{t \to \infty} \| \exp(-t\varrho \circ F) \|_{W^{p,k}(\gamma)} = 0 \quad \text{for all } p, k \in \mathbb{N}.$$

PROOF. The quantity $\|D_H^k \exp(-t\varrho \circ F)\|_{\mathcal{H}_k}$ is easily estimated by the product of $\exp(-t\varrho \circ F)$ and a sum of products of $|t \partial^{(l)} \varrho \circ F|$ and $\|D_H^m F\|_{\mathcal{H}_m}$, where $l = (l_1, \ldots, l_d)$, $l_i, m \leqslant k$. By Hölder's inequality, it suffices to show that the integral of $t^n \exp(-t\varrho \circ F) |\partial^{(l)} \varrho \circ F|^n$ against the measure γ tends to zero as $t \to \infty$ for every n. This follows by the Lebesgue dominated convergence theorem, for we have the pointwise convergence to zero and the following uniform estimate (ensured by the inequality $e^{-s} \leqslant n! s^{-n}$):

$$t^n \exp(-t\varrho \circ F) |\partial^{(l)} \varrho \circ F|^n \leqslant n! |(\partial^{(l)} \varrho / \varrho) \circ F|^n,$$

where the right-hand side is γ-integrable by Proposition 9.7.4. □

9.7.6. Proposition. *Let the hypotheses of Theorem 9.3.5 be fulfilled. Then $\lim_{\varepsilon > 0} C_{p,k}(\{\varrho \circ F \leqslant \varepsilon\}) = 0$ for all p, k. Hence the set $(\varrho \circ F)^{-1}(0)$ is slim.*

PROOF. Since $\exp(-\varepsilon^{-1} \varrho \circ F) \geqslant e^{-1}$ on the set $\{\varrho \circ F \leqslant \varepsilon\}$, it remains to apply Lemma 9.7.5. □

By using these results, the following important fact was established in Hirsch, Song [548].

9.7.7. Theorem. *Under the hypotheses of Theorem 9.3.5, let ϱ be a continuous density of $\gamma \circ F^{-1}$. Then the set $\{\varrho > 0\}$ is connected.*

The following result in the Gaussian case was obtained in Ben Arous, Léandre [126], Hirsch, Song [547]; it is also proved in Nualart [848, p. 108]. We give a different proof.

9.7.8. Proposition. *Suppose that in Theorem 9.3.3 the measure μ is nonnegative, the hypotheses are fulfilled for all m and r, and that for the continuous version of the density we have $\varrho(x_0) = 0$ for some x_0. Then $\partial^{(\alpha)} \varrho(x_0) = 0$ for all α. In particular, this is true in the situation of Theorem 9.3.5.*

PROOF. It follows from our reasoning in §9.3 that $\partial^{(\alpha)} \varrho$ is the density of the measure $(\eta \cdot \mu) \circ F^{-1}$, where the function $\eta \in \bigcap_{1 \leqslant p < \infty} L^p(\mu)$ is such that $\partial_{h_{i_1}} \cdots \partial_{h_{i_m}} \eta \in \bigcap_{1 \leqslant p < \infty} L^p(\mu)$ for all $i_1, \ldots, i_m$ and all $m \in \mathbb{N}$. Therefore, $\partial^{(\alpha)} \varrho = (\eta_\alpha \cdot \mu) \circ F^{-1} = g_\alpha \cdot (\mu \circ F^{-1}) = g_\alpha \cdot \varrho$, where $g_\alpha \in \bigcap_{1 \leqslant p < \infty} L^p(\varrho \, dx)$. Now our claim follows by Corollary 3.4.11. □

Now, following Nourdin, Peccati [839], Nourdin, Viens [844], we derive a formula for induced densities which gives some information on behavior of densities on support.

Let $f \in W^{2,1}(\gamma)$ be mean zero and let $\sigma(f)$ be the σ-field generated by f. The corresponding conditional expectation will be denoted by $\mathbb{E}^{\sigma(f)}$. Set
$$\eta := -(D_H f, D_H L^{-1} f)_H,$$
where L^{-1} is defined on $1^\perp$ in $L^2(\gamma)$. There is a Borel function ψ_f on $\mathbb{R}^1$ such that
$$\mathbb{E}^{\sigma(f)} \eta = \psi_f(f).$$

9.7.9. Lemma. *One can always find a nonnegative function for ψ_f.*

PROOF. Let φ be a bounded Borel function with bounded support. Set
$$\Phi(t) = \int_0^t \varphi(s) \, ds, \ t \geqslant 0, \quad \Phi(t) = -\int_t^0 \varphi(s) \, ds, \ t < 0.$$
Clearly, Φ is a bounded Lipschitzian function, so $\Phi(f) \in W^{2,1}(\gamma)$, and we have $D_H \Phi(f) = \Phi'(f) D_H f$, whence
$$\mathbb{E}[f \Phi(f)] = \mathbb{E}[\Phi(f) L L^{-1} f] = -\mathbb{E}[\Phi(f) \delta(D_H L^{-1} f)] \qquad (9.7.1)$$
$$= -\mathbb{E}[(D_H \Phi(f), D_H L^{-1} f)_H] = -\mathbb{E}[\varphi(f)(D_H f, D_H L^{-1} f)_H].$$

Let $\varphi \geqslant 0$. Then $t\Phi(t) \geqslant 0$, so the left-hand side is nonnegative. In particular, this is true if φ is the indicator function of a bounded Borel set, which shows that $\mathbb{E}^{\sigma(f)}\eta \geqslant 0$. This yields our claim. □

Relation (9.7.1) leads to the following important identity:
$$\mathbb{E}[\varphi(f)\psi_f(f)] = \mathbb{E}[\varphi(f)\eta] = \mathbb{E}[f\Phi(f)]; \qquad (9.7.2)$$
whence for $\mu := \gamma \circ f^{-1}$ we obtain
$$\int_{\mathbb{R}^1} \Phi'(t)\psi_f(t)\,\mu(dt) = \int_{\mathbb{R}^1} t\Phi(t)\,\mu(dt). \qquad (9.7.3)$$
This identity can be written as the equation
$$(\psi_f \cdot \mu)' = -t \cdot \mu \qquad (9.7.4)$$
in the sense of distributions. Introducing the measure $\sigma := \psi_f \cdot \mu$ we find
$$\sigma' = -\frac{t}{\psi_f} \cdot \sigma, \qquad (9.7.5)$$
where t/ψ_f is integrable with respect to σ. This equality shows that the measure σ is Fomin differentiable, hence it has an absolutely continuous density θ (even if μ does not). An analogous construction for multidimensional mappings is considered in Airault, Malliavin, Viens [29]; see also Remark 11.1.12.

9.7.10. Theorem. *Suppose that the measure $\gamma \circ f^{-1}$ is absolutely continuous. Then its density ϱ_f admits the following representation:*
$$\varrho_f(t) = \frac{\|f\|_{L^1(\gamma)}}{2\psi_f(t)} \exp\left(-\int_0^t \frac{s}{\psi_f(s)}\,ds\right) \qquad (9.7.6)$$
for a.e. t in the support of $\gamma \circ f^{-1}$.

PROOF. We recall (see Proposition 9.7.1) that the topological support of $\gamma \circ f^{-1}$ is a closed interval $[a,b]$ (possibly unbounded). Since the integral of f vanishes, we have $0 \in (a,b)$. The density θ vanishes outside $[a,b]$. Since $\psi_f \geqslant 0$ and $\sigma \geqslant 0$, it follows from (9.7.4) that θ is increasing on $[a,0]$ and decreasing on $[0,b]$. Then we conclude that $\theta(t) > 0$ for all $t \in (a,b)$ and arrive at the relation
$$\frac{\theta'(t)}{\theta(t)} = -\frac{t}{\psi_f(t)}, \quad t \in (a,b),$$
whence we find
$$\theta(t) = \theta(0) \exp\left(-\int_0^t \frac{s}{\psi_f(s)}\,ds\right).$$
Since the integral of f vanishes, the integral of $|f|$ equals $2\theta(0)$. □

9.7.11. Corollary. *Suppose that $\gamma \circ f^{-1}$ has no atoms and that the function $s \mapsto s/\psi_f(s)$ is locally Lebesgue integrable in the interior of the support of $\gamma \circ f^{-1}$. Then $\gamma \circ f^{-1}$ admits a density expressed by (9.7.6).*

PROOF. Due to the local integrability of $t/\psi_f(t)$ on (a,b) it follows from (9.7.5) that θ on (a,b) is proportional to the exponent of the integral of the function $-t/\psi_f(t)$, hence is strictly positive on (a,b). Now (9.7.4) shows that μ has density θ/ψ_f. □

9.8. Comments and exercises

The following books are entirely or partly devoted to the Malliavin calculus and applications of Malliavin's method: Bell [**106**], [**108**], Bogachev [**191**], Da Prato [**307**], Davydov, Lifshitz, Smorodina [**336**], Fang [**414**], Huang, Yan [**564**], Ikeda, Watanabe [**571**], Malliavin [**767**], Nualart [**848**], Sanz-Sole [**982**], Shigekawa [**1017**], Üstünel [**1149**], Üstünel, Zakai [**1164**], Watanabe [**1178**]. Surveys of general character on the Malliavin calculus (including different approaches and typical applications) are given in Anulova et al. [**74**], Bernard [**144**], Bogachev [**189**], [**190**], Bogachev, Smolyanov [**233**], Daletskiĭ [**317**], Ikeda, Watanabe [**570**], Itô [**581**]–[**584**], Kusuoka [**681**], [**683**]–[**685**], Kusuoka, Stroock [**687**]–[**689**], Michel, Pardoux [**813**], Norris [**837**], Nualart [**849**], Nualart, Zakai [**858**], Ocone [**864**], Stroock [**1082**]–[**1085**], Veretennikov [**1171**], [**1172**] Williams [**1193**].

In relation to the Malliavin calculus, see also Bakhtin, Mattingly [**85**], Baudoin, Hairer [**101**], Berger [**143**], Cruzeiro, Zambrini [**302**], Donati-Martin, Pardoux [**353**], Ewald [**403**], Gong, Ma [**493**], Ikeda, Shigekawa, Taniguchi [**569**], Ikeda, Watanabe [**581**], Imkeller [**572**], Krée [**642**], [**643**], Komatsu [**627**], Kruk, Russo, Tudor [**648**], Kulik [**654**], Malliavin [**753**], [**754**], [**757**], [**764**], [**765**], Mattingly, Pardoux [**788**], Meyer [**805**], Osswald [**869**], Taniguchi [**1107**], Üstünel [**1143**]–[**1147**], Zhou [**1223**]. Some additional references to works published before 1990 can be found in [**233**].

In addition to the aforementioned works, the regularity of induced measures and smooth disintegrations are studied in Bally [**92**], Bismut [**154**], Bogachev [**184**], Daletskiĭ, Steblovskaya [**329**], Elliott, Kohlmann [**385**], Estrade, Pontier, Florchinger [**399**], Fang [**408**], Krée [**641**], Lanjri Zadi, Nualart [**696**], Lescot [**716**]–[**718**], Picard [**886**], Picard, Savona [**888**], Ren [**944**], Sintes-Blanc [**1039**], Steblovskaya [**1073**], Watanabe [**1181**], [**1182**].

Malliavin's method for diverse classes of non-Gaussian measures (jump processes, processes with independent increments, Lévy measures, etc.) are discussed in Applebaum [**75**], Bass, Cranston [**100**], Bichteler, Gravereaux, Jacod [**148**], Bismut [**151**], [**155**], Byczkowski, Graczyk [**256**], Carlen, Pardoux [**263**], Davydov, Smorodina [**336**], Dermoune [**341**], Dermoune, Krée [**342**], Di Nunno [**346**], Di Nunno et al. [**347**], Di Nunno, Øksendal, Proske [**348**], Elliott, Kohlmann [**384**], Elliott, Tsoi [**386**], [**387**], Franz, Privault, Schott [**450**], Fournier [**448**], Fournier, Giet [**449**], Graczyk [**501**], Gravereaux [**502**], Hiraba [**543**], Ishikawa, Kulik [**651**], [**655**], Kunita [**578**], Léandre [**699**], Norris [**838**], Osswald [**870**], Privault [**924**], Smorodina [**1060**], [**1061**], Solé, Utzet, Vives [**1070**], Wu [**1194**], Yablonski [**1196**]. Product measures are considered in Osswald [**868**].

The absolute continuity of distributions, requiring much weaker conditions, is studied in Bouleau, Hirsch [**245**]–[**247**], Imkeller, Nualart [**574**], Kusuoka [**675**], Shigekawa [**1009**].

Hargé [**517**]–[**519**] considered some analogs of the approximate continuity on the Wiener space.

An analog of Theorem 9.2.8 for decompositions in the Chebyshev–Hermit polynomials was obtained in Shevlyakov [**1007**].

Connections with white noise calculus developed in Hida [**537**], Hida et al. [**538**], Kuo [**671**], are also discussed in Potthoff [**913**], [**914**].

Stochastic calculus related to the fractional Brownion motion is studied in Biagini et al. [**146**], Decreusefond, Üstünel [**339**], Hu [**562**], Nualart [**848**], Nualart, Saussereau [**853**].

Differentiability along vector fields was introduced implicitly in the Malliavin calculus and in the construction of the extended stochastic integral of Skorohod (a closed idea was used in Hitsuda [**549**]); see Skorohod [**1047**], and also Kabanov, Skorohod [**587**]. Explicitly (including the concept of the vector logarithmic derivative) this was done by Yu.L. Daletskiĭ (see [**317**] and also Daletskiĭ, Maryanin [**320**], where general measures on nonlinear manifolds were considered). Norin [**832**], [**833**] considered differentiability along vector fields on linear spaces in relation to the construction of extended stochastic integrals for differentiable measures (see a detailed account in Norin [**834**]). Differentiability along vector fields was investigated in Bogachev [**176**], [**179**], [**189**], Bogachev, Smolyanov [**233**], Gaveau, Moulinier [**470**], Dalecky, Fomin [**319**]. Vector fields of differentiability arise naturally in extensions of the Malliavin calculus to non-Gaussian processes, whose distributions have no constant vectors of differentiability and so do not allow us to construct vector fields of differentiability in the same simple manner as in the Gaussian case. Divergence of Banach space valued vector fields is studied in Maas, Neerven [**747**], Mayer-Wolf, Zakai [**793**], [**794**].

Diverse results and further references related to the classes considered in §9.5 can be found in Airault, Malliavin [**20**], Bogachev [**180**], [**183**], [**184**], Bogachev, Smolyanov [**233**], Davydov [**333**], Davydov, Lifshits [**335**], Fang [**408**], Gaveau, Moulinier [**469**], Prat [**916**], [**917**], Smorodina, Lifshits [**1069**], Uglanov [**1136**]. On smoothness of stopping times and local times, see Airault, Malliavin, Ren [**25**], Airault, Ren, Zhang [**30**], Ren, Zhang [**949**].

The study of infinite dimensional oscillatory integrals in relation to smoothness of distributions was initiated in the papers Gaveau, Vauthier [**472**], Malliavin [**752**], [**756**], [**760**], Gaveau, Moulinier [**469**] for Gaussian measures and in the paper Uglanov [**1134**] for general smooth measures (see also Uglanov [**1141**], §3.2). This study was continued by many authors: concerning the case of smooth measures, see Bogachev [**177**]–[**180**], [**182**]–[**184**], Bogachev, Smolyanov [**233**], Daletsky, Steblovskaya [**330**], Steblovskaya [**1073**], [**1075**], and also Albeverio, Steblovskaya [**56**]; concerning the Gaussian case, see Ben Arous [**124**], Hara, Ikeda [**516**], Ikeda, Manabe [**568**], Malliavin [**758**], [**759**], Matsumoto, Taniguchi [**787**], Moulinier

[**818**], Nualart, Steblovskaya [**854**], Prat [**916**], Ren [**941**], Sugita, Taniguchi [**1094**], [**1095**], Taniguchi [**1110**]–[**1114**], Ueki [**1127**]. In [**184**], infinite dimensional oscillatory integrals dependent on a finite dimensional parameter were used for the proof of the rapid decreasing of the characteristic functionals of multidimensional mappings.

On applications of Malliavin's method to estimating the integrals of $\exp(-tF)$ and finding asymptotics of more general integrals, see Kusuoka, Osajima [**686**], Kusuoka, Stroock [**692**], Takanobu, Watanabe [**1099**].

A construction of surface measures generated by quasi-invariant or differentiable measures on infinite dimensional spaces was suggested by Skorohod [**1044**], [**1046**]. Another construction was developed by Uglanov [**1132**]–[**1136**], [**1139**]–[**1142**] (see also Efimova, Uglanov [**380**]), where his method was applied in particular to the study of smoothness of distributions of some functionals on spaces with smooth measures (the main goal of the construction was application to partial differential equations on infinite dimensional spaces). A geometric approach, on which this method is based, requires rather stringent restrictions of differential and topological character on the considered surfaces (in particular, the function determining the surface must be continuous; in addition, some conditions of continuity are imposed on its derivative). Uglanov's method was applied in Yakhlakov [**1197**]. A completely different approach was suggested in Airault, Malliavin [**20**] for Gaussian measures (earlier, for Gaussian measures, diverse results connected with surface measures were obtained in Goodman [**495**], Kuo [**666**], Hertle [**533**], [**534**] and implicitly in Stengle [**1081**], where in place of surface measures conditional expectations were used). In spite of the restriction on the class of measures, the principal feature of this work was that one considered measures on level sets of Sobolev functions without any topological conditions such as continuity. The results on the existence of surface measures were derived from the results on the regularity of distributions of functionals, unlike the reverse order in Uglanov's approach. In Bogachev [**181**], a construction of surface measures for not necessarily Gaussian smooth measures was developed on the basis of the Malliavin calculus. This approach was further substantially developed in the works Pugachev [**928**]–[**930**], [**931**], [**935**], which we follow in §9.8. Infinite dimensional variational problems connected with smooth measures and surface measures are studied in Dalecky, Steblovskaya [**325**], [**326**], Weizsäcker, Smolyanov [**1188**] and in a more general form in Uglanov [**1141**]. Infinite dimensional analogs of Hausdorff measures are studied in Feyel, de La Pradelle [**426**], Hino [**542**].

Convergence of finite dimensional images of measures in variation without assumption of nondegeneracy of derivatives of convergent measures is studied in Aleksandrova, Bogachev, Pilipenko [**59**], which we follow in §9.6. In the one-dimensional case this problem was considered in Davydov [**334**]. Convergence of densities of measures induced by nondegenerate mappings is investigated in Ren, Watanabe [**948**]. Example 9.6.16 covers Theorem 2.1 and Theorem 2.2 in Agrachev et al. [**5**]. Some applications are discussed in Breton [**253**], Kulik [**656**], [**658**]. Proposition 9.7.2 is well known for Gaussian measures (see Nualart [**848**]). On supports of induced measures, positivity and lower estimates of densities, see Aida, Kusuoka, Stroock [**10**], Chaleyat-Maurel, Nualart [**278**], Chaleyat-Maurel, Sanz-Sole [**279**], Hirsch, Song [**547**], [**548**], Kohatsu-Higa [**620**], Shirikyan [**1032**].

Let us mention the following open question. Let γ be a Radon Gaussian measure on a Hilbert space X and let f be a continuous polynomial on X such that its gradient has no zeros. Is it true that $\gamma \circ f^{-1}$ has a smooth density?

Exercises

9.8.1. Let f be a function on $[0,1]$. Prove that for every set E of Lebesgue measure zero belonging to the set of points of differentiability of f the set $f(E)$ has measure zero as well.

HINT: See [**193**, Chapter 5].

9.8.2. Let f be a function on $[0,1]$ and let E be the set of points at which f has a zero derivative. Prove that the set $f(E)$ has Lebesgue measure zero.

HINT: See [**193**, Chapter 5].

9.8.3. Prove that the assertion in Example 9.2.1 remains valid for every absolutely continuous function f.

HINT: See [**193**, Chapter 5].

9.8.4. Give an example of an infinitely differentiable function f on the real line for which the function
$$F(w) = \int_0^1 f(w_t)\,dt$$
on the space $C[0,1]$ with the Wiener measure has an unbounded density of distribution.

9.8.5. Can a Gaussian measure γ be differentiable along a Borel vector v for which $v(x) \notin H(\gamma)$ for all x?

9.8.6. Extend Theorem 9.2.9 to not necessarily Gaussian differentiable measures.

9.8.7. Provide the details of justification of Example 9.2.6 in the case of a nonconstant diffusion coefficient σ.

HINT: Observe that by the Girsanov theorem it suffices to consider the case $b = 0$; show that the collection of vectors $\partial_h \Phi(x)$, where h is Lipschitzian on $[0,1]$ and $h(0) = 0$, is $*$-dense in X for P-a.e. x; to this end take h such that $h'(s) = v$ on $[0, c]$ and $h'(0) = 0$ otherwise and use that the expectation of $\partial_h \Phi$ equals the integral of Av, where A is the integral of the expectation of $\sigma(\xi_s)$ over $[0, c]$ (see the equation for $\partial_h \Phi$ in the proof of Theorem 8.2.6).

9.8.8. Let ξ_1 and ξ_2 be independent standard Gaussian random variables. Show that $\xi_1^2 - \xi_2^2$ has an unbounded density of distribution. Show that $\xi_1 - \xi_2^2$ has a density of distribution of the class C_b^∞.

9.8. Comments and exercises

9.8.9. Let γ be a centered Radon Gaussian measure on a locally convex space X and let $F = (F_1, \ldots, F_n)$, where each function F_i belongs to $\mathcal{X}_0 \oplus \mathcal{X}_1 \oplus \mathcal{X}_2$. Suppose that all nontrivial linear combinations of the operators $D_H^2 F_1, \ldots, D_H^2 F_n$ have infinite dimensional ranges. Prove that $\gamma \circ F^{-1}$ has a density of the class $\mathcal{S}(\mathbb{R}^n)$.

9.8.10. Let γ be a centered Radon Gaussian measure on a locally convex space and let Q be a γ-measurable quadratic form on X. Show that a necessary and sufficient condition for the existence of a bounded density of distribution of Q is the existence of a two-dimensional subspace L in $H(\gamma)$ on which the form Q is positive or negative definite (moreover, in this case the density of distribution of Q is automatically of bounded variation).

HINT: If such a subspace L exists, then we consider the conditional measures on the sets $x + L$, which are Gaussian measures that differ by their means, and show that the restrictions of Q to $x + L$ have densities with uniformly bounded variations; alternatively, one can apply Malliavin's method to the vector field $\partial_{e_1} Q e_1 + \partial_{e_2} Q e_2$, where e_1 and e_2 span L and are orthogonal in $H(\gamma)$. If there is no such L, then, as noted in §1.4, the function Q can be represented as $\alpha_1 l_1^2 - \alpha_2 l^2 + c$, where l_1 and l_2 are γ-measurable linear functionals orthogonal in $L^2(\gamma)$, and α_1 and α_2 are numbers of different sign.

9.8.11. Show that the quadratic form
$$Q(w) = \int_0^1 \int_0^t q(t,s) \, dw_s \, dw_t, \quad q \in L^2([0,1]^2),$$
has a bounded density of distribution on $C[0,1]$ with the Wiener measure P_W precisely when there exists a two-dimensional plane $L \subset C_0^\infty[0,1]$ such that Q on L is either strictly positive definite or strictly negative definite.

HINT: Use the previous exercise and the fact that Q on $H(P_W)$ is given by
$$Q(h) = \int_0^1 \int_0^t q(t,s) \, h'(s) \, h'(t) \, ds \, dt.$$

9.8.12. Let $f(t) = t^k + c_{k-1} t^{k-1} + \cdots + c_0$ be a polynomial. Prove that Lebesgue measure of the set $\{t : |f(t)| \leq \varepsilon\}$ does not exceed $2k\varepsilon^{1/k}$.

9.8.13. (Fang [407]) Let γ be a centered Radon Gaussian measure on a locally convex space X and $F \in W^\infty(X, \mathbb{R}^d)$. Prove that there exists a sequence of mappings $F_n = (F_n^1, \ldots, F_n^d)$ of the class $W^\infty(X, \mathbb{R}^d)$ with
$$\left[\det(D_H F_n^i, D_H F_n^j)_{i,j \leq d}\right]^{-1} \in L^p(\gamma)$$
for all $p < \infty$ convergent to F in all Sobolev norms.

9.8.14. Let γ be a centered Radon Gaussian measure on a locally convex space X and let $F_n \in W^\infty(X, \mathbb{R}^d)$ be such that for all $p, r \in \mathbb{N}$ we have
$$\sup_n \|F_n\|_{W^{p,r}(\gamma, \mathbb{R}^d)} < \infty, \quad \sup_n \|\Delta_n^{-1}\|_{L^p(\gamma)} < \infty,$$
where Δ_n is the determinant of the Malliavin matrix for F_n. Prove that the densities of the measures $\gamma \circ F_n^{-1}$ are uniformly bounded in every Sobolev norm $\|\cdot\|_{p,k}$.

9.8.15. Let the measure γ be the same as in the previous exercise, $F \in W^{2,2}(\gamma)$, $D_H F \neq 0$ a.e. and $(|LF| + \|D_H^2 F(x)\|_{\mathcal{H}}) |D_H F|_H^{-2} \in L^1(\gamma)$. Prove that the measure

$\gamma \circ F^{-1}$ has a density ϱ of bounded variation (in particular, bounded). If
$$|D_H F|_H^{-1} \in L^4(\gamma),$$
then
$$\|\varrho\|_{L^\infty} \leqslant c_1 \|LF\|_{L^2(\gamma)} \| |D_H F|_H^{-1} \|_{L^4(\gamma)} \leqslant c_2 \|F\|_{2,2} \| |D_H F|_H^{-1} \|_{L^4(\gamma)},$$
where c_1 and c_2 are constants independent of F.

HINT: See Bogachev [**191**, Example 6.9.4].

9.8.16. Let μ be a Radon probability measure on a locally convex space X infinitely differentiable along vectors from a densely embedded separable Hilbert space H. Let $f \in W^\infty(\mu)$ and $E := \bigcup_{n=1}^\infty \{x \colon D_H^n f(x) \neq 0\}$. Prove that the measure $\mu|_E \circ f^{-1}$ is absolutely continuous.

HINT: If $f \in C^k(\mathbb{R}^1)$, then the set $\{f' = 0\} \cap \{f^{(k)} \neq 0\}$ does not contain its limit points.

9.8.17. Show that Proposition 9.7.2 may fail for mappings to $\mathbb{R}^2$ by considering the following example from Nualart [**848**, p. 107]: the mapping $F = (F_1, F_2)$ on $\mathbb{R}^2$ with the standard Gaussian measure is given by
$$F_1(x_1, x_2) = (\xi_1 + 3\pi/2)\cos(2\xi_2 + \pi), \quad F_2(x_1, x_2) = (\xi_1 + 3\pi/2)\sin(2\xi_2 + \pi),$$
where $\xi_1 = \arctan x_1$, $\xi_2 = \arctan x_2$

9.8.18. Let ξ_t be the diffusion process in $\mathbb{R}^d$ governed by the stochastic differential equation
$$d\xi_t = \sigma(\xi_t)dw_t + b(\xi_t)dt, \quad \xi_0 = 0,$$
where $\sigma = (\sigma^{ij})$ and $b = (b^i)$ are C_b^∞-mappings and σ^{-1} is uniformly bounded. Let μ^ξ denote the distribution of this process in the space $X = C_0([0,T], \mathbb{R}^d)$ and let $T\colon X \to \mathbb{R}^n$ be a continuous linear surjection. Show that the induced measure $\mu^\xi \circ T^{-1}$ admits a density of the class C_b^∞. In particular, the random vector
$$\eta(\omega) = \int_0^T \xi_t(\omega)\,dt$$
has a smooth distribution density.

9.8.19. Let γ be a centered Radon Gaussian measure on a locally convex space X and let polynomials $f_n \in \mathcal{P}_d(\gamma)$ be such that the sequence of measures $\gamma \circ f_n^{-1}$ is uniformly tight. Prove that $\{f_n\}$ is bounded in $L^2(\gamma)$.

HINT: Assuming that $c_n := \|f_n\|_2 \to +\infty$ as $n \to \infty$, show that the polynomials $g_n := f_n/c_n$ converge to zero in measure and obtain a contradiction with the fact that $\|g_n\|_2 = 1$.

9.8.20. Let μ be a probability measure on $\mathbb{R}^2$ with a continuous density $\varrho > 0$. Show that its projection on $\mathbb{R}^1$ has a density that is locally separated from zero.

9.8.21. Let μ be a probability measure on $\mathbb{R}^2$ with a smooth bounded density $\varrho > 0$. Is it true that its projection on $\mathbb{R}^1$ admits a continuous density?

9.8.22. Let μ be a probability measure on $\mathbb{R}^n$ with a density of the class $\mathcal{S}(\mathbb{R}^n)$. Show that its projection on $\mathbb{R}^{n-1}$ admits a density of the class $\mathcal{S}(\mathbb{R}^{n-1})$.

CHAPTER 10

Infinite dimensional transformations

This chapter is devoted to infinite dimensional transformations of differentiable measures. Mostly we are concerned with various properties of infinite dimensional images of smooth measures under smooth transformations. In particular, we study transformations taking a given measure into an equivalent one. Such problems arise in measure theory as well as in applications in the theory of random processes, mathematical physics, the theory of representations of groups, and other areas.

10.1. Linear transformations of Gaussian measures

Transformations of Gaussian measures are discussed in detail in the books Bogachev [**191**] and Üstünel, Zakai [**1164**], but in order to make our exposition self-contained we include the key results without proofs. Throughout this section we assume that γ is a centered Radon Gaussian measure on a locally convex space X, $H = H(\gamma)$ is its Cameron–Martin space with its Hilbert norm $|\cdot|_H$, and R_γ is its covariance operator.

For any operator $A \in \mathcal{L}(H)$ let $\widehat{A}$ denote the γ-measurable linear extension of A described in §1.4. The symbols $\mathcal{H}(H)$ or $\mathcal{H}$ denote the class of Hilbert–Schmidt operators on H.

10.1.1. Theorem. (i) *Let $T\colon X \to X$ be a γ-measurable linear mapping and $\gamma \circ T^{-1} = \gamma$. Let T_0 be a proper linear modification of T and let U be the restriction of T_0 to H. Then $U \in \mathcal{L}(H)$ and U^* is an isometry (i.e., U^* preserves distances). If U is injective, then U is an orthogonal operator.*

(ii) *Conversely, for every $U \in \mathcal{L}(H)$ such that U^* is an isometry, there exists a γ-measurable proper linear mapping T that preserves the measure γ and coincides with U on H.*

In infinite dimensional spaces, an operator T may preserve the measure γ without being injective on $H(\gamma)$. For example, let γ be the countable product of the standard Gaussian measures on the real line. Then the mapping $T\colon \mathbb{R}^\infty \to \mathbb{R}^\infty$, $Tx = (x_2, \ldots, x_n, \ldots)$, takes γ into γ, but is not injective on l^2. Hence the isometry U^* may fail to be a surjection: it is an isometry between H and $U^*(H)$.

10.1.2. Definition. *A measurable linear automorphism of the space X is a γ-measurable linear mapping T with the following properties:*

(i) there is a set Ω such that $\gamma(\Omega) = 1$ and T maps Ω one-to-one on Ω and $T(X\backslash\Omega) \subset X\backslash\Omega$;

(ii) for every $B \in \mathcal{B}(X)$ both sets $T(B)$ and $T^{-1}(B)$ are γ-measurable and $\gamma\bigl(T^{-1}(B)\bigr) = \gamma\bigl(T(B)\bigr) = \gamma(B)$.

10.1.3. Proposition. *Let $T\colon X \to X$ be a γ-measurable linear mapping. The following conditions are equivalent:*

(i) *the mapping T is a measurable linear automorphism;*

(ii) *the mapping T takes all sets of measure zero to sets of measure zero and its proper linear version is an orthogonal operator on $H(\gamma)$;*

(iii) *the mapping T takes all measurable sets to measurable sets and one has $\gamma(B) = \gamma\bigl(T^{-1}(B)\bigr) = \gamma\bigl(T(B)\bigr)$ for all $B \in \mathcal{B}(X)$.*

We recall that a mapping has Lusin's property (N) or satisfies Lusin's condition (N) if it takes all measure zero sets to measure zero sets.

10.1.4. Corollary. *Suppose that a γ-measurable linear mapping T satisfies Lusin's condition (N), its proper linear version is injective on $H(\gamma)$ and $\gamma \circ T^{-1} = \gamma$. Then T is a measurable linear automorphism.*

10.1.5. Corollary. *Let T be a measurable linear automorphism and let a mapping S be such that $S \circ T = T \circ S = I$. Then S is a measurable linear automorphism.*

We observe that not every measurable linear mapping T with $\gamma \circ T^{-1} = \gamma$ takes all sets of measure zero to sets of measure zero. For example, if γ is the countable power of the standard Gaussian measure on the real line, then one can take a linear subspace $X \subset \mathbb{R}^\infty$ such that $\gamma(X) = 1$ and an algebraic complement of X will have a Hamel basis $\{v_\alpha\}$ of cardinality of the continuum. Now we can redefine the identity operator on this algebraic complement mapping $\{v_\alpha\}$ onto $\mathbb{R}^\infty$. This will give a linear version of the identity operator which maps a set of measure zero onto the whole space. One can show that if $\gamma \circ T^{-1} = \gamma$ and $H(\gamma)$ is infinite dimensional, then T has a version which maps some measure zero set onto a set of full measure, but there is also a version which takes all sets of measure zero to sets of measure zero.

10.1.6. Proposition. *Let $A, B \in \mathcal{L}(H)$ and let $\widehat{B}$ transform the measure γ into an equivalent measure. Then $\widehat{AB} = \widehat{A}\widehat{B}$ γ-a.e.*

We shall say that an operator $A \in \mathcal{L}(H)$ has property (E) if A is invertible and $AA^* - I \in \mathcal{H}$.

10.1.7. Lemma. (i) *An operator $A \in \mathcal{L}(H)$ has property (E) precisely when $A = U(I+K)$, where U is an orthogonal operator and K is a symmetric Hilbert–Schmidt operator such that $I + K$ is invertible.*

(ii) *If $A \in \mathcal{L}(H)$ has property (E), then A^* and A^{-1} have this property as well. In addition, the composition of two operators with property (E) has this property.*

(iii) Let $A \in \mathcal{L}(H)$ and $A(H) = H$. Then $AA^* - I$ is a Hilbert–Schmidt operator if and only if $A = (I + S)W$, where S is a symmetric Hilbert–Schmidt operator, the operator $I + S$ is invertible and W^* is an isometry.

10.1.8. Theorem. (i) Let $T\colon X \to X$ be a γ-measurable linear mapping, let T_0 be its proper linear version, and let $\gamma \circ T^{-1} \sim \gamma$. Then $A := T_0|_H$ maps H continuously onto H and $AA^* - I \in \mathcal{H}$.

(ii) Conversely, for any operator $A \in \mathcal{L}(H)$ satisfying the conditions $A(H) = H$ and $AA^* - I \in \mathcal{H}$, there is a γ-measurable proper linear mapping T for which $T|_H = A$ and $\gamma \circ T^{-1} \sim \gamma$.

10.1.9. Corollary. Let T be a γ-measurable linear mapping. The following conditions are equivalent:

(i) a linear version of T maps H into H and has property (E) and T satisfies Lusin's condition (N);

(ii) there is a set Ω with $\gamma(\Omega) = 1$ such that T maps Ω one-to-one onto Ω, $T(X\backslash\Omega) \subset X\backslash\Omega$ and $\gamma \circ T^{-1} \sim \gamma$.

In this case there is a γ-measurable linear mapping S that is inverse to T, i.e., $TS = ST = I$.

10.1.10. Corollary. Suppose that a proper linear version of a γ-measurable linear mapping T has property (E) on H and f is an H-Lipschitzian function. Then $f \circ T$ is also H-Lipschitzian and $D_H(f \circ T)(x) = T^* D_H f(Tx)$. An analogous assertion is true for mappings f to any separable Hilbert space.

Now we give formulas for the Radon–Nikodym densities of equivalent Gaussian measures. Recall the concept of a regularized Fredholm–Carleman determinant for operators of the form $I + K$, $K \in \mathcal{H}$. The main idea is seen in the case where the operator K is diagonal and has eigenvalues k_i. Then the product $\det K := \prod_{i=1}^{\infty}(1 + k_i)$ may diverge if K has no trace. However, as one can easily verify, the product $\det_2 K := \prod_{i=1}^{\infty}(1 + k_i)e^{-k_i}$ converges. Here we have $\det_2 K = \det K \exp(-\operatorname{trace} K)$ if K is a nuclear operator.

Let $K \in \mathcal{H}$ be a finite dimensional operator with range $K(H)$. Set

$$\det{}_2(I + K) := \det\bigl(I + K|_{K(H)}\bigr) \exp\bigl(-\operatorname{trace} K|_{K(H)}\bigr).$$

Then the following Carleman inequality (see, e.g., Gohberg, Krein [488, Ch. IV, §2]) is fulfilled:

$$|\det{}_2(I + K)| \leqslant \exp\left[\frac{1}{2}\|K\|_{\mathcal{H}}^2\right]. \tag{10.1.1}$$

For finite dimensional operators A and B, letting $I + C = (I + A)(I + B)$, one has

$$\det{}_2(I + A) \det{}_2(I + B) = \det{}_2(I + C) \exp(\operatorname{trace} AB). \tag{10.1.2}$$

Now we can extend the function $\det_2$ to all Hilbert–Schmidt operators. If $K \in \mathcal{H}$ and the operator $I + K$ is not invertible (which corresponds to the existence of an eigenvalue -1 for K), then we put $\det_2(I + K) := 0$.

10.1.11. Proposition. *Let $K \in \mathcal{H}$ and let the operator $I+K$ be invertible. Then for every sequence of finite dimensional operators K_n convergent to K in the Hilbert–Schmidt norm, the sequence $\det{}_2(I + K_n)$ converges to a limit denoted by $\det{}_2(I+K)$ and independent of our choice of the approximating sequence. The function $K \mapsto \det{}_2(I+K)$ on the space $\mathcal{H}$ is locally uniformly continuous on the set of operators whose spectra do not contain -1. Moreover, the function $\det{}_2$ satisfies (10.1.1) and (10.1.2).*

If $\{e_n\}$ is an arbitrary orthonormal basis in H, then for every $K \in \mathcal{H}$ we have

$$\det{}_2(I+K) = \lim_{n\to\infty} \det\bigl(\delta_{ij} + (Ke_i, e_j)\bigr)_{i,j=1}^n \exp\Bigl[-\sum_{i=1}^n (Ke_i, e_i)\Bigr].$$

Let $K \in \mathcal{H}$ and let the operator $T = I + K$ on H be invertible. Set

$$\Lambda_K(x) := \bigl|\det{}_2(I+K)\bigr| \exp\Bigl[\delta K(x) - \frac{1}{2}|Kx|_H^2\Bigr].$$

10.1.12. Theorem. *Let $S = (I + \widehat{K})^{-1}$. Then*

$$\frac{d(\gamma \circ T^{-1})}{d\gamma}(x) = \frac{1}{\Lambda_K(Sx)}, \quad \frac{d(\gamma \circ S^{-1})}{d\gamma}(x) = \Lambda_K(x). \qquad (10.1.3)$$

10.1.13. Theorem. *Two centered Radon Gaussian measures μ and ν on X are equivalent precisely when $H(\mu)$ and $H(\nu)$ coincide as sets and there exists an invertible operator $C \in \mathcal{L}\bigl(H(\mu)\bigr)$ such that $CC^* - I \in \mathcal{H}\bigl(H(\mu)\bigr)$ and*

$$|h|_{H(\nu)} = |C^{-1}h|_{H(\mu)} \quad \text{for all } h \in H(\mu). \qquad (10.1.4)$$

If $C - I \in \mathcal{H}\bigl(H(\mu)\bigr)$, then

$$\frac{d\nu}{d\mu}(x) = \frac{1}{\Lambda_{C-I}\bigl(\widehat{C}^{-1}x\bigr)}. \qquad (10.1.5)$$

Finally, if $\mu \sim \nu$, then there exists a symmetric operator C on $H(\mu)$ such that $C - I \in \mathcal{H}\bigl(H(\mu)\bigr)$ and (10.1.5) is fulfilled.

10.1.14. Corollary. *Let μ and ν be two equivalent centered Radon Gaussian measures on X. Then there exist an orthonormal basis $\{e_n\}$ in $H(\mu)$ and a sequence $\{\lambda_n\}$ of real numbers distinct from -1 such that $\sum_{n=1}^\infty \lambda_n^2 < \infty$ and for every sequence of independent standard Gaussian random variables ξ_n on a probability space (Ω, P) one has the equality*

$$\mu = P \circ \Bigl(\sum_{n=1}^\infty \xi_n e_n\Bigr)^{-1} \quad \text{and} \quad \nu = P \circ \Bigl(\sum_{n=1}^\infty (1+\lambda_n)\xi_n e_n\Bigr)^{-1}.$$

10.1.15. Corollary. *Two centered Radon Gaussian measures μ and ν on X are equivalent precisely when there exists an invertible symmetric nonnegative operator T on $H(\mu)$ such that $T - I \in \mathcal{H}\bigl(H(\mu)\bigr)$ and*

$$(f,g)_{L^2(\nu)} = (TR_\mu f, R_\mu g)_{H(\mu)}, \quad \forall f, g \in X^*. \qquad (10.1.6)$$

10.1. Linear transformations of Gaussian measures

An equivalent condition: the norms $\|f\|_{L^2(\mu)}$ and $\|f\|_{L^2(\nu)}$ are equivalent on X^* and the quadratic form $(f,f)_{L^2(\nu)} - (f,f)_{L^2(\mu)}$ on the space X_μ^* is generated by a Hilbert–Schmidt operator on X_μ^*.

10.1.16. Corollary. *Let μ and ν be two equivalent Radon Gaussian measures on X. Then $d\nu/d\mu = \exp F$, where F is a μ-measurable second order polynomial of the form*

$$F(x) = c + \sum_{n=1}^{\infty} c_n \xi_n(x) + \sum_{n=1}^{\infty} \alpha_n \big(\xi_n(x)^2 - 1\big) \quad \mu\text{-a.e.}, \qquad (10.1.7)$$

where $c \in \mathbb{R}^1$, $\sum_{n=1}^{\infty} c_n^2 < \infty$, $\sum_{n=1}^{\infty} \alpha_n^2 < \infty$, $\alpha_n < 1/2$, $\{\xi_n\}$ is an orthonormal basis in X_γ^*, and both series converge a.e. and in $L^2(\mu)$. Conversely, if F has such a form, then $\exp F \in L^1(\mu)$ and the measure with density $\|\exp F\|_{L^1(\mu)}^{-1} \exp F$ with respect to μ is Gaussian.

In the case where X is a separable Hilbert space and μ and ν have covariance operators K_μ and K_ν, we obtain $H(\mu) = \sqrt{K_\mu}(X)$. Assuming that K_μ and K_ν have dense ranges (which can be achieved by passing to the closure of $H(\mu)$ in X), we write C in the form $C = \sqrt{K_\nu}\sqrt{K_\mu}^{-1}$. On the other hand, $C = \sqrt{K_\mu} C_0 \sqrt{K_\mu}^{-1}$, where $C_0 = \sqrt{K_\mu}^{-1}\sqrt{K_\nu} \in \mathcal{L}(X)$ is an invertible operator. Here we have $C - I \in \mathcal{H}(H(\mu))$ precisely when $C_0 - I \in \mathcal{H}(X)$. Therefore, the equivalence of the measures μ and ν is characterized by the continuity and invertibility of the operator $\sqrt{K_\mu}^{-1}\sqrt{K_\nu}$ and the inclusion $\sqrt{K_\mu}^{-1}\sqrt{K_\nu} - I \in \mathcal{H}(X)$. The latter condition can be written as

$$\sqrt{K_\mu}^{-1} K_\nu \sqrt{K_\mu}^{-1} - I \in \mathcal{H}(X) \qquad (10.1.8)$$

since $A - I \in \mathcal{H}(X)$ precisely when $AA^* - I \in \mathcal{H}(X)$. Certainly, here one can interchange μ and ν.

In the considered Hilbert case there is a sufficient (but not necessary) condition for equivalence that does not require finding square roots of the covariance operators.

10.1.17. Example. Suppose that $H(\mu) = H(\nu)$ and $K_\nu = (I+Q)K_\mu$, where $Q \in \mathcal{H}(X)$ and the operator is $I + Q$ invertible. Then $\mu \sim \nu$.

10.1.18. Example. A Gaussian measure ν on $L^2[0,1]$ is equivalent to the Wiener measure P^W if and only if $a_\nu \in H(P^W)$ and its covariance operator R_ν is an integral operator with a kernel K_ν of the following form:

$$K_\nu(t,s) = \min(t,s) + \int_0^t \int_0^s Q(u,v)\, du\, dv,$$

where $Q \in L^2\big([0,1]^2\big)$ is a symmetric function such that it generates the integral operator without eigenvalue -1. In this case for a.e. (t,s) one has the equality $Q(t,s) = \partial_t \partial_s K_\nu(t,s)$.

10.1.19. Example. Let $\tau \in C^1[0,1]$, $\tau'(t) > 0$, $\tau(0) = 0$, $\tau(1) = 1$. Set
$$Tx(t) = x(\tau(t))/\sqrt{\tau'(t)}.$$
The measure $\nu = P^W \circ T^{-1}$, i.e., the distribution of the process $w_{\tau(t)}/\sqrt{\tau'(t)}$, is equivalent to the Wiener measure P^W precisely when the function τ' is absolutely continuous and $\tau'' \in L^2[0,1]$.

Let us find the Radon–Nikodym density of the measure induced by the Ornstein–Uhlenbeck process ξ on $[0,1]$ with $\xi_0 = 0$ with respect to the Wiener measure P^W. Girsanov's theorem gives at once the equality
$$\frac{d\mu^\xi}{dP^W}(w) = \exp\left(-\frac{1}{2}\int_0^1 w_t\, dw_t - \frac{1}{8}\int_0^1 w_t^2\, dt\right), \tag{10.1.9}$$
where by the Itô formula we obtain
$$\int_0^1 w_s\, dw_s = \frac{1}{2}(w_1^2 - 1).$$
However, we shall derive the same expression from the previous results. The measure μ^ξ on $C[0,1]$ is the image of P^W under the linear operator T defined by the equation
$$Tx(t) = x(t) - \frac{1}{2}\int_0^t Tx(s)\, ds.$$
This equation is uniquely solvable. The inverse operator S is given by the formula
$$Sx(t) = x(t) + \frac{1}{2}\int_0^t x(s)\, ds.$$
It is readily seen that $Q = S - I$ is a nuclear operator on $H = H(P^W)$ and its complexification has no eigenvalues. In addition,
$$4|Qx|_H^2 = \|x\|_{L^2[0,1]}^2 \quad \text{and} \quad \delta Q(x) = -2^{-1}\sum_{n=1}^\infty (Qx, e_n)_H \widehat{e}_n(x),$$
which can be written as
$$-\frac{1}{2}\sum_{n=1}^\infty (x, e'_n)_{L^2[0,1]} \int_0^1 e'_n(s)\, dx(s) = -\frac{1}{2}\int_0^1 x(s)\, dx(s),$$
where $\{e_n\}$ is any orthonormal basis in H (i.e., $\{e'_n\}$ is a basis in $L^2[0,1]$). Now formula (10.1.9) follows from (10.1.5).

10.2. Nonlinear transformations of Gaussian measures

In this section we also assume that γ is a centered Radon Gaussian measure on a locally convex space X and $H = H(\gamma)$ is its Cameron–Martin space with the Hilbert norm $|\cdot|_H$. We shall consider images of the measure γ under measurable mappings $T\colon X \to X$ of the following special form:
$$T(x) = x + F(x), \quad \text{where } F\colon X \to H.$$

10.2. Nonlinear transformations of Gaussian measures

Typical results of this section state that the mapping $I + F$ transforms γ into an equivalent measure if $F \in W^{2,1}(\gamma, H)$ satisfies certain technical conditions and $I + D_H F$ is invertible on H. All proofs can be found in [**191**]. We start with a lemma of independent interest.

10.2.1. Lemma. *Let $T = I + F \colon X \to X$, where $F \colon X \to H$ is a Borel mapping such that for every $h \in H$ we have*

$$|F(x+h) - F(x)|_H \leqslant \lambda |h|_H \quad \gamma\text{-a.e.}, \tag{10.2.1}$$

where $\lambda < 1$. Then there exists a set Ω of full γ-measure which is mapped by T one-to-one onto itself and the inverse mapping S has the form $S = I + G$, where a Borel mapping $G \colon X \to H$ satisfies condition (10.2.1) with constant $\frac{\lambda}{1-\lambda}$ in place of λ. In addition, an inverse mapping of the indicated form is unique up to a redefinition on a set of measure zero.

Let us introduce the following notation for mappings F of the class $W^{2,1}_{\mathrm{loc}}(\gamma, H)$:

$$\Lambda_F(x) := \left|\det{}_2\bigl(I + D_H F(x)\bigr)\right| \exp\!\left[\delta F(x) - \frac{1}{2}|F(x)|_H^2\right].$$

10.2.2. Theorem. *Let $F \colon X \to H$ be a γ-measurable mapping such that for every $h \in H$ we have*

$$|F(x+h) - F(x)|_H \leqslant \lambda |h|_H \quad \gamma\text{-a.e.},$$

where $\lambda < 1$. Suppose that $D_H F(x)$ is a Hilbert–Schmidt operator for a.e. x and γ-a.e. one has $\|D_H F(x)\|_{\mathcal{H}} \leqslant M < \infty$. Then

(i) there exists a set Ω of full measure which is mapped by T one-to-one onto itself and the inverse mapping S has the form $S = I + G$, where G is Lipschitzian along H with constant $\lambda(1-\lambda)^{-1}$ and $\|D_H G\|_{\mathcal{H}} \leqslant M(1-\lambda)^{-1}$;

(ii) one has $\gamma \circ T^{-1} \ll \gamma$, $\gamma \circ S^{-1} \ll \gamma$, and

$$\frac{d(\gamma \circ T^{-1})}{d\gamma}(x) = \Lambda_G(x) = \frac{1}{\Lambda_F(T^{-1}(x))}, \quad \frac{d(\gamma \circ S^{-1})}{d\gamma} = \Lambda_F.$$

10.2.3. Definition. *Denote by $\mathcal{H}\text{-}\mathcal{C}^1_{\mathrm{loc}}$ the class of γ-measurable mappings $F \colon X \to H$ for which there exists a γ-measurable function τ such that $\tau(x) > 0$ a.e. and for a.e. x the mapping $h \mapsto F(x+h)$ is Fréchet differentiable along H at every point of the ball $U_{\tau(x)} = \{h \in H \colon |h|_H < \tau(x)\}$, the corresponding derivative is a Hilbert–Schmidt operator, and the mapping $h \mapsto D_H F(x+h)$, $U_{\tau(x)} \to \mathcal{H}$, is continuous.*

10.2.4. Theorem. *Let $T(x) = x + F(x)$, where $F \colon X \to H$ is a mapping of the class $\mathcal{H}\text{-}\mathcal{C}^1_{\mathrm{loc}}$. Set*

$$M = \bigl\{x \colon \det{}_2\bigl(I + D_H F(x)\bigr) \neq 0\bigr\}.$$

Then there exists a partition of M into disjoint measurable sets M_n such that on each M_n we have $T = T_n = I + F_n$, where, for every n, the mapping F_n belongs to $W^{2,1}(\gamma, H)$ and is bounded and Lipschitzian along H, T_n is *bijective and takes γ into an equivalent measure. In addition, for every bounded measurable function g one has the equality*

$$\int_X g(T_n(x)) \Lambda_{F_n}(x) \gamma(dx) = \int_X g(x) \gamma(dx). \tag{10.2.2}$$

Furthermore, the set $T^{-1}(x) \cap M$ for a.e. x is of at most countable cardinality $N(x, M)$ and for every bounded measurable function f one has the equality

$$\int_{T(M_n)} f(x) \gamma(dx) = \int_{M_n} f(T(x)) \Lambda_F(x) \gamma(dx).$$

Finally, $\gamma|_M \circ T^{-1} \ll \gamma$ and

$$\frac{d(\gamma|_M \circ T^{-1})}{d\gamma}(x) = \sum_{y \in T^{-1}(x) \cap M} \frac{1}{\Lambda_F(y)}.$$

It is unknown whether the assumption of continuity of $D_H F$ along H can be omitted in this theorem.

10.2.5. Corollary. *If T in Theorem 10.2.4 is bijective and M has full measure, then $\gamma \circ T^{-1} \sim \gamma$ and for $S = T^{-1}$ we have*

$$\frac{d(\gamma \circ T^{-1})}{d\gamma} = \frac{1}{\Lambda_F \circ T^{-1}}, \quad \frac{d(\gamma \circ S^{-1})}{d\gamma} = \Lambda_F.$$

10.2.6. Corollary. *Let $Tx = x + F(x)$, where a γ-measurable mapping $F\colon X \to H$ is such that for γ-a.e. x the mapping $h \mapsto F(x+h)$ is Fréchet differentiable, $D_H F(x) \in \mathcal{H}$ a.e. and the mapping $h \mapsto D_H F(x+h)$, $H \to \mathcal{H}$, is continuous. If the operator $D_H F(x)$ has no eigenvalue -1 for all x from a measurable set B, then the measure $\gamma|_B \circ T^{-1}$ is absolutely continuous with respect to γ.*

10.2.7. Example. *Let γ be a centered Gaussian measure on a separable Banach space X, let $H = H(\gamma)$, and let $F\colon X \to H$ be a continuously Fréchet differentiable mapping, where H is equipped with the norm $|\cdot|_H$. Suppose that $I + F'(x)$ maps H one-to-one onto H for each x. Then we have $\gamma \circ (I + F)^{-1} \sim \gamma$.*

Let us apply the presented results for the proof of the Girsanov theorem on the distribution of the diffusion process defined by the stochastic differential equation

$$d\xi_t = dw_t + b(\xi_t)dt, \quad \xi_0 = 0.$$

For simplicity we consider the one-dimensional case. Let $b \in C_b^\infty(\mathbb{R}^1)$. The measure μ^ξ on the space $C[0,1]$ is the image of the Wiener measure P^W under the mapping $T(w)(t) = \xi_t(w)$ determined from the integral equation

$$T(w)(t) = w(t) + \int_0^t b(T(w)(s)) \, ds.$$

10.2. Nonlinear transformations of Gaussian measures

This equation is uniquely solvable on $[0,1]$ since for every continuous function φ the mapping

$$V(x)(t) = \varphi(t) + \int_a^t b(x(s))\, ds$$

is a contraction of $C[a, a_1]$ if $|a_1 - a| \sup |b'| < 1$. The inverse mapping $S = T^{-1}$ is given by the formula

$$S(x)(t) = x(t) - \int_0^t b(x(s))\, ds.$$

It is clear that $G := S - I$ takes values in $H = H(P^W)$ and is infinitely Fréchet differentiable. For every x the operator $D_H G(x)$ is nuclear. One has the equality

$$\partial_h G(x)(t) = -\int_0^t b'(x(s)) h(s)\, ds, \quad h \in C[0,1].$$

It is verified directly that the complexification of the operator $D_H G(x)$ has no eigenvalues since the linear equation $\lambda h'(t) = -b'(x(t)) h(t)$, $h(0) = 0$, has only zero solution in the complexification of $W_0^{2,1}[0,1]$. Hence we have $\det_2(I + D_H G(x)) = 1$ and $\operatorname{trace}_H D_H G(x) = 0$. In addition,

$$|G(x)|_H^2 = \int_0^1 |b(x(s))|^2\, ds.$$

Let $\{e_n\}$ be an orthonormal basis in H. Then $\{e_n'\}$ is an orthonormal basis in $L^2[0,1]$, whence we obtain

$$b(x(s)) = \sum_{n=1}^\infty (b \circ x, e_n')_{L^2[0,1]} e_n'(s)$$

in $L^2[0,1]$. Therefore, $\delta G(x)$ coincides with

$$\sum_{n=1}^\infty (G(x), e_n)_H \widehat{e}_n(x) = -\sum_{n=1}^\infty \int_0^1 b(x(s)) e_n'(s)\, ds \int_0^1 e_n'(s)\, dx(s)$$

$$= -\int_0^1 b(x(s))\, dx(s).$$

Now the formula for $\Lambda_G(x)$ gives the same expression as the Girsanov theorem:

$$\frac{d\mu^\xi}{dP^W}(x) = \exp\left(\int_0^1 b(x(t))\, dx(t) - \frac{1}{2} \int_0^1 |b(x(t))|^2\, dt\right).$$

In the case of a bounded Borel drift b one can find a uniformly bounded sequence of mappings $b_j \in C_b^\infty(\mathbb{R}^1)$ convergent a.e. to b. Then the corresponding Radon–Nikodym densities are uniformly integrable, which yields $\mu^\xi \ll P^W$. Finally, one can easily derive the desired formula for the density.

10.3. Transformations of smooth measures

The results analogous to those obtained above are also known for some other classes of measures. However, the corresponding conditions and formulations turn out to be either complicated even for linear transformations or require more stringent restrictions on measures and transformations. Let us consider transformations of a Radon probability measure μ on a locally convex space X differentiable along vectors of a separable Hilbert space H, continuously embedded into X, under an invertible Borel mapping of the form $Tx = x + F(x)$, where $F\colon X \to H$. Suppose first that $X = \mathbb{R}^n$ and $\mu = \varrho\, dx$ has the logarithmic gradient $\beta \in W^{1,1}_{\mathrm{loc}}(\mathbb{R}^n, \mathbb{R}^n)$.

10.3.1. Proposition. *If T is one-to-one, locally Lipschitzian and one has $\det DT(x) \ne 0$ a.e., then, letting $S = T^{-1}$, we have $\mu \circ S^{-1} \sim \mu$, $\mu \circ T^{-1} \sim \mu$, and*

$$\frac{d(\mu \circ S^{-1})}{d\mu} = \Lambda_F, \qquad \frac{d(\mu \circ T^{-1})}{d\mu}(x) = \bigl[\Lambda_F(T^{-1}(x))\bigr]^{-1}, \qquad (10.3.1)$$

$$\Lambda_F := |\det{}_2 DT| \exp\left(\delta_\mu F + \int_0^1 (1-t)\bigl(D\beta(x + tF(x))F(x), F(x)\bigr)\, dt\right),$$

where $\delta_\mu F = \operatorname{div} F + (F, \beta)$.

PROOF. Let $S = T^{-1}$. By the change of variables formula the measure $\mu \circ S^{-1}$ has a density

$$\Lambda_F(x) = |\det DT(x)| \exp\Bigl[\ln \varrho(x + F(x)) - \ln \varrho(x)\Bigr].$$

with respect to μ. Writing the difference of logarithms by Taylor's formula with the remainder in the Lagrange form and taking into account the equality $\nabla \ln \varrho = \beta$, we obtain the desired formula for Λ_F. $\square$

Let us consider the infinite dimensional case. Let $\{e_n\}$ be an orthonormal basis in H and let P_n be the projector in H on the linear span H_n of the vectors $e_1, \ldots, e_n$. Let us decompose X into a direct topological sum $X = H_n \oplus Y_n$. Suppose that a Radon probability measure μ on X is differentiable along all e_n and, for all $i = 1, \ldots, n$ and $y \in Y_n$, the functions $z \mapsto \beta^\mu_{e_i}(y + z)$ belong to the class $W^{1,1}_{\mathrm{loc}}(H_n)$. Then the matrices $\Phi_n(x) := \bigl(\partial_{e_i}\beta^\mu_{e_j}(x)\bigr)_{i,j \le n}$ exist μ-a.e. For the measure on $\mathbb{R}^n$ with density $\exp(-V)$ this matrix is $-V''$. In particular, if the function V is convex, then this matrix is nonpositive.

Let $\sup_x |F(x)|_H = M_0 < \infty$ and let

$$|F(x+h) - F(x)|_H \le \lambda |h|_H, \qquad x \in X, h \in H,$$

where $\lambda < 1$. Then there exists a mapping $S = T^{-1}$ that is Lipschitzian along H with constant $\lambda(1-\lambda)^{-1}$.

10.3.2. Theorem. *Let numbers $C \ge 0$ and $M \ge 0$ be such that*

$$\|D_H F(x)\|_\mathcal{H} \le C < \infty, \qquad \|\Phi_n\|_{\mathcal{L}(H_n)} \le M \quad \forall n.$$

10.3. Transformations of smooth measures

Then we have $\mu \circ T^{-1} \ll \mu$ and $\mu \circ S^{-1} \ll \mu$. If there exists a bounded mapping $\Psi\colon X \to \mathcal{L}(H)$ for which one has $\partial_{e_i}\beta^\mu_{e_j}(x) = \bigl(\Psi(x)e_i, e_j\bigr)_H$ a.e., then we have $\mu \circ T^{-1} \sim \mu \sim \mu \circ S^{-1}$ and

$$\frac{d(\mu \circ S^{-1})}{d\mu} = \Lambda_F, \quad \frac{d(\mu \circ T^{-1})}{d\mu}(x) = \bigl[\Lambda_F\bigl(T^{-1}(x)\bigr)\bigr]^{-1}, \quad (10.3.2)$$

where we set

$$\Lambda_F := |\det{}_2(I + D_H F)| \exp(\delta_\mu F + \eta_F),$$

$$\eta_F := \int_0^1 (1-t)\bigl(\Psi\bigl(x + tF(x)\bigr)F(x), F(x)\bigr)_H\, dt.$$

PROOF. For $F_n := P_n F$ in place of F the first assertion follows from the previous proposition by using differentiable conditional measures μ^y on $H_n + y$, where $y \in Y_n$. Since $\beta^{\mu^y}_{e_i}(z) = \beta^\mu_{e_i}(y + z)$, $z \in H_n$, the hypotheses of this proposition are fulfilled. In addition, the functions Λ_{F_n} are uniformly integrable. Indeed, let us take $q > 1$ such that $q\lambda < 1$. Clearly, the same is true also for λF_n in place of F_n. We have $|\eta_{qF_n}| \leq q^2 M M_0^2$ and

$$|\det{}_2(I + qD_H F_n)| \leq \exp\bigl(\|qD_H P_n F\|_{\mathcal{H}}^2/2\bigr) \leq \exp(q^2 C^2 / 2).$$

Let $T_{n,q} := I + qF_n$. Then $T_{n,q}^{-1} = I + G_n$ and

$$\bigl(I + qD_H F_n \circ (I + G_n)\bigr) \circ (I + D_H G_n) = I.$$

Setting $R := qD_H F_n \circ (I + G_n)$, we find from the previous equality that

$$D_H G_n = (I + R)^{-1} - I = -R(I + R)^{-1},$$

whence $\|D_H G_n\|_{\mathcal{H}} \leq qC(1-q\lambda)^{-1}$. It follows from (10.1.2) that

$$|\det{}_2(I + qD_H F_n)| \geq \varepsilon > 0.$$

Since the integral of the function $|\det{}_2(I + qD_H F_n)|\exp(q\delta_\mu F_n + \eta_{qF_n})$ equals 1, the obtained estimates show that the functions $\exp(q\delta_\mu F_n)$ have uniformly bounded integrals. This gives the boundedness in $L^q(\mu)$ of the functions

$$\Lambda_{F_n} = d\bigl(\mu\circ(I + G_n)^{-1}\bigr)/d\mu.$$

This reasoning also applies to $I - qF_n$. Hence the sequence of functions $\exp(-\delta_\mu F_n)$ is bounded in $L^q(\mu)$. Therefore, the sequence of functions $\Lambda_{F_n}^{-1}$ is bounded in $L^q(\mu)$ as well. By the equality $\mu\circ(I+G_n)^{-1} = \Lambda_{F_n}\cdot\mu$ and the change of variables formula we have

$$\int_X \bigl[\Lambda_{F_n}\circ(I+F_n)^{-1}\bigr]^{-q}\, d\mu = \int_X \Lambda_{F_n}^{1-q}\, d\mu.$$

Thus, the functions $d\bigl(\mu\circ(I+F_n)^{-1}\bigr)/d\mu$ are uniformly integrable as well. Since $I + F_n \to I + F$ a.e. and $I + G_n \to I + G$ a.e., the first assertion follows from Theorem 1.2.20. For the proof of the formula for the density we observe that, as $n \to \infty$, we have $\det{}_2(I + D_H F_n) \to \det{}_2(I + D_H F)$ and $\delta_\mu F_n \to \delta_\mu F$ in measure μ. So we should verify convergence $\eta_{F_n} \to \eta_F$ in measure, for which it suffices to apply the last assertion of Theorem 1.2.20 to the measure $\mu\otimes\lambda$, where λ is Lebesgue measure on $[0,1]$. □

From this result one can derive an analog of Theorem 10.2.4 (see Kulik, Pilipenko [**659**]). However, one should keep in mind that the uniform boundedness of $\Psi(x)$ is a very strong restriction on μ. So it would be of interest to find other sufficient conditions for the absolute continuity of transformations of various classes of non-Gaussian measures.

10.4. Absolutely continuous flows

Let v be a vector field on a locally convex space X with a Radon measure μ. Let us consider the differential equation

$$x'(t) = v(x(t)), \quad x(0) = x. \qquad (10.4.1)$$

For applications it is important to know whether there exists a flow of transformations $\{U_t\}$, i.e., $U_t \circ U_s = U_{t+s}$, such that the family $\{U_t(x)\}$ is a solution to (10.4.1). Even in the case where v is a bounded continuous field on the real line, there might be no flow resolving (10.4.1). A sufficient condition for the existence of a flow of solutions is the global Lipschitzness of v (in the case of a Banach space). However, in applications one has to deal with vector fields that do not satisfy this restrictive condition. Then the situation becomes more complicated even if v is a smooth vector field on $\mathbb{R}^n$. Another important problem is to find conditions under which the flow U_t preserves the measure μ or transforms it into equivalent measures. Both problems are important also in the infinite dimensional case. The first results here belong to Cruzeiro [**293**], [**294**], [**296**]. Her investigations were continued in Peters [**882**], [**883**] (Gaussian measures) and Bogachev, Mayer-Wolf [**210**], [**211**], [**212**] (smooth measures); see also further references in the comments. Let β_v^μ denote the logarithmic derivative of the measure μ along the vector field v.

10.4.1. Proposition. *Let μ be a differentiable probability measure on $\mathbb{R}^n$ with a locally strictly positive density and let v be a vector field of the class $W^{1,1}_{\mathrm{loc}}(\mathbb{R}^n, \mathbb{R}^n)$ such that for some $c > 0$ we have*

$$\int_{\mathbb{R}^n} \exp\bigl(c|\nabla v(x)|\bigr)\,\mu(dx) + \int_{\mathbb{R}^n} \exp\bigl(c|\beta_v^\mu(x)|\bigr)\,\mu(dx) < \infty.$$

Then there exists a flow $\{U_t(x)\}$ defined for all $t \in \mathbb{R}^1$ and μ-a.e. x such that for all t one has the equality

$$U_t(x) = x + \int_0^t v\bigl(U_s(x)\bigr)\,ds \quad \mu\text{-a.e.}$$

In addition, $\mu \circ U_t^{-1} = k_t \cdot \mu$, where for sufficiently small t we have

$$k_t(x) = \exp \int_0^t \beta_v^\mu\bigl(U_{-s}(x)\bigr)\,ds.$$

10.4. Absolutely continuous flows

10.4.2. Theorem. *Let γ be a centered Radon Gaussian measure on a locally convex space and let $v\colon X \to H = H(\gamma)$ be a Borel mapping of the class $G_{\mathrm{oper}}^{p,1}(\mu, H)$, where $p \geqslant 1$. Suppose that β_v^γ exists and*

$$\exp(\lambda \|D_H v\|_{\mathcal{L}(H)}) + \exp(\lambda |\beta_v^\gamma|) \in L^1(\gamma), \quad \forall \lambda \geqslant 0.$$

Then there exist transformations $U_t \colon X \to X$ for which

$$U_t(x) = x + \int_0^t v(U_s(x))\, ds \quad \text{for all } t \text{ for } \gamma\text{-a.e. } x.$$

In addition, the measures γ and $\gamma \circ U_t^{-1}$ are equivalent for all t and

$$k_t(x) = \frac{d(\gamma \circ U_t^{-1})}{d\gamma}(x) = \exp \int_0^t \beta_v^\gamma(U_{-s}(x))\, ds.$$

Analogous assertions are valid for the equation

$$U_t(x) = Q_t x + \int_0^t Q_{t-s} v(U_s(x))\, ds$$

on a separable Banach space X, where $\{Q_t\}$ is a measurable family of continuous linear operators on X such that $Q_t|_H$ is a semigroup of orthogonal operators (see Bogachev, Mayer-Wolf [**211**], Peters [**883**]). There are analogous results for flows generated by vector fields dependent on time. The case of a Gaussian measure is considered in the book Üstünel, Zakai [**1164**] and the paper Ambrosio, Figalli [**64**], to which we return below.

Let H be a separable Hilbert space continuously and densely embedded in a locally convex space X and let $j\colon X^* \to H$ be the canonical embedding. Suppose that a Radon probability measure μ is differentiable along $j(X^*)$, $\{e_i\}$ is an orthonormal basis in H, and for every e_i there exists a number $\varepsilon_i > 0$ such that $\exp(\varepsilon_i |\beta_{e_i}^\mu|) \in L^1(\mu)$. Suppose also that $e_i = j(l_i)$, where a sequence $\{l_i\} \subset X^*$ generates $\mathcal{B}(X)$. Set $P_n x := \sum_{i=1}^n l_i(x) e_i$.

10.4.3. Theorem. *Suppose that $v\colon X \to H$ is a Borel mapping of the class $G_{\mathrm{oper}}^{2,1}(\mu, H)$ and that for some $c > 0$ and all $\lambda > 0$ we have*

$$|v|_H^{1+c} + \exp(\lambda \|D_H v\|_{\mathcal{L}(H)}) + \exp(\lambda |\beta_v^\mu|) \in L^1(\mu).$$

Then the conclusion of Theorem 10.4.2 *remains valid if*

$$\sup_n \int_X \exp(\lambda |\beta_{P_n v}^\mu(x)|)\, \mu(dx) < \infty, \quad \forall \lambda.$$

The latter condition can be efficiently verified for product measures (for example, for Gaussian measures), but in the general case it is rather restrictive.

We observe that in the case of a smooth field, for the existence of a flow it is not enough to have the boundedness of the divergence of the field. For example, let γ be the standard Gaussian measure on $\mathbb{R}^2$ and let v be defined by

$$v\colon x = (x_1, x_2) \mapsto \left(x_1^2, (2x_1 - x_1^3) e^{x_2^2/2} \int_{x_2}^\infty e^{-s^2/2}\, ds\right).$$

Then $v \in C^\infty$ and $|v|$, $\|\nabla v\| \in L^{1-\varepsilon}(\gamma)$ for all $\varepsilon \in (0,1)$, moreover,
$$\delta_\gamma v(x) = \operatorname{div} v(x) - (v(x), x) = 0$$
since $\beta^\gamma(x) = -x$. For every initial point x, the solution explodes in finite time.

In the paper Ambrosio, Figalli [**64**], the existence of an absolutely continuous flow on a space with a Gaussian measure γ is proven without assumption of the exponential integrability of $D_H v$. We shall state this result in a somewhat less general but simpler formulation (in particular, for v independent of time).

10.4.4. Theorem. *Suppose that $v \in W^{q,1}(\gamma, H)$, where $q > 1$, and that $\exp\bigl(c \min(\delta_\gamma v, 0)\bigr) \in L^1(\gamma)$ for some $c > 0$. Let $T \leqslant c/p'$. Then, there exist mappings U_t, $t \in [0,T]$, satisfying the same equation as in Theorem 10.4.2 such that $\gamma \circ U_t^{-1} = k_t \cdot \gamma$ and $\|k_t\|_{L^s(\mu)} \leqslant \bigl\|\exp\bigl(T\min(\delta_\gamma v, 0)\bigr)\bigr\|_{L^s(\gamma)}$ whenever one has $s \leqslant c/T$.*

Concerning the uniqueness problem for solutions to such equations, see Ambrosio, Figalli [**64**] and Bogachev, Mayer-Wolf [**211**], [**212**].

Bell [**106**], [**107**], Daletskiĭ, Sokhadze [**324**], and Dalecky, Fomin [**319**, Ch. VIII] investigated the situation where a flow generated by a vector field along which the measure is differentiable is given in advance. For smooth measures μ, one can consider transformations of the form $T(x) = x + v(x)$ included in the family $T_t(x) = x + tv(x)$, where $0 \leqslant t \leqslant 1$. Typically, such a family is not a flow, but one can modify the reasoning for flows and obtain a condition for the equivalence of the images of μ. This is done in Bell [**107**], although, so far the conditions on μ and v are very restrictive and are not efficient.

10.5. Negligible sets

In this section, we discuss several concepts of a "zero-set" in infinite dimensional locally convex spaces. Here there is no reasonable substitute for Lebesgue measure (as well as any preference, say, in the choice of a distinguished nondegenerate Gaussian measure among the continuum of mutually singular measures). So one has to define negligible sets without making use of any specific fixed measure, but by means of certain families of measures or in terms of suitable partitions. One such definition belongs to Christensen [**288**], who called a Borel set A in a Banach space X universally zero if there exists a Borel probability measure μ such that $\mu(A + x) = 0$ for all x. Another definition was suggested by Aronszajn [**79**], who introduced the following class $\mathfrak{A}^\mathcal{B}$ of exceptional Borel sets.

For every vector e in a locally convex space X we denote by $\mathfrak{A}^\mathcal{B}_e$ the class of all Borel sets A such that $\operatorname{mes}(t\colon x + te \in A) = 0$ for every $x \in X$, where mes is Lebesgue measure. For every sequence $\{e_n\}$ in X, let $\mathfrak{A}^\mathcal{B}\{e_n\}$ be the class of all sets of the form $A = \bigcup_n A_n$, where $A_n \in \mathfrak{A}^\mathcal{B}_{e_n}$ for all n.

10.5. Negligible sets

10.5.1. Definition. *A set A is called exceptional if it belongs to the class $\mathfrak{A}^\mathcal{B}\{e_n\}$ for every sequence $\{e_n\}$ with a dense linear span in X. The class of all exceptional sets will be denoted by $\mathfrak{A}^\mathcal{B}$.*

The corresponding class is smaller than Christensen's class, but both coincide with the class of Borel sets of zero Lebesgue measure in finite dimensional spaces. Then Phelps [**885**] introduced the class of Gaussian null sets.

10.5.2. Definition. *A Borel set A is called a Gaussian null set if it is zero for every nondegenerate Radon Gaussian measure on X. Let $\mathcal{G}^\mathcal{B}$ be the class of all such sets in X.*

It is easily verified (which was done in [**885**]) that $\mathfrak{A}^\mathcal{B} \subset \mathcal{G}^\mathcal{B}$. In [**166**], the author introduced the following definition.

10.5.3. Definition. *A set $A \in \mathcal{B}(X)$ is called negligible if it has measure zero for every measure differentiable along vectors from a sequence with a dense linear span. Let $\mathcal{P}^\mathcal{B}$ be the class of all such sets.*

The same class of sets is obtained if in place of densely differentiable measures we take densely continuous measures. This follows from the fact that every measure continuous along vectors a_n is absolutely continuous with respect to some measure differentiable along all a_n (see Theorem 5.3.2). In place of the class of Gaussian measures in Definition 10.5.2 one could take a larger class of measures of product-type (see Chapter 4) or the subclass of the latter class consisting of cubic measures considered in Mankiewicz [**776**].

10.5.4. Definition. (i) *A measure ν on a locally convex space X is called cubic if it is the image of the countable power of Lebesgue measure on $[0,1]$ under a mapping $T\colon [0,1]^\infty \to X$ of the form $Tx = c + \sum_{n=1}^\infty x_n a_n$, where $c \in X$, vectors a_n are linearly independent and have a dense linear span and $\sum_{n=1}^\infty p(a_n) < \infty$ for every continuous seminorm p.*
(ii) *A Borel set A is called a cubic zero set if $\nu(A) = 0$ for every cubic measure ν on X. The class of all such sets is denoted by $\mathcal{MK}^\mathcal{B}$.*

The coincidence of $\mathcal{MK}^\mathcal{B}$ and $\mathcal{P}^\mathcal{B}$ is almost obvious: since cubic measures are densely continuous, one has $\mathcal{P}^\mathcal{B} \subset \mathcal{MK}^\mathcal{B}$; on the other hand, as is clear from Theorem 5.3.2, for every densely continuous measure μ, one can find a cubic measure ν such that $\mu \ll \nu * |\mu|$, so we have $\nu * |\mu|(A) = 0$ for every $A \in \mathcal{MK}^\mathcal{B}$ since any shift of a cubic measure is a cubic measure, hence $\nu(A - x) = 0$ for all x. Cubic zero sets play an important role in the proof of the theorem below.

The classes introduced so far may be extended in such a way that in the finite dimensional case they will embrace all Lebesgue zero sets (not necessarily Borel). To this end, we denote by $\mathcal{L}$ the class of all sets in X that are measurable with respect to every measure on X differentiable along vectors from a sequence with a dense linear span. Then the above definitions extend naturally to $\mathcal{L}$ and lead to broader classes $\mathfrak{A}_e$, $\mathfrak{A}\{e_n\}$, $\mathfrak{A}$, $\mathcal{G}$, $\mathcal{P}$, and $\mathcal{MK}$, in the definitions of which we now allow sets from $\mathcal{L}$ and not only Borel sets. In the author's papers [167]–[169], it was shown that $\mathcal{G}^\mathcal{B} = \mathcal{P}^\mathcal{B}$ and $\mathfrak{A} = \mathcal{G} = \mathcal{P}$. This gave the inclusion $\mathcal{G}^\mathcal{B} \subset \mathfrak{A}$, however, it was unclear whether sets A_n in the corresponding decomposition can be chosen in $\mathcal{B}(X)$ (and not merely in $\mathcal{L}$), so that the problem of the coincidence of $\mathcal{G}^\mathcal{B}$ and $\mathfrak{A}^\mathcal{B}$ was not resolved. This problem had remained open for quite a long time until Csörnyei [304] gave a positive answer in the case of a separable Banach space. The final result is as follows.

10.5.5. Theorem. *One has $\mathfrak{A}^\mathcal{B} \subset \mathcal{G}^\mathcal{B} = \mathcal{P}^\mathcal{B}$ (if X is a separable Banach space, then $\mathfrak{A}^\mathcal{B} = \mathcal{G}^\mathcal{B}$) and $\mathfrak{A} = \mathcal{G} = \mathcal{P}$.*

PROOF. The equality $\mathcal{G}^\mathcal{B} = \mathfrak{A}^\mathcal{B}$ for separable Banach spaces, proved in [304], is the most difficult part of this theorem. In addition to the cited paper, it is proved in detail in the book Benyamini, Lindenstrauss [138, Theorem 6.32], so we shall not reproduce this proof. The inclusions $\mathcal{P}^\mathcal{B} \subset \mathcal{G}^\mathcal{B}$ and $\mathcal{P} \subset \mathcal{G}$ are trivial. The inclusion $\mathfrak{A}^\mathcal{B} \subset \mathcal{P}^\mathcal{B}$ is clear from the fact that any measure μ continuous along a vector e_n vanishes on the class $\mathfrak{A}^\mathcal{B}_{e_n}$ due to the existence of absolutely continuous conditional measures on the straight lines $x + \mathrm{I\!R}^1 e_n$. For the proof of the inclusion $\mathcal{G}^\mathcal{B} \subset \mathcal{P}^\mathcal{B}$ we consider a probability measure μ differentiable along vectors $\{e_n\}$ with a dense linear span. Then there exists a nondegenerate Radon Gaussian measure γ concentrated on a separable Hilbert space E continuously embedded into $D(\mu)$. If $B \in \mathcal{G}^\mathcal{B}$, then $\gamma(B - x) = 0$ for all x, whence $\mu * \gamma(B) = \gamma * \mu(B) = 0$. Hence $\mu(B - y) = 0$ for γ-a.e. $y \in E$, whence $\mu(B) = 0$ by the continuity of the function $y \mapsto \mu(B - y)$ on $D(\mu)$. The same reasoning gives the inclusion $\mathcal{G} \subset \mathcal{P}$. Finally, let us show that $\mathcal{P} \subset \mathfrak{A}$. Let $A \in \mathcal{P}$ and let $\{e_n\}$ have a dense linear span. We may assume that the vectors e_n are linearly independent. Denote by E_n the linear span of $e_1, \ldots, e_n$. The set $E = \bigcup_{n=1}^\infty E_n$ is a linear space. Let Γ be a linear subspace algebraically complementing E to X. Set $A_n := A \cap (\Gamma + E_{n-1})$. Then $A_n \in \mathcal{P}$ and $A = \bigcup_{n=1}^\infty A_n$. For every $x \in X$, the set $A_n \cap (x + \mathrm{I\!R}^1 e_n)$ contains at most one point. □

Let us give examples in which decompositions of sets into components from $\mathfrak{A}^\mathcal{B}_{e_n}$ are easy to construct.

10.5.6. Example. (i) Let A be a Borel set of Lebesgue measure zero in $\mathrm{I\!R}^n$. Set

$$B_1 := \{x \in A : \lambda(t : x + te_1 \in A) > 0\}, \quad A_1 := A \backslash B_1,$$

$$B_k := \{x \in B_{k-1} : \lambda(t : x + te_k \in A) > 0\}, \quad A_k := B_{k-1} \backslash B_k, \ k > 1,$$

where λ is Lebesgue measure on the real line and $\{e_k\}$ is a basis in $\mathrm{I\!R}^n$. Then we have $A = A_1 \cup \cdots \cup A_n$ and $A_k \in \mathfrak{A}^\mathcal{B}_{e_k}$.

(ii) Suppose that E is an n-dimensional subspace in a locally convex space X, $e_1, \ldots, e_n$ is a basis in E and λ_E is Lebesgue measure on E which arises when we identify E and $\mathrm{I\!R}^n$ by using the given basis. Let $A \subset X$ be a

Borel set such that $\lambda_E\bigl(E\cap(A-x)\bigr)=0$ for all $x\in X$. Then $A=A_1\cup\cdots\cup A_n$, where the sets $A_k\in\mathfrak{A}^{\mathcal{B}}_{e_k}$ are defined in the same manner as in (i).

PROOF. The sets B_k and A_k are Borel (see Lemma 1.2.14). In the case of $\mathrm{I\!R}^n$ we identify $\mathrm{I\!R}^{n-1}$ with the orthogonal complement of e_1. By Fubini's theorem the set $C:=\bigl\{x\in\mathrm{I\!R}^{n-1}\colon\lambda(B_1\cap(x+\mathrm{I\!R}^1 e_1)>0\bigr\}$ has measure zero. By using induction on n, we may assume that for C our assertion is already proven and $C=C_2\cup\cdots\cup C_n$, where sets $C_k\in\mathfrak{A}^{\mathcal{B}}_{e_k}$ are obtained by the indicated formula for C in place of A and for the basis $e_2,\ldots,e_n$ in $\mathrm{I\!R}^{n-1}$. Now we can take $A_k:=B_1\cap C_k$, $k\geqslant 2$. Assertion (ii) follows from what we have already proved since the sets A_k are Borel in the general case and the relationships $A=A_1\cup\cdots\cup A_n$ and $A_k\in\mathfrak{A}^{\mathcal{B}}_{e_k}$ can be verified in each subspace $x+E$. □

Similar classes $\mathfrak{A}^G$ and $\mathcal{G}^G$ arise if in place of $\mathcal{L}$ we take the class $\mathcal{L}^G$ of all sets measurable with respect to every nondegenerate Radon Gaussian measure on X. Then by the same reasoning as above one has $\mathfrak{A}^G=\mathcal{G}^G$. No examples are known distinguishing the classes $\mathcal{L}\subset\mathcal{L}^G$ (or the classes $\mathcal{G}^G$ and $\mathcal{P}$). An analogous question arises also for the class of cubic measures in place of the class of Gaussian measures.

10.5.7. Corollary. *For any set $A\in\mathcal{L}$, the following conditions are equivalent*:
(i) $\gamma(A)=0$ *for every nondegenerate Radon Gaussian measure γ*;
(ii) $\gamma(A)=0$ *for every nondegenerate Radon Gaussian measure on X with Hilbert support*;
(iii) $\mu(A)=0$ *for every measure μ continuous along vectors from a sequence with a dense linear span.*

Let us recall that a stable measure μ is called completely asymmetric if it cannot be represented in the form $\mu=\mu_1*\mu_2$, where μ_1 and μ_2 are stable measures of the same order as μ and μ_1 is symmetric, but is not a Dirac measure (see §4.2).

10.5.8. Corollary. *Let μ be a stable of order α Radon measure on a locally convex space X that is not completely asymmetric and is positive on all nonempty open sets. Then μ vanishes on all Borel negligible sets.*

PROOF. We have $\mu=\nu*\mu_0$, where μ_0 is the symmetric stable measure that is the symmetrization of μ, i.e., $\mu_0=\mu(2^{-1/\alpha}\cdot)*\mu(-2^{-1/\alpha}\cdot)$. The measure μ_0 is also positive on all nonempty open sets. It suffices to show that μ_0 vanishes on $\mathcal{P}^{\mathcal{B}}$. This is seen from the fact that by Theorem 4.2.1 the measure μ_0 has the form
$$\mu_0(B)=\int_T \gamma_t(B)\,\sigma(dt),$$
where T is some topological space with a Radon probability measure σ and each γ_t is a nondegenerate Radon Gaussian measure. □

The following result obtained in [**167**]–[**169**] gives a solution of Aronszajn's problem of characterization of measures vanishing on exceptional sets. We recall that a *densely differentiable* measure is a measure differentiable along vectors from a sequence with a dense linear span. Similarly, *densely continuous* measures are defined.

10.5.9. Theorem. (i) *A Radon measure μ on X vanishes on all sets from $\mathfrak{A}^\mathcal{B}\{e_n\}$ if and only if it is continuous along all vectors e_n.*

(ii) *A Radon measure μ vanishes on all sets from $\mathcal{P}^\mathcal{B}$ if and only if it belongs to the closure of the linear space generated by densely differentiable measures in the topology of convergence on Borel sets.*

PROOF. (i) If the measure μ is continuous along all e_n, then it has absolutely continuous conditional measures on the straight lines $x + \mathrm{IR}^1 e_n$. Hence it vanishes on $\mathfrak{A}^\mathcal{B}_{e_n}$. For the proof of the converse let us show that if the measure μ vanishes on $\mathfrak{A}^\mathcal{B}_h$, then it is continuous along h. We may assume that $\mu \geqslant 0$ passing to $|\mu|$. Let γ be the image of the standard Gaussian measure on the real line under the mapping $t \mapsto th$. We verify that $\mu \ll \gamma * \mu$. Let $A \in \mathcal{B}(X)$ and $\gamma * \mu(A) = 0$. The set $B := \{x \colon \gamma(A-x) > 0\}$ is Borel by Lemma 1.2.14 and $\mu(B) = 0$. Set $C := A \backslash B$. For any $x \in C$ we have $\gamma(A - x) = 0$, i.e., $\lambda(t \colon th + x \in C) = 0$, where λ is Lebesgue measure. We observe that the last equality holds for all $x \in X$ and not only for $x \in C$. This is clear from the fact that if the set $\{t \colon th + x \in C\}$ is not empty, then it can be written in the form $\{t \colon th + x_0 \in C\} + \alpha$, where $x_0 \in C$ and $\alpha \in \mathrm{IR}^1$ are chosen in such a way that $\alpha h + x = x_0 \in C$. Thus, $C \in \mathcal{A}^\mathcal{B}_h$ and $\mu(C) = 0$, whence $\mu(A) = 0$, i.e., $\mu \ll \gamma * \mu$. This yields the continuity of μ along h.

(ii) It is clear that any measure in the indicated closure vanishes on $\mathcal{P}^\mathcal{B}$. Conversely, suppose that a Radon measure μ vanishes on $\mathcal{P}^\mathcal{B}$. If μ does not belong to the aforementioned closure, then there exists a continuous linear functional l on the space of all Radon measures with the topology of convergence on Borel sets such that $l(\mu) > 0$, but $l(\nu) = 0$ for every densely differentiable measure ν. This functional is represented as the integral of some Borel function f with finitely many values. It is easy to show that once the integral of f is zero with respect to every densely differentiable measure, it vanishes also with respect to every densely continuous measure. This implies that the measure of the set $\{f \neq 0\}$ is zero for every densely continuous measure, hence for $|\mu|$, i.e., one has $l(\mu) = 0$. □

Naturally the question arises whether it suffices to take only the sequential closure in (ii), i.e., whether every measure vanishing on $\mathcal{P}^\mathcal{B}$ can be obtained as a limit of a sequence of linear combinations of densely differentiable measures. This question is connected with Aronszajn's problem on the existence of exceptional measures. A nonzero measure μ is called *exceptional in the sense of Aronszajn* if it vanishes on $\mathfrak{A}^\mathcal{B}$, but, for every sequence $\{e_n\}$ with a dense linear span, it has a set of full measure in $\mathfrak{A}^\mathcal{B}\{e_n\}$. This problem was solved positively in the author's paper [**167**]. Aronszajn proved himself that every measure μ vanishing on $\mathcal{A}^\mathcal{B}$ has the form $\mu = \mu_0 + \sum_{n=1}^\infty \mu_n$, where all measures μ_i are mutually singular, the measure μ_0 is either exceptional or zero, and every measure μ_n with $n \geqslant 1$ vanishes on $\mathfrak{A}^\mathcal{B}_{a_n}$ for some $a_n \in X$, i.e., is continuous along a_n as shown above.

10.5.10. Theorem. (i) *If a nonzero Radon measure μ is such that $\mu_x \perp \mu$ for all $x \neq 0$ and μ vanishes on $\mathfrak{A}^\mathcal{B}$, then μ is an exceptional measure.*

(ii) *There exist exceptional measures. In particular, the distribution of any non-Gaussian homogeneous stable process with independent increments is an exceptional measure.*

(iii) *Every exceptional measure belongs to the closure described in Theorem 10.5.9(ii), but does not belong to the sequential closure.*

PROOF. (i) Let $\mu = \mu_0 + \sum_{n=1}^\infty \mu_n$ be the decomposition given by the aforementioned theorem of Aronszajn. Our condition shows that $\mu_n = 0$ for all $n \geqslant 1$. Hence $\mu = \mu_0$ is an exceptional measure. (ii) For constructing an exceptional measure it suffices to find a measure of the form

$$\mu(B) = \int_T \gamma_t(B)\, \sigma(dt)$$

on a separable Hilbert space X, where each γ_t is a nondegenerate Gaussian measure on X and σ is some probability measure on the space T such that $\mu_x \perp \mu$ for all $x \neq 0$. An artificial example of this kind was constructed in [**167**] (see Example 4.5.1(i)), but one can use a natural example (see Example 4.2.4 or Example 4.5.1(ii)): the distribution μ^ξ in $L^2[0,1]$ of any homogeneous stable of order $\alpha \in (0,2)$ process with independent increments. We have seen in Example 4.2.4 that $\mu_x^\xi \perp \mu^\xi$ if $x \neq 0$. The fact that μ^ξ is a mixture of nondegenerate Gaussian measures follows from Theorem 4.2.1. (iii) It is clear that any exceptional measure μ belongs to the aforementioned closure. However, it cannot belong to the sequential closure. Indeed, suppose that μ is a limit (on every Borel set) of a sequence of measures of the form $\nu_{n,1} + \cdots + \nu_{n,k_n}$, where every measure $\nu_{n,m}$ is densely continuous. Then the measure μ cannot be mutually singular with every such measure $\nu_{n,m}$. Hence there exists a nonzero measure ν such that $\nu \ll \nu_{n,m}$ and $\nu \ll \mu$. However, this is impossible since the measure ν is at the same time exceptional (being absolutely continuous with respect to an exceptional measure) and densely continuous, so it vanishes on some class $\mathfrak{A}^\mathcal{B}\{e_n\}$. □

10.5.11. Remark. (i) In order for a set to be negligible, it suffices that all densely differentiable symmetric measures vanish on this set (since every densely differentiable measure is absolutely continuous with respect to some symmetric densely differentiable measure). However, there exist Borel sets that are not negligible, but are zero sets for all symmetric Gaussian measures. For example, let K be an injective compact selfadjoint operator in an infinite dimensional Hilbert space H and let $A = K(H) + h$, where

$h \notin K(H)$. The Borel set A is not negligible, but $\gamma(A) = 0$ for every symmetric Gaussian measure γ since $\gamma(A) = \gamma(-A)$ and $A \cap (-A) = \varnothing$, but $\gamma(A)$ is either 1 or 0 by the zero-one law.

(ii) We have defined negligible sets by means of an uncountable family of measures. It is easily verified that in the infinite dimensional case it is not enough to consider a countable family of measures. Moreover, even if we take a countable family of Radon probability measures m_n (say, on an infinite dimensional separable Hilbert space X) and declare a Borel set A to be negligible if $m_n(A - x) = 0$ for all n and all x, then the class thus obtained will not coincide with $\mathcal{P}^{\mathcal{B}}$. Indeed, let $m = \sum_{n=1}^{\infty} 2^{-n} m_n$. Let us take a compact operator with a range Y such that $m(Y) = 1$. There is a compact operator with a dense range Z such that $Y \cap Z = 0$ (Exercise 1.6.5). Then Z does not belong to $\mathcal{P}^{\mathcal{B}}$, although $m(Z - x) = 0$ for all $x \in X$ since $Y \cap (Z - x)$ contains at most one point, but a Dirac measure taken for m obviously does not produce the class $\mathcal{P}^{\mathcal{B}}$.

(iii) Although nondegenerate Gaussian measures and densely differentiable measures produce the same class of negligible sets, on every infinite dimensional separable Banach space there exists a densely differentiable probability measure that is mutually singular with every Gaussian measure (Remark 5.3.7). In addition, one can construct a cubic measure that is mutually singular with all Gaussian measures (Exercise 5.5.3). Nevertheless, the classes of Borel Gaussian null sets and Borel cubic null sets coincide.

The classes of negligible or exceptional sets introduced above are invariant with respect to affine isomorphisms of X and possess many useful properties of finite dimensional Lebesgue measure zero sets (see the results below on the differentiability of Lipschitzian functions). However, they are not stable with respect to nonlinear diffeomorphisms. For example, if f is a real analytic diffeomorphism of the real line that is not affine, then the mapping $F \colon x \mapsto f \circ x$ on $C[0, 1]$ is a diffeomorphism under which the image of some negligible set is not negligible, which follows from the next result obtained in Tolmachev [**1118**].

10.5.12. Theorem. *Let μ^ξ be the measure on $C[0, 1]$ generated by the diffusion process ξ satisfying the stochastic differential equation*

$$d\xi_t = \sigma(\xi_t) dw_t + b(\xi_t) dt, \quad \xi_0 = x_0,$$

where σ is real-analytic and nonconstant and b is bounded continuous. Then μ^ξ is concentrated on a Borel negligible set.

PROOF. Suppose we are given a sequence of finite partitions π_n of the interval $[a, b]$ by points $a = t_0^{(n)} < t_1^{(n)} < \cdots < t_{k(n)}^{(n)} = b$ such that $\max_{i \leqslant k(n)} |t_i^{(n)} - t_{i-1}^{(n)}|$ tends to 0 as $n \to \infty$. The quadratic variation of the function x on $[a, b]$ along the sequence of partitions $\{\pi_n\}$ of the interval

10.5. Negligible sets

$[a, b]$ is the quantity

$$Q^{a,b}_{\{\pi_n\}}(x) = \lim_{n\to\infty} Q^{a,b}_{\pi_n}(x), \quad Q^{a,b}_{\pi_n}(x) := \sum_{i=1}^{k(n)} |x(t_i^{(n)}) - x(t_{i-1}^{(n)})|^2$$

if this limit exists and is finite. For example, if x is Lipschitzian, then $Q^{a,b}_{\{\pi_n\}}(x) = 0$. Moreover, in this case $Q^{a,b}_{\{\pi_n\}}(y + x) = Q^{a,b}_{\{\pi_n\}}(y)$ if $Q^{a,b}_{\{\pi_n\}}(y)$ exists. It is known (see, e.g., [**477**, V. 2, Ch. I, §1, p. 43, Ch. III, §1, p. 229]) that, for every interval $[a, b] \subset [0, 1]$, for a.e. ω the trajectory $t \mapsto \xi(t, \omega)$ has finite quadratic variation along the sequence of partitions π_n obtained by dividing $[a, b]$ into 2^n equal intervals and one has

$$Q^{a,b}_{\{\pi_n\}}(\xi(\cdot, \omega)) = \int_a^b \sigma^2(\xi(u, \omega))\, du.$$

We shall consider only points a, b of the form $m2^{-n}$. Thus, for μ^ξ-a.e. x simultaneously for all such points we have

$$Q^{a,b}_{\{\pi_n\}}(x) = \int_a^b \sigma^2(x(u))\, du. \tag{10.5.1}$$

The set of all such points x will be denoted by X_0. We show that for all $x \in X_0$ and $h \in C_0[0, 1]$, $h \not\equiv 0$, the set

$$M_{x,h} := \{\lambda \colon x + \lambda h \in X_0\}$$

is at most countable. First we observe that if $x, h \in C[0, 1]$ and a, b are such that the quantity $Q^{a,b}_{\{\pi_n\}}(x + \lambda h)$ is defined at least for three different values of λ, then it is defined for all λ and the function

$$\lambda \mapsto P(\lambda) := Q^{a,b}_{\{\pi_n\}}(x + \lambda h)$$

is a second order polynomial. Indeed,

$$Q^{a,b}_{\pi_n}(x + \lambda h) = \sum_{i=1}^{k(n)} |x(t_i^{(n)}) - x(t_{i-1}^{(n)})|^2 + \lambda^2 \sum_{i=1}^{k(n)} |h(t_i^{(n)}) - h(t_{i-1}^{(n)})|^2$$

$$+ 2\lambda \sum_{i=1}^{k(n)} \big(x(t_i^{(n)}) - x(t_{i-1}^{(n)})\big)\big(h(t_i^{(n)}) - h(t_{i-1}^{(n)})\big).$$

It remains to use the fact that if a sequence of second order polynomials has a finite limit at three different points, then it converges everywhere to a second order polynomial. Suppose now that the set $M_{x,h}$ is uncountable. Then the quantity $P(\lambda)$ is defined for all λ. Set

$$F_{a,b}(\lambda) = \int_a^b \sigma^2(x(u) + \lambda h(u))\, du.$$

It is easily verified that the function $F_{a,b}$ is real-analytic and one has

$$F'_{a,b}(\lambda) = \int_a^b 2\sigma'(x(u) + \lambda h(u))\sigma(x(u) + \lambda h(u))h(u)\, du.$$

Therefore, $F_{a,b}(\lambda) = P(\lambda)$ for all λ. By the boundedness of $F_{a,b}$ this is only possible if both functions are constant. Thus, we have $F'_{a,b} = 0$ for all a, b of the above form. Hence $\sigma'(x(u) + \lambda h(u))\sigma(x(u) + \lambda h(u))h(u) = 0$. The latter is only possible if $\sigma' \equiv 0$ since $h(t_0) \neq 0$ for some t_0. □

Now, if g is a real-analytic function such that $0 < g' \leq C$, $|g''| \leq C$, then letting $\sigma(z) := g'(g^{-1}(z))$, we obtain by Itô's formula that the process $\xi_t := g(w_t)$ satisfies the hypotheses of the previous theorem. Hence the mapping $x \mapsto g \circ x$ takes a full measure set M for the Wiener measure (which is not negligible) to the negligible set X_0. The inverse mapping (given by $x \mapsto f \circ x$, $f = g^{-1}$) takes the negligible set X_0 to the set M that is not negligible. In the case where g' and g'' are not bounded, similar reasoning applies in a sufficiently small ball.

The following example is borrowed from [169].

10.5.13. Example. Let X be a separable Hilbert space X. There exist two nuclear selfadjoint operators A and B with dense ranges in X for which $A(X) \cap B(X) = 0$ (see Exercise 1.6.5). Let $F(x) = x + (x, x)Ax$. Then F is a diffeomorphism in some neighborhood of the origin V, the set $V \cap B(X)$ is not negligible, but the set $F(B(X))$ is negligible. In particular, under the inverse diffeomorphism F^{-1} the image of some negligible set is not negligible.

PROOF. Since $F'(0) = I$, by the inverse mapping theorem F is a diffeomorphism between some neighborhood of the origin V and $W = F(V)$. It is readily seen that the set $F(B(X))$ is Borel. We show that it is negligible. Even more is true: every straight line $y + \mathbb{R}^1 h$ intersects it in at most two points distinct from the origin. Indeed, let $y_1 = x_1 + (x_1, x_1)Ax_1$, $y_2 = x_2 + (x_2, x_2)Ax_2$ and $y = x + (x, x)Ax$ be distinct and belong to a straight line, where $x_1, x_2, x \in B(X)$. We may assume that y belongs to the interval joining y_1 and y_2, i.e., $y = \alpha y_1 + (1 - \alpha)y_2$, $\alpha \in (0, 1)$. Since $A(X) \cap B(X) = 0$, we have

$x = \alpha x_1 + (1 - \alpha)x_2$ and $(x, x)Ax = \alpha(x_1, x_1)Ax_1 + (1 - \alpha)(x_2, x_2)Ax_2$.

Hence $(x, x)x = \alpha(x_1, x_1)x_1 + (1 - \alpha)(x_2, x_2)x_2$. If $x \neq 0$, then we arrive at the equality $(x, x) = (x_1, x_1) = (x_2, x_2)$, which is impossible for points of the same interval. □

It was a long standing problem (raised in Yost [1209]) whether every Lipschitzian image of a Borel negligible set is zero in the sense of Christensen. Counterexamples were constructed in Matoušková [785], Lindenstrauss, Matoušková, Preiss [726].

10.6. Infinite dimensional Rademacher's theorem

Here we present several basic facts related to the Rademacher theorem in the infinite dimensional case. The classical Rademacher theorem asserts that any Lipschitzian mapping $F \colon \mathbb{R}^n \to \mathbb{R}^k$ is almost everywhere

Fréchet differentiable. This result has no straightforward infinite dimensional generalizations. The main reason is not the absence of infinite dimensional analogs of Lebesgue measure, but merely the existence of Lipschitzian mappings between Hilbert spaces without points of Fréchet differentiability. Nevertheless, there are many works devoted to diverse generalizations of the Rademacher theorem to the infinite dimensional case. The point is that in the finite dimensional case this theorem can be reformulated in many equivalent ways that admit infinite dimensional generalizations. Here we discuss one such possibility by using differentiable measures.

There are simple examples of Lipschitzian everywhere Gâteaux differentiable mappings from a Hilbert space H to an infinite dimensional Hilbert space E without points of Fréchet differentiability. However, the case of finite dimensional E had been a long standing problem before Preiss [918] finally proved that every real Lipschitzian function on an Asplund space (in particular, on any Hilbert space) is Fréchet differentiable at the points of some everywhere dense set. It is still unknown whether this is true if $\dim E > 1$. However, even for real functions, the positive result of Preiss does not give a generalization of the Rademacher theorem since, for every probability Borel measure μ on l^2, there exists a Lipschitzian convex function f which is not Fréchet differentiable μ-a.e.

10.6.1. Example. For every n, there exists a compact ellipsoid $K_n \subset l^2$ such that $\mu(K_n) > 1 - 1/n$. Set

$$f_n(x) := \operatorname{dist}(x, K_n) \quad \text{and} \quad f(x) := \sum_{n=1}^{\infty} 2^{-n} f_n(x).$$

Then the function f is convex, Lipschitzian with constant 1 and is not Fréchet differentiable μ-a.e.

PROOF. The functions f_n are convex, Lipschitzian with constant 1 and satisfy the inequality $f_n(x) \leqslant |x|$. Hence f has the same properties. We observe that f_n is not Fréchet differentiable at the points of K_n. Indeed, if $x_0 \in K_n$ and the Fréchet derivative $f'_n(x_0)$ exists, then $f'_n(x_0) = 0$ since $f_n(x_0) = 0$, i.e., x_0 is a minimum point. Hence in some ball $B(x_0, r)$ of radius $r > 0$ we have $f_n(x) \leqslant |x - x_0|/8$. By the compactness of K_n the ball $B(x_0, r)$ contains a ball $B(z, r/4)$ without common points with K_n (Exercise 1.6.14). Then $f(z) \geqslant r/4$ contrary to the estimate $f(z) \leqslant r/8$. Now we use the following fact: if continuous functions φ and ψ are convex and the sum $\varphi + \psi$ is Fréchet differentiable at x_0, then φ and ψ are Fréchet differentiable at x_0 (see the discussion before Proposition 4.7 in Benyamini, Lindenstrauss [138, p. 86]). We apply this fact to $\varphi = f_n$ and $\psi = \sum_{k \neq n} 2^{-k} f_k$. □

It is clear that in the same manner one can construct an example in every infinite dimensional Banach space. A considerably more subtle example was constructed in Matoušek, Matoušková [783] (see also Matoušková [784]).

10.6.2. Theorem. *There exists an equivalent norm p on the space l^2 for which the set of points of Fréchet differentiability belongs to the class of negligible sets $\mathfrak{A}^\mathcal{B} = \mathcal{P}^\mathcal{B} = \mathcal{G}^\mathcal{B}$.*

See Benyamini, Lindenstrauss [**138**, Example 6.46] for a somewhat simpler construction of a convex function with the same property.

We observe that even the convolution with a Gaussian (or smooth) measure, which improves the differentiability, does not help in this case. Let us consider the following example from Bogachev, Priola, Tolmachev [**213**].

10.6.3. Theorem. *Let X be an infinite dimensional separable Banach space and let μ be a Borel probability measure on X such that the norm is μ-integrable. Then, there exists a convex Lipschitzian function f on X such that at μ-almost every point $x \in X$ the convolution $f * \mu$ is not Fréchet differentiable along X.*

PROOF. There is a balanced convex compact set K such that $\mu(K) > 0$ and $\mu(\bigcup_{n=1}^\infty nK) = 1$ (see Theorem 1.2.6). Set $A := 2K$. We take a sequence of elements $v_n \in X$ such that $\|v_n\| \to 0$ and the distance between A and $A + v_n$ is greater than $\|v_n\|/8$. This is possible since for every n there exists a vector x_n with $\|x_n\| = 1/n$ separated from $2A$ by a closed hyperplane of the form $\{x: l_n(x) = c_n\}$, $l_n \in X^*$, i.e., $l_n(a) \leqslant c_n$ whenever $a \in 2A$ and $l_n(x_n) > c_n$. Let us set $Z_n := \operatorname{Ker} l_n$ and take any element $z_n \in Z_n$ with $\|x_n - z_n\| \leqslant 2\operatorname{dist}(x_n, Z_n)$, where $\operatorname{dist}(x, M) := \inf\{\|x - m\|: m \in M\}$ for every set M. Then $v_n := 2(x_n - z_n)$ is a suitable vector. Indeed, we have

$$\operatorname{dist}(v_n, 2A) \geqslant \operatorname{dist}(v_n, x_n + Z_n) = \operatorname{dist}(x_n - 2z_n, Z_n)$$

$$= \operatorname{dist}(x_n, Z_n) \geqslant \frac{1}{2}\|x_n - z_n\| = \frac{1}{4}\|v_n\|.$$

Hence for any vectors $a_1, a_2 \in A$ we obtain $\|a_1 - a_2 - v_n\| \geqslant \|v_n\|/4$. We observe that $\|v_n\| \leqslant 4/n$. Let us consider the function

$$f(x) = \operatorname{dist}(x, A),$$

which is convex by the convexity of A and is Lipschitzian with constant 1. We show that the convolution $f * \mu$ is not Fréchet differentiable at the points of K. It is clear that $f * \mu$ is Lipschitzian with constant 1 and convex. Suppose that the function $f * \mu$ is Fréchet differentiable at some point $x \in K$. Then, for every μ-measurable set E, the function

$$F_E(z) = \int_E f(z + y)\, \mu(dy)$$

is also Fréchet differentiable at x. This follows from the fact that F_E and $F_{X \setminus E}$ are convex and continuous and the sum of two convex continuous functions is Fréchet differentiable at x if and only if each of them is Fréchet differentiable at x, as noted above. Now we take the set $E = K$ and put $F := F_K$. Since $x \in K$, we have $x + y \in A$ for all $y \in K$, hence $F(x) = 0$. Therefore, since x is a minimum point of F, we have $DF(x) = 0$, where

$DF(x)$ is the Fréchet derivative of F at the point x. In order to obtain a contradiction it suffices to show that $F(x+h)$ is not $o(|h|)$, i.e., there exists a sequence of vectors h_n with $\|h_n\| \to 0$ and $F(x+h_n) \geqslant c\|h_n\|$ for some $c > 0$ independent of n. It suffices to obtain an estimate

$$f(x+y+h_n) \geqslant c_1 \|h_n\|$$

for every $y \in K$ with some constant $c_1 > 0$. Then we can take $c = c_1\mu(K)$. Let $h_n = v_n$ be the vectors constructed above and let $B(a,r)$ denote the closed ball of radius r in X centered at a. Let $y \in K$. We observe that

$$B(x+y+v_n, \|v_n\|/8) \cap A = \varnothing.$$

Indeed, if $u \in A$, then $\|x+y+v_n-u\| > \|v_n\|/8$ since $x+y \in A$. Therefore, one has $f(x+y+h_n) \geqslant \|h_n\|/8$. Hence we can take $c_1 = 1/8$. Thus, the set of points of Fréchet differentiability of $f * \mu$ does not intersect the set K. Finally, replacing K by nK in the described construction, we obtain the corresponding function f_n such that f_n is Lipschitzian with constant 1 and convex and the set of Fréchet differentiability of the function $f_n * \mu$ does not intersect nK. The function $f := \sum_{n=1}^{\infty} 2^{-n} f_n$ is also convex and Lipschitzian with constant 1. As noted above, the function $f * \mu$ is not Fréchet differentiable on the union of nK (which is a set of full μ-measure) since $2^{-n} f_n * \mu$ and $\sum_{k \neq n} 2^{-k} f_k * \mu$ are Lipschitzian and convex. $\square$

It is unknown whether there exists a Lipschitzian function f on a separable Hilbert space X such that its convolution $f * \mu$ with some nondegenerate Gaussian measure μ on X is Fréchet differentiable only on a Gaussian null set. It is also interesting whether a Lipschitzian function f can have this property for every Gaussian measure μ (i.e., not only for some μ).

The situation with Gâteaux differentiability is more favorable. For example, every locally Lipschitzian function on a separable normed space X is Gâteaux differentiable outside some "exceptional" set for many possible choices of the concept "exceptional" (for example, for the classes mentioned above this fact was established in Christensen [**289**], Mankiewicz [**776**], Aronszajn [**79**], and Phelps [**885**] and became a motivation for introducing the respective classes).

10.6.4. Theorem. *Let X be a separable Banach space. Let $F\colon X \to Y$ be a locally Lipschitzian mapping with values in a Banach space Y with the Radon–Nikodym property. Then F is Hadamard differentiable (hence Gâteaux differentiable) outside some Borel negligible (hence exceptional) set.*

PROOF. It suffices to show that the set of points where the Lipschitzian mapping $F\colon X \to Y$ is not Gâteaux differentiable belongs to $\mathcal{P}^{\mathcal{B}}$. Let $\{h_n\}$ be a linearly independent sequence with a dense linear span, let μ be a Radon probability measure continuous along all h_n, and let H_n be the linear span of $h_1, \ldots, h_n$. It is easy to verify that the set B_n of all points x where there exists a derivative of the mapping F along H_n is Borel (Exercise 10.8.5). Since the sets $x + H_n$ can be equipped with conditional measures continuous along $h_1,\ldots,h_n$, the finite dimensional Rademacher theorem yields that $\mu(B_n) = 1$. Hence $B := \bigcap_{n=1}^{\infty} B_n$ has full μ-measure. For every $x \in B$, on the linear span of $\{h_n\}$ we obtain a linear operator $T(x)$ with $T(x)(h) = \partial_h F(x)$ such that $\|T(x)(h)\| \leqslant \Lambda \|h\|$, where Λ is a Lipschitz constant for F. Hence the operator $T(x)$ is bounded. Now it is easy to show that $T(x)$ serves as the Gâteaux derivative for F at the point x. Indeed, given $h \in X$ and $\varepsilon > 0$ we can find a vector h_0 in the linear span of $\{h_n\}$ such that $\|h - h_0\| \leqslant \varepsilon$. There is $\delta > 0$ such that $\|F(x+th_0) - F(x) - tT(x)(h_0)\| \leqslant \varepsilon$ if $|t| \leqslant \delta$. Then for such t the quantity $\|F(x+th) - F(x) - tT(x)(h)\|$ is estimated by

$$\|F(x+th) - F(x+th_0)\| + \|F(x+th_0) - F(x) - tT(x)(h_0)\| + \|tT(x)(h - h_0)\|,$$

which does not exceed $2\Lambda\delta\varepsilon + \varepsilon$. Since F is Lipschitzian, it is Hadamard differentiable at x. $\square$

10.6.5. Corollary. *In the above theorem, for every compactly embedded normed space $E \subset X$ the Fréchet derivative $D_E F(x)$ along E exists for all points x outside some Borel negligible set.*

In particular, given a densely continuous measure μ such that $E \subset C(\mu)$, the Fréchet derivative $D_E F(x)$ exists μ-a.e.

However, once we admit the differentiability along a smaller subspace, it would be natural to impose the Lipschitz condition also only along this subspace. This point of view is consistent with the basic constructions of the Malliavin calculus and the theory of differentiable measures. The proof presented above gives the following assertion.

10.6.6. Theorem. *Suppose that X is a locally convex space, $E \subset X$ is a continuously embedded Banach space, Y is a Banach space with the Radon–Nikodym property, and $F\colon X \to Y$ is a Borel mapping such that for some number C and all $x \in X$ one has*

$$\|F(x+h) - F(x)\|_Y \leqslant C\|h\|_E, \quad \forall\, h \in E. \tag{10.6.1}$$

Let μ be a Radon measure on X continuous along vectors h_n with the linear span dense in E. Then F has the Gâteaux derivative $D_E F(x)$ along E μ-a.e.

If μ is quasi-invariant along all h_n, then it suffices to have (10.6.1) for μ-a.e. x. In this case, there exists a modification of F satisfying (10.6.1) for all x. In particular, these assertions are true if μ is a Gaussian measure and $E = H(\mu)$.

PROOF. The justification of the first assertion is completely analogous to the proof of the previous theorem. The second assertion is based on the following observation: if a mapping f between Banach spaces E and Y is such that for some $C > 0$ for all points x in a dense set one has the estimate $\|f(x+h) - f(x)\|_Y \leqslant C\|h\|_E$ for all $h \in E$, then f is Lipschitzian with constant C. Indeed, let $y, h \in E$ and $\varepsilon > 0$. We find a point $x \in E$ with the

indicated property such that $\|x - y - h/2\|_E \leqslant \varepsilon/2$. Then
$$\|f(y+h) - f(y)\|_Y \leqslant \|f(y+h) - f(x)\|_Y + \|f(x) - f(y)\|_Y$$
$$\leqslant C(\|y+h-x\|_E + \|x-y\|_E) \leqslant C\|h\|_E + C\varepsilon,$$

whence our claim follows. By hypothesis, the set Ω_0 of all points x for which (10.6.1) is true has full measure. By the quasi-invariance of μ we obtain a full measure set $\Omega := \bigcap_{n,k}(\Omega_0 + r_k h_n)$, where $\{r_k\}$ is the set of all rational numbers. By the above observation, for every $x \in \Omega$ the mapping $h \mapsto F(x+h)$ from E to Y is Lipschitzian with constant C (which gives a modification that is Lipschitzian along E if we redefine F by zero outside Ω). Now an obvious modification of our reasoning from the previous theorem completes the proof. $\square$

10.6.7. Corollary. *Under the assumptions of Theorem 10.6.6, for every normed space B compactly embedded into the space E, the Fréchet derivative $D_B F$ exists μ-a.e.*

The question arises about Fréchet differentiability of the mapping F along the subspace E. As the following example from Bogachev, Mayer-Wolf [**209**] shows, the last corollary is not valid, in general, for the whole space E.

10.6.8. Example. Let $X = \mathbb{R}^\infty$, $\mu = \prod_{n=1}^\infty \mu_n$, where μ_n are identical standard Gaussian measures on $\mathbb{R}^1$, $E = l^2$, $F \colon X \to l^2$, $F(x) = (f_n(x_n))$, where f_n is an 2^{1-n}-periodic function on the real line such that
$$f_n(t) = t \text{ if } t \in [0, 2^{-n}], \quad f_n(t) = 2^{1-n} - t \text{ if } t \in [2^{-n}, 2^{1-n}].$$
Then $E = H(\mu) = Q(\mu)$ and F is Lipschitzian along E, but at no point is Fréchet differentiable along E.

It is not difficult to modify this example in order to make the mapping F everywhere Gâteaux differentiable along E. It was conjectured in Enchev, Stroock [**397**] that an analogous example exists also with a real function on a space with a Gaussian measure. Such an example has recently been constructed in Bogachev, Priola, Tolmachev [**213**].

10.6.9. Theorem. *Let γ the countable power of the standard Gaussian measure on $\mathbb{R}^1$ and $H = H(\gamma) = l^2$. There exists a Borel function f on the space $\mathbb{R}^\infty$ that is Lipschitzian along H with constant 1, but the set of points where f is Fréchet differentiable along H has γ-measure zero.*

Moreover, such a function exists for every centered Radon Gaussian measure μ on a locally convex space X such that its Cameron–Martin space is infinite dimensional.

PROOF. Let γ_n be the standard Gaussian measure with density ϱ_n on $\mathbb{R}^n$ and let $0 < \varepsilon < 1/2$. We show that there is a set $A_n \subset \mathbb{R}^n$ with the following properties: there exist a natural number M_n, points $a_{n,i}$ and

positive numbers $r_{n,i}$ such that
$$A_n = \bigcup_{i=1}^{M_n} B(a_{n,i}, r_{n,i}) \setminus B(a_{n,i}, \varepsilon r_{n,i}),$$
where closed balls $B(a_{n,i}, r_{n,i})$ are pairwise disjoint and $\gamma_n(A_n) > 1 - 9\varepsilon 2^{-n}$. Indeed, we take a cube Q with $\gamma_n(Q) > 1 - \varepsilon 4^{-n}$. Let us find $r \in (0, 2^{-n})$ such that
$$\sup_{x \in B(a,r)} \varrho_n(x) \leqslant 2 \inf_{x \in B(a,r)} \varrho_n(x) \qquad (10.6.2)$$
for every ball $B(a,r) \subset Q$. Then we find finitely many pairwise disjoint balls $B(a_{n,i}, r_{n,i}) \subset Q$, $i = 1, \ldots, M_n$, with $0 < r_{n,i} < r$ such that the γ_n-measure of their union is greater than $1 - \varepsilon 4^{-n}$. Let λ_n be Lebesgue measure on $\mathbb{R}^n$. For every ball $B(a, \delta) \subset Q$ with $\delta < r$ we have
$$\gamma_n(B(a, \varepsilon\delta)) \leqslant 2\varepsilon^n \gamma_n(B(a, \delta)),$$
which follows from (10.6.2) and the equality $\lambda_n(B(a, \varepsilon\delta)) = \varepsilon^n \lambda_n(B(a, \delta))$. Therefore,
$$\gamma_n\left(\bigcup_{i=1}^{M_n} B(a_{n,i}, r_{n,i}) \setminus B(a_{n,i}, \varepsilon r_{n,i})\right) \geqslant (1 - 4\varepsilon^n) \gamma_n\left(\bigcup_{i=1}^{M_n} B(a_{n,i}, r_{n,i})\right)$$
$$> (1 - 4\varepsilon^n)(1 - \varepsilon 4^{-n}) > 1 - \varepsilon 4^{-n} - 4\varepsilon^n > 1 - 9\varepsilon 2^{-n}.$$

Let us represent $\mathbb{R}^\infty$ as $\mathbb{R}^1 \times \mathbb{R}^2 \times \mathbb{R}^3 \times \cdots$ and the measure γ as the product of the measures γ_n. Let $\varepsilon > 0$ be fixed. For every natural number n we take the set $A_n \subset \mathbb{R}^n$ constructed above and define a Borel set in $\mathbb{R}^\infty$ by
$$A := \prod_{n=1}^{\infty} A_n.$$
Then we have
$$\gamma(A) \geqslant \prod_{n=1}^{\infty} (1 - 9\varepsilon 2^{-n}) \geqslant 1 - 9\varepsilon.$$
Finally, we define a Borel function f on $\mathbb{R}^\infty$ by the formula
$$f(x) := \mathrm{dist}_H(x, A) := \inf\{|x - y|_H : \ y \in A\}$$
if there is at least one element $y \in A$ with $x - y \in H$. Otherwise we put $f(x) = 0$. The fact that f is Borel measurable follows from the equality
$$\{x : f(x) < c\} = (A + cU) \cup (X \setminus (A + H)), \quad c > 0,$$
where U is the open unit ball in H. The sets $A + cU$ and $A + H$ are Borel since $A + cU$ is the union of the sets $A + cV_n$, where V_n is the closed ball of radius $1 - 1/n$ in H, and these sets are compact in X. We have
$$|f(x+h) - f(x)| \leqslant |h|_H, \quad h \in H,$$
i.e., the function f is Lipschitzian along H. Let us show that f is not Fréchet differentiable along H at the points of the set A. Indeed, suppose that a

point $a \in A$ is such that the function f is Fréchet differentiable along H at a. Since $f(a) = 0$ and $f \geqslant 0$, one has $D_H f(a) = 0$. This will give a contradiction if we show that for every $\delta > 0$ there exists a vector $h \in H$ with $|h|_H < \delta$ such that

$$\{x\colon |a + h - x|_H \leqslant \varepsilon |h|_H\} \cap A = \varnothing$$

because in that case $f(a+h) \geqslant \varepsilon |h|_H$ contrary to the fact that the Fréchet derivative of f along H vanishes at a. We have $a = (a_n)$, where $a_n \in A_n$. Let us choose n such that $2^{-n} < \delta$. Then $a_n \in B(a_{n,i}, r_{n,i}) \backslash B(a_{n,i}, \varepsilon r_{n,i})$ for some $i \leqslant M_n$. It remains to take the vector h whose nth component in our representation of $\mathbb{R}^\infty$ equals $a_{n,i} - a_n$ and all other components equal zero. Then $|h|_H \leqslant r_{n,i} < 2^{-n} < \delta$. In addition, the ball of radius $\varepsilon |h|_H$ in the metric of H centered at $a + h$ does not intersect A since otherwise the ball in $\mathbb{R}^n$ of radius $\varepsilon r_{n,i}$ centered at $a_{n,i}$ would intersect the set A_n, which is impossible by our construction.

The previous step gives a function f which is Lipschitzian along H with constant 1, but is not Fréchet differentiable at the points of the set A with $\gamma(A) \geqslant 1 - 9\varepsilon$. Now we construct such a set for every $\varepsilon = 1/k$, $k \in \mathbb{N}$, and denote the corresponding function by f_k. Let us represent $X = \mathbb{R}^\infty$ as $X = \prod_{k=1}^\infty X_k$, where each X_k is a copy of the space $\mathbb{R}^\infty$, and equip X with the standard Gaussian product measure γ (which equals the countable product of the standard Gaussian product measures $\gamma^{(k)}$ on the factors X_k). Let π_k denote the projection operator on the kth factor. The Cameron–Martin space H of the measure γ coincides with l^2 and equals the Hilbert sum of the Cameron–Martin spaces H_k of the measures $\gamma^{(k)}$. The function

$$f(x) := \sum_{k=1}^\infty 2^{-k} f_k\big(\pi_k(x)\big)$$

is Lipschitzian along H with constant 1. It is not Fréchet differentiable along H γ-a.e. Indeed, for every k let $E_k := A_k \times \prod_{n \neq k} X_n$. We can write $f = 2^{-k} f_k + g_k$, where $g_k := \sum_{n \neq k} 2^{-n} f_n$. The function g_k does not depend on $\pi_k(x)$. Since f_k is not Fréchet differentiable along H_k at the points of the set A_k, the function f is not Fréchet differentiable along H_k (hence also along H) at the points of the set E_k. It remains to observe that we have $\gamma(E_k) = \gamma^{(k)}(A_k) \geqslant 1 - 9k^{-1}$.

The case of a general locally convex space X equipped with a centered Radon Gaussian measure μ whose Cameron–Martin space $H(\mu)$ is infinite dimensional reduces to the considered case. Indeed, we know that there exists an injective Borel measurable linear mapping T defined on a Borel linear subspace $E \subset \mathbb{R}^\infty$ with $\gamma(E) = 1$ such that $E_0 := T(E)$ is a Borel and Souslin subspace in X with $\mu(E_0) = 1$, μ coincides with the image of γ under T and T is an isometry between l^2 and $H(\mu)$. The mapping $S := T^{-1}\colon E_0 \to E$ is Borel measurable (see [**193**, Theorem 6.8.6]). Having constructed our function f on $\mathbb{R}^\infty$, we obtain a Borel measurable function $f_0 = f \circ S$ on E_0 with the required properties. Set $f_0(x) = 0$ outside E_0. □

This example gives a partial answer to Question (iii) in [**191**, p. 266] concerning the set of points of Fréchet H-differentiability of $\mathrm{dist}_H(x, B)$.

In relation to the existence (by Theorem 10.6.2) of a convex Lipschitzian function on $X = l^2$ whose set of points of Fréchet differentiability along the whole space X is negligible, an analogous question arises for the differentiability along H. Namely, suppose that a continuous function f on a separable Hilbert space (or on $\mathbb{R}^\infty$) is Lipschitzian along the Cameron–Martin space H of some nondegenerate Gaussian measure γ. Can it occur that the set of points where the function f is Fréchet differentiable along H is a Gaussian null set? Finally, it would be interesting to know whether in Theorem 10.6.9 one can find a function f with the additional property that it is convex along H. In particular, it is not clear whether one can find f of the form $\mathrm{dist}_H(x, A)$ for some Borel set A that is convex along H.

10.7. Triangular and optimal transformations

Here we briefly discuss two interesting classes of transformations of measures that have been intensively investigated in the last years. These two classes do not almost overlap, but have several common features: (a) in the one-dimensional case, these transformations are monotone functions, (b) any measure that is not too degenerate can be transformed into any other measure by means of transformations from both classes, (c) transformations of both classes are related to interesting nonlinear functional inequalities such as transport inequalities, logarithmic Sobolev inequalities, and others.

We shall start with triangular mappings, which are simpler. Suppose we are given spaces X_k in a finite or countable number. Let $X = \prod_k X_k$. A mapping $T = (T_1, T_2, \ldots) \colon X \to X$ is called *triangular* if we have

$$T_k(x) = T_k(x_1, \ldots, x_k) \quad \text{for all } k, \text{ where } x = (x_i).$$

If $X_k = \mathbb{R}^1$ for all k, then a triangular mapping is called *increasing* if every function $x_k \mapsto T_k(x_1, \ldots, x_k)$ is increasing. Analogously, we define increasing triangular mappings defined on subsets of $\mathbb{R}^\infty$. We observe that no monotonicity with respect to other variables is required. Basic properties of triangular mappings are presented in Bogachev [**193**, §10.10(vii)]. The word "triangular" in the name is explained by the fact that the derivative of a differentiable triangular mapping on $\mathbb{R}^n$ is given by a triangular matrix. In spite of their rather special form, triangular mappings possess rich possibilities in diverse problems related to transformations of measures. For example, as we shall see below, the countable power of Lebesgue measure on $[0,1]$ can be transformed into an arbitrary Borel probability measure on $[0,1]^\infty$ by a Borel increasing triangular transformation. Unlike general measurable isomorphisms, triangular transformations are constructed efficiently.

10.7. Triangular and optimal transformations

In the case of an absolutely continuous measure on $\mathbb{R}^n$ they can be given by explicit (although rather involved for $n > 2$) formulas.

Let us recall that every Borel probability measure μ on the product of two Souslin spaces X_1 and X_2 admits conditional Borel probability measures μ_{x_1} on X_2, where $x_1 \in X_1$, such that, for every Borel set B in the space $X_1 \times X_2$, the function $x_1 \mapsto \mu_{x_1}(B^{x_1})$, where $B^{x_1} = \{x_2 \in X_2 \colon (x_1, x_2) \in B\}$, is Borel on X_1 and

$$\mu(B) = \int_{X_1} \mu_{x_1}(B^{x_1})\, \mu_1(dx_1),$$

where μ_1 is the projection of the measure μ on X_1. We observe that if we are given a Borel measure μ on the product of three Souslin spaces X_1, X_2, and X_3, then its conditional measures on X_2 are also conditional measures of its projection on $X_1 \times X_2$.

10.7.1. Theorem. (i) *Let X_1 and X_2 be Souslin spaces and let μ and ν be Borel probability measures on $X_1 \times X_2$. Suppose that the projection of μ on the first factor and the conditional measures μ_{x_1}, $x_1 \in X_1$, on X_2 have no atoms. Then, there exists a Borel triangular mapping $T \colon X_1 \times X_2 \to X_1 \times X_2$ such that $\mu \circ T^{-1} = \nu$.*

(ii) *Let $X = \prod_{n=1}^{\infty} X_n$, where each X_n is a Souslin space. Let μ be a Borel probability measure on X such that its projections on all $\prod_{j=1}^{n} X_j$ and the conditional measures on X_n have no atoms for all n. Then, for every Borel probability measure ν on X, there exists a triangular Borel mapping $T \colon X \to X$ such that $\mu \circ T^{-1} = \nu$.*

Let us explain the method of constructing triangular mappings in the case of absolutely continuous measures on $[0,1]^2$. First we find an increasing function $T_1 \colon [0,1] \to [0,1]$ that transforms the projection of μ to the projection of ν. This is done by an explicit formula with the aid of the distribution functions of both projections (in the present case, by means of densities of our measures). Then, for each fixed x_1, we transform the conditional measure μ_{x_1} into the conditional measure $\nu_{T_1(x_1)}$ by an increasing function $T_2(x_1, \cdot) \colon [0,1] \to [0,1]$ (which is also done explicitly by using densities of the given measures). In the multidimensional case the construction continues inductively, and its pleasant feature is that the components constructed at the nth step are unchanged at the subsequent steps.

Increasing Borel triangular transformations of measures are called *canonical triangular mappings*. A canonical triangular transformation of the measure μ into the measure ν will be denoted by $T_{\mu,\nu}$ (as we shall see, under broad assumptions, such a transformation is unique up to a modification). In the case where the measures μ and ν are defined on all of $\mathbb{R}^n$ an analogous construction yields a triangular increasing Borel mapping $T_{\mu,\nu} = (T_1, \ldots, T_n)$ with values in $\mathbb{R}^n$ defined on some Borel set $\Omega \subset \mathbb{R}^n$ of full μ-measure. Then every function T_k as a function of variables $x_1, \ldots, x_k$ is defined on some Borel set in $\mathbb{R}^k$ whose intersections with the straight lines parallel to the kth coordinate line are intervals. This is obvious from our inductive construction and the one-dimensional case, in which the composition $G_{\nu_1} \circ F_{\mu_1}$ is defined either on the whole real line or on a ray or on an interval (if the function G_{ν_1} has no finite limits at the points 0 and 1, and the measure μ_1 is concentrated on an interval). For example, if μ is Lebesgue measure on $[0,1]$ regarded on the whole real line and ν is the standard Gaussian measure, then the mapping $T_{\mu,\nu}$ is defined on the interval $(0,1)$, but has no increasing extension to $\mathbb{R}^1$. If a measure ν on $\mathbb{R}^n$ has bounded support, then the mapping $T_{\mu,\nu}$ is defined on the whole space $\mathbb{R}^n$. The same is true for every measure ν if the projection of the measure μ on the first coordinate line and its conditional measures on the remaining coordinate lines are not concentrated on intervals. For example, this is fulfilled if the measure μ is equivalent to Lebesgue measure since we take a strictly positive Borel version of its density. Since conditional measures are defined uniquely up to a set of measure zero, canonical triangular mappings are defined up to modifications. However, the uniqueness of canonical mappings holds in a broader class of transformations.

10.7.2. Lemma. (i) *Let μ and ν be two Borel probability measures on $\mathbb{R}^n$ having atomless projections on the first coordinate line and atomless conditional measures on all other coordinate lines. Then the mapping $T_{\mu,\nu}$ is injective on some Borel set of full μ-measure. The same is true for measures on $\mathbb{R}^\infty$.*

(ii) *Let μ be a Borel probability measure on $\mathbb{R}^\infty$. If two increasing triangular Borel mappings $T = (T_n)_{n=1}^\infty$ and $S = (S_n)_{n=1}^\infty$ are such that $\mu \circ T^{-1} = \mu \circ S^{-1}$ and for every n the mapping $(T_1, \ldots, T_n)$ is injective on a Borel set of full measure with respect to the projection of μ on $\mathbb{R}^n$, then $T(x) = S(x)$ for μ-a.e. x.*

In particular, if the projections of the measures μ and ν on the spaces $\mathbb{R}^n$ are absolutely continuous, then there exists the canonical triangular mapping $T_{\mu,\nu}$ and it is unique up to μ-equivalence in the class of increasing Borel triangular mappings transforming μ into ν.

Canonical triangular mappings depend continuously on both measures.

10.7.3. Theorem. *Suppose that Borel probability measures μ_j on $\mathbb{R}^\infty$ converge in variation to a measure μ and Borel probability measures ν_j on $\mathbb{R}^\infty$ converge in variation to a measure ν. Suppose also that the measures μ_j and μ satisfy the hypotheses of Theorem 10.7.1. Then the canonical triangular mappings T_{μ_j,ν_j}, defined in an arbitrary way to Borel mappings of the whole space outside their initial domains of definition, converge in measure μ to the mapping $T_{\mu,\nu}$.*

Therefore, there is a subsequence in $\{T_{\mu_j,\nu_j}\}$ convergent to $T_{\mu,\nu}$ almost everywhere with respect to μ. In this formulation this theorem extends to countable products of arbitrary Souslin spaces (see Aleksandrova [58]), and if the factors are metrizable, then the assertion about convergence in measure μ of the whole sequence remains valid. As [**193**, Example 10.4.24] shows, there might be no convergence almost everywhere for the whole sequence T_{μ_j, ν_j}.

For increasing triangular mappings, the following change of variables formula holds.

10.7.4. Lemma. *Let $T = (T_1, \ldots, T_n) \colon \mathbb{R}^n \to \mathbb{R}^n$ be an increasing Borel triangular mapping. Suppose that the functions $x_i \mapsto T_i(x_1, \ldots, x_i)$ are absolutely continuous on all intervals for a.e. $(x_1, \ldots, x_{i-1}) \in \mathbb{R}^{i-1}$. Set by definition $\det DT := \prod_{i=1}^n \partial_{x_i} T_i$. Then, for every Borel function φ integrable on the set $T(\mathbb{R}^n)$, the function $\varphi \circ T \det DT$ is integrable over $\mathbb{R}^n$ and one has*

$$\int_{T(\mathbb{R}^n)} \varphi(y) \, dy = \int_{\mathbb{R}^n} \varphi(T(x)) \det DT(x) \, dx. \tag{10.7.1}$$

If the mapping T is defined only on a Borel set $\Omega \subset \mathbb{R}^n$ and every function T_i is defined on a Borel set in $\mathbb{R}^i$ whose intersections by the straight lines parallel to the ith coordinate line are intervals, and the indicated condition is fulfilled for the intervals of these sections, then the same assertion is true with Ω in place of $\mathbb{R}^n$.

Let us give a simple sufficient condition on two measures μ and ν ensuring the absolute continuity of the ith component of $T_{\mu,\nu}$ with respect to x_i.

10.7.5. Example. Let a probability measure μ be absolutely continuous and let a probability measure ν be equivalent to Lebesgue measure. Then the canonical triangular mapping $T_{\mu,\nu}$ satisfies the condition of the previous lemma.

If the measure ν is not equivalent to Lebesgue measure, then the ith component of the canonical triangular mapping may be discontinuous. For example, the canonical mapping of Lebesgue measure on $[0,1]$ to the measure ν whose density equals 2 on $[0, 1/4] \cup [3/4, 1]$ and equals 0 on $(1/4, 3/4)$ has a jump. Nevertheless, the indicated change of variables formula remains true without the assumption of the absolute continuity made in the lemma if T is the canonical mapping of absolutely continuous measures (certainly, not every increasing Borel triangular mapping has such a form).

10.7.6. Proposition. *Let μ and ν be two probability measures on $\mathbb{R}^n$ with densities ϱ_μ and ϱ_ν with respect to Lebesgue measure. Then, the canonical triangular mapping $T_{\mu,\nu} = (T_1, \ldots, T_n)$ satisfies the equality*

$$\varrho_\mu(x) = \varrho_\nu(T_{\mu,\nu}(x)) \det DT_{\mu,\nu}(x) \quad \text{for } \mu\text{-a.e. } x, \tag{10.7.2}$$

where $\det DT_{\mu,\nu} := \prod_{i=1}^n \partial_{x_i} T_i$ exists almost everywhere by the monotonicity of T_i in x_i.

Let us stress once again that the partial derivative in the formulation is the usual partial derivative existing almost everywhere, but not the derivative in the sense of generalized functions (which has a singular component in the case of a function that is not absolutely continuous).

It is proved in Zhdanov [**1220**], Zhdanov, Ovsienko [**1221**] that the mapping $T_{\mu,\nu}$ belongs to some Sobolev class if both measures μ and ν have strictly positive densities belonging to suitable Sobolev classes.

We shall say that a Borel probability measure μ with a twice differentiable density $\exp(-\Phi_n)$ on $\mathbb{R}^n$ is uniformly convex with constant $C > 0$ if Φ_n is a convex function and one has the estimate $D^2\Phi_n(x) \geqslant C \cdot I$. A Borel probability measure μ on $\mathbb{R}^\infty$ is called uniformly convex with constant $C > 0$ if its projections on the spaces $\mathbb{R}^n$ are uniformly convex with constant C.

The next result is obtained in Bogachev, Kolesnikov, Medvedev [**204**].

10.7.7. Theorem. *Let a probability measure μ on $\mathbb{R}^n$ be uniformly convex with constant C (for example, let μ be the standard Gaussian measure). Let ν be an absolutely continuous probability measure on $\mathbb{R}^n$ such that for $f := d\nu/d\mu$ we have $f \ln f \in L^1(\mu)$. Then, there exists a Borel increasing triangular mapping T such that $\nu = \mu \circ T^{-1}$ and*

$$\int_{\mathbb{R}^n} |x - T(x)|^2 \, \mu(dx) \leqslant \frac{2}{C} \int_{\mathbb{R}^n} f(x) \ln f(x) \, \mu(dx).$$

In the case of the standard Gaussian measure one has $C = 1$.

The next theorem is proven in Bogachev, Kolesnikov [**198**].

10.7.8. Theorem. *Let a Borel probability measure μ on $X := \mathbb{R}^\infty$ be uniformly convex with constant $C > 0$ and $H = l^2$. Let $\nu \ll \mu$ be a probability measure and let $f := d\nu/d\mu$.*

(i) *If $f \ln f \in L^1(\mu)$, then the canonical triangular mapping $T_{\mu,\nu}$ has the following property:*

$$\int_X |T_{\mu,\nu}(x) - x|_H^2 \, \mu(dx) \leqslant \frac{2}{C} \int_X f \ln f \, d\mu.$$

(ii) *If μ has the form $\mu_1 \otimes \mu'$, where μ' is a measure on the product of the remaining straight lines, then there exists a Borel triangular mapping T of the form $T(x) = x + F(x)$ with $F\colon X \to H$ such that $\nu = \mu \circ T^{-1}$.*

(iii) *If μ is equivalent to the measure μ_{e_1}, where $e_1 = (1, 0, 0, \ldots)$, then there exists a Borel mapping T of the form $T(x) = x + F(x)$ such that $F\colon X \to H$ and $\nu = \mu \circ T^{-1}$.*

Conditions (ii) and (iii) are satisfied for countable products of identical uniformly convex measures on the real line, in particular, for products of the standard Gaussian measure. It is not clear whether in assertions (ii) and (iii) one can take for T the canonical triangular mappings $T_{\mu,\nu}$.

10.7.9. Corollary. *Let γ be a Radon Gaussian measure on a locally convex space X and let μ be a Radon probability measure with $\mu \ll \gamma$. Then, there exists a Borel mapping $F\colon X \to H(\gamma)$ such that $\mu = \gamma \circ (I + F)^{-1}$.*

10.7. Triangular and optimal transformations

Let us proceed to optimal mappings. The general setting of the problem is this. Suppose we are given two probability spaces $(X, \mathcal{A}, \mu)$ and $(Y, \mathcal{B}, \nu)$ and a nonnegative $\mathcal{A}\otimes\mathcal{B}$-measurable function h on $X\times Y$ called a cost function. The so-called Monge problem is to find a measurable transformation $T\colon X \to Y$ taking μ to ν and minimizing the quantity

$$C(T) := \int_X h\bigl(x, T(x)\bigr)\,\mu(dx).$$

This nonlinear problem is not always solvable even for nice functions h and nice spaces. More general is the Kantorovich problem of finding the so-called optimal plan, i.e., a measure σ on $X\times Y$ such that its projections on X and Y are the measures μ and ν, respectively, and σ gives minimum to

$$K(\mu, \nu) := \inf \int_{X\times Y} h(x,y)\,\eta(dx\,dy),$$

where inf is taken over all probability measures η on $X\times Y$ whose projections are μ and ν. The Kantorovich problem is linear and is solvable under much broader assumptions. For example, if X and Y are compact and the function h is continuous, then the Kantorovich problem is the minimization problem for a continuous linear function on the convex compact set of probability measures (the space of measures is equipped with the weak topology). Any solution to the Monge problem also gives a solution to the Kantorovich problem: we take for σ the image of μ under the mapping $x \mapsto \bigl(x, T(x)\bigr)$. There is extensive literature on these problems; see Ambrosio [60], Rachev, Rüschendorf [937], Villani [1174], [1175]. We shall mention only some of the simplest facts directly related to the main subject of this book. According to the Brenier theorem [252] (see also [1174, p. 66]), for every pair of Borel probability measures μ and ν on $\mathbb{R}^d$ with finite second moments such that μ is absolutely continuous, there exists a convex function V on $\mathbb{R}^d$ (finite on a convex set of full μ-measure) such that the mapping $T = \nabla V$ transforms μ into ν and is a unique optimal transport between μ and ν for the quadratic cost function $h(x,y) = |x-y|^2$. McCann [797] (see also [1174, pp. 96, 133]) made precise the Brenier theorem as follows: For any two Borel probability measures μ and ν on $\mathbb{R}^d$ such that μ is absolutely continuous (no moment condition is imposed), there exists a mapping T that is the gradient of some convex function V and transforms μ into ν. Such a mapping is unique (any two such mappings coincide μ-a.e.). If both measures are absolutely continuous and are given by densities ϱ_μ and ϱ_ν, then almost everywhere the Monge–Ampère equation

$$\det D_{\mathrm{ac}}^2 V \varrho_\nu\bigl(\nabla V(x)\bigr) = \varrho_\mu(x)$$

is fulfilled. Optimal mappings are rarely triangular: the Jacobi matrix of a triangular mapping is triangular, but for an optimal mapping it is symmetric, hence both conditions are fulfilled only for diagonal matrices, which is possible only for mappings $T = (T_k)$ with $T_k(x) = T_k(x_k)$. The investigation of the Monge–Ampère equation on infinite dimensional spaces with

Gaussian measures was initiated in Feyel, Üstünel [**430**] and continued in Bogachev, Kolesnikov [**200**], Feyel, Üstünel [**431**]; close problems of optimal transportation are considered in Shao [**991**]. We mention two results in this direction. The first one is an important existence theorem obtained in [**430**]. Let X be a locally convex space and let γ be a centered Radon Gaussian measure on X with the Cameron–Martin space H. Without loss of generality, one may assume that $X = \mathbb{R}^\infty$ and γ is the countable power of the standard Gaussian measure on the real line; then $H = l^2$. Suppose that we are given a probability measure of the form $g \cdot \gamma$ such that the following quantity is finite:

$$W_H(\gamma, g \cdot \gamma)^2 = \inf_{m \in \mathcal{P}(\gamma, g \cdot \gamma)} \int_{X \times X} |x_1 - x_2|_H^2 \, m(dx_1, dx_2),$$

where $\mathcal{P}(\gamma, g \cdot \gamma)$ is the set of all Radon probability measures on $X \times X$ whose projections to the first and second factors equal γ and $g \cdot \gamma$, respectively. A constructive sufficient condition in order that $W_H(\gamma, g \cdot \gamma)$ be finite is that the entropy of g is finite, i.e., the function $g \ln g$ is γ-integrable.

10.7.10. Theorem. *There exists a unique γ-measurable mapping $T \colon X \to X$, called the optimal transport, that transforms γ into $g \cdot \gamma$ and satisfies the equality*

$$W_H(\gamma, g \cdot \gamma)^2 = \int_X |T(x) - x|_H^2 \, d\gamma.$$

Moreover, $T = I + \nabla_H \Phi$, where $\Phi \in W^{2,1}(\gamma)$ is a 1-convex function. If $g > 0$ γ-a.e. and $\ln g \in L^1(\gamma)$, then T has an inverse mapping S, i.e., $T \circ S(x) = S \circ T(x) = x$ for γ-a.e. x. In addition, S is the optimal transport of the measure $g \cdot \gamma$ to γ and $S = I + \nabla_H \Psi$, where a function $\Psi \in W^{2,1}(\gamma)$ is 1-convex.

The Monge–Ampère equation in the described situation was derived in Bogachev, Kolesnikov [**201**], [**200**]. Let us introduce some additional notation. If the measure $\partial_k \Phi \cdot \gamma$ is Skorohod differentiable along h, then we have the measure

$$\Phi_{hk} := d_h(\partial_k \Phi \cdot \gamma) - \partial_k \Phi \cdot d_h \gamma.$$

The density of the absolutely continuous part of Φ_{kh} (with respect to γ) will be denoted by Φ_{kh}^{ac}. If $\Phi \in W^{2,2}(\gamma)$, then

$$\Phi_{hk} = \partial_h \partial_k \Phi \cdot \gamma.$$

Let $\{e_i\}$ be an orthonormal basis in H. If $\ln g \in L^1(\gamma)$ and $g \ln g \in L^1(\gamma)$, then one can show that for all vectors $h, k \in H$ there exists bounded Radon measures Ψ_{hk} and Φ_{hk}. With this notation, the following result holds.

10.7.11. Theorem. (i) *Let $\ln g \in L^1(\gamma)$ and $g \ln g \in L^1(\gamma)$. Then, there exist $\mathcal{H}$-valued mappings $D_{\mathrm{ac}}^2 \Psi$ and $D_{\mathrm{ac}}^2 \Phi$ with matrix elements $\Phi_{e_i e_j}^{\mathrm{ac}}$*

and $\Psi^{\mathrm{ac}}_{e_i e_j}$ and a subsequence $\{n_k\}$ such that γ-a.e. there exist finite limits

$$\mathcal{L}_0 \Psi = \lim_{m\to\infty} \frac{1}{m} \sum_{k=1}^{m} \sum_{i=1}^{n_k} \big(\Psi^{\mathrm{ac}}_{e_i e_i} - x_i \partial_{e_i} \Psi\big),$$

$$\mathcal{L}_0 \Phi = \lim_{m\to\infty} \frac{1}{m} \sum_{k=1}^{m} \sum_{i=1}^{n_k} \big(\Phi^{\mathrm{ac}}_{e_i e_i} - x_i \partial_{e_i} \Phi\big).$$

Furthermore,

$$g = \det{}_2(I + D^2_{\mathrm{ac}}\Psi)\exp\big(\mathcal{L}_0\Psi - |\nabla_H \Psi|^2_H/2\big),$$

$$1/g(T) = \det{}_2(I + D^2_{\mathrm{ac}}\Phi)\exp\big(\mathcal{L}_0\Phi - |\nabla_H \Phi|^2_H/2\big).$$

(ii) *Let* $g > c > 0$ *and* $g \ln g \in L^1(\gamma)$. *Then* $\mathcal{L}_0 \Psi = \mathcal{L}_{\mathrm{ac}} \Psi$ *and*

$$g = \det{}_2(I + D^2_{\mathrm{ac}}\Psi)\exp\big(\mathcal{L}_{\mathrm{ac}}\Psi - |\nabla_H \Psi|^2_H/2\big).$$

(iii) *Let* $0 < g < C$ *and* $\ln g \in L^1(\gamma)$. *Then* $\mathcal{L}_0 \Phi = \mathcal{L}_{\mathrm{ac}} \Phi$ *and*

$$1/g(T) = \det{}_2(I + D^2_{\mathrm{ac}}\Phi)\exp\big(\mathcal{L}_{\mathrm{ac}}\Phi - |\nabla_H \Phi|^2_H/2\big).$$

All of these equalities hold almost everywhere.

10.8. Comments and exercises

The results presented in §§10.1–10.3 go back to the pioneering works Cameron, Martin [**259**], [**260**] and Cameron, Fagen [**258**], where in the case of the classical Wiener space shifts, linear mappings and then more general nonlinear transformations $T\omega = \omega + F(\omega)$ were consequently studied. One of the basic assumptions in this circle of problems is that F takes values in the Cameron–Martin space. The next important step was accomplished by G. Maruyama, Yu.V. Prohorov, A.V. Skorohod, and I.V. Girsanov in their studies of distributions of diffusion processes. Such distributions are images of the Wiener measure under transformations defined implicitly by integral equations. For example, for the equation $d\xi_t = dw_t + b(\xi_t)dt$ the corresponding transformation T is determined from the equation

$$T(x)(t) = x(t) + \int_0^t b\big(x(s)\big)\,ds,$$

which enables us to write it in the form $Tx = x + F(x)$, where F takes values in $H(P^W)$. Though, for nonsmooth b the mapping F will not be differentiable, but this is compensated by a special form of dependence of F on x. It was shown by Maruyama [**782**] that the distribution of a one-dimensional diffusion with unit diffusion coefficient is equivalent to the Wiener measure and a formula was derived for the corresponding Radon–Nikodym derivative; however, for the suggested approach it was essential that the considered diffusion was one-dimensional. Prohorov [**927**] obtained a multidimensional generalization in the case of a constant diffusion coefficient. This result was extended to nonconstant diffusion coefficients by

A.V. Skorohod [**1041**] and I.V. Girsanov [**480**]. Later nonlinear transformations of the form $I + F$ for Gaussian measures on Hilbert or Banach spaces were investigated in Gross [**503**], Baklan, Shatashvili [**86**], Skorohod [**1043**], and other works. Theorem 10.2.4, one of the central results in this circle of problems, was obtained in Kusuoka [**673**] in the case of separable Banach spaces; under somewhat less general conditions a close result had been established in the papers Ramer [**939**], [**940**], which were of great importance for finding more general formulations. There are many works on this subject, see, in particular, Airault [**17**], Badrikian [**84**], Baudoin, Thieullen [**102**], Buckdahn [**254**], Buckdahn, Föllmer [**255**], Enchev [**396**], Feyel, Üstünel, Zakai [**432**], Kusuoka [**678**], Liptser, Shiryaev [**728**], Luo [**740**], Mazziotto, Millet [**796**], Skorohod [**1046**], [**1048**], Üstünel [**1151**], Üstünel, Zakai [**1154**]–[**1160**], [**1162**], [**1163**], [**1165**], Yano [**1200**], Zakai, Zeitouni [**1216**]. A detailed discussion of the principal results is given in the books Bogachev [**191**] and Üstünel, Zakai [**1164**]. For applications of nonlinear transformations of Gaussian measures, see also Albeverio, Yoshida [**57**], Sanz-Solé M., Malliavin [**984**].

On transformations of smooth measures, see Belopol'skaya, Daletskiĭ [**120**], Daletskiĭ, Steblovskaya [**328**], [**330**], Kulik, Pilipenko [**659**], Smolyanov, Weizsäcker [**1056**], [**1057**], Smolyanov, Weizsäcker, Vittich [**1058**], and Steblovskaya [**1074**], [**1076**]. Additional results related to this subject will be discussed in Chapter 11.

Transformations of measures by flows in finite and infinite dimensions are studied in Aida [**7**], Ambrosio [**61**], Ambrosio, Crippa [**62**], Ambrosio, Figalli [**64**], Ambrosio, Gigli, Savaré [**66**], Bogachev, Mayer-Wolf [**211**], [**212**], Brayman [**249**]–[**251**], Cipriano, Cruzeiro [**276**], Cruzeiro [**293**]–[**296**], Daletskiĭ, Steblovskaya [**327**], DiPerna, Lions [**352**], Fang, Luo [**415**], Mayer-Wolf [**791**], Kunita [**660**], [**661**], Luo [**739**], Peters [**882**], [**883**], Pilipenko [**900**], [**901**], [**902**], [**903**], Tolmachev, Khitruk [**1120**], Üstünel, Zakai [**1164**].

Certain infinite dimensional negligible sets are discussed also in Hunt, Sauer, Yorke [**565**], Johnson, Skoug [**586**]. For Gaussian measures, the result of Theorem 10.6.6 was obtained in Kusuoka [**674**].

On infinite dimensional versions of the Rademacher theorem, see also Ambrosio, Durand-Cartagena [**63**], Kulik [**657**], Röckner, Schied [**961**], Schied [**988**].

According to the well-known Sard theorem, for any infinitely differentiable mapping $F\colon \mathbb{R}^n \to \mathbb{R}^k$, the image of the set of critical points, that is, points where the derivative of F is not surjective, has measure zero. There is no infinite dimensional analog of the Sard theorem in this sense. In the papers Katznelson, Malliavin [**597**] and Malliavin, Katznelson [**770**], an example is constructed of a mapping $F\colon C[0,1] \to \mathbb{R}^n$, where $C[0,1]$ is equipped with the Wiener measure P^W, such that F is infinitely Fréchet differentiable, but the image (under the mapping F) of the restriction of P^W to the set $\{x\colon F'(x) = 0\}$ has a smooth density. In particular, unlike the finite dimensional case, the condition in Theorem 9.2.4 is far from being necessary. However, some infinite dimensional version of the Sard theorem for real functions was obtained in Lescot [**719**], where it was shown that if a real function f on a space with a Gaussian measure γ belongs to the Gevrey class (defined in [**719**]), then $f(\{D_H f = 0\})$ has measure zero. In Getzler [**475**], Kusuoka [**680**], Üstünel, Zakai [**1162**], [**1164**], analogous questions are considered for infinite dimensional mappings of the form $T(x) = x + F(x)$, where $F \in W^{2,1}(\gamma, H)$. For example, in the last two papers the proof of the following fact is given. If the mappings $h \mapsto F(x+h)$ are differentiable and the mappings $h \mapsto D_H F(x+h)$ with values in $\mathcal{H}(H)$ are continuous, then the image of the set $\{x\colon \det_2(I + D_H F(x)) = 0\}$ has γ-measure zero.

On triangular transformations, see also Knothe [**616**], Talagrand [**1106**]. Our exposition follows the recent works Aleksandrova [**58**], Bogachev, Kolesnikov [**198**], Bogachev, Kolesnikov, Medvedev [**204**]. Optimal mappings are applied for proving various useful inequalities, see the books Villani [**1174**], [**1175**] and the papers Bogachev, Kolesnikov [**199**], Cordero-Erausquin, Nazaret, Villani [**292**], Kolesnikov [**624**]. A new class of transformations of measures related to optimal transports is introduced in Bogachev, Kolesnikov [**202**], [**203**].

Exercises

10.8.1. Construct an example of a measurable mapping $F\colon X \to H$, where H is the Cameron–Martin space of a Radon Gaussian measure on a locally convex space X, such that F is smooth along H, $\|D_H F\|_{\mathcal{L}(H)} \leqslant 1/2$, but the measure $\gamma \circ (I+F)^{-1}$ is singular with respect to γ.

HINT: Take the countable power of the standard Gaussian measure and look for $F = (F_n)$ with values $H = l^2$, where $F_n(x) = f_n(x_n)$ with suitable smooth functions f_n.

10.8.2. Show that there exists a nonlinear diffeomorphism of the plane preserving the standard Gaussian measure.

HINT: Using polar coordinates observe that the standard Gaussian measure is invariant under any Borel transformation which is a rotation in every circle $|x| = r$.

10.8.3. Let f be a homeomorphism of $\mathrm{I\!R}^1$ preserving the standard Gaussian measure γ. Prove that either $f(x) = x$ or $f(x) = -x$.

HINT: Show that $f(0) = 0$ using that $[0, +\infty)$ is the only ray of γ-measure $1/2$; assuming that $f(1) > 0$ and $f(x) \neq x$ for some x, obtain a contradiction using that $[0, x]$ and $[0, y]$ have different γ-measures if $x \neq y$.

10.8.4. Construct two nondegenerate centered Gaussian measures μ and ν on l^2 such that none of them is the image of the other one under a continuous linear mapping.

HINT: Show that if a bounded linear operator A takes μ to ν, then A takes $H(\mu)$ onto $H(\nu)$; so it suffices to choose $H(\mu)$ and $H(\nu)$ in such a way that none of them could be mapped onto the other by a bounded linear operator on l^2; by Exercise 1.6.22 this will be the case if the eigenvalues of the covariance operator of μ are 3^{-n} and the eigenvalues of the covariance operator of ν are 2^{-n} for even n and 4^{-n} for odd n.

10.8.5. Suppose that X is a locally convex space, $E \subset X$ is a finite dimensional subspace, Y is a Banach space, and $F\colon X \to Y$ is a Borel mapping such that
$$\|F(x+h) - F(x)\|_Y \leqslant C|h| \quad \text{for all } x \in X,\ h \in E,$$
where $|\cdot|$ is some norm on E. Prove that the set of all points at which F is Gâteaux differentiable along E is Borel in X.

HINT: Use the fact that the set of points of convergence of a sequence of Borel functions is Borel measurable.

10.8.6. Find an example of an infinitely differentiable strictly increasing function on the real line under which the image of the standard Gaussian measure has an infinitely differentiable density, but is not a differentiable measure.

10.8.7. Let μ be a probability measure on $\mathbb{R}^n$ differentiable along all vectors. Find a sufficient condition on a differentiable mapping F ensuring that the measure $\mu \circ F^{-1}$ is differentiable along all vectors.

10.8.8. Let γ be the standard Gaussian measure on $\mathbb{R}^n$ and let $F\colon \mathbb{R}^\infty \to \mathbb{R}^\infty$ be infinitely differentiable. Find a sufficient condition in order that the measure $\gamma \circ F^{-1}$ be analytic along all vectors.

10.8.9. Let p be a continuous seminorm on a separable Banach space X such that p is not identically zero. Show that the sets $\{p = c\}$ are negligible.

HINT: Show that for any dense countable set $\{a_n\} \subset X$ the set $\{p = c\}$ can be represented as a countable union of sets E_n such that the sets $\{t\colon x + ta_n \in E_n\}$ are finite.

10.8.10. Let f be a nonconstant continuous polynomial on a separable Banach space X. Show that the sets $\{f = c\}$ are negligible.

10.8.11. Let μ be a densely continuous Radon measure on a Banach space X and let F be a Fréchet differentiable mapping on X such that its derivative is invertible at every point. Show that the measure $\mu \circ F^{-1}$ has absolutely continuous finite dimensional projections.

10.8.12. Use Example 10.5.13 to show that there exist a nondegenerate centered Gaussian measure γ on a separable Hilbert space X and a polynomial diffeomorphism F of X such that the measure $\gamma \circ F^{-1}$ is mutually singular with every measure possessing a nonzero vector of continuity.

10.8.13. Let γ be a centered Radon Gaussian measure, let H be its Cameron–Martin space, and let Q be a γ-measurable polynomial mapping with values in the space $\mathcal{H}(H)$. Set $f(x) = \det_2(I + Q(x))$. Show that either f is a constant a.e. or its distribution is absolutely continuous.

HINT: Consider the functions $t \mapsto f(x+th)$, $h \in H$.

10.8.14. Let γ be a centered Radon Gaussian measure, let H be its Cameron–Martin space, and let F be an H-valued mapping whose components F_n in some orthonormal basis in H are measurable second order polynomials orthogonal to all first order polynomials. Show that $\gamma \circ (I+F)^{-1} \ll \gamma$.

HINT: Use the previous exercise.

CHAPTER 11

Measures on manifolds

This chapter is concerned with differentiable measures on general measurable spaces and on measurable spaces equipped with certain differential structures enabling us to consider differentiations along vector fields. In the case of an abstract measurable space without any additional structure one can introduce various topologies on the space of measures $\mathcal{M}$ and study the continuity or differentiability of families of measures μ_u parameterized by elements u of some space U with a manifold structure (in a suitable sense). This leads to the mapping $u \mapsto \mu_u$ from U to $\mathcal{M}$ and enables one to consider its differential properties. The simplest and most important case is a family of measures parameterized by points of the real line or an interval. In the previous chapters, we dealt with the partial case of this situation where the measures μ_t were the shifts of a measure μ by vectors th. A more general case is a family of the images of a measure μ under transformations T_t. Certainly, one can deal with families of other types. Here, even more than in the case of linear spaces, the principal role is played not by a general theory, but rather by examples which it covers. So the first two sections should be regarded not as a basis of a general theory (which, perhaps, does not exist at all for the whole circle of the problems under consideration), but a rather limited collection of universal concepts and constructions which in every concrete situation are filled in with an appropriate content so that the results obtained in these concrete situations are not corollaries of some general facts, but realizations of a programme inspired by these universal concepts and constructions. The situation is similar to the one in the theory of partial differential equations, in which an effective theory is developed for very special classes of equations (elliptic, parabolic, hyperbolic, etc.), and for each class this theory is highly individual. The principal examples of infinite dimensional nonlinear manifolds with measures studied so far are infinite products of finite dimensional manifolds, spaces of mappings from one finite dimensional manifold into another (for example, the space of mappings from an interval or a circle to a nonlinear finite dimensional manifold), more special spaces of this kind consisting of mappings with various additional properties (for example, the space of homeomorphisms or diffeomorphisms, the space of volume preserving transformations), the space of operators on a Hilbert space and its various subsets, and also diverse spaces of measures including the special, but important for applications, case of spaces of configurations. All these examples will be considered below, but the reader should be warned right now that a considerable portion of results will be presented without proofs since, in many cases, technical details of justifications are technically involved.

11.1. Measurable manifolds and Malliavin's method

As we have already noted above, the integration by parts formulae discussed in Chapter 6 are naturally extended to the case of differentiability along vector fields. A vector field on a linear space is merely a measurable mapping of the space into itself. The situation is more difficult if we deal with an infinite dimensional "manifold". The point is that a standard concept of an infinite dimensional differentiable manifold, where charts are neighborhoods in a Banach or locally convex space, turns out to be too rigid for most of applications. For example, the infinite dimensional torus (the countable power of the circle) is a compact topological space with a natural differentiable structure, but is has no local charts homeomorphic to open sets in locally convex spaces since only finite dimensional linear spaces are locally compact. Thus, we need a more general concept of a "differentiable manifold". Such a concept was implicitly introduced in the Malliavin calculus. In fact, the Malliavin operator L (or the corresponding operator "caré du champ") imports some differentiable structure. This was employed already in the paper Stroock [**1082**]; an abstract approach applicable to arbitrary measurable spaces was suggested in Smorodina [**1059**], [**1060**], Bogachev [**179**]. In the framework of the theory of Dirichlet forms this idea was developed in Bouleau, Hirsch [**247**].

Denote by $C_p^\infty(\mathbb{R}^n)$ the class of all smooth functions on $\mathbb{R}^n$ with derivatives of at most polynomial growth.

11.1.1. Definition. *A collection $(X, \mathcal{B}, \mu, \mathcal{E})$ is called a measurable manifold if $(X, \mathcal{B}, \mu)$ is a measurable space with a measure $\mu \geq 0$ equipped with an algebra $\mathcal{E} \subset \bigcap_{p<\infty} L^p(\mu)$ that is dense in each $L^p(\mu)$, consists of $\mathcal{B}$-measurable functions and satisfies the following condition:*

$$\varphi(f_1, \ldots, f_n) \in \mathcal{E} \quad \text{for all} \quad f_1, \ldots, f_n \in \mathcal{E} \quad \text{and} \quad \varphi \in C_p^\infty(\mathbb{R}^n).$$

11.1.2. Example. Let $(X_n, \mathcal{B}_n, \mu_n, \mathcal{E}_n)$ be measurable manifolds with probability measures μ_n. Then their product $X = \prod_n X_n$ equipped with the product-measure $\mu = \bigotimes_n \mu_n$ is a measurable manifold with the algebra $\mathcal{E}$ of all functions of the form $f = \varphi(f_1 \circ \pi_1, \ldots, f_n \circ \pi_n)$, where $\varphi \in C_b^\infty(\mathbb{R}^n)$, $f_i \in \mathcal{E}_i$, and π_i is the natural projection on X_i.

Sometimes the following stronger condition is useful.

11.1.3. Definition. *The algebra $\mathcal{E}$ satisfies condition (C) if, for every open set $U \subset \mathbb{R}^n$, every function $\psi \in C^\infty(U)$ and every mapping $F = (f_1, \ldots, f_n) \colon X \to U$ such that $f_i \in \mathcal{E}$ and $\partial^{(\alpha)}\psi(F) \in \bigcap_{p<\infty} L^p(\mu)$, we have $\psi(F) \in \mathcal{E}$.*

11.1.4. Example. In the situation of Theorem 8.2.4, the class $D^\infty(\mu)$ satisfies condition (C). In particular, if γ is a Gaussian measure, then the Sobolev class $W^\infty(\gamma) = H^\infty(\gamma) = D^\infty(\gamma)$ satisfies condition (C).

PROOF. It suffices to use the characterization of these classes in terms of the directional differentiability. □

11.1.5. Definition. *A linear mapping* $v\colon \mathcal{E} \to \mathcal{E}$ *is called a smooth vector field on* X *if, setting* $\partial_v f := v(f)$, *for all* $f_i \in \mathcal{E}$, $\varphi \in C_p^\infty(\mathbb{R}^n)$, *one has the equality*

$$\partial_v[\varphi(f_1,\ldots,f_n)] = \sum_{i=1}^n \partial_{x_i}\varphi(f_1,\ldots,f_n)\partial_v f_i.$$

11.1.6. Definition. *A measure* λ *on* $(X,\mathcal{B})$ *is called differentiable along a vector field* v *if* $\mathcal{E} \subset L^1(\lambda)$ *and there exists a measure* $d_v\lambda$ *such that for all bounded* $f \in \mathcal{E}$ *one has*

$$\int_X \partial_v f(x)\, \lambda(dx) = -\int_X f(x)\, d_v\lambda(dx). \tag{11.1.1}$$

If $d_v\lambda \ll \lambda$, *then the corresponding Radon–Nikodym density is denoted by the symbol* β_v^λ *(or just by* β_v*) and is called a logarithmic derivative of the measure* λ *along* v.

Let us observe that if the class $\mathcal{E}$ separates measures on $\mathcal{B}$, then a logarithmic derivative is unique.

One can extend the concepts of a vector field and the respective differentiability in order to admit not necessarily smooth fields $v\colon \mathcal{E} \to L^0(\lambda)$. Namely, in (11.1.1) one should take only f with $\partial_v f \in L^1(\lambda)$.

A mapping $F = (f_1,\ldots,f_n)$, where $f_i \in \mathcal{E}$, is called nondegenerate if there exist smooth vector fields $v_1,\ldots,v_n$ such that

$$\det\bigl((\partial_{v_i} f_j)_{i,j=1}^n\bigr)^{-1} \in \mathcal{E}.$$

Under some additional assumptions on X one can define an analog of logarithmic gradients. Namely, suppose that X is a measurable Riemannian manifold in the sense that a suitable linear space of smooth vector fields on X is equipped with a measurable family of inner products $(\,\cdot\,,\cdot\,)_x$, $x \in X$. Suppose also that a divergence $\operatorname{div} v$ is defined. Denote by TX the completion of the linear space of smooth vector fields on X with respect to the inner product

$$(u,v)_0 = \int_X \bigl(u(x),v(x)\bigr)_x \mu(dx).$$

11.1.7. Definition. *A mapping* $\beta^\mu\colon X \to TX^*$ *is called a logarithmic gradient of a measure* μ *if for every smooth vector field* v *on* X *one has*

$$\langle \beta^\mu, v\rangle = \beta_v^\mu - \operatorname{div} v \quad \mu\text{-a.e.}$$

If a gradient ∇ is also defined on $\mathcal{E}$ (i.e., $\nabla f \in TX$), then we define the Laplace–Beltrami operator Δ by the equality

$$\Delta f = \mathrm{div}\nabla f.$$

The following identity holds:

$$\int_X (\nabla f, \nabla g)_x\, \mu(dx) = -\int_X f[\Delta g + \langle \beta^\mu, \nabla g\rangle]\, \mu(dx).$$

Indeed,

$$\int_X (\nabla f, \nabla g)_x\, \mu(dx) = -\int_X f\beta^\mu_{\nabla g}\, \mu(dx),$$
$$\int_X f\langle \beta^\mu, \nabla g\rangle\, \mu(dx) = \int_X f[\beta^\mu_{\nabla g} - \mathrm{div}\nabla g]\, \mu(dx).$$

11.1.8. Definition. *A Malliavin operator on a probability space (Ω, P) is a symmetric linear operator L on $L^2(P)$ defined on a dense linear subspace $\mathcal{R}$ in $\bigcap_{r<\infty} L^r(P)$ and taking values in $\bigcap_{r<\infty} L^r(P)$ such that the following conditions are fulfilled:*

1) $f(\varphi_1,\ldots,\varphi_n) \in \mathcal{R}$ *for all* $f \in C_b^\infty(\mathbb{R}^n)$, $\varphi_i \in \mathcal{R}$;
2) $\Gamma(f,f) \geqslant 0$ *a.e. for all* $f \in \mathcal{R}$, *where*

$$\Gamma(f,g) := \frac{1}{2}[L(fg) - fLg - gLf];$$

3) *for all* $\varphi \in C_b^\infty(\mathbb{R}^n)$ *and* $f_1,\ldots,f_n \in \mathcal{R}$ *one has*

$$L\bigl(\varphi(f_1,\ldots,f_n)\bigr) = \sum_{i=1}^n \frac{\partial \varphi}{\partial x_i}(f_1,\ldots,f_n)Lf_i + \sum_{i,j=1}^n \frac{\partial^2 \varphi}{\partial x_i \partial x_j}(f_1,\ldots,f_n)\Gamma(f_i,f_j).$$

A typical example of a Malliavin operator is the Ornstein–Uhlenbeck operator.

It follows from 3) that $1 \in \mathcal{R}$ and $L1 = 0$. By the symmetry of L one has

$$\int_\Omega Lf(x)\, P(dx) = 0 \quad \forall f \in \mathcal{R}.$$

Along with the symmetry of L this yields the following identity:

$$\int_\Omega f(x)Lg(x)\, P(dx) = -\int_\Omega \Gamma(f,g)(x)\, P(dx). \tag{11.1.2}$$

Operators of this type were introduced in Stroock [**1082**] in the form of generators of symmetric diffusion semigroups. Concerning operators Γ ("opérateurs carré du champ"), see Bakry, Emery [**89**]. Their relation to differentiation along vector fields is described by the following result from Bogachev [**179**].

11.1.9. Proposition. (i) *Let L be a Malliavin operator, $\mathcal{E} = \mathcal{R} \cap L^\infty$. Then $(\Omega, P, \mathcal{E})$ is a measurable manifold. In addition, if $f_1,\ldots,f_n \in \mathcal{R}$, then the mappings*

$$f \mapsto \Gamma(f,f_i), \quad \mathcal{E} \to \bigcap_{r<\infty} L^r(P),$$

11.1. Measurable manifolds and Malliavin's method

define vector fields v_i (in particular, the Malliavin matrix $\Gamma(f_i, f_j)$ coincides with $\partial_{v_i} f_j$), the measure P is differentiable along v_i and $d_{v_i} P = L f_i \cdot P$, i.e., we have $\beta_{v_i}^P = L f_i$.

(ii) *Conversely, suppose that a measure μ on a measurable manifold $(X, \mu, \mathcal{E})$ is differentiable along certain fields $v_1, \ldots, v_n$ and $d_{v_i} \mu = \beta_{v_i}^\mu \cdot \mu$, where we have $\beta_{v_i}^\mu \in \bigcap_{p<\infty} L^p(\mu)$. Then the formula*

$$Lf = \sum_{i=1}^n (\partial_{v_i}^2 f + \beta_{v_i}^\mu \partial_{v_i} f)$$

defines a Malliavin operator.

PROOF. (i) Since the class $\mathcal{E}$ consists of bounded functions, compositions with C^∞-functions are the same as compositions with C_b^∞-functions. Let $\varphi \in C_b^\infty(\mathbb{R}^k)$, $g_1, \ldots, g_k \in \mathcal{E}$, $h \in \mathcal{R}$. The field $f \mapsto \Gamma(f, h)$ will be denoted by v. Then, applying condition 3) from Definition 11.1.8 to the function $y_{k+1} \varphi(y_1, \ldots, y_k)$ on $\mathbb{R}^{k+1}$, we find that

$$L\big(h\varphi(g_1, \ldots, g_k)\big) = \varphi(g_1, \ldots, g_k) Lh + h \sum_{i=1}^k \partial_{x_i} \varphi(g_1, \ldots, g_k) L g_i$$
$$+ h \sum_{i,j=1}^k \partial_{x_i} \partial_{x_j} \varphi(g_1, \ldots, g_k) \Gamma(g_i, g_j) + 2 \sum_{i=1}^k \partial_{x_i} \varphi(g_1, \ldots, g_k) \Gamma(h, g_i).$$

Hence

$$\partial_v [\varphi(g_1, \ldots, g_k)] = \frac{1}{2} \big[L\big(h\varphi(g_1, \ldots, g_k)\big) - hL\varphi(g_1, \ldots, g_k) - \varphi(g_1, \ldots, g_k) Lh \big]$$
$$= \sum_{i=1}^k \partial_{x_i} \varphi(g_1, \ldots, g_k) \Gamma(h, g_i) = \sum_{i=1}^k \partial_{x_i} \varphi(g_1, \ldots, g_k) \partial_v g_i.$$

Therefore, v is a vector field. Clearly, $\partial_v f \in L^r(P)$ for any $f \in \mathcal{E}$ and $r < \infty$. By construction we have $\partial_{v_i} f_j = \Gamma(f_i, f_j)$. Finally, for any $f \in \mathcal{E}$ we have

$$\int_\Omega \partial_{v_i} f \, P(dx) = \frac{1}{2} \int_\Omega [L(f f_i) - f L f_i - f_i L f] \, P(dx) = - \int_\Omega f L f_i \, P(dx)$$

since the operator L is symmetric and the integral of $L(f f_i)$ vanishes. Therefore, we obtain $d_{v_i} P = L f_i \cdot P$.

(ii) It suffices to prove our assertion for a single vector field $v \colon \mathcal{E} \to \mathcal{E}$. We have $L \colon \mathcal{E} \to \bigcap_{p<\infty} L^p(\mu)$. For any $f \in \mathcal{E}$ one has

$$\Gamma(f, f) = \frac{1}{2} L(f^2) - f L f (\partial_v f)^2 + f \partial_v^2 f + f \partial_v f \beta_v^\mu - f \partial_v^2 f + f \partial_v f \beta_v^\mu$$
$$= (\partial_v f)^2 \geq 0.$$

Given $f_1, \ldots, f_k \in \mathcal{E}$ and $\varphi \in C_b^\infty(\mathbb{R}^k)$, we have

$$L\varphi(f_1,\ldots,f_k) = \partial_v \sum_{i=1}^k \partial_{x_i}\varphi(f_1,\ldots,f_k)\partial_v f_i + \beta_v^\mu \sum_{i=1}^k \partial_{x_i}\varphi(f_1,\ldots,f_k)\partial_v f_i$$

$$= \sum_{i=1}^k \Big[\partial_{x_i}\varphi(f_1,\ldots,f_k)\partial_v^2 f_i + \beta_v^\mu \partial_{x_i}\varphi(f_1,\ldots,f_k)\partial_v f_i\Big]$$

$$+ \sum_{i,j=1}^k \partial_{x_j}\partial_{x_i}\varphi(f_1,\ldots,f_k)\partial_v f_i \partial_v f_j.$$

This ensures condition 3) from Definition 11.1.8 since

$$\partial_v f_i \partial_v f_j = L(f_i f_j) - f_i L f_j - f_j L f_i = \Gamma(f_i, f_j)$$

due to the equality

$$L(f_i f_j) = f_i \partial_v^2 f_j + f_j \partial_v^2 f_i + \partial_v f_i \partial_v f_j.$$

It remains to verify that L is symmetric. Let $f, g \in \mathcal{E}$. Integrating by parts we obtain

$$\int_X f L g \, \mu(dx) = - \int_X \partial_v f \partial_v g \, \mu(dx),$$

which coincides with the integral of gLf. □

11.1.10. Example. If μ is an atomless probability measure on a standard measurable space $(\Omega, \mathcal{B})$, then there exists a measurable structure $\mathcal{E}$ on $(\Omega, \mathcal{B})$ such that, for every centered square-integrable random variable ξ on (Ω, μ), one can find a measurable vector field $v(\xi)$ on Ω along which the measure μ is differentiable and $\beta_{v(\xi)}^\mu = \xi$ a.e.

PROOF. It suffices to consider an isomorphism ψ of measurable spaces $(\Omega, \mathcal{B}, \mu)$ and $\mathbb{R}^1$ with the Borel σ-algebra and the standard Gaussian measure γ. Then one can export the differentiable structure of $\mathbb{R}^1$ to Ω by using this isomorphism. Namely, let $\mathcal{E}$ consist of all functions $f = \psi \circ h$, where $\psi \in C_b^\infty(\mathbb{R}^1)$. Given a vector field u on the real line, we set $\partial_v f := (\psi' u) \circ h$. If $f_i = \psi \circ h$, $i = 1, \ldots, n$, $\varphi \in C_p^\infty(\mathbb{R}^n)$, then $F = \varphi(f_1, \ldots, f_n) \in \mathcal{E}$ and $\partial_v F = \sum_{i=1}^n \partial_{x_i}\varphi(f_1, \ldots, f_n)\partial_v f_i$. Finally, if γ is differentiable along u, then the equality $\gamma = P \circ h^{-1}$ yields

$$\int_\Omega \partial_v f(\omega) \, P(d\omega) = \int_{-\infty}^{+\infty} \psi'(x) u(x) \, \gamma(dx) = - \int_{-\infty}^{+\infty} \psi(x) \beta_u^\gamma(x) \, \gamma(dx)$$

$$= - \int_\Omega f(\omega) \beta_u^\gamma(h(x)) \, P(d\omega).$$

Hence $\beta_v^\mu = \beta_u^\gamma \circ h$. It remains to choose u such that $\beta_u^\gamma = \xi \circ h^{-1}$, which is possible since $\eta := \xi \circ h^{-1}$ belongs to $L^2(\gamma)$ and has zero mean. Note that u

can be found explicitly from the relation $u'(x) - xu(x) = \eta(x)$, whence we find

$$u(x) = e^{x^2/2} \int_{-\infty}^{x} \eta(t) e^{-t^2/2} \, dt.$$

One can show directly (Exercise 11.6.6) that $u \in L^2(\gamma)$. Alternatively, one can observe that there is $\zeta \in W^{2,2}(\gamma)$ such that $\zeta''(x) - x\zeta'(x) = \eta(x)$ since the Ornstein–Uhlenbeck operator has bounded inverse on the orthogonal complement to 1, hence we can take $u = \zeta'$. This shows that actually one has $u \in W^{2,1}(\gamma)$. □

In this example, the imported differentiable structure on Ω may be very exotic, such that this construction cannot be considered as satisfactory. Usually there is some original "smooth" structure on Ω (for example, the space of smooth cylindrical functions if Ω is the path space), and it would be natural to seek vector fields on this original structure. For example, it would be interesting to answer the following question. Let M_t be a continuous martingale on $[0, 1]$, let Ω be the space of trajectories with the algebra $\mathcal{F}C^\infty$, and let η_t be an adapted square-integrable process. Is it always possible to find a measurable vector field v on Ω along which the measure μ on Ω induced by M_t is differentiable with respect to $\mathcal{F}C^\infty$ and

$$\beta_v^\mu = \int_0^1 \eta_t \, dM_t \text{ a.e.?}$$

As follows from Example 9.1.1, the answer is positive if M_t is a diffusion process.

One of the principal results of the Malliavin calculus extends to measurable manifolds in the following way (the proof, completely analogous to the reasoning from Chapter 9, is given in Bogachev, Smolyanov [233]).

11.1.11. Theorem. *If $F = (f_1, \ldots, f_n)$ is a nondegenerate mapping, then the measure $\mu \circ F^{-1}$ has a density $p \in \mathcal{S}(\mathbb{R}^n)$.*

11.1.12. Remark. Given a Malliavin operator L on a measurable manifold $(X, \mu, \mathcal{E})$, one can define natural analogs of the function ψ_f and equation (9.7.4) considered at the end of §9.7. Let $f \in \mathcal{E}$ be such that there exists a function $g \in \mathcal{E}$ for which $Lg = f$ (or a function with this property from a suitable analog of the second Sobolev class). Set

$$\eta := -\Gamma(f, g).$$

There is a Borel function ψ_f on $\mathbb{R}^1$ such that

$$\mathbb{E}^{\sigma(f)} \eta = \psi_f(f),$$

where $\mathbb{E}^{\sigma(f)}$ denotes the conditional expectation with respect to the σ-algebra generated by f. Let $\Phi \in C^\infty(\mathbb{R}^1)$ be such that $\varphi := \Phi'$ has compact support and $\Phi(0) = 0$. Similarly to (9.7.1) we have

$$\mathbb{E}[f\Phi(f)] = \mathbb{E}[\Phi(f)Lg] = -\mathbb{E}\big[\Gamma\big(\Phi(f), g\big)\big] \qquad (11.1.3)$$
$$= -\mathbb{E}\big[\Phi'(f)\Gamma(f,g)\big] = \mathbb{E}\big[\Phi'(f)\eta\big] = \mathbb{E}[\varphi(f)\psi_f(f)].$$

Therefore, equalities (9.7.3)–(9.7.5) hold also in this more general case.

In the case of a mapping $F = (f_1, \ldots, f_n)$ to $\mathbb{R}^n$, where $f_i \in \mathcal{E}$, assuming that there exist functions $g_i \in \mathcal{E}$ with $Lg_i = f_i$, we set

$$\eta_{ij} := -\Gamma(f_i, g_j).$$

Let $\mathbb{E}^{\sigma(F)}$ denote the conditional expectation with respect to the σ-algebra generated by F. There exist Borel functions ψ_{ij} on $\mathbb{R}^n$ such that

$$\mathbb{E}^{\sigma(F)}\eta_{ij} = \psi_{ij}(F).$$

Now in place of (11.1.3) we obtain

$$\mathbb{E}[f_i\Phi(F)] = \mathbb{E}[\Phi(F)Lg_i] = -\mathbb{E}\big[\Gamma\big(\Phi(F), g_i\big)\big] \qquad (11.1.4)$$
$$= -\sum_{j=1}^n \mathbb{E}\big[\partial_{x_j}\Phi(F)\Gamma(f_j, g_i)\big] = \sum_{j=1}^n \mathbb{E}\big[\partial_{x_j}\Phi(F)\eta_{ji}\big] \qquad (11.1.5)$$
$$= \sum_{j=1}^n \mathbb{E}[\partial_{x_j}\Phi(F)\psi_{ji}(F)],$$

which yields the following system of equations on $\mathbb{R}^n$:

$$\sum_{j=1}^n \partial_{x_j}\big[\psi_{ji} \cdot (\mu \circ F^{-1})\big] = x_i \cdot \mu \circ F^{-1}, \quad i = 1, \ldots, n.$$

The following definition goes back to Malliavin [**762**]; close motives can be found in the works Nualart, Üstünel [**855**], Nualart, Üstünel, Zakai [**857**], Üstünel, Zakai [**1152**], [**1153**] on analysis on the Wiener space; an abstract approach presented below was suggested in Bogachev [**179**].

11.1.13. Definition. *Let $\mathcal{A}$ be a subalgebra in $\mathcal{E}$ and let $\mathbb{E}^{\mathcal{A}}$ be the conditional expectation corresponding to the σ-algebra $\sigma(\mathcal{A})$. We call $\mathbb{E}^{\mathcal{A}}$ smooth if $\mathbb{E}^{\mathcal{A}}(\mathcal{E}) \subset \mathcal{E}$.*

Formally, this definition differs from the definition given in [**762**], where the equality $\mathbb{E}^{\mathcal{A}}(\mathcal{E}) = \mathcal{A}$ is required. However, in the situation of the above definition one can replace the initial subalgebra $\mathcal{A} \subset \mathcal{E}$ by the intersection $\mathcal{A}_1 = \mathcal{E} \cap L^2(\mathcal{A})$, where $L^2(\mathcal{A}) := \mathbb{E}^{\mathcal{A}}\big(L^2(\mu)\big)$, which yields the equalities

$$\sigma(\mathcal{A}) = \sigma(\mathcal{A}_1), \quad \mathbb{E}^{\mathcal{A}_1}(\mathcal{E}) = \mathbb{E}^{\mathcal{A}}(\mathcal{E}) = \mathcal{A}_1.$$

If $\mathcal{E}$ is a Fréchet space continuously embedded into $L^2(\mu)$, we can make $\mathcal{A}_1$ closed in $\mathcal{E}$ by taking its closure in $\mathcal{E}$.

11.1.14. Remark. It would be interesting to obtain conditions sufficient for the equality $\mathcal{A} = \mathcal{E} \cap L^2(\mathcal{A})$ assuming that $\mathcal{A}$ is closed. In the general case this equality may be false as the following simple example shows. Let

$$\mathcal{E} = C_b(\mathbb{R}^1), \quad \mathcal{A} = \{f \in \mathcal{E}: \exists \lim_{|t| \to \infty} f(t)\}.$$

Then $\mathcal{A}$ is closed in $\mathcal{E}$, $\sigma(\mathcal{E}) = \sigma(\mathcal{A}) = \mathcal{B}(\mathbb{R}^1)$, but $\mathbb{E}^{\mathcal{A}}\mathcal{E} = \mathcal{E}$ is broader than $\mathcal{A}$, where μ is the standard Gaussian measure.

11.1.15. Remark. Suppose that $\mathcal{E}$ is equipped with a structure of a Fréchet space (or a barrelled space) such that the natural embedding $\mathcal{E} \to L^2(\mu)$ is continuous and that $\mathcal{A}$ is closed in $\mathcal{E}$. Then the mapping $\mathbb{E}^{\mathcal{A}}: \mathcal{E} \to \mathcal{E}$ is automatically continuous provided that $\mathbb{E}^{\mathcal{A}}$ is smooth (the mapping $\mathbb{E}^{\mathcal{A}}: \mathcal{E} \to \mathcal{A}$ is continuous provided that $\mathbb{E}^{\mathcal{A}}$ is smooth in the sense of Malliavin). Indeed, since the mapping $\mathbb{E}^{\mathcal{A}}: \mathcal{E} \to L^2(\mu)$ is continuous, we can apply the closed graph theorem. In particular, this is fulfilled in the case considered in Malliavin [**762**], hence in Definition 4.2 in [**762**] the mapping $\mathbb{E}^{\mathcal{A}}$ is automatically continuous.

An important example of a smooth conditional expectation is given by the σ-algebra $\mathcal{A}$ generated by any nondegenerate mapping $F = (f_1, \ldots, f_n)$. This was proved for functionals on the Wiener space in Üstünel, Zakai [**1153**] (see the proof of Proposition 2.10 in [**1153**]) under the assumption of $\mathcal{A}$-measurability of $\partial_{v_i} f_j$ and stated in Malliavin [**762**] without that assumption. According to the next proposition, the same is true in a more general case (the proof delegated to Exercise 11.6.4 can be found in Bogachev [**186**], where, however, the assumption of the $\mathcal{A}$-measurability of $\partial_{v_i} f_j$ was mistakenly omitted in the formulation).

11.1.16. Proposition. *Let a mapping $F = (f_1, \ldots, f_n)$ be nondegenerate and let $\mathcal{A} = \sigma(F)$. Then $\mathbb{E}^{\mathcal{A}}\mathcal{E} \subset \mathcal{E}$ provided that $\mathcal{E}$ satisfies condition* (C) *and the functions $\partial_{v_i} f_j$ are $\mathcal{A}$-measurable.*

As we shall now see, in many cases the assumption of $\mathcal{A}$-measurability of $\partial_{v_i} f_j$ can be omitted indeed.

We need the following observation. Let $F = (f_1, \ldots, f_n)$, where $f_i \in \mathcal{E}$, be a nondegenerate mapping and let $\mathcal{A}$ be the σ-field generated by F. Suppose that $\varphi \in \mathcal{E}$. We know that the measures $\mu \circ F^{-1}$ and $(\varphi \cdot \mu) \circ F^{-1}$ have smooth densities ϱ and ϱ_φ, respectively. We observe that $\varrho_\varphi = \eta \cdot \varrho$, where $\eta \in L^1(\mu \circ F^{-1})$ is a Borel function. Therefore,

$$\frac{\varrho_\varphi}{\varrho} \circ F \in L^1(\mu),$$

where $\varrho_\varphi/\varrho := 0$ on the set where ϱ vanishes. Note also that $\varrho_\varphi(F)/\varrho(F)$ coincides with the conditional expectation of φ with respect to $\mathcal{A}$. Indeed, for any function of the form $\psi \circ F$, where ψ is a bounded Borel function on $\mathbb{R}^n$, we have

$$\int_X \frac{\varrho_\varphi\bigl(F(x)\bigr)}{\varrho\bigl(F(x)\bigr)}\,\psi\bigl(F(x)\bigr)\mu(dx) = \int_{\mathbb{R}^n} \varrho_\varphi(y)\psi(y)\,dy = \int_X \psi\bigl(F(x)\bigr)\varphi(x)\,\mu(dx).$$

Therefore, since $\mathbb{E}^{\mathcal{A}}$ is a contraction on $L^p(\mu)$, one has

$$\frac{\varrho_\varphi}{\varrho} \in L^p(\mu\circ F^{-1}) \quad \text{if } \varphi \in L^p(\mu).$$

Let μ be a Radon probability measure on a locally convex space X infinitely differentiable along all vectors from a continuously embedded separable Hilbert space H such that the corresponding logarithmic derivatives belong to all $L^p(\mu)$ with $p < \infty$.

11.1.17. Proposition. *Let $f_1, \ldots, f_n \in W_H^\infty(\mu)$ and let $F = (f_1, \ldots, f_n)$ be a nondegenerate mapping and let $\mathcal{A}$ be the σ-field generated by F. Then, for any $\varphi \in W_H^\infty(\mu)$, one has $\mathbb{E}^{\mathcal{A}}\varphi \in W_H^\infty(\mu)$.*

PROOF. Let us set $\varrho_\varepsilon := \varrho + \varepsilon$, $\varepsilon > 0$. Then $\varrho_\varphi/\varrho_\varepsilon \in C^\infty(\mathbb{R}^n)$ and we have

$$D_H\left(\frac{\varrho_\varphi(F)}{\varrho_\varepsilon(F)}\right) = \varrho_\varepsilon^{-1}(F)[\varrho_\varphi'(F)]\circ D_H F + \varrho_\varepsilon^{-2}(F)\varrho_\varphi(F)[\varrho'(F)]\circ D_H F.$$

For every fixed $p < \infty$, both terms on the right are uniformly bounded in $L^p(\mu)$ with respect to $\varepsilon \in (0,1)$ because $\varrho^{-1}(F)\varrho_\varphi(F)$, $\varrho^{-1}(F)\varrho_\varphi'(F)$, and $\varrho^{-1}(F)\varrho'(F)$ belong to all $L^p(\mu)$. Repeatedly differentiating the obtained identity, using that $\varrho^{-1}(F)\partial^{(\alpha)}\varrho_\varphi(F) \in L^p(\mu)$ for all $p < \infty$ and letting $\varepsilon \to 0$, we finally obtain the inclusion $\varrho_\varphi(F)/\varrho(F) \in W_H^{p,r}(\mu)$ for all $p < \infty$ and all $r \in \mathbb{N}$. □

A vector field v on X is called $\mathcal{A}$-basic (see [**762**]) if the function $\partial_v f$ is measurable with respect to $\mathcal{A}$ for every $\mathcal{A}$-measurable function $f \in \mathcal{E}$.

11.1.18. Proposition. *Let a field v be $\mathcal{A}$-basic and let $\mathbb{E}^{\mathcal{A}}$ be smooth. Then the following Malliavin identity holds:*

$$\partial_v(\mathbb{E}^{\mathcal{A}}f) = \mathbb{E}^{\mathcal{A}}(\partial_v f) + \mathbb{E}^{\mathcal{A}}(f\beta_v^\mu) - (\mathbb{E}^{\mathcal{A}}f)(\mathbb{E}^{\mathcal{A}}\beta_v^\mu), \quad f \in \mathcal{E}.$$

PROOF. By our condition the function $\partial_v(\mathbb{E}^{\mathcal{A}}f)$ belongs to the class $\mathcal{E}$ and is $\mathcal{A}$-measurable. Since $\mathcal{E}$ is dense in $L^2(\mu)$ and $\mathbb{E}^{\mathcal{A}}(\mathcal{E}) \subset \mathcal{E}$, the set of $\mathcal{A}$-measurable functions from $\mathcal{E}$ is dense in the space of $\mathcal{A}$-measurable functions in $L^2(\mu)$. For every $\mathcal{A}$-measurable function $\varphi \in \mathcal{E}$, we have

$$\int_X \partial_v(\mathbb{E}^{\mathcal{A}}f)\varphi\,\mu(dx) = -\int_X \mathbb{E}^{\mathcal{A}}f\partial_v\varphi\,\mu(dx) - \int_X \mathbb{E}^{\mathcal{A}}f\varphi\,\beta_v^\mu\,\mu(dx)$$

$$= \int_X \bigl[\mathbb{E}^{\mathcal{A}}(\partial_v f)\varphi + \mathbb{E}^{\mathcal{A}}(f\beta_v^\mu) + (\mathbb{E}^{\mathcal{A}}f)(\mathbb{E}^{\mathcal{A}}\beta_v^\mu)\bigr]\varphi\,\mu(dx)$$

because
$$\int_X \mathbb{E}^{\mathcal{A}} f \partial_v \varphi \, \mu(dx) = \int_X f \partial_v \varphi \, \mu(dx)$$
$$= -\int_X (\partial_v f + f \beta_v^\mu) \varphi \, \mu(dx) = -\int_X \bigl[\mathbb{E}^{\mathcal{A}} (\partial_v f + f \beta_v^\mu)\bigr] \varphi \, \mu(dx).$$

This relationship gives the required identity. □

In the papers Nualart, Üstünel [**855**], Nualart, Üstünel, Zakai [**857**], Üstünel, Zakai [**1152**], [**1153**] in the case of the Wiener space, some classes of σ-algebras are discussed in relation to the following interesting question. Suppose that two smooth functionals f and g have the gradients $D_H f$ and $D_H g$ orthogonal in $L^2(\gamma, H)$. Is it true that f and g are independent? In general, this is false (see Exercise 11.6.3). The converse is also of interest: if f and g are independent, then are $D_H f$ and $D_H g$ orthogonal in $L^2(\gamma, H)$? This is also false (see Nualart, Üstünel, Zakai [**856**] and Exercise 11.6.3). However, under additional assumptions there are positive assertions.

11.2. Differentiable families of measures

Let $(X, \mathcal{B})$ be a measurable space and let $\mathcal{M}(X)$ be the Banach space of all real measures on $\mathcal{B}$ with the variation norm. Suppose that $\mathcal{M} \subset \mathcal{M}(X)$ is a linear subspace equipped with some topology τ consistent with its linear structure. The most important example is the case where X is a completely regular topological space with its Borel σ-algebra $\mathcal{B}(X)$ (or the Baire σ-algebra $\mathcal{B}_0(X)$), $\mathcal{M} = \mathcal{M}(X)$ is the space of all Borel (or Baire) measures on X, and τ is one of the following topologies:

(i) the topology τ_s of setwise convergence;

(ii) the topology τ_v generated by the variation norm;

(iii) the weak topology τ_w of the Banach space $\mathcal{M}(X)$ equipped with the variation norm;

(iv) the topology $\tau_{\mathcal{F}}$ generated by the duality with some linear space $\mathcal{F}$ of bounded real functions on X satisfying the condition

$$\|\mu\| = \sup_{f \in \mathcal{F},\, \|f\| \leqslant 1} \int_X f(x)\, \mu(dx)$$

for all $\mu \in \mathcal{M}$, where $\|f\| := \sup_x |f(x)|$.

Typical examples of $\mathcal{F}$ are: $C_b(X)$ with its sup-norm, the space $\mathcal{B}_b(X)$ of all bounded $\mathcal{B}$-measurable functions with its sup-norm, and the space $\mathcal{FC}_b^\infty$, where X is a locally convex space. Convergence of a sequence $\mu_n \to \mu$ in $\tau_{\mathcal{F}}$ is equivalent to

$$\int_X f(x)\, \mu_n(dx) \to \int_X f(x)\, \mu(dx)$$

for every element $f \in \mathcal{F}$. It is clear that τ_s is a special case of $\tau_{\mathcal{F}}$, where $\mathcal{F}$ is the linear space generated by the indicator functions of measurable sets.

11.2.1. Definition. *A family of measures $t \mapsto \mu(t) \in \mathcal{M}$, $t \in (a,b)$, is called τ-differentiable at a point $t_0 \in (a,b)$ if a limit*
$$\mu'(t_0) = \lim_{s \to 0} \frac{\mu(t_0+s) - \mu(t_0)}{s}$$
exists in the topology τ.

The definition of continuity is completely analogous. Higher order differentiability is defined inductively.

Similarly, we define the differentiability of a family of measures $\mu(y)$ indexed by elements y of a differentiable manifold Y, provided that some concept of differentiability of mappings between Y and $\mathcal{M}$ is chosen (for example, if Y is a normed space, this can be Gâteaux differentiability, Hadamard differentiability, etc.). Clearly, the derivative may depend on our choice of τ.

A typical example of a family of measures arises if we consider a fixed measure μ and the family of its images $\mu(t)$ with respect to certain measurable transformations $T_t \colon X \to X$. In Chapter 3, we studied the special case where X is a locally convex space and $\mu(t) = \mu_{th}$, $h \in X$.

11.2.2. Remark. Note the following obvious property of any τ_s-differentiable family of measures $\mu(t)$: for every set $E \in \mathcal{B}$, the family of measures
$$\mu_E(t) \colon B \mapsto \mu(t)(B \cap E)$$
is also τ_s-differentiable and $\mu'_E(t)(B) = \mu'(t)(B \cap E)$. In the case of the family of translations $\mu(\cdot + th)$ on a linear space, the obtained family will not be in general a family of translations of the restriction of μ to E since $(B + th) \cap E$ may substantially differ from $(B \cap E) + th$.

In addition, for every measurable mapping $F \colon X \to Y$ to a measurable space $(Y, \mathcal{E})$, the family of measures $\nu(t) := \mu(t) \circ F^{-1}$ is τ_s-differentiable. The same is true for $\tau_\mathcal{F}$-differentiability if on Y we consider some class $\mathcal{F}_Y$ of functions f such that $f \circ F \in \mathcal{F}$.

Let us discuss connections between differentiabilities in different topologies. Let I be a fixed interval of the real line and let $\{\mu(t)\}$, $t \in I$, be a family of measures on $\mathcal{B}$, where $X \neq \varnothing$.

11.2.3. Lemma. *Let $\mathcal{M} = \mathcal{M}(X)$. The following conditions are equivalent:*

(i) *the family $\{\mu(t)\}$ is τ_w-differentiable at $t_0 \in I$;*
(ii) *the family $\{\mu(t)\}$ is $\tau_{\mathcal{B}_b}$-differentiable at $t_0 \in I$;*
(iii) *the family $\{\mu(t)\}$ is τ_s-differentiable at $t_0 \in I$;*
(iv) *for every $A \in \mathcal{B}$, the function $t \mapsto \mu(t)$ is differentiable at $t_0 \in I$.*

In addition, any of these conditions along with the additional condition that $\mathcal{F}$ is continuously embedded into $\mathcal{B}_b(X)$ yields that the family $\{\mu(t)\}$ is $\tau_\mathcal{F}$-differentiable at t_0.

PROOF. Clearly, condition (iv) follows from any of conditions (i)–(iii). If it is fulfilled, then by the Nikodym theorem (see Chapter 1), there exists a measure ν on $\mathcal{B}$ such that, for every $A \in \mathcal{B}$, we have

$$\nu(A) = \lim_{s \to 0} \frac{\mu(t_0 + s)(A) - \mu(t_0)(A)}{s}.$$

By the Nikodym theorem, for every sequence $s_n \to 0$, the sequence of measures $s_n^{-1}(\mu(t_0 + s_n) - \mu(t_0))$ is bounded in variation. In addition, this sequence of measures converges to ν in the topology τ_w (see [**193**, Corollary 4.7.27]). Hence we obtain τ_w-differentiability, which implies also (ii) and (iii) since the topology τ_w is the strongest of the three indicated topologies. The last assertion of the lemma is obvious. □

11.2.4. Proposition. *Let $\{\mu(t)\}$ be a τ_s-continuous family. Then, there exists a probability measure ν on $\mathcal{B}$ such that $\mu(t) \ll \nu$ for all t and*

$$\lim_{\nu(A) \to 0} \mu(t)(A) = 0$$

uniformly in t from every compact set in I. In addition, for every such measure ν, one can find a version of the Radon–Nikodym density $f(t) = d\mu(t)/d\nu$, for which the mapping $(t, x) \mapsto f(t)(x)$ will be $\mathcal{B}(I) \otimes \mathcal{B}$-measurable.

PROOF. Let $J \subset I$ be a compact set. By the Nikodym theorem, the set $K = \{\mu(t) \colon t \in J\}$ is bounded in variation. This set is compact in the topology τ_s. Hence it is τ_w-compact (see [**193**, Theorem 4.7.27]). In addition, there exists a probability measure ν dominating all measures $\mu(t)$, $t \in J$, i.e., $\mu(t) \ll \nu$ for all t. Then the same is true for the whole I. Finally, letting $f(t)$ be the Radon–Nikodym density of the measure $\mu(t)$ with respect to ν, we obtain a random process $(t, x) \mapsto f(t)(x)$ on $(X, \mathcal{B}, \nu)$ which is continuous in probability. By the well-known result (see, e.g., Meyer [**804**, §IV.3] or Neveu [**827**, §III.4]), there exists a jointly measurable modification of this process. □

11.2.5. Proposition. *Suppose that $\mathcal{F}$ is a closed subspace in $\mathcal{B}_b(X)$ equipped with sup-norm and $\{\mu(t)\}$ is $\tau_\mathcal{F}$-differentiable. Then the following assertions are true.*

(i) *The family $\{\mu(t)\}$ is τ_v-continuous. In particular, the conclusion of Proposition 11.2.4 is valid. If, in addition, the family $\{\mu(t)\}$ is τ_s-differentiable, then the measure ν from Proposition 11.2.4 dominates $\mu'(t)$ for each t.*

(ii) *If the family $\{\mu'(t)\}$ is τ_v-continuous (for example, if the family $\{\mu(t)\}$ is twice $\tau_\mathcal{F}$-differentiable), then the family $\{\mu(t)\}$ is τ_v-differentiable.*

In addition, the τ_v-continuity of $\{\mu(t)\}$ still holds if in place of the completeness of $\mathcal{F}$ we are given that the function $t \mapsto \|\mu'(t)\|$ is locally integrable.

PROOF. If $\mathcal{F}$ is closed, then by the Banach–Steinhaus theorem, the family of measures $\mu'(t)$ with t from a fixed interval, is bounded with respect to the norm in $\mathcal{F}^*$, which gives the boundedness in variation due to our condition on $\mathcal{F}$. By the mean value theorem we obtain the estimate $\|\mu(t + h) - \mu(t)\| \leq C|h|$, which yields the continuity of $\{\mu_t\}$ in variation.

In the case of a local integrability of $\|\mu'(t)\|$ for any $h > 0$ we obtain the estimate
$$\|\mu(t+h) - \mu(t)\| \leqslant \int_t^{t+h} \|\mu'(s)\|\,ds,$$
which also yields the continuity. For $h < 0$ the reasoning is similar. The identity
$$h^{-1}\int_X f(x)\,[\mu(t_0+h) - \mu(t_0)](dx) - \int_X f(x)\,\mu'(t_0)(dx)$$
$$= h^{-1}\int_0^h \int_X f(x)\,[\mu'(t_0+sh) - \mu'(t_0)](dx)\,ds$$
for all $f \in \mathcal{F}$ yields the estimate
$$\|h^{-1}[\mu(t_0+h) - \mu(t_0)] - \mu'(t_0)\| \leqslant \|\mu'(t_0+sh) - \mu'(t_0)\|,$$
which in the case of a τ_v-continuous family of measures $\mu'(t)$ obviously yields τ_v-differentiability. So (ii) is proven as well. $\square$

11.2.6. Theorem. *Suppose that the family $\{\mu(t)\}$ is either τ_s-differentiable or $\tau_\mathcal{F}$-differentiable and $\mu'(t) \ll \mu(t)$ for Lebesgue almost all $t \in I$.*

Suppose also that the function $t \mapsto \|\mu'(t)\|$ is Lebesgue integrable on I. Then the following assertions are true.

(i) *For all $a, t \in I$ we have*
$$\mu(t) - \mu(a) = \int_a^t \mu'(s)\,ds, \tag{11.2.1}$$
where the integral on the right is an $\mathcal{M}(X)$-valued Bochner integral.

(ii) *There exist a probability measure ν on $\mathcal{B}$ and two $\mathcal{B}(I)\otimes\mathcal{B}$-measurable functions f and g such that $\mu(t) = f(t,\,\cdot\,)\cdot\nu$ for all t, $\mu'(t) = g(t,\,\cdot\,)\cdot\nu$ for almost all t and*
$$f(t,x) - f(a,x) = \int_a^t g(s,x)\,ds \tag{11.2.2}$$
for all $t \in I$ and all $x \in X$, i.e., $g(t,x) = \partial_t f(t,x)$.

(iii) *For almost every $t \in I$, the family $\{\mu(t)\}$ is τ_v-differentiable at the point t and $\mu'(t) \ll \mu(t)$.*

In addition, in case (i), if $\mu(t) \geqslant 0$ for all t, then one has $\mu'(t) \ll \mu(t)$ for all t.

PROOF. (i) The previous proposition yields the existence of a common dominating measure for all $\mu'(t)$ (in the second case mentioned in the formulation this proposition gives τ_v-continuity of our family, whence we obtain the existence of a common dominating measure for $\mu(t)$). Hence $\mathcal{M}(X)$ has a separable subspace containing almost all measures $\mu'(t)$. Then the mapping $s \mapsto \mu'(s)$ is Bochner integrable.

(ii) According to Proposition 11.2.4, there exist a probability measure ν on X dominating all measures $\mu(t)$ and a $\mathcal{B}(I)\otimes\mathcal{B}$-measurable function F_0 such that $\mu(t) = f_0(t,\,\cdot\,)\cdot\nu$. It is not difficult to verify that the measure

$m(dt\,dx) := \mu'(t)\,dt$ on $\mathcal{B}(I)\otimes\mathcal{B}$ is absolutely continuous with respect to the measure $\lambda\otimes\nu$, where λ is Lebesgue measure on I. Denote by g the corresponding Radon–Nikodym density, where we choose its $\mathcal{B}(I)\otimes\mathcal{B}$-measurable version. For ν-a.e. x the function $t \mapsto g(t,x)$ is Lebesgue integrable on I. Moreover, as one can easily verify, the set of all points x for which this is not true belongs to $\mathcal{B}$ (this set consists of the points x for which the integral of $\min(|g|,n)$ in t increases to infinity as $n \to \infty$, and the indicated integral is a $\mathcal{B}$-measurable function of x). Redefining g by zero for such x, we obtain a $\mathcal{B}(I)\otimes\mathcal{B}$-measurable version of g that is integrable in t for each x. Since we have passed to a version, the relationship $\mu'(t) = g(t,\,\cdot\,)\cdot\nu$ is fulfilled only for almost all t. The function
$$f(t,x) := f_0(a,x) + \int_a^t g(s,x)\,ds$$
is $\mathcal{B}(I)\otimes\mathcal{B}$-measurable. We show that we have obtained a version of the function f_0. Indeed, for any fixed t, for every function $\varphi \in \mathcal{F}$ we have
$$\int_X \varphi(x)\,\mu(t)(dx) = \int_X \varphi(x)\,\mu(a)(dx) + \int_a^t \int_X \varphi(x)\,\mu'(s)(dx)\,ds$$
$$= \int_X \varphi(x)\,\mu(a)(dx) + \int_a^t \int_X \varphi(x)g(s,x)\,\nu(dx)\,ds$$
$$= \int_X \varphi(x)\Big(f_0(a,x) + \int_a^t g(s,x)\,ds\Big)\,\nu(dx) = \int_X \varphi(x)f(t,x)\,\nu(dx).$$

Hence, for every fixed t, we have $\mu(t) = f(t,\,\cdot\,)\cdot\nu$.

(iii) Equality (11.2.2) holds not only pointwise, but also in the sense of the Bochner integral for mappings with values in $L^1(\nu)$. For such integrals, an analog of the classical Lebesgue differentiation theorem is valid (see Dunford, Schwartz [**376**, III.12.8, Theorem 8, p. 217]), that is, for almost every fixed t, the equality $\lim_{h\to 0} h^{-1}[f(t+h,\,\cdot\,) - f(t,\,\cdot\,)] = g(t,\,\cdot\,)$ holds in the norm of $L^1(\nu)$. This means τ_v-differentiability of $\mu(t)$ at the point t, and the corresponding derivative coincides with the $\tau_{\mathcal{F}}$-derivative. For the proof of the relationship $\mu'(t) \ll \mu(t)$ for a.e. t in the case of τ_s-differentiability we set
$$Z_1 := \{(t,x)\colon f(t,x)=0\} \quad \text{and} \quad Z_2 := \{(t,x)\colon g(t,x)=0\}.$$
For fixed x we have $g(t,x) = \partial_t f(t,x)$ for almost all t. Hence $\{t\colon (t,x)\in Z_2\}$ is contained in $\{t\colon (t,x)\in Z_1\}$ up to a set of Lebesgue measure zero. By Fubini's theorem the set $Z_1\backslash Z_2$ has $\lambda\otimes\nu$-measure zero. The same theorem shows that for λ-almost every t the set $\{x\colon (t,x)\in Z_2\}$ is contained in $\{x\colon (t,x)\in Z_1\}$ up to a set of ν-measure zero. For such t we obtain the relationship $\mu'(t) \ll \mu(t)$. This is always true in the case $\mu(t) \geqslant 0$: if $\mu(t_0)(B) = 0$, then t_0 is a point of minimum, whence $\mu'(t_0)(B) = 0$. □

If $\mu'(t) \ll \mu(t)$, then the Radon–Nikodym density of the measure $\mu'(t)$ with respect to $\mu(t)$ is denoted by $\beta(t)$ and called the *logarithmic derivative* at the point t.

If in the above theorem $\mu(t) \geqslant 0$, then

$$\beta(t)(x) = \frac{\partial}{\partial t} \ln f(t, x) \qquad (11.2.3)$$

for $\lambda \otimes \nu$-almost every point (t, x) for which $f(t, x) > 0$. Indeed, for any fixed x we have $\partial f(t, x)/\partial t = g(t, x)$ for a.e. t. On the other hand, for almost every fixed t we have the equality $g(t, x) = \beta(t)(x) f(t, x)$ for ν-a.e. x since

$$f(t) \cdot \nu = \mu'(t) = \beta(t) \cdot \mu(t) = \beta(t) f(t, \cdot) \cdot \nu.$$

Therefore, for $\lambda \otimes \nu$-a.e. (t, x) we have the equality

$$f(t, x) - f(a, x) = \int_a^t \beta(s)(x) f(s, x) \, ds, \qquad (11.2.4)$$

which yields (11.2.3). This reasoning leads to the following assertion, analogous to the fact that in the linear case the exponential integrability of the logarithmic derivative implies the equivalence of all shifts of the given measure.

11.2.7. Theorem. *Let the family $\{\mu(t)\}$ be τ_s-differentiable or $\tau_{\mathcal{F}}$-differentiable and let the mapping $(t, x) \mapsto \beta(t)(x)$ be $\mathcal{B}(I) \otimes \mathcal{B}$-measurable. Suppose that for some $a, b \in I$ we have*

$$\int_a^b \|\mu'(t)\| \, dt < \infty \quad \text{and} \quad \int_a^b |\beta(t)(x)| \, dt < \infty \quad |\mu(a)| + |\mu(b)|\text{-a.e.}$$

Then the measures $\mu(t)$ with $a \leqslant t \leqslant b$ are equivalent and

$$\frac{d\mu(t)}{d\mu(a)}(x) = \exp\left(\int_a^t \beta(s)(x) \, ds\right).$$

PROOF. We have to establish the equality

$$f(b, x) = f(a, x) \exp\left(\int_a^b \beta(s)(x) \, ds\right)$$

for ν-a.e. x. For ν-a.e. fixed x we have either the integrability of the function $t \mapsto |\beta(t)(x)|$ on $[a, b]$ or the equality $f(b, x) = f(a, x) = 0$. Hence the desired equality follows from the following assertion: if $f \in C[a, b]$ and $g \in L^1[a, b]$ are such that

$$f(t) = \int_a^t f(s) g(s) \, ds,$$

then

$$f(t) = f(a) \exp\left(\int_a^t g(s) \, ds\right).$$

For justification it suffices to observe that the right-hand side satisfies the same linear integral equation as f. □

11.2. Differentiable families of measures

The integrability of the logarithmic derivative in this theorem is essential even in the case of the family of shifts of a smooth measure on the real line: it suffices to take a smooth density with compact support.

Trivial examples show that in assertion (ii) of Theorem 11.2.6 the measure $\mu'(t)$ may be singular with respect to $\mu(t)$ for some exceptional t from a set of Lebesgue measure zero (let $\mu(t) = t\delta$, $t \in (-1,1)$, where δ is Dirac's measure; then $\mu(0) = 0$ and $\mu'(0) = \delta$). The above theorem states that this cannot occur if $\mu(t) \geq 0$. The reason is simple: if a differentiable nonnegative function φ vanishes at some point t_0, then this is a point of local minimum, hence $\varphi'(t_0) = 0$. The absolute continuity of $\mu'(t)$ with respect to $\mu(t)$ turns out to be connected with the differentiability of the Hahn–Jordan decomposition of the measures $\mu(t)$, which holds automatically in the case of Fomin differentiability on linear spaces. The situation is different for general families. Let us consider the following example.

11.2.8. Example. Let $\mu(t)$ be the measure on $[0,1]$ given by a density $f(t)$ defined as follows: $f(t)(x) = t\sin(1/t)\sin(x/t)$ if $t \neq 0$, $f(0) = 0$. Then it is easily verified that this family is τ_s-differentiable, but its positive part $\mu(t)^+$ is not τ_s-differentiable at zero.

As follows from assertion (iii) of the previous theorem and the next theorem, under reasonable assumptions the failure of τ_s-differentiability of $\mu(t)^+$ cannot occur for points t from a set of positive measure.

11.2.9. Theorem. *The following conditions are equivalent:*
(i) *the family* $\{\mu(t)\}$ *is* τ_v*-differentiable at* t_0 *and* $\mu(t_0)' \ll \mu(t_0)$;
(ii) *the positive and negative parts* $\mu(t)^+$ *and* $\mu(t)^-$ *are* τ_v*-differentiable at the point* t_0.

In this case, $(\mu(t)^+)' = \mu'(t)(\,\cdot \cap X_t^+)$ *and* $(\mu(t)^-)' = \mu'(t)(\,\cdot \cap X_t^-)$, *where* X_t^+ *and* X_t^- *are the sets from the Hahn–Jordan decomposition of* $\mu(t)$.

PROOF. The implication (ii)$\Rightarrow$(i) is clear from the facts proven above. Suppose that (i) is fulfilled. We may assume that $t_0 = 0$. We slightly modify the reasoning from the proof of Theorem 3.3.1. Set $A_t := X_t^+ \backslash X_0^+$, $B_t := X_0^+ \backslash X_t^+ = X_t^- \backslash X_0^-$. Then

$$t^{-1}[|\mu(0)| + |\mu(t)|](A_t) = t^{-1}[\mu(t)(A_t) - \mu(0)(A_t)] \leq \|t^{-1}[\mu(t) - \mu(0)]\|,$$

whence

$$\limsup_{t \to 0} \frac{[|\mu(0)| + |\mu(t)|](A_t)}{t} \leq \|\mu'(0)\|. \tag{11.2.5}$$

Let us show that

$$\lim_{t \to 0} t^{-1}|\mu(t)|(A_t) = 0. \tag{11.2.6}$$

It suffices to verify this equality for every sequence $t_n \to 0$. Suppose that we have $t_n^{-1}|\mu(t_n)|(A_{t_n}) \geq \varepsilon > 0$ for all n. Passing to a subsequence, we may assume that $\sum_{n=1}^{\infty} t_n < \infty$. Set $E_m := \bigcup_{n=m}^{\infty} A_{t_n}$. By inequality (11.2.5)

we have the estimate
$$|\mu(0)|(E_m) \leq \sum_{n=m}^{\infty} Ct_n.$$
Hence $\lim_{m\to\infty} |\mu(0)|(E_m) = 0$. Since $|\mu'(0)| \ll \mu(0)$, one has $|\mu'(0)|(E_m) \to 0$. Therefore, there exists m such that $|\mu'(0)|(E_m) < \varepsilon$. Set $E := E_m$ and consider the family of measures $\mu_E(t) := \mu(t)(\,\cdot\, \cap E)$. As observed in Remark 11.2.2, the family $\mu_E(t)$ is differentiable at zero and we have the equality $\mu'_E(0) = \mu'(0)(\,\cdot\, \cap E)$. The sets $X_t^+ \cap E$ and $X_t^- \cap E$ give the Hahn–Jordan decomposition for $\mu_E(t)$. Hence, whenever $n \geq m$, similarly to (11.2.5) we obtain the estimate
$$\limsup_{n\to\infty} t_n^{-1} |\mu_E(t_n)|(A_{t_n}) \leq \|\mu'_E(0)\| = |\mu'_E(0)|(E) < \varepsilon,$$
which contradicts our choice of ε. Thus, (11.2.6) is proven. Replacing $\mu(t)$ by $-\mu(t)$, we obtain also (11.2.6) for B_t in place of A_t. It is straightforward to verify that
$$\mu(t)^+ - \mu(0)^+ = \mu(t)(X_t^+ \cap \,\cdot\,) - \mu(0)(X_0^+ \cap \,\cdot\,)$$
$$= \mu(t)(X_0^+ \cap \,\cdot\,) - \mu(0)(X_0^+ \cap \,\cdot\,) + \mu(t)(A_t \cap \,\cdot\,) - \mu(t)(B_t \cap \,\cdot\,).$$
Since $\lim_{t\to\infty} t^{-1}[\|\mu(t)|(A_t) + |\mu(t)|(B_t)] = 0$ according to what has been proven, one has $\|t^{-1}[\mu(t)^+ - \mu(0)^+] - \mu'(0)(X_0^+ \cap \,\cdot\,)\| = 0$, as required. □

11.2.10. Corollary. *If $\{\mu(t)\}$ is τ_s-differentiable, then $\{\mu(t)^+\}$ and $\{\mu(t)^-\}$ are τ_v-differentiable at t for almost all t.*

11.2.11. Definition. *Let $p \in [1, +\infty)$. The family of nonnegative measures $\mu(t)$ is called L^p-differentiable at a point t_0 if there exists a nonnegative measure ν such that $\mu(t) = f(t) \cdot \nu$ for all t and the mapping $\Phi \colon I \to L^p(\nu)$, $t \mapsto f(t)^{1/p}$, is differentiable at t_0. In the case $p = 2$, the quantity $\|d\sqrt{f(t)}/dt\|^2_{L^2(\nu)}$ is called Fisher's information number at the point t.*

A family of signed measures is called L^p-differentiable if its positive and negative parts have this property.

11.2.12. Proposition. (i) *If for some $p > 1$ a family of measures $\{\mu(t)\}$ is L^p-differentiable at a point t_0, then it is τ_v-differentiable at this point and its logarithmic derivative $\beta(t_0)$ belongs to $L^p(\mu(t_0))$.*

(ii) *Conversely, if a family of measures $\{\mu(t)\}$ is τ_s-differentiable (or $\tau_{\mathcal{F}}$-differentiable with $\mu'(t) \ll \mu(t)$ for a.e. t) and the function $\|\beta(t)\|_{L^p(\mu(t))}$ is locally integrable on I, then this family is L^p-differentiable at almost all points t.*

(iii) *If in case (i) we have $\mu(t) \geq 0$ for all t, then*
$$\Phi'(t_0) = p^{-1} f(t_0)^{1/p} \beta(t_0),$$

11.2. Differentiable families of measures

and in case (ii) for all $a, b \in I$ we have

$$f(b)^{1/p} - f(a)^{1/p} = p^{-1}\int_a^b f(t)^{1/p}\beta(t)\,dt$$

ν-a.e. and as an equality for $L^p(\nu)$-valued Bochner integrals.

PROOF. Let $\mu(t) \geq 0$. Then (i) and the first part of (iii) follow from the fact that the mapping $\Psi \colon f \mapsto |f|^p$ from $L^p(\nu)$ to $L^1(\nu)$ is Fréchet differentiable and we have $D\Psi(f)(h) = p|f|^{p-1}h$ (Exercise 11.6.2). Since $\Phi'(t) \in L^p(\nu)$, one has

$$\int_X |\beta(t)|^p\,d\mu(t) = \int_X |\beta(t)|^p f(t)\,d\nu = p^p\int_X |\Phi'(t)|^p\,d\nu < \infty,$$

i.e., we have the inclusion $\beta(t) \in L^p(\mu(t))$. In case (ii) the integrability of the function $t \mapsto \|f(t)^{1/p}\beta(t)\|_{L^p(\nu)} = \|\beta(t)\|_{L^p(\mu(t))}$ gives the Bochner integrability of the mapping $t \mapsto f(t)^{1/p}\beta(t)$ with values in $L^p(\nu)$. Moreover, (11.2.4) yields the equality

$$f(b)^{1/p} - f(a)^{1/p} = p^{-1}\int_a^b f(t)^{1/p}\beta(t)\,dt$$

ν-a.e. and also in the sense of $L^p(\nu)$-valued Bochner integrals. The case of signed measures reduces to the considered one. □

Let us consider the family of measures $\mu(t) = \mu \circ T_t^{-1}$, where μ is a probability measure on X and $\{T_t\}_{t\in\mathbb{R}^1}$ is a group of measurable (in both directions) bijections of the space X. If this family is τ_s or τ_v-continuous, then all measures $\mu(t)$ are absolutely continuous with respect to the probability measure $\lambda = \sum_{n=1}^\infty 2^{-n}\mu(r_n)$, where $\{r_n\}$ is the set of all rational numbers. Then $\lambda \circ T_t^{-1} \ll \lambda$ for all t. Hence $\lambda \circ T_t^{-1} = \varrho_t \cdot \lambda$. It is readily seen that the mapping $t \mapsto \varrho_t$ from $\mathbb{R}^1$ to $L^1(\lambda)$ is continuous.

11.2.13. Theorem. *If, for some t_0, the family $\{\mu(t)\}$ is either τ_s-differentiable at t_0 or $\tau_{\mathcal{F}}$-differentiable at t_0 with $\mu'(t_0) \ll \mu(t_0)$ and $f \circ T_t \in \mathcal{F}$ for all $f \in \mathcal{F}$, then the family $\{\mu(t)\}$ is τ_v-differentiable on $\mathbb{R}^1$, $\beta(t)$ exists for all t and $\beta(t)(x) = \beta(0)(T_{-t}(x))$ $\mu(t)$-a.e. If the function $t \mapsto \beta(0)(T_{-t}x)$ is locally integrable for μ-a.e. x, then for all b we have*

$$\frac{d\mu(b)}{d\mu(0)}(x) = \exp\left(\int_0^b \beta(0)(T_{-t}x)\,dt\right).$$

The family $\{\mu(t)\}$ is L^p-differentiable precisely when $\beta(0) \in L^p(\mu)$. This is also equivalent to the estimate $\|f_t^{1/p} - f_0^{1/p}\|_{L^p(\lambda)} \leq C|t|$, where $\mu_t = f_t \cdot \lambda$.

PROOF. The differentiability at all points is clear from the equality

$$\int_X \varphi\,[\mu(t+h) - \mu(t)](dx) = \int_X \varphi \circ T_{t-t_0}\,[\mu(t_0+h) - \mu(t_0)](dx),$$

which shows also that $\mu'(t) = \mu'(t_0) \circ T_{t-t_0}^{-1}$. Hence we obtain $\mu'(t) \ll \mu(t)$ for all t. In addition, $\beta(t) = \beta(0) \circ T_{-t}$. Now the equivalence of the measures $\mu(t)$ and the formula for their densities follow from Theorem 11.2.7. If the given family is L^p-differentiable, then $\beta(0) \in L^p(\mu(0))$. Conversely, under this condition we have $\beta(t) = \beta(0) \circ T_{-t} \in L^p(\mu(0) \circ T_t^{-1}) = L^p(\mu(t))$ and $\|\beta(t)\|_{L^p(\mu(t))} = \|\beta(0)\|_{L^p(\mu(0))}$ for all t. It remains to apply Proposition 11.2.12. Finally, the equivalence of L^p-differentiability to the estimate indicated in the theorem follows by Lemma 1.1.1 and the fact that $L^p(\lambda)$ with $1 < p < \infty$ has the Radon–Nikodym property. In order to apply this lemma one should take $h = f_0^{1/p}$ and the group of operators $\widetilde{T}_t$ on $L^p(\lambda)$ defined by the following formula: $\widetilde{T}_t\varphi := \varphi \circ T_{-t} \varrho_t^{1/p}$. The change of variables formula and the equality $(\varrho_t \cdot \lambda) \circ T_{-t}^{-1} = \lambda$ yield $\|\widetilde{T}_t\varphi\|_{L^p(\lambda)} = \|\varphi\|_{L^p(\lambda)}$. In order to verify the group property, we observe that $\varrho_{t+s} = \varrho_s \circ T_{-t} \varrho_t$ since $\lambda \circ T_{t+s}^{-1} = (\varrho_s \cdot \lambda) \circ T_t^{-1} = (\varrho_s \circ T_{-t} \varrho_t) \cdot \lambda$. Therefore, for all $s, t \geq 0$ we have the equalities $\widetilde{T}_t \widetilde{T}_s \varphi = \varphi \circ T_{-t-s} \varrho_s^{1/p} \circ T_{-t} \varrho_t^{1/p} = \widetilde{T}_{t+s}\varphi$. □

It follows from the proof and Lemma 1.1.1 that L^p-differentiability automatically yields the continuous differentiability of the mapping $t \mapsto f_t^{1/p}$.

Let us discuss the differentiability of conditional measures. Suppose that a family of measures $\mu(t) \geq 0$ is τ_s-differentiable on $(X, \mathcal{B})$, ν is a probability measure on $\mathcal{B}$ such that $\mu(t) \ll \nu$ for all t, $\pi \colon X \to Y$ is a measurable mapping to a measurable space $(Y, \mathcal{E})$, $\nu_Y := \nu \circ \pi^{-1}$, and on the sets $\pi^{-1}(y)$ there exist conditional measures ν_y, $y \in Y$, uniquely determined ν_Y-a.e. (e.g., let X and Y be Souslin spaces). Then, for any measure $\sigma \ll \nu$, there exist ν_Y-a.e. uniquely determined measures σ_y on $\pi^{-1}(y)$ such that

$$\sigma(B) = \int_Y \sigma_y(B)\,\nu_Y(dy), \quad B \in \mathcal{B}.$$

11.2.14. Theorem. (i) *There exist τ_s-differentiable measures $\mu(t)_y$ on $\pi^{-1}(y)$ such that*

$$\mu(t)(B) = \int_Y \mu(t)_y(B)\,\nu_Y(dy), \quad B \in \mathcal{B}.$$

(ii) *Let $X = X_1 \times X_2$, where $(X_1, \mathcal{B}_1)$ and $(X_2, \mathcal{B}_2)$ are measurable spaces, $Y = X_1$, $\mathcal{B} = \mathcal{B}_1 \otimes \mathcal{B}_2$, and let π be the projection on X_1. If $t \mapsto \|\mu'(t)\| \in L^1_{\mathrm{loc}}(\mathbb{R}^1)$ and $\mu(t) \sim \mu(0)$ for all t, then there exist conditional measures $\mu(t)_{x_1}$ on X_2 with respect to $\mu(t)_{X_1}$ such that for every fixed x_1 and for every $B \in \mathcal{B}_2$ the function $t \mapsto \mu(t)_{x_1}(B)$ is almost everywhere differentiable. If $\mu(t) = \mu(0) \circ T_t^{-1}$, where $T_t(x_1, x_2) = (x_1, S_t(x_2))$ and $\{S_t\}$ is a group of measurable (in both directions) bijections of X_2, then one can choose conditional measures μ_{x_1} such that the family $\mu_{x_1} \circ S_t^{-1}$ will be τ_s-differentiable.*

11.2. Differentiable families of measures

PROOF. (i) Since $\mu'(s) \ll \nu$ and

$$\mu(t) - \mu(0) = \int_Y [\mu(t)_y - \mu(0)_y]\, \nu_Y(dy) = \int_Y \int_0^t \big(\mu'(s)\big)_y\, ds\, \nu_Y(dy),$$

where the measures $\big(\mu'(s)\big)_y \ll \nu_y$ are concentrated on $\pi^{-1}(y)$, we obtain that $\mu(t)_y - \mu(0)_y$ is the integral of $\big(\mu'(s)\big)_y$ over $[0,t]$ for ν_Y-a.e. y. (ii) We take a function $f(t,\,\cdot\,)$ on X according to Theorem 11.2.6. In the case $\mu(t) \sim \mu(0)$ we can take $\nu = \mu(0)$, which gives $\mu(t) \sim \nu$. Then

$$\mu(t) = f(t,\,\cdot\,)\cdot\nu = \int_{X_1} \varrho(t,x_1)^{-1} f(t,x_1,\,\cdot\,)\cdot\nu_{x_1}\, \mu(t)_{X_1}(dx_1),$$

where $\varrho(t,\,\cdot\,)$ is the density of the measure $\mu(t)_{X_1}$ with respect to the measure ν_{X_1} and $\varrho(t,x_1) > 0$ for $\mu(t)_{X_1}$-a.e. x_1. Using Theorem 11.2.6, we take the version of $\varrho(t,x_1)$ that is locally absolutely continuous in t. Let g be the function from (11.2.2). There is a set $\Omega \subset X_1$ of full ν_{X_1}-measure such that for $x_1 \in \Omega$ the function $(s,x_2) \mapsto g(s,x_1,x_2)$ is integrable over the sets $[-R,R] \times X_2$ with respect to $ds \otimes \nu_{x_1}$. Whenever $x_1 \notin \Omega$ or $\varrho(t,x_1) = 0$ we put $\mu(t)_{x_1} = \nu_{X_2}$. For all other x_1 we put $\mu(t)_{x_1} := \varrho(t,x_1)^{-1} f(t,x_1,\,\cdot\,)\cdot\nu_{x_1}$. This gives versions of conditional measures. Whenever $x_1 \in \Omega$ we have

$$\varrho(t,x_1)\mu(t)^{x_1}(B) = \int_B f(0,x_1,x_2)\,\nu_{x_1}(dx_2) + \int_0^t \int_B g(s,x_1,x_2)\,\nu_{x_1}(dx_2)\,ds,$$

which gives the local absolute continuity of the left-hand side. Hence, as soon as $\varrho(\tau,x_1) > 0$, the function $t \mapsto \mu(t)_{x_1}(B)$ is absolutely continuous in a neighborhood of τ. The set $Z := \{(t,x_1)\colon \varrho(t,x_1) = 0\}$ has measure zero with respect to the measure $ds \otimes \nu_{X_1}$, which is easily seen from Fubini's theorem and the relationship $\mu_{X_1}(t) \sim \nu_{X_1}$. Hence, for ν_{X_1}-a.e. x_1, we have $\varrho(t,x_1) > 0$ a.e., which completes the proof of the first assertion. In the second case the family of measures

$$\sigma_{x_1}(t) := f(0,x_1,\,\cdot\,)\cdot\nu_{x_1} + \int_0^t g(s,x_1,\,\cdot\,)\cdot\nu_{x_1}\, ds$$

is τ_s-differentiable and represents $\mu(t)$ as the integral of $\sigma_{x_1}(t)$ with respect to the measure ν_{X_1}. We observe that $\mu(t)_{X_1} = \mu(0)_{X_1}$ since T_t does not change projections on X_1. Set $\varrho := d\mu(0)_{X_1}/d\nu_{X_1}$, $\mu(t)_{x_1} := \varrho(x_1)^{-1}\sigma_{x_1}(t)$ if $\varrho(x_1) > 0$, $\mu(t)_{x_1} := \nu_{X_2}$ if $\varrho(x_1) = 0$. These measures are conditional for $\mu(t)$. On the other hand, taking conditional measures $\mu(0)_{x_1}$ for $\mu(0)$, we obtain for $\mu(t)$ the conditional measures $\mu(0)_{x_1} \circ S_t^{-1}$. Hence, for μ_{X_1}-a.e. x_1, we have $\mu(t)_{x_1} = \mu(0)_{x_1} \circ S_t^{-1}$ for all t. To this end, it suffices to obtain this equality for all rational t and then apply Proposition 1.3.2. □

Suppose that $\{T_t\}_{t \in \mathbb{R}^1}$ and $\{S_t\}_{t \in \mathbb{R}^1}$ are two groups of bimeasurable bijections of spaces X_1 and X_2 extended to X by the formulas $T_t(x_1,x_2) = \big(T_t(x_1),x_2\big)$, $S_t(x_1,x_2) = \big(x_1,S_t(x_2)\big)$. Let the families $\mu_1(t) := \mu \circ T_t^{-1}$ and $\mu_2(t) := \mu \circ S_t^{-1}$ be τ_s-differentiable. The proof of the following result is analogous.

11.2.15. Theorem. *There exist conditional measures μ_{x_1} for μ such that for each x_1 the family $\{\mu_{x_1} \circ S_t^{-1}\}$ is τ_s-differentiable and for every fixed $B \in \mathcal{B}(X_2)$ the functions $t \mapsto \mu_{T_{-t}(x_1)}(B)$ are a.e. differentiable.*

11.3. Current and loop groups

A current group is the space of mappings $\mathcal{F}(M,G)$ from a manifold M to a Lie group G; $\mathcal{F}(S^1, G)$ is called a loop group. Usually more narrow subspaces $C(M,G)$ and $C^\infty(M,G)$ in $\mathcal{F}(M,G)$ are considered consisting of continuous and smooth mappings, respectively; $\mathcal{F}(M,G)$, $C(M,G)$, and $C^\infty(M,G)$ have a natural group structure: $(f \cdot g)(m) := f(m) \cdot g(m)$. The reader familiar with the notion of a Haar measure on a group G can imagine how useful an analogous object on an infinite dimensional group $C(M,G)$ could be. One cannot achieve a complete analogy, but it is possible to construct measures preserving some features of the Haar measure. Here we consider one of the technically simplest cases: the group $W(G)$ of continuous mappings $\omega \in C(I, G)$ from the interval $I = [0,1]$ to G such that $\omega(0) = e$, where e is the group unit in G. In addition, we shall assume that the group G is connected and is an n-dimensional Lie subgroup in the group of invertible linear operators $GL(\mathbb{R}^d)$ with some $d \geq n$ and possesses a left- and right-invariant Haar measure m. These assumptions are not too restrictive. The Lie algebra of the group G will be denoted by $\mathfrak{g}$. The embedding of G into $GL(\mathbb{R}^d)$ generates an embedding of the algebra $\mathfrak{g}$ into the algebra of d-dimensional matrices $\mathfrak{gl}_d$. Hence $\mathfrak{g}$ can be equipped with the inner product $\langle \xi, \eta \rangle := \operatorname{trace} \xi\eta$, which is invariant with respect to the adjoint action of G ($\operatorname{Ad}(g) \colon \xi \mapsto g\xi g^{-1}$).

An important role is played below by the following analog of the classical Cameron–Martin space of the Wiener measure:

$$H(G) := \left\{ h \in W(G) \colon \text{the mapping } h \text{ is absolutely continuous,} \right.$$

$$\left. \|h\|_H^2 := \int_0^1 |h(s)^{-1} h'(s)|^2 \, ds < \infty \right\}.$$

One can verify that $h(s)^{-1} h'(s) \in \mathfrak{g}$.

Let $\xi_1, \ldots, \xi_n$ be an orthonormal basis in $\mathfrak{g}$. The Laplace operator on $C^\infty(G)$ is defined by the equality $\Delta f(g) := \frac{1}{2} \sum_{i=1}^n \partial_i^2 f(g)$, where

$$\partial_i f(g) := \frac{d}{dt} f\big(g \exp(t\xi_i)\big)\big|_{t=0}.$$

The factor $1/2$ is included only for simplification of some formulae below connected with the Wiener process; Δ does not depend on the basis. As in the case of a linear space, one can show that there exists a positive smooth function p on $(0,1) \times G \times G$ such that

$$\frac{\partial p}{\partial t} = \Delta, \quad \lim_{t \to 0} \int_G p(t,x,y) \varphi(y) \, m(dy) = \varphi(x)$$

11.3. Current and loop groups

for all $x \in G$ and $\varphi \in C_b(G)$. The function p serves as an analog of the Gaussian kernel $p(t, x, y) = (2\pi t)^{-n/2} \exp[-|x-y|^2/(2t)]$ in the linear case. The Wiener measure μ on $W(G)$ is defined by analogy with the Wiener measure on $C[0, 1]$. Let $0 \leqslant s_1 < s_2 < \cdots < s_k \leqslant 1$. If $s_1 > 0$, then let $\mu^{s_1,\ldots,s_n}$ denote the probability measure on G^k with the density

$$p_{s_1,\ldots,s_k}(x_1,\ldots,s_k) := p(s_1, e, x_1)p(s_2 - s_1, x_1, x_2)$$
$$\times p(s_3 - s_2, x_2, x_3) \cdots p(s_k - s_{k-1}, x_{k-1}, x_k)$$

with respect to the measure m^k. If $s_1 = 0$, then the measure $\mu^{s_1,\ldots,s_n}$ is concentrated on $e \times G^{k-1}$ and is given there by the density $p_{s_2,\ldots,s_k}(x_2,\ldots,x_k)$. It is straightforward to verify that we obtain consistent probability measures. By the Kolmogorov theorem, we obtain a probability measure μ on $\mathcal{F}(I, G)$ with the indicated finite dimensional distributions. As in the case of the Wiener measure on $C[0,1]$, this measure is concentrated on $W(G)$ (more precisely, the set $W(G)$ has outer measure 1 and so the constructed measure can be restricted to $W(G)$ by the standard procedure). Justification of all these assertions is technically rather involved, but does not meet any principal difficulties. Another way of constructing the measure μ at once on $W(G)$ is to obtain it as the distribution of a continuous process defined by the stochastic differential equation in Itô's form

$$dx(t) = x(t)dB(t) + \frac{1}{2}x(t)\sum_{i=1}^{n}\xi_i^2 dt, \quad x(0) = e,$$

where $B(t)$ is a standard Wiener process in the Lie algebra $\mathfrak{g}$; for example, one can set $B(t) := \sum_{i=1}^{n} w_i(t)\xi_i$, where $w_1,\ldots,w_n$ are independent real Wiener processes. Clearly, we obtain at once a process $x(t)$ with values in $\mathfrak{gl}_n$, but it should be verified that in fact we obtain a process in G. In order to show that the distribution of this process coincides with the measure μ, it suffices to verify that its transition probabilities satisfy the parabolic equation written above. This can be done by using Itô's formula.

Let μ_h denote the shift of the measure μ by h, i.e., its image under the mapping $x \mapsto x \cdot h$, where $h \in H(G)$. The measure μ_h is the distribution of the process $x_h(t) := x(t) \cdot h(t)$. One can verify that it satisfies the stochastic equation

$$dx_h(t) = x(t)dB(t)h(t) + \frac{1}{2}x(t)\sum_{i=1}^{n}\xi_i^2 h(t)dt + x(t)h'(t)dt$$
$$= x_h(t)h(t)^{-1}dB(t)h(t) + \frac{1}{2}x_h(t)h(t)^{-1}\sum_{i=1}^{n}\xi_i^2 h(t)dt + x_h(t)h(t)^{-1}h'(t)dt,$$

where $x_h(0) = e$. Set

$$Y_t(x) := \int_0^t x(s)^{-1} dx(s) - \frac{1}{2}\int_0^t \sum_{i=1}^{n} \xi_i^2 ds.$$

11.3.1. Theorem. *The measure μ_h is equivalent to the measure μ and is given with respect to μ by the density*

$$\varrho_h(x) := \exp\left(\int_0^1 \langle h(s)^{-1}h'(s), dY_s(x)\rangle - \frac{1}{2}\|h\|_H^2\right). \tag{11.3.1}$$

PROOF. The main observation is that the process $\widetilde{B}(t) = h(t)^{-1}B(t)h(t)$ is also Wiener. This is clear from the fact that the operators $\xi \mapsto h(t)^{-1}\xi h(t)$ in $\mathfrak{gl}_n$ are unitary, and for the Wiener process $B(t)$ any process of the form $U(t)B(t)$, where $U(t)$ are unitary operators continuously dependent on t, is also Wiener. In addition, by using the invariance of the inner product in $\mathfrak{g}$ with respect to the action of G one can derive the identity $h(t)^{-1}\sum_{i=1}^n \xi_i^2 h(t) = \sum_{i=1}^n \xi_i^2$. Therefore, the equation for x_h can be written in the form

$$dx_h(t) = x_h(t)d\widetilde{B}(t) + \frac{1}{2}x_h(t)\sum_{i=1}^n \xi_i^2 dt + x_h(t)h(t)^{-1}h'(t)dt.$$

By the Girsanov theorem (see Theorem 1.5.6) one has $\mu \sim \mu_h$ and $d\mu_h/d\mu$ is given by (11.3.1), where the stochastic integral can be also written via $dx(s)$ by means of the given expression for Y_t. □

We shall mention only one more example of a similar kind studied in Driver [**360**] and related to more general spaces M. Let M be a compact connected finite dimensional Riemannian manifold without boundary, $W(M) = C([0,1], M)$, and let ∇ be a covariant derivative on the tangent space TM of the manifold M consistent with the Riemannian metric g on M. Let us fix a point $a \in M$. Denote by P the Wiener measure on the space $W(M)$ concentrated on the set $W_a(M)$ of paths with $w(a) = a$. For every C^1-function $h\colon [0,1] \to T_aM$ with $h(0) = 0$, one can define a vector field on $W_a(M)$ by $X^h(w)(s) = H(w)(s)h(s)$, where $H(w)(s)$ denotes the stochastic parallel transport along w (with respect to P) on the interval $[0,s]$ (see Driver [**360**], Watanabe, Ikeda [**571**, Ch. V, §5]). Given a vector field X^h, it is natural to try to construct the associated flow, i.e., a family $\{T_t\}$ of measurable transformations of $W(M)$ such that

$$\frac{d}{dt}T_t(w) = X^h(T_t(w)), \quad T_0(w) = w \quad \text{for } P\text{-a.e. } w. \tag{11.3.2}$$

11.3.2. Theorem. *Let ∇ have a skew-symmetric torsion (for example, let it be a Levy–Civita connection) and $X = W_a(M)$. Then*
  (i) *equation (11.3.2) has a unique solution $\{T_t\}$, which is a flow;*
  (ii) *the measures $P_t := P \circ T_t^{-1}$ are equivalent to P;*
  (iii) *the measure P is differentiable along the flow $\{T_t\}$ and for sufficiently small positive ε one has*

$$\int_X \exp\left(\varepsilon\big[\beta_h(0)(w)/\|h'\|\big]^2\right) P(dw) \leqslant K,$$

where K is a constant, $\|h'\| = \||h'(s)|\|_{L^2[0,1]}$ and $\beta_h(0)$ is the logarithmic derivative of the flow at zero.

In particular, the following integration by parts formula holds:
$$\int_X \partial_h f g \, P(dx) = -\int_X f[\partial_h g + g\beta_h(0)] \, P(dx), \quad \partial_h f = \frac{d}{dt} T_t f \Big|_{t=0}.$$

The details of the construction and justification turn out to be considerably more complicated than in the case of a group, although at the level of ideas they are similar.

11.4. Poisson spaces

Let M be a connected smooth complete Riemannian manifold of dimension d equipped with a σ-finite measure σ having a density ϱ with respect to the Riemannian volume on M, where $\sqrt{\varrho} \in W^{1,2}_{\text{loc}}(M)$. For the purposes of this section it is enough to assume that $M = \mathbb{R}^d$. The configuration space $\Gamma = \Gamma_M$, or the space of configurations in M, is the space of Borel measures γ on M with values in $\mathbb{Z}_+ \cup \{+\infty\}$ such that, for every compact set $K \subset M$, we have $\gamma(K) < \infty$. Points $\gamma \in \Gamma$ have the form $\gamma = \sum_i k_i \delta_{x_i}$, where δ_{x_i} is Dirac's measure at the point $x_i \in M$, numbers k_i are nonnegative integers, and the set $\{x_i\}$ is locally finite. Configurations are locally finite collections of points from M, which can have finite multiplicities. A representation of configurations in the form of measures is convenient for several reasons, one of which concerns a topology or metric on Γ. Namely, we equipped Γ with the vague topology, i.e., the topology generated by all functions of the form
$$\langle \varphi, \gamma \rangle := \int_M \varphi(x) \, \gamma(dx), \quad \varphi \in C_0(M).$$
It is not difficult to show that this topology is metrizable. Moreover, one can introduce a complete separable metric defining this topology. The measure σ on M is not involved in our construction of the topology, but it is needed for defining the so-called Poisson measure $\pi := \pi_\sigma$ on the space Γ with intensity σ. This measure is determined by the requirement that $\gamma \mapsto \gamma(A)$ is a Poisson random variable with intensity $\sigma(A)$ and, for any disjoint Borel sets $B_1, \ldots, B_k$, the random variables $\gamma(B_1), \ldots, \gamma(B_k)$ are independent. For example, for every two disjoint bounded Borel sets $A, B \subset M$, one has
$$\pi\{\gamma \colon |\gamma \cap A| = m, \ |\gamma \cap B| = n\} = \frac{\sigma(A)^m}{m!} \frac{\sigma(B)^n}{n!} e^{-\sigma(A) - \sigma(B)},$$
where $|A|$ is the cardinality of the set A. One can verify that the set of configurations with multiple points has zero π-measure. So we shall consider the space of configurations without multiple points, i.e., now we can identify configurations with countable locally finite subsets of the manifold M. In other words, measures γ can be identified with their supports. Let us emphasize that Γ is not a linear space (measures γ assume integer values),

although it is naturally embedded into the linear space of all locally finite Borel measures on M.

Let $F\colon M \to M$ be a Borel mapping such that $F(x) = x$ outside some compact set $\Omega \subset M$. This mapping generates a transformation $T\colon \Gamma \to \Gamma$ by the formula $T(\gamma) = \{F(x)\colon x \in \gamma\}$, i.e., each configuration γ is taken to the configuration $F(\gamma)$, which is also locally finite since the set $\gamma \cap \Omega$ is finite. In the representation of γ by means of a locally finite measure $\gamma = \sum_i k_i \delta_{x_i}$ the element $F(\gamma)$ is merely $\gamma \circ F^{-1}$, the image of the measure γ under F, and has the form $F(\gamma) = \sum_i k_i \delta_{F(x_i)}$.

11.4.1. Theorem. *Let $\sigma \circ F^{-1} \ll \sigma$ and $\Lambda := d(\sigma \circ F^{-1})/d\sigma$. Then $\pi \circ T^{-1} \ll \pi$ and $d(\pi \circ T^{-1})/d\pi(\gamma) = \prod_{x \in \gamma \cap \Omega} \Lambda(x)$.*

PROOF. On the right we have a measurable function. The space $\Gamma = \Gamma_M$ is the product of two configuration spaces Γ_Ω and $\Gamma_{M\setminus\Omega}$, and the Poisson measure π_σ is the product of the Poisson measures π_1 and π_2 on Γ_Ω and $\Gamma_{M\setminus\Omega}$, respectively. The mapping T transforms only the first factor and is identical on the second one. By the boundedness of Ω the space Γ_Ω splits into the countable family of T-invariant sets Γ_n consisting of configurations in Ω with n distinct points. The set Γ_n can be regarded as the factor-space of Ω^n/S_n, $n = 0, 1, \ldots$, where Ω^n/S_n is the factorization of Ω^n with respect to the action of the group of permutations S_n, provided that we identify elements Ω^n which differ only by a permutation of points (the power M^n consists of ordered collections $(x_1, \ldots, x_n)$, but a configuration $\{x_1, \ldots, x_n\}$ is a disordered collection). More precisely, we should also delete from this factor-space all collections that have multiple points, but this is not essential since the σ^n-measure of such collections is zero. The set Γ_0 is a single point and $\pi_1(\Gamma_0) = e^{-\sigma(\Omega)}$. The restriction of the Poisson measure π_1 to Ω^n/S_n is the image of the measure $e^{-n}(\sigma|_\Omega)^n/n!$ on Ω^n under the canonical projection $\Omega^n \to \Omega^n/S_n$. On every set Γ_n, the measure π_1 is transformed by T into an equivalent measure with the density $\varrho(\gamma) = \prod_{x_i \in \gamma} \Lambda(x_i)$ with respect to π_1, which is clear from the change of variables formula for the transformation of σ^n. □

A considerably more general result is proved in Pugachev [**936**].

11.5. Diffeomorphism groups

For many applications it is of interest to have measures on groups of diffeomorphisms of manifolds similar to Haar measures. There is no exact analog of a Haar measure on them, for example, there is no nonzero locally finite measure equivalent to all shifts. However, it is possible to construct measures equivalent to the shifts by elements of dense subgroups. We shall consider one of the simplest examples: the so-called Shavgulidze measure on the group $\operatorname{Diff}^1_+([0,1])$ of all C^1-diffeomorphisms of the interval $[0,1]$ leaving the endpoints fixed. We equip this group with the metric from $C^1[0,1]$. Let $\operatorname{Diff}^2_+([0,1])$ denote the subgroup of C^2-diffeomorphisms. Let P_W be

11.5. Diffeomorphism groups

the Wiener measure on the subspace $C_0[0,1] \subset C[0,1]$ of functions x with $x(0) = 0$. Let us consider the mapping

$$S\colon \operatorname{Diff}^1_+([0,1]) \to C_0[0,1], \quad S(\varphi)(t) = \ln \varphi'(t) - \ln \varphi'(0).$$

This mapping is a homeomorphism and

$$S^{-1}(x)(t) = Jx(t)/Jx(1), \quad Jx(t) := \int_0^t \exp x(s)\, ds.$$

For every $\varphi \in \operatorname{Diff}^2_+([0,1])$ put $L_\varphi(\psi) := \varphi \circ \psi$. It is easy to see that

$$S \circ L_\varphi \circ S^{-1} = T_{S(\varphi)},$$

where for every ξ from the space $C_0^1[0,1] := C^1[0,1] \cap C_0[0,1]$ we set

$$T_\xi(x)(t) := x(t) + \xi\bigl(Jx(t)/Jx(1)\bigr).$$

Set $\mu(B) := P_W\bigl(S(B)\bigr)$ for every $B \in \mathcal{B}\bigl(\operatorname{Diff}^1_+([0,1])\bigr)$, i.e., we transport the Wiener measure to $\operatorname{Diff}^1_+([0,1])$.

11.5.1. Theorem. *For each $\varphi \in \operatorname{Diff}^2_+([0,1])$ we have $\mu \circ L_\varphi^{-1} \sim \mu$.*

PROOF. It is clear from the previous equalities that we have to verify the equivalence of the measure P_W and its image under the mapping T of the form $T(x) = x + F(x)$, where $F(x)(t) = \xi\bigl(Jx(t)/Jx(1)\bigr)$. The mapping T is a homeomorphism and is continuously Fréchet differentiable. It is obvious that F takes values in the space $H = W_0^{2,1}[0,1]$. For applying a general result on transformations of Gaussian measures, we show that the operator $D_H F(x)$ has no eigenvalues. Let $v \in H$ and $D_H F(x)v = \lambda v$, i.e., one has equality $\partial_v F(x) = \lambda v$. This yields the integral equation

$$\lambda v(t) = Jx(1)^{-2} \xi'\bigl(Jx(t)/Jx(1)\bigr)$$
$$\times \left[Jx(1) \int_0^t v(s) \exp x(s)\, ds - Jx(t) \int_0^1 v(s) \exp x(s)\, ds \right], \quad v(0) = 0.$$

Let $\lambda \neq 0$ (the case $\lambda = 0$ is obvious). If the integral of $v \exp x$ over $[0,1]$ vanishes, the condition $v(0) = 0$ easily yields that $v = 0$. If the indicated integral does not vanish, we may assume that it equals $Jx(1)$. Then the function

$$u(t) := v(t)/\xi'\bigl(Jx(t)/Jx(1)\bigr)$$
$$= \bigl(\lambda Jx(1)\bigr)^{-1} \int_0^t v(s) \exp x(s)\, ds - \bigl(\lambda Jx(1)\bigr)^{-1} \int_0^t \exp x(s)\, ds$$

satisfies the equation

$$u(t) = c \int_0^t \alpha(s) u(s) \exp x(s)\, ds - c \int_0^t \exp x(s)\, ds, \quad u(0) = 0,$$

where $\alpha(t) := \xi'\big(Jx(t)/Jx(1)\big)$ and c is a constant such that $u(1) = 0$. This equation is solved explicitly by the method of variation of constants: its solution has the form

$$u(t) = -c \exp z_2(t) \int_0^t \exp z_1(s)\, ds,$$

where the functions z_1 and z_2 are explicitly expressed via x and α. Hence $u(1) \neq 0$ (if $c \neq 0$), which is a contradiction. $\square$

The absence of eigenvalues of $D_H F(x)$ simplifies the formula known from §10.2 for the Radon–Nikodym density $d(\mu \circ T^{-1})/d\mu = 1/\Lambda_F \circ T^{-1}$, where the function Λ_F is defined by $\Lambda_F = \exp\big(\delta F - |F|_H^2/2\big)$, and gives the following expression:

$$\Lambda_F(x) = \exp\left(-\frac{1}{Jx(1)} \int_0^1 \xi'\Big(\frac{Jx(t)}{Jx(1)}\Big) e^{x(t)}\, dx(t) - \frac{1}{2}|F(x)|_H^2\right)$$

$$= \exp\left(-\frac{1}{Jx(1)} \int_0^1 \xi'\Big(\frac{Jx(t)}{Jx(1)}\Big) e^{x(t)}\, dx(t) - \frac{1}{2Jx(1)^2}\int_0^1 \Big|\xi'\Big(\frac{Jx(t)}{Jx(1)}\Big)\Big|^2 e^{2x(t)}\, dt\right),$$

where the first integral in the exponent is stochastic. In the case $\xi \in C_0^2[0,1]$, due to Exercise 11.6.5 we have

$$-\frac{1}{Jx(1)} \int_0^1 \xi'\Big(\frac{Jx(t)}{Jx(1)}\Big) e^{x(t)}\, dx(t)$$

$$= \frac{1}{Jx(1)^2} \int_0^1 \xi''\Big(\frac{Jx(t)}{Jx(1)}\Big) e^{2x(t)}\, dt + \frac{\xi'(0)}{Jx(1)} - \frac{\xi'(1)}{Jx(1)} e^{x(1)}.$$

For $\varphi \in \mathrm{Diff}_+^3([0,1])$ this gives the formula $d(\mu \circ L_\varphi^{-1})/d\mu = 1/\Lambda_\varphi \circ L_\varphi^{-1}$,

$$\Lambda_\varphi(z) = \Lambda_F\big(S(z)\big) = \exp\Big(\frac{\varphi''(0)}{\varphi'(0)} z'(0) - \frac{\varphi''(1)}{\varphi'(1)} z'(1)\Big)$$

$$\times \exp \int_0^1 \left[\frac{\varphi'''(z(t))}{\varphi'(z(t))} - \frac{3}{2}\Big|\frac{\varphi''(z(t))}{\varphi'(z(t))}\Big|^2\right] |z'(t)|^2\, dt.$$

Similarly, on the group $\mathrm{Diff}_+^1(S^1)$ of orientation preserving diffeomorphisms of S^1 one constructs a probability measure μ quasi-invariant with respect to the action of the group $\mathrm{Diff}_+^2(S^1)$. To this end, on the hyperplane $E := \{x \in C_0[0,1]\colon x(1) = 1\}$ we take the measure P_1 which is equal to the image of P_W under the mapping $\Pi x(t) = x(t) + x(1)t$. The mapping

$$S\colon \mathrm{Diff}_+^1(S^1) \to S^1 \times E, \quad Sz(t) = \big(z(0), \ln z'(t) - \ln z'(0)\big),$$

where S^1 is parameterized by the interval $[0,1)$, is a homeomorphism. Let us set $\mu := P_1 \circ \Psi^{-1}$, $\Psi = S^{-1}$. By using the previous theorem, it is not difficult to show the quasi-invariance of μ with respect to the left action of the group $\mathrm{Diff}_+^2(S^1)$. For shifts by the elements of $\mathrm{Diff}_+^3(S^1)$, the corresponding Radon–Nikodym density is given by the same formula as above, but without

11.5. Diffeomorphism groups

the first factor with boundary values. The presented formulae yield the differentiability of the measure μ with respect to the shift.

Shavgulidze [**999**], [**1001**], [**1002**] suggested also a very interesting construction of quasi-invariant and smooth measures μ on the groups of diffeomorphisms of multidimensional manifolds. This construction is technically more involved, so we only mention the main steps (the details can be found in [**1002**] and Khafizov [**603**], where a detailed proof of the differentiability of μ is given). Let Ω be a bounded domain in $\mathbb{R}^n$ with smooth boundary and let $\mathrm{Diff}^k(\Omega)$ be the space of all C^k-diffeomorphisms of the closure of Ω.

11.5.2. Lemma. *Let $m, k \in \mathbb{N}$, where $k \geqslant 3m + 1$. Then, for every index $i = 1, \ldots, n$, there exist real numbers $c_0, \ldots, c_m$ such that the differential operator Q defined on $\mathrm{Diff}^k(\Omega)$ by $Q(f) = \sum_{j=0}^{m} c_j \partial_i^j \big[(f')^{-1} \partial_i^{k-j} f\big]$ has the following properties*:

(i) $Q(f) = (f')^{-1} \partial_i^k f +$ *terms of lower order in* ∂_i;

(ii) *for all φ and f from $\mathrm{Diff}^k(\Omega)$ the expression $Q(\varphi \circ f) - Q(f)$ as a differential operator in f has order less than $k - m$.*

Let us fix $k, m = 2l \in \mathbb{N}$ with $2m > n$ and $2k \geqslant 3m - 2$. Let L_k be the space of mappings $f \in W^{2,2k+m}(\Omega, \mathbb{R}^n)$ vanishing at the boundary of Ω along with their first k derivatives (belonging to $W^{2,k}(\mathbb{R}^n, \mathbb{R}^n)$ when defined by zero outside Ω). The space L_k has the natural Hilbert norm. The affine space $I + L_k$, where I is the identity mapping on Ω, is equipped with the topology imported from L_k. One can verify that $G_k := (I + L_k) \bigcap \mathrm{Diff}^{2k}(\Omega)$ is open in $I + L_k$ with the induced topology. Let $H := W_0^{2,m}(\Omega, \mathbb{R}^n)$ be the subspace of the Hilbert space $W^{2,m}(\Omega, \mathbb{R}^n)$ consisting of the mappings vanishing at $\partial \Omega$ along with their first m derivatives. It is known that H is closed and that the operator $(-1)^m \Delta^m$ has an inverse T, which is a nonnegative Hilbert–Schmidt operator. Hence there exists a centered Gaussian measure γ on H with the covariance operator T^2. The idea of Shavgulidze is to transport γ to $\mathrm{Diff}^{2k}(\Omega)$ by means of nonlinear differential operators constructed in Lemma 11.5.2. The lemma gives an operator $Q\colon G_k \to W^{2,m}(\Omega, \mathbb{R}^n)$ such that the principal part of $Q(f)$ is $(f')^{-1} L f$, where $L f = \sum_{i=1}^{n} \partial_i^{2k} f / \partial x_i^{2k}$, and for every $\varphi \in \mathrm{Diff}^{2k}(\Omega)$ the expression $Q(\varphi \circ f) - Q(f)$ contains only derivatives of f of order at most $2k - m$. The mapping Q is differentiable and its derivative at I is L. It is known that L is a linear isomorphism between L_k and $W^{2,m}(\Omega, \mathbb{R}^n)$. By the inverse mapping theorem, there exists a neighborhood W of the point I in G_k such that $Q\colon W \to Q(W) \subset W^{2,m}(\Omega, \mathbb{R}^n)$ is a diffeomorphism. We take a probability measure $\nu = \varrho \cdot \gamma$, where ϱ is a smooth function on H whose support is a ball in $Q(W)$. Set $\mu_0(A) = \nu\big(Q(A \cap W)\big)$. Let G be the subgroup in $\mathrm{Diff}^{2k}(\Omega)$ consisting of all g such that $g - I \in W_0^{2,2k+2m}(\Omega, \mathbb{R}^n)$. Let $\{g_i\}$ be a countable dense subset of G with $g_i \in \mathrm{Diff}^{2k+2m+2}(\Omega)$. Finally, we set $\mu(A) = \sum_{i=1}^{\infty} \mu_0(g_i A)$, where $c_i > 0$ and $\sum_{i=1}^{\infty} c_i = 1$.

11.5.3. Theorem. *The measure μ on $\mathrm{Diff}^{2k}(\Omega)$ is left-quasi-invariant with respect to the action of the subgroup G and the measure $\mu * \mu_1$, where $\mu_1(A) := \mu(A^{-1})$, is left- and right-quasi-invariant with respect to the action of G. In addition, for a suitable choice of c_i, the measure μ is differentiable along the left action of the one-parameter semigroup $\{\varphi_t\} \subset G$ such that the mapping $t \mapsto \varphi_t$ from $\mathbb{R}^1$ to $C^{2k+2m+1}(\overline{\Omega})$ is continuously differentiable.*

The main idea of justification is that due to nice properties of Q the mapping $T_\varphi = Q \circ L_\varphi \circ Q^{-1}$, where $L_\varphi(f) = \varphi \circ f$, has the following special form:

$$T_\varphi f - f = Q(\varphi \circ Q^{-1} f) - Q(Q^{-1} f) = T^{1/2} \Delta^{m/2} P_\varphi(Q^{-1} f),$$

P_φ is a differential operator of some order strictly less than $2k$ such that we have $\Delta^{m/2} P_\varphi g \in W_0^{2,m}(\Omega)$. For φ close to I, the mapping T_φ satisfies the hypotheses of Corollary 10.2.6.

11.6. Comments and exercises

Spaces with measures equipped with differential structures were considered by many authors in different contexts; see Bogachev [**176**], [**179**], [**189**], Bogachev, Smolyanov [**233**], Bouleau, Hirsch [**247**], Chentsov [**283**], Davydov, Lifshits, Smorodina [**336**], Pilipenko [**899**], Smorodina [**1059**], Stroock [**1082**]. Families of measures dependent on a parameter arise in many applications, in particular, in the theory of random processes and mathematical statistics. Chentsov [**283**] studied manifolds of measures. The results of §11.2 in different degrees of generality can be found in many works; in particular, Theorems 11.2.6, 11.2.7, 11.2.9 in the presented form are borrowed from Smolyanov, Weizsäcker [**1055**] (see also Weizsäcker, Smolyanov [**1187**]). Connections between the exponential integrability of logarithmic derivatives and quasi-invariance in the nonlinear case were investigated in Daletskiĭ, Sokhadze [**324**], Bell [**107**], and also in the works on flows of transformations (see §10.4 and the corresponding comments).

Quasi-invariant and smooth measures on loop groups and path spaces, transition probabilities of diffusions in such spaces, and related questions of analysis have been studied by many authors; let us especially note the works Malliavin, Malliavin [**750**] and Driver [**361**]–[**365**]; see also Aida [**9**], Airault, Malliavin [**21**], [**22**], Andersson, Driver [**70**], Baxendale [**104**], Bell [**110**]–[**113**], Belopol'skaya, Daletskiĭ [**120**], Cecil, Driver [**274**], Cruzeiro, Malliavin [**297**], [**299**], [**300**], Cruzeiro, Zhang [**303**], Driver, Gordina [**366**], [**367**], Driver, Lohrenz [**368**], Driver, Röckner [**369**], Driver, Srimurthy [**370**], Driver, Thalmaier [**371**], Elworthy, Le Jan, Li [**390**], [**391**], Elworthy, Li [**392**]–[**395**], Fang [**410**]–[**413**], Fang, Malliavin [**416**], Fang, Shao [**419**], Getzler [**476**], Gordina [**498**], Gross [**507**], [**508**], Hsu [**555**]–[**560**], Inahama [**576**], [**577**], Kontsevich, Suhov [**632**], Léandre [**706**]–[**708**], Léandre, Norris [**710**], Li [**721**], [**722**], Lyons, Qian [**741**], Malliavin [**763**], [**767**], [**768**], Márquez-Carreras, Tindel [**781**], Melcher [**799**], Nekludov [**821**], Neretin

[823], Pickrell [892], Rudowicz [977], Sadasue [978], Shigekawa [1014], Sidorova [1034], Sidorova et al. [1035], Smolyanov [1052], Srimurthy [1072], Werner [1192]. Our presentation in §11.3 is based on the paper Gordina [498].

Measures and analysis on the Poisson space and other spaces of configurations were studied by many authors; see Albeverio, Daletskii, Lytvynov [38], Albeverio et al. [37], Albeverio, Kondratiev, Röckner [46], Albeverio, Smorodina [55], Bogachev, Daletskii [165], Bogachev, Pugachev, Röckner [215], [214], Bouleau, Denis [244], Carlen, Pardoux [263], Davydov, Lifshits, Smorodina [336], Dermoune [341], Dermoune, Krée, Wu [342], Elliott, Tsoi [387], Ismagilov [579], Ishikawa, Kunita [578], Ismagilov [580], Kachanovsky [588], Kondratiev, Lytvynov, Us [629], Lifshits, Shmeleva [724], Neretin [825], [824], Nicaise [828], Picard [887], Privault [921]–[923], Pugachev [933], [936], Ren, Röckner, Zhang [947], Röckner, Schied [961], Smorodina [1060]–[1067], [1068], Wu [1194], Vershik, Gel'fand, Graev [1173]. Concerning the distribution of the gamma process, see Renesse, Yor, Zambotti [951], Tsilevich, Vershik, Yor [1124]. For probabilistic foundations, see Kingman [606]. Measures interesting from the point of view of differential structures were considered also on other infinite dimensional manifolds; see Airault, Bogachev [18], Airault, Malliavin, Thalmaier [27], [28], Borodin, Olshansky [241], Driver, Gordina [366], [367], Paycha [877], [878], Pickrell [889], [891]. Fang, Shao, Sturm [420], Lott, Villani [736], Sturm [1086], and Renesse, Sturm [950] discuss the geometry of the spaces of measures corresponding to the Kantorovich–Rubinshtein metric. These spaces are interesting nonlinear manifolds possessing measures quasi-invariant with respect to broad classes of transformations and differentiable along many vector fields.

Submanifolds of the Wiener space and certain nonlinear manifolds connected with the Wiener space are studied in Aida [6]–[8], Airault [16], Airault, Malliavin, Ren [26], Airault, Van Biesen [31], [32], Gaveau [468], Hu, Üstünel, Zakai [563], Kuo [662], Kusuoka [679], [680], [682], Shigekawa [1015], Taniguchi [1109], Van Biesen [1168], Zakai [1215].

Differential forms on the Wiener space were first considered in Malliavin [755], and then in Arai, Mitoma [76], Kusuoka [682], Shigekawa [1009]. Differential forms of finite codegree n were introduced in Smolyanov [1051] as measures on a Hilbert space X with values in the space of n-linear functionals on X. Such forms are in duality with usual n-dimensional differential forms on X. For these forms, there is an analog of the Stokes formula; see Weizsäcker, Léandre, Smolyanov [1186], Weizsäcker, Smolyanov [1189].

Analysis on manifolds of the type of countable products is developed in Albeverio, Daletskii, Kondratiev [36], Yoshida [1208]. On measures on countable products of finite dimensional groups and groups of operators, see Bendikov [127], Bendikov, Saloff-Coste [128], [129], Gordina [496], [497], Hora [554], and Shimomura [1026]. On measures on groups in relation to the theory of representations, see Airault, Malliavin [**23**], Albeverio, Kosyak [**48**], Goldin, Menikoff, Sharp [**489**], Kosyak [**635**], [**636**]. On groups of homeomorphisms and diffeomorphisms, see Airault, Malliavin [**24**], Cruzeiro, Malliavin [**301**], Gliklikh [**483**], Gordina, Lescot [**499**], Kuzmin [**695**], Malliavin [**769**], Neretin [**826**], Pickrell [**890**], Shavgulidze [**1002**], [**1003**], [**1004**], [**1005**], Shimomura [**1031**].

Exercises

11.6.1. Justify Example 11.2.8.

11.6.2. Let μ be a nonnegative measure and $p \in (1, +\infty)$. Prove that the mapping $F\colon x \mapsto |x|^p$ from $L^p(\mu)$ to $L^1(\mu)$ is Fréchet differentiable and one has the equality $DF(x)h = p|x|^{p-1}h$.

11.6.3. (i) Show that the functions $f(x,y) = \max(x,y)$ and $g(x,y) = \min(x,y)$ are not independent on $\mathbb{R}^2$ with the standard Gaussian measure, but almost everywhere one has the equality $(\nabla f(x), \nabla g(x)) = 0$.

(ii) (A.V. Shaposhnikov) Let γ be the standard Gaussian measure on $\mathbb{R}^3$,
$$f(x,y,z) = \frac{x+yz}{\sqrt{1+z^2}}, \quad g(x,y,z) = \frac{-xz+y}{\sqrt{1+z^2}}.$$
Show that f and g are independent standard Gaussian random variables, hence $\xi = f^2$ and $\eta = g^2$ are also independent, but $(\nabla \xi, \nabla \eta) = -4f^2g^2(1+z^2)^{-2}$, which has a negative expectation.

11.6.4. Prove Proposition 11.1.16.

11.6.5. Let $\eta(\,\cdot\,, \,\cdot\,)\colon [0,1]\times C[0,1] \to \mathbb{R}^1$ be a Borel function having a bounded partial derivative $\partial_t \eta$ such that, for every t, the function $x \mapsto \eta(t,x)$ is measurable with respect to the σ-algebra $\mathcal{F}_t$ generated by the functions $x \mapsto x(s)$, $s \leqslant t$, on $C[0,1]$. Prove that for every function $f \in C^1(\mathbb{R}^1)$, one has
$$\int_0^1 \eta(t,x)f'(x(t))\,dx(t) = -\int_0^1 \partial_t \eta(t,x)f(x(t))\,dt + \eta(1,x)f(x(1)) - \eta(0,x)f(0),$$
where the first integral is stochastic with respect to the Wiener process $x(t)$.

11.6.6. Suppose that η is a measurable function on $\mathbb{R}^1$ such that
$$\int_0^{+\infty} |\eta(t)|^2 e^{-t^2/2}\,dt < \infty.$$
Show that
$$\int_0^{+\infty} e^{x^2/2} \left(\int_x^{+\infty} \eta(t) e^{-t^2/2}\,dt \right)^2 dx < \infty.$$

HINT: Use integration by parts and the estimates
$$\int_x^{+\infty} e^{-t^2/2}\,dt \leqslant \frac{C}{x} e^{-x^2/2}, \quad \int_0^x e^{t^2/2}\,dt \leqslant \frac{C}{x} e^{x^2/2}, \quad x > 0.$$

11.6.7. Let μ be a Borel probability measure on the d-dimensional torus $\mathbb{T}^d$ whose logarithmic gradient associated with the standard manifold structure on $\mathbb{T}^d$ vanishes. Show that μ is the normalized Lebesgue measure on $\mathbb{T}^d$. Extend this assertion to the infinite dimensional torus $\mathbb{T}^\infty$.

CHAPTER 12

Applications

In this chapter we consider a number of typical applications of the Malliavin calculus and differentiable measures. Some important applications have already been discussed above, for example, the absolute continuity of distributions of functionals of random processes and regularity of the corresponding densities, and also smooth approximations in Banach spaces. Here we shall be concerned with more special applications in the most diverse fields such as partial differential equations, differential geometry, stochastic analysis, and mathematical physics.

12.1. A probabilistic approach to hypoellipticity

An impressive application of Malliavin's method of proving smoothness of the induced measures is a probabilistic approach to hypoellipticity, i.e., techniques of proving smoothness of solutions of degenerate elliptic and parabolic equations by probabilistic tools. The chief goal of the fundamental work [751] entitled "Stochastic variational calculus and hypoelliptic operators" was to apply analysis on the Wiener space to the proof of hypoellipticity of second order differential operators associated with diffusion processes. Later the method was successfully applied to other problems related to the study of fundamental solutions of parabolic equations; in particular, it was effectively employed to the proof of the so-called index theorems. In this section, we briefly present basic ideas of these applications. A differential operator L with infinitely differentiable coefficients is called *hypoelliptic* if for every $u \in \mathcal{D}'$ with $Lu \in C^\infty(\Omega)$, where Ω is an open domain, one has $u \in C^\infty(\Omega)$. This is equivalent to the fact that L preserves singular supports of distributions. A trivial example of a hypoelliptic operator is the differentiation on the real line. On the other hand, the differentiation operator $\partial/\partial x_1$ on the plane is not hypoelliptic. According to Weyl's theorem, the Laplacian Δ on $\mathbb{R}^d$ is hypoelliptic. Moreover, any uniformly elliptic second order operator with smooth coefficients is hypoelliptic. The assumption of nondegeneracy cannot be omitted. For example, the δ-function is a solution to the equation $x^3 F'' = 0$. However, it was observed already by Kolmogorov that by adding a first order term to a degenerate second order elliptic operator one sometimes can obtain a hypoelliptic operator (see Ikeda, Watanabe [571, Example 8.1, Ch. V]). This phenomenon was investigated by Hörmander, who obtained rather general sufficient conditions for hypoellipticity (see Hörmander [**550**], [**551**]). Malliavin suggested a probabilistic approach to this circle of problems, the essence of which is establishing smoothness of finite dimensional images of measures. An important role in this approach is played by connections between second order elliptic operators and diffusion processes. Namely, for any matrix-valued mapping $\sigma\colon \mathbb{R}^d \to L(\mathbb{R}^d)$ and vector field $b\colon \mathbb{R}^d \to \mathbb{R}^d$ with the components $\sigma^{ij}, b^i \in C_b^\infty$, there exists a unique diffusion process $\xi = (\xi_{t,x})$ satisfying the stochastic differential equation in Itô's form

$$\xi_{t,x} = \sigma(\xi_{t,x})dw_t + b(\xi_{t,x})dt, \quad \xi_{0,x} = x.$$

Denote by $P(t,x,B) := P(\xi_{t,x} \in B)$, $B \in \mathcal{B}(\mathbb{R}^d)$, the transition probabilities of the process ξ and by $\{P_t\}_{t \geq 0}$ the corresponding semigroup, which acts on functions by the formula

$$P_t f(x) = \int_{\mathbb{R}^d} f(y)\, P(t,x,dy)$$

and acts on measures by the formula

$$P_t^* \mu(B) = \int_{\mathbb{R}^d} P(t,x,B)\, \mu(dx).$$

The classical Kolmogorov theorem states that this semigroup has generator

$$Lf = \sum_{i,j=1}^d a^{ij}(x)\frac{\partial^2 f}{\partial x_i \partial x_j}(x) + \sum_{i=1}^d b^i(x)\frac{\partial f}{\partial x_i}(x),$$

where $A(x) = \bigl(a^{ij}(x)\bigr)_{i,j=1}^d = 2^{-1}\sigma(x)\sigma^*(x)$. For every fixed x, the transition probabilities $P(t,x,\cdot)$ satisfy the equation

$$\frac{\partial P(t,x,\cdot)}{\partial t} = L^* P(t,x,\cdot),$$

where L^* is the adjoint operator to L in the sense of the equality

$$\int f\, d(L^*\mu) = \int Lf\, d\mu.$$

The meaning of the term "generator" is that the indicated semigroup is continuous in suitable spaces (for example, in $C_b(\mathbb{R}^d)$ in the case of bounded b, in some smaller subspaces in more general cases, and sometimes in certain spaces $L^1(\mu)$) and its generator coincides with L on $C_0^\infty(\mathbb{R}^d)$.

Suppose that the transition probabilities $P(t,x,\cdot)$ have twice differentiable densities $p(t,x,y)$. Then these densities as functions of (t,y) satisfy the direct Kolmogorov equation

$$\frac{\partial p(t,x,y)}{\partial t} = \sum_{i,j=1}^d \frac{\partial^2}{\partial y_i \partial y_j}\bigl(a^{ij}(y)p(t,x,y)\bigr) - \sum_{i=1}^d \frac{\partial}{\partial y_i}\bigl(b^i(y)p(t,x,y)\bigr),$$

and as functions of (t,x) they satisfy the backward Kolmogorov equation
$$\frac{\partial p}{\partial t} = Lp.$$
However, in general, there might be no transition densities (this phenomenon arises in the case of degeneracy of the diffusion matrix). Thus, Hörmander's theory gives a sufficient condition for the smoothness of transition probabilities. P. Malliavin [**751**] discovered that one can prove the differentiability of transition probabilities by using a suitable generalization of geometric measure theory on the Wiener space. In subsequent years Malliavin's method was developed by many authors. The results on hypoellipticity obtained in this way are sometimes the best possible. In addition, the Malliavin calculus turned out to be an efficient tool in the study of asymptotic behavior of transition densities. Here we briefly discuss some of these results.

Let $V_0, \ldots, V_n$ be C_0^∞-vector fields on $\mathbb{R}^d$ of the form
$$V_i(x) = \sum_{j=1}^{d} v_i^j(x) \frac{\partial}{\partial x_j}.$$
Let us consider the differential operator
$$L = \frac{1}{2} \sum_{i=1}^{n} V_i^2 + V_0.$$
The operator L generates the diffusion process $X(t,x)$ satisfying the stochastic differential equation in the Stratonovich form
$$dX(t,x,w) = \sum_{i \leq n} V_i\big(X(t,x,w)\big) \circ dw_i(t) + V_0\big(X(t,x,w)\big) dt$$
with initial condition $X(0,x,w) = x$. According to Theorem 8.2.6, for all x and t we have $X(t,x,\cdot) \in W^\infty(P_W, \mathbb{R}^d)$. Set
$$M(t,x,w) = \Big(\big(DX^i(t,x,w), DX^j(t,x,w)\big)_H\Big)_{i,j=1}^{d}.$$

12.1.1. Theorem. (i) *If for all $t > 0$ and $p > 1$ we have*
$$\sup_x \int |\det M(t,x,w)|^{-p} P(dw) < \infty,$$
then $P(t,x,dy) = p(t,x,y)dy$, where $p(t,\cdot,\cdot) \in C_b^\infty(\mathbb{R}^d \times \mathbb{R}^d)$, $t > 0$.
(ii) *If, in addition, for every $p > 1$ one has*
$$\limsup_{t \to 0} t \ln \sup_x \int |\det M(t,x,w)|^{-p} P(dw) = 0,$$
then the operator L is hypoelliptic.

For the proof, see Kusuoka [**681**], Kusuoka, Stroock [**688**].

Now we turn to conditions for hypoellipticity expressed in terms of the coefficients of L. Denote by $\mathcal{A}$ the Lie algebra of vector fields generated by $V_1, \ldots, V_n$ and by $\mathcal{A}_0$ the Lie algebra generated by $V_0, \ldots, V_n$. Let $\mathcal{J}$

be the ideal generated by $\mathcal{A}$ in $\mathcal{A}_0$. For any multi-index $\alpha = (\alpha_1, \ldots, \alpha_k)$, where $\alpha_i \in \{0, 1, \ldots, n\}$, we define vector fields V_i^α inductively as follows. If $|\alpha| = 0$, then $V_i^\alpha = V_i$. Then we let
$$V_i^{(\alpha_1, \ldots, \alpha_{k+1})} = \left[V_{\alpha_{k+1}}, V_i^{(\alpha_1, \ldots, \alpha_k)} \right],$$
where $[U, V]$ is the commutator of the fields V and U. Let
$$v_k(x) = \inf \Big\{ \sum_{i=1}^n \sum_{\|\alpha\| \leqslant k-1} \big(V_i^\alpha(x), z\big)^2; \ z \in \mathbb{R}^d, |z| = 1 \Big\}, \quad k \in \mathbb{N},$$
where $\|\alpha\|$ denotes $|\alpha|$ plus the cardinality of the set of those indices i for which $\alpha_i = 0$.

12.1.2. Theorem. (i) *Let* $\dim \mathcal{A}_0(x) = n$ *for all* $x \in \mathbb{R}^d$. *Then the operator L is hypoelliptic.*

(ii) *Suppose that* $\dim \mathcal{J}(x) = n$ *for all* $x \in \mathbb{R}^d$. *Then the operator $\frac{\partial}{\partial t} - L^*$ is hypoelliptic and the transition probabilities of $X(t, x)$ have smooth densities.*

(iii) *Let* $\dim \mathcal{A}(x) = n$ *for all* $x \in \mathbb{R}^d$. *Then the operator $\frac{\partial}{\partial t} - L^*$ is hypoelliptic and the transition probabilities of $X(t, x)$ have strictly positive smooth densities.*

(iv) *Let* $\dim \mathcal{J}(x_0) = n$ *for some* $x_0 \in \mathbb{R}^d$. *Then the transition probabilities of $X(t, x)$ have smooth densities.*

12.1.3. Theorem. *If* $\inf\{v_k(x) \colon x \in \mathbb{R}^d\} > 0$ *for some* $k \geqslant 1$, *then L is hypoelliptic. In particular, this is true if the Lie algebra $[V_0, \ldots, V_n]$ generated by the vector fields $V_0, \ldots, V_n$ spans $\mathbb{R}^d$ at every point $x \in \mathbb{R}^d$.*

One of the most general results obtained so far in this direction is the following theorem from Kusuoka [**681**] (see also Kusuoka, Stroock [**688**]). Denote by $WF(u)$ the wave front of the distribution u (see Hörmander [**550**, v. 1, §8.1]). Let S^{d-1} be the unit sphere in $\mathbb{R}^d$.

12.1.4. Theorem. *Let*
$$u_k(x, z) = \sum_{i=1}^n \sum_{\|\alpha\| \leqslant k} \big(V_i^\alpha(x), z\big)^2, \quad (x, z) \in \mathbb{R}^d \times S^{d-1}.$$

Suppose that a point $(x_0, z_0) \in \mathbb{R}^d \times S^{d-1}$ has a neighborhood $U \times V$ such that there exist a function $\psi \in C^\infty(U)$ and a continuous strictly increasing function $h \colon [0, \infty) \to [0, \infty)$ for which

(i) $h(0) = 0$, $\lim\limits_{t \to 0} t^\varepsilon \ln h(t) = 0$ *for every* $\varepsilon > 0$,

(ii) $\inf\{u_k(x, z), \ z \in V\} \geqslant h\big(\psi(x)^2\big)$ *for every* $x \in U$,

(iii) *there exist integer numbers* $m \geqslant 0$, $i_1, \ldots, i_m \in \{0, \ldots, n\}$ *such that* $V_{i_1} \cdots V_{i_m} \psi(x_0) \neq 0$.

Let $c \in C^\infty(\mathbb{R}^d)$, $u \in \mathcal{D}'(\mathbb{R}^d)$ and $(x_0, z_0) \notin WF\big((L + c)u\big)$. Then we have $(x_0, z_0) \notin WF(u)$.

12.1. A probabilistic approach to hypoellipticity

Detailed proofs of the main results can be found in the books Bell [**106**], [**108**], Ikeda, Watanabe [**571**], Nualart [**848**], Shigekawa [**1017**], and Watanabe [**1178**]; so we do not repeat these proofs.

Interesting applications of the Malliavin calculus to the study of asymptotic behavior of transition probabilities can be found in Ben Arous [**125**], Ikeda [**567**], Ikeda, Shigekawa, Taniguchi [**569**], Kusuoka [**677**], Léandre [**700**]–[**702**], [**705**], [**709**], Uemura [**1128**], and Watanabe [**1179**], [**1180**]. The formulation of the next result (see [**125**], [**701**]) on asymptotic behavior of transition densities in small times requires some additional notation.

Let $d_m(x)$ denote the dimension of the vector subspace in $\mathbb{R}^d$ generated by all vectors $V_i^\alpha(x)$ with $|\alpha| \leqslant m$. Set $D(x) = \sum_{k=1}^\infty k\big(d_k(x) - d_{k-1}(x)\big)$.

12.1.5. Theorem. *There exist $c_0 > 0$ and $c_k \in \mathbb{R}^1$ such that for every $m \in N$ we have*

$$p(t,x,x) = t^{-D/2}\left(\sum_{k=0}^{m} c_k t^k + O(t^{D+1})\right) \quad \text{as } t \to 0.$$

In the case where the matrix A is positive definite, the well-known Varadhan theorem states that $\lim_{t\to 0} 2t \ln p(t,x,y) = -d^2(x,y)$, where d is the distance in the Riemannian metric generated by the tensor $G = (G^{ij}) = A^{-1}$. The proofs of the following two results are given in Léandre [**700**]–[**702**].

12.1.6. Theorem. *Let $\dim A(x) = n$ for every x. Then, uniformly on every compact set in $\mathbb{R}^d \times \mathbb{R}^d$, we have*

$$\liminf_{t\to 0} 2t \ln p(t,x,y) \geqslant -d^2(x,y),$$

where $d(\cdot,\cdot)$ is the Riemannian metric on $\mathbb{R}^d$ associated with the operator L.

12.1.7. Theorem. *Suppose that for some k we have*

$$\inf\left\{\sum_{Y \in L_k} \langle Y(x), v\rangle^2,\ x, v \in \mathbb{R}^d, |v| = 1\right\} > 0,$$

where L_k denotes the Lie algebra obtained by forming the Lie brackets up to order k by means of the vector fields $V_1, \ldots, V_n$. Then

$$\limsup_{t\to 0} 2t \ln p(t,x,y) \leqslant -d^2(x,y).$$

The same is true for every partial derivative $p^{(r)}(t,x,y)$.

To the same circle of ideas belong applications of the Malliavin calculus to the so-called index theorems. The reader should be warned from the very beginning that expression of the type "probabilistic proofs of the Atyah–Singer theorem" frequently used in the literature are somewhat misleading. There exist no purely probabilistic proofs of such theorems, but only some steps in the known analytic proofs can be accomplished with the help of probabilistic reasoning. This approach is not simpler than the usual ones, and approximately the same topological background is required along with a number of difficult preliminary analytical results. However, this approach is a good illustration of effectiveness of probabilistic methods in the study of heat kernels. In addition, probabilistic representations of various arising quantities (such as solutions of the heat equation, traces, etc.) may be of independent interest. A detailed discussion of these issues would require considerable background in differential topology and the theory of elliptic operators (see the recent book Gilkey [**479**]). For this reason we confine ourselves to introductory notes. A detailed presentation of this direction in relation to the Malliavin calculus can be found in Bismut [**153**], [**156**], Duncan [**375**], Ikeda [**567**], Léandre [**703**], [**704**], Shigekawa, Ueki [**1019**], Shigekawa, Ueki, Watanabe [**1020**], Ueki [**1126**]; see also Akiyama [**33**], Popescu [**912**] (see also the survey Elworthy [**389**]). Apparently, the paper Duncan [**374**] was one of the first works which used a probabilistic approach for the index theorems. Let us recall some topological notions (see, for example, Palais [**873**]). Let M be a smooth manifold and let E and F be two bundles over M (we assume that all these objects are of the class C^∞). A differential operator $d\colon C^\infty(E) \to C^\infty(F)$ is a linear mapping that is a differential operator in local coordinates. By choosing suitable sections which form bases at all points x, one can describe d by means of matrices $P(x,d) = \bigl(P_{ij}(x, \partial/\partial x_1, \ldots, \partial/\partial x_n)\bigr)$, where P_{ij} is a polynomial in partial derivatives. The order r of the operator d is defined as the maximal degree of the polynomials P_{ij}. By taking the terms of order r, replacing $\partial/\partial x_j$ by $i\xi_j$ and multiplying this by $(-1)^{r/2}$, we obtain a matrix-valued mapping $Q(x,\xi)$ on the tangent bundle, called the principal symbol of the operator d. The operator d is called elliptic if $\dim E_x = \dim F_x = p$ for all x and the matrix $Q(x,\xi)$ is invertible for all x and all nonzero ξ. It is known that in this case the spaces $\ker d$ and $\operatorname{coker} d = C^\infty(F)/\mathcal{R}(d)$, where $\mathcal{R}(d)$ is the range of d, are finite dimensional. The analytic index $i_a(d)$ is defined by the equality

$$i_a(d) = \dim \ker d - \dim \operatorname{coker} d.$$

It is also known that the index $i_a(d)$ is stable under topological deformations. I.M. Gelfand raised the problem of characterizing $i_a(d)$ in topological terms. A positive answer was given by M. Atiyah and I. Singer. They proved the equality

$$i_a(d) = i_t(d),$$

where $i_t(d)$ is the topological index of d defined as the result of the action of some rational class of cohomologies $\operatorname{ch} d$ (dependent on d) on the class of cohomologies $\mathcal{T}$ of the complexified tangent bundle of M. Moreover, it was proved that the index d is local in the sense that it can be obtained as the integral over M of some function ϱ_d constructed by means of d. By using a special algebraic procedure the general case reduces to the study of the Dirac operator D acting on sections of the spinor bundle. Another important topological ingredient is the equality

$$\operatorname{Ind} D^+ = \int_M \operatorname{Trace}_s Q(\varepsilon, x, x)\, dx,$$

where Trace_s is the so-called super trace and Q is the heat kernel. The role of probability theory is to provide asymptotic decompositions of the heat kernel, which enable one to derive an explicit expression for the index. A very special case of the index theorem is the Gauss–Bonnet–Chern theorem; a stochastic approach to it is discussed in Gilkey [**479**], Ikeda [**567**]. This theorem states that for any compact oriented manifold M of even dimension $m = 2l$ with the Euler characteristic

$$\chi(M) = \sum_{j=0}^m (-1)^j \dim H_j(M),$$

where $H_j(M)$ is the space of harmonic j-forms on M, there exists a function C on M (called Chern's polynomial) satisfying the equality

$$\chi(M) = \int_M C(x)\, m(dx),$$

where m is the Riemannian volume on M. The starting point of the stochastic approach is a representation of $\chi(M)$ in the form of the integral over M of a function expressed via the fundamental solution Q of the equation $\partial u/\partial t = \frac{1}{2} \square u$. In this way, transition probabilities of a diffusion are involved, and one can study asymptotics of Q by probabilistic methods.

12.2. Equations for measures

We proceed to applications of differentiable measures to nondegenerate infinite dimensional diffusions. These applications are based on analytic tools of the theory of elliptic and parabolic equation for measures and the theory of semigroups and have the following three aspects:

1) constructions of infinite dimensional diffusions,

2) the study of differential properties of their transition probabilities and stationary distributions,

3) the study of the uniqueness problem for stationary distributions and transition probabilities and equations for them (respectively, elliptic and parabolic).

All these problems are of considerable interest also in the finite dimensional case, which we now briefly discuss in order to have some landmarks in the infinite dimensional situation. Unlike the previous section, where smooth coefficients have been considered, but some degeneracy of the diffusion matrices have been allowed, here the main specifics are connected with possible singularity of the drift coefficients, whereas the diffusion coefficients will be sufficiently regular and nondegenerate (the most typical problems arise even for the unit diffusion coefficient). Thus, suppose that on $\mathbb{R}^d$ we are given a Borel mapping $\sigma = (\sigma^{ij})_{i,j \leqslant d}$ with values in the space of positive symmetric operators and a Borel vector field $b = (b^i)_{i \leqslant d}$. If these mappings are Lipschitzian, then, according to what has been said in Chapter 1, there exists a diffusion process $\xi_{t,x}$ satisfying the stochastic differential equation

$$d\xi_{t,x} = \sigma(\xi_{t,x})dW_t + b(\xi_{t,x})dt, \quad \xi_{0,x} = x,$$

where W_t is a Wiener process in $\mathbb{R}^d$. The nondegeneracy of σ yields that the transition probabilities of this diffusion have densities satisfying the Kolmogorov equations described in the previous section. Let us recall that the generator of the transition semigroup of the diffusion ξ has the form

$$L\varphi = \sum_{i,j \leqslant d} a^{ij} \partial_{x_i} \partial_{x_j} \varphi + \sum_{i \leqslant d} b^i \partial_{x_i} \varphi, \quad A := (a^{ij}) := \frac{1}{2}\sigma\sigma^*.$$

Suppose also that $\xi_{t,x}$ possesses an *invariant* (or *stationary*) probability measure μ, i.e., $P_t^* \mu = \mu$ for all $t \geqslant 0$, where $\{P_t\}_{t \geqslant 0}$ is the semigroup introduced above. One can show that μ has a density ϱ with respect to Lebesgue measure and satisfies the elliptic equation

$$L^* \mu = 0, \qquad (12.2.1)$$

which is understood in the sense of the identity

$$\int_{\mathbb{R}^d} L\varphi(x)\,\mu(dx) = 0, \quad \varphi \in C_0^\infty(\mathbb{R}^d). \qquad (12.2.2)$$

In addition, the semigroup $\{P_t\}_{t \geqslant 0}$ extends to a strongly continuous Markov semigroup $\{T_t\}_{t \geqslant 0}$ in $L^1(\mu)$ (and also in all $L^p(\mu)$ with $p \in [1, +\infty)$). Under the stated assumptions, there are no other invariant probability measures. In addition, there are no other solutions of the equation (12.2.1) in the class of probability measures. A convenient sufficient condition for the existence of a probability invariant measure is given by the Hasminskii theorem [**521**]: it requires the existence of a function $V \in C^2(\mathbb{R}^d)$ (a Lyapunov function) such that, as $|x| \to \infty$, we have $V(x) \to +\infty$ and $LV(x) \to -\infty$. For example, if A is uniformly bounded, then it suffices to have $(b(x), x) \to -\infty$ as $|x| \to +\infty$ (then we can take $V(x) = (x, x)$). Generalizations of the Hasminskii condition were obtained in Bogachev, Röckner [**221**]. Certainly, not every diffusion has an invariant probability measure. For example, the Wiener process has no nonzero finite invariant measures. Elliptic equation (12.2.1) for stationary distributions appeared in the celebrated paper of Kolmogorov [**625**], written before the discovery of stochastic differential equations. This equation proved to be an efficient tool in the study of diffusion processes with singular coefficients, when stochastic equations either have no solutions or these equations and their solutions must be defined in a certain generalized sense, the result of which is a loss of the main advantages of stochastic equations. An important advantage of equation (12.2.1) is that it is at least meaningful under very broad assumptions: it is only required that the coefficients a^{ij} and b^i be locally integrable *with respect to the solution* μ; with respect to Lebesgue measure they may be very singular.

12.2. Equations for measures

12.2.1. Example. Let $\mu = \varrho\, dx$, where $\varrho \in C^\infty(\mathbb{R}^d)$ is a probability density (possibly having zeros). Set $A := I$ and $b(x) := \nabla\varrho(x)/\varrho(x)$ if $\varrho(x) > 0$ and $b(x) = 0$ if $\varrho(x) = 0$. Then the function $|b|$ is locally integrable with respect to μ, but not necessarily with respect to Lebesgue measure if ϱ has zeros; for example, if $\varrho(x) = (2\pi)^{-1/2} x^2 \exp(-x^2/2)$ on the real line, then $b(x) = 2x^{-1} - x$. The measure μ satisfies equation (12.2.1) since

$$\int_{\mathbb{R}^d} [\Delta\varphi\,\varrho + (\nabla\varphi, \nabla\varrho)]\, dx = 0$$

for all $\varphi \in C_0^\infty(\mathbb{R}^d)$ due to the integration by parts formula.

In this example the solution itself is smooth, in spite of possible singularity of the drift. However, unlike usual elliptic equations, such equations may have nondifferentiable solutions. For example, if $b = 0$, then, for every strictly positive probability density ϱ, the measure $\mu = \varrho\, dx$ satisfies (12.2.1) with the coefficient $A(x) = \varrho(x)^{-1} I$.

In the study of equation (12.2.1) the following result from Bogachev, Krylov, Röckner [**206**] is frequently used.

12.2.2. Theorem. (i) *Suppose that a Borel measure μ on $\mathbb{R}^d$ (possibly, signed) satisfies equation (12.2.1), where the coefficients are locally integrable with respect to μ. Then the locally finite measure $(\det A)^{1/d} \cdot \mu$ is absolutely continuous. In particular, the restriction of μ to the set $\{\det A \neq 0\}$ is absolutely continuous.*

(ii) *Suppose that A is continuous, the operators $A(x)$ are invertible, for some $p > d$ we have $a^{ij} \in W^{p,1}_{\mathrm{loc}}(\mathbb{R}^d)$ and either $b^i \in L^p_{\mathrm{loc}}(\mathbb{R}^d)$ or $b^i \in L^p_{\mathrm{loc}}(\mu)$. Then μ has a continuous density ϱ in the Sobolev class $W^{p,1}_{\mathrm{loc}}(\mathbb{R}^d)$.*

Let us mention yet another result from the paper Bogachev, Krylov, Röckner [**205**], [**219**] (for its development, see [**207**], [**230**], [**231**] and Metafune, Pallara, Rhandi [**802**]), related to μ-square integrable drifts.

12.2.3. Proposition. *Suppose that a probability measure μ on $\mathbb{R}^d$ satisfies equation (12.2.1), where the mappings A and A^{-1} are uniformly bounded, A is uniformly Lipschitzian, and a Borel vector field b satisfies the condition $|b| \in L^2(\mu)$. Then*
(i) *the measure μ has a density $\varrho \in W^{1,1}(\mathbb{R}^d)$ such that*

$$\int_{\mathbb{R}^d} \frac{|\nabla\varrho(x)|}{\varrho(x)}\, dx \leqslant \alpha^{-1} \int_{\mathbb{R}^d} |b(x) + b_0(x)|^2\, \mu(dx), \qquad (12.2.3)$$

where $\alpha = \sup_x \|A^{-1/2}(x)\|$ and $b_0^i(x) := \sum_{j=1}^d \partial_{x_j} a^{ij}(x)$;

(ii) *the mapping $\nabla\varrho/\varrho$ coincides μ-a.e. with the orthogonal projection of the vector field $A^{-1}(b - b_0)$ on the closure of the class $\{\nabla f \colon f \in C_0^\infty(\mathbb{R}^d)\}$ in the space $L^2(\mu, \mathbb{R}^d)$ equipped with the inner product*

$$(u, v)_2 = \int_{\mathbb{R}^d} (Au(x), v(x))\, \mu(dx).$$

In particular, if $A = I$, then $b_0 = 0$ and $\nabla \varrho/\varrho$ is the orthogonal projection of b on the closure in $L^2(\mu, \mathbb{R}^d)$ of the space of the gradients of functions from $C_0^\infty(\mathbb{R}^d)$.

This result has infinite dimensional analogs presented below. Note that α does not depend on the dimension d, in particular, if $A = I$, then $\alpha = 1$.

12.2.4. Theorem. *If in the previous theorem $|b| \in L^p(\mu)$ with $p > d$, then $\varrho \in W^{p,1}(\mathbb{R}^d)$ and hence $\varrho \in L^\infty(\mathbb{R}^d)$.*

For a proof, see Metafune, Pallara, Rhandi [802] and Bogachev, Krylov, Röckner [207], [208].

Unlike equations with smooth coefficients on bounded domains, for equation (12.2.1) with $d > 1$ there is no uniqueness theorem in the class of probability measures even if $A = I$ and the coefficient b is infinitely differentiable. Examples are constructed in Bogachev, Röckner, Stannat [229] and Shaposhnikov [995], [996]. If we do not require the local boundedness of b, then there is an example also for $d = 1$: the measure with a probability density $\varrho \in C^\infty(\mathbb{R}^1)$ vanishing only at the origin satisfies our equation with $A(x) = 1$ and $b(x) = \varrho'(x)/\varrho(x)$, $b(0) := 0$. This equation is satisfied also by the measure with the density $\varrho_1(x) = c\varrho(x)$ for $x > 0$, $\varrho_1(x) = 0$ for $x < 0$ (where c is a normalization constant). There are several sufficient conditions for the uniqueness of solution of (12.2.1) in the class of probability measures. For example, the existence of a Lyapunov function mentioned above as a sufficient condition for the existence of solutions ensures the uniqueness as well. Another sufficient condition in the case $A = I$ is the inclusion $|b|^2 \in L^2(\mu)$. It turns out that the uniqueness of solution to (12.2.1) is closely related to the properties of the semigroup corresponding to L, which we now explain.

Let us recall that elliptic equation (12.2.1) appeared as an equation for invariant measures of the semigroup. For this reason its solutions are called *infinitesimally invariant* measures for L. However, our previous results made no assumptions on the existence of semigroups. Naturally the questions arise about connections between infinitesimally invariant measures and properly invariant measures of semigroups. If an operator semigroup $(T_t)_{t \geq 0}$ acts on the space of bounded functions or on $L^1(\mu)$, then a measure μ is called *invariant* for $(T_t)_{t \geq 0}$ provided that

$$\int_{\mathbb{R}^d} T_t f \, d\mu = \int_{\mathbb{R}^d} f \, d\mu$$

for all bounded Borel functions f. In the case of a semigroup on $L^1(\mu)$ it suffices to have this identity for all $f \in C_0^\infty(\mathbb{R}^d)$. Certainly, first of all the question arises about the existence of semigroups corresponding in some sense to the operator L. There are precise answers to these questions. Let a probability measure μ on $\mathbb{R}^d$ satisfy equation (12.2.1) for the coefficients of which the condition of part (ii) of Theorem 12.2.2 is fulfilled. Then, there exists at least one strongly continuous semigroup $(T_t^\mu)_{t \geq 0}$ on $L^1(\mu)$ such that the domain of definition of its generator contains the class $C_0^\infty(\mathbb{R}^d)$

12.2. Equations for measures

and on this class the generator coincides with the operator L. By using this semigroup one can construct some diffusion process for which it will serve as the transition semigroup. It turns out, remarkably, that if μ is a unique probability solution of (12.2.1), then there are no other semigroups on $L^1(\mu)$ with the described properties and μ is a true invariant measure for $(T_t^\mu)_{t\geq 0}$. If equation (12.2.1) has a probability solution different from μ, then, for every probability solution ν (in particular, for μ), there exists more than one strongly continuous semigroup $(T_t^\nu)_{t\geq 0}$ on $L^1(\nu)$ whose generator coincides with L on $C_0^\infty(\mathbb{R}^d)$, and no measure ν is a true invariant measure for any semigroup $(T_t^\nu)_{t\geq 0}$ (and these semigroups have no invariant measures at all). Thus, the infinitesimal invariance turns out to be a broader concept than invariance with respect to semigroups.

Summing up, we can say that probability solutions to equations (12.2.1) serve as a starting point for constructing and investigating diffusion processes, which are not covered by the standard approach in the case of singularities of drift coefficients. So this approach has become popular also in the infinite dimensional case, where additional problems arise even for smooth drift coefficients.

In the case of an infinite dimensional locally convex space X one can introduce an analog of equation (12.2.1). Let μ be a Radon probability measure on X and let $\mathcal{F}$ be a dense linear subspace in $L^1(\mu)$, on which a linear mapping $L\colon \mathcal{F} \to L^1(\mu)$ is defined. The equation

$$L^*\mu = 0 \tag{12.2.4}$$

will be understood as the identity

$$\int_X L\varphi(x)\,\mu(dx) = 0, \quad \varphi \in \mathcal{F}.$$

Furthermore, we consider operators of the form

$$L\varphi(x) = \sum_{i,j} a^{ij}(x)\partial_{e_i}\partial_{e_j}\varphi(x) + \sum_i b^i(x)\partial_{e_i}\varphi(x)$$

defined on the linear space $\mathcal{F}$ of functions of the form

$$\varphi(x) = \varphi_0\bigl(l_1(x),\ldots,l_n(x)\bigr), \quad \varphi_0 \in C_b^\infty(\mathbb{R}^n),$$

where $\{l_i\}$ is some fixed sequence of functionals in X^*, for example, coordinate functionals in the case of l^2 or $\mathbb{R}^\infty$, $\{e_i\} \subset X$ is a sequence such that $l_i(e_j) = \delta_{ij}$, and the functions a^{ij} and b^i on X are Borel measurable. It is clear that we have $L(\mathcal{F}) \subset L^1(\mu)$ if $a^{ij}, b^i \in L^1(\mu)$. Set $A = (a^{ij})$ and $b = (b^i)$, although we do not assume that $A(x)$ and $b(x)$ correspond to some operator and vector in X.

As compared to the well-studied finite dimensional case, in infinite dimensions, only some partial results have been obtained so far. Let us mention some of them with the simplest formulations. Let γ be a centered Radon Gaussian measure on a locally convex space X and $H = H(\gamma)$. In

the following theorem we consider the coefficient A of the form
$$A(x) = I + K(x),$$
where $K\colon X \to \mathcal{L}(H)$ and the operators $K(x)$ are symmetric. Let us consider solutions of equation (12.2.4) with the operator
$$Lf = \sum_{i,j}(Ae_i, e_j)_H \partial_{e_i}\partial_{e_j} f + {}_{X^*}\langle f', B\rangle_X,$$
where $\{e_i\}$ is a fixed orthonormal basis in H. For $\mathcal{F}$ one can take here the space of smooth (of the class C_b^∞) functions of finitely many coordinates.

12.2.5. Theorem. *Let μ be a Radon probability measure on X satisfying equation (12.2.4), where*
$$b(x) = -x + v(x), \quad v\colon X \to H, \quad |v(\,\cdot\,)|_H \in L^2(\mu).$$
Suppose that $K(x) = S(x)T$, where T is a symmetric nuclear operator on H with $\|T\| \leqslant 1$ and the mapping $x \mapsto S(x)$ satisfies the following condition:
$$\|S(x)\|_{\mathcal{L}(H)} \leqslant q < 1, \quad \|S(x+h) - S(x)\|_{\mathcal{L}(H)} \leqslant C|h|_H$$
for all $x \in X$ and $h \in H$. Then
  (i) *the measure μ has the form $\mu = f^2 \cdot \gamma$, where $f \in W^{1,2}(\gamma)$;*
  (ii) *the logarithmic gradient β_H^μ exists;*
  (iii) *if $K = 0$, then $\beta_H^\mu(x) = -x + u(x)$, where u is the orthogonal projection of v on the closure in $L^2(\mu, H)$ of the set of mappings $\nabla_H \varphi$, where $\varphi \in \mathcal{F}C_b^\infty$.*

Under the indicated condition, there exists a diffusion process ξ_t in X satisfying the equation
$$d\xi_t = \sqrt{A(\xi_t)}dw_t + \frac{1}{2}b(\xi_t)dt, \quad \xi_0 = x, \qquad (12.2.5)$$
where w_t is a Wiener process associated with H in the following sense: if $l \in X^*$ and $|j_H(l)|_H = 1$, then $l(w_t)$ is a real Wiener process. If $A = I$ and $b(x) = -x$, then we obtain the *Ornstein–Uhlenbeck process*. As in the one-dimensional case, the infinite dimensional Ornstein–Uhlenbeck process is expressed through the Wiener process by the formula indicated in §1.5; its transition semigroup is $\{T_{t/2}\}$, where $\{T_t\}$ is the Ornstein–Uhlenbeck semigroup. According to Theorem 12.2.5, all invariant measures of the diffusion ξ_t are absolutely continuous with respect to the measure γ, which is a unique invariant probability measure of the Ornstein–Uhlenbeck process. Thus, invariant measures of a diffusion whose coefficients are sufficiently small perturbations of the coefficients of the Ornstein–Uhlenbeck process, are absolutely continuous with respect to γ. In the case $A = I$ this gives a positive solution to the problem posed by Shigekawa [**1011**] (this problem was solved in Bogachev, Röckner [**219**]). In view of Theorem 12.2.5, we obtain that if $A = I$, $b(x) = -x + v(x)$ and $\sup_x |v(x)|_H < \infty$, then equation (12.2.4) has a unique solution in the class of probability measures (the existence of a solution in this case was established in [**1011**]; see more general

conditions for existence in Hino [**540**] and Bogachev, Röckner [**222**]). Generalizations to nonconstant diffusion coefficients are obtained in Albeverio, Bogachev, Röckner [**35**], Bogachev, Krylov, Röckner [**205**], Hino [**540**].

It would be interesting to investigate small perturbations of non-Gaussian symmetric diffusions (this concept is discussed in the next section). In this direction, the following result was obtained in Albeverio, Bogachev, Röckner [**35**].

12.2.6. Theorem. *Let a Radon probability measure μ satisfy the logarithmic Sobolev inequality (see §8.8), be quasi-invariant along $j(X^*)$ and have a logarithmic gradient $\beta := \beta_H^\mu$. Let $b(x) = \beta_H^\mu(x) + v(x)$, where $v\colon X \to H$ is a Borel mapping, $\sup_x |v(x)|_H < \infty$, and there exists an orthonormal basis $\{e_n\}$ in H with $e_n = j(l_n)$, $l_n \in X^*$, such that $P_n\beta$ is a function of $P_n x$, where $P_n x := \sum_{i=1}^n l_i(x)e_i$. Then equation (12.2.4) has a unique solution in the class of Radon probability measures.*

Certainly, equations of type (12.2.5) have solutions under much broader assumptions. Is it true that the transition probabilities $P(t,x,\cdot)$ of such diffusions possess nonzero vectors of differentiability? An interesting counterexample was constructed by Tolmachev [**1119**]. In his example, coefficients are smooth and A is uniformly invertible. Let us formulate two theorems from [**1119**]. Let X be an infinite dimensional separable Hilbert space.

12.2.7. Theorem. *Given $\varepsilon > 0$, there exist a Hilbert space H densely embedded into X by a Hilbert–Schmidt operator, an infinitely Fréchet differentiable mapping σ_0 on X with values in the space $\mathcal{N}(X)$ of nuclear operators on X, and an infinitely Fréchet differentiable mapping $v\colon X \to X$ with $\|\sigma_0(x)\|_{\mathcal{N}(X)} \leqslant \varepsilon$ and $\|v(x)\| \leqslant \varepsilon$ and bounded derivatives such that there is a unique probability measure μ invariant for the diffusion ξ_t generated by the stochastic differential equation*

$$d\xi_t = \sigma(\xi_t)dW_t + b(\xi_t)dt$$

with $\sigma(x) = I + \sigma_0(x)$, $b(x) = -\frac{1}{2}x + v(x)$, where W_t is the Wiener process associated with H, but μ has no nonzero vectors of continuity. In addition, the transition probabilities $P_t(x,\cdot)$ of ξ_t have no vectors of continuity as well.

12.2.8. Theorem. *Given $\varepsilon > 0$, there exist a Hilbert space H densely embedded into X by a Hilbert–Schmidt operator, an infinitely Fréchet differentiable mapping σ_0 with values in the space $\mathcal{N}(X)$ of nuclear operators on X, and an infinitely Fréchet differentiable mapping $v\colon X \to X$ with $\|\sigma_0(x)\|_{\mathcal{N}(X)} \leqslant \varepsilon$ and $\|v(x)\| \leqslant \varepsilon$ and bounded derivatives such that there are two different probability measures ν_1 and ν_2 invariant for the diffusion ξ_t generated by the above stochastic differential equation with $\sigma(x) = I + \sigma_0(x)$, $b(x) = -\frac{1}{2}x + v(x)$ such that ν_1 is equivalent to a Gaussian measure and possesses a vector logarithmic derivative β_H^ν, but ν_2 has no nonzero vectors of continuity.*

In both theorems σ_0 and v can be made zero outside some ball. It remains an open problem whether such examples exist in the case $A = I$. So far only in some very special cases positive results have been obtained. In Piech [**897**], [**898**], the coefficient A has the form $I + K(x)$, where $K(x)$ are sufficiently small nuclear operators and $b\colon X \to H$ is Lipschitzian. Under some additional conditions, it is shown that the measures $P(t, x, \,\cdot\,)$ are absolutely continuous with respect to the corresponding transition probabilities of the Wiener process and are differentiable along a dense subspace. Analogous conditions were used in Shavgulidze [**998**], Belopol'skaya, Dalecky [**120**]. Unfortunately, the formulations in all of these works are too technical. If we have $A = I$ and $b(x) = -x + v(x)$, where $v\colon X \to H$ and $\sup_x |v(x)|_H < \infty$, then by an infinite dimensional version of the Girsanov theorem, the measures $P(t, x, \,\cdot\,)$ are absolutely continuous with respect to transition probabilities of the Ornstein–Uhlenbeck process $X(t)$ (for which one has $b(x) = -x$). Under some technical assumptions, their Radon–Nikodym derivatives belong to certain Sobolev classes. Various results of this kind are obtained in Asadian [**80**], [**81**], Bogachev, Da Prato, Röckner [**194**], Bogachev, Krylov, Röckner [**205**], Bonaccorsi, Fuhrman [**236**], [**237**], Bonaccorsi, Zambotti [**238**], Da Prato, Debussche [**308**], Da Prato, Debussche, Goldys [**309**], Es-Sarhir [**398**], Fuhrman [**458**], [**459**], [**460**], Gaveau, Moulinier [**470**], Gozzi [**500**], Hino [**540**], Moulinier [**817**], Peszat, Zabczyk [**879**], and Simão [**1036**]–[**1038**]. The differentiability of $P(t, x, \,\cdot\,)$ with respect to x was studied in Belopol'skaya, Dalecky [**120**], Dalecky, Fomin [**319**], Da Prato, Elworthy, Zabczyk [**310**], Da Prato, Zabczyk [**312**]–[**314**].

12.3. Logarithmic gradients and symmetric diffusions

Here we discuss connections between logarithmic gradients of measures and the so-called symmetrizable diffusions. A diffusion $(\xi_t)_{t \geq 0}$ in a topological space X with the transition semigroup $(T_t)_{t \geq 0}$ is called *symmetrizable* if there exists a Borel probability measure μ on X such that for all bounded Borel functions f and g one has

$$\int_X T_t f(x)\, g(x)\, \mu(dx) = \int_X f(x)\, T_t g(x)\, \mu(dx).$$

In this case the diffusion $(\xi_t)_{t \geq 0}$ is called μ-symmetric. It is clear that μ is a an invariant measure for the diffusion $(\xi_t)_{t \geq 0}$ with the additional property that the generator L of the semigroup $(T_t)_{t \geq 0}$ on $L^2(\mu)$ is symmetric on $L^2(\mu)$. In fact, for introducing all of these concepts there is no need to consider the diffusion itself (for this reason no precise definition of diffusions is given here), we only need a measure μ and a semigroup $(T_t)_{t \geq 0}$ on $L^2(\mu)$.

In his celebrated work [**625**] Kolmogorov studied the following problem. Let $(\xi_t)_{t \geq 0}$ be a diffusion in a finite dimensional Riemannian manifold X (in Kolmogorov's case it was compact) with generator $L = (\Delta + b)/2$, where b is a smooth vector field on X. When is the process $\eta_t = \xi_{T-t}$ governed by the same equation, i.e., has the same generator L? The answer found by

Kolmogorov says: if and only if b is the gradient of a function. Earlier, this question had been considered by Schrödinger [990] in the one-dimensional case. In [625], Kolmogorov considered only solutions of the Fokker–Planck–Kolmogorov equations as densities of transition probabilities: in those days no stochastic integration was developed (the corresponding stochastic equation would have the form $d\xi_t = dw_t + 2^{-1}b(\xi_t)dt$). Moreover, the case of compact X was mostly considered, when invariant probability measures always exist. In this case the property studied by Kolmogorov is equivalent to the symmetrizability of $(\xi_t)_{t\geq 0}$. Thus, Kolmogorov's result is a criterion of symmetrizability of a diffusion in a compact manifold. In the noncompact case, additional conditions are required in order to ensure the existence of invariant probability measures. The most convenient conditions of this kind were discovered by Hasminskii and mentioned in the previous section. The difference between existence of an invariant measure $\mu = \varrho\,dx$ and symmetrizability of the diffusion $(\xi_t)_{t\geq 0}$ with generator $L = (\Delta + b)/2$ can be seen from Proposition 12.2.3(ii), which leads to the following result.

12.3.1. Proposition. *Let a probability measure μ on $\mathbb{R}^d$ satisfy equation $L^*\mu = 0$, where $L\varphi = \Delta\varphi + (\nabla\varphi, b)$ and $|b| \in L^2(\mu)$. The operator L on domain $C_0^\infty(\mathbb{R}^d)$ in $L^2(\mu)$ is symmetric precisely when b coincides μ-a.e. with the logarithmic gradient of μ.*

Thus, nonsymmetrizability of a diffusion is related with a nongradient component in its drift coefficient.

Note also that in the case where the diffusion $(\xi_t)_{t\geq 0}$ is not symmetric and the reversed process $(\eta_t)_{t\geq 0}$ is a diffusion with a (different) drift b^*, the drift coefficients b and b^* are related by the duality equation

$$b + b^* = 2\nabla\varrho/\varrho,$$

where ϱ is a density of μ (see Föllmer [437]).

In the last few years considerable progress has been achieved in the study of analogous problems, both in finite and infinite dimensional cases; see, for example, Albeverio, Röckner [54], Bogachev, Röckner [217], [219], [222], Bogachev, Röckner, Zhang [232], Da Prato, Zabczyk [314], Deuschel, Föllmer [344], Föllmer [436], [437], Föllmer, Wakolbinger [439], Fritz [451], Haussmann, Pardoux [524], Kirillov [611], Kusuoka, Stroock [691], Long [734], Millet, Nualart, Sanz [814], [815], and Vintschger [1176]. In particular, the following result from [217], [219] (its first assertion was proved in [54]) shows that in the infinite dimensional case logarithmic gradients are precisely drift coefficients of symmetric diffusions. It can be regarded as a probabilistic counterpart of Theorem 8.4.1.

12.3.2. Theorem. *Let γ be a centered Gaussian measure on a separable Banach space with a dense Cameron–Martin space $H = H(\gamma)$ and let μ be a Borel probability measure on X such that $X^* \subset L^2(\mu)$, the logarithmic gradient β_H^μ exists and $\|\beta_H^\mu(\,\cdot\,)\|_X \in L^2(\mu)$. Then, there exists an X-valued*

Wiener process $(w_t)_{t\geqslant 0}$ with covariance $\mathbb{E}[l(w_t)k(w_t)] = \bigl(j(l), j(k)\bigr)_H$ for which the stochastic differential equation

$$d\xi_t = dw_t + \frac{1}{2}b(\xi_t)dt, \ \xi_0 = x, \qquad (12.3.1)$$

where $b = \beta_H^\mu$, has a weak solution $(\xi_t)_{t\geqslant 0}$ for μ-a.e. $x \in X$ that is a μ-symmetric diffusion with an invariant measure μ.

Conversely, if there exists a μ-symmetric diffusion $(\xi_t)_{t\geqslant 0}$ satisfying the above equation for μ-a.e. $x \in X$ and $\langle k, b\rangle \in L^2(\mu)$ for all $k \in X^*$, then μ is differentiable along all vectors in $j(X^*)$, β_H^μ exists and $\beta_H^\mu = b$ μ-a.e.

It is worth noting that Theorem 12.3.2 gives some characterization of mappings $\beta\colon X \to X$ that are logarithmic gradients of measures on X. It would be interesting to find other necessary or sufficient conditions for an infinite dimensional mapping b to be the drift coefficient of a symmetric diffusion. On this subject see Bogachev, Röckner [**219**], [**222**], Kirillov [**611**]. The following question arises: if μ is a probability measure on $\mathbb{R}^n$ satisfying the equation $L^*\mu = 0$, where $A = I$ and $b = \nabla V$, $V \in C^1(\mathbb{R}^n)$, is it true that $b = \beta^\mu$? S.V. Shaposhnikov has recently constructed a simple counterexample for $n = 2$ (see also Bogachev, Kirillov, Shaposhnikov [**197**]): let

$$b(x,y) = \nabla V(x,y), \quad V(x,y) = -\bigl[\ln(1+x^2) + \ln(1+y^2) + x + x^3/3 + y + y^3/3\bigr];$$

then μ with density $c(1+x^2)^{-1}(1+y^2)^{-1}$ is a probability solution.

12.4. Dirichlet forms and differentiable measures

Let H be a separable Hilbert space continuously and densely embedded into a locally convex space X, let $j(X^*) \subset H \subset X$ be a standard triple, and let μ be a Radon probability measure on X. The *classical Dirichlet form* $\mathcal{E}$ is defined on $\mathcal{FC}_b^\infty$ by the equality

$$\mathcal{E}(f, f) = \int \bigl(\nabla_H f(x), \nabla_H f(x)\bigr)_H \mu(dx). \qquad (12.4.1)$$

Let us recall that a nonnegative quadratic form Q on a dense linear subspace L in a Hilbert space E is called *closed* if L is complete with respect to the inner product $(x, x)_Q = (x, x)_E + Q(x, x)$. The form Q is called *closable* if it extends to a closed form. Closability means that if a sequence $\{x_n\}$ is fundamental in $(\cdot, \cdot)_Q$ and $(x_n, x_n)_E \to 0$, then $(x_n, x_n)_Q \to 0$. A general representation of a closable form is this: $Q(x, x) = (-L_\mu x, x)_E$, where L_μ is a nonpositive symmetric operator with a dense domain of definition $D(L_\mu) \subset E$. For $Q = \mathcal{E}$ given by (12.4.1) we have

$$(f, f)_Q = \int_X |f(x)|^2 \mu(dx) + \int_X \bigl(\nabla_H f(x), \nabla_H f(x)\bigr)_H \mu(dx).$$

12.4.1. Definition. (i) *A closed nonnegative quadratic form $\bigl(\mathcal{E}, D(\mathcal{E})\bigr)$ with domain $D(\mathcal{E})$ dense in $L^2(\mu)$ is called a Dirichlet form if we have*

$$u^\sharp := u^+ \wedge 1 \in D(\mathcal{E}) \quad \text{and} \quad \mathcal{E}(u^\sharp, u^\sharp) \leqslant \mathcal{E}(u, u) \quad \text{for all } u \in D(\mathcal{E}).$$

12.4. Dirichlet forms and differentiable measures

(ii) *A Dirichlet form is called irreducible if the equality* $\mathcal{E}(u,u) = 0$ *implies* $u = \mathrm{const}\ \mu$-*a.e.*

Now a few words are in order about partial Dirichlet forms. We shall assume that we are given a linear subspace K in X^* separating the points in X. For notational simplicity we identify K with the subspace $j_H(X^*)$ in X; in place of $j(k)$ we shall write k. Let μ be a Radon probability measure on X. Let us fix a vector $k \in X$ such that there exists a logarithmic derivative $\beta_k^\mu \in L^2(\mu)$. Let $D(\mathcal{E}_k^\mu)$ be the space of all functions $u \in L^2(\mu)$ such that there exists $f_k \in L^2(\mu)$ with

$$\int_X u\partial_k v\,\mu(dx) = -\int_X f_k v\,\mu(dx) - \int_X uv\beta_k^\mu\,\mu(dx), \quad \forall v \in \mathcal{FC}_b^\infty.$$

In this case f_k is uniquely determined by u, and we set

$$\partial_k^\mu u := f_k.$$

For $u, v \in D(\mathcal{E}_k^\mu)$ put

$$\mathcal{E}_k^\mu(u,v) := \int_X \partial_k^\mu u\,\partial_k^\mu v\,\mu(dx).$$

In order to show that $(\mathcal{E}_k^\mu, D(\mathcal{E}_k^\mu))$ is a Dirichlet form, we need the characterization from the next proposition. Let us recall that for every closed subspace X_k in X such that $X = \mathrm{I\!R}^1 k \oplus X_k$ and $\pi_k \colon X \to X_k$ is the canonical projection, there exist conditional measures μ^y on the straight lines $y + \mathrm{I\!R}^1 k$, $y \in X_k$, which possess locally absolutely continuous densities f_k^y with respect to the corresponding Lebesgue measures (induced by the mappings $t \mapsto tk + y$), and for every Borel function u on X and $\nu_k := \mu\circ\pi_k^{-1}$ we have

$$\int_X u(x)\,\mu(dx) = \int_{X_k}\int_{y+\mathrm{I\!R}^1 k} u(y+z)\,\mu^y(dz)\,\nu_k(dy)$$
$$= \int_X \int_{-\infty}^\infty u(y+sk)f_k^y(s)\,ds\,\nu_k(dy).$$

12.4.2. Proposition. *Let* $u \in L^2(\mu)$. *Then* $u \in D(\mathcal{E}_k^\mu)$ *if and only if for* ν_k-*almost all* $y \in X_k$ *the following is true: the function* $t \mapsto u(y+tk)$ *coincides Lebesgue a.e. on the set* $\{t\colon f_k^y(t) \neq 0\}$ *with a function* $\widetilde{u}^y$ *that is locally absolutely continuous on this set and*

$$\int_X \int_{y+\mathrm{I\!R}^1 k} |\partial_t \widetilde{u}^y(t)|^2\,\mu^y(dt)\,\nu_k(dy) < \infty.$$

In this case $\partial_k^\mu u(y+tk) = d\widetilde{u}^y/dt$. *In addition,* $(\mathcal{E}_k^\mu, D(\mathcal{E}_k^\mu))$ *is a Dirichlet form on* $L^2(\mu)$.

PROOF. This assertion is easily deduced from the results in Chapter 3 on differentiable conditional measures (Exercise 12.7.2). □

12.4.3. Corollary. Let $D(\mathcal{E}_K^\mu)$ be the class of all $u \in L^2(\mu)$ with the property that $u \in D(\mathcal{E}_k^\mu)$ for all $k \in K$ and there exists a $(\mathcal{B}(X), \mathcal{B}(H))$-measurable mapping $\nabla_H u \in L^2(\mu, H)$ such that, for every fixed $k \in K$, we have $(\nabla_H u, k)_H = \partial_k^\mu u$ a.e. Let

$$\mathcal{E}_K^\mu(u,v) = \int_X (\nabla_H u, \nabla_H v)_H \, \mu(dx), \quad u,v \in D(\mathcal{E}_K^\mu).$$

Then $(\mathcal{E}_K^\mu, D(\mathcal{E}_K^\mu))$ is a Dirichlet form on $L^2(\mu)$.

12.4.4. Remark. There exists an orthonormal basis $\{k_n\}$ in H with $k_n \in K$. Hence if $u \in D(\mathcal{E}_K^\mu)$ and $\partial_{k_n}^\mu u = 0$ μ-a.e. for all n, then μ-a.e. we have $\nabla_H u = 0$.

The following lemma is the main part of the proof of Theorem 12.5.2 stated below.

12.4.5. Lemma. Let $\varrho \in L^1(\mu)$, $\varrho \geqslant 0$, $\nu := \varrho \cdot \mu$, and let $k \in K$ be such that there exists $\beta_k^\nu \in L^2(\nu)$ and $\sqrt{\varrho} \beta_k^\mu \in L^2(\mu)$. Then we have $\sqrt{\varrho} \in D(\mathcal{E}_k^\mu)$ and $\partial_k^\mu \sqrt{\varrho} = \sqrt{\varrho}(\beta_k^\nu - \beta_k^\mu)/2$.

PROOF. First we consider the case of the real line. Then the measures μ and ν possess absolutely continuous densities p and q, respectively. It is clear that $q = \varrho p$, hence the function ϱ is locally absolutely continuous on the set $U_p = \{p > 0\}$. We observe that on the set $U_q = \{q > 0\}$ we have $q'/q = p'/p + \varrho'/\varrho$. The integrability conditions are written as follows: $q'/q \in L^2(qdx)$ and $\sqrt{\varrho} p'/p \in L^2(pdx)$. The first inclusion can be written as $\sqrt{\varrho} q'/q \in L^2(pdx)$ since $p\varrho(q')^2/q^2 = (q')^2/q$. On the open set U_q we have

$$(\sqrt{\varrho})' = \frac{1}{2}\frac{\varrho'}{\sqrt{\varrho}} = \frac{1}{2}\frac{1}{\sqrt{\varrho}}\left(\frac{q'p - qp'}{p^2}\right) = \frac{\sqrt{\varrho}}{2}\left(\frac{q'}{q} - \frac{p'}{p}\right).$$

It remains to observe that $\sqrt{\varrho} p'/p \in L^2(pdx)$ and $\sqrt{\varrho} q'/q \in L^2(pdx)$, as indicated above. In particular,

$$\int_{\mathbb{R}^1} |(\sqrt{\varrho})'(t)|^2 p(t)\,dt \leqslant \frac{1}{2}\int_{\mathbb{R}^1} \frac{q'(t)^2}{q(t)^2}q(t)\,dt + \frac{1}{2}\int_{\mathbb{R}^1} \varrho(t)\frac{p'(t)^2}{p(t)^2}p(t)\,dt. \tag{12.4.2}$$

In the general case, let Y be a closed hyperplane complementing the one-dimensional space $\mathbb{R}^1 k$. Denote by μ_0 and ν_0 the projections of the measures μ and ν on Y. We know that the measures μ and ν possess differentiable conditional measures μ^y and ν^y on the straight lines $y + \mathbb{R}^1 k$, $y \in Y$. It follows from our assumptions that on μ_0-almost all straight lines $y + \mathbb{R}^1 k$ we arrive at the condition of the one-dimensional case considered above. Then estimate (12.4.2) shows that Proposition 12.4.2 is applicable. □

In relation to form (12.4.1) the following questions arise.
A) The closability of $\mathcal{E}$.
B) The essential selfadjointness of L_μ on $\mathcal{FC}_b^\infty$.
C) The properties of the generated semigroup $T_t = e^{tL_\mu}$ in the closable case, where $\mathcal{E}(f,f) = -(L_\mu f, f)$.

12.4.6. Theorem. *Let $\{e_n\}$ be an orthonormal basis in H such that one has $e_n \in j(X^*)$. Suppose that the measure μ is differentiable along all vectors e_n and $\beta_{e_n}^\mu \in L^2(\mu)$. Then the form $\mathcal{E}$ is closable.*

PROOF. We have to prove the following: if a sequence of functions f_n from $\mathcal{FC}_b^\infty$ converges to zero in $L^2(\mu)$ and is Cauchy in the norm $\|\cdot\|_Q$, then one has $(f_n, f_n)_Q \to 0$. The fundamentality in the norm $\|\cdot\|_Q$ yields a mapping $\Lambda \in L^2(\mu, H)$ to which the mappings $\nabla_H f_n$ converge in $L^2(\mu, H)$. Let us verify that $\Lambda = 0$ μ-a.e. To this end, it suffices to show that $(\Lambda, e_i)_H = 0$ for each i. It is clear that $(\Lambda, e_i)_H$ is an $L^2(\mu)$-limit of the functions $\partial_{e_i} f_n$. For every function $\varphi \in \mathcal{FC}_b^\infty$ we have

$$\int_X (\Lambda(x), e_i)_H \varphi(x)\, \mu(dx) = \lim_{n\to\infty} \int_X \partial_{e_i} f_n(x) \varphi(x)\, \mu(dx)$$

$$= -\lim_{n\to\infty} \int_X f_n(x)[\partial_{e_i}\varphi(x) + \varphi(x)\beta_{e_i}^\mu(x)]\, \mu(dx) = 0$$

by convergence $f_n \to 0$ in $L^2(\mu)$. Since the function φ is arbitrary, our claim is proven. □

We observe that the partial forms $D(\mathcal{E}_k^\mu)$ can be defined without condition of differentiability along k. Then the closability of the forms $D(\mathcal{E}_{k_n}^\mu)$ for sufficiently large collection $\{k_n\}$ implies the closability of the full Dirichlet form. For example, in the case $X = \mathbb{R}^2$, the closability of the gradient Dirichlet form follows from the closability of the partial forms

$$\mathcal{E}_i(\varphi, \varphi) := \int_{\mathbb{R}^2} |\partial_{x_i}\varphi|^2\, \mu(dx), \quad \varphi \in C_0^\infty(\mathbb{R}^2),\ i = 1, 2.$$

Pugachev [**932**], [**934**] constructed an example showing that the closability of the full Dirichlet form on $\mathbb{R}^2$ does not imply the closability of the partial forms. The problem of existence of such examples had remained open for a long time. In this relation we observe that the following problem posed by Fukushima more than thirty years ago is still open: Does there exist a singular measure μ on $\mathbb{R}^2$ for which the gradient Dirichlet form (12.4.1) is closable? There is no such measures on the real line, so any measure on the plane with closable partial Dirichlet forms must be absolutely continuous.

Now we discuss Problem B under the assumptions of Theorem 12.4.6. We observe that the form $\mathcal{E}$ corresponds to the operator

$$L_\mu \varphi = \Delta_H \varphi + {}_{X^*}\langle \varphi', \beta_H^\mu\rangle_X, \quad \varphi \in \mathcal{FC}_b^\infty,$$

where $\Delta_H \varphi := \sum_{n=1}^\infty \partial_{e_n}^2 \varphi$ and φ' is the Gâteaux derivative. Indeed, let

$$\varphi(x) = \psi(l_1(x), \ldots, l_n(x)), \quad \text{where } \psi \in C_b^\infty(\mathbb{R}^n) \text{ and } l_i \in X^*.$$

Hence

$$\nabla_H \varphi = \sum_{i=1}^n \partial_{x_i}\psi(l_1, \ldots, l_n) j(l_i), \quad \varphi' = \sum_{i=1}^n \partial_{j(l_i)}\varphi\, l_i.$$

We may assume that the vectors $j(l_i)$ are orthonormal in H. Thus, one has

$$\int_X \bigl(\nabla_H \varphi(x), \nabla_H \varphi(x)\bigr)_H \mu(dx) = \int_X \sum_{i=1}^n \bigl(\partial_{j(l_i)} \varphi(x)\bigr)^2 \mu(dx)$$

$$= -\int_X \varphi(x) \Bigl(\sum_{i=1}^n \partial^2_{j(l_i)} \varphi(x) + \sum_{i=1}^n \partial_{j(l_i)} \varphi(x) \beta^\mu_{j(l_i)}(x)\Bigr) \mu(dx)$$

$$= -\int \varphi(x) \bigl(\Delta_H \varphi(x) + \langle \varphi'(x), \beta^\mu_H(x)\rangle\bigr) \mu(dx).$$

12.4.7. Theorem. *Let X be a Hilbert space.*

(i) *If β^μ_H has a continuous Fréchet derivative bounded on bounded sets, then L_μ is essentially selfadjoint on $C^2_0(X)$.*

(ii) *Suppose that there exist mappings b_n having continuous second order Fréchet derivatives bounded by a power of the norm at infinity, $b_n \to \beta^\mu_H$ in $L^2(\mu, X)$, and for some numbers C_n and C one has*

$$\bigl(b'_n(x)h, h\bigr)_X \leqslant C|h|^2_X, \quad \bigl(b_n(x), x\bigr)_X \leqslant C_n(1 + |x|^2_X), \quad \forall x, h \in X.$$

Then the operator L_μ is essentially selfadjoint on $C^2_b(X)$.

For proofs and other results in this direction, see Albeverio, Kondratiev, Röckner [**45**], Kondratiev, Tsycalenko [**630**], [**631**]. The essential selfadjointness of Dirichlet and Schrödinger operators on infinite dimensional spaces and related problems of uniqueness (such as Markov uniqueness, etc.) are studied in Eberle [**377**], Frolov [**455**], [**456**], Liskevich, Röckner [**729**], Liskevich et al. [**730**], Liskevich, Semenov [**731**], [**732**], Long, Simão [**735**], Röckner, Zhang [**963**], Song [**1071**], and Takeda [**1100**]–[**1102**].

12.5. The uniqueness problem for invariant measures

In Chapter 7 we discussed the problem of uniqueness of a measure with a given logarithmic gradient and in §12.3 we described connections between logarithmic gradients and drift coefficients of diffusion processes. Now we are ready to present an example of a linear equation (12.3.1) with a continuous linear operator b such that several different centered Gaussian measures are solutions to the equation $L^*\mu = 0$. At the same time this gives an example of a nondegenerate Gaussian diffusion with several different invariant probabilities. Such an example is impossible in $\mathbb{R}^n$, but becomes possible in infinite dimensions, which was first noticed by J. Zabczyk [**1210**], who constructed an example of a diffusion $d\xi_t = B\xi_t dt + v\, dw_t$ with a fixed vector v and an unbounded linear operator B for which the transition probabilities are positive on nonempty open sets, but there are many invariant probability measures for $(\xi_t)_{t \geqslant 0}$. Below we describe a simple example of a diffusion governed by the equation $d\xi_t = dw_t + 2^{-1} B\xi_t\, dt$ with a bounded linear operator B and possessing distinct invariant probability measures that are nondegenerate Gaussian measures. In addition, the process $(\xi_t)_{t \geqslant 0}$ is symmetric with respect to these Gaussian measures.

12.5. The uniqueness problem for invariant measures

12.5.1. Example. By Theorem 7.6.7, there exist centered Gaussian measures μ_1 and μ_2 on a separable Hilbert space X, a Hilbert space H densely embedded into X by a Hilbert–Schmidt operator, and an operator $B \in \mathcal{L}(X)$ such that B is the logarithmic gradient along H for both measures. Let w_t be the Wiener process in X associated with H (see §12.2). Then the stochastic equation

$$d\xi_t = dw_t + 2^{-1} B\xi_t\, dt, \quad \xi_0 = x,$$

has a unique strong solution ξ_t, for which the measures μ_1 and μ_2 are invariant, and the process ξ_t is symmetric with respect to both measures μ_1 and μ_2.

PROOF. Since the operator B is bounded, the indicated equation has a unique solution which is given by the following formula (see Da Prato, Zabczyk [**312**, Theorem 5.4]):

$$\xi_t = e^{tB/2}x + w_t + \frac{1}{2}\int_0^t Be^{(t-s)B/2}w_s\, ds = e^{tB/2}x + \int_0^t e^{(t-s)B/2}\, dw_s, \quad t \geqslant 0.$$

According to [**312**, Theorem 5.2], for any $t > 0$, the distribution P_t of the random vector ξ_t is a Gaussian measure with mean $e^{tB}x$ and covariance

$$(Q_t v, v)_X = \int_0^t \bigl(e^{(t-s)B/2} Q e^{(t-s)B^*/2}v, v\bigr)_X ds,$$

where Q is the covariance of the distribution of the random vector w_1. Since the function under the integral above is continuous and the operator Q is injective, the operators Q_t are also injective, i.e., P_t has full support. Since in the present case the transition semigroup maps the linear span of functions $\exp(il)$, where $l \in X^*$, to the same linear span, we obtain that every measure satisfying the equation $L^*\mu = 0$, where $L\varphi = \Delta_H\varphi + \langle\varphi', B\rangle$, is actually an invariant measure of the process ξ_t. Thus, both measures μ_1 and μ_2 are invariant measures, and then the process ξ_t is symmetric with respect to each of them by Theorem 12.3.2. $\square$

Now we continue our consideration which we began in §7.6 and, following the paper Bogachev, Röckner [**219**], give a description of all Radon probability measures μ for which $\mathcal{G}_{K,ac}^{\beta\mu} = \{\mu\}$ or $\mathcal{G}_K^{\beta\mu} = \{\mu\}$ (see the notation in §7.6). Though, the obtained description is very implicit and is expressed in terms of the properties of Dirichlet forms (so that in fact we are talking of an easy reformulation of the indicated equalities, which in some cases may be more convenient for verification). We shall use notation from §12.4. Let us recall that a Dirichlet form $\mathcal{E}$ on $L^2(\mu)$ is irreducible if the equality $\mathcal{E}(f,f) = 0$ implies that $f = \text{const}$.

12.5.2. Theorem. (i) *One has* $\mathcal{G}_{K,ac}^{\beta\mu} = \{\mu\}$ *if and only if the Dirichlet form* $\bigl(\mathcal{E}_K^\mu, D(\mathcal{E}_K^\mu)\bigr)$ *is irreducible.*

(ii) *One has* $\mathcal{G}_K^{\beta\mu} = \{\mu\}$ *if and only if for all* $\nu \in \mathcal{G}_K^{\beta\mu}$ *the Dirichlet forms* $\bigl(\mathcal{E}_K^\nu, D(\mathcal{E}_K^\nu)\bigr)$ *are irreducible.*

12.5.3. Remark. Since irreducibility is so crucial for the uniqueness problem, we mention here some of its characterizations in terms of the semigroup $(T_t)_{t>0}$ and the generator $(L_\mu, D(L_\mu))$ of the form $(\mathcal{E}_K^\mu, D(\mathcal{E}_K^\mu))$ on the space $L^2(\mu)$. The following conditions are equivalent:

(i) the Dirichlet form $(\mathcal{E}_K^\mu, D(\mathcal{E}_K^\mu))$ is irreducible;

(ii) the semigroup $(T_t)_{t>0}$ is irreducible, i.e., if a function $g \in L^2(\mu)$ is such that $T_t(gf) = gT_tf$ for all $t > 0$, $f \in L^2(\mu)$, then $g = \mathrm{const}$;

(iii) if $g \in L^2(\mu)$ is such that $T_tg = g$ for all $t > 0$, then $g = \mathrm{const}$;

(iv) for each $g \in L^2(\mu)$ the functions T_tg converge in $L^2(\mu)$ to the integral of g as $t \to +\infty$.

(v) if $u \in D(L_\mu)$ and $L_\mu u = 0$, then $u = \mathrm{const}$.

The proof is delegated to Exercise 12.7.3.

12.6. Existence of Gibbs measures

Here we consider an interesting application to statistical mechanics: a construction of Gibbs distributions, i.e., measures with given conditional distributions. To this end, we present a version of Theorem 7.5.4 on existence of measures with given logarithmic derivatives for the case of the space $X = \mathbb{R}^S$, where S is a countable set. Suppose that we are given an infinite symmetric matrix $J = (J_{s,t})_{s,t \in S}$ with nonnegative elements $J_{s,t}$, and a family $q = (q_s)_{s \in S}$ of positive numbers with $\sum_{s \in S} q_s < \infty$. We shall write $Jq \leqslant Cq$, where $C \in [0, \infty)$, if $\sum_{s \in S} q_s J_{s,t} \leqslant Cq_t$ for all $t \in S$. Denote by $l^1(q)$ the weighted l^1-space of all collections $x = (x_s)_{s \in S}$ for which

$$\|x\|_{l^1(q)} = \sum_{s \in S} q_s |x_s| < \infty.$$

We observe that the condition $Jq \leqslant Cq$ is fulfilled if

$$J_{t,s} = b_{t,s}c_{t,s}, \quad \text{where } b_{t,u}q_u \leqslant C_1 q_t \text{ and } \sum_{u \in S} c_{t,u} \leqslant C_2 \text{ for all } t, u \in S.$$

This holds for $S = \mathbb{Z}^d$ if $J_{n,j} = a(n-j)$, where a is an even nonnegative function such that $a(n) \leqslant \mathrm{const}\, q_n^2$ and $q_j q_{n-j} \leqslant \mathrm{const}\, q_n$ for all $n, j \in \mathbb{Z}^d$. For example, the latter is true if $q_n \sim |n|^{-r}$, $r > d$ and $a(n) \leqslant \mathrm{const}|n|^{-2r}$. Suppose that J defines a bounded operator on $l^1(q)$, i.e., for all $x \in l^1(q)$ and some $\lambda \geqslant 0$ one has

$$\sum_{s,t \in S} q_t J_{t,s} |x_s| \leqslant \lambda \sum_{t \in S} q_t |x_t|.$$

The minimal possible λ is the operator norm $\|J\|_{\mathcal{L}(l^1(q))}$. It is clear that the condition $Jq \leqslant Cq$ implies the estimate $\|J\|_{\mathcal{L}(l^1(q))} \leqslant C$. In order to have an example of suitable J and q we take the integer lattice $S = \mathbb{Z}^d$ in $\mathbb{R}^d$ and for $n \in \mathbb{Z}^d$ put $q_n = (|n|+1)^{-r}$, where $r > d$, i.e., $\sum_{n \in S} q_n < \infty$. For example, if

$$\|J\|_p^2 = \sup_{n \in \mathbb{Z}^d} \sum_{j \in \mathbb{Z}^d} (1+|j|)^{2p} J_{n,j+n}^2 < \infty, \quad \forall p \in \mathbb{N}, \qquad (12.6.1)$$

12.6. Existence of Gibbs measures

then $\|J\|_{\mathcal{L}(l^1(q))} < \infty$. Indeed, $J_{n,j} \leq \|J\|_r(1+|n-j|)^{-r}$ and

$$\sum_{j \in S} J_{n,j} q_j \leq C_1(|n/2|+1)^{-r} + \sum_{|j|<|n|/2} J_{n,j} q_j \leq C_1(|n/2|+1)^{-r}$$

$$+ C_2(|n/2|+1)^{-r} \leq C_3 q_n.$$

Let us introduce the weighted Banach space

$$X_0 = \Big\{ x \in \mathbb{R}^S \colon |x|_0 = \Big(\sum_{s \in S} q_s |x_s|^\alpha\Big)^{1/\alpha} < \infty \Big\},$$

where $\alpha \geq 2$ and $q = (q_s)_{s \in S} \in l_1$. We assume that

$$\|J\|_{\mathcal{L}(l^1(q))} \leq \lambda, \quad \sum_{t \in S} J_{s,t} \leq \lambda.$$

Suppose that we are given a family of functions $b = (b_s)_{s \in S}$ on X_0 continuous on all balls in X_0 with respect to the topology from X and satisfying the conditions $|b^s(x)| \leq C_s + C_s |x|_0^{p_s}$ and

$$x_s b^s(x) \leq c - (\lambda + \varepsilon)|x_s|^\alpha + \sum_{t \in S} J_{s,t} |x_t|^\alpha \tag{12.6.2}$$

with some positive numbers c, ε, C_s, p_s.

12.6.1. Theorem. *Suppose that S is the union of increasing finite sets S_k, where $k \in \mathbb{N}$, and that we are given continuously differentiable functions G_k on $\mathbb{R}^{S_k}$ such that $b^i = \partial_{x_i} G_k$ on $\mathbb{R}^{S_k}$ for all $i \in S_k$. Then, there exists a probability measure μ on X_0 such that for every n the function b^n is the logarithmic derivative of μ along e_n.*

12.6.2. Remark. Let $J_{n,j} = a(n-j)$, where a is an even nonnegative function, $q_n \sim |n|^{-r}$, $r > d$ and $a(n) \leq \text{const}|n|^{-2r}$. Then in the above theorem the estimates on $|b^s(x)|$ can be omitted, but in that case the assertions will be weaker: the function b^n will be the logarithmic derivative of μ along the vector e_n with respect to $\text{Lip}_0(X_0)$.

Let us apply this theorem to the classical spin system on the lattice $\mathbb{Z}^d$ with the configuration space $\mathbb{R}^S$, $S = \mathbb{Z}^d$, possessing the formal energy functional

$$E(x) = \sum_{n \in S} V_n(x_n) + \sum_{n,j \in S} W_{n,j}(x_n, x_j),$$

where $W_{n,j}(x_n, x_j) = W_{n,j}(x_j, x_n)$ and $W_{n,n} = 0$. We shall assume that the functions V_n and $W_{n,j}$ are continuously differentiable and satisfy the following estimates:

$$|W_{n,j}(x_n, x_j)| \leq J_{n,j}(1 + |x_n|^\alpha + |x_j|^\alpha), \tag{12.6.3}$$

$$|\partial_{x_n} W_{n,j}(x_n, x_j)| \leq J_{n,j}(1 + |x_n|^{\alpha-1} + |x_j|^{\alpha-1}), \tag{12.6.4}$$

$$x_n \partial_{x_n} V_n(x_n) \leq C - M|x_n|^\alpha, \tag{12.6.5}$$

where $J_{n,j} \geq 0$, $C, M > 0$.

12.6.3. Example. Suppose that we have (12.6.3), (12.6.4), (12.6.5), where $\alpha \geqslant 2$ and $J = (J_{n,j})_{n,j \in S}$ is a symmetric matrix such that there exists a family of positive numbers $q = (q_s)_{s \in S}$ for which $\sum_{s \in S} q_s < \infty$ and $\|J\|_{\mathcal{L}(l^1(q))} \leqslant \lambda < M/3$, $\sum_{n \in S} J_{n,j} \leqslant \lambda$ for all j. Then, there exists a probability measure μ concentrated on X_0 such that μ is differentiable along all standard unit vectors $e_n \in \mathbb{R}^S$ with respect to the class $\mathrm{Lip}_0(X_0)$ and

$$\beta_{e_n}^\mu(x) = \partial_{x_n} V_n(x_n) + \sum_{j \in S} \partial_{x_n} W_{n,j}(x_n, x_j).$$

In particular, the regular conditional measures for μ on the straight lines of the form $\mathbb{R}^1 e_n + y$, $y \in \Pi_n := \{x \colon x_n = 0\}$, possess continuously differentiable densities $p(x_n | x_n^c)$, for which

$$\partial_{x_n} p(x_n | x_n^c) / p(x_n | x_n^c) = \beta_{e_n}^\mu(x), \quad x = (x_n, x_n^c),$$

i.e., one has the equalities

$$p(x_n | x_n^c) = \Theta(x_n^c) \exp\left[V_n(x_n) + \sum_{j \in S} W_{n,j}(x_n, x_j)\right], \qquad (12.6.6)$$

where $\Theta(x_n^c)$ are norming constants.

PROOF. The function

$$b_n(x) = \partial_{x_n} V_n(x_n) + \sum_{j \in S} \partial_{x_n} W_{n,j}(x_n, x_j)$$

is continuous on all balls in X_0 since the series converges uniformly on every ball in X_0 due to (12.6.4) and our condition on J. We have $M = 3\lambda + \varepsilon$, where $\varepsilon > 0$. The inequality $z \leqslant 1 + z^\alpha/\alpha$ for $z \geqslant 0$ yields $|x_n| |x_j|^{\alpha-1} \leqslant |x_n|^\alpha/\alpha + |x_j|^\alpha$, whence $|x_n| + |x_n|^\alpha + |x_n| |x_j|^{\alpha-1} \leqslant 1 + 2|x_n|^\alpha + |x_j|^\alpha$. By the estimate $\sum_{j \in S} J_{n,j} \leqslant \lambda$ we obtain

$$x_n b_n(x) \leqslant C - (3\lambda + \varepsilon)|x_n|^\alpha + \sum_{j \in S} J_{n,j}(1 + 2|x_n|^\alpha + |x_j|^\alpha)$$

$$\leqslant C + \lambda - (\lambda + \varepsilon)|x_n|^\alpha + \sum_{j \in S} J_{n,j} |x_j|^\alpha,$$

i.e., (12.6.2) is fulfilled. Let $S_k := \{s \in S \colon |s| \leqslant k\}$, $k \in \mathbb{N}$. The restriction of $(b_i)_{i \in S_k}$ to $\mathbb{R}^{S_k}$ is the gradient of the continuously differentiable function $E_k(x) := \sum_{s \in S_k} V(x_s) + \sum_{n,j \in S} W_{n,j}(x_n, x_j)$. Hence Theorem 12.6.1 applies. We observe that the series in (12.6.6) converges uniformly on all balls in X_0 according to (12.6.3). □

12.7. Comments and exercises

Applications of the Malliavin calculus to hypoellipticity and related problems are discussed in Anulova et al. [74], Bally [90], [91], Bell [109], Bell, Mohammed [114]–[116], Bernard [144], Bismut [149]–[152], Cattiaux [267]–[269], Cattiaux, Mesnager [272], Chen, Zhou [282], Coquio, Gravereaux [291], Diebolt [349], Duc, Nualart, Sanz [372],

Heya [**535**], [**536**], Ikeda, Watanabe [**570**], Kusuoka [**681**], [**683**], [**685**], Kusuoka, Stroock [**687**], Ocone [**864**], Stroock [**1082**]–[**1085**], Takeuchi [**1103**], [**1104**], Taniguchi [**1108**], Vârsan [**1169**], Veretennikov [**1171**], [**1172**], Watanabe [**1183**], Williams [**1193**]. For results on smoothness of conditional densities with applications to nonlinear filtering and partial hypoellipticity, see Bismut, Michel [**157**], Chaleyat-Maurel [**277**], Elliott, Kohlmann [**382**], Florchinger [**434**], [**435**], Kusuoka, Stroock [**690**], Michel [**811**], [**812**], Michel, Pardoux [**813**], Nualart, Zakai [**859**], Ocone [**864**], Pardoux [**874**], Üstünel [**1145**]; applications in financial mathematics are discussed in the books Carmona, Tehranchi [**266**], Malliavin, Thalmaier [**774**] and the papers Bally, Caramellino, Zanette [**94**], Bavouzet-Morel, Messaoud [**103**], Biagini et al. [**147**], Bouchard, Ekeland, Touzi [**242**], Davis, Johansson [**332**], Detemple, Garcia, Rindisbacher [**343**], El-Khatib, Privault [**381**], Ewald [**401**], [**402**], Ewald, Zhang [**404**], Forster, Lütkebohmert, Teichmann [**444**], Fournié et al. [**446**], [**447**], Gobet, Kohatsu-Higa [**486**], Imkeller [**573**], Kohatsu-Higa, Montero [**621**], León et al. [**715**], Montero, Kohatsu-Higa [**816**], Petrou [**884**], Temam [**1116**].

Applications to integral equations and stochastic partial differential equations are discussed in the works on the Malliavin calculus cited in Chapter 9 and also in Ankirchner, Imkeller, Dos Reis [**71**], Bally, Pardoux [**96**], Carmona, Nualart [**264**], Hirsch [**544**], Márquez-Carreras [**779**], Márquez-Carreras, Sarrà [**780**], Nualart [**850**], Nualart, Quer-Sardanyons [**852**], Ocone [**865**], Pardoux, Zhang [**875**], Rovira, Sanz-Solé [**972**], Sanz-Solé [**982**], [**983**], Sanz-Solé, Torrecilla-Tarantino [**985**], Suzuki [**1096**].

Delay equations are considered in Bell, Mohammed [**114**], [**116**], Vârsan [**1170**], Yan, Mohammed [**1199**].

For diverse applications of the Malliavin calculus, see Airault, Malliavin, Viens [**29**], Bally, Bavouzet, Messaoud [**93**], Bally, Gyöngy, Pardoux [**95**], Caballero, Fernández, Nualart [**257**], Decreusefond [**337**], Di Nunno, Øksendal, Proske [**348**], Eckmann, Hairer [**378**], Flandoli, Tudor [**433**], Gloter, Gobet [**484**], Gobet [**485**], Gobet, Munos [**487**], Guérin, Méléard, Nualart [**509**], Hayashi [**525**], Kohatsu-Higa [**619**], Kohatsu-Higa, Pettersson [**622**], Kohatsu-Higa, Yasuda [**623**], Kusuoka [**672**], Kusuoka, Yoshida [**694**], Nourdin, Peccati [**839**]–[**841**], Nourdin, Peccati, Reinert [**842**], Nourdin, Peccati, Réveillac [**843**], Nualart, Ortiz-Latorre [**851**], Privault, Réveillac [**925**], Privault, Wei [**926**], Sakamoto, Yoshida [**979**], Uchida, Yoshida [**1125**], Watanabe [**1185**], Yoshida [**1201**]–[**1206**].

Sobolev classes and gradients in the framework of Dirichlet forms are discussed in Bouleau [**243**], Bouleau, Hirsch [**247**], Fukushima, Oshima, Takeda [**466**].

Concerning infinite dimensional elliptic and parabolic equations connected with measures and diffusion processes, in addition to the already mentioned works see Airault [**15**], Albeverio, Høegh-Krohn [**41**], [**42**], Albeverio, Høegh-Krohn, Streit [**43**], Antoniouk, Antoniouk [**72**], [**73**], Barbu et al. [**98**], Belopol'skaya, Daletsky [**121**], Belyaev [**123**], [**122**], Bentkus

[**130**], [**131**], Berezanskiĭ [**139**], Berezansky, Kondratiev [**140**], Bogachev, Da Prato, Röckner [**195**], Cattiaux, Fradon [**270**], Cattiaux, Léonard [**271**], Cerrai [**275**], Cruzeiro [**295**], Da Prato [**305**], [**306**], Da Prato, Zabczyk [**315**], Daletskiĭ [**316**], Elson [**388**], Föllmer [**438**], Gross [**504**], Hino [**539**], Kirillov [**607**], [**608**], [**611**], [**612**], Kondratiev [**628**], Kuo [**667**], [**670**], Lobuzov [**733**], Ługiewicz, Zegarliński [**737**], Malliavin [**761**], Norin [**831**], Piech [**893**]–[**898**], Röckner [**959**], [**960**], Shigekawa [**1018**], Uglanov [**1129**], [**1132**], Zabczyk [**1211**].

Elliptic and parabolic equations for measures arising as equations for stationary distributions and transition probabilities of diffusion processes are studied in Albeverio, Bogachev, Röckner [**35**], Bogachev, Krylov, Röckner [**205**]–[**207**], Bogachev, Röckner [**217**], [**219**]–[**224**], Bogachev, Röckner, Shaposhnikov [**226**], Bogachev, Röckner, Stannat [**229**], Bogachev, Röckner, Wang [**230**], [**231**], Bogachev, Röckner, Zhang [**232**], Cruzeiro, Malliavin [**298**], Metafune, Pallara, Rhandi [**802**]. A survey of this subject is given in Bogachev, Krylov, Röckner [**208**]. The series of papers Kirillov [**607**]–[**615**] is devoted to applications of logarithmic gradients in mathematical physics. In relation to differential operators acting on differentiable measures we recall that according to the well-known theorem of Peetre, any linear operator $L\colon C^\infty(\mathbb{R}^n) \to C^\infty(\mathbb{R}^n)$ with the property $\operatorname{supp} Lf \subset \operatorname{supp} f$ is a differential operator of a locally finite order. A partial generalization of this result to infinite dimensional Hilbert spaces (for operators acting on smooth functions or smooth measures) was obtained in Smolyanov [**1049**].

Belopol'skaya [**119**], Belopol'skaya, Dalecky [**120**], Belopol'skaya [**117**], [**118**] investigate equations which are satisfied for the logarithmic gradients of transition probabilities of diffusion processes and for the logarithmic gradients of densities of their stationary distributions. Some nonlinear parabolic and elliptic equations can be solved in this way.

Exercises

12.7.1. Prove Proposition 12.3.1.

12.7.2. Provide the details of the proof of Proposition 12.4.2.

12.7.3. Justify Remark 12.5.3.

12.7.4. Let μ be a Radon probability measure on a locally convex space X, let H be a densely embedded separable Hilbert space, and let a measure μ be differentiable along all vectors H and have the property that for every $h \in H$ there exists $c_h > 0$ such that $\exp(c_h|\beta_h^\mu|) \in L^1(\mu)$. Let $\{e_n\}$ be a basis in H and H_n the linear span of $e_1, \ldots, e_n$. Prove that every function $f \in W^\infty(\mu)$ has a version such that the function $h \mapsto f(x+h)$ is infinitely differentiable on all H_n for each x.

HINT: Use the existence of positive continuous densities of the conditional measures on the subspace $x + H_n$ and the fact that the assertion is true in the finite dimensional case.

References

[1] Accardi L., Bogachev V.I. *The Ornstein–Uhlenbeck process associated with the Lévy Laplacian and its Dirichlet form.* Probab. Math. Stat. 1997. V. 17, №1. P. 95–114. [*275*]^{1,2}

[2] Adams R.A. Sobolev spaces. Academic Press, New York, 1975; 268 p. [*48, 66*]

[3] Adams R.A., Fournier J.J.F. Sobolev spaces. 2nd ed. Academic Press, New York, 2003; xiii+305 p. [*50, 56, 57, 66*]

[4] Agafontsev B.V., Bogachev V.I. *Asymptotic properties of polynomials in Gaussian random variables.* Dokl. Ross. Akad. Nauk. 2009. V. 429, №1. P. 151–154 (in Russian); English transl.: Dokl. Math. 2009. V. 80, №3. P. 806–809. [*278*]

[5] Agrachev A., Kuksin S., Sarychev A., Shirikyan A. *On finite-dimensional projections of distributions for solutions of randomly forced 2D Navier–Stokes equations.* Ann. Inst. H. Poincaré Probab. Statist. 2007. V. 43. P. 399–415. [*325*]

[6] Aida S. D^∞-*cohomology groups and* D^∞-*maps on submanifolds of the Wiener space.* J. Funct. Anal. 1992. V. 107, №2. P. 289–301. [*399*]

[7] Aida S. *Certain gradient flows and submanifolds in Wiener spaces.* J. Funct. Anal. 1993. V. 112, №2. P. 346–372. [*366, 399*]

[8] Aida S. *On the Ornstein–Uhlenbeck operators on Wiener–Riemann manifolds.* J. Funct. Anal. 1993. V. 116, №1. P. 83–110. [*399*]

[9] Aida S. *Sobolev spaces over loop groups.* J. Funct. Anal. 1995. V. 127, №1. P. 155–172. [*398*]

[10] Aida S., Kusuoka S., Stroock D. *On the support of Wiener functionals.* Asymptotic problems in probability theory: Wiener functionals and asymptotics (Sanda/Kyoto, 1990), pp. 3–34. Pitman Research Notes Math. Ser., 284, Longman Sci. Tech., Harlow, 1993. [*326*]

[11] Aida S., Masuda T., Shigekawa I. *Logarithmic Sobolev inequalities and exponential integrability.* J. Funct. Anal. 1994. V. 126, №1. P. 83–101. [*252, 274*]

[12] Aida S., Shigekawa I. *Logarithmic Sobolev inequalities and spectral gaps: perturbation theory.* J. Funct. Anal. 1994. V. 126, №2. P. 448–475. [*252, 274*]

[13] Aida S., Stroock D. *Moment estimates derived from Poincaré and logarithmic Sobolev inequalities.* Math. Research Letters. 1994. V. 1, №1. P. 75–86. [*252, 274*]

[14] Aimar H., Forzani L., Scotto R. *On Riesz transforms and maximal functions in the context of Gaussian harmonic analysis.* Trans. Amer. Math. Soc. 2007. V. 359, №5. P. 2137–2154. [*238*]

[15] Airault H. *Projection of the infinitesimal generator of a diffusion.* J. Funct. Anal. 1989. V. 85, №2. P. 353–391. [*425*]

[16] Airault H. *Differential calculus on finite codimensional submanifold of the Wiener space.* J. Funct. Anal. 1991. V. 100. P. 291–316. [*399*]

[17] Airault H. *Formules de Cameron–Martin–Girsanov–Kusuoka–Buckdahn. Application à l'etude des retractes.* C. R. Acad. Sci. Paris, sér. 1. 1993. T. 316, №7. P. 727–731. [*366*]

---

[1] Numbers in italics in square brackets indicate the page numbers where the corresponding work is cited.

[2] Article titles are given in italics to distinguish them from book titles.

[18] Airault H., Bogachev V. *Realization of Virasoro unitarizing measures on the set of Jordan curves.* C. R. Math. Acad. Sci. Paris. 2003. T. 336, №5. P. 429–434. [*399*]

[19] Airault H., Bogachev V.I., Lescot P. *Finite-dimensional sections of functions from fractional Sobolev classes on infinite-dimensional spaces.* Dokl. Ross. Akad. Nauk. 2003. V. 391, №3. P. 320–323 (in Russian); English transl.: Dokl. Math. 2003. V. 68, №1. P. 71–74. [*274*]

[20] Airault H., Malliavin P. *Intégration géometrique sur l'espaces de Wiener.* Bull. Sci. Math. (2). 1988. V. 112, №1. P. 3–52. [*303, 324, 325*]

[21] Airault H., Malliavin P. *Integration on loop groups II. Heat equation for the Wiener measure.* J. Funct. Anal. 1992. V. 104, №1. P. 71–109. [*398*]

[22] Airault H., Malliavin P. *Integration by parts formulas and dilatation vector fields on elliptic probability spaces.* Probab. Theory Relat. Fields. 1996. V. 106, №4. P. 447–494. [*398*]

[23] Airault H., Malliavin P. *Unitarizing probability measures for representations of Virasoro algebra.* J. Math. Pures Appl. (9). 2001. V. 80, №6. P. 627–667. [*400*]

[24] Airault H., Malliavin P. *Quasi-invariance of Brownian measures on the group of circle homeomorphisms and infinite-dimensional Riemannian geometry.* J. Funct. Anal. 2006. V. 241, №1. P. 99–142. [*400*]

[25] Airault H., Malliavin P., Ren J. *Smoothness of stopping times of diffusion processes.* J. Math. Pures Appl. (9). 1999. V. 78, №10. P. 1069–1091. [*324*]

[26] Airault H., Malliavin P., Ren J. *Geometry of foliations on the Wiener space and stochastic calculus of variations.* C. R. Math. Acad. Sci. Paris. 2004. T. 339, №9. P. 637–642. [*399*]

[27] Airault H., Malliavin P., Thalmaier A. *Support of Virasoro unitarizing measures.* C. R. Math. Acad. Sci. Paris. 2002. V. 335, №7. P. 621–626. [*399*]

[28] Airault H., Malliavin P., Thalmaier A. *Canonical Brownian motion on the space of univalent functions and resolution of Beltrami equations by a continuity method along stochastic flows.* J. Math. Pures Appl. (9). 2004. V. 83, №8. P. 955–1018. [*399*]

[29] Airault H., Malliavin P., Viens F. *Stokes formula on the Wiener space and n-dimensional Nourdin–Peccati analysis.* J. Funct. Anal. 2010. V. 258, №5. P. 1763–1783. [*322, 425*]

[30] Airault H., Ren J., Zhang X. *Smoothness of local times of semimartingales.* C. R. Acad. Sci. Paris, sér. I Math. 2000. T. 330, №8. P. 719–724. [*324*]

[31] Airault H., Van Biesen J. *Géometrie riemannienne en codimension finie sur l'espace de Wiener.* C. R. Acad. Sci. Paris, sér. 1. 1990. T. 311, №2. P. 125–130. [*246, 399*]

[32] Airault H., Van Biesen J. *Processus d'Ornstein–Uhlenbeck sur une sous variété de codimension finie de l'espace de Wiener.* Bull. Sci. Math. (2). 1991. V. 115, №2. P. 195–210. [*399*]

[33] Akiyama H. *Geometric aspects of Malliavin's calculus on vector bundles.* J. Math. Kyoto Univ. 1986. V. 26, №4. P. 673–696. [*406*]

[34] Alberti G. *A Lusin type theorem for gradients.* J. Funct. Anal. 1991. V. 100, №1. P. 110–118. [*250, 257*]

[35] Albeverio S., Bogachev V.I., Röckner M. *On uniqueness of invariant measures for finite and infinite dimensional diffusions.* Comm. Pure Appl. Math. 1999. V. 52. P. 325–362. [*413, 426*]

[36] Albeverio S., Daletskii A., Kondratiev Y. *Stochastic analysis on product manifolds: Dirichlet operators on differential forms.* J. Funct. Anal. 2000. V. 176, №2. P. 280–316. [*399*]

[37] Albeverio S., Daletskii A., Kondratiev Y., Lytvynov E. *Laplace operators in de Rham complexes associated with measures on configuration spaces.* J. Geom. Phys. 2003. V. 47, №2-3. P. 259–302. [*399*]

[38] Albeverio S., Daletskii A., Lytvynov E. *Laplace operators on differential forms over configuration spaces.* J. Geom. Phys. 2001. V. 37, №1-2. P. 15–46. [*399*]

[39] Albeverio S., Daletsky Yu.L., Kondratiev Yu.G., Streit L. *Non-Gaussian infinite-dimensional analysis.* J. Funct. Anal. 1996. V. 138, №2. P. 311–350. [*276*]

[40] Albeverio S., Fukushima M., Hansen W., Ma Z.M., Röckner M. *An invariance result for capacities on Wiener space.* J. Funct. Anal. 1992. V. 106, №1. P. 35–49. [*276*]

[41] Albeverio S., Høegh-Krohn R. *Dirichlet forms and diffusion processes on rigged Hilbert spaces.* Z. Wahr. theor. verw. Geb. 1977. B. 40. S. 1–57. [*70, 71, 111, 155, 190, 223, 425*]

[42] Albeverio S., Høegh-Krohn R. *Hunt processes and analytic potential theory on rigged Hilbert spaces.* Ann. Inst. H. Poincaré (B). 1977. V. 13. P. 269–291. [*190, 223, 425*]

[43] Albeverio S., Høegh-Krohn R., Streit L. *Energy forms, Hamiltonians, and distorted Brownian paths.* J. Math. Phys. 1977. V. 18, №5. P. 907–917. [*425*]

[44] Albeverio S., Kondratiev Yu., Pasurek T., Röckner M. *Existence of and a priori estimates for Euclidean Gibbs states.* Tr. Mosk. Mat. Obs. 2006. V. 67. P. 3–103 (in Russian); transl.: Trans. Moscow Math. Soc. 2006. P. 1–85. [*224*]

[45] Albeverio S., Kondratiev Yu.G., Röckner M. *Dirichlet operators via stochastic analysis.* J. Funct. Anal. 1995. V. 128, №1. P. 102–138. [*274, 420*]

[46] Albeverio S., Kondratiev Yu.G., Röckner M. *Analysis and geometry on configuration spaces: the Gibbsian case.* J. Funct. Anal. 1998. V. 157, №1. P. 242–291. [*399*]

[47] Albeverio S., Kondratiev Yu.G., Röckner M., Tsikalenko T.V. *A priori estimates for symmetrizing measures and their applications to Gibbs states.* J. Funct. Anal. 2000. V. 171, №2. P. 366–400. [*224*]

[48] Albeverio S., Kosyak A. *Quasiregular representations of the infinite-dimensional Borel group.* J. Funct. Anal. 2005. V. 218, №2. P. 445–474. [*400*]

[49] Albeverio S., Kusuoka S., Röckner M. *On partial integration in infinite dimensional Dirichlet forms.* J. London Math. Soc. 1990. V. 42. P. 122–136. [*103*]

[50] Albeverio S., Liang S., Zegarlinski B. *Remark on the integration by parts formula for the ϕ_3^4-quantum field model.* Infin. Dim. Anal. Quantum Probab. Relat. Top. 2006. V. 9, №1. P. 149–154. [*132*]

[51] Albeverio S., Ma Z.M., Röckner M. *Partitions of unity in Sobolev spaces over infinite-dimensional state spaces.* J. Funct. Anal. 1997. V. 143, №1. P. 247–268. [*275*]

[52] Albeverio S., Röckner M. *Classical Dirichlet forms on topological vector spaces: construction of an associated diffusion process.* Probab. Theory Relat. Fields. 1989. V. 83. P. 405–434. [*275*]

[53] Albeverio S., Röckner M. *Classical Dirichlet forms on topological vector spaces: closability and a Cameron–Martin formula.* J. Funct. Anal. 1990. V. 88, №2. P. 395–437. [*172, 229, 275*]

[54] Albeverio S., Röckner M. *Stochastic differential equations in infinite dimensions: solutions via Dirichlet forms.* Probab. Theory Relat. Fields. 1991. V. 89, №3. P. 347–386. [*236, 275, 415*]

[55] Albeverio S., Smorodina N.V. *A distributional approach to multiple stochastic integrals and transformations of the Poisson measure.* Acta Appl. Math. 2006. V. 94. P. 1–19; – II. ibid. 2008. V. 102. P. 319–343. [*399*]

[56] Albeverio S., Steblovskaya V. *Asymptotics of infinite-dimensional integrals with respect to smooth measures. I.* Infin. Dim. Anal. Quantum Probab. Relat. Top. 1999. V. 2, №4. P. 529–556. [*324*]

[57] Albeverio S., Yoshida M.W. *H-C^1 maps and elliptic SPDEs with polynomial and exponential perturbations of Nelson's Euclidean free field.* J. Funct. Anal. 2002. V. 196, №2. P. 265–322. [*366*]

[58] Aleksandrova D.E. *Convergence of triangular transformations of measures.* Teor. Verojatn. i Primen. 2005. V. 50, №1. P. 145–150 (in Russian); English transl.: Theory Probab. Appl. 2005. V. 50, №1. P. 113–118. [*360, 367*]

[59] Aleksandrova D.E., Bogachev V.I., Pilipenko A.Yu. *On the convergence of induced measures in variation.* Mat. Sbornik. 1999. V. 190, N 9. P. 3–20 (in Russian); English transl.: Sb. Math. 1999. V. 190, №9–10. P. 1229–1245. [*325*]

[60] Ambrosio L. *Lecture notes on optimal transport problems.* Lecture Notes in Math. 2003. V. 1812. P. 1–52. [*363*]

[61] Ambrosio L. *Transport equation and Cauchy problem for non-smooth vector fields.* Lecture Notes in Math. 2008. V. 1927. P. 1–41. [*366*]

[62] Ambrosio L., Crippa G. *Existence, uniqueness, stability and differentiability properties of the flow associated to weakly differentiable vector fields.* UMI Lecture Notes, V. 5, pp. 3–57. Springer, Berlin, 2008. [*366*]

[63] Ambrosio L., Durand-Cartagena E. *Metric differentiability of Lipschitz maps defined on Wiener spaces.* Rend. Circ. Mat. Palermo (2). 2009. V. 58, №1. P. 1–10. [*366*]

[64] Ambrosio L., Figalli A. *On flows associated to Sobolev vector fields in Wiener spaces: an approach à la DiPerna–Lions.* J. Funct. Anal. 2009. V. 256, №1. P. 179–214. [*341, 366*]

[65] Ambrosio L., Fusco N., Pallara D. Functions of bounded variation and free discontinuity problems. Clarendon Press, Oxford University Press, New York, 2000; xviii+434 p. [*53, 66*]

[66] Ambrosio L., Gigli N., Savaré G. Gradient flows in metric spaces and in the space of probability measures. Birkhäuser, Basel, 2005; viii+333 p. [*366*]

[67] Ambrosio L., Miranda M., Maniglia S., Pallara D. *Towards a theory of BV functions in abstract Wiener spaces.* Physica D: Nonlin. Phenom. 2010. V. 239. [*275*]

[68] Ambrosio L., Miranda M. (Jr.), Maniglia S., Pallara D. *BV functions in abstract Wiener spaces.* J. Funct. Anal. 2010. V. 258, №3. P. 785–813. [*275*]

[69] Ambrosio L., Tilli P. Topics on analysis in metric spaces. Oxford University Press, Oxford, 2004; viii+133 p. [*274*]

[70] Andersson L., Driver B.K. *Finite-dimensional approximations to Wiener measure and path integral formulas on manifolds.* J. Funct. Anal. 1999. V. 165, №2. P. 430–498. [*398*]

[71] Ankirchner S., Imkeller P., Dos Reis G. *Classical and variational differentiability of BSDEs with quadratic growth.* Electron. J. Probab. 2007. V. 12, №53. P. 1418–1453. [*425*]

[72] Antoniouk A.V., Antoniouk A.V. *Smoothing properties of semigroups for Dirichlet operators of Gibbs measures.* J. Funct. Anal. 1995. V. 127, №2. P. 390–430. [*132, 425*]

[73] Antoniouk A.V., Antoniouk A.V. *Nonlinear estimates approach to the regularity properties of diffusion semigroups.* Nonlinear analysis and applications: to V. Lakshmikantham on his 80th birthday. V. 1, 2, pp. 165–226. Kluwer Acad. Publ., Dordrecht, 2003. [*425*]

[74] Anulova S.V., Veretennikov A.Yu., Krylov N.V., Liptser R.Sh., Shiryaev A.N. Stochastic calculus. Current problems in mathematics. Fundamental directions, V. 45. P. 5–253. Itogi Nauki i Tekhniki, Akad. Nauk SSSR, Vsesoyuz. Inst. Nauchn. i Tekhn. Inform., Moscow, 1989 (in Russian); English transl.: Stochastic calculus. Probability theory, III. Encyclopaedia Math. Sci., 45, Springer, Berlin, 1998; 253 p. [*323, 424*]

[75] Applebaum D. *Universal Malliavin calculus in Fock and Lévy–Itô spaces.* Commun. Stoch. Anal. 2009. V. 3, №1. P. 119–141. [*323*]

[76] Arai A., Mitoma I. *De Rham–Hodge–Kodaira decomposition in infinite dimension.* Math. Ann. 1991. B. 291, №1. S. 51–73. [*399*]

[77] Araki H. *Hamiltonian formalism and the canonical commutation relations in quantum field theory.* J. Math. Phys. 1960. V. 1. P. 492–504. [*104*]

[78] Araki H. *On representations of the canonical commutation relations.* Comm. Math. Phys. 1971. V. 20. P. 9–25. [*104*]

[79] Aronszajn N. *Differentiability of Lipchitzian mappings between Banach spaces.* Studia Math. 1976. V. 57, №2. P. 147–190. [*342, 353*]

[80] Asadian F. *Regularity of measures induced by solutions of infinite dimensional stochastic differential equations.* Rocky Mount. J. Math. 1997. V. 27, №2. P. 387–423. [*414*]

[81] Asadian F. *Subspaces of differentiability of measures generated by stochastic differential equations in Hilbert space.* Dyn. Contin. Discrete Impuls. Syst., Ser. A, Math. Anal. 2009. V. 16, №S1, Suppl. P. 98–103. [*414*]
[82] Averbuh V.I., Smolyanov O.G., Fomin S.V., *Generalized functions and differential equations in linear spaces. I, Differentiable measures.* Trudy Moskovsk. Matem. Ob. 1971. V. 24. P. 133–174 (in Russian); English transl.: Trans. Moscow Math. Soc. 1971. V. 24. P. 140–184. [*ix, 103, 155, 187*]
[83] Averbuh V.I., Smolyanov O.G., Fomin S.V., *Generalized functions and differential equations in linear spaces. II, Differentiable operators and their Fourier transforms.* Trudy Moskovsk. Matem. Ob. 1972. V. 27. P. 248–262 (in Russian); English transl.: Trans. Moscow Math. Soc. 1972. V. 27. P. 255–270. [*ix, 103*]
[84] Badrikian A. *Transformation of Gaussian measures.* Ann. Math. Blaise Pascal. 1996. Numero Special. P. 13–58. [*366*]
[85] Bakhtin Yu., Mattingly J.C. *Malliavin calculus for infinite-dimensional systems with additive noise.* J. Funct. Anal. 2007. V. 249, №2. P. 307–353. [*323*]
[86] Baklan V.V., Shatashvili A.D., *Transformations of Gaussian measures by nonlinear mappings in Hilbert space.* Dopovidi Akad. Nauk. Ukrain. RSR. 1965. №9. P. 1115–1117 (in Russian). [*366*]
[87] Bakry D. *Transformations de Riesz pour les semigroupes symétriques.* Lecture Notes in Math. 1985. V. 1123. P. 130–174 [*274*]
[88] Bakry D. *L'hypercontractivité et son utilization en théorie des semigroupes.* Lecture Notes in Math. 1994. V. 1581. P. 1–114. [*274*]
[89] Bakry D., Emery M. *Diffusions hypercontractives.* Lecture Notes in Math. 1985. V. 1123. P. 177–206. [*274, 372*]
[90] Bally V. *On the connection between the Malliavin covariance matrix and Hörmander's condition.* J. Funct. Anal. 1991. V. 96, №2. P. 219–255. [*424*]
[91] Bally V. *A stochastic ellipticity assumption.* Rev. Roumaine Math. Pures Appl. 1992. V. 37, №5. P. 353–361. [*424*]
[92] Bally V. *Lower bounds for the density of locally elliptic Itô processes.* Ann. Probab. 2006. V. 34, №6. P. 2406–2440. [*323*]
[93] Bally V., Bavouzet M.-P., Messaoud M. *Integration by parts formula for locally smooth laws and applications to sensitivity computations.* Ann. Appl. Probab. 2007. V. 17, №1. P. 33–66. [*425*]
[94] Bally V., Caramellino L., Zanette A. *Pricing and hedging American options by Monte Carlo methods using a Malliavin calculus approach.* Monte Carlo Methods Appl. 2005. V. 11, №2. P. 97–133. [*425*]
[95] Bally V., Gyöngy I., Pardoux E. *White noise driven parabolic SPDEs with measurable drifts.* J. Funct. Anal. 1994. V. 120. P. 484–510. [*425*]
[96] Bally V., Pardoux E. *Malliavin calculus for white noise driven parabolic SPDEs.* Potential Anal. 1998. V. 9, №1. P. 27–64. [*425*]
[97] Bally V., Saussereau B. *A relative compactness criterion in Wiener–Sobolev spaces and application to semi-linear stochastic PDEs.* J. Funct. Anal. 2004. V. 210, №2. P. 465–515. [*254*]
[98] Barbu V., Bogachev V.I., Da Prato G., Röckner M. *Weak solutions to the stochastic porous media equation via Kolmogorov equations: the degenerate case.* J. Funct. Anal. 2006. V. 237, №1. P. 54–75. [*425*]
[99] Barthe F., Cordero-Erausquin D., Fradelizi M. *Shift inequalities of Gaussian type and norms of barycenters.* Studia Math. 2001. V. 146, №3. P. 245–259. [*107*]
[100] Bass R.F., Cranston M. *The Malliavin calculus for pure jump processes and application for local time.* Ann. Probab. 1986. V. 14, №2. P. 490–532. [*323*]
[101] Baudoin F., Hairer M. *A version of Hörmander's theorem for the fractional Brownian motion.* Probab. Theory Relat. Fields. 2007. V. 139, №3-4. P. 373–395. [*323*]
[102] Baudoin F., Thieullen M. *Pinning class of the Wiener measure by a functional: related martingales and invariance properties.* Probab. Theory Relat. Fields. 2003. V. 127, №1. P. 1–36. [*366*]

[103] Bavouzet-Morel M.-P., Messaoud M. *Computation of Greeks using Malliavin's calculus in jump type market models.* Electron. J. Probab. 2006. V. 11, №10. P. 276–300. [*425*]

[104] Baxendale P. *Brownian motions in the diffeomorphism group. I.* Compos. Math. 1984. V. 53. P. 19–50. [*398*]

[105] Bell D. *A quasi-invariance theorem for measures on Banach spaces.* Trans. Amer. Math. Soc. 1985. V. 290, №2. P. 851–855. [*172*]

[106] Bell D. The Malliavin calculus. Wiley and Sons, NY, 1987; 105 p. [*323, 342, 405*]

[107] Bell D. *Transformations of measure on an infinite-dimensional vector space.* Seminar on Stochastic Processes (Vancouver, 1990), Progr. Probab. V. 24. P. 15–25. Birkhäuser Boston, Boston, 1991. [*342, 398*]

[108] Bell D.R. Degenerate stochastic differential equations and hypoellipticity. Longman, Harlow, 1995; xii+114 p. [*323, 405*]

[109] Bell D.R. *Stochastic differential equations and hypoelliptic operators.* Real and stochastic analysis. P. 9–42. Birkhäuser Boston, Boston, 2004. [*424*]

[110] Bell D. *Divergence theorems in path space.* J. Funct. Anal. 2005. V. 218, №1. P. 130–149. [*398*]

[111] Bell D. *Divergence theorems in path space. II. Degenerate diffusions.* C. R. Math. Acad. Sci. Paris. 2006. T. 342, №11. P. 869–872. [*398*]

[112] Bell D. *Quasi-invariant measures on the path space of a diffusion.* C. R. Math. Acad. Sci. Paris. 2006. T. 343, №3. P. 197–200. [*398*]

[113] Bell D. *Divergence theorems in path space. III. Hypoelliptic diffusions and beyond.* J. Funct. Anal. 2007. V. 251, №1. P. 232–253. [*398*]

[114] Bell D., Mohammed S.E.A. *The Malliavin calculus and stochastic delay equations.* J. Funct. Anal. 1991. V. 99, №1. P. 75–99. [*424, 425*]

[115] Bell D., Mohammed S.E.A. *Hypoelliptic parabolic operators with exponential degeneracies.* C. R. Acad. Sci. Paris, sér. 1. 1993. T. 317, №11. P. 1059–1064. [*424*]

[116] Bell D.R., Mohammed S.E.A. *Smooth densities for degenerate stochastic delay equations with hereditary drift.* Ann. Probab. 1995. V. 23, №4. P. 1875–1894. [*424, 425*]

[117] Belopol'skaya Ya.I. *Smooth measures and nonlinear equations of mathematical physics.* J. Math. Sci. 1996. V. 80, №6. P. 2226–2235. [*426*]

[118] Belopol'skaya Ya.I. *Burgers equation on a Hilbert manifold and the motion of incompressible fluid.* Methods Funct. Anal. Topology. 1999. V. 5, №4. P. 15–27. [*426*]

[119] Belopol'skaya Ya.I. *Smooth diffusion measures and their transformations.* Zap. Nauchn. Sem. S.-Peterburg. Otdel. Mat. Inst. Steklov. (POMI). 1999. V. 260. P. 31–49 (in Russian); English transl.: J. Math. Sci. (New York). 2002. V. 109, №6. P. 2047–2060. [*426*]

[120] Belopol'skaya Ya.I., Dalecky Yu.L. Stochastic equations and differential geometry. Transl. from the Russian. Kluwer Academic Publ., Dordrecht, 1990; xvi+260 p. (Russian ed.: Kiev, 1990). [*366, 398, 414, 426*]

[121] Belopol'skaya Ya.I., Daletsky Yu.L. *Smoothness of transition probabilities of Markov processes described by stochastic differential equations in Hilbert space.* Dopov./Dokl. Akad. Nauk Ukraini. 1994. №9. P. 40–45 (in Russian). [*425*]

[122] Belyaev A.A. *Integral representation of functions harmonic in a domain of Hilbert space.* Vestnik Moskov. Univ. Ser. I Mat. Mekh. 1981. №6. P. 44–47 (Russian); English transl.: Moscow Univ. Math. Bull. 1981. V. 36. №6. P. 54–58. [*425*]

[123] Belyaev A.A. *The mean value theorem for harmonic functions in a domain of Hilbert space.* Vestnik Moskov. Univ. Ser. I Mat. Mekh. 1982. №5. P. 32–35 (in Russian); Moscow Univ. Math. Bull. 1982. V. 37. №5. P. 38–42. [*425*]

[124] Ben Arous G. *Methodes de Laplace et de la phase stationaire sur l'espace de Wiener.* Stochastics. 1988. V. 25. P. 125–153. [*324*]

[125] Ben Arous G. *Devéloppement asymptotique du noyau de la chaleur hypoelliptique sur la diagonale.* Ann. Inst. Fourier. 1989. V. 39, №1. P. 73–99. [*405*]

[126] Ben Arous G., Léandre R. *Annulation plate du noyau de la chaleur.* C. R. Acad. Sci. Paris, sér. I Math. 1991. T. 312, №6. P. 463–464. [*321*]

[127] Bendikov A. *Potential theory on infinite-dimensional abelian groups.* De Gruyter, Berlin, 1995; vi+184 p. [*399*]

[128] Bendikov A., Saloff-Coste L. *On the hypoellipticity of sub-Laplacians on infinite dimensional compact groups.* Forum Math. 2003. V. 15, №1. P. 135–163. [*399*]

[129] Bendikov A., Saloff-Coste L. *Spaces of smooth functions and distributions on infinite-dimensional compact groups.* J. Funct. Anal. 2005. V. 218, №1. P. 168–218. [*399*]

[130] Bentkus V.Ju. *Existence and uniqueness of the solution of the Poisson equation for generalized measures on an infinite-dimensional space.* Mat. Zametki. 1976. V. 20, №6. 825–834 (in Russian); English transl.: Math. Notes. 1976. V. 20, №5–6. P. 1020–1025. [*426*]

[131] Bentkus V.Yu. *Ellipticity of an infinite-dimensional iterated Laplace operator. I, II.* Litovsk. Mat. Sb. 1979. V. 19, №4. P. 13–28; ibid., 1980. V. 20, №1. P. 3–13 (in Russian); English transl.: Lithuanian Math. J. 1979. V. 19, №4. P. 455–465; 1980. V. 20, №1. P. 1–7. [*71, 111, 426*]

[132] Bentkus V.Yu. *Analyticity of Gaussian measures.* Teor. Verojatn. i Primen. 1982. V. 27, №1. P. 147–154 (in Russian); English transl.: Theor. Probab. Appl. 1982. V. 27, №1. P. 155–161. [*71, 111*]

[133] Bentkus V.Yu. *Differentiable functions in spaces c_0 and R^k.* Litovsk. Mat. Sb. 1983. V. 23, №4. P. 26–36 (in Russian); English transl.: Lithuanian Math. J. 1983. V. 23, №2. P. 146–155. [*155*]

[134] Bentkus V.Yu. *Estimates of the closeness of sums of independent random elements in the space $C[0,1]$.* Litovsk. Mat. Sb. 1983. V. 23, №1. P. 7–16 (in Russian); English transl.: Lithuanian Math. J. 1983. V. 23, №1. P. 9–18. [*149*]

[135] Bentkus V., Juozulynas A., Paulauskas V. *Bounds for stable measures of convex shells and stable approximations.* Ann. Probab. 2000. V. 28, №3. P. 1280–1300. [*110*]

[136] Bentkus V.Yu., Račkauskas A.Yu. *On the rate of convergence in the central limit theorem in infinite-dimensional spaces.* Litovsk. Mat. Sb. 1981. V. 21, №4. P. 9–18 (in Russian); English transl.: Lith. Math. J. 1981. V. 21, №4. P. 271–276. [*149*]

[137] Bentkus V.Yu., Rachkauskas A.Yu. *Estimates of the rate of convergence of sums of independent random variables in a Banach space. I, II.* Litovsk. Mat. Sb. 1982. V. 22, №3. P. 12–28; V. 22, №4. P. 8–20 (in Russian); English transl.: Lith. Math. J. 1982. V. 22, №3. P. 222–234; 1982. V. 22, №4. P. 344–353. [*149*]

[138] Benyamini Y., Lindenstrauss J. *Geometric nonlinear functional analysis.* Amer. Math. Soc., Providence, Rhode Island, 2000; 488 p. [*5, 37, 344, 351*]

[139] Berezanskiĭ Yu.M. *Selfadjoint operators in spaces of functions of infinitely many variables.* Transl. from the Russian. Amer. Math. Soc., Providence, Rhode Island, 1986; xiv+383 p. (Russian ed.: Kiev, 1978). [*276, 426*]

[140] Berezansky Y.M., Kondratiev Y.G. *Spectral methods in infinite-dimensional analysis.* V. 1, 2. Transl. from the Russian. Kluwer Academic Publ., Dordrecht, 1995; xviii+576 p., viii+432 p. (Russian ed.: Kiev, 1988). [*276, 426*]

[141] Berezhnoi V.E. *On the equivalence of norms in the space of γ-measurable polynomials.* Vestnik Moskov. Univ. Ser. I Mat. Mekh. 2004. №4. P. 54–56 (in Russian); English transl.: Moscow Univ. Math. Bull. 2004. V. 59. №4. P. 37–39. [*251*]

[142] Berezhnoy V.E. *On the equivalence of integral norms on the space of measurable polynomials with respect to a convex measure.* Theory Stoch. Process. 2008. V. 14, №1. P. 7–10. [*251*]

[143] Berger M.A. *A Malliavin-type anticipative stochastic calculus.* Ann. Probab. 1988. V. 16, №1. P. 231–245. [*323*]

[144] Bernard P. *Une presentation de calcul des variations stochastique (calcul de Malliavin) pour les non-specialistes.* Ann. Sci. Univ. Clermont-Ferrand II Math. 1986. №23. P. 13–45. [*323, 424*]

[145] Besov O.V., Il'in V.P., Nikolskiĭ S.M. *Integral representations of functions and imbedding theorems.* V. I, II. Translated from the Russian. Winston & Sons, Washington; Halsted Press, New York – Toronto – London, 1978, 1979; viii+345 p., viii+311 p. (Russian ed.: Moscow, 1975). [*66*]

[146] Biagini F., Hu Y., Øksendal B., Zhang T. *Stochastic calculus for fractional Brownian motion and applications.* Springer-Verlag, London, 2008; xii+329 p. [*324*]

[147] Biagini F., Øksendal B., Sulem A., Wallner N. *An introduction to white-noise theory and Malliavin calculus for fractional Brownian motion. Stochastic analysis with applications to mathematical finance.* Proc. R. Soc. Lond. Ser. A Math. Phys. Eng. Sci. 2004. V. 460, №2041. P. 347–372. [*425*]

[148] Bichteler K., Gravereaux J.B., Jacod J. *Malliavin calculus for processes with jumps.* Gordon and Breach, 1987; x+161 p. [*323*]

[149] Bismut J.-M. *Martingales, the Malliavin calculus and hypoellipticity under general Hørmander's conditions.* Z. Wahr. theor. verw. Geb. 1981. B. 56, №4. S. 469–505. [*281, 424*]

[150] Bismut J.-M. *An introduction to the stochastic calculus of variations.* Stochastic differential systems (Bad Honnef, 1982), Lecture Notes in Control and Information Sci. V. 43. P. 33–72. Springer, Berlin–New York, 1982. [*424*]

[151] Bismut J.-M. *Calcul des variations stochastique et processus de sauts.* Z. Wahr. theor. verw. Geb. 1983. B. 63, №2. S. 147–235. [*323, 424*]

[152] Bismut J.-M. *Large deviations and the Malliavin calculus.* Birkhäuser Boston, Boston, 1984; viii+216 p. [*424*]

[153] Bismut J.-M. *The Atiyah–Singer theorems: a probabilistic approach. I. The index theorem.* J. Funct. Anal. 1984. V. 57, №1. P. 56–99; *II. The Lefschetz fixed point formulas.* J. Funct. Anal. 1984. V. 57, №3. P. 329–349. [*406*]

[154] Bismut J.-M. *Last exit decompositions and regularity at the boundary of transition probabilities.* Z. Wahr. theor. verw. Geb. 1985. B. 69, №1. S. 65–98. [*323*]

[155] Bismut J.-M. *Jump processes.* Proc. Taniguchi Intern. Symp. on Stochastic Analysis (Katata/Kyoto, 1982, K. Itô ed.), P. 53–104. North-Holland, Amsterdam, 1985. [*323*]

[156] Bismut J.-M. *Probability and geometry.* Lecture Notes in Math. 1986. V. 1206. P. 1–60. [*406*]

[157] Bismut J.-M., Michel D. *Diffusions conditionelles. I. Hypoellipticité partielle.* J. Funct. Anal. 1981. V. 44, №2. P. 174–211; *II. Générateur conditionel. Application au filtrage.* J. Funct. Anal. 1982. V. 45, №2. P. 274–292. [*425*]

[158] Blanchere S., Chafai D., Fougeres F., Gentil I., Malrien F., Roberto C., Scheffer G. *Sur les inégalités de Sobolev logarithmiques.* Panoramas et Synthèses, Soc. Math. France, 2000; 213 p. [*274*]

[159] Bobkov S.G. *The size of singular component and shift inequalities.* Ann. Probab. 1999. V. 27, №1. P. 416–431. [*107*]

[160] Bobkov S.G. *Isoperimetric and analytic inequalities for log-concave probability measures.* Ann. Probab. 1999. V. 27, №4. P. 1903–1921. [*131*]

[161] Bobkov S.G. *Remarks on the growth of L^p-norms of polynomials.* Lecture Notes in Math. 2000. V. 1745. P. 27–35. [*131*]

[162] Bobkov S.G., Götze F. *Exponential integrability and transportation cost related to logarithmic Sobolev inequalities.* J. Funct. Anal. 1999. V. 163, №1. P. 1–28. [*274*]

[163] Bobkov S.G., Ledoux M. *From Brunn–Minkowski to Brascamp–Lieb and to logarithmic Sobolev inequalities.* Geom. Funct. Anal. 2000. V. 10, №5. P. 1028–1052. [*274*]

[164] Bochkarev S.V. *A method of averaging in the theory of orthogonal series and some problems in the theory of bases.* Trudy Mat. Inst. Steklov. 1978. V. 146. 87 p. (in Russian); English transl.: Proc. Steklov Inst. Math. 1980. №3, vi+92 p. [*148*]

[165] Bogachev L.V., Daletskii A.Yu. *Poisson cluster measures: quasi-invariance, integration by parts and equilibrium stochastic dynamics.* J. Funct. Anal. 2009. V. 256. P. 432–478. [*399*]

[166] Bogachev V.I. *Negligible sets and differentiable measures in Banach spaces*. Vestn. Mosk. Univ., Ser. I. 1982. №3. P. 47–52 (Russian); English transl.: Moscow Univ. Math. Bull. 1982. V. 37, №3. P. 54–59. [*95, 103, 134, 155, 196, 343*]

[167] Bogachev V.I. *Three problems of Aronszajn from measure theory*. Funk. Anal. i Pril. 1984. V. 18, №3. P. 75–76 (in Russian); English transl.: Funct. Anal. Appl. 1984, №3. V. 18. P. 242–244. [*187, 344, 346, 347*]

[168] Bogachev V.I. *Negligible sets in locally convex spaces*. Matem. Zametki. 1984. V. 36, №1. P. 51–64 (in Russian); English transl.: Math. Notes. 1984. V. 36. P. 519–526. [*344, 346*]

[169] Bogachev V.I. *Some results on differentiable measures*. Mat. Sb. 1985. V. 127, №3. P. 336–351 (in Russian); English transl.: Math. USSR Sbornik. 1986. V. 55, №2. P. 335–349. [*109, 131, 134, 155, 344, 346, 350*]

[170] Bogachev V.I. *Locally convex spaces with the CLT property and supports of measures*. Vestnik Moskovsk. Univ. 1986. №6. P. 16–20 (in Russian); English transl.: Moscow Univ. Math. Bull. 1986. V. 41, №6. P. 19–23. [*145*]

[171] Bogachev V.I. *Indices of asymmetry of stable measures*. Matem. Zametki. 1986. V. 40, №1. P. 127–138 (in Russian); English transl.: Math. Notes. 1986. V. 40. P. 569–575. [*108, 109*]

[172] Bogachev V.I. *On links between between differential properties of measures and their supports*. In: "Differential equations, harmonic analysis and their applications", P. 65–66. Moscow State Univ., Moscow, 1987. [*155*]

[173] Bogachev V.I. *On analytic properties of distributions of diffusion processes*. In: "The First All-Union School on Ergodic Theory of Markov Processes", p. 9, Kyzyl, 1987 (in Russian). [*124*]

[174] Bogachev V.I. *On Skorokhod differentiability of measures*. Teor. Verojatn. i Primen. 1988. V. 38, №2. P. 349–354 (in Russian); English transl.: Theory Probab. Appl. 1988. V. 33, №2. P. 330–334. [*91, 92*]

[175] Bogachev V.I. *Subspaces of differentiability of smooth measures on infinite-dimensional spaces*. Dokl. Akad. Nauk SSSR. 1988. V. 299, №1. P. 18–22 (in Russian); English transl.: Sov. Math. Dokl. 1988. V. 37, №1. P. 304–308. [*131, 134, 155*]

[176] Bogachev V.I. *Distributions of smooth functionals on probability manifolds and nonlinear transformations of measures*. In: "Extremal problems, functional analysis and their applications", pp. 56–59, Moscow State Univ., Moscow, 1988 (in Russian). [*324, 398*]

[177] Bogachev V.I. *Distributions of polynomial functionals and oscillatory integrals*. Proc. Conference "On methods of algebra and analysis", pp. 42–43, Tartu University, Tartu, 1988 (in Russian). [*324*]

[178] Bogachev V.I. *Distributions of smooth functionals of stochastic processes and infinite dimensional oscillating integrals*. Proc. of the Fifth Internat. Vilnius Conference on Probab. Theory and Math. Statistics, V. 3, pp. 76–77, Vilnius, 1989 (in Russian). [*324*]

[179] Bogachev V.I. *Differential properties of measures on infinite dimensional spaces and the Malliavin calculus*. Acta Univ. Carolinae, Math. et Phys. 1989. V. 30, №2. P. 9–30. [*103, 124, 155, 324, 370, 372, 398*]

[180] Bogachev V.I. *Analytic functionals of stochastic processes and infinite dimensional oscillatory integrals*. Probab. Theory and Math. Stat., Proc. 5th Vilnius Conf. on Probab. Theory and Math. Stat. V. 1. P. 152–163. "Mokslas"/VSP, Vilnius, Utrecht, 1990. [*324*]

[181] Bogachev V.I. *Smooth measures, the Malliavin calculus and approximation in infinite dimensional spaces*. Acta Univ. Carolinae, Math. et Phys. 1990. V. 31, №2. P. 9–23. [*103, 155, 299, 325*]

[182] Bogachev V.I. *Polynômes et intégrales oscillantes sur les espaces de dimension infinie*. C. R. Acad. Sci. Paris. 1990. T. 311, №12. P. 807–812. [*324*]

[183] Bogachev V.I. *The distributions of analytic functionals of random processes.* Dokl. Akad. Nauk SSSR. 1990. V. 312, №6. P. 1291–1296 (in Russian); English transl.: Sov. Math. Dokl. 1990. V. 41, №3. P. 517–522. [*324*]

[184] Bogachev V.I. *Functionals of random processes and infinite-dimensional oscillatory integrals connected with them.* Izvest. Akad. Nauk SSSR. 1992. V. 156, №2. P. 243–278 (in Russian); English transl.: Russian Sci. Izv. Math. 1993. V. 40, №2. P. 235–266. [*124, 126, 292, 297, 299, 323, 324*]

[185] Bogachev V.I. *Infinite dimensional integration by parts and related problems.* Preprint SFB 256 №235. Univ. Bonn, 1992. P. 1–35. [*161, 187, 224*]

[186] Bogachev V.I. *Remarks on integration by parts in infinite dimensions.* Acta Univ. Carolinae, Math. et Phys. 1993. V. 34, №2. P. 11–29. [*161, 187, 224, 377*]

[187] Bogachev V.I. *Remarks on invariant measures and reversibility of infinite dimensional diffusions.* Probability theory and mathematical statistics (St. Petersburg, 1993). P. 119–132. Gordon and Breach, Amsterdam, 1996. [*224*]

[188] Bogachev V.I. *Vector logarithmic derivatives of measures and their applications.* Séminaire d'Initiation à l'Analyse 1995/1996, ed. by G. Choquet, G. Godefroy, M. Rogalski, J. Saint Raymond, Paris, 29 p. [*223*]

[189] Bogachev V.I. *Differentiable measures and the Malliavin calculus.* Scuola Normale Superiore di Pisa, Preprint №16 (1995), 197 p.; J. Math. Sci. (New York). 1997. V. 87, №4. P. 3577–3731. [*xiv, 103, 223, 323, 324, 398*]

[190] Bogachev V.I. *The Malliavin calculus.* Lectures held at the University of Minnesota, Winter 1997, Minneapolis, 1997; 57 p. [*323*]

[191] Bogachev V.I. *Gaussian measures.* Amer. Math. Soc., Providence, Rhode Island, 1998; 433 p. [*23, 35, 146, 147, 220, 223, 237, 239, 251, 252, 255, 259, 265, 267, 272, 273, 278, 296, 301, 323, 328, 329, 335, 358, 366*]

[192] Bogachev V.I. *Extensions of H-Lipschitzian mappings with infinite-dimensional range.* Infin. Dim. Anal., Quantum Probab. Relat. Top. 1999. V. 2, №3. P. 1–14. [*250*]

[193] Bogachev V.I. *Measure theory.* V. 1, 2. Springer, Berlin, 2007; xvii+500 p., xiii+575 p. [*6, 10, 11, 13, 17, 18, 19, 20, 71, 91, 96, 107, 113, 122, 128, 131, 158, 207, 246, 266, 309, 326, 357, 361, 381*]

[194] Bogachev V.I., Da Prato G., Röckner M. *Regularity of invariant measures for a class of perturbed Ornstein–Uhlenbeck operators.* Nonlin. Diff. Equ. Appl. 1996. V. 3, №2. P. 261–268. [*414*]

[195] Bogachev V.I., Da Prato G., Röckner M. *On parabolic equations for measures.* Comm. Partial Diff. Eq. 2008. V. 33, №3. P. 397–418. [*426*]

[196] Bogachev V.I., Goldys B.G. *Second derivatives of convex functions in the sense of A.D. Aleksandrov in infinite-dimensional measure spaces.* Mat. Zametki. 2006. V. 79, №4. P. 488–504 (in Russian); English transl.: Math. Notes. 2006. V. 79, №3-4. P. 454–467. [*187*]

[197] Bogachev V.I., Kirillov A.I., Shaposhnikov S.V. *Invariant measures of diffusions with gradient drifts.* Dokl. Ross. Akad. Nauk. 2010 (in Russian); English transl.: Dokl. Math. 2010. [*416*]

[198] Bogachev V.I., Kolesnikov A.V. *Nonlinear transformations of convex measures.* Teor. Verojatn. i Primen. 2005. V. 50, №1. P. 27–51. (in Russian); English transl.: Theory Probab. Appl. 2006. V. 50, №1. P. 34–52. [*362, 367*]

[199] Bogachev V.I., Kolesnikov A.V. *Integrability of absolutely continuous transformations of measures and applications to optimal mass transportation.* Teor. Verojatn. i Primen. 2005. V. 50, №3. P. 433–456 (in Russian); English transl.: Theory Probab. Appl. 2006. V. 50, №3. P. 367–385. [*367*]

[200] Bogachev V.I., Kolesnikov A.V. *On the Monge–Ampère equation in infinite dimensions.* Infin. Dim. Anal. Quantum Probab. Relat. Top. 2005. V. 8, №4. P. 547–572. [*364*]

[201] Bogachev V.I., Kolesnikov, A.V. *On the Monge–Ampère equation on Wiener space.* Dokl. Ross. Akad. Nauk. 2006. V. 406, №1. P. 7–11 (in Russian); English transl.: Dokl. Math. 2006. V. 73, №1. P. 1–5. [*364*]

[202] Bogachev V.I., Kolesnikov A.V. *Transformations of measures by Gauss maps.* Dokl. Ross. Akad. Nauk. 2008. V. 422, №4. P. 446–449 (in Russian); English transl.: Dokl. Math. 2008. V. 78. №2. P. 720–723. [*367*]

[203] Bogachev V.I., Kolesnikov A.V. *Mass transport generated by a flow of Gauss maps.* J. Funct. Anal. 2009. V. 256, №3. P. 940–957. [*367*]

[204] Bogachev V.I., Kolesnikov A.V., Medvedev K.V. *Triangular transformations of measures.* Matem. Sbornik. 2005. V. 196, №3. P. 3–30 (in Russian); English transl.: Sbornik Math. 2005. V. 196, №3. P. 309–335. [*362, 367*]

[205] Bogachev V.I., Krylov N.V., Röckner M. *Regularity of invariant measures: the case of non-constant diffusion part.* J. Funct. Anal. 1996. V. 138, №1. P. 223–242. [*409, 413, 414, 426*]

[206] Bogachev V.I., Krylov N.V., Röckner M. *On regularity of transition probabilities and invariant measures of singular diffusions under minimal conditions.* Comm. Partial Diff. Eq. 2001. V. 26, №11–12. P. 2037–2080. [*409, 426*]

[207] Bogachev V.I., Krylov N.V., Röckner M. *Elliptic equations for measures: regularity and global bounds of densities.* J. Math. Pures Appl. 2006. V. 85, №6. P. 743–757. [*409, 410, 426*]

[208] Bogachev V.I., Krylov N.V., Röckner M. *Elliptic and parabolic equations for measures.* Uspehi Mat. Nauk. 2009. V. 64, №6. P. 5–116 (in Russian); English transl.: Russian Math. Surveys. 2009. V. 64, №6. P. 973–1078. [*410, 426*]

[209] Bogachev V.I., Mayer-Wolf E. *Remarks on Rademacher's theorem in infinite dimension.* Potential Anal. 1996. V. 5, №1. P. 23–30. [*355*]

[210] Bogachev V.I., Mayer-Wolf E. *The flows generated by vector fields of Sobolev type and the corresponding transformations of probability measures.* Dokl. Ross. Akad. Nauk. 1998. V. 358, №4. P. 442–446 (in Russian); English transl.: Dokl. Math. 1998. V. 57, №1. P. 64–68. [*340*]

[211] Bogachev V.I., Mayer-Wolf E. *Absolutely continuous flows generated by Sobolev class vector fields in finite and infinite dimensions.* J. Funct. Anal. 1999. V. 167, №1. P. 1–68. [*170, 233, 340, 341, 342, 366*]

[212] Bogachev V., Mayer-Wolf E. *Dynamical systems generated by Sobolev class vector fields in finite and infinite dimensions.* J. Math. Sci. (New York). 1999. V. 94, №3. P. 1394–1445. [*340, 342, 366*]

[213] Bogachev V.I., Priola E., Tolmachev N.A. *On Fréchet differentiability of Lipschitzian functions on spaces with Gaussian measures.* Dokl. Ross. Akad. Nauk. 2007. V. 414, №2. P. 151–155 (in Russian); English transl.: Dokl. Math. 2007. V. 75, №3. P. 353–357. [*352, 355*]

[214] Bogachev V.I., Pugachev O.V., Röckner M. *Surface measures and tightness of Sobolev capacities on Poisson space.* Dokl. Ross. Akad. Nauk. 2002. V. 386, №1. P. 7–10 (in Russian); English transl.: Dokl. Math. 2002. V. 66, №2. P. 157–160. [*399*]

[215] Bogachev V.I., Pugachev O.V., Röckner M. *Surface measures and tightness of (r,p)-capacities on Poisson space.* J. Funct. Anal. 2002. V. 196. P. 61–86. [*399*]

[216] Bogachev V.I., Röckner M. *Les capacités gaussiennes sont portées par des compacts metrisables.* C. R. Acad. Sci. Paris. 1992. T. 315, №2. P. 197–202. [*273, 276*]

[217] Bogachev V.I., Röckner M. *Hypoellipticity and invariant measures for infinite dimensional diffusions.* C. R. Acad. Sci. Paris. 1994. T. 318, №6. P. 553–558. [*224, 415, 426*]

[218] Bogachev V.I., Röckner M. *Mehler formula and capacities for infinite dimensional Ornstein–Uhlenbeck processes with general linear drift.* Osaka J. Math. 1995. V. 32, №2. P. 237–274. [*275, 276*]

[219] Bogachev V.I., Röckner M. *Regularity of invariant measures on finite and infinite dimensional spaces and applications.* J. Funct. Anal. 1995. V. 133, №1. P. 168–223. [*212, 223, 224, 409, 412, 415, 421, 426*]

[220] Bogachev V.I., Röckner M. *Elliptic equations for infinite dimensional probability distributions and Lyapunov functions.* C. R. Acad. Sci. Paris, sér. 1. 1999. T. 329. P. 705–710. [*224*]

[221] Bogachev V.I., Röckner M. *A generalization of Khasminskii's theorem on the existence of invariant measures for locally integrable drifts.* Teor. Verojatn. i Primen. 2000. V. 45, №3. P. 417–436; correction: ibid. 2001. V. 46, №3. P. 600 (in Russian); English transl.: Theory Probab. Appl. 2000. V. 45, №3. P. 363–378. [*408, 426*]

[222] Bogachev V.I., Röckner M. *Elliptic equations for measures on infinite-dimensional spaces and applications.* Probab. Theory Relat. Fields. 2001. V. 120, №4. P. 445–496. [*132, 224, 413, 415, 426*]

[223] Bogachev V.I., Röckner M. *Invariant measures of diffusion processes: regularity, existence, and uniqueness problems.* Stochastic partial differential equations and applications (G. Da Prato et al., eds.). P. 69–87. Marcel Dekker, New York, 2002. [*426*]

[224] Bogachev V.I., Röckner M. *On L^p-uniqueness of symmetric diffusion operators on Riemannian manifolds.* Mat. Sb. 2003. V. 194, №7. P. 15–24 (in Russian); English transl.: Sb. Math. 2003. V. 194, №7. P. 969–978. [*426*]

[225] Bogachev V.I., Röckner M., Schmuland B. *Generalized Mehler semigroups and applications.* Probab. Theor. Relat. Fields. 1996. V. 105, №2. P. 193–225. [*275*]

[226] Bogachev V.I., Röckner M., Shaposhnikov S.V. *Estimates of densities of stationary distributions and transition probabilities of diffusion processes.* Teor. Verojatn. i Primen. 2007. V. 52, №2. P. 240–270 (in Russian); English transl.: Theory Probab. Appl. 2008. V. 52, №2. P. 209–236. [*426*]

[227] Bogachev V.I., Röckner M., Shaposhnikov S.V. *Positive densities of transition probabilities of diffusion processes.* Teor. Verojatn. i Primen. 2008. V. 53, №2. P. 213–239 (in Russian); English transl.: Theory Probab. Appl. 2009. V. 53, №2. P. 194–215. [*172*]

[228] Bogachev V.I., Röckner M., Shaposhnikov S.V. *Lower estimates of densities of solutions of elliptic equations for measures.* Dokl. Ross. Akad. Nauk. 2009. V. 426, №2. P. 156–161 (in Russian); English transl.: Dokl. Math. 2009. V. 79, №3. P. 329–334. [*172*]

[229] Bogachev V.I., Röckner M., Stannat W. *Uniqueness of solutions of elliptic equations and uniqueness of invariant measures of diffusions.* Mat. Sb. 2002. V. 197, №7. P. 3–36 (in Russian); English transl.: Sb. Math. 2002. V. 193, №7. P. 945–976. [*410, 426*]

[230] Bogachev V.I., Röckner M., Wang F.-Y. *Elliptic equations for invariant measures on finite and infinite dimensional manifolds.* J. Math. Pures Appl. 2001. V. 80. P. 177–221. [*409, 426*]

[231] Bogachev V.I., Röckner M., Wang F.-Yu. *Invariance implies Gibbsian: some new results.* Comm. Math. Phys. 2004. V. 248. P. 335–355. [*132, 409, 426*]

[232] Bogachev V.I., Röckner M., Zhang T.S. *Existence and uniqueness of invariant measures: an approach via sectorial forms.* Appl. Math. Optim. 2000. V. 41. P. 87–109. [*415, 426*]

[233] Bogachev V.I., Smolyanov O.G. *Analytic properties of infinite dimensional distributions.* Uspehi Mat. Nauk. 1990. V. 45, №3. P. 3–83 (in Russian); English transl.: Russian Math. Surveys. 1990. V. 45, №3. P. 1–104. [*103, 111, 123, 131, 145, 155, 167, 299, 323, 324, 375, 398*]

[234] Bogachev V.I., Smolyanov O.G. Real and functional analysis: a university course. Regular and Chaotic Dynamics, Moscow – Izhevsk, 2009; 724 p. (in Russian). [*3, 147, 235*]

[235] Bojarski B., Iwaniec T. *Analytical foundations of the theory of quasiconformal mappings in R^n.* Annales Acad. Sci. Fennicae, Ser. A. I. Math. 1983. V. 8. P. 257–324. [*54*]

[236] Bonaccorsi S., Fuhrman M. *Regularity results for infinite dimensional diffusions. A Malliavin calculus approach.* Atti Accad. Naz. Lincei Cl. Sci. Fis. Mat. Natur. Rend. Lincei (9) Mat. Appl. 1999. V. 10, №1. P. 35–45. [*414*]

[237] Bonaccorsi S., Fuhrman M. *Integration by parts and smoothness of the law for a class of stochastic evolution equations.* Infin. Dim. Anal. Quantum Probab. Relat. Top. 2004. V. 7, №1. P. 89–129. [*414*]

[238] Bonaccorsi S., Zambotti L. *Integration by parts on the Brownian meander.* Proc. Amer. Math. Soc. 2004. V. 132, №3. P. 875–883. [*414*]

[239] Borell C. *Convex measures on locally convex spaces.* Ark. Math. 1974. V. 12. P. 239–252. [*113, 131*]

[240] Borell C. *A note on conditional probabilities of a convex measure.* Vector space measures and applications, II (Proc. Conf., Univ. Dublin, Dublin, 1977). P. 68–72. Lecture Notes in Phys. V. 77. Springer, Berlin – New York, 1978. [*114*]

[241] Borodin A., Olshansky G. *Infinite random matrices and ergodic measures.* Comm. Math. Phys. 2001. V. 223. P. 87–123. [*399*]

[242] Bouchard B., Ekeland I., Touzi N. *On the Malliavin approach to Monte Carlo approximation of conditional expectations.* Finance Stoch. 2004. V. 8, №1. P. 45–71. [*425*]

[243] Bouleau N. *Differential calculus for Dirichlet forms: the measure-valued gradient preserved by image.* J. Funct. Anal. 2005. V. 225. P. 63–73. [*425*]

[244] Bouleau N., Denis L. *Energy image density property and the lent particle method for Poisson measures.* J. Funct. Anal. 2009. V. 257. P. 1144–1174. [*399*]

[245] Bouleau N., Hirsch F. *Proprietés d'absolue continuité dans les espaces de Dirichlet et application aux équations différentielles stochastiques.* Lecture Notes in Math. 1986. V. 1204. P. 131–161. [*324*]

[246] Bouleau N., Hirsch F. *Formes de Dirichlet générales et densité des variables aléatoires réelles sur l'espace de Wiener.* J. Funct. Anal. 1986. V. 69, №2. P. 229–259. [*324*]

[247] Bouleau N., Hirsch F. *Dirichlet forms and analysis on Wiener space.* De Gruyter, Berlin – New York, 1991; x+325 p. [*273, 275, 324, 370, 398, 425*]

[248] Brascamp H., Lieb E.H. *On extensions of Brunn–Minkowski and Prékopa–Leindler theorems, including inequalities for Log concave functions, and with applications to diffusion equations.* J. Funct. Anal. 1976. V. 22, №4. P. 366–389. [*122, 131*]

[249] Brayman V.B. *On existence of a solution for differential equation with interaction in abstract Wiener space.* Theory Stoch. Process. 2004. V. 10, №3-4. P. 9–20. [*366*]

[250] Brayman V.B. *On the existence and uniqueness of the solution of a differential equation with interaction governed by generalized function in abstract Wiener space.* Theory Stoch. Process. 2005. V. 11, №3-4. P. 29–41. [*366*]

[251] Brayman V.B. *On existence of a solution for differential equation with interaction.* Random Oper. Stoch. Equat. 2006. V. 14, №4. P. 325–334. [*366*]

[252] Brenier Y. *Polar factorization and monotone rearrangement of vector–valued functions.* Comm. Pure Appl. Math. 1991. V. 44. P. 375–417. [*363*]

[253] Breton J.-C. *Convergence in variation of the joint laws of multiple stable stochastic integrals.* Probab. Math. Statist. 2008. V. 28, №1. P. 21–40. [*326*]

[254] Buckdahn R. *Anticipative Girsanov transformations.* Probab. Theory Relat. Fields. 1991. V. 89, №2. P. 211–238. [*366*]

[255] Buckdahn R., Föllmer H. *A conditional approach to the anticipating Girsanov transformation.* Probab. Theory Relat. Fields. 1993. V. 95, №3. P. 311–390. [*366*]

[256] Byczkowski T., Graczyk P. *Malliavin calculus for stable processes on Heisenberg group.* Probab. Math. Stat. 1992. V. 13, №2. P. 277–292. [*323*]

[257] Caballero M., Fernández B., Nualart D. *Estimation of densities and applications.* J. Theoret. Probab. 11 (1998), no. 3, 831–851. [*425*]

[258] Cameron R.H., Fagen R.E. *Nonlinear transformations of Volterra type in Wiener space.* Trans. Amer. Math. Soc. 1953. V. 7, №3. P. 552–575. [*365*]

[259] Cameron R.H., Martin W.T. *Transformation of Wiener integral under translation.* Ann. Math. 1944. V. 45. P. 386–396. [*365*]

[260] Cameron R.H., Martin W.T. *Transformations of Wiener integrals under a general class transformation.* Trans. Amer. Math. Soc. 1945. V. 58. P. 184–219. [*365*]

[261] Caraman P. *Module and p-module in an abstract Wiener space.* Rev. Roumaine Math. Pures Appl. 1982. V. 27, №5. P. 551–599. [*276*]

[262] Caraman P. *Quasiconformal mappings in abstract Wiener spaces.* Lecture Notes in Math. 1983. P. 23–49. [*276*]

[263] Carlen E., Pardoux E. *Differential calculus and integration by parts on Poisson space.* Stochastics, Algebra and Analysis in Classical and Quantum Mechanics (S. Albeverio et al. eds.). P. 63–73. Kluwer Acad. Publ., Dordrecht, 1990. [*323, 399*]

[264] Carmona R., Nualart D. *Random non-linear wave equations: smoothness of the solutions.* Probab. Theory Relat. Fields. 1988. V. 79, №4. P. 469–508. [*425*]

[265] Carmona R.A., Nualart D. *Traces of random variables on Wiener space and the Onsager–Machlup functional.* J. Funct. Anal. 1992. V. 107, №2. P. 402–438. [*275*]

[266] Carmona R.A., Tehranchi M.R. Interest rate models: an infinite dimensional stochastic analysis perspective. Springer-Verlag, Berlin, 2006; xiv+235 p. [*425*]

[267] Cattiaux P. *Hypoellipticité et hypoellipticité partielle pour les diffusion avec une condition frontière.* Ann. Inst. H. Poincaré. 1986. V. 22, №1. P. 67–112. [*424*]

[268] Cattiaux P. *Regularité au bord pour les densités et les densités conditionnelles d'une diffusion reflechie hypoelliptique.* Stochastics. 1987. V. 20, №4. P. 309–340. [*424*]

[269] Cattiaux P. *Calcul stochastique et opérateurs dégénérés du second ordre. I. Résolvantes, théorème de Hörmander et applications.* Bull. Sci. Math. (2). 1990. V. 114, №4. P. 421–462; — *II. Problème de Dirichlet.* Ibid. 1991. V. 115, №1. P. 81–122. [*424*]

[270] Cattiaux P., Fradon M. *Entropy, reversible diffusion processes, and Markov uniqueness.* J. Funct. Anal. 1996. V. 138, №1. P. 243–272. [*64, 426*]

[271] Cattiaux P., Léonard C. *Minimization of the Kullback information of diffusion processes.* Ann. Inst. H. Poincaré. 1994. V. 30, №1. P. 83–132; correction: ibid., 1995. V. 31, №4. P. 705–707. [*426*]

[272] Cattiaux P., Mesnager L. *Hypoelliptic non-homogeneous diffusions.* Probab. Theory Relat. Fields. 2002. V. 123, №4. P. 453–483. [*424*]

[273] Cattiaux P., Roelly S., Zessin H. *Une approche gibbsienne des diffusions browniennes infini-dimensionnelles.* Probab. Theory Relat. Fields. 1996. V. 104, №2. P. 147–179. [*224*]

[274] Cecil M., Driver B.K. *Heat kernel measure on loop and path groups.* Infin. Dimens. Anal. Quantum Probab. Relat. Top. 2008. V. 11, №2. P. 135–156. [*398*]

[275] Cerrai S. Second order PDE's in finite and infinite dimension. A probabilistic approach. Lecture Notes in Math. V. 1762. Springer-Verlag, Berlin, 2001; x+330 p. [*426*]

[276] Cipriano F., Cruzeiro A.-B. *Flows associated with irregular $\mathbb{R}^d$-vector fields.* J. Diff. Eq. 2005. V. 219, №1. P. 183–201. [*366*]

[277] Chaleyat-Maurel M. *Robustesse du filtre et calcul des variations stochastique.* J. Funct. Anal. 1986. V. 68, №1. P. 55–71. [*425*]

[278] Chaleyat-Maurel M., Nualart D. *Points of positive density for smooth functionals.* Electron. J. Probab. 1998. V. 3, №1. P. 1–8. [*326*]

[279] Chaleyat-Maurel M., Sanz-Sole M. *Positivity of the density for the stochastic wave equation in two spatial dimensions.* ESAIM Probab. Stat. 2003. V. 7. P. 89–114. [*326*]

[280] Chatterji S.D., Mandrekar V. *Quasi-invariance of measures under translation.* Math. Z. 1977. B. 154, №1. S. 19–29. [*155*]

[281] Cheeger J. *Differentiability of Lipschitz functions on metric measure spaces.* Geom. Funct. Anal. 1999. V. 9. P. 428–517. [*274*]

[282] Chen M.F., Zhou X.Y. *Applications of Malliavin calculus to stochastic differential equations with time-dependent coefficients.* Acta Math. Appl. Sinica (English Ser.). 1991. V. 7, №3. P. 193–216. [*424*]

[283] Chentsov (Čencov) N.N. Statistical decision rules and optimal inference. Translated from the Russian. Amer. Math. Soc., Providence, Rhode Island, 1982; viii+499 p. (Russian ed.: Moscow, 1972). [*398*]

[284] Chojnowska-Michalik A., Goldys B. *On regularity properties of nonsymmetric Ornstein–Uhlenbeck semigroup in L^p spaces.* Stochastics Stoch. Rep. 1996. V. 59, №3-4. P. 183–209. [*275*]

[285] Chojnowska-Michalik A., Goldys B. *Generalized Ornstein–Uhlenbeck semigroups: Littlewood–Paley–Stein inequalities and the P.A. Meyer equivalence of norms.* J. Funct. Anal. 2001. V. 182, №2. P. 243–279. [*275*]

[286] Chojnowska-Michalik A., Goldys B. *Symmetric Ornstein–Uhlenbeck semigroups and their generators.* Probab. Theory Related Fields. 2002. V. 124, №4. P. 459–486. [*275*]

[287] Chow P.-L., Menaldi J.-L. *Infinite-dimensional Hamilton–Jacobi–Bellman equations in Gauss–Sobolev spaces.* Nonlinear Anal. 1997. V. 29, №4. P. 415–426. [*275*]

[288] Christensen J.P.R. *On sets of Haar measure zero in abelian Polish groups.* Israel J. Math. 1972. V. 13. P. 255–260. [*342*]

[289] Christensen J.P.R. *Measure theoretic zero sets in infinite dimensional spaces and applications to differentiability of Lipschitz mappings. II.* Publ. Dep. Math. Lyon. 1973. V. 10. P. 29–39. [*353*]

[290] Clark J.M. *The representation of functionals of Brownian motion by stochastic integrals.* Ann. Math. Stat. 1970. V. 41. P. 1282–1295; correction: ibid., 1971. V. 42. P. 1778. [*275*]

[291] Coquio A., Gravereaux J.-B. *Calcul de Malliavin et régularité de la densité d'une probabilité invariante d'une chaine de Markov.* Ann. Inst. H. Poincaré, Probab. Statist. 1992. V. 28, №4. P. 431–478. [*424*]

[292] Cordero-Erausquin D., Nazaret B., Villani C. *A mass-transportation approach to sharp Sobolev and Gagliardo–Nirenberg inequalities.* Adv. Math. 2004. V. 182, №2. P. 307–332. [*367*]

[293] Cruzeiro A.-B. *Équations différentielles sur l'espace de Wiener et formules de Cameron–Martin non-linéaires.* J. Funct. Anal. 1983. V. 54, №2. P. 206–227. [*340, 366*]

[294] Cruzeiro A.-B. *Unicité de solutions d'équations différentielles sur l'espace de Wiener.* J. Funct. Anal. 1984. V. 58, №3. P. 335–347. [*340, 366*]

[295] Cruzeiro A.-B. *Processus sur l'espace de Wiener associés à des opérateurs élliptiques à coefficients dans certains espaces de Sobolev.* J. Funct. Anal. 1987. V. 72, №2. P. 346–367. [*366, 426*]

[296] Cruzeiro A.-B. *Flows in infinite dimensions and associated transformations of Gaussian measures.* Stochastic methods in mathematics and physics (Karpacz, 1988). P. 290–301. World Sci. Publ., Teaneck, New Jersey, 1989. [*340, 366*]

[297] Cruzeiro A.-B., Malliavin P. *Renormalized differential geometry on path space: structural equation, curvature.* J. Funct. Anal. 1996. V. 139, №1. P. 119–181. [*398*]

[298] Cruzeiro A.-B., Malliavin P. *Nonperturbative construction of invariant measure through confinement by curvature.* J. Math. Pures Appl. (9). 1998. V. 77, №6. P. 527–537. [*426*]

[299] Cruzeiro A.-B., Malliavin P. *Non-existence of infinitesimally invariant measures on loop groups.* J. Funct. Anal. 2008. V. 254, №7. P. 1974–1987. [*398*]

[300] Cruzeiro A.-B., Malliavin P. *Stochastic calculus of variations on complex line bundle and construction of unitarizing measures for the Poincaré disk.* J. Funct. Anal. 2009. V. 256, №2. P. 385–408. [*398*]

[301] Cruzeiro A.-B., Malliavin P. *Renormalized stochastic calculus of variations for a renormalized infinite-dimensional Brownian motion.* Stochastics. 2009. V. 81, №3-4. P. 385–399. [*400*]

[302] Cruzeiro A.-B., Zambrini J.-C. *Malliavin calculus and Euclidean quantum mechanics. I.* J. Funct. Anal. 1991. V. 96, №1. P. 62–95. [*323*]

[303] Cruzeiro A.B., Zhang X. *Finite dimensional approximation of Riemannian path space geometry.* J. Funct. Anal. 2003. V. 205, №1. P. 206–270. [*398*]

[304] Csörnyei M. *Aronszajn null and Gaussian null sets coincide.* Israel J. Math. 1999. V. 111. P. 191–201. [*344*]

[305] Da Prato G. *Kolmogorov equations for stochastic PDEs.* Birkhäuser Verlag, Basel, 2004; x+182 p. [*426*]

[306] Da Prato G. *An introduction to infinite-dimensional analysis.* Springer-Verlag, Berlin, 2006; x+209 p. [*426*]

[307] Da Prato G. *Introduction to stochastic analysis and Malliavin calculus.* Edizioni della Normale, Scuola Normale Superiore di Pisa, Pisa, 2007; xvi+187 p. [*323*]

[308] Da Prato G., Debussche A. *Absolute continuity of the invariant measures for some stochastic PDEs.* J. Statist. Phys. 2004. V. 115, №1-2. P. 451–468. [*414*]

[309] Da Prato G., Debussche A., Goldys B. *Some properties of invariant measures of non symmetric dissipative stochastic systems.* Probab. Theory Relat. Fields. 2002. V. 123, №3. P. 355–380. [*254, 414*]

[310] Da Prato G., Elworthy D., Zabczyk J. *Strong Feller property for stochastic semilinear equations.* Stoch. Anal. Appl. 1995. V. 13, №1. P. 35–46. [*414*]

[311] Da Prato G., Malliavin P., Nualart D. *Compact families of Wiener functionals.* C. R. Acad. Sci. Paris. 1992. T. 315. P. 1287–1291. [*253*]

[312] Da Prato G., Zabczyk J. *Stochastic differential equations in infinite dimensions.* Cambridge Univ. Press, Cambridge, 1992; xviii+454 p. [*414, 421*]

[313] Da Prato G., Zabczyk J. *Regular densities of invariant measures in Hilbert spaces.* J. Funct. Anal. 1995. V. 130, №2. P. 427–449. [*414*]

[314] Da Prato G., Zabczyk J. *Ergodicity for infinite-dimensional systems.* Cambridge University Press, Cambridge, 1996; xii+339 p. [*414, 415*]

[315] Da Prato G., Zabczyk J. *Second order partial differential equations in Hilbert spaces.* Cambridge University Press, Cambridge, 2002; xvi+379 p. [*426*]

[316] Daletskiĭ (Daleckiĭ) Yu.L. *Infinite-dimensional elliptic operators and parabolic equations connected with them.* Uspehi Matem. Nauk. 1967. V. 22, №4. P. 3–54 (in Russian); English transl.: Russian Math. Surv. 1967. V. 22, №4. P. 1–53. [*103, 426*]

[317] Daletskiĭ Yu.L. *Stochastic differential geometry.* Uspehi Matem. Nauk. 1983. V. 38, №3. P. 87–111 (in Russian); English transl.: Russian Math. Surv. 1983. V. 38, №3. P. 97–125. [*323, 324*]

[318] Daletskiĭ Yu.L. *Biorthogonal analogue of the Hermite polynomials and the inversion of the Fourier transform with respect to a non-Gaussian measure.* Funkts. Anal. Prilozh. 1991. V. 25, №2. P. 68–70 (in Russian); English transl.: Funct. Anal. Appl. 1991. V. 25, №2. P. 138–140. [*276*]

[319] Dalecky (Daletskiĭ) Yu.L., Fomin S.V. *Measures and differential equations in infinite-dimensional space.* Transl. from the Russian. With additional material by V.R. Steblovskaya, Yu.V. Bogdansky and N.Yu. Goncharuk. Kluwer Acad. Publ., Dordrecht, 1991; xvi+337 p. (Russian ed.: Moscow, 1983). [*103, 276, 324, 342, 414*]

[320] Daletskiĭ Yu.L., Maryanin B.D. *Smooth measures on infinite-dimensional manifolds.* Dokl. Akad. Nauk SSSR. 1985. V. 285, №6. P. 1297–1300 (in Russian); English transl.: Sov. Math. Dokl. 1985. V. 32. P. 863–866. [*324*]

[321] Daletskiĭ Yu.L., Paramonova S.N. *Stochastic integrals with respect to a normally distributed additive set function.* Dokl. Akad. Nauk SSSR. 1973. V. 208, №3. P. 512–515 (in Russian); English transl.: Sov. Math. Dokl. 1973. V. 14. P. 96–100. [*274*]

[322] Daletskiĭ Yu.L., Paramonova S.N. *On a formula from the theory of Gaussian measures and on the estimation of stochastic integrals.* Teor. Verojatn. i Primen. 1974. V. 19, №4. P. 844–849 (in Russian); English. transl.: Theory Probab. Appl. 1974. V. 19, №4. P. 812–817. [*274*]

[323] Daletskiĭ Yu.L., Paramonova S.N. *Integration by parts with respect to measures in function spaces. I, II.* Teor. Verojatn. Mat. Stat. 1977. V. 17. P. 51–60; 1978. V. 18. P. 37–45 (in Russian); English transl.: Theory Probab. Math. Stat. 1979. V. 17. P. 55–65; V. 18. P. 39–46. [*274*]

[324] Daletskiĭ Yu.L., Sokhadze G.A. *Absolute continuity of smooth measures.* Funkt. Anal. Prilozh. 1988. V. 22, №2. P. 77–78 (in Russian); English transl.: Funct. Anal. Appl. 1988. V. 22, №2. P. 149–150. [*342, 398*]

[325] Dalecky Yu.L., Steblovskaya V.R. *Smooth measures: absolute continuity, stochastic integrals, variational problems.* Probability theory and mathematical statistics (Kiev, 1991). P. 52–69. World Sci. Publ., River Edge, New Jersey, 1992. [*325*]

[326] Dalecky Yu.L., Steblovskaya V.R. *On infinite-dimensional variational problems.* Stoch. Anal. Appl. 1996. V. 14, №1. P. 47–71. [*325*]

[327] Daletskiĭ Yu.L., Steblovskaya V.R. *On absolutely continuous and invariant evolution of smooth measure in Hilbert space.* C. R. Acad. Sci. Paris, sér. I Math. 1996. T. 323, №7. P. 823–827. [*366*]

[328] Daletskiĭ Yu.L., Steblovskaya V.R. *On transformations of smooth measure related to parabolic and hyperbolic differential equations in infinite dimensions.* Stoch. Anal. Appl. 1998. V. 16, №5. P. 945–963. [*366*]

[329] Daletskiĭ Yu.L., Steblovskaya V.R. *Measures with smooth finite-dimensional projections.* Skorokhod's ideas in probability theory. P. 132–146. Inst. Math., Kiev, 2000. [*323*]

[330] Daletsky Yu.L., Steblovskaya V.R. *The smooth measures in Banach spaces and their smooth mappings.* Probability theory and mathematical statistics, Vol. I (Vilnius, 1989), pp. 245–257. "Mokslas", Vilnius, 1990. [*324, 366*]

[331] Davis M.H.A. *Functionals of diffusion process as stochastic integrals.* Math. Proc. Camb. Soc. 1980. V. 87, №1. P. 157–166. [*275*]

[332] Davis M.H.A., Johansson M.P. *Malliavin Monte Carlo Greeks for jump diffusions.* Stoch. Process. Appl. 2006. V. 116, №1. P. 101–129. [*425*]

[333] Davydov Yu.A. *On distributions of multiple Wiener–Itô integrals.* Teor. Verojatn. i Primen. 1990. V. 35, №1. P. 51–62 (in Russian); English transl.: Theory Probab. Appl. 1990. V. 35, №1. P. 27–37. [*324*]

[334] Davydov Yu.A. *On convergence in variation of one-dimensional image measures.* Zap. Nauchn. Semin. POMI. 1992. V. 194. P. 48–58 (in Russian); English transl.: J. Math. Sci. (New York). 1995. V. 75, №5. P. 1903–1909. [*325*]

[335] Davydov Yu.A., Lifshits M.A. *Fibering method in some probabilistic problems.* Itogi Nauki Tekh., Ser. Teor. Verojatn., Mat. Stat., Teor. Kibern. 1984. V. 22. P. 61–157 (in Russian); English transl.: J. Sov. Math. 1985. V. 31. P. 2796–2858. [*127, 324*]

[336] Davydov Yu.A., Lifshits M.A., Smorodina N.V. *Local properties of distributions of stochastic functionals.* Translated from the Russian. Amer. Math. Soc., Providence, Rhode Island, 1998; xiii+184 p. (Russian ed.: Moscow, 1995). [*147, 296, 323, 398, 399*]

[337] Decreusefond L. *Perturbation analysis and Malliavin calculus.* Ann. Appl. Probab. 1998. V. 8, №2. P. 496–523. [*425*]

[338] Decreusefond L., Hu Y.Z., Üstünel A.-S. *Une inégalité d'interpolation sur l'espace de Wiener.* C. R. Acad. Sci. Paris Sér. I Math. 1993. T. 317, №11. P. 1065–1067. [*274*]

[339] Decreusefond L., Üstünel A.S. *Stochastic analysis of the fractional Brownian motion.* Potential Anal. 1999. V. 10, №2. P. 177–214. [*324*]

[340] Denis L. *Convergence quasi-partout pour les capacités définies par un semi-groupe sous-markovien.* C. R. Acad. Sci. Paris Sér. I Math. 1992. T. 315, №10. P. 1033–1036. [*276*]

[341] Dermoune A. *Distributions sur l'espace de P. Lévy et calcul stochastique.* Ann. Inst. H. Poincaré. Probab. et Statist. 1990. V. 26, №1. P. 101–119. [*399*]

[342] Dermoune A., Krée P., Wu L. *Calcul stochastique non adapté par rapport à la mesure aléatoire de Poisson.* Lecture Notes in Math. 1988. V. 1321. P. 477–484. [*323, 399*]

[343] Detemple J., Garcia R., Rindisbacher M. *Representation formulas for Malliavin derivatives of diffusion processes.* Finance Stoch. 2005. V. 9, №3. P. 349–367. [*425*]

[344] Deuschel J., Föllmer H. *Time reversal and smoothing of infinite dimensional diffusion processes.* Proc. I Internat. Ascona–Como Meeting "Stoch. processes in Classical and Quantum Systems" (S. Albeverio, G. Casati, and D. Merlini eds.). P. 179–186. Lecture Notes in Phys. V. 262. Springer, Berlin, 1986. [*132, 415*]

[345] Deville R., Godefroy G., Zizler V. Smoothness and renormings in Banach spaces. Longman Scientific & Technical, Harlow; John Wiley & Sons, New York, 1993; xii+376 p. [*155, 303*]

[346] Di Nunno G. *Random fields: non-anticipating derivative and differentiation formulas.* Infin. Dim. Anal. Quantum Probab. Relat. Top. 2007. V. 10, №3. P. 465–481. [*323*]

[347] Di Nunno G., Meyer-Brandis T., Øksendal B., Proske F. *Malliavin calculus and anticipative Itô formulae for Lévy processes.* Infin. Dim. Anal. Quantum Probab. Relat. Top. 2005. V. 8, №2. P. 235–258. [*323*]

[348] Di Nunno G., Øksendal B., Proske F. Malliavin calculus for Lévy processes with applications to finance. Springer-Verlag, Berlin, 2009; xiv+413 p. [*323, 425*]

[349] Diebolt J. *Regularité de la loi de la solution d'une équation différentielle excitée par un processus gaussien physiquement réalisable.* C. R. Acad. Sci. Paris, sér. 1. 1987. T. 305, №4. P. 143–146. [*424*]

[350] Diestel J. Geometry of Banach spaces – selected topics. Lecture Notes in Math. V. 485. Springer-Verlag, Berlin – New York, 1975; xi+282 p. [*266*]

[351] Diestel J., Uhl J.J. Vector measures. Amer. Math. Soc., Providence, 1977; xiii+322 p. [*5*]

[352] DiPerna R.J., Lions P.L. *Ordinary differential equations, transport theory and Sobolev spaces.* Invent. Math. 1989. V. 98, №3. P. 511–547. [*366*]

[353] Donati-Martin C., Pardoux E. *EDPS réfléchies et calcul de Malliavin.* Bull. Sci. Math. 1997. V. 121, №5. P. 405–422. [*323*]

[354] Dorogovtsev A.A. *Elements of stochastic differential calculus.* (Russian) In: "Mathematics today, 1988", P. 105–131. Vyscha Shkola, Kiev, 1988. [*240*]

[355] Dorogovtsev A.A. Stochastic analysis and random maps in Hilbert space. VSP, Utrecht, 1994; iv+110 p. (Russian ed.: Kiev, 1992). [*240*]

[356] Dorogovtsev A.A. Anticipating stochastic equations. Inst. Math. Akad. Nauk Ukr., Kiev, 1996; 152 p. (in Russian). [*240*]

[357] Dorogovtsev A.A. *Measurable functionals and finitely absolutely continuous measures on Banach spaces.* Ukr. Mat. Zh. 2000. V. 52, №9. P. 1194–1204 (in Russian); English transl.: Ukr. Math. J. 2000. V. 52, №9. P. 1366–1379. [*251*]

[358] Doss H., Royer G. *Processus de diffusion associés aux mesures de Gibbs sur R^{Z^d}.* Z. Wahr. theor. verw. Geb. 1979. B. 46. S. 125–158. [*132*]

[359] Dragičević O., Volberg A. *Bellman functions and dimensionless estimates of Littlewood–Paley type.* J. Operator Theory. 2006. V. 56, №1. P. 167–198. [*275*]

[360] Driver B. *A Cameron–Martin type quasi-invariance theorem for the Brownian motion on a compact manifold.* J. Funct. Anal. 1992. V. 110, №2. P. 272–376. [*392*]

[361] Driver B.K. *Integration by parts for heat kernel measures revisited.* J. Math. Pures Appl. (9). 1997. V. 76, №8. P. 703–737. [*398*]

[362] Driver B.K. *Integration by parts and quasi-invariance for heat kernel measures on loop groups.* J. Funct. Anal. 1997. V. 149, №2. P. 470–547; correction: ibid. 1998. V. 155, №1. P. 297–301. [*398*]

[363] Driver B.K. *Analysis of Wiener measure on path and loop groups.* Finite and infinite dimensional analysis in honor of Leonard Gross (New Orleans, 2001). P. 57–85. Contemp. Math. V. 317. Amer. Math. Soc., Providence, Rhode Island, 2003. [*398*]

[364] Driver B.K. *Heat kernels measures and infinite dimensional analysis.* Heat kernels and analysis on manifolds, graphs, and metric spaces (Paris, 2002). P. 101–141. Contemp. Math. V. 338. Amer. Math. Soc., Providence, Rhode Island, 2003. [*398*]

[365] Driver B.K. *Curved Wiener space analysis.* Real and stochastic analysis. P. 43–198. Trends Math., Birkhäuser Boston, Boston, 2004. [*398*]

[366] Driver B., Gordina M. *Heat kernel analysis on infinite-dimensional Heisenberg groups.* J. Funct. Anal. 2008. V. 255. P. 2395–2461. [*398*]

[367] Driver B., Gordina M. *Integrated Harnack inequalities on Lie groups.* J. Diff. Geom. 2009. V. 83, №3. P. 501–550. [*398*]

[368] Driver B.K., Lohrenz T. *Logarithmic Sobolev inequalities for pinned loop groups.* J. Funct. Anal. 1996. V. 140, №2. P. 381–448. [*398*]

[369] Driver B., Röckner M. *Construction of diffusions on path and loop spaces of compact Riemannian manifolds.* C. R. Acad. Sci. Paris, sér. 1. 1992. T. 315, №5. P. 603–608. [*398*]

[370] Driver B.K., Srimurthy V.K. *Absolute continuity of heat kernel measure with pinned Wiener measure on loop groups.* Ann. Probab. 2001. V. 29, №2. P. 691–723. [*398*]

[371] Driver B.K., Thalmaier A. *Heat equation derivative formulas for vector bundles.* J. Funct. Anal. 2001. V. 183, №1. P. 42–108. [*398*]

[372] Duc N.M., Nualart D., Sanz M. *Application of Malliavin calculus to a class of stochastique differential equations.* Probab. Theory Relat. Fields. 1990. V. 84, №4. P. 549–571. [*424*]

[373] Dudley R.M., Kanter M. *Zero-one laws for stable measures.* Proc. Amer. Math. Soc. 1974. V. 45, №2. P. 245–252; correction: ibid. 1983. V. 88, №4. P. 689–690. [*108*]

[374] Duncan T.E. *The heat equation, the Kac formula and some index formulas.* Partial Diff. Equations and Geometry, Lect. Notes in Pure and Appl. Math. V. 48. P. 57–76. Dekker, New York, 1979. [*406*]

[375] Duncan T.E. *Stochastic calculus on manifolds with applications.* Recent Advances in Stoch. Calculus (J.S. Baras, V. Mirelli eds.). P. 105–140. Springer, New York, 1990. [*406*]

[376] Dunford N., Schwartz J.T. *Linear operators, I. General Theory.* Interscience, New York, 1958; xiv+858 p. [*235, 383*]

[377] Eberle A. *Uniqueness and non-uniqueness of singular diffusion operators.* Lecture Notes in Math. V. 1718. Springer, Berlin, 1999; 262 p. [*420*]

[378] Eckmann J.-P., Hairer M. *Uniqueness of the invariant measure for a stochastic PDE driven by degenerate noise.* Comm. Math. Phys. 2001. V. 219, №3. P. 523–565. [*425*]

[379] Edwards R.E. *Functional analysis. Theory and applications.* Holt, Rinehart and Winston, New York – London, 1965; xiii+781 p. [*3, 135, 136*]

[380] Efimova E.I., Uglanov A.V. *Green's formula on a Hilbert space.* Mat. Sb. 1982. V. 119, №2. P. 225–232 (in Russian); English transl.: Math. USSR Sb. 1984. V. 47. P. 215–222. [*325*]

[381] El-Khatib Y., Privault N. *Computations of Greeks in a market with jumps via the Malliavin calculus.* Finance Stoch. 2004. V. 8, №2. P. 161–179. [*425*]

[382] Elliott R.J., Kohlmann M. *The existence of smooth densities for the prediction filtering and smoothing problems.* Acta Appl. Math. 1989. V. 14, №3. P. 269–286. [*425*]

[383] Elliott R.J., Kohlmann M. *Martingale representation and the Malliavin calculus.* Appl. Math. Optim. 1989. V. 20, №1. P. 105–112. [*275*]

[384] Elliott R.J., Kohlmann M. *Integration by parts and densities for jump processes.* Stochastics Stoch. Rep. 1989. V. 27, №2. P. 83–97. [*323*]

[385] Elliott R.J., Kohlmann M. *Integration by parts, homogeneous chaos expansions and smooth densities.* Ann. Probab. 1989. V. 17, №1. P. 194–207. [*323*]

[386] Elliott R.J., Tsoi A.H. *Integration by parts for the single jump process.* Statistics and Probab. Lett. 1991. V. 12, №5. P. 363–370. [*323*]

[387] Elliott R.J., Tsoi A.H. *Integration by parts for Poisson processes.* J. Multivar. Anal. 1993. V. 44, №2. P. 179–190. [*323, 399*]

[388] Elson C.M. *An extension of Weyl's lemma to infinite dimensions.* Trans. Amer. Math. Soc. 1974. V. 194. P. 301–324. [*426*]

[389] Elworthy K.D. *Geometric aspects of stochastic analysis.* In: Development of mathematics 1950–2000. P. 437–484. Birkhäuser, Basel, 2000. [*406*]

[390] Elworthy K.D., Le Jan Y., Li X.-M. *Integration by parts formulae for degenerate diffusion measures on path spaces and diffeomorphism groups.* C. R. Acad. Sci. Paris Sér. I Math. 1996. V. 323, №8. P. 921–926. [*398*]

[391] Elworthy K.D., Le Jan Y., Li X.-M. *On the geometry of diffusion operators and stochastic flows.* Lecture Notes in Math. V. 1720. Springer-Verlag, Berlin, 1999; ii+118 p. [*398*]

[392] Elworthy K.D., Li X.-M. *A class of integration by parts formulae in stochastic analysis. I.* In: Itô's stochastic calculus and probability theory. P. 15–30. Springer, Tokyo, 1996. [*398*]

[393] Elworthy K.D., Li X.-M. *Some families of q-vector fields on path spaces.* Infin. Dim. Anal. Quantum Probab. Relat. Top. 2003. V. 6, suppl. P. 1–27. [*398*]

[394] Elworthy K.D., Li X.-M. *Itô maps and analysis on path spaces.* Math. Z. 2007. B. 257. S. 643–706. [*398*]

[395] Elworthy K.D., Li X.-M. *An L^2 theory for differential forms on path spaces. I.* J. Funct. Anal. 2008. V. 254, №1. P. 196–245. [*398*]

[396] Enchev O. *Non linear transformation on the Wiener space.* Ann. Probab. 1993. V. 21, №4. P. 2169–2188. [*366*]

[397] Enchev O., Stroock D. *On Rademacher's theorem on Wiener space.* Ann. Probab. 1993. V. 21, №1. P. 25–33. [*355*]

[398] Es-Sarhir A. *Sobolev regularity of invariant measures for generalized Ornstein–Uhlenbeck operators.* Infin. Dim. Anal. Quantum Probab. Relat. Top. 2006. V. 9, №4. P. 595–606. [*414*]

[399] Estrade A., Pontier M., Florchinger P. *Filtrage avec observation discontinue sur une variété. Existence d'une densité régulière.* Stochastics Stoch. Rep. 1996. V. 56, №1-2. P. 33–51. [*323*]

[400] Evans C., Gariepy R.F. *Measure theory and fine properties of functions.* CRC Press, Boca Raton – London, 1992; viii+268 p. [*52, 57, 66, 175*]

[401] Ewald C.-O. *Local volatility in the Heston model: a Malliavin calculus approach.* J. Appl. Math. Stoch. Anal. 2005. №3. P. 307–322. [*425*]

[402] Ewald C.-O. *The Malliavin gradient method for the calibration of stochastic dynamical models.* Appl. Math. Comput. 2006. V. 175, №2. P. 1332–1352. [*425*]

[403] Ewald C.-O. *A note on the Malliavin derivative operator under change of variable.* Statist. Probab. Lett. 2008. V. 78, №2. P. 173–178. [*323*]

[404] Ewald C.-O., Zhang A. *A new technique for calibrating stochastic volatility models: the Malliavin gradient method.* Quant. Finance. 2006. V. 6, №2. P. 147–158. [*425*]

[405] Fabes E., Gutierrez C.E., Scotto R. *Weak-type estimates for the Riesz transforms associated with Gaussian measure.* Rev. Mat. Iberoamericana. 1994. V. 10. P. 229–281. [*238, 275*]

[406] Fang S. *Une inégalité isopérimétrique sur l'espace de Wiener.* Bull. Sci. Math. (2). 1988. V. 112, №3. P. 345–355. [*306*]

[407] Fang S. *Pseudo-théorème de Sard pour les applications réelles et connexité sur l'espace de Wiener.* Bull. Sci. Math. 1989. V. 113, №4. P. 483–492. [*320, 327*]

[408] Fang S. *Non-dégénéréscence des pseudo-normes de Sobolev sur l'espace de Wiener.* Bull. Sci. Math. (2). 1991. V. 115, №2. P. 223–234. [*323, 324*]

[409] Fang S. *On derivatives of holomorphic functions on a complex Wiener space.* J. Math. Kyoto Univ. 1994. V. 34, №3. P. 637–640. [*275*]

[410] Fang S. *Une inégalité du type de Poincaré sur un espace de chemins.* C. R. Acad. Sci. Paris, sér. 1. 1994. T. 318, №3. P. 257–260. [*398*]

[411] Fang S. *Rotations et quasi-invariance sur l'espace des chemins.* Potential Anal. 1995. V. 4, №1. P. 67–77. [*398*]

[412] Fang S. *Integration by parts for heat measures over loop groups.* J. Math. Pures Appl. (9). 1999. V. 78, №9. P. 877–894. [*398*]

[413] Fang S. *Integration by parts formula and logarithmic Sobolev inequality on the path space over loop groups.* Ann. Probab. 1999. V. 27, №2. P. 664–683. [*398*]

[414] Fang S. *Introduction to Malliavin calculus.* Beijing, 2004; x+153 p. [*323*]

[415] Fang S., Luo D. *Transport equations and quasi-invariant flows on the Wiener space.* Bull. Sci. Math. 2010. V. 134, №3. P. 295–328. [*366*]

[416] Fang S., Malliavin P. *Stochastic analysis on the path space of a Riemannian manifold: I. Markovian stochastic calculus.* J. Funct. Anal. 1993. V. 118. P. 249–274. [*398*]

[417] Fang S.Z., Ren J.G. *Quelques proprietés des fonctions holomorphes sur un espace de Wiener complexe.* C. R. Acad. Sci. Paris, sér. 1. 1992. T. 315, №4. P. 447–450. [*275*]

[418] Fang S., Ren J. *Sur le squelette et les dérivées de Malliavin des fonctions holomorphe sur espace de Wiener complexe.* J. Math. Kyoto Univ. 1993. V. 33, №3. P. 749–764. [*275*]

[419] Fang S., Shao J. *Optimal transport for Monge–Kantorovich problem on loop groups.* J. Funct. Anal. 2007. V. 248, №1. P. 225–257. [*398*]

[420] Fang S., Shao J., Sturm K.-Th. *Wasserstein space over the Wiener space.* Probab. Theory Related Fields. 2010. V. 146, №3-4. P. 535–565. [*399*]

[421] Federer H. *Geometric measure theory.* Springer-Verlag, New York, 1969; xiv+676 p. [*52, 53, 56, 308*]

[422] Feissner G.F. *Hypercontractive semigroups and Sobolev's inequality.* Trans. Amer. Math. Soc. 1975. V. 210. P. 51–62. [*274*]

[423] Fernique X. *Fonctions aléatoires gaussiennes, vecteurs aléatoires gaussiens.* Université de Montréal, Centre de Recherches Mathématiques, Montréal, 1997; iv+217 p. [*35*]

[424] Feyel D., La Pradelle A. de. *Espaces de Sobolev gaussiens.* Ann. Inst. Fourier. 1989. V. 39, №4. P. 875–908. [*274*]

[425] Feyel D., La Pradelle A. de. *Capacités gaussiens.* Ann. Inst. Fourier. 1991. V. 41, №1. P. 49–76. [*276*]

[426] Feyel D., La Pradelle A. de. *Hausdorff measures on the Wiener space.* Potential. Anal. 1992. V. 1, №2. P. 177–189. [*274, 325*]

[427] Feyel D., La Pradelle A. de. *Harmonic analysis in infinite dimension.* Potential Anal. 1993. V. 2, №1. P. 23–36. [*274*]

[428] Feyel D., La Pradelle A. de. *Opérateurs linéaires gaussiennes.* Potential Anal. 1994. V. 3, №1. P. 89–105. [*272*]

[429] Feyel D., Üstünel A.S. *The notion of convexity and concavity on Wiener space.* J. Funct. Anal. 2000. V. 176, №2. P. 400–428. [*180, 275*]

[430] Feyel D., Üstünel A.S. *Monge-Kantorovitch measure transportation and Monge–Ampère equation on Wiener space.* Probab. Theory Relat. Fields. 2004. V. 128, №3. P. 347–385. [*364*]

[431] Feyel D., Üstünel A.S. *Solution of the Monge–Ampère equation on Wiener space for general log-concave measures.* J. Funct. Anal. 2006. V. 232, №1. P. 29–55. [*364*]

[432] Feyel D., Üstünel A.S., Zakai M. *The realization of positive random variables via absolutely continuous transformations of measure on Wiener space.* Probab. Surv. 2006. V. 3. P. 170–205. [*366*]

[433] Flandoli F., Tudor C.A. *Brownian and fractional Brownian stochastic currents via Malliavin calculus.* J. Funct. Anal. 2010. V. 258, №1. P. 279–306. [*425*]

[434] Florchinger P. *Malliavin calculus with time dependent coefficients and applications to nonlinear filtering.* Probab. Theory Relat. Fields. 1990. V. 86, №2. P. 203–223. [*425*]

[435] Florchinger P. *Existence of a smooth density for the filter in nonlinear filtering with infinite-dimensional noise.* Systems Control Lett. 1991. V. 16, №2. P. 131–137. [*425*]

[436] Föllmer H. *Time reversal on Wiener space.* Lecture Notes in Math. 1986. V. 1158. P. 119–129. [*415*]

[437] Föllmer H. *Random fields and diffusion processes.* Lecture Notes in Math. 1988. V. 1362. P. 101–202. [*132, 415*]

[438] Föllmer H. *Martin boundaries on Wiener space.* Diffusion Processes and Related Problems (M. Pinsky ed.). V. 1. P. 3–16. Birkhäuser, 1989. [*426*]

[439] Föllmer H., Wakolbinger A. *Time reversal of infinite dimensional diffusions.* Stoch. Process. Appl. 1986. V. 22, №1. P. 59–77. [*415*]

[440] Fomin S.V. *Differentiable measures in linear spaces.* Proc. Int. Congress of Mathematicians, sec. 5. P. 78–79. Izdat. Moskov. Univ., Moscow, 1966 (in Russian). [*ix, 69, 103*]

[441] Fomin S.V. *Differentiable measures in linear spaces.* Uspehi Matem. Nauk. 1968. V. 23, №1. P. 221–222 (in Russian). [*ix, 69, 103*]

[442] Fomin S.V. *Generalized functions of infinitely many variables and their Fourier trasnforms.* Uspehi Mat. Nauk. 1968. V. 23, №2. P. 215–216 (in Russian). [*ix, 69, 103*]

[443] Fomin S.V. *Some new problems and results in non-linear functional analysis.* Vestnik Moskov. Univ. Ser. I Mat. Mekh. 1970. V. 25, №2. P. 57–65 (in Russian); English transl.: Moscow Univ. Math. Bull. 1970. V. 25, №1-2. P. 81–86. [*ix, 69, 95, 103*]

[444] Forster B., Lütkebohmert E., Teichmann J. *Absolutely continuous laws of jump-diffusions in finite and infinite dimensions with applications to mathematical finance.* SIAM J. Math. Anal. 2009. V. 40, №5. P. 2132–2153. [*425*]

[445] Forzani L., Scotto R., Sjögren P., Urbina W. *On the L^p boundedness of the non-centered Gaussian Hardy–Littlewood maximal function.* Proc. Amer. Math. Soc. 2002. V. 130, №1. P. 73–79. [*238*]

[446] Fournié E., Lasry J.-M., Lebuchoux J., Lions P.-L. *Applications of Malliavin calculus to Monte-Carlo methods in finance. II.* Finance Stoch. 2001. V. 5, №2. P. 201–236. [*425*]

[447] Fournié E., Lasry J.-M., Lebuchoux J., Lions P.-L., Touzi N. *Applications of Malliavin calculus to Monte Carlo methods in finance.* Finance Stoch. 1999. V. 3, №4. P. 391–412. [*425*]

[448] Fournier N. *Malliavin calculus for parabolic SPDEs with jumps.* Stoch. Process. Appl. 2000. V. 87, №1. P. 115–147. [*323*]

[449] Fournier N., Giet J.-S. *Existence of densities for jumping stochastic differential equations.* Stoch. Process. Appl. 2006. V. 116, №4. P. 643–661. [*323*]

[450] Franz U., Privault N., Schott R. *Non-Gaussian Malliavin calculus on real Lie algebras.* J. Funct. Anal. 2005. V. 218, №2. P. 347–371. [*323*]

[451] Fritz J. *Stationary measures of stochastic gradient systems, infinite lattice models.* Z. Wahr. theor. verw. Geb. 1982. B. 59, №4. S. 479–490. [*132, 415*]

[452] Frolov N.N. *Embedding theorems for spaces of functions of countably many variables, I.* Proceedings Math. Inst. of Voronezh Univ., Voronezh University. 1970. № 1. P. 205–218 (in Russian). [*253, 274*]

[453] Frolov N.N. *Embedding theorems for spaces of functions of countably many variables and their applications to the Dirichlet problem.* Dokl. Akad. Nauk SSSR. 1972. V. 203, №1. P. 39–42 (in Russian); English transl.: Soviet Math. 1972. V. 13, №2. P. 346–349. [*253, 274*]

[454] Frolov N.N. *On a coercivity inequality for an elliptic operator in an infinite number of independent variables.* Matem. Sb. 1973. V. 90, P. 402–413 (in Russian); English transl.: Math. USSR Sbornik. 1973. V. 19. P. 395–406. [*274*]

[455] Frolov N.N. *Essential self-adjointness of an infinite-dimensional operator.* Mat. Zametki. 1978. V. 24, №2. P. 241–248 (in Russian); English transl.: Math. Notes. 1979. V. 24. P. 630–634. [*274, 420*]

[456] Frolov N.N. *Self-adjointness of the Schrödinger operator with an infinite number of variables.* Sib. Mat. Zh. 1981. V. 22, №1. P. 198–204 (in Russian); English transl.: Sib. Math. J. 1981. V. 22. P. 147–152. [*274, 420*]

[457] Frolov N.N. *Embedding theorems for spaces of functions of a countable number of variables and their applications.* Sibirsk. Mat. Zh. 1981. V. 22, №4. P. 199–217 (in Russian); English transl.: Siberian Math. J. 1981. V. 22, №4. P. 638–652. [*253, 274*]

[458] Fuhrman M. *Smoothing properties of stochastic equations in Hilbert spaces.* Nonlinear Diff. Eq. Appl. 1996. V. 3. P. 445–464. [*414*]

[459] Fuhrman M. *Regularity properties of transition probabilities in infinite dimensions.* Stochastics Stoch. Rep. 2000. V. 69, №1-2. P. 31–65. [*414*]

[460] Fuhrman M. *Logarithmic derivatives of invariant measure for stochastic differential equations in Hilbert spaces.* Stochastics Stoch. Rep. 2001. V. 71, №3-4. P. 269–290. [*414*]

[461] Fukushima M. Dirichlet forms and Markov processes. North-Holland, Amsterdam, 1980; x+196 p. [*276*]

[462] Fukushima M. *A note on capacities in infinite dimensions.* Lecture Notes in Math. 1988. V. 1299. P. 80–85. [*273*]

[463] Fukushima M. *BV functions and distorted Ornstein–Uhlenbeck processes over the abstract Wiener space.* J. Funct. Anal. 2000. V. 174. P. 227–249. [*275*]

[464] Fukushima M., Hino M. *On the space of BV functions and a related stochastic calculus in infinite dimensions.* J. Funct. Anal. 2001. V. 183, №1. P. 245–268. [*275*]

[465] Fukushima M., Kaneko H. *On (r,p)-capacities for general Markovian semigroups.* Infinite dimensional analysis and stochastic processes (S. Albeverio, ed.). P. 41–47. Pitman, Boston, 1985. [*275*]

[466] Fukushima M., Oshima Y., Takeda M. Dirichlet forms and symmetric Markov processes. De Gruyter, Berlin – New York, 1994; 392 p. [*60, 276, 425*]

[467] García-Cuerva J., Mauceri G., Meda S., Sjögren P., Torrea J.L. *Functional calculus for the Ornstein–Uhlenbeck operator.* J. Funct. Anal. 2001. V. 183, №2. P. 413–450. [*275*]

[468] Gaveau B. *Analyse sur certaines sous-varietés de l'espace de Wiener.* C. R. Acad. Sci. Paris, sér. 1. 1985. T. 301, №19. P. 881–884. [*399*]

[469] Gaveau B., Moulinier J.-M. *Intégrales oscillantes stochastiques: estimation asymptotique de fonctionnelles charactéristiques.* J. Funct. Anal. 1983. V. 54, №2. P. 161–176. [*296, 324*]

[470] Gaveau B., Moulinier J.M. *Régularité des mesures et perturbation stochastiques de champs des vecteurs sur des espaces de dimension infinie.* Publ. Res. Inst. Sci. Kyoto Univ. 1985. V. 21, №3. P. 593–616. [*324, 414*]

[471] Gaveau B., Trauber P. *L'intégral stochastique comme opérateur de divergence dans l'espace fonctionnel.* J. Funct. Anal. 1982. V. 46, №2. P. 230–238. [*261*]

[472] Gaveau B., Vauthier J. *Intégrales oscillantes stochastiques et l'équation de Pauli.* J. Funct. Anal. 1981. V. 44, №3. P. 388–400. [*324*]

[473] Gel'fand I.M., Vilenkin N.Ya. Generalized functions. V. 4. Applications of harmonic analysis. Translated from the Russian. Academic Press, New York – London, 1964; xiv+384 p. (Russian ed.: Moscow, 1961). [*104*]

[474] Georgii H.-O. Gibbs measures and phase transitions. Walter de Gruyter, Berlin, 1988; xiv+525 p. [*132*]

[475] Getzler E. *Degree theory for Wiener maps.* J. Funct. Anal. 1988. V. 68, №3. P. 388–403. [*367*]

[476] Getzler E. *Dirichlet forms on loop spaces.* Bull. Sci. Math. (2). 1989. V. 113, №2. P. 151–174. [*398*]

[477] Gihman I.I., Skorohod A.V. The theory of stochastic processes. I–III. Translated from the Russian. Corrected printing of the first edition. Springer, Berlin, 2004, 2007; viii+574 p., vii+441 p., vii+387 p. [*35, 103, 123, 132, 170, 349*]

[478] Gihman I.I., Skorohod A.V. Stochastic differential equations and their applications. Naukova Dumka, Kiev, 1982; 611 p. (in Russian). [*35*]

[479] Gilkey P.B. Asymptotic formulae in spectral geometry. Chapman & Hall/CRC, Boca Raton, Florida, 2004; viii+304 p. [*406, 407*]

[480] Girsanov I.V. *On transforming a certain class of stochastic processes by absolutely continuous substitution of measures.* Teor. Verojatn. i Primen. 1960. V. 5, №3. P. 314–330 (in Russian); English transl.: Theory Probab. Appl. 1960. V. 5. P. 285–301. [*366*]

[481] Girsanov I.V., Mityagin B.S. *Quasi-invariant measures in topological linear spaces.* Nauchn. Dokl. Vyssh. Shkoly, Fiz.-Mat. Nauki. 1959. №2. P. 5–9 (in Russian). [*103*]

[482] Giusti E. Minimal surfaces and functions of bounded variation. Birkhäuser Verlag, Basel, 1984; xii+240 p. [*66*]

[483] Gliklikh Y. *Global analysis in mathematical physics. Geometric and stochastic methods.* Springer-Verlag, New York, 1997; xvi+213 p. [*400*]

[484] Gloter A., Gobet E. *LAMN property for hidden processes: the case of integrated diffusions.* Ann. Inst. H. Poincaré. Probab. Stat. 2008. V. 44, №1. P. 104–128. [*425*]

[485] Gobet E. *Local asymptotic mixed normality property for elliptic diffusion: a Malliavin calculus approach.* Bernoulli. 2001. V. 7, №6. P. 899–912. [*425*]

[486] Gobet E., Kohatsu-Higa A. *Computation of Greeks for barrier and look-back options using Malliavin calculus.* Electron. Comm. Probab. 2003. V. 8. P. 51–62. [*425*]

[487] Gobet E., Munos R. *Sensitivity analysis using Itô–Malliavin calculus and martingales, and application to stochastic optimal control.* SIAM J. Control Optim. 2005. V. 43, №5. P. 1676–1713. [*425*]

[488] Gohberg I.C., Krein M.G. *Introduction to the theory of linear nonselfadjoint operators.* (English) Translated from the Russian. Amer. Math. Soc., Rhode Island, Providence, 1969; xv+378 p. (Russian ed.: Moscow, 1965). [*331*]

[489] Goldin G.A., Menikoff R., Sharp D.H. *Particle statistics from induced representations of a local current group.* J. Math. Phys. 1980. V. 21, №4. P. 650–664. [*400*]

[490] Gol'dshteĭn V.M., Reshetnyak Yu.G. *Quasiconformal mappings and Sobolev spaces.* Translated and revised from the 1983 Russian original. Kluwer Academic Publ., Dordrecht, 1990; xx+371 p. (Russian ed.: Introduction to the theory of functions with generalized derivatives and quasiconformal mappings. Moscow, 1983). [*50, 52, 54, 56, 66, 312*]

[491] Goldys B., Gozzi F., van Neerven J.M.A.M. *On closability of directional gradients.* Potential Anal. 2003. V. 18, №4. P. 289–310. [*275*]

[492] Gomilko A.M., Dorogovtsev A.A. *Localization of the extended stochastic integral.* Mat. Sb. 2006. V. 197, №9. P. 19–42 (in Russian); English transl.: Sb. Math. 2006. V. 197, №9. P. 1273–1295. [*257*]

[493] Gong F.-Z., Ma Z.-M. *Invariance of Malliavin fields on Itô's Wiener space and on abstract Wiener space.* J. Funct. Anal. 1996. V. 138, №2. P. 449–476. [*323*]

[494] Goodman V. *Quasi-differentiable functions on Banach spaces.* Proc. Amer. Math. Soc. 1971. V. 30. P. 367–370. [*155*]

[495] Goodman V. *A divergence theorem for Hilbert space.* Trans. Amer. Math. Soc. 1972. V. 164. P. 411–426. [*325*]

[496] Gordina M. *Heat kernel analysis and Cameron–Martin subgroup for infinite-dimensional groups.* J. Funct. Anal. 2000. V. 171, №1. P. 192–232. [*399*]

[497] Gordina M. *Holomorphic functions and the heat kernel measure on an infinite-dimensional complex orthogonal group.* Potent. Anal. 2000. V. 12. P. 325–357. [*399*]

[498] Gordina M. *Quasi-invariance for the pinned Brownian motion on a Lie group.* Stoch. Process. Appl. 2003. V. 104, №2. P. 243–257. [*399*]

[499] Gordina M., Lescot P. *Riemannian geometry of* $\mathrm{Diff}(S^1)/S^1$. J. Funct. Anal. 2006. V. 239, №2. P. 611–630. [*400*]

[500] Gozzi F. *Smoothing properties of nonlinear transition semigroups: case of Lipschitz nonlinearities.* J. Evol. Equ. 2006. V. 6, №4. P. 711–743. [*414*]

[501] Graczyk P. *Malliavin calculus for stable processes on homogeneous groups.* Studia Math. 1991. V. 100, №3. P. 183–205. [*323*]

[502] Gravereaux J.-B. *Calcul de Malliavin et probabilité invariante d'une chaine de Marcov.* Ann. Inst. H. Poincaré. 1988. V. 24, №2. P. 159–188. [*323*]

[503] Gross L. *Integration and nonlinear transformations in Hilbert space.* Trans. Amer. Math. Soc. 1960. V. 94. P. 404–440. [*366*]

[504] Gross L. *Potential theory on Hilbert space.* J. Funct. Anal. 1967. V. 1, №2. P. 123–181. [*x, 103, 426*]

[505] Gross L. *Logarithmic Sobolev inequalities.* Amer. J. Math. 1975. V. 97, №4. P. 1061–1083. [*250, 274*]

[506] Gross L. *Logarithmic Sobolev inequalities and contractive properties of semigroups.* Lecture Notes in Math. 1993. V. 1563. P. 54–82. [*274*]

[507] Gross L. *Uniqueness of ground states for Schödinger operators over loop groups.* J. Funct. Anal. 1993. V. 112, №2. P. 373–441. [*398*]

[508] Gross L. *Heat kernel analysis on Lie groups.* Stochastic analysis and related topics, VII (Kusadasi, 1998). P. 1–58. Progr. Probab., V. 48. Birkhäuser Boston, Boston, 2001. [*398*]

[509] Guérin H., Méléard S., Nualart E. *Estimates for the density of a nonlinear Landau process.* J. Funct. Anal. 2006. V. 238, №2. P. 649–677. [*425*]

[510] Guerquin M. *Non-hilbertian structure of the Wiener measure.* Colloq. Math. 1973. V. 28. P. 145–146. [*155*]

[511] Gutiérrez C. *On the Riesz transform for Gaussian measures.* J. Funct. Anal. 1994. V. 120, №1. P. 107–134. [*275*]

[512] Gutiérrez C.E., Segovia C., Torrea J. *On higher Riesz transforms for Gaussian measures.* J. Fourier Anal. Appl. 1996. V. 2, №6. P. 583–596. [*238, 275*]

[513] Hajłasz P. *Change of variables formula under minimal assumptions.* Colloq. Math. 1993. V. 64, №1. P. 93–101. [*56, 68*]

[514] Hajłasz P., Koskela P. *Sobolev met Poincaré.* Mem. Amer. Math. Soc. 2000. V. 145, №688; x+101 p. [*274*]

[515] Hamza M. *Détermination des formes de Dirichlet sur $\mathbb{R}^n$.* Thèse 3e cycle. Univ. Orsay, 1975. [*61*]

[516] Hara K., Ikeda N. *Quadratic Wiener functionals and dynamics on Grassmannians.* Bull. Sci. Math. 2001. V. 125. P. 481–528. [*324*]

[517] Hargé G. *Regularité de certaines fonctionnelles sur l'espace de Wiener.* C. R. Acad. Sci. Paris, sér. 1. 1993. T. 316, №6. P. 593–596. [*324*]

[518] Hargé G. *Continuité approximative de certaines fonctionnelles sur l'espace de Wiener.* J. Math. Pures Appl. (9). 1995. V. 74, №1. P. 59–93. [*324*]

[519] Hargé G. *Limites approximatives sur l'espace de Wiener.* Potential Anal. 2002. V. 16, №2. P. 169–191. [*324*]

[520] Hariya Y. *Integration by parts formulae for Wiener measures restricted to subsets in $\mathbb{R}^d$.* J. Funct. Anal. 2006. V. 239. P. 594–610. [*132*]

[521] Has'minskiĭ R.Z. *Stochastic stability of differential equations.* Transl. from the Russian. Sijthoff & Noordhoff, Alphen aan den Rijn—Germantown, Md., 1980. xvi+344 p. (Russian ed.: Moscow, 1969). [*408*]

[522] Haussmann U.G. *Functionals of Itô processes as stochastic integrals.* SIAM J. Control Optim. 1978. V. 16, №2. P. 252–269. [*275*]

[523] Haussmann U.G. *On the integral representation of functionals of Itô processes.* Stochastics. 1979. V. 3, №1. P. 17–28. [*275*]

[524] Haussmann U.G., Pardoux E. *Time reversal of diffusions.* Ann. Probab. 1986. V. 14, №4. P. 1188–1205. [*415*]

[525] Hayashi M. *Asymptotic expansions for functionals of a Poisson random measure.* J. Math. Kyoto Univ. 2008. V. 48, №1. P. 91–132. [*425*]

[526] Hebey E. *Nonlinear analysis on manifolds: Sobolev spaces and inequalities.* New York University, Courant Institute of Mathematical Sciences, New York; Amer. Math. Soc., Providence, Rhode Island, 1999; x+309 p. [*66*]

[527] Hegerfeld G. *On canonical commutation relations and infinite dimensional measures.* J. Math. Phys. 1972. V. 13, №1. P. 45–50. [*104*]

[528] Hegerfeld G. *Gårding domains and analytic vectors for quantum fields.* J. Math. Phys. 1972. V. 13, №6. P. 821–827. [*104*]

[529] Hegerfeld G. *Probability measures on distribution spaces and quantum field theoretical models.* Rep. Math. Phys. 1975. V. 7, №3. P. 403–409. [*104*]

[530] Hegerfeld G., Melsheimer O. *The form of representations of the canonical commutation relations for Bose fields and connection with finitely many degrees of freedom.* Comm. Math. Phys. 1969. V. 12. P. 304–323. [*104*]

[531] Heinonen J. *Lectures on analysis on metric spaces.* Springer, New York, 2001; x+140 p. [*274*]

[532] Heinonen J. *Nonsmooth calculus.* Bull. Amer. Math. Soc. 2007. V. 44, №2. P. 163–232. [*274*]

[533] Hertle A. *Gaussian surface measures and the Radon transform on separable Banach spaces.* Lecture Notes in Math. 1980. V. 794. P. 513–531. [*325*]

[534] Hertle A. *Gaussian plane and spherical means in separable Hilbert spaces.* Lecture Notes in Math. 1982. V. 945. P. 314–335. [*325*]

[535] Heya N. *The absolute continuity of a measure induced by infinite dimensional stochastic differential equations.* J. Math. Sci. Univ. Tokyo. 2005. V. 12, №1. P. 77–104. [*425*]

[536] Heya N. *Hypoelliptic stochastic differential equations in infinite dimensions.* J. Math. Sci. Univ. Tokyo. 2005. V. 12, №3. P. 399–416. [*425*]

[537] Hida T. *Analysis of Brownian functionals.* Carleton Math. Lecture Notes. №13. Carleton Univ., Ottawa, 1975; iv+61 p. [*324*]

[538] Hida T., Kuo H., Pothoff J., Streit L. White noise. An infinite-dimensional calculus. Kluwer Acad. Publ., Dordrecht, 1993; xiv+516 p. [*276, 324*]

[539] Hino M. *Spectral properties of Laplacians on an abstract Wiener space with a weighted Wiener measure.* J. Funct. Anal. 1997. V. 147, №2. P. 485–520. [*426*]

[540] Hino M. *Existence of invariant measures for diffusion processes on a Wiener space.* Osaka J. Math. 1998. V. 35, №3. P. 717–734. [*413, 414*]

[541] Hino M. *Integral representation of linear functionals on vector lattices and its application to BV functions on Wiener space.* Stochastic analysis and related topics in Kyoto. P. 121–140. Adv. Stud. Pure Math., 41, Math. Soc. Japan, Tokyo, 2004. [*275*]

[542] Hino M. *Sets of finite perimeter and the Hausdorff–Gauss measure on the Wiener space.* J. Funct. Anal. 2010. V. 258, №5. P. 1656–1681. [*275, 325*]

[543] Hiraba S. *Existence and smoothness of transition density for jump-type Markov processes: applications of Malliavin calculus.* Kodai Math. J. 1992. V. 15, №1. P. 29–49. [*323*]

[544] Hirsch F. *Proprieté d'absolue continuité pour les équations différentielles stochastiques dependant du passé.* J. Funct. Anal. 1988. V. 76, №1. P. 193–216. [*425*]

[545] Hirsch F. *Theory of capacity on the Wiener space.* Stochastic analysis and related topics, V (Silivri, 1994). P. 69–98. Progr. Probab. V. 38. Birkhäuser Boston, Boston, 1996. [*276*]

[546] Hirsch F. *Lipschitz functions and fractional Sobolev spaces.* Potential Anal. 1999. V. 11, №4. P. 415–429. [*274*]

[547] Hirsch F., Song S. *Criteria of positivity for the density of the law of a Wiener functional.* Bull. Sci. Math. 1997. V. 121, №4. P. 261–273. [*326, 321*]

[548] Hirsch F., Song S. *Properties of the set of positivity for the density of a regular Wiener functional.* Bull. Sci. Math. 1998. V. 122, №1. P. 1–15. [*321, 326*]

[549] Hitsuda M. *Formula for Brownian partial derivative.* Proc. Second Japan–USSR Symp. on Probab. Theory. V. 2. P. 111–114. Inst. Math. Kyoto Univ., Kyoto, 1972. [*324*]

[550] Hörmander L. The analysis of linear partial differential operators. I–IV. Springer-Verlag, Berlin – New York, 1983, 1985; ix+391 p., viii+391 p., viii+525 p., vii+352 p. [*402, 404*]

[551] Hörmander L. *Hypoelliptic second order differential equations.* Acta Math. 1967. V. 119. P. 147–171. [*402*]

[552] Holley R., Stroock D. *Diffusions on an infinite dimensional torus.* J. Funct. Anal. 1981. V. 42, №1. P. 29–63. [*132*]

[553] Hora A. *On Banach space of functions associated with a homogeneous additive process.* Publ. RIMS, Kyoto Univ. 1988. V. 24. P. 739–757. [*104, 107, 131*]

[554] Hora A. *Quasi-invariant measures on commutative Lie groups of infinite product type.* Math. Z. 1991. B. 206, №2. S. 169–192. [*399*]

[555] Hsu E.P. *Quasi-invariance of the Wiener measure on the path space over a compact Riemannian manifold.* J. Funct. Anal. 1995. V. 134, №2. P. 417–450. [*398*]

[556] Hsu E.P. *Integration by parts in loop spaces.* Math. Ann. 1997. V. 309, №2. P. 331–339. [*398*]

[557] Hsu E.P. *Analysis on path and loop spaces.* Probability theory and applications (Princeton, 1996). P. 277–347. IAS/Park City Math. Ser., 6, Amer. Math. Soc., Providence, Rhode Island, 1999. [*398*]

[558] Hsu E.P. *Quasi-invariance of the Wiener measure on path spaces: noncompact case.* J. Funct. Anal. 2002. V. 193, №2. P. 278–290. [*398*]

[559] Hsu E.P. Stochastic analysis on manifolds. American Math. Soc., Providence, Rhode Island, 2002; xiv+281 p. [*398*]

[560] Hsu E.P. *Characterization of Brownian motion on manifolds through integration by parts. Stein's method and applications.* Lecture Notes Ser. Inst. Math. Sci. Nat. Univ. Singapore, 5. P. 195–208. Singapore Univ. Press, Singapore, 2005. [*398*]

[561] Hu J., Ren J. *Infinite dimensional quasi continuity, path continuity and ray continuity of functions with fractional regularity.* J. Math. Pures Appl. (9). 2001. V. 80, №1. P. 131–152. [*274*]

[562] Hu Y. *Integral transformations and anticipative calculus for fractional Brownian motions.* Mem. Amer. Math. Soc. 2005. V. 175, №825; viii+127 p. [*324*]

[563] Hu Y., Üstünel A.S., Zakai M. *Tangent processes on Wiener space.* J. Funct. Anal. 2002. V. 192, №1. P. 234–270. [*399*]

[564] Huang Z., Yan J. *Introduction to infinite dimensional stochastic analysis.* Kluwer Acad. Publ., Dordrecht; Science Press, Beijing, 2000; xii+296 p. [*276, 323*]

[565] Hunt B.R., Sauer T., Yorke J.A. *Prevalence: a translation-invariant "almost-everywhere" on infinite-dimensional spaces.* Bull. Amer. Math. Soc. 1992. V. 27, №2. P. 217–238; addendum: ibid. 1993. V. 28, №2. P. 306–307. [*366*]

[566] Ibragimov I.A., Rozanov Y.A. Gaussian random processes. Translated from the Russian. Springer-Verlag, New York – Berlin, 1978; x+275 p. (Russian ed.: Moscow, 1970). [*35*]

[567] Ikeda N. *Probabilistic methods in the study of asymptotics.* Lecture Notes in Math. 1990. V. 1427. P. 196–325. [*405, 406, 407*]

[568] Ikeda N., Manabe S. *Asymptotic formulae for stochastic oscillatory integrals.* In: Asymptotic problems in probability theory: Wiener functionals and asymptotics (Sanda/Kyoto, 1990). P. 136–155. Pitman Res. Notes Math. Ser. V. 284, Longman Sci. Tech., Harlow, 1993. [*324*]

[569] Ikeda N., Shigekawa I., Taniguchi S. *The Malliavin calculus and long time asymptotics of certain Wiener integrals.* Proc. Centre Math. Anal. Austral. Nat. Univ. 1985. V. 9. P. 46–113. [*323, 405*]

[570] Ikeda N., Watanabe S. *Malliavin calculus of Wiener functionals and its applications.* Pitman Res. Notes Math. Ser. V. 150. P. 132–178. Longman Scientific, Harlow, 1986. [*323, 425*]

[571] Ikeda N., Watanabe S. Stochastic differential equations and diffusion processes. North-Holland, Amsterdam, 1989; xvi+555 p. [*31, 35, 232, 274, 323, 392, 401, 405*]

[572] Imkeller P. *Enlargement of the Wiener filtration by an absolutely continuous random variable via Malliavin's calculus.* Probab. Theory Relat. Fields. 1996. V. 106, №1. P. 105–135. [*323*]

[573] Imkeller P. *Malliavin's calculus in insider models: additional utility and free lunches.* Math. Finance. 2003. V. 13, №1. P. 153–169. [*425*]

[574] Imkeller P., Nualart D. *Integration by parts on Wiener space and the existence of occupation densities.* Ann. Probab. 1994. V. 22, №1. P. 469–493. [*324*]

[575] Inahama Y. *Meyer's inequality of the type $L(\ln L)^\alpha$ on abstract Wiener spaces.* J. Math. Kyoto Univ. 1999. V. 39, №1. P. 87–113. [*274*]

[576] Inahama Y. *Logarithmic Sobolev inequality on free loop groups for heat kernel measures associated with the general Sobolev spaces.* J. Funct. Anal. 2001. V. 179, №1. P. 170–213. [*398*]

[577] Inahama Y. *Logarithmic Sobolev inequality for H_0^s-metric on pinned loop groups.* Infin. Dim. Anal. Quantum Probab. Relat. Top. 2004. V. 7, №1. P. 1–26. [*398*]

[578] Ishikawa Y., Kunita H. *Malliavin calculus on the Wiener-Poisson space and its application to canonical SDE with jumps.* Stoch. Process. Appl. 2006. V. 116, №12. P. 1743–1769. [*323, 399*]

[579] Ismagilov R.S. *On unitary representations of the group of diffeomorphisms of the space R^n, $n > 2$*. Mat. Sb. 1975. V. 98. P. 55–71 (in Russian); English transl.: Math. USSR, Sb. 1975. V. 27. P. 51–65. [*399*]

[580] Ismagilov R.S. *Representations of infinite-dimensional groups*. Amer. Math. Soc., Providence, Rhode Island, 1996; x+197 p. [*399*]

[581] Itô K. *Malliavin calculus on a Segal space*. Lecture Notes in Math. 1987. V. 1322. P. 50–72. [*323*]

[582] Itô K. *On Malliavin calculus*. Probability theory (Singapore, 1989). P. 47–72. De Gruyter, Berlin, 1992. [*323*]

[583] Itô K. *An elementary approach to Malliavin fields*. Pitman Research Notes in Math. Sci. V. 284 (D. Elworthy and N. Ikeda, eds.). P. 35–89. Longman, 1993. [*323*]

[584] Itô K. *A measure-theoretic approach to Malliavin calculus*. New trends in stochastic analysis (Charingworth, 1994). P. 220–287. World Sci. Publ., River Edge, New Jersey, 1997. [*323*]

[585] Janson S. *Gaussian Hilbert spaces*. Cambridge Univ. Press, Cambridge, 1997; x+340 p. [*240*]

[586] Johnson G.W., Skoug D.L. *Scale-invariant measurability in Wiener space*. Pacific J. Math. 1979. V. 83, №1. P. 157–176. [*366*]

[587] Kabanov Ju.M., Skorohod A.V. *Extended stochastic integrals*. Proceedings of the School and Seminar on the Theory of Random Processes (Druskininkai, 1974), Part I, pp. 123–167. Inst. Fiz. i Mat. Akad. Nauk Litovsk. SSR, Vilnius, 1975 (in Russian). [*324*]

[588] Kachanovsky N.A. *A generalized Malliavin derivative connected with the Poisson- and gamma-measures*. Methods Funct. Anal. Topol. 2003. V. 9, №3. P. 213–240. [*399*]

[589] Kakutani S. *On equivalence of infinite product measures*. Ann. Math. 1948. V. 49. P. 214–224. [*107*]

[590] Kallenberg O. *Foundations of modern probability*. 2nd ed. Springer-Verlag, New York, 2002; xx+638 p. [*35, 125*]

[591] Karatzas I., Ocone D., Li J. *An extension of Clark's formula*. Stochastics Stoch. Rep. 1991. V. 37, №3. P. 127–131. [*275*]

[592] Kashin B.S., Saakyan A.A. *Orthogonal series*. Translated from the Russian. Amer. Math. Soc., Providence, Rhode Island, 1989; xii+451 p. (Russian ed.: Moscow, 1984). [*148*]

[593] Kats M.P. *The second derivative of twice differentiable measure might not be countably additive*. Vestnik Moskov. Univ. Ser. I Mat. Mekh. 1972. V. 27, №5. P. 35–43 (in Russian); English transl.: Moscow Univ. Math. Bull. 1972. V. 27, №5. P. 27–34. [*151*]

[594] Kats M.P. *Conditions for the countable additivity of the derivatives of differentiable measures*. Vestnik Moskov. Univ. Ser. I Mat. Mekh. 1972. V. 27, №6. P. 18–25 (in Russian); English transl.: Moscow Univ. Math. Bull. 1972. V. 27, №6. P. 82–88. [*151*]

[595] Kats M.P. *Countable additivity of the derivatives of sufficiently smooth measures*. Uspehi Mat. Nauk. 1973. V. 28, №3. P. 183–184 (in Russian). [*151*]

[596] Kats M.P. *Quasi-invariance and differentiability of measures*. Uspehi Mat. Nauk. 1978. V. 33, №3. P. 175 (in Russian); English transl.: Russ. Math. Surv. 1978. V. 33, №3. P. 159. [*156*]

[597] Katznelson Y., Malliavin P. *Un contre-exemple au théorème de Sard en dimension infinie*. C. R. Acad. Sci. Paris, sér. 1. 1988. T. 306, №1. P. 37–41. [*366*]

[598] Kauhanen J. *Failure of the condition N below $W^{1,n}$*. Ann. Acad. Sci. Fenn. Math. 2002. V. 27, №1. P. 141–150. [*55*]

[599] Kazumi T., Shigekawa I. *Measures of finite (r,p)-energy and potentials on a separable metric space*. Lecture Notes in Math. 1992. V. 1526. P. 415–444. [*270, 272, 275*]

[600] Keith S. *A differentiable structure for metric measure spaces*. Adv. Math. 2004. V. 183, №2. P. 271–315. [*274*]

[601] Khafizov M.U. *The differentiability space of a product measure.* Vestn. Mosk. Univ., Ser. I. 1989. №2. P. 81–84 (in Russian); English transl.: Moscow Univ. Math. Bull. 1989. V. 44, №2. P. 105–108. [*131, 141*]

[602] Khafizov M.U. *Some new results on differentiable measures.* Vestn. Mosk. Univ., Ser. I. 1990. №4. P. 63–66 (in Russian); English transl.: Moscow Univ. Math. Bull. 1990. V. 45, №4. P. 34–36. [*103, 155, 187*]

[603] Khafizov M.U. *A quasi-invariant smooth measure on the diffeomorphisms group of a domain.* Mat. Zametki. 990. V. 48, №2. P. 134–142 (in Russian); English transl.: Math. Notes. 1990. V. 48, №3-4. P. 968–974. [*397*]

[604] Khafizov M.U. *The measure derivative has unbounded variation.* Izv. Vyssh. Uchebn. Zaved., Mat. 1992. №12. P. 57–58 (in Russian); English transl.: Russ. Math. 1992. V. 36, №12. P. 57–58. [*196*]

[605] Khafizov M.U. *On the continuity space of product measure.* Mat. Zametki. 1993. V. 54, №5. P. 119–128 (in Russian); English transl.: Math. Notes. 1993. V. 54, №5. P. 1159–1164. [*131, 136, 156*]

[606] Kingman J.F.C. *Poisson processes.* Clarendon Press, Oxford, 1993; viii+104 p. [*399*]

[607] Kirillov A.I. *Two mathematical problems of canonical quantization. I, II, III, IV.* Teoret. Mat. Fiz. 1991. V. 87, №1. P. 22–33; 1991. V. 87, №2. P. 163–172; 1992. V. 91, №3. P. 377–395; 1992. V. 93, №2. P. 249–263 (in Russian); English transl.: Theoret. Math. Phys. 1991. V. 87, №1. P. 345–353; 1991. V. 87, №2. P. 447–454; 1992. V. 91, №3. P. 591–603; 1992. V. 93, №2. P. 1251–1261. [*205, 223, 426*]

[608] Kirillov A.I. *Brownian motion with drift in a Hilbert space and its application in integration theory.* Teor. Verojat. i Primen. 1993. V. 38, №3. P. 629–634 (in Russian); Theory Probab. Appl. 1993. V. 38, №3. P. 529–533. [*224, 426*]

[609] Kirillov A.I. *On the determination of measures on function spaces by means of numerical densities and function integrals.* Mat. Zametki. 1993. V. 53, №5. P. 152–155 (in Russian); English transl.: Math. Notes. 1993. V. 53, №5-6. P. 555–557. [*224, 426*]

[610] Kirillov A.I. *A field of sine-Gordon type in space-time of arbitrary dimension: the existence of the Nelson measure.* Teoret. Mat. Fiz. 1994. V. 98. №1. P. 12–28 (in Russian); English transl.: Theoret. Math. Phys. 1994. V. 98. №1. P. 8–19. [*224, 426*]

[611] Kirillov A.I. *Infinite-dimensional analysis and quantum theory as semimartingale calculi.* Uspekhi Mat. Nauk. 1994. V. 49, №3. P. 43–92 (in Russian); English transl.: Russian Math. Surveys. 1994. V. 49, №3. P. 43–95. [*223, 224, 415, 416, 426*]

[612] Kirillov A.I. *On the reconstruction of measures from their logarithmic derivatives.* Izv. Ross. Akad. Nauk Ser. Mat. 1995. V. 59, №1. P. 121–138 (in Russian); English transl.: Izv. Math. 1995. V. 59, №1. P. 121–139. [*224, 426*]

[613] Kirillov A.I. *A field of sine-Gordon type in space-time of arbitrary dimension. II. Stochastic quantization.* Teoret. Mat. Fiz. 1995. V. 105, №2. P. 179–197 (in Russian); English transl.: Theoret. Math. Phys. 1995. V. 105, №2. P. 1329–1345. [*224, 426*]

[614] Kirillov A.I. *Stochastic quantization using a modified Langevin equation.* Teoret. Mat. Fiz. 1998. V. 115, №1. P. 46–55 (in Russian); English transl.: Theoret. Math. Phys. 1998. V. 115, №1. P. 410–417. [*224, 426*]

[615] Kirillov A.I. *Generalized differentiable product measures.* Mat. Zametki. 1998. V. 63, №1. P. 37–55 (in Russian); English transl.: Math. Notes. 1998. V. 63, №1-2. P. 33–49. [*224, 426*]

[616] Knothe H. *Contributions to the theory of convex bodies.* Michigan Math. J. 1957. V. 4. P. 39–52. [*367*]

[617] Knutova E.M. *Asymptotic properties of Student's conditional distributions in a Hilbert space.* Vestn. Samar. Gos. Univ. Mat. Mekh. Fiz. Khim. Biol. 2001. №4. P. 42–55 (in Russian). [*132*]

[618] Kobanenko, K.N. *On the extension of generalized Lipschitz mappings.* Mat. Zametki. 1998. V. 63, №5. P. 789–791 (in Russian); English transl.: Math. Notes. 1998. V. 63, №5. P. 693–695. *[250]*

[619] Kohatsu-Higa A. *Weak approximations. A Malliavin calculus approach.* Math. Comp. 2001. V. 70, №233. P. 135–172. *[425]*

[620] Kohatsu-Higa A. *Lower bounds for densities of uniformly elliptic random variables on Wiener space.* Probab. Theory Relat. Fields. 2003. V. 126. P. 421–457. *[326]*

[621] Kohatsu-Higa A., Montero M. *Malliavin calculus in finance.* Handbook of computational and numerical methods in finance. P. 111–174. Birkhäuser Boston, Boston, 2004. *[425]*

[622] Kohatsu-Higa A., Pettersson R. *Variance reduction methods for simulation of densities on Wiener space.* SIAM J. Numer. Anal. 2002. V. 40, №2. P. 431–450. *[425]*

[623] Kohatsu-Higa A., Yasuda K. *Estimating multidimensional density functions using the Malliavin–Thalmaier formula.* SIAM J. Numer. Anal. 2009. V. 47, №2. P. 1546–1575. *[425]*

[624] Kolesnikov A.V. *Convexity inequalities and optimal transport of infinite-dimensional measures.* J. Math. Pures Appl. 2004. V. 83, №11. P. 1373–1404. *[367]*

[625] Kolmogorov A.N. *Zur Umkehrbarkeit der statistischen Naturgesetze.* Math. Ann. 1937. B. 113. S. 766–772. *[408]*

[626] Kolmogorov A.N., Fomin S.V. Introductory real analysis. V. 1. Metric and normed spaces. V. 2. Measure. The Lebesgue integral. Hilbert space. Transl. from the 2nd Russian ed. Corr. repr. Dover, New York, 1975; xii+403 p. (Russian ed.: Elements of the theory of functions and functional analysis, Moscow, 1954, 1960). *[1]*

[627] Komatsu T. *On the Malliavin calculus for SDE's on Hilbert spaces.* Acta Appl. Math. 2003. V. 78, №1-3. P. 223–232. *[323]*

[628] Kondrat'ev Yu.G. *Dirichlet operators and smoothness of the solutions of infinite-dimensional elliptic equations.* Dokl. Akad. Nauk SSSR. V. 282, №2. P. 269–273 (in Russian); English transl.: Sov. Math. Dokl. 1985. V. 31. P. 461–464. *[426]*

[629] Kondratiev Y.G., Lytvynov E.W., Us G.F. *Analysis and geometry on R_+-marked configuration space.* Methods Funct. Anal. Topology. 1999. V. 5, №1. P. 29–64. *[399]*

[630] Kondratiev Yu.G., Tsycalenko T.V. *Dirichlet operators and differential equations.* Selecta Math. Sov. 1991. V. 10. P. 345–397. *[420]*

[631] Kondratiev Yu.G., Tsycalenko T.V. *Infinite-dimensional Dirichlet operators I: Essential self-adjointness and associated elliptic equations.* Potential Anal. 1993. V. 2. P. 1–21. *[420]*

[632] Kontsevich M.L., Suhov Yu.M. *On Malliavin measures, SLE, and CFT.* Tr. Matem. Inst. im. V.A. Steklova. 2007. V. 258. P. 107–153 (ArXiv Math-Ph. 0609056). *[398]*

[633] Körezlioğlu H., Üstünel A.S. *A new class of distributions on Wiener spaces.* Lecture Notes in Math. 1990. V. 1444. P. 106–121. *[276]*

[634] Koskela P. *Extensions and imbeddings.* J. Funct. Anal. 1998. V. 159. P. 369–383. *[56]*

[635] Kosyak A.V. *Criteria for irreducibility and equivalence of regular Gaussian representations of groups of finite upper-triangular matrices of infinite order.* Selecta Math. Soviet. 1992. V. 11, №3. P. 241–291. *[400]*

[636] Kosyak A.V. *Irreducible regular Gaussian representations of the groups of the interval and circle diffeomorphisms.* J. Funct. Anal. 1994. V. 125, №2. P. 493–547. *[400]*

[637] Krée M. *Propriété de trace en dimension infinie, d'espaces du type Sobolev.* C. R. Acad. Sci., sér. A. 1974. T. 297. P. 157–164. *[274]*

[638] Krée M. *Propriété de trace en dimension infinie, d'espaces du type Sobolev.* Bull. Soc. Math. France. 1977. V. 105, №2. P. 141–163. *[274]*

[639] Krée M., Krée P. *Continuité de la divergence dans les espace de Sobolev relatifs à l'espace de Wiener.* C. R. Acad. Sci., sér. 1 math. 1983. T. 296, №20. P. 833–836. *[274]*

[640] Krée P. *Calcul d'intégrales et de dérivées en dimension infinie.* J. Funct. Anal. 1979. V. 31, №2. P. 150–186. [*274*]

[641] Krée P. *Regularité C^∞ des lois conditionelles par rapport à certaines variables aléatoires.* C. R. Acad. Sci. Paris, sér. 1 math. 1983. T. 296, №4. P. 223–225. [*323*]

[642] Krée P. *Calcul vectoriel des variations stochastiques par rapport à une mesure de probabilité $H - C^\infty$ fixée.* C. R. Acad. Sci. Paris, sér. 1. 1985. T. 300, №15. P. 557–560. [*323*]

[643] Krée P. *La théorie des distributions en dimension quelconque et l'intégration stochastique.* Lecture Notes in Math. 1988. V. 1316. P. 170–233. [*323*]

[644] Kruglov V.M. Supplementary chapters of probability theory. Vysš. Škola, Moscow, 1984; 264 p. (in Russian). [*35*]

[645] Krugova E.P. *On the integrability of logarithmic derivatives of measures.* Mat. Zametki. 1993. V. 53, №5. P. 76–86 (in Russian); English transl.: Math. Notes. 1993. V. 53, №5-6. P. 506–512. [*163*]

[646] Krugova E.P. *On differentiability of convex measures.* Matem. Zametki. 1995. V. 57, №6. P. 51–61 (in Russian); English transl.: Math. Notes. 1995. V. 58, №6. P. 1294–1301. [*115, 115, 116, 131*]

[647] Krugova E.P. *On translates of convex measures.* Mat. Sb. 1997. V. 188, №2. P. 57–66 (in Russian); English transl.: Sb. Math. 1997. V. 188, №2. P. 227–236. [*118, 131*]

[648] Kruk I., Russo F., Tudor C.A. *Wiener integrals, Malliavin calculus and covariance measure structure.* J. Funct. Anal. 2007. V. 249, №1. P. 92–142. [*323*]

[649] Kufner A. Weighted Sobolev spaces. John Wiley, New York, 1985; 116 p. [*66*]

[650] Kufner A., John O., Fučik S. Function spaces. Noordhoff International Publ., Leyden; Academia, Prague, 1977; xv+454 p. [*66*]

[651] Kulik A.M. *Admissible formal differentiations for pure discontinuous Markov processes.* Theory Stoch. Process. 2000. V. 6, №1-2. P. 90–96. [*323*]

[652] Kulik A.M. *Filtering and finite-dimensional characterization of logarithmically convex measures.* Ukrain. Mat. Zh. 2002. V. 54, №3. P. 323–331 (in Russian); English transl.: Ukrainian Math. J. 2002. V. 54, №3. P. 398–408. [*131*]

[653] Kulik A.M. *Log-Sobolev inequality, exponential integrability and large deviation estimates for $C(\alpha,\beta)$ log-concave measures.* Random Oper. Stoch. Eq. 2002. V. 10, №2. P. 105–122. [*131, 274*]

[654] Kulik A.M. *Malliavin calculus for functionals with generalized derivatives and some applications for stable processes.* Ukr. Math. J. 2002. V. 54, №2. P. 266–279. [*323*]

[655] Kulik A.M. *Malliavin calculus for Lévy processes with arbitrary Levy measures.* Teor. Ĭmovīr. Mat. Stat. 2005. №72. P. 67–83 (in Ukrainian); English transl.: Theory Probab. Math. Statist. 2006. №72. P. 75–92. [*323*]

[656] Kulik A.M. *On a convergence in variation for distributions of solutions of SDE's with jumps.* Random Oper. Stoch. Eq. 2005. V. 13, №3. P. 297–312. [*326*]

[657] Kulik A.M. *Markov uniqueness and Rademacher theorem for smooth measures on infinite-dimensional space under successful filtration condition.* Ukrain. Mat. Zh. 2005. V. 57, №2. P. 170–186; English tranls.: Ukrain. Math. J. 2005. V. 57, №2. P. 200–220. [*366*]

[658] Kulik A.M. *Exponential ergodicity of the solutions to SDE's with a jump noise.* Stochastic Process. Appl. 2009. V. 119, №2. P. 602–632. [*326*]

[659] Kulik A.M., Pilipenko A.Yu. *Nonlinear transformations of smooth measures on infinite-dimensional spaces.* Ukr. Mat. Zh. 2000. V. 52, №9. P. 1226–1250 (in Russian); English transl.: Ukrain. Math. J. 2000. V. 52, №9. P. 1403–1431. [*340, 366*]

[660] Kunita H. Stochastic flows and stochastic differential equations. Cambridge Univ. Press, Cambridge, 1990; xiv+346 p. [*366*]

[661] Kunita H. *Stochastic flows acting on Schwartz distributions.* J. Theoret. Probab. 1994. V. 7, №2. P. 247–278. [*366*]

[662] Kuo H. *Integration theory in infinite dimensional manifolds.* Trans. Amer. Math. Soc. 1971. V. 159. P. 57–78. [*399*]

[663] Kuo H. *Differential measures.* Chinese J. Math. 1974. №2. P. 188–198. [*103*]

[664] Kuo H. *Integration by parts for abstract Wiener measures.* Duke Math. J. 1974. V. 41. P. 373–379. [*103*]

[665] Kuo H. *On Gross differentiation on Banach spaces.* Pacif. J. Math. 1975. V. 59. P. 135–145. [*103*]

[666] Kuo H.-H. *Gaussian measures in Banach spaces.* Springer-Verlag, Berlin – New York, 1975; vi+224 p. [*35, 155, 325*]

[667] Kuo H. *Potential theory associated with Uhlenbeck–Ornstein process.* J. Funct. Anal. 1976. V. 21, №1. P. 63–75. [*103, 426*]

[668] Kuo H. *The chain rule for differentiable measures.* Studia Math. 1978. V. 63, №2. P. 145–155. [*103*]

[669] Kuo H. *Differential calculus for measures on Banach spaces.* Lecture Notes in Math. 1978. V. 644. P. 270–285. [*103*]

[670] Kuo H. *Brownian motion, diffusions and infinite dimensional calculus.* Lecture Notes in Math. 1988. V. 1316. P. 130–169. [*103, 426*]

[671] Kuo H. *White noise distribution theory.* CRC Press, Boca Raton, Florida, 1996; xii+378 p. [*276, 324*]

[672] Kusuoka S. (Seiichiro) *Existence of densities of solutions of stochastic differential equations by Malliavin calculus.* J. Funct. Anal. 2010. V. 258, №3. P. 758–784. [*425*]

[673] Kusuoka S. (Shigeo) *The nonlinear transformation of Gaussian measure on Banach space and its absolute continuity I, II.* J. Fac. Sci. Univ. Tokyo, sec. 1A. 1982. V. 29, №3. P. 567–598; 1983. V. 30, №1. P. 199–220. [*366*]

[674] Kusuoka S. *Dirichlet forms and diffussion processes on Banach spaces.* J. Fac. Sci. Univ. Tokyo, sec. 1A. 1982. V. 29, №1. P. 79–95. [*366*]

[675] Kusuoka S. *Analytic functionals of Wiener processes and absolute continuity.* Lecture Notes in Math. 1982. V. 923. P. 1–46. [*324*]

[676] Kusuoka S. *On the absolute continuity of the law of a system of multiple Wiener integral.* J. Fac. Sci. Univ. Tokyo, sec. 1A. 1983. V. 30, №1. P. 191–198. [*287*]

[677] Kusuoka S. *The generalized Malliavin calculus based on Brownian sheet and Bismut's expansion for large deviation.* Lecture Notes in Math. 1984. V. 1158. P. 141–157. [*405*]

[678] Kusuoka S. *Some remarks on Getzler's degree theorem.* Lecture Notes in Math. 1988. V. 1299. P. 239–249. [*366*]

[679] Kusuoka S. *Degree theorem in certain Riemann–Wiener manifolds.* Lecture Notes in Math. 1988. V. 1322. P. 93–108. [*399*]

[680] Kusuoka S. *On the foundations of Wiener–Riemannian manifolds.* Stochastic Anal., Path Integration and Dynamics, Pitman Research Notes in Math. Series. 1989. V. 200. P. 130–164. [*399*]

[681] Kusuoka S. *The Malliavin calculus and its applications.* Sugaku Exp. 1990. V. 3, №2. P. 127–144. [*323, 403, 404, 425*]

[682] Kusuoka S. *Analysis on Wiener spaces I, nonlinear maps.* J. Funct. Anal. 1991. V. 98, №1. P. 122–168; — *II, Differential forms.* Ibidem. 1992. V. 103, №2. P. 229–274. [*399*]

[683] Kusuoka S. *More recent theory of Malliavin calculus.* Sugaku Expositions. 1992. V. 5, №2. P. 127–144. [*323, 425*]

[684] Kusuoka S. *Stochastic analysis as infinite-dimensional analysis.* Sugaku Exp. 1997. V. 10, №2. P. 183–194. [*323*]

[685] Kusuoka S. *Malliavin calculus revisited.* J. Math. Sci. Univ. Tokyo. 2003. V. 10, №2. P. 261–277. [*323, 425*]

[686] Kusuoka S., Osajima Y. *A remark on the asymptotic expansion of density function of Wiener functionals.* J. Funct. Anal. 2008. V. 255, №9. P. 2545–2562. [*325*]

[687] Kusuoka S., Stroock D.W. *Applications of the Malliavin calculus, I.* Stochastic Analysis, Proc. Taniguchi Intern. Symp. Katata and Kyoto (K. Itô ed.). P. 271–306. North-Holland, Amsterdam, 1984. [*323, 425*]

[688] Kusuoka S., Stroock D.W. *Applications of the Malliavin calculus, II.* J. Fac. Sci. Univ. Tokyo. 1985. V. 32, №1. P. 1–76. [*323, 403, 404, 425*]

[689] Kusuoka S., Stroock D.W. *Applications of the Malliavin calculus, III.* J. Fac. Sci. Univ. Tokyo. 1987. V. 34. P. 391–442. [*323, 425*]
[690] Kusuoka S., Stroock D.W. *The partial Malliavin calculus and its application to nonlinear filtering.* Stochastics. 1984. №12. P. 83–142. [*425*]
[691] Kusuoka S., Stroock D.W. *Some boundedness properties of certain stationary diffusion semigroups.* J. Funct. Anal. 1985. V. 60, №2. P. 243–264. [*415*]
[692] Kusuoka S., Stroock D.W. *Precise asymptotics of certain Wiener functionals.* J. Funct. Anal. 1991. V. 99, №1. P. 1–74. [*325*]
[693] Kusuoka S., Taniguchi S. *Pseudo-convex domains in almost complex abstract Wiener spaces.* J. Funct. Anal. 1993. V. 117, №1. P. 62–117. [*275*]
[694] Kusuoka S., Yoshida N. *Malliavin calculus, geometric mixing, and expansion of diffusion functionals.* Probab. Theory Relat. Fields. 2000. V. 116, №4. P. 457–484. [*425*]
[695] Kuzmin P.A. *On circle diffeomorphisms with discontinuous derivatives and quasi-invariance subgroups of Malliavin–Shavgulidze measures.* J. Math. Anal. Appl. 2007. V. 330, №1. P. 744–750. [*400*]
[696] Lanjri Zadi N., Nualart D. *Smoothness of the law of the supremum of the fractional Brownian motion.* Electron. Comm. Probab. 2003. V. 8. P. 102–111. [*323*]
[697] Larsson-Cohn L. *On the constants in the Meyer inequality.* Monatsh. Math. 2002. V. 137, №1. P. 51–56. [*274*]
[698] Lascar B. *Propriétés locales d'espaces de type Sobolev en dimension infinie.* Comm. Partial Dif. Equations. 1976. V. 1, №6. P. 561–584. [*274*]
[699] Léandre R. *Régularité de processus de sauts dégénérés. I, II.* Ann. Inst. H. Poincaré. 1985. V. 21, №2. P. 125–146; 1988. V. 24, №2. P. 209–236. [*323*]
[700] Léandre R. *Minoration en temps petit de la densité d'une diffusion dégénérée.* J. Funct. Anal. 1987. V. 74, №2. P. 399–414. [*405*]
[701] Léandre R. *Majoration en temps petit de la densité d'une diffusion dégénérée.* Probab. Theory Relat. Fields. 1987. V. 74, №2. P. 289–294. [*405*]
[702] Léandre R. *Quantative and geometric applications of the Malliavin calculus.* Contemp. Math. 1988. V. 73. P. 173–196. [*405*]
[703] Léandre R. *Sur le théorème d'Atiyah–Singer.* Probab. Theory Relat. Fields. 1988. V. 80, №1. P. 199–237. [*406*]
[704] Léandre R. *Sur le théorème de l'indice des familles.* Lecture Notes in Math. 1988. V. 1321. P. 348–419. [*406*]
[705] Léandre R. *Développement asymptotique de la densité d'une diffusion dégénérée.* Forum Math. 1992. V. 4, №1. P. 45–75. [*405*]
[706] Léandre R. *Integration by parts formulas and rotationally invariant Sobolev calculus on free loop spaces.* J. Geom. Phys. 1993. V. 11. P. 517–528. [*398*]
[707] Léandre R. *Invariant Sobolev calculus on the free loop space.* Acta Appl. Math. 1997. V. 46, №3. P. 267–350. [*398*]
[708] Léandre R. *Logarithmic Sobolev inequalities for differentiable paths.* Osaka J. Math. 2000. V. 37, №1. P. 139–145. [*398*]
[709] Léandre R. *Malliavin calculus of Bismut type in semi-group theory.* Far East J. Math. Sci. 2008. V. 30, №1. P. 1–26. [*405*]
[710] Léandre R., Norris J. *Integration by parts and Cameron–Martin formulas for the free path space of a compact Riemannian manifold.* Lecture Notes in Math. 1997. V. 1655. P. 16–23. [*398*]
[711] Ledoux M. *On an integral criterion for hypercontractivity of diffusion semigroups and extremal functions.* J. Funct. Anal. 1992. V. 105, №2. P. 444–465. [*274*]
[712] Ledoux M. The concentration of measure phenomenon. Amer. Math. Soc., Providence, Rhode Island, 2001; x+181 p. [*274*]
[713] Lee Y.-J. *Sharp inequalities and regularity of heat semigroup on infinite-dimensional spaces.* J. Funct. Anal. 1987. V. 71, №1. P. 69–87. [*275*]
[714] Leindler L. *On a certain converse of Hölder's inequality. II.* Acta Sci. Math. (Szeged). 1972. V. 33. P. 217–223. [*131*]

[715] León J.A., Solé J.L., Utzet F., Vives J. *On Lévy processes, Malliavin calculus and market models with jumps.* Finance Stoch. 2002. V. 6, №2. P. 197–225. [*425*]

[716] Lescot P. *Un critère de régularité des lois pour certaines fonctionnelles de Wiener.* C. R. Acad. Sci. Paris, sér. 1. 1993. T. 316, №1. P. 1313–1318. [*323*]

[717] Lescot P. *Desintégration de la mesure de Wiener sous certains fonctionnelles dégénérées.* C. R. Acad. Sci. Paris, sér. 1. 1993. T. 317, №1. P. 93–95. [*323*]

[718] Lescot P. *Un théorème de désintégration en analyse quasi-sure.* Lecture Notes in Math. 1993. V. 1557. P. 256–275. [*323*]

[719] Lescot P. *Sard's theorem for hyper-Gevrey functionals on the Wiener space.* J. Funct. Anal. 1995. V. 129, №1. P. 191–220. [*367*]

[720] Lescot P., Malliavin P. *A pseudo-differential symbolic calculus for fractional derivatives on the Wiener space.* C. R. Acad. Sci. Paris, sér. 1. 1996. T. 322, №5. P. 475–479. [*246*]

[721] Li X.D. *Sobolev spaces and capacities theory on path spaces over a compact Riemannian manifold.* Probab. Theory Related Fields. 2003. V. 125, №1. P. 96–134. [*398*]

[722] Li X.D. *Asymptotic behavior of divergences and Cameron–Martin theorem on loop spaces.* Ann. Probab. 2004. V. 32, №3B. P. 2409–2445. [*398*]

[723] Lifshits M.A. *Gaussian random functions.* Translated from the Russian. Kluwer Academic Publ., Dordrecht, 1995; xii+333 p. (Russian ed.: Kiev, 1995). [*35*]

[724] Lifshits M.A., Shmileva E.Yu. *Poisson measures that are quasi-invariant with respect to multiplicative transformations.* Teor. Verojatn. i Primen. 2001. V. 46, №4. P. 697–712 (in Russian); English transl.: Theory Probab. Appl. 2003. V. 46, №4. P. 652–666. [*399*]

[725] Linde W. *Probability in Banach spaces – stable and infinitely divisible distributions.* Wiley, 1986; 195 p. [*109*]

[726] Lindenstrauss J., Matoušková E., Preiss D. *Lipschitz image of a measure-null set can have a null complement.* Israel J. Math. 2000. V. 118. P. 207–219. [*350*]

[727] Lindenstrauss J., Zafriri L. *Classical Banach spaces.* V. 1,2. Springer-Verlag, Berlin, 1977, 1979; xiii+188 p., x+243 p. [*135*]

[728] Liptser R.S., Shiryaev A.N. *Statistics of random processes.* V. 1, 2. Springer, Berlin, 2001; xv+427 p., xv+402 p. (Russian ed.: Moscow, 1974). [*366*]

[729] Liskevich V., Röckner M. *Strong uniqueness for certain infinite-dimensional Dirichlet operators and applications to stochastic quantization.* Ann. Scuola Norm. Sup. Pisa Cl. Sci. (4). 1998. V. 27, №1. P. 69–91. [*420*]

[730] Liskevich V., Röckner M., Sobol Z., Us O. L^p-*uniqueness for infinite-dimensional symmetric Kolmogorov operators: the case of variable diffusion coefficients.* Ann. Scuola Norm. Sup. Pisa Cl. Sci. (4). 2001. V. 30, №2. P. 285–309. [*420*]

[731] Liskevich V.A., Semenov Yu.A. *Conditions for the selfadjointness of Dirichlet operators.* Ukrain. Mat. Zh. 1990. V. 42, №2. P. 284–289 (in Russian); English transl.: Ukrain. Math. J. 1990. V. 42, №2. P. 253–257. [*420*]

[732] Liskevich V.A., Semenov Ju.A. *Dirichlet operators: a priori estimates and the uniqueness problem.* J. Funct. Anal. 1992. V. 109, №1. P. 199–213. [*420*]

[733] Lobuzov A.A. *First boundary-value problem for a parabolic equation in an abstract Wiener space.* Mat. Zametki. 1981. V. 30, №2. P. 221–233 (in Russian); English transl.: Math. Notes. 1982. V. 30. P. 592–599. [*426*]

[734] Long H. *Necessary and sufficient conditions for the symmetrizability of differential operators over infinite dimensional state spaces.* Forum Math. 2000. V. 12, №2. P. 167–196. [*415*]

[735] Long H., Simão I. *A note on the essential self-adjointness of Ornstein–Uhlenbeck operators perturbed by a dissipative drift and a potential.* Infin. Dim. Anal. Quantum Probab. Relat. Top. 2004. V. 7, №2. P. 249–259. [*420*]

[736] Lott J., Villani C. *Weak curvature conditions and functional inequalities.* J. Funct. Anal. 2007. V. 245. P. 311–333. [*399*]

[737] Ługiewicz P., Zegarliński B. *Coercive inequalities for Hörmander type generators in infinite dimensions.* J. Funct. Anal. 2007. V. 247, №2. P. 438–476. [*426*]

[738] Lunardi A. On the Ornstein–Uhlenbeck operator in L^2 spaces with respect to invariant measures. Trans. Amer. Math. Soc. 1997. V. 349, №1. P. 155–169. [*275*]

[739] Luo D. Well-posedness of Fokker–Planck type equations on the Wiener space. Infin. Dim. Anal. Quantum Probab. Relat. Top. 2010. [*366*]

[740] Luo S. A nonlinear deformation of Wiener space. J. Theor. Probab. 1998. V. 11, №2. P. 331–350. [*366*]

[741] Lyons T., Qian Z. Stochastic Jacobi fields and vector fields induced by varying area on path spaces. Probab. Theory Relat. Fields. 1997. V. 109, №4. P. 539–570. [*398*]

[742] Lyons T., Röckner M. A note on tightness of capacities associated with Dirichlet forms. Bull. London Math. Soc. 1992. V. 24. P. 181–189. [*276*]

[743] Ma Z.M., Röckner M. An introduction to the theory of (non-symmetric) Dirichlet forms. Springer, Berlin, 1992; 209 p. [*276*]

[744] Maas J. Malliavin calculus and decoupling inequalities in Banach spaces. J. Math. Anal. Appl. 2010. V. 363, №2. P. 383–398. [*275*]

[745] Maas J. Analysis of infinite dimensional diffusions. Dissertation, Delft Technical University, Delft, 2009; ix+265 p. [*275*]

[746] Maas J., van Neerven J. On analytic Ornstein–Uhlenbeck semigroups in infinite dimensions. Arch. Math. (Basel). 2007. V. 89, №3. P. 226–236. [*275*]

[747] Maas J., van Neerven J. A Clark–Ocone formula in UMD Banach spaces. Electron. Comm. Probab. 2008. V. 13. P. 151–164. [*275, 324*]

[748] Maas J., van Neerven J. Boundedness of Riesz transforms for elliptic operators on abstract Wiener spaces. J. Funct. Anal. 2009. V. 257, №8. P. 2410–2475. [*275*]

[749] Malgrange B. Ideals of differentiable functions. Oxford University Press, London, 1966; 106 p. [*293*]

[750] Malliavin M., Malliavin P. Integration on loop groups I. Quasi-invariant measures. J. Funct. Anal. 1990. V. 93, №1. P. 207–237; — II. Heat equation for the Wiener measure. Ibid. 1992. V. 104, №1. P. 71–109; — III. Asymptotic Peter–Weyl orthogonality. Ibid. 1992. V. 108, №1. P. 13–46. [*398*]

[751] Malliavin P. Stochastic calculus of variation and hypoelliptic operators. Proc. Intern. Symp. on Stoch. Diff. Eq. (Res. Inst. Math. Sci., Kyoto Univ., Kyoto, 1976). P. 195–263. Wiley, New York – Chichester – Brisbane, 1978. [*x, 274, 403*]

[752] Malliavin P. Sur certaines intégrales stochastiques oscillantes. C. R. Acad. Sci., sér. 1. 1982. T. 295, №3. P. 295–300. [*324*]

[753] Malliavin P. Implicit functions in finite corank on the Wiener space. Proc. Taniguchi Intern. Symp. on Stochastic Anal., Katata and Kyoto (K. Itô ed.). P. 369–386. North-Holland, Amsterdam, 1984. [*323*]

[754] Malliavin P. Analyse différentielle sur l'espace de Wiener. Proc. Intern. Congr. Math. (Warszawa, Aug. 16-24, 1983). V. 2. P. 1089–1096. PWN, Warszawa, 1984. [*323*]

[755] Malliavin P. Calcul des variations, intégrales stochastiques et complex de de Rham sur l'espace de Wiener. C. R. Acad. Sci. Paris, sér. 1. 1984. T. 299, №8. P. 347–350. [*399*]

[756] Malliavin P. Intégrales stochastiques oscillantes et une formule de Feynman–Kac positive. C. R. Acad. Sci., sér. 1. 1985. T. 300, №5. P. 141–143. [*324*]

[757] Malliavin P. Analyticité transverse d'opérateurs hypoelliptiques C^3 sur des fibres principaux. Spectre équivariant et courbure. C. R. Acad. Sci. Paris, sér. 1. 1985. T. 301, №16. P. 767–770. [*323*]

[758] Malliavin P. Analyticité réelle des lois de certaines fonctionnelles additives. C. R. Acad. Sci. Paris, sér. 1. 1986. T. 302, №2. P. 73–78. [*324*]

[759] Malliavin P. Estimation du rayon d'analyticité transverse et calcul des variations. C. R. Acad. Sci., sér. 1. 1986. T. 302, №9. P. 359–362. [*324*]

[760] Malliavin P. Minoration de l'état fondamental de l'équation de Schrödinger du magnétisme et calcul des variations stochastique. C. R. Acad. Sci. Paris, sér. 1. 1986. T. 302, №13. P. 481–486. [*324*]

[761] Malliavin P. *Hypoellipticity in infinite dimension*. Diffusion Processes and Related Problems, Evanston, 1989 (M. Pinsky, ed.). V. 1. P. 17–32. Birkhäuser Boston, Boston, 1989. [*426*]

[762] Malliavin P. *Smooth σ-fields*. Stochastic Analysis (E. Mayer-Wolf, E. Merzbach, A. Schwartz, eds.). P. 371–382. Academic Press, New York – Boston, 1991. [*376, 377*]

[763] Malliavin P. *Naturality of quasi-invariance of some measures*. Stochastic analysis and applications (Lisbon, 1989; A.-B. Cruzeiro, ed.), Progr. Probab. V. 26. P. 144–154. Birkhäuser Boston, Boston, 1991. [*398*]

[764] Malliavin P. *Infinite dimensional analysis*. Bull. Sci. Math. 1993. V. 117, №1. P. 63–90. [*273, 323*]

[765] Malliavin P. *Universal Wiener space*. Barcelona Seminar on Stochastic Analysis, Progress in Probab. V. 32 (D. Nualart and M. Sanz eds.). P. 77–102. Birkhäuser, Basel, 1993. [*323*]

[766] Malliavin P. *Filtered Wiener space versus abstract Wiener space*. The Legacy of Norbert Wiener: A Centennial Symposium (Cambridge, MA, 1994), pp. 185–197, Proc. Sympos. Pure Math., 60, Amer. Math. Soc., Providence, Rhode Island, 1997. [*275*]

[767] Malliavin P. Stochastic analysis. Springer-Verlag, Berlin, 1997; xii+343 p. [*246, 323, 398*]

[768] Malliavin P. *Heat measures and unitarizing measures for Berezinian representations on the space of univalent functions in the unit disk*. Perspectives in analysis. P. 253–268. Springer, Berlin, 2005. [*398*]

[769] Malliavin P. *Invariant or quasi-invariant probability measures for infinite dimensional groups. I. Non-ergodicity of Euler hydrodynamic*. Japan. J. Math. 2008. V. 3, №1. P. 1–17; *II. Unitarizing measures or Berezinian measures*. ibid. P. 19–47. [*400*]

[770] Malliavin P., Katznelson Y. *Image des points critiques d'une application régulière*. Lecture Notes in Math. 1987. V. 1322. P. 85–92. [*366*]

[771] Malliavin P., Nualart D. *Quasi sure analysis of stochastic flows and Banach space valued smooth functionals on the Wiener space*. J. Funct. Anal. 1993. V. 112, №2. P. 287–317. [*273*]

[772] Malliavin P., Nualart D. *Quasi sure analysis and Stratonovich anticipative SDE's*. Probab. Theory Relat. Fields. 1993. V. 96, №1. P. 45–55. [*273*]

[773] Malliavin P., Nualart E. *Density minoration of a strongly non-degenerated random variable*. J. Funct. Anal. 2009. V. 256, №12. P. 4197–4214. [*172*]

[774] Malliavin P., Thalmaier A. Stochastic calculus of variations in mathematical finance. Springer-Verlag, Berlin, 2006; xii+142 p. [*425*]

[775] Malý J., Martio O. *Lusin's condition (N) and mappings of the class $W^{1,n}$*. J. Reine. Angew. Math. 1995. B. 458. S. 19–36. [*55*]

[776] Mankiewicz P. *On the differentiability of Lipschitz mappings in Fréchet spaces*. Studia Math. 1973. V. 45. P. 15–29. [*343, 353*]

[777] Marcus D. *Non-stable laws with all projections stable*. Z. Wahr. theor. verw. Geb. 1983. B. 64. S. 139–156. [*108*]

[778] Markushevich A.I. Theory of analytic functions. V. 2. Moscow, 1968; 624 p. (in Russian); English transl.: Theory of functions of a complex variable. V. I, II, III. 2nd English ed. Chelsea Publ., New York, 1977; xxii+1238 p. [*98*]

[779] Márquez-Carreras D. *On the asymptotics of the density in perturbed SPDE's with spatially correlated noise*. Infin. Dimens. Anal. Quantum Probab. Relat. Top. 2006. V. 9, №2. P. 271–285. [*425*]

[780] Márquez-Carreras D., Sarrà M. *Behaviour of the density in perturbed SPDE's with spatially correlated noise*. Bull. Sci. Math. 2003. V. 127, №4. P. 348–367. [*425*]

[781] Márquez-Carreras D., Tindel S. *On exponential moments for functionals defined on the loop group*. Stochastic Anal. Appl. 2003. V. 21, №6. P. 1333–1352. [*398*]

[782] Maruyama G. *On the transition probability functionals of the Markov process*. Natural Sci. Rep. Ochanomizu Univ. 1954. V. 5, №1. P. 10–20. [*365*]

[783] Matoušek J., Matoušková E. *A highly non-smooth norm on Hilbert space.* Israel J. Math. 1999. V. 112. P. 1–27. [*351*]

[784] Matoušková E. *An almost nowhere Fréchet smooth norm on superreflexive spaces.* Studia Math. 1999. V. 133, №1. P. 93–99. [*351*]

[785] Matoušková E. *Lipschitz images of Haar null sets.* Bull. London Math. Soc. 2000. V. 32, №2. P. 235–244. [*350*]

[786] Matoušková E., Zajiček L. *Second order differentiability and Lipschitz smooth points of convex functionals.* Czech. Math. J. 1998. V. 48, №4. P. 617–640. [*187*]

[787] Matsumoto H., Taniguchi S. *Wiener functionals of second order and their Lévy measures.* Electron. J. Probab. 2002. V. 7, №14. P. 1–30. [*324*]

[788] Mattingly J.C., Pardoux E. *Malliavin calculus for the stochastic 2D Navier–Stokes equation.* Comm. Pure Appl. Math. 2006. V. 59, №12. P. 1742–1790. [*323*]

[789] Mauceri G., Meda S. *BMO and H^1 for the Ornstein–Uhlenbeck operator.* J. Funct. Anal. 2007. V. 252, №1. P. 278–313. [*275*]

[790] Mauceri G., Meda S., Sjögren P. *Sharp estimates for the Ornstein–Uhlenbeck operator.* Ann. Sc. Norm. Super. Pisa Cl. Sci. (5). 2004. V. 3, №3. P. 447–480. [*238, 275*]

[791] Mayer-Wolf E. *Preservation of measure continuity under conditioning.* J. Funct. Anal. 1993. V. 115, №1. P. 227–246. [*366*]

[792] Mayer-Wolf E. *Covariance inequalities induced by biorthogonal operators.* J. Funct. Anal. 1996. V. 140, №1. P. 170–193. [*224*]

[793] Mayer-Wolf E., Zakai M. *The divergence of Banach space valued random variables on Wiener space.* Probab. Theory Relat. Fields. 2005. V. 132, №2. P. 291–320; erratum: ibid. 2008. V. 140, №3-4. P. 631–633. [*324*]

[794] Mayer-Wolf E., Zakai M. *The Clark–Ocone formula for vector valued Wiener functionals.* J. Funct. Anal. 2005. V. 229, №1. P. 143–154; corrigendum: ibid. 2008. V. 254, №7. P. 2020–2021. [*324*]

[795] Maz'ja V.G. Sobolev spaces. Transl. from the Russian. Springer-Verlag, Berlin, 1985; xix+486 p. (Russian ed.: Leningrad, 1985). [*66*]

[796] Mazziotto G., Millet A. *Absolute continuity of the law of an infinite-dimensional Wiener functional with respect to the Wiener probability.* Probab. Theory Relat. Fields. 1990. V. 85, №3. P. 403–411. [*366*]

[797] McCann R.J. *Existence and uniqueness of monotone measure-preserving maps.* Duke Math. J. 1995. V. 80. P. 309–323. [*363*]

[798] Medvedev K.V. *Certain properties of triangular transformations of measures.* Theory Stoch. Process. 2008. V. 14, №1. P. 95–99. [*132*]

[799] Melcher T. *Malliavin calculus for Lie group-valued Wiener functions.* Infin. Dimens. Anal. Quantum Probab. Relat. Top. 2009. V. 12, №1. P. 67–89. [*398*]

[800] Metafune G. *L^p-spectrum of Ornstein–Uhlenbeck operators.* Ann. Scuola Norm. Sup. Pisa Cl. Sci. (4). 2001. V. 30, №1. P. 97–124. [*275*]

[801] Metafune G., Pallara D., Priola E. *Spectrum of Ornstein–Uhlenbeck operators in L^p spaces with respect to invariant measures.* J. Funct. Anal. 2002. V. 196, №1. P. 40–60. [*275*]

[802] Metafune G., Pallara D., Rhandi A. *Global regularity of invariant measures.* J. Funct. Anal. 2005. V. 223. P. 396–424. [*409, 410, 426*]

[803] Metafune G., Prüss J., Rhandi A., Schnaubelt R. *The domain of the Ornstein–Uhlenbeck operator on an L^p-space with invariant measure.* Ann. Sc. Norm. Super. Pisa Cl. Sci. (5). 2002. V. 1, №2. P. 471–485. [*275*]

[804] Meyer P.-A. Probability and potentials. Blaisdell, Waltham – Toronto – London, 1966; xiii+266 p. [*266, 381*]

[805] Meyer P.-A. *Variations de solutions d'une E.D.S. d'après J.-M. Bismut.* Lecture Notes in Math. 1982. V. 921. P. 151–165. [*323*]

[806] Meyer P.-A. *Note sur les processus d'Ornstein–Uhlenbeck.* Lecture Notes in Math. 1982. V. 920. P. 95–133. [*274*]

[807] Meyer P.-A. *Quelques résultats analytiques sur semigroupe d'Ornstein–Uhlenbeck en dimension infinie.* Lecture Notes in Control and Inform. Sci. 1983. V. 49. P. 201–214. [*274*]

[808] Meyer P.-A. *Transformation de Riesz pour les lois gaussiennes.* Lecture Notes in Math. 1984. V. 1059. P. 179–193. [*274*]

[809] Meyer P.-A., Yan J.A. *A propos des distributions sur l'espace de Wiener.* Lecture Notes in Math. 1987. V. 1247. P. 8–26. [*276*]

[810] Meyer P.-A., Yan J.A. *Distributions sur l'espace de Wiener (suite) d'après I. Kubo et Y. Yokoi.* Lecture Notes in Math. 1989. V. 1372. P. 382–392. [*276*]

[811] Michel D. *Régularité des lois conditionnelles en théorie du filtrage non-linéaire et calcul des variation stochastique.* J. Funct. Anal. 1981. V. 41, №1. P. 8–36. [*425*]

[812] Michel D. *Conditional laws and Hörmander's condition.* Proc. Taniguchi Intern. Conf. on Stoch. Anal., Katata and Kyoto, 1982 (K. Itô ed.). P. 387–408. Kinokuniaya/North-Holland, 1984. [*425*]

[813] Michel D., Pardoux E. *An introduction to Malliavin calculus and some of its applications.* Recent Advances in Stoch. Calculus, Distinguished Lect. Series on Stoch. Calculus, Progress in Automation and Information Systems (J.S. Baras, V. Mirelli, eds.). P. 65–104. Springer, New York, 1990. [*323, 425*]

[814] Millet A., Nualart D., Sanz M. *Integration by parts and time reversal for diffusion processes.* Ann. Probab. 1989. V. 17, №1. P. 208–238. [*415*]

[815] Millet A., Nualart D., Sanz M. *Time reversal for infinite-dimensional diffusions.* Probab. Theory Relat. Fields. 1989. V. 82, №3. P. 315–347. [*415*]

[816] Montero M., Kohatsu-Higa A. *Malliavin calculus applied to finance.* Phys. A. 2003. V. 320, №1-4. P. 548–570. [*425*]

[817] Moulinier J.-M. *Absolue continuité de probabilités de transition par rapport à une mesure gaussienne dans une espace de Hilbert.* J. Funct. Anal. 1985. V. 64, №2. P. 257–295. [*414*]

[818] Moulinier J.-M. *Fonctionnelles oscillantes stochastiques et hypoellipticité.* Bull. Sci. Math. (2). 1985. V. 109, №1. P. 37–60. [*296, 325*]

[819] Mushtari D.Kh. Probabilities and topologies on linear spaces. Kazan Mathematics Foundation, Kazan', 1996; xiv+233 p. [*35*]

[820] van Neerven J.M.A.M. *Second quantization and the L^p-spectrum of nonsymmetric Ornstein–Uhlenbeck operators.* Infin. Dim. Anal. Quantum Probab. Relat. Top. 2005. V. 8, №3. P. 473–495. [*275*]

[821] Neklyudov M.Yu. *The derivative of the Wiener measure on trajectories in a compact Lie group.* Mat. Zametki. 2004. V. 75, №5. P. 789–792 (in Russian); English transl.: Math. Notes. 2004. V. 75, №5-6. P. 734–738. [*398*]

[822] Nelson E. *The free Markov field.* J. Funct. Anal. 1973. V. 12. P. 211–227. [*274*]

[823] Neretin Ju.A. *Some remarks on quasi-invariant actions of loop groups and the group of diffeomorphisms of the circle.* Comm. Math. Phys. 1994. V. 164. P. 599–626. [*399*]

[824] Neretin Yu.A. Categories of symmetries and infinite-dimensional groups. Transl. from the Russian. Clarendon Press, Oxford, 1996; xiv+417 p. [*399*]

[825] Neretin Yu.A. *On the correspondence between the boson Fock space and the space L^2 with respect to Poisson measure.* Mat. Sb. 1997. V. 188, №11. P. 19–50 (Russian); English transl.: Sb. Math. 1997. V. 188, №11. P. 1587–1616. [*399*]

[826] Neretin Yu.A. *Fractional diffusions and quasi-invariant actions of infinite-dimensional groups.* "Prostran. Petel i Gruppy Diffeomorf." Tr. Mat. Inst. Steklova. 1997. V. 217. P. 135–181 (Russian); English transl.: Proc. Steklov Inst. Math. 1997. V. 217, №2. P. 126–173. [*400*]

[827] Neveu J. Bases mathématiques du calcul des probabilités. Masson et Cie, Paris, 1964; xiii+203 p. English transl.: Mathematical foundations of the calculus of probability. Holden-Day, San Francisco, 1965; 231 p. [*381*]

[828] Nicaise F. *Anticipative direct transformations on the Poisson space.* Ann. Inst. H. Poincaré Probab. Statist. 2003. V. 39, №4. P. 557–592. [*399*]

[829] Nikol'skii S.M. *Approximation of functions of several variables and imbedding theorems.* Transl. from the Russian. Springer-Verlag, New York – Heidelberg, 1975; viii+418 p. (Russian ed.: Moscow, 1977). [*66*]

[830] Nishimura T. *Exterior product bundle over complex abstract Wiener space.* Osaka J. Math. 1992. V. 29, №2. P. 233–245. [*275*]

[831] Norin N.V. *Heat potential on Hilbert space.* Mat. Zametki. 1984. V. 35, №4. P. 531–548 (in Russian); English transl.: Math. Notes. 1984. V. 35. P. 279–288. [*426*]

[832] Norin N.V. *An extended stochastic integral for non-Gaussian measures in locally convex spaces.* Usp. Mat. Nauk. 1986. V. 41, №3. P. 199–200 (in Russian); English transl.: Russ. Math. Surv. 1986. V. 41, №3. P. 229–230. [*324*]

[833] Norin N.V. *Stochastic integrals and differentiable measures.* Teor. Verojatn. i Primen. 1987. V. 32, №1. P. 114–124 (in Russian); English transl.: Theory Probab. Appl. 1987. V. 32. P. 107–116. [*324*]

[834] Norin N.V. *The extended stochastic integral in linear spaces with differentiable measures and related topics.* World Sci. Publ., River Edge, New Jersey, 1996; xii+257 p. [*103, 324*]

[835] Norin N.V., Smolyanov O.G. *Some results on logarithmic derivatives of measures on a locally convex space.* Mat. Zametki. V. 54, №6. P. 135–138 (in Russian); English transl.: Math. Notes. 1993. V. 54, №6. P. 1277–1279. [*131, 224*]

[836] Norin N.V., Smolyanov O.G. *Logarithmic derivatives of measures and Gibbs distributions.* Dokl. Ross. Akad. Nauk. 1997. V. 354, №4. P. 456–460 (in Russian); English transl.: Dokl. Math. 1997. V. 55, №3. P. 398–401. [*132*]

[837] Norris J.R. *Simplified Malliavin calculus.* Lecture Notes in Math. 1986. V. 1204. P. 101–130. [*323*]

[838] Norris J.R. *Integration by parts for jump processes.* Lecture Notes in Math. 1988. V. 1321. P. 271–315. [*323*]

[839] Nourdin I., Peccati G. *Stein's method on Wiener chaos.* Probab. Theory Relat. Fields. 2009. V. 145, №1. P. 75–118. [*321, 425*]

[840] Nourdin I., Peccati G. *Stein's method and exact Berry–Esséen asymptotics for functionals of Gaussian fields.* Ann. Probab. 2009. V. 37, №6. P. 2231–2261. [*425*]

[841] Nourdin I., Peccati G. *Stein's method meets Malliavin calculus: a short survey with new estimates.* Recent Advances in Stochastic Dynamics and Stochastic Analysis (J. Duan, S. Luo, and C. Wang eds.), World Scientific, Hackensack, New Jersey, 2010. [*425*]

[842] Nourdin I., Peccati G., Reinert G. *Second order Poincaré inequalities and CLTs on Wiener space.* J. Funct. Anal. 2009. V. 257. P. 593–609. [*425*]

[843] Nourdin I., Peccati G., Réveillac A. *Multivariate normal approximation using Stein's method and Malliavin calculus.* Ann. Inst. H. Poincaré. 2010. V. 46, №1. P. 45–58. [*425*]

[844] Nourdin I., Viens F. *Density formula and concentration inequalities with Malliavin calculus.* Probab. Theory Relat. Fields. 2009. V. 145, №1. P. 75–118. [*321*]

[845] Nowak E. *Distance en variation entre une mesure de Gibbs et sa translatée.* C. R. Acad. Sci. Paris. 1998. V. 326. P. 239–242. [*132*]

[846] Nowak E. *Absolute continuity between a Gibbs measure and its translate.* Theory Probab. Appl. 2005. V. 49, №4. P. 713–724. [*132*]

[847] Nualart D. *Analysis on Wiener space and anticipating stochastic calculus.* Lecture Notes in Math. 1998. V. 1690. P. 123–227. [*31*]

[848] Nualart D. *The Malliavin calculus and related topics.* 2nd ed. Springer-Verlag, Berlin, 2006; xiv+382 p. [*31, 232, 240, 260, 261, 262, 323, 324, 326, 328, 405*]

[849] Nualart D. *Malliavin calculus and its applications.* Amer. Math. Soc., Providence, Rhode Island, 2009; viii+85 p. [*323*]

[850] Nualart D. *Application of Malliavin calculus to stochastic partial differential equations.* Lecture Notes in Math. 2009. V. 1962. P. 73–109. [*425*]

[851] Nualart D., Ortiz-Latorre S. *Central limit theorems for multiple stochastic integrals and Malliavin calculus.* Stoch. Process. Appl. 2008. V. 118, №4. P. 614–628. [*425*]

[852] Nualart D., Quer-Sardanyons L. *Existence and smoothness of the density for spatially homogeneous SPDEs.* Potential Anal. 2007. V. 27, №3. P. 281–299. [*425*]

[853] Nualart D., Saussereau B. *Malliavin calculus for stochastic differential equations driven by a fractional Brownian motion.* Stochastic Process. Appl. 2009. V. 119, №2. P. 391–409. [*324*]

[854] Nualart D., Steblovskaya V. *Asymptotics of oscillatory integrals with quadratic phase function on Wiener space.* Stochastics Stoch. Rep. 1999. V. 66, №3-4. P. 293–309. [*325*]

[855] Nualart D., Üstünel A.S. *Geometric analysis of conditional independence on Wiener space.* Probab. Theory Relat. Fields. 1991. V. 89, №4. P. 407–422. [*376, 379*]

[856] Nualart D., Üstünel A.S., Zakai M. *Some remarks on independence and conditioning on Wiener space.* Lecture Notes in Math. 1990. V. 1444. P. 122–127. [*379*]

[857] Nualart D., Üstünel A.S., Zakai M. *Some relations among classes of σ-fields on the Wiener spaces.* Probab. Theory Relat. Fields. 1990. V. 85, №1. P. 119–129. [*376, 379*]

[858] Nualart D., Zakai M. *Generalized stochastic integrals and the Malliavin calculus.* Probab. Theory Relat. Fields. 1986. V. 73, №2. P. 255–280. [*275, 281, 323*]

[859] Nualart D., Zakai M. *The partial Malliavin calculus.* Lecture Notes in Math. 1989. V. 1372. P. 362–381. [*425*]

[860] Nualart D., Zakai M. *A summary of some identities of the Malliavin calculus.* Lecture Notes in Math. 1989. V. 1390. P. 192–196. [*264*]

[861] Nualart E. *Exponential divergence estimates and heat kernel tail.* C. R. Math. Acad. Sci. Paris. 2004. T. 338, №1. P. 77–80. [*172*]

[862] Obata N. *White noise calculus and Fock space.* Lecture Notes in Math. V. 1577. Springer-Verlag, Berlin, 1994; x+183 p. [*276*]

[863] Ocone D. *Malliavin's calculus and stochastic integral representation of functionals of diffusion processes.* Stochastics. 1984. V. 12, №3-4. P. 161–185. [*275*]

[864] Ocone D. *A guide to the stochastic calculus of variations.* Lecture Notes in Math. 1988. V. 1316. P. 1–80. [*323, 425*]

[865] Ocone D. *Stochastic calculus of variations for stochastic partial differential equations.* J. Funct. Anal. 1988. V. 79, №2. P. 288–331. [*425*]

[866] Ocone D.L., Karatzas I. *A generalized Clark representation formula, with application to optimal portfolios.* Stochastics Stoch. Rep. 1991. V. 34, №3-4. P. 187–220. [*275*]

[867] Ogawa S. *Quelques propriétés de l'intégrale stochastique du type noncausal.* Japan J. Appl. Math. 1984. V. 1, №2. P. 405–416. [*262*]

[868] Osswald H. *Malliavin calculus for product measures on $\mathbb{R}^N$ based on chaos.* Stochastics. 2005. V. 77, №6. P. 501–514. [*323*]

[869] Osswald H. *Malliavin calculus on extensions of abstract Wiener spaces.* J. Math. Kyoto Univ. 2008. V. 48, №2. P. 239–263. [*323*]

[870] Osswald H. *A smooth approach to Malliavin calculus for Lévy processes.* J. Theoret. Probab. 2009. V. 22, №2. P. 441–473. [*323*]

[871] Paclet Ph. *Espaces de Dirichlet en dimension infinie.* C. R. Acad. Sci. Paris, sér. A-B. 1979. T. 288, №21. P. A981–A983. [*274*]

[872] Paclet Ph. *Capacité fonctionnelle sur un espace de Hilbert.* C. R. Acad. Sci. Paris, sér. A-B. 1979. T. 289, №5. P. A337–A340. [*274*]

[873] Palais R.S. *Seminar on the Atiyah–Singer index theorem.* With contributions by M.F. Atiyah, A. Borel, E.E. Floyd, R.T. Seeley, W. Shih and R. Solovay. Princeton University Press, Princeton, New Jersey, 1965; x+366 p. [*406*]

[874] Pardoux E. *Filtrage non linéaire et equations aux dérivées partielles stochastiques associées.* Lecture Notes in Math. 1991. V. 1464. P. 67–163. [*425*]

[875] Pardoux E., Zhang T.S. *Absolute continuity of the law of the solution of a parabolic SPDE.* J. Funct. Anal. 1993. V. 112, №2. P. 447–458. [*425*]

[876] Paulauskas V., Račkauskas A. *Approximation theory in the central limit theorem. Exact results in Banach spaces.* Transl. from the Russian. Kluwer Academic Publ., Dordrecht, 1987; xviii+156 p. [*149, 155, 296*]

[877] Paycha S. *Mesures et determinants en dimension infinie dans le modele des cordes de Polyakov.* C. R. Acad. Sci. Paris, sér. 1. 1989. T. 309, P. 201–204. [*399*]

[878] Paycha S. *Bosonic strings and measures on infinite dimensional manifolds.* Stochastics, algebra and analysis in classical quantum dynamics, Proc. 4th French–German Encounter Math. Phys. P. 189–203. CIRM, Reidel, Marseille, 1990. [*399*]

[879] Peszat S., Zabczyk J. *Strong Feller property and irreducibility for diffusions on Hilbert spaces.* Ann. Probab. 1995. V. 23, №1. P. 157–172. [*414*]

[880] Peszat S. *On a Sobolev space of functions of infinite number of variables.* Bull. Polish Acad. Sci. Math. 1993. V. 41, №1. P. 55–60. [*254, 274*]

[881] Peszat S. *Sobolev spaces of functions on an infinite-dimensional domain.* Stochastic processes and related topics (Siegmundsberg, 1994). P. 103–116. Gordon and Breach, Yverdon, 1996. [*254, 274*]

[882] Peters G. *Flows on the Wiener space generated by vector fields with low regularity.* C. R. Acad. Sci. Paris, sér. 1. 1995. T. 320, №8. P. 1003–1008. [*340, 366*]

[883] Peters G. *Anticipating flows on the Wiener space generated by vector fields of low regularity.* J. Funct. Anal. 1996. V. 142, №1. P. 129–192. [*340, 341, 366*]

[884] Petrou E. *Malliavin calculus in Lévy spaces and applications to finance.* Electron. J. Probab. 2008. V. 13, №27. P. 852–879. [*425*]

[885] Phelps R.R. *Gaussian null sets and differentiability of Lipschitz map on Banach spaces.* Pacif. J. Math. 1978. V. 77, №2. P. 523–531. [*343, 353*]

[886] Picard J. *On the existence of smooth densities for jump processes.* Probab. Theory Relat. Fields. 1996. V. 105, №4. P. 481–511. [*323*]

[887] Picard J. *Formules de dualité sur l'espace de Poisson.* Ann. Inst. H. Poincaré Probab. Statist. 1996. V. 32, №4. P. 509–548. [*399*]

[888] Picard J., Savona C. *Smoothness of harmonic functions for processes with jumps.* Stoch. Process. Appl. 2000. V. 87, №1. P. 69–91. [*323*]

[889] Pickrell D. *Measures on infinite-dimensional Grassmann manifolds.* J. Funct. Anal. 1987. V. 70, №2. P. 323–356. [*399*]

[890] Pickrell D. *On YM_2 measures and area-preserving diffeomorphisms.* J. Geom. Phys. 1996. V. 19, №4. P. 315–367. [*400*]

[891] Pickrell D. *Invariant measures for unitary groups associated to Kac–Moody Lie algebras.* Mem. Amer. Math. Soc. 2000. V. 146, №693; x+125 p. [*399*]

[892] Pickrell D. *An invariant measure for the loop space of a simply connected compact symmetric space.* J. Funct. Anal. 2006. V. 234, №2. P. 321–363. [*399*]

[893] Piech M.A. *A fundamental solution of the parabolic equation on Hilbert space.* J. Funct. Anal. 1969. V. 3. P. 85–114. [*426*]

[894] Piech M.A. *A fundamental solution of the parabolic equation on Hilbert space II: the semigroup property.* Trans. Amer. Math. Soc. 1970. V. 150. P. 257–286. [*426*]

[895] Piech M.A. *Some regularity properties of diffusion processes on abstract Wiener spaces.* J. Funct. Anal. 1971. V. 8. P. 153–172. [*426*]

[896] Piech M.A. *The Ornstein–Uhlenbeck semigroup in an infinite dimensional L^2 setting.* J. Funct. Anal. 1975. V. 18. P. 271–285. [*155, 426*]

[897] Piech M.A. *Smooth functions on Banach spaces.* J. Math. Anal. Appl. 1977. V. 57, №1. P. 56–67. [*414, 426*]

[898] Piech M.A. *Differentiability of measures associated with parabolic equation on infinite dimensional spaces.* Trans. Amer. Math. Soc. 1979. V. 253. P. 191–209. [*414, 426*]

[899] Pilipenko A.Yu. *On the properties of an operator of stochastic differentiation constructed on a group.* Ukr. Mat. Zh. 1996. V. 48. №4. P. 566–571 (in Russian); English transl.: Ukr. Math. J. 1996. V. 48, №4. P. 623–630. [*398*]

[900] Pilipenko A.Yu. *Transformations of measures in infinite-dimensional spaces by a flow generated by a stochastic differential equation.* Mat. Sb. 2003. V. 194, №4.

P. 85–106 (in Russian); English transl.: Sb. Math. 2003. V. 194, №3-4. P. 551–573. [*366*]

[901] Pilipenko A.Yu. *Transfer of absolute continuity by a flow generated by a stochastic equation with reflection.* Ukrain. Mat. Zh. 2006. V. 58, №12. P. 1663–1673 (in Russian); English transl.: Ukrainian Math. J. 2006. V. 58, №12. P. 1891–1903. [*366*]

[902] Pilipenko A.Yu. *Transformation of Gaussian measure by infinite-dimensional stochastic flow.* Random Oper. Stoch. Eq. 2006. V. 14, №3. P. 275–290. [*366*]

[903] Pilipenko A.Yu. *Liouville's theorem and its generalizations.* In "Mathematics today'07". № 13. P. 47–77. Kiev, 2007 (in Russian). [*366*]

[904] Pisier G. *Riesz transforms: a simpler analytic proof of P.-A. Meyer's inequality.* Lecture Notes in Math. 1988. V. 1321. P. 485–501. [*274*]

[905] Pitcher T.S. *Likelihood ratios for diffusion processes with shifted mean value.* Trans. Amer. Math. Soc. 1961. V. 101. P. 168–176. [*103, 123, 124, 281*]

[906] Pitcher T.S. *The admissible mean values of stochastic process.* Trans. Amer. Math. Soc. 1963. V. 108. P. 538–546. [*103, 281*]

[907] Pitcher T.S. *Likelihood ratios for stochastic processes related by groups of transformations. I, II.* Illinois J. Math. 1963. V. 7, №3. P. 369–414; 1964. V. 8, №2. P. 271–279. [*103, 281*]

[908] Pitcher T.S. *Parameter estimation for stochastic processes.* Acta Math. 1964. V. 121, №1. P. 1–40. [*103, 281*]

[909] Ponomarev S.P. *Property N of homeomorphisms of the class W_p.* Sib. Mat. Zh. 1987. V. 28, №2. P. 140–148 (in Russian); English transl.: Sib. Math. J. 1987. V. 28. P. 291–298. [*55*]

[910] Ponomarev S.P. *Submersions and preimages of sets of measure zero.* Sib. Mat. Zh. 1987. V. 28, №1. P. 199–210 (in Russian); English transl.: Sib. Math. J. 1987. V. 28. P. 153–163. [*284, 310*]

[911] Ponomarev S.P. *The N^{-1}-property of mappings, and Luzin's (N) condition.* Mat. Zametki. 1995. V. 58, №3. P. 411–418 (in Russian); English transl.: Math. Notes. 1995. V. 58, №3-4. P. 960–965. [*55*]

[912] Popescu I. *Morse inequalities, a function space integral approach.* J. Funct. Anal. 2006. V. 235, №1. P. 1–68. [*406*]

[913] Potthoff J. *On the connection of the white-noise and Malliavin calculi.* Proc. Japan Acad. Ser. A. 1986. V. 62, №2. P. 43–45. [*324*]

[914] Potthoff J. *White-noise approach to Malliavin's calculus.* J. Funct. Anal. 1987. V. 71, №2. P. 207–217. [*324*]

[915] Potthoff J. *On Meyer's equivalence.* Nagoya Math. J. 1988. V. 111. P. 99–109. [*274*]

[916] Prat J.-J. *Équation de Schrödinger: analyticité transverse de la densité de la loi d'une fonctionnelle additive.* Bull. Sci. Math. (2). 1991. V. 115, №2. P. 133–176. [*324, 325*]

[917] Prat J.-J. *Gevrey regularity of C^3-functionals and stochastic calculus of variation.* Stochastics Stoch. Rep. 1993. V. 44, №3-4. P. 167–211. [*324*]

[918] Preiss D. *Differentiability of Lipschitz functions on Banach spaces.* J. Funct. Anal. 1990. V. 91, №2. P. 312–345. [*351*]

[919] Prékopa A. *Logarithmic concave measures with application to stochastic programming.* Acta Sci. Math. 1971. V. 32. P. 301–316. [*131*]

[920] Prékopa A. *On logarithmic concave measures and functions.* Acta Sci. Math. 1973. V. 34. P. 335–343. [*131*]

[921] Privault N. *Girsanov theorem for anticipative shifts on Poisson space.* Probab. Theory Relat. Fields. 1996. V. 104. P. 61–76. [*399*]

[922] Privault N. *Calcul des variations stochastique pour la mesure de densité uniforme.* Potential Anal. 1997. V. 7, №2. P. 577–601. [*399*]

[923] Privault N. *Connections and curvature in the Riemannian geometry of configuration spaces.* J. Funct. Anal. 2001. V. 185, №2. P. 367–403. [*399*]

[924] Privault N. *Stochastic analysis in discrete and continuous settings: with normal martingales.* Lecture Notes in Math., Springer, Berlin, 2009; xvi+282 p. [*323*]

[925] Privault N., Réveillac A. *Stein estimation for the drift of Gaussian processes using the Malliavin calculus.* Ann. Statist. 2008. [*425*]

[926] Privault N., Wei X. *A Malliavin calculus approach to sensitivity analysis in insurance.* Insurance Math. Econom. 2004. V. 35, №3. P. 679–690. [*425*]

[927] Prohorov Yu.V. *Convergence of random processes and limit theorems in probability theory.* Teor. Verojatn. i Primen. 1956. V. 1, №2. P. 177–238 (in Russian); English transl.: Theory Probab. Appl. 1956. V. 1. P. 157–214. [*35, 365*]

[928] Pugachev O.V. *Surface measures in infinite-dimensional spaces.* Mat. Zametki. 1998. V. 63, №1. P. 106–114 (in Russian); English transl.: Math. Notes. 1998. V. 63, №1. P. 94–101. [*299, 325*]

[929] Pugachev O.V. *The Gauss–Ostrogradskii formula in infinite-dimensional space.* Mat. Sb. 1998. V. 189, №5. P. 115–128 (in Russian); English transl.: Sb. Math. 1998. V. 189, №5. P. 757–770. [*299, 325*]

[930] Pugachev O.V. *Construction of non-Gaussian surface measures by the Malliavin method.* Mat. Zametki. 1999. V. 65, №3. P. 377–388 (in Russian); English transl.: Math. Notes. 1999. V. 65, №3-4. P. 315–325. [*299, 325*]

[931] Pugachev O.V. *Tightness of Sobolev capacities in infinite dimensional spaces.* Infin. Dim. Anal. Quantum Probab. Relat. Top. 1999. V. 2, №3. P. 427–440. [*272, 299, 325*]

[932] Pugachev O.V. *On the closability of classical Dirichlet forms in the plane.* Dokl. Ross. Akad. Nauk. 2001. V. 380, №3. P. 315–318 (in Russian); English transl.: Dokl. Math. 2001. V. 64, №2. P. 197–200. [*61, 419*]

[933] Pugachev O.V. *Sobolev capacities of configurations with multiple points in Poisson space.* Mat. Zametki. 2004. V. 76, №6. P. 874–882 (in Russian); English transl.: Math. Notes. 2004. V. 76, №6. P. 816–823. [*399*]

[934] Pugachev O.V. *On closability of classical Dirichlet forms.* J. Funct. Anal. 2004. V. 207, №2. P. 330–343. [*61, 419*]

[935] Pugachev O.V. *Capacities and surface measures in locally convex spaces.* Teor. Verojatn. i Primen. 2008. V. 53, №1. P. 178–188 (in Russian); English transl.: Theory Probab. Appl. 2009. V. 53, №1. P. 179–190. [*276, 299, 325*]

[936] Pugachev O.V. *Quasi-invariance of Poisson distributions with respect to transformations of configurations.* Dokl. Ross. Akad. Nauk. 2008. V. 420, №4. P. 455–458 (in Russian); English transl.: Dokl. Math. 2008. V. 77, №3. P. 420–423. [*394, 399*]

[937] Rachev S.T., Rüschendorf L. *Mass transportation problems.* V. 1. Springer, New York, 1998; 508 p. [*363*]

[938] Radó T., Reichelderfer P.V. *Continuous transformations in analysis.* Springer-Verlag, Berlin, 1955; vi+441 p. [*312*]

[939] Ramer R. *On nonlinear transformations of Gaussian measures.* J. Funct. Anal. 1974. V. 15. P. 166–187. [*366*]

[940] Ramer R. *Integration on infinite dimensional manifolds.* Ph.D. Dissertation. Amsterdam Univ., 1974; 155 p. [*366*]

[941] Ren J. *Formule de Stokes et estimation explicite d'une intégrale stochastique oscillante.* Bull. Sci. Math. (2). 1988. V. 112, №4. P. 453–471. [*325*]

[942] Ren J. *Analyse quasi-sure des équations différentielles stochastiques.* Bull. Sci. Math. (2). 1990. V. 114, №2. P. 187–213. [*273*]

[943] Ren J. *Topologie p-fine sur l'espace de Wiener et théorème des fonctions implicites.* Bull. Sci. Math. (2). 1990. V. 114, №2. P. 99–114. [*273*]

[944] Ren J. *On smooth martingales.* J. Funct. Anal. 1994. V. 120, №1. P. 72–81. [*273, 323*]

[945] Ren J., Röckner M. *Ray Hölder-continuity for fractional Sobolev spaces in infinite dimensions and applications.* Probab. Theory Relat. Fields. 2000. V. 117, №2. P. 201–220. [*274*]

[946] Ren J., Röckner M. *A remark on sets in infinite dimensional spaces with full or zero capacity.* Stochastic analysis: classical and quantum. P. 177–186. World Sci. Publ., Hackensack, New Jersey, 2005. [*276*]

[947] Ren J., Röckner M., Zhang X. *Kusuoka–Stroock formula on configuration space and regularities of local times with jumps.* Potential Anal. 2007. V. 26, №4. P. 363–396. [*399*]

[948] Ren J., Watanabe S. *A convergence theorem for probability densities and conditional expectations of Wiener functionals.* Dirichlet forms and stochastic processes (Beijing, 1993). P. 335–344. De Gruyter, Berlin, 1995. [*325*]

[949] Ren J., Zhang X. *Regularity of local times of random fields.* J. Funct. Anal. 2007. V. 249, №1. P. 199–219. [*324*]

[950] von Renesse M.-K., Sturm K.-T. *Entropic measure and Wasserstein diffusion.* Ann. Probab. 2009. V. 37, №3. P. 1114–1191. [*399*]

[951] von Renesse M.-K., Yor M., Zambotti L. *Quasi-invariance properties of a class of subordinators.* Stoch. Proc. Appl. 2008. V. 118, №1. P. 2038–2057. [*399*]

[952] Reshetnyak Yu.G. *Property N for the space mappings of class $W_{n,loc}^1$.* Sib. Mat. Zh. 1987. V. 28, №5. P. 149–153 (in Russian); English transl.: Sib. Math. J. 1987. V. 28, №5. P. 810–813. [*55*]

[953] Reshetnyak Yu.G. Space mappings with bounded distortion. Transl. from the Russian. Amer. Math. Soc., Providence, Rhode Island, 1989; xvi+362 p. (Russian ed.: Moscow, 1982). [*54, 55, 66*]

[954] Reshetnyak Yu.G. *Sobolev classes of functions with values in a metric space.* Sibirsk. Mat. Zh. 1997. V. 38, №3. P. 657–675 (in Russian); English transl.: Siberian Math. J. 1997. V. 38, №3. P. 567–583. [*274*]

[955] Reshetnyak Yu.G. *Sobolev classes of functions with values in a metric space. II.* Sibirsk. Mat. Zh. 2004. V. 45, №4. P. 855–870 (in Russian); English transl.: Siberian Math. J. 2004. V. 45, №4. P. 709–721. [*274*]

[956] Reshetnyak Yu.G. *On the theory of Sobolev classes of functions with values in a metric space.* Sibirsk. Mat. Zh. 2006. V. 47, №1. P. 146–168 (in Russian); English transl.: Siberian Math. J. 2006. V. 47, №1. P. 117–134. [*274*]

[957] Rockafellar R.T. Convex analysis. Princeton University Press, Princeton, New Jersey, 1997; xviii+451 p. [*112, 115*]

[958] Röckner M. *Dirichlet forms on infinite dimensional state space and applications.* Stochastic Analysis and Related Topics P. 131–186. H. Körezlioglu and A.S. Üstünel, eds. Birkhäuser, Boston, 1992. [*239, 276*]

[959] Röckner M. *On the parabolic Martin boundary of the Ornstein–Uhlenbeck operator on Wiener space.* Ann. Probab. 1992. V. 20, №2. P. 1063–1085. [*426*]

[960] Röckner M. *L^p-analysis of finite and infinite dimensional diffusion operators.* Lecture Notes in Math. 1998. V. 1715. P. 65–116. [*426*]

[961] Röckner M., Schied A. *Rademacher's theorem on configuration spaces and applications.* J. Funct. Anal. 1999. V. 169, №2. P. 325–356. [*366, 399*]

[962] Röckner M., Schmuland B. *Tightness of general $C_{1,p}$ capacities on Banach spaces.* J. Funct. Anal. 1992. V. 108, №1. P. 1–12. [*272*]

[963] Röckner M., Zhang T.S. *Uniqueness of generalized Schrödinger operators and applications.* J. Funct. Anal. 1992. V. 105, №1. P. 187–231; — II. Ibid. 1994. V. 119, №2. P. 455–467. [*64, 223, 420*]

[964] Roelly S., Zessin H. *Une caractérisation des mesures de Gibbs sur $C(0,1)^{\mathbb{Z}^d}$ par le calcul des variations stochastiques.* Ann. Inst. H. Poincaré. 1993. V. 29, №3. P. 327–338. [*224*]

[965] Romanov V.A. *On continuous and totally discontinuous measures in linear spaces.* Dokl. Akad. Nauk SSSR. V. 227, №3. P. 569–570 (in Russian); English transl.: Sov. Math. Dokl. 1976. V. 17. P. 472–474. [*103*]

[966] Romanov V.A. *Decomposition of a measure in linear space into a sum of H-continuous and completely H-discontinuous measures.* Vestn. Mosk. Univ., Ser. I.

1976. V. 31, №4. P. 63–66 (in Russian); English transl.: Moscow Univ. Math. Bull. 1976. V. 31, №3-4. P. 99–102. [*103*]
[967] Romanov V.A. *Limits of differentiable measures in Hilbert space.* Ukrain. Mat. Zh. 1981. V. 33, №2. P. 215–219 (in Russian). [*103*]
[968] Romanov V.A. *Passage to the limit with measures in a Hilbert space with respect to various types of convergence.* Ukrain. Mat. Zh. 1984. V. 36, №1. P. 69–73 (in Russian); English transl.: Ukrainian Math. J. 1984. V. 36, №1. P. 63–66. [*103*]
[969] Romanov V.A. *The structure of sets of continuity directions of measure-products and their series in Hilbert space.* Mat. Zametki. 1990. V. 47, №1. P. 166–168 (in Russian). [*103*]
[970] Romanov V.A. *On the nonequivalence of three definitions of continuous directions for vector measures.* Mat. Zametki. 1995. V. 57, №2. P. 310–312 (in Russian); English transl.: Math. Notes. 1995. V. 57, №1-2. P. 220–222. [*103*]
[971] Romanov V.A. *On the nonequivalence of various definitions of differentiability directions for vector measures.* Mat. Zametki. 2002. V. 72, №4. P. 528–534 (in Russian); English transl.: Math. Notes. 2002. V. 72, №3-4. P. 489–494. [*103*]
[972] Rovira C., Sanz-Solé M. *Stochastic Volterra equations in the plane: smoothness of the law.* Stoch. Anal. Appl. 2001. V. 19, №6. P. 983–1004. [*425*]
[973] Royer G. Initiation to logarithmic Sobolev inequalities. Amer. Math. Soc., Providence, Rhode Island, 2007; 119 p. [*274*]
[974] Rozanov Yu.A. Infinite-dimensional Gaussian distributions. Trudy Matem. Steklov Inst. 1968. V. 108. P. 1–161 (in Russian); English transl.: Proc. Steklov Inst. Math. V. 108. American Math. Soc., Providence, Rhode Island, 1971. [*35*]
[975] Rudenko A.V. *Approximation of densities of absolutely continuous components of measures in a Hilbert space by the Ornstein–Uhlenbeck semigroup.* Ukrain. Mat. Zh. 2004. V. 56, №12. P. 1654–1664 (in Russian); English transl.: Ukrainian Math. J. 2004. V. 56, №12. P. 1961–1974. [*277*]
[976] Rudin W. Functional analysis. 2nd ed. McGraw-Hill, Boston, 1991; xv+424 p. [*1, 97, 99*]
[977] Rudowicz R. *Admissible translations of the Brownian motion on a Lie group.* Probab. Math. Statist. 1991. V. 12, №2. P. 311–317. [*399*]
[978] Sadasue G. *Equivalence-singularity dichotomy for the Wiener measures on path groups and loop groups.* J. Math. Kyoto Univ. 1995. V. 35, №4. P. 653–662. [*399*]
[979] Sakamoto Y., Yoshida N. *Expansion of perturbed random variables based on generalized Wiener functionals.* J. Multivariate Anal. 1996. V. 59, №1. P. 34–59. [*425*]
[980] Saks S. Theory of the integral. Warszawa, 1937; xv+343 p. [*157, 158, 159*]
[981] Samoilenko Yu.S. Spectral theory of families of selfadjoint operators. Transl. from the Russian. Kluwer Academic Publ., Dordrecht, 1991; xvi+293 p. (Russian ed.: Kiev, 1984). [*104*]
[982] Sanz-Solé M. Malliavin calculus. With applications to stochastic partial differential equations. EPFL Press, Lausanne, 2005; viii+162 p. [*323, 425*]
[983] Sanz-Solé M. *Properties of the density for a three-dimensional stochastic wave equation.* J. Funct. Anal. 2008. V. 255, №1. P. 255–281. [*425*]
[984] Sanz-Solé M., Malliavin P. *Smoothness of the functional law generated by a nonlinear SPDE.* Chin. Ann. Math. Ser. B. 2008. V. 29, №2. P. 113–120. [*366*]
[985] Sanz-Solé M., Torrecilla-Tarantino I. *Probability density for a hyperbolic SPDE with time dependent coefficients.* ESAIM Probab. Stat. 2007. V. 11. P. 365–380. [*425*]
[986] Savinov E.A. *Logarithmic derivatives of symmetric distributions in a sequence space and their probability properties.* Vestn. Samar. Gos. Univ. Estestvennonauchn. Ser. 2004. Special Issue. P. 36–49 (in Russian). [*132*]
[987] Scheutzow M., von Weizsäcker H. *Which moments of a logarithmic derivative imply quasiinvariance?* Doc. Math. 1998. V. 3. P. 261–272. [*188*]
[988] Schied A. *Geometric analysis for symmetric Fleming–Viot operators: Rademacher's theorem and exponential families.* Potential Anal. 2002. V. 17, №4. P. 351–374. [*366*]

[989] Schmuland B. *Tightness of Gaussian capacities on subspaces.* C. R. Math. Rep. Acad. Sci. Canada. 1992. V. 14, №4. P. 125–130. [*276*]

[990] Schrödinger E. *Über die Umkehrung der Naturgesetze.* Sitzungsber. Preuss. Akad. Wiss. Berlin, Phys.-Math. Kl. 1931. B. 12. S. 144–153. [*415*]

[991] Shao J. *Hamilton–Jacobi semi-groups in infinite dimensional spaces.* Bull. Sci. Math. 2006. V. 130, №8. P. 720–738. [*364*]

[992] Shaposhnikov A.V. *Differentiability of measures in the sense of Skorokhod and related properties.* Dokl. Akad. Nauk, Ross. Akad. Nauk. 2009. V. 429, №2. P. 163–167 (in Russian); English transl.: Dokl. Math. 2009. V. 80, №3. P. 818–822. [*99*]

[993] Shaposhnikov A.V. *A Lusin type theorem for vector fields on the Wiener space.* Dokl. Akad. Nauk, Ross. Akad. Nauk. 2010. [*250*]

[994] Shaposhnikov S.V. *Positiveness of invariant measures of diffusion processes.* Dokl. Akad. Nauk, Ross. Akad. Nauk. 2007. V. 415, №2. P. 174–179 (in Russian); English transl.: Dokl. Math. 2007. V. 76, №1. P. 533–538. [*172*]

[995] Shaposhnikov S.V. *The nonuniqueness of solutions to elliptic equations for probability measures.* Dokl. Akad. Nauk, Ross. Akad. Nauk. 2008. V. 420, №3. P. 320–323 (in Russian); English transl.: Dokl. Math. 2008. V. 77, №3. P. 401–403. [*410*]

[996] Shaposhnikov S.V. *On nonuniqueness of solutions to elliptic equations for probability measures.* J. Funct. Anal. 2008. V. 254. P. 2690–2705. [*410*]

[997] Shatskikh S.Ya. *Some properties of logarithmic derivatives of elliptically contoured measures.* Vestn. Samar. Gos. Univ. Mat. Mekh. Fiz. Khim. Biol. 2001. №4. P. 109–114 (Russian). [*132*]

[998] Shavgulidze E.T. *The direct Kolmogorov equation for measures in the Hilbert scale of spaces.* Vestn. Mosk. Univ., Ser. I. 1978. №3. P. 19–28 (in Russian); English transl.: Moscow Univ. Math. Bull. 1978. V. 33, №3. P. 17–25. [*414*]

[999] Shavgulidze E.T. *An example of a measure quasi-invariant under the action of the diffeomorphism group of the circle.* Funk. Anal. i Pril. 1978. V. 12, №3. P. 55–60 (in Russian); English transl.: Funct. Anal. Appl. 1979. V. 12. P. 203–207. [*397*]

[1000] Shavgulidze E.T. *Hahn–Jordan decomposition for smooth measures.* Mat. Zametki. 1981. V. 30, №3. P. 439–442 (in Russian); English transl.: Math. Notes. 1982. V. 30. P. 710–712. [*153*]

[1001] Shavgulidze E.T. *On a measure that is quasi-invariant with respect to the action of a group of diffeomorphisms of a finite-dimensional manifold.* Dokl. Akad. Nauk SSSR. 1988. V. 303, №4. P. 811–814 (in Russian); English transl.: Sov. Math. Dokl. 1989. V. 38, №3. P. 622–625. [*397*]

[1002] Shavgulidze E.T. *Quasiinvariant measures on groups of diffeomorphisms.* Tr. Mat. Inst. Steklova. 1997. V. 217. P. 189–208 (in Russian); English transl.: in "Loop spaces and groups of diffeomorphisms" (ed. A.G. Sergeev et al.). Proc. Steklov Inst. Math. 1997. V. 217. P. 181–202. [*397, 400*]

[1003] Shavgulidze E.T. *Quasi-invariant measures with respect to diffeomorphism groups on spaces of curves and surfaces.* Vestnik Moskov. Univ. Ser. I Mat. Mekh. 1999. №6. P. 19–25 (in Russian); English transl.: Moscow Univ. Math. Bull. 1999. V. 54, №6. P. 19–24. [*400*]

[1004] Shavgulidze E.T. *Some properties of quasi-invariant measures on groups of diffeomorphisms of the circle.* Russ. J. Math. Phys. 2000. V. 7, №4. P. 464–472. [*400*]

[1005] Shavgulidze E.T. *Properties of the convolution operation for quasi-invariant measures on groups of diffeomorphisms of a circle.* Russ. J. Math. Phys. 2001. V. 8, №4. P. 495–498. [*400*]

[1006] Shepp L.A. *Distinguishing sequence of random variables from a translate of itself.* Ann. Math. Statist. 1965. V. 36, №4. P. 1107–1112. [*156*]

[1007] Shevlyakov A.Yu. *On distributions of square integrable functionals of Gaussian measures.* Teor. Sluchajnykh Protsessov (Theory of Random Processes). 1985. №13. P. 104–110 (in Russian). [*324*]

[1008] Shigekawa I. *Absolute continuity of probability laws of Wiener functionals.* Proc. Japan Acad., ser. A. 1978. V. 54, №8. P. 230–233. [*281*]

[1009] Shigekawa I. *Derivatives of Wiener functionals and absolute continuity of induced measures.* J. Math. Kyoto Univ. 1980. V. 20, №2. P. 263–289. [*281, 324*]
[1010] Shigekawa I. *De Rham–Hodge–Kodaira's decomposition on an abstract Wiener space.* J. Math. Kyoto Univ. 1986. V. 26, №2. P. 191–202. [*274, 399*]
[1011] Shigekawa I. *Existence of invariant measures of diffusions on an abstract Wiener space.* Osaka J. Math. 1987. V. 24, №1. P. 37–59. [*412*]
[1012] Shigekawa I. *Sobolev spaces over the Wiener space based on an Ornstein–Uhlenbeck operator.* J. Math. Kyoto Univ. 1992. V. 32, №4. P. 731–748. [*274*]
[1013] Shigekawa I. *An example of regular (r,p)-capacities and essential self-adjointness of a diffusion operator.* J. Math. Kyoto Univ. 1995. V. 35, №4. P. 639–651. [*276*]
[1014] Shigekawa I. *Differential calculus on a based loop group.* New trends in stochastic analysis (Charingworth, 1994). P. 375–398. World Sci. Publ., River Edge, New Jersey, 1997. [*399*]
[1015] Shigekawa I. *Vanishing theorem of the Hodge-Kodaira operator for differential forms on a convex domain of the Wiener space.* Infin. Dim. Anal. Quantum Probab. Relat. Top. 2003. V. 6, suppl. P. 53–63. [*399*]
[1016] Shigekawa I. *Orlicz norm equivalence for the Ornstein–Uhlenbeck operator.* Stochastic analysis and related topics in Kyoto. P. 301–317. Adv. Stud. Pure Math., 41, Math. Soc. Japan, Tokyo, 2004. [*274*]
[1017] Shigekawa I. *Stochastic analysis.* Amer. Math. Soc., Providence, Rhode Island, 2004; xii+182 p. [*50, 232, 240, 274, 323, 405*]
[1018] Shigekawa I. *Schrödinger operators on the Wiener space.* Commun. Stoch. Anal. 2007. V. 1, №1. P. 1–17. [*426*]
[1019] Shigekawa I., Ueki N. *A stochastic approach to the Riemann–Roch theorem.* Osaka J. Math. 1988. V. 25, №4. P. 759–784. [*406*]
[1020] Shigekawa I., Ueki N., Watanabe S. *A probabilistic proof of the Gauss–Bonnet–Chern theorem for manifolds with boundary.* Osaka J. Math. 1989. V. 26, №4. P. 897–930. [*406*]
[1021] Shigekawa I., Yoshida N. *Littlewood–Paley–Stein inequality for a symmetric diffusion.* J. Math. Soc. Japan. 1992. V. 44, №2. P. 251–280. [*274*]
[1022] Shimomura H. *An aspect of quasi-invariant measures on R^∞.* Publ. RIMS, Kyoto Univ. 1976. V. 11, №3. P. 749–773. [*136, 155*]
[1023] Shimomura H. *Ergodic decomposition of quasi-invariant measures.* Publ. RIMS Kyoto Univ. 1978. V. 14, №1. P. 359–381. [*155*]
[1024] Shimomura H. *Remark to the previous paper "Ergodic decomposition of quasi-invariant measures".* Publ. RIMS Kyoto Univ. 1983. V. 19, №1. P. 203–205. [*155*]
[1025] Shimomura H. *Some results on quasi-invariant measures on infinite-dimensional space.* J. Math. Kyoto Univ. 1981. V. 21, №4. P. 703–713. [*155*]
[1026] Shimomura H. *Quasi-invariant measures on the orthogonal group over the Hilbert space.* Publ. RIMS Kyoto Univ. 1984. V. 20, №1. P. 107–117. [*399*]
[1027] Shimomura H. *An aspect of differentiable measures on R^∞.* Publ. RIMS, Kyoto Univ. 1987. V. 23, №5. P. 791–811. [*103*]
[1028] Shimomura H. *On the distributions of logarithmic derivative of differentiable measures on $\mathbf{R}$.* Publ. RIMS, Kyoto Univ. 1989. V. 25, №1. P. 75–78. [*104, 107*]
[1029] Shimomura H. *A characterization of Borel measures which vanish on the sets of capacity zero on an abstract Wiener space.* Preprint. 1992. [*270, 272*]
[1030] Shimomura H. *Canonical representations generated by translationally quasi-invariant measures.* Publ. Res. Inst. Math. Sci. 1996. V. 32, №4. P. 633–669. [*104*]
[1031] Shimomura H. *Quasi-invariant measures on the group of diffeomorphisms and smooth vectors of unitary representations.* J. Funct. Anal. 2001. V. 187, №2. P. 406–441. [*400*]
[1032] Shirikyan A. *Qualitative properties of stationary measures for three-dimensional Navier–Stokes equations.* J. Funct. Anal. 2007. V. 249, №2. P. 284–306. [*326*]
[1033] Shiryaev A.N. *Probability.* Translated from the Russian. Springer-Verlag, New York, 1996; xvi+623 p. (3nd Russian ed.: Moscow, 2004). [*107*]

[1034] Sidorova N.A. *The Smolyanov surface measure on trajectories in a Riemannian manifold.* Infin. Dim. Anal. Quantum Probab. Relat. Top. 2004. V. 7, №3. P. 461–471. [*399*]

[1035] Sidorova N.A., Smolyanov O.G., von Weizsacker H., Wittich O. *The surface limit of Brownian motion in tubular neighborhoods of an embedded Riemannian manifold.* J. Funct. Anal. 2004. V. 206, №2. P. 391–413. [*399*]

[1036] Simão I. *Regular fundamental solution for a parabolic equation on an infinite-dimensional space.* Stoch. Anal. Appl. 1993. V. 11, №2. P. 235–247. [*414*]

[1037] Simão I. *Regular transition densities for infinite-dimensional diffusions.* Stoch. Anal. Appl. 1993. V. 11, №3. P. 309–336. [*414*]

[1038] Simão I. *Smoothness of the transition densities corresponding to certain stochastic equations on a Hilbert space.* Comm. Appl. Anal. 2006. V. 10, №2-3. P. 253–268. [*414*]

[1039] Sintes-Blanc A. *On the density of some Wiener functionals: an application of Malliavin calculus.* Publ. Mat. 1992. V. 36, №2B. P. 981–987. [*323*]

[1040] Sjögren P. *Operators associated with the Hermite semigroup – a survey.* J. Fourier Anal. Appl. 1997. V. 3, Special Issue. P. 813–823. [*238, 275*]

[1041] Skorohod A.V. *On the differentiability of measures corresponding to Markov processes.* Teor. Verojatn. i Primen. 1960. V. 5, №1. P. 45–53 (in Russian); English transl.: Theory Probab. and Appl. 1960. V. 5. P. 40–49. [*366*]

[1042] Skorohod A.V. *Random processes with independent increments.* Nauka, Moscow, 1964; 280 p. (in Russian). [*127*]

[1043] Skorohod A.V. *Nonlinear transformations of stochastic measures in functional spaces.* Dokl. Akad. Nauk SSSR. 1966. V. 168, №6. P. 1269–1271 (in Russian); English transl.: Soviet Math. 1966. V. 7. P. 838–840. [*366*]

[1044] Skorohod A.V. *Surface integrals and the Green formula in Hilbert space.* Teor. Verojatnost. i Mat. Statist. 1970. №2. P. 172–175 (in Russian). [*325*]

[1045] Skorohod A.V. *Admissible shifts of measures in Hilbert space.* Teor. Verojatnost. i Primenen. 1970. V. 15. P. 577–598 (in Russian); English transl.: Theor. Probability Appl. 1970. V. 15. P. 557–580. [*103, 170*]

[1046] Skorohod A.V. *Integration in Hilbert space.* Transl. from the Russian. Springer-Verlag, Berlin – New York, 1974; xii+177 p. (Russian ed.: Moscow, 1974). [*70, 103, 132, 170, 325, 366*]

[1047] Skorohod A.V. *On a generalization of a stochastic integral.* Teor. Verojatn. i Primen. 1975. V. 20. P. 223–238 (in Russian); English transl.: Theory Probab. Appl. 1975. V. 20. P. 219–233. [*324*]

[1048] Skorokhod A.V. *Asymptotic methods in the theory of stochastic differential equations.* Transl. from the Russian. Amer. Math. Soc., Providence, Rhode Island, 1989; xvi+339 p. (Russian ed.: Kiev, 1987). [*366*]

[1049] Smolyanov O.G. *Linear differential operators in spaces of measures and functions on a Hilbert space.* Uspehi matem. nauk. 1973. V. 28, №5. P. 251–252 (in Russian). [*426*]

[1050] Smolyanov O.G. *Analysis on topological vector spaces and its applications.* Izdat. Moskov. gos. univ., Moscow, 1979; 80 p. [*103*]

[1051] Smolyanov O.G. *De Rham currents and Stokes' formula in a Hilbert space.* Dokl. Akad. Nauk SSSR. 1986. V. 286, №3. P. 554–558 (in Russian); English transl.: Sov. Math. Dokl. 1986. V. 33. P. 140–144. [*399*]

[1052] Smolyanov O.G. *Smooth measures on loop groups.* Dokl. Akad. Nauk. 1995. V. 345, №4. P. 455–458 (Russian); English transl.: [*399*]

[1053] Smolyanov O.G., Shavgulidze E.T. *Continual integrals.* Izdat. Moskovsk. Univ., Moscow, 1989; 150 p. (in Russian). [*276*]

[1054] Smolyanov O.G., Uglanov A.V. *Every Hilbert subspace of a Wiener space has measure zero.* Matem. Zamet. 1973. V. 14, №3. P. 369–374 (in Russian); English transl.: Math. Notes. 1973. V. 14. P. 772–774. [*155*]

[1055] Smolyanov O.G., von Weizsäcker H. *Differentiable families of measures.* J. Funct. Anal. 1993. V. 118, №2. P. 454–476. [*398*]

[1056] Smolyanov O.G., von Weizsäcker H. *Change of measures and their logarithmic derivatives under smooth transformations.* C. R. Acad. Sci. Paris, sér. I Math. 1995. T. 321, №1. P. 103–108. [*366*]

[1057] Smolyanov O.G., von Weizsäcker H. *Smooth probability measures and associated differential operators.* Infin. Dim. Anal. Quantum Probab. Relat. Top. 1999. V. 2, №1. P. 51–78. [*366*]

[1058] Smolyanov O.G., von Weizsäcker H., Vittich O. *Analytic continuation of the change of variables formula for smooth measures.* Dokl. Ross. Akad. Nauk. 2008. V. 419, №4. P. 466–470 (in Russian); English transl.: Dokl. Math. 2008. V. 77, №2. P. 265–268. [*366*]

[1059] Smorodina N.V. *A differential calculus on measurable spaces and smoothness conditions for distribution densities of random variables.* Dokl. Akad. Nauk SSSR. 1987. V. 282, №5. P. 1053–1057; (in Russian); English transl.: Sov. Math. Dokl. 1987. V. 35. P. 184–188. [*323, 370, 398*]

[1060] Smorodina N.V. *Differential calculus in configuration space and stable measures. I.* Teor. Verojatn. i Primen. 1988. V. 33, №3. P. 522–534 (in Russian); English transl.: Theory Probab. Appl. 1988. V. 33, №3. P. 488–499. [*323, 370, 399*]

[1061] Smorodina N.V. *The Gauss–Ostrogradskij formula for the configuration space.* Teor. Verojatn. i Primen. 1990. V. 35, №4. P. 727–739 (in Russian); English transl.: Theory Probab. Appl. 1991. V. 35, №4. P. 725–736. [*399*]

[1062] Smorodina N.V. *Asymptotic expansion of the distribution of a homogeneous functional of a strictly stable vector.* Teor. Verojatn. i Primen. 1996. V. 41, №1. P. 133–163 (in Russian); English transl.: Theory Probab. Appl. 1996. V. 41, №1. P. 91–115. [*399*]

[1063] Smorodina N.V. *Asymptotic expansion of the distribution of a homogeneous functional of a strictly stable vector. II.* Teor. Verojatn. i Primen. 1999. V. 44, №2. P. 458–465 (in Russian); English transl.: Theory Probab. Appl. 2000. V. 44, №2. P. 419–427. [*399*]

[1064] Smorodina N.V. *Multiple stochastic integrals and "non-Poisson" transformations of the gamma measure.* Zap. Nauchn. Sem. S.-Peterburg. Otdel. Mat. Inst. Steklov. (POMI). 2005. V. 328. Verojatn. i Stat. 9. P. 191–220 (in Russian); English transl.: J. Math. Sci. (New York). 2006. V. 139, №3. P. 6608–6624. [*399*]

[1065] Smorodina N.V. *Invariant and quasi-invariant transformations of of measures corresponding to stable processes with independent increments.* Zap. Nauchn. Semin. POMI. 2006. V. 339. P. 135–150 (in Russian); English transl.: J. Math. Sci. (New York). 2007. V. 145, №2. P. 4914–4922. [*399*]

[1066] Smorodina N.V. *Transformations of measures generated by Lévy jump processes.* Zap. Nauchn. Sem. S.-Peterburg. Otdel. Mat. Inst. Steklov. (POMI). Verojatn. i Stat. 11. 2007. V. 341. P. 174–188 (in Russian); English transl.: J. Math. Sci. (New York). 2007. V. 147, №4. P. 6950–6957. [*399*]

[1067] Smorodina N.V. *Measure preserving transformations of multidimensional stable Lévy processes.* Zap. Nauchn. Sem. S.-Peterburg. Otdel. Mat. Inst. Steklov. (POMI). 2007. V. 351. P. 242–252 (in Russian); English transl.: J. Math. Sci. (New York). 2008. V. 152, №6. P. 934–940. [*399*]

[1068] Smorodina N.V. *The invariant and quasi-invariant transformations of the stable Lévy processes.* Acta Appl. Math. 2007. V. 97, №1-3. P. 239–250. [*399*]

[1069] Smorodina N.V., Lifshits M.A. *On the distribution of the norm of a stable vector.* Teor. Verojatn. i Primen. 1989. V. 34, №2. P. 304–313 (in Russian); English transl.: Theory Probab. Appl. 1989. V. 34, №2. P. 266–274. [*324*]

[1070] Solé J.L., Utzet F., Vives J. *Canonical Lévy process and Malliavin calculus.* Stoch. Process. Appl. 2007. V. 117, №2. P. 165–187. [*323*]

[1071] Song S. *A study of Markovian maximality, change of probability measure and regularity.* Potential Anal. 1994. V. 3, №4. P. 391–422. [*420*]

[1072] Srimurthy V.K. *On the equivalence of measures on loop space.* Probab. Theory Relat. Fields. 2000. V. 118, №4. P. 522–546. [*399*]

[1073] Steblovskaya V.R. *Smoothness of finite-dimensional images of measures.* Ukr. Mat. Zh. 1989. V. 41, №2. P. 261–265 (in Russian); English transl.: Ukr. Math. J. 1989. V. 41, №2. P. 233–237. [*323, 324*]

[1074] Steblovskaya V.R. *On transformations of measures related to second order differential equations.* Stochastic analysis and applications (Powys, 1995). P. 408–418. World Sci. Publ., River Edge, New Jersey, 1996. [*366*]

[1075] Steblovskaya V. *On a method of asymptotic expansions for infinite dimensional integrals with respect to smooth measures.* Methods Funct. Anal. Topol. 2000. V. 6, №1. P. 3–13. [*324*]

[1076] Steblovskaya V. *Finite dimensional images of smooth measures.* Methods Funct. Anal. Topol. 2004. V. 10, №3. P. 64–76. [*366*]

[1077] Steerneman T. *On the total variation and Hellinger distance between signed measures.* Proc. Amer. Math. Soc. 1983. V. 88, №4. P. 684–690. [*8*]

[1078] Stein E.M. *Topics in harmonic analysis related to the Littlewood–Paley theory.* Princeton University Press, Princeton, 1970; 149 p. [*237*]

[1079] Stein E. *Singular integrals and differentiability properties of functions.* Princeton University Press, Princeton, 1970; xiv+290 p. [*56, 57, 66*]

[1080] Stein E., Weiss G. *Introduction to Fourier analysis on Euclidean spaces.* Princeton University Press, Princeton, 1971; x+297 p. [*66*]

[1081] Stengle G. *A divergence theorem for Gaussian stochastic process expectations.* J. Math. Anal. Appl. 1968. V. 21. P. 537–546. [*325*]

[1082] Stroock D. *The Malliavin calculus, a functional analytic approach.* J. Funct. Anal. 1981. V. 44, №2. P. 212–257. [*281, 323, 370, 372, 398, 425*]

[1083] Stroock D. *The Malliavin calculus and its application to second order parabolic differential equation. I, II.* Math. Systems Theory. 1981. V. 14, №1. P. 25–65; 1981. V. 14, №2. P. 141–171. [*323, 425*]

[1084] Stroock D. *The Malliavin calculus and its applications.* Lecture Notes in Math. 1981. V. 851. P. 394–432. [*323, 425*]

[1085] Stroock D. *Some applications of stochastic analysis to partial differential equations.* Lecture Notes in Math. 1983. V. 976. P. 268–381. [*323, 425*]

[1086] Sturm K.-T. *On the geometry of metric measure spaces. I, II.* Acta Math. 2006. V. 196, №1. P. 65–131, 133–177. [*399*]

[1087] Sudakov V.N. *Linear sets with quasi-invariant measure.* Dokl. Akad. Nauk SSSR. 1959. V. 127, №3. P. 524–525 (in Russian). [*103*]

[1088] Sudakov V.N. *On the characterization of quasi-invariance of measures in Hilbert space.* Usp. Mat. Nauk. 1963. V. 18, №1. P. 188–190 (in Russian). [*103*]

[1089] Sudakov V.N. *Geometric problems of the theory of infinite-dimensional probability distributions.* Trudy Mat. Inst. Steklov. 1976. V. 141. P. 1–190 (in Russian); English transl.: Proc. Steklov Inst. Math. 1979. №2. P. 1–178. [*132*]

[1090] Sugita H. *Sobolev space of Wiener functionals and Malliavin calculus.* J. Math. Kyoto Univ. 1985. V. 25, №1. P. 31–48. [*274*]

[1091] Sugita H. *On a characterization of the Sobolev spaces over an abstract Wiener space.* J. Math. Kyoto Univ. 1985. V. 25, №4. P. 717–725. [*274*]

[1092] Sugita H. *Positive Wiener functionals and potential theory over abstract Wiener spaces.* Osaka J. Math. 1988. V. 25, №3. P. 665–698. [*270*]

[1093] Sugita H. *Properties of holomorphic Wiener functions — skeleton, contraction, and local Taylor expansion.* Probab. Theory Relat. Fields. 1994. V. 100, №1. P. 117–130. [*275*]

[1094] Sugita H., Taniguchi S. *Oscillatory integrals with quadratic phase function on a real abstract Wiener space.* J. Funct. Anal. 1998. V. 155, №1. P. 229–262. [*325*]

[1095] Sugita H., Taniguchi S. *A remark on stochastic oscillatory integrals with respect to a pinned Wiener measure.* Kyushu J. Math. 1999. V. 53, №1. P. 151–162. [*325*]

[1096] Suzuki M. *Applications of the Malliavin calculus to McKean equations.* J. Fac. Sci. Univ. Tokyo. 1990. V. 37, №1. P. 129–149. [*425*]

[1097] Sytaja G.N. *The admissible translations of weighted Gaussian measures.* Teor. Verojatnost. i Primenen. 1969. V. 14, №3. P. 527–531 (in Russian); Theor. Probability Appl. 1969. V. 14, №3. P. 506–509. [*132*]

[1098] Sztencel R. *On the lower tail of stable seminorm.* Bull. Acad. Polon Sci. 1984. V. 32, №11–12. P. 231–234. [*109*]

[1099] Takanobu S., Watanabe S. *Asymptotic expansion formulas of the Schilder type for a class of conditional Wiener functional integrations.* Asymptotic problems in probability theory: Wiener functionals and asymptotics (Sanda/Kyoto, 1990). P. 194–241. Pitman Res. Notes Math. Ser., 284, Longman Sci. Tech., Harlow, 1993. [*325*]

[1100] Takeda M. *On the uniqueness of Markovian self-adjoint extension of diffusion operators on infinite dimensional space.* Osaka J. Math. 1985. V. 22. P. 733–742. [*420*]

[1101] Takeda M. *On the uniqueness of the Markovian self-adjoint extension.* Lecture Notes in Math. 1987. V. 1250. P. 319–325. [*420*]

[1102] Takeda M. *The maximum Markovian self-adjoint extensions of generalized Schrödinger operators.* J. Math. Soc. Japan. 1992. V. 44. P. 113–130. [*420*]

[1103] Takeuchi A. *The Malliavin calculus for SDE with jumps and the partially hypoelliptic problem.* Osaka J. Math. 2002. V. 39, №3. P. 523–559. [*425*]

[1104] Takeuchi A. *Malliavin calculus for degenerate stochastic functional differential equations.* Acta Appl. Math. 2007. V. 97, №1-3. P. 281–295. [*425*]

[1105] Talagrand M. *Mesures gaussiennes sur un espace localement convexe.* Z. Wahrscheinlichkeitstheorie verw. Geb. 1983. B. 64. S. 181–209. [*29*]

[1106] Talagrand M. *Transportation cost for Gaussian and other product measures.* Geom. Funct. Anal. 1996. V. 6. P. 587–600. [*367*]

[1107] Taniguchi S. *Malliavin's stochastic calculus of variation for manifold-valued Wiener functionals and its applications.* Z. Wahr. theor. verw. Geb. 1983. B. 65, №2. S. 269–290. [*323*]

[1108] Taniguchi S. *Applications of Malliavin's calculus to time-dependent systems of heat equations.* Osaka J. Math. 1985. V. 22, №2. P. 307–320. [*425*]

[1109] Taniguchi S. *On Ricci curvatures of hypersurfaces in abstract Wiener spaces.* J. Funct. Anal. 1996. V. 136, №1. P. 226–244. [*399*]

[1110] Taniguchi S. *On the exponential decay of oscillatory integrals on an abstract Wiener space.* J. Funct. Anal. 1998. V. 154, №2. P. 424–443. [*325*]

[1111] Taniguchi S. *Stochastic oscillatory integrals with quadratic phase function and Jacobi equations.* Probab. Theory Relat. Fields. 1999. V. 114, №3. P. 291–308. [*325*]

[1112] Taniguchi S. *On Wiener functionals of order 2 associated with soliton solutions of the KdV equation.* J. Funct. Anal. 2004. V. 216, №1. P. 212–229. [*325*]

[1113] Taniguchi S. *On the quadratic Wiener functional associated with the Malliavin derivative of the square norm of Brownian sample path on interval.* Electron. Comm. Probab. 2006. V. 11. P. 1–10. [*325*]

[1114] Taniguchi S. *On the Jacobi field approach to stochastic oscillatory integrals with quadratic phase function.* Kyushu J. Math. 2007. V. 61, №1. P. 191–208. [*325*]

[1115] Tarieladze V.I. *On a topological description of characteristic functionals.* Dokl. Akad. Nauk SSSR. 1987. V. 295, №6. P. 1320–1323 (in Russian); English transl.: Soviet Math. Dokl. 1988. V. 36, №1. P. 205–208. [*96*]

[1116] Temam E. *Analysis of error with Malliavin calculus: application to hedging.* Math. Finance. 2003. V. 13, №1. P. 201–214. [*425*]

[1117] Timan A.F. Theory of approximation of functions of a real variable. Transl. from the Russian. Pergamon Press, Oxford, 1963; xii+631 p. (Russian ed.: Moscow, 1960). [*287*]

[1118] Tolmachev A.N. *A property of distributions of diffusion processes.* Matem. Zamet. 1993. V. 54, №3. P. 106–113 (in Russian); English transl.: Math. Notes. 1993. V. 54. P. 946–950. [*348*]

[1119] Tolmachev N.A. *On the smoothness and singularity of invariant measures and transition probabilities of infinite-dimensional diffusions.* Teor. Verojatn. i Primen.

1998. V. 43, №4. P. 798–808 (in Russian); English transl.: Theory Probab. Appl. 1998. V. 43, №4. P. 655–664. [*413*]

[1120] Tolmachev N.A., Khitruk F.A. *Transformations of Gaussian measures by stochastic flows.* Dokl. Ross. Akad. Nauk. 2004. V. 399, №4. P. 454–459 (in Russian); English transl.: Dokl. Math. 2004. V. 70, №3. P. 918–923. [*366*]

[1121] Tortrat A. *Lois $e(\lambda)$ dans les espaces vectoriels et lois stables.* Z. Wahr. theor. verw. Geb. 1976. B. 37, №2. P. 175–182. [*109*]

[1122] Trèves J.F. *Linear partial differential equations with constant coefficients: existence, approximation and regularity of solutions.* Gordon and Breach, New York, 1966; x+534 p. [*xiii*]

[1123] Triebel H. *Interpolation theory. Function spaces. Differential operators.* Deutscher Verlag des Wissenschaften, Berlin, 1978; 528 p. [*66, 243*]

[1124] Tsilevich N., Vershik A., Yor M. *An infinite-dimensional analogue of the Lebesgue measure and distinguished properties of the gamma process.* J. Funct. Anal. 2001. V. 185, №1. P. 274–296. [*399*]

[1125] Uchida M., Yoshida N. *Information criteria for small diffusions via the theory of Malliavin–Watanabe.* Stat. Inference Stoch. Process. 2004. V. 7, №1. P. 35–67. [*425*]

[1126] Ueki N. *A stochastic approach to the Poincaré–Hopf theorem.* Probab. Theory Relat. Fields. 1989. V. 82, №2. P. 271–293. [*406*]

[1127] Ueki N. *Asymptotic expansion of stochastic oscillatory integrals with rotation invariance.* Ann. Inst. H. Poincare Probab. Statist. 1999. V. 35, №4. P. 417–457. [*325*]

[1128] Uemura H. *On a short time expansion of the fundamental solution of heat equations by the method of Wiener functionals.* J. Math. Kyoto Univ. 1987. V. 27, №3. P. 417–431. [*405*]

[1129] Uglanov A.V. *The heat equation for measures in a rigged Hilbert space.* Vestnik Moskov. Univ. Ser. I Mat. Meh. 1971. V. 26, №1. P. 52–60 (Russian); English transl.: Moscow Univ. Math. Bull. 1972. V. 27 (1972), №5. P. 10–18. [*426*]

[1130] Uglanov A.V. *Differentiable measures in a rigged Hilbert space.* Vestnik Mosk. Univ. 1972. №5. P. 14–24 (in Russian); English transl.: Mosc. Univ. Math. Bull. 1972. V. 27, №5-6. P. 10–18. [*103, 276*]

[1131] Uglanov A.V. *A result on differentiable measures on a linear space.* Mat. Sb. 1976. V. 29. P. 242–247 (in Russian); English transl.: Math. USSR Sb. 1976. V. 29. P. 217–222. [*103*]

[1132] Uglanov A.V. *Surface integrals and differential equations on an infinite-dimensional space.* Dokl. Akad. Nauk SSSR. 1979. V. 247, №6. P. 1331–1335 (in Russian); English transl.: Sov. Math. Dokl. 1979. V. 20. P. 917–920. [*325, 426*]

[1133] Uglanov A.V. *Surface integrals in a Banach space.* Mat. Sb. 1979. V. 110, №2. P. 189–217 (in Russian); English transl.: Math. USSR Sb. 1981. V. 38. P. 175–199. [*325*]

[1134] Uglanov A.V., *On the division of generalized functions of an infinite number of variables by polynomials.* Dokl. Akad. Nauk SSSR. 1982. V. 264, №5. P. 1096–1099 (in Russian); English transl.: Soviet Math. Dokl. 1982. V. 25, №3. P. 843–846. [*276, 324, 325*]

[1135] Uglanov A.V. *The Newton–Leibniz formula on Banach spaces and approximation of functions of an infinite-dimensional argument.* Izv. Akad. Nauk SSSR, Ser. Mat. 1987. V. 51, №1. P. 152–170 (in Russian); English transl.: Math. USSR Izv. 1988. V. 30. P. 145–161. [*155, 306, 325*]

[1136] Uglanov A.V. *Smoothness of distributions of functionals of random processes.* Teor. Verojatn. i Primen. 1988. V. 33, №3. P. 535–544 (in Russian); English transl.: Theory Probab. Appl. 1988. V. 33. P. 500–508. [*324, 325*]

[1137] Uglanov A.V. *A quotient of smooth measures is a smooth function.* Izv. Vyssh. Uchebn. Zaved., Mat. 1989. №9. P. 72–76 (in Russian); English transl.: Sov. Math. 1989. V. 33, №9. P. 71–75. [*161, 188*]

[1138] Uglanov A.V. *Factorization of smooth measures.* Mat. Zametki. 1990. V. 48, №5. P. 121–127 (in Russian); English transl.: Math. Notes. 1990. V. 48, №5. P. 1158–1162. [*103*]

[1139] Uglanov A.V. *Surface integrals in Fréchet spaces.* Mat. Sb. 1998. V. 189, №11. P. 139–157 (in Russian); English transl.: Sb. Math. 1998. V. 189, №11. P. 1719–1737. [*325*]

[1140] Uglanov A.V. *The infinite-dimensional Lagrange problem.* Dokl. Ross. Akad. Nauk. 2000. V. 372, №6. P. 740–742 (Russian); English transl.: Dokl. Math. 2000. V. 61, №3. P. 430–432. [*325*]

[1141] Uglanov A.V. Integration on infinite-dimensional surfaces and its applications. Kluwer Acad. Publ., Dordrecht, 2000; 262 p. [*103, 324, 325, 325*]

[1142] Uglanov A.V. *Surface integrals in locally convex spaces.* Tr. Mosk. Mat. Ob-va. 2001. V. 62. P. 262–285 (in Russian); English transl.: Trans. Mosc. Math. Soc. 2001. P. 249–270. [*325*]

[1143] Üstünel A.S. *Applications of integration by parts formula for infinite-dimensional semimartingales.* J. Multivar. Anal. 1986. V. 18, №2. P. 287–299. [*323*]

[1144] Üstünel A.S. *Representation of distributons on Wiener space and stochastique calculus of variations.* J. Funct. Anal. 1987. V. 70, №1. P. 126–139. [*276, 323*]

[1145] Üstünel A.S. *Some comments on the filtering of diffusions and the Malliavin calculus.* Lecture Notes in Math. 1988. V. 1316. P. 247–267. [*323, 425*]

[1146] Üstünel A.S. *Extension of the Itô calculus via the Malliavin calculus.* Stochastics. 1988. V. 23, №3. P. 353–375. [*323*]

[1147] Üstünel A.S. *The Itô formula for anticipative processes with non-monotonous time scale via the Malliavin calculus.* Probab. Theory Relat. Fields. 1988. V. 79, №2. P. 249–269. [*323*]

[1148] Üstünel A.S. *Intégrabilité exponentielle de fonctionelles de Wiener.* C. R. Acad. Sci. Paris, sér. 1. 1992. T. 315, №9. P. 997–1000. [*274*]

[1149] Üstünel A.S. An introduction to analysis on Wiener space. Lecture Notes in Math. V. 1610. Springer-Verlag, Berlin, 1995; x+95 p. [*274, 276, 323*]

[1150] Üstünel A.S. *Some exponential moment inequalities for the Wiener functionals.* J. Funct. Anal. 1996. V. 136, №1. P. 154–170. [*274*]

[1151] Üstünel A.S. *Entropy, invertibility and variational calculus of adapted shifts on Wiener space.* J. Funct. Anal. 2009. V. 257, №11. P. 3655–3689. [*366*]

[1152] Üstünel A.S., Zakai M. *On independence and conditioning on Wiener space.* Ann. Probab. 1989. V. 17, №4. P. 141–154. [*376, 379*]

[1153] Üstünel A.S., Zakai M. *On the structure of independence on Wiener space.* J. Funct. Anal. 1990. V. 90, №1. P. 113–137. [*376, 377*]

[1154] Üstünel A.S., Zakai M. *Transformations of Wiener measure under anticipative flows.* Probab. Theory Relat. Fields. 1992. V. 93, №1. P. 91–136. [*366*]

[1155] Üstünel A.S., Zakai M. *Applications of the degree theorem to absolute continuity on Wiener space.* Probab. Theory Relat. Fields. 1993. V. 95, №4. P. 509–520. [*366*]

[1156] Üstünel A.S., Zakai M. *Transformation of the Wiener measure under non-invertible shifts.* Probab. Theory Relat. Fields. 1994. V. 99, №4. P. 485–500. [*366*]

[1157] Üstünel A.S., Zakai M. *The composition of Wiener functionals with non absolutely continuous shifts.* Probab. Theory Relat. Fields. 1994. V. 98, №1. P. 163–184. [*366*]

[1158] Üstünel A.S., Zakai M. *Random rotations of the Wiener path.* Probab. Theory Relat. Fields. 1995. V. 103, №3. P. 409–429. [*366*]

[1159] Üstünel A.S., Zakai M. *Measures induced on Wiener space by monotone shifts.* Probab. Theory Relat. Fields. 1996. V. 105, №4. P. 545–563. [*366*]

[1160] Üstünel A.S., Zakai M. *Degree theory on Wiener space.* Probab. Theory Relat. Fields. 1997. V. 108, №2. P. 259–279. [*366*]

[1161] Üstünel A.S., Zakai M. *The construction of filtrations on abstract Wiener space.* J. Funct. Anal. 1997. V. 143, №1. P. 10–32. [*275*]

[1162] Üstünel A.S., Zakai M. *The Sard inequality on Wiener space.* J. Funct. Anal. 1997. V. 149, №1. P. 226–244. [*366*]

[1163] Üstünel A.S., Zakai M. *On the uniform integrability of the Radon–Nikodým densities for Wiener measure.* J. Funct. Anal. 1998. V. 159, №2. P. 642–663. [*366*]

[1164] Üstünel A.S., Zakai M. Transformation of measure on Wiener space. Springer-Verlag, Berlin, 2000; xiv+296 p. [*274, 323, 329, 341, 367*]

[1165] Üstünel A.S., Zakai M. *The invertibility of adapted perturbations of identity on the Wiener space.* C. R. Math. Acad. Sci. Paris. 2006. T. 342, №9. P. 689–692. [*366*]

[1166] Vakhania N.N., Tarieladze V.I. *Covariance operators of probability measures in locally convex spaces.* Teor. Verojatn. i Primen. 1978. V. 23. P. 3–26 (in Russian); English transl.: Theor. Probab. Appl. 1978. V. 23. P. 1–21. [*142*]

[1167] Vakhania N.N., Tarieladze V.I., Chobanyan S.A. Probability distributions in Banach spaces. Translated from the Russian. Kluwer Academic Publ., 1991; xxvi+482 p. (Russian ed.: Moscow, 1985). [*35, 106, 138, 156, 203, 220, 270*]

[1168] Van Biesen J. *The divergence on submanifolds of the Wiener space.* J. Funct. Anal. 1993. V. 113, №2. P. 426–461. [*399*]

[1169] Vârsan C. *On the regularity of the probabilities associated with diffusions.* Rev. Roum. Math. Pures Appl. 1987. V. 32, №2. P. 159–177. [*425*]

[1170] Vârsan C. *On the regularity of the probabilities associated with a class of stochastic equations with delay.* Rev. Roum. Math. Pures Appl. 1988. V. 33, №6. P. 529–536. [*425*]

[1171] Veretennikov A.Yu. *Probabilistic approach to hypoellipticity.* (Russian) Uspekhi Mat. Nauk. 1983. V. 38, №3. P. 113–125 (in Russian); English transl.: Russian Math. Surveys. 1983. V. 38, №3. P. 127–140. [*323, 425*]

[1172] Veretennikov A.Yu. *Probabilistic problems in the theory of hypoellipticity.* Izv. Akad. Nauk SSSR Ser. Mat. 1984. V. 48, №6. P. 1151–1170 (in Russian); English transl.: Izvestiya Math. 1985. V. 25, №3. P. 455–473. [*323, 425*]

[1173] Vershik A.M., Gel'fand I.M., Graev M.I. *Representations of the group of diffeomorphisms.* Uspehi Mat. Nauk. 1975. V. 30. №6. P. 3–50 (in Russian); English transl.: Russ. Math. Surv. 1975. V. 30, №6. P. 1–50. [*399*]

[1174] Villani C. Topics in optimal transportation. Amer. Math. Soc., Rhode Island, 2003; 370 p. [*363, 367*]

[1175] Villani C. Optimal transport, old and new. Springer, New York, 2009; xxii+973 p. [*363, 367*]

[1176] von Vintschger R. *A duality equation for diffusions on an abstract Wiener space.* Stochastic Anal., Path Integration and Dynamics. Pitman Research Notes in Math. Ser. V. 200. P. 215–222. Longman Sci. Tech., Harlow; John Wiley, New York, 1989. [*415*]

[1177] Vodop'yanov S.K. *Topological and geometric properties of mappings with an integrable Jacobian in Sobolev classes. I.* Sibirsk. Mat. Zh. 2000. V. 41, №1. P. 23–48 (in Russian); English transl.: Siberian Math. J. 2000. V. 41, №1. P. 19–39. [*274*]

[1178] Watanabe S. Stochastic differential equations and Malliavin calculus. Tata Inst. of Fundamental Research, Bombay; Springer-Verlag, Berlin, 1984; iii+111 p. [*274, 323, 405*]

[1179] Watanabe S. *Analysis of Wiener functionals (Malliavin calculus) and its applications to heat kernels.* Ann. Probab. 1987. V. 15, №1. P. 1–39. [*405*]

[1180] Watanabe S. *Short time asymptotic problems in Wiener functional integration theory. Applications to heat kernels and index theorems.* Lecture Notes in Math. 1990. V. 1444. P. 1–62. [*405*]

[1181] Watanabe S. *Donsker's δ-function in the Malliavin calculus.* Stochastic Analysis. P. 495–502. Academic Press, Boston, 1991. [*323*]

[1182] Watanabe S. *Disintegration problems in Wiener functional integrations.* Probability theory (Singapore, 1989). P. 181–198. De Gruyter, Berlin, 1992. [*323*]

[1183] Watanabe S. *Stochastic analysis and its applications.* Sugaku Expositions. 1992. V. 5, №1. P. 51–69. [*425*]

[1184] Watanabe S. *Fractional order Sobolev spaces on Wiener space.* Probab. Theory Relat. Fields. 1993. V. 95. P. 175–198. [*243, 274*]

[1185] Watanabe S. *Itô calculus and Malliavin calculus.* Stochastic analysis and applications, pp. 623–639, Abel Symp., 2, Springer, Berlin, 2007. [*425*]

[1186] von Weizsäcker H. (fon Vaĭtszekker Kh.), Léandre R., Smolyanov O.G. *Algebraic properties of infinite-dimensional differential forms of finite codegree.* Dokl. Ross. Akad. Nauk. 1999. V. 369, №6. P. 727–731 (in Russian); English transl.: Dokl. Math. 1999. V. 60, №3. P. 412–415. [*399*]

[1187] von Weizsäcker H. (fon Vaĭtszekker Kh.), Smolyanov O.G. *Smooth curves in spaces of measures and shifts of differentiable measures along vector fields.* Dokl. Ross. Akad. Nauk. 1994. V. 339, №5. P. 584–587 (in Russian); English transl.: Russian Acad. Sci. Dokl. Math. 1995. V. 50, №5. P. 476–481. [*398*]

[1188] von Weizsäcker H. (fon Vaĭtszekker Kh.), Smolyanov O.G. *Noether theorems for infinite-dimensional variational problems.* Dokl. Ross. Akad. Nauk. 1998. V. 361, №5. P. 583–586 (in Russian); English transl.: Dokl. Math. 1998. V. 58, №1. P. 91–94. [*325*]

[1189] von Weizsäcker H. (fon Vaĭtszekker Kh.), Smolyanov O.G. *Differential forms on infinite-dimensional spaces and an axiomatic approach to the Stokes formula.* Dokl. Ross. Akad. Nauk. 1999. V. 367, №2. P. 151–154 (in Russian); English transl.: Dokl. Math. 1999. V. 60, №1. P. 22–25. [*399*]

[1190] von Weizsäcker H. (fon Vaĭtszekker Kh.), Smolyanov O.G. *Relations between smooth measures and their logarithmic gradients and derivatives.* Dokl. Ross. Akad. Nauk. 1999. V. 369, №2. P. 158–162 (in Russian); English transl.: Dokl. Math. 1999. V. 60, №3. P. 337–340. [*223*]

[1191] Wentzell A.D. *A course in the theory of stochastic processes.* Translated from the Russian. McGraw-Hill International Book, New York, 1981; x+304 p. (Russian ed.: Moscow, 1975). [*35, 125*]

[1192] Werner W. *The conformally invariant measure on self-avoiding loops.* J. Amer. Math. Soc. 2008. V. 21, №1. P. 137–169. [*399*]

[1193] Williams D. *To begin at the beginning.* Lecture Notes in Math. 1981. V. 851. P. 1–55. [*323, 425*]

[1194] Wu L.M. *Construction de l'opérateur de Malliavin sur l'espace de Poisson.* Lecture Notes in Math. 1987. V. 1247. P. 100–114. [*323, 399*]

[1195] Wu L.M. *Un traitement unifié de la représentation des fonctionnelles de Wiener.* Lecture Notes in Math. 1990. V. 1426. P. 166–187. [*275*]

[1196] Yablonski A. *The calculus of variations for processes with independent increments.* Rocky Mountain J. Math. 2008. V. 38, №2. P. 669–701. [*323*]

[1197] Yakhlakov V.Yu. *Surface measures on surfaces of finite codimension in a Banach space.* Mat. Zametki. 1990. V. 47, №4. P. 147–156 (in Russian); English transl.: Math. Notes. 1990. V. 47, №4. P. 414–421 (1990). [*325*]

[1198] Yamasaki Y., Hora A. *Differentiable shifts for measures on infinite dimensional spaces.* Publ. RIMS, Kyoto Univ. 1987. V. 23. P. 275–296. [*75, 103, 104, 141*]

[1199] Yan F., Mohammed S. *A stochastic calculus for systems with memory.* Stoch. Anal. Appl. 2005. V. 23, №3. P. 613–657. [*425*]

[1200] Yano K. *A generalization of the Buckdahn–Föllmer formula for composite transformations defined by finite dimensional substitution.* J. Math. Kyoto Univ. 2002. V. 42, №4. P. 671–702. [*366*]

[1201] Yoshida Nakahiro. *Asymptotic expansions of maximum likelihood estimators for small diffusions via the theory of Malliavin–Watanabe.* Probab. Theory Related Fields. 1992. V. 92, №3. P. 275–311. [*425*]

[1202] Yoshida Nakahiro. *Asymptotic expansion of Bayes estimators for small diffusions.* Probab. Theory Related Fields. 1993. V. 95, №4. P. 429–450. [*425*]

[1203] Yoshida Nakahiro. *Asymptotic expansions for perturbed systems on Wiener space: maximum likelihood estimators.* J. Multivar. Anal. 1996. V. 57, №1. P. 1–36. [*425*]

[1204] Yoshida Nakahiro. *Malliavin calculus and asymptotic expansion for martingales.* Probab. Theory Relat. Fields. 1997. V. 109, №3. P. 301–342. [*425*]

[1205] Yoshida Nakahiro. *Malliavin calculus and martingale expansion.* Bull. Sci. Math. 2001. V. 125, №6-7. P. 431–456. [*425*]

[1206] Yoshida Nakahiro. *Conditional expansions and their applications.* Stochastic Process. Appl. 2003. V. 107, №1. P. 53–81. [*425*]

[1207] Yoshida Nobuo. *A large deviation principle for (r,p)-capacities on the Wiener space.* Probab. Theory Related Fields. 1993. V. 94, №4. P. 473–488. [*276*]

[1208] Yoshida Nobuo. *The Littlewood–Paley–Stein inequality on an infinite dimensional manifold.* J. Funct. Anal. 1994. V. 122, №2. P. 402–427. [*399*]

[1209] Yost D. *If every donut is a teacup, then every Banach space is a Hilbert space.* Seminar on Functional Anal. V. 1. P. 127–148. Univ. de Murcia, 1987. [*350*]

[1210] Zabczyk J. *Structural properties and limit behaviour of linear stochastic systems in Hilbert spaces.* Math. Control Theory. Banach Center Publ., V. 14. P. 591–609. Warsaw, 1981. [*420*]

[1211] Zabczyk J. *Parabolic equations on Hilbert spaces.* Lecture Notes in Math. 1998. V. 1715. P. 116–213. [*426*]

[1212] Zakai M. *Malliavin calculus.* Acta Appl. Math. 1985. V. 3, №2. P. 175–207. [*281*]

[1213] Zakai M. *Malliavin derivatives and derivatives of functionals of the Wiener process with respect to a scale parameter.* Ann. Probab. 1985. V. 13, №2. P. 609–615. [*259*]

[1214] Zakai M. *Stochastic integration, trace and the skeleton of Wiener functionals.* Stochastics Stoch. Rep. 1985. V. 32. P. 93–108. [*275*]

[1215] Zakai M. *Rotations and tangent processes on Wiener space.* Lecture Notes in Math. 2005. V. 1857. P. 205–225. [*399*]

[1216] Zakai M., Zeitouni O. *When does Ramer formula look like Girsanov formula?* Ann. Probab. 1992. V. 20, №3. P. 1436–1440. [*366*]

[1217] Zambotti L. *Integration by parts formulae on convex sets of paths and applications to SPDEs with reflection.* Probab. Theory Relat. Fields. 2002. V. 123, №4. P. 579–600. [*132*]

[1218] Zambotti L. *Integration by parts on δ-Bessel bridges, $\delta > 3$ and related SPDEs.* Ann. Probab. 2003. V. 31, №1. P. 323–348. [*132*]

[1219] Zambotti L. *Integration by parts on the law of the reflecting Brownian motion.* J. Funct. Anal. 2005. V. 223, №1. P. 147–178. [*132*]

[1220] Zhdanov R.I. *Continuity and differentiability of triangular mappings.* Dokl. Ross. Akad. Nauk. 2010 (in Russian); English transl.: Dokl. Math. 2010. [*362*]

[1221] Zhdanov R.I., Ovsienko Yu.V. *Estimates of Sobolev norms of triangular mappings.* Vestn. Mosk. Univ., Ser. I. 2007, №1. P. 3–6 (in Russian); English transl.: Moscow Univ. Math. Bull. 2007. V. 62, №1. P. 1–4. [*362*]

[1222] Zhikov V.V. *Weighted Sobolev spaces.* Mat. Sb. 1998. V. 189, №8. P. 27–58 (in Russian); English transl.: Sb. Math. 1998. V. 189, №8. P. 1139–1170. [*62*]

[1223] Zhou X.Y. *Infinite-dimensional Malliavin calculus and its application.* Acta Math. Appl. Sinica (English Ser.). 1992. V. 8, №2. P. 97–114. [*323*]

[1224] Ziemer W. *Weakly differentiable functions.* Springer-Verlag, New York – Berlin, 1989; xvi+308 p. [*54, 66*]

[1225] Zinn J. *Admissible translates of stable measures.* Studia Math. 1976. V. 54, №3. P. 245–257. [*110*]

[1226] Zolotarev V.M. *One-dimensional stable distributions.* Translated from the Russian. Amer. Math. Soc., Providence, Rhode Island, 1986; x+284 p. (Russian ed.: Moscow, 1983). [*109*]

Subject Index

Symbols:[1]

$\mathcal{A}_\mu$, 6
apDf, 52
apD$_i f$, 53
$\mathfrak{A}$, 344
$\mathfrak{A}_e$, 344
$\mathfrak{A}\{e_n\}$, 344
$\mathfrak{A}^\mathcal{B}$, 343
$\mathfrak{A}_e^\mathcal{B}$, 342
$\mathfrak{A}^\mathcal{B}\{e_n\}$, 343
$BV(\mathbb{R}^n)$, 51
$\mathcal{B}(X)$, 6
$\mathcal{B}a(X)$, 6
β_H^μ, 190
β_h^μ, 79
β_v^μ, 181
$\beta_h^{\mu,\mathrm{loc}}$, 184
C_0^∞, 2
$C_b(T)$, 2
$C(\mu)$, 133
$C_\mathcal{F}$, 265
$C_{p,r}$, 272
c_0, 2
$D(\mu)$, 133
$D_C(\mu)$, 133
$D_H^k f$, 229, 230
$D^{p,r}(\mu)$, 230, 239
$D^\infty(\mu)$, 230
$D^{\infty,r}(\mu)$, 230
$D^{p,\infty}(\mu)$, 230
$D^{p,r}(\mu, E)$, 230
$\mathcal{D}(\mathbb{R}^n)$, 2
$\mathcal{D}'(\mathbb{R}^n)$, 2
$\mathfrak{D}(v, \mu)$, 181

---

[1] The symbols are ordered alphabetically according to their names, with the exception of the entries beginning with mathematical symbols.

$d_h\mu$, 69
$d_v\mu$, 181
$d_h^m\mu$, 70
$d_{h_1} \ldots d_{h_n}\mu$, 70
$d\nu/d\mu$, 8
$\det_2(I + K)$, 331
dist (x, E), 4
dist$_H(x, A)$, 277, 358
δv, 254
δ_a, 23
$\delta_\mu v$, 181
$\mathbb{E}^\mathcal{B}$, 22
$\mathcal{E}^{p,\alpha}(\mu)$, 243
$\mathcal{F}C_b^\infty(X)$, 4, 227
$G^{p,1}(\mu)$, 62, 233, 239
$G^{p,r}(\mu)$, 233, 239
$G_2(H, \mu, X)$, 203
$G_{\mathrm{oper}}^{p,1}(\mu, E)$, 233
$\mathcal{G}$, 344
$\mathcal{G}^\mathcal{B}$, 343
γ_h, 26
$H(\gamma)$, 25
$H(\mu)$, 133
$H(\mu, \nu)$, 8
$H_p(\mu)$, 133
$H^{p,r}(\mu)$, 235, 239
$H^\infty(\mu)$, 236
$\mathcal{H}$, 3
$\mathcal{H}(H, E)$, 3
$\mathcal{H}_n$, 3
$\widehat{h}$, 24
I_A, 6
j_H, 189
$L^p(\mu)$, 7
$L^p_{\mathrm{loc}}(\Omega)$, 39
$L^\infty(\mu)$, 7
$L^p(\mu, X)$, 7
$\mathcal{L}(X)$, 3
$\mathcal{L}(X, Y)$, 3
$\mathcal{L}^0(\mu)$, 7

$\mathcal{L}^p(\mu)$, 7
$\mathcal{L}^\infty(\mu)$, 7
$\mathcal{L}^p(\mu, X)$, 7
$\mathcal{L}_{(1)}(H)$, 4
$\mathcal{MK}$, 344
$\mathcal{MK}^\mathcal{B}$, 343
$\widetilde{\mu}$, 16
μ_h, 69
μ^+, μ^-, 6
$\mu \sim \nu$, 8
$\mu \perp \nu$, 8
$\mu \circ f^{-1}$, 9
$\mu \otimes \nu$, 16
$\mu \ast \nu$, 16
$\nu \ll \mu$, 8
P_W, 23
$\mathcal{P}$, 344
$\mathcal{P}^\mathcal{B}$, 343
$Q(\mu)$, 133
$\mathbb{R}^\infty$, 2
$\mathcal{S}(\mathbb{R}^n)$, 2
$\mathcal{S}'(\mathbb{R}^n)$, 2
$\sigma(X)$, 6
$\mathrm{trace}_H A$, 183
$W^{p,r}_{\mathrm{loc}}(\gamma)$, 241
$W^{p,r}_{\mathrm{loc}}(\gamma, E)$, 241
$W^{p,1}(\mu)$, 58, 228
$W^\infty(\mu)$, 228
$W^{\infty,r}(\mu)$, 228
$W^{p,\infty}(\mu)$, 228
$W^{p,k}(\Omega)$, 39
$W^{p,k}(\Omega, \mathbb{R}^d)$, 40
$W^{p,k}_{\mathrm{loc}}(\Omega, \mathbb{R}^d)$, 40
$W^{p,r}(\mathbb{R}^n)$, 65, 239
X^*, 2
X^*_γ, 24
$\mathcal{X}_k$, 29, 251

$|\cdot|_H$, 2, 25
$(\cdot, \cdot)_H$, 2, 25
$\|\cdot\|_{\mathcal{L}(X,Y)}$, 3
$\|f\|_p$, 7
$\|f\|_\infty$, 7
$\|f\|_{p,k}$, 40
$\|f\|_{p,r}$, 228
$\|f\|_{L^p(\mu)}$, 7
$\|f\|_{p,1,\mu}$, 58
$|\mu|$, 6
$\|\mu\|$, 6
$\nabla_H F$, 5, 228, 233
$\nabla^n_H F$, 228, 230
$\partial_h f$, 78, 157, 227
$\partial_v \varphi$, 181

A

Albeverio–Høegh Krohn
  differentiability, 70
absolute continuity of measures, 8
absolutely continuous mapping, 4
absolutely convex hull, 2
absolutely convex set, 2
absolutely summing mapping, 4
absolutely summing operator, 4
approximate Jacobian, 52
approximate derivative, 52, 53
approximately differentiable
  mapping, 52
atomless measure, 7

B

Baire σ-algebra, 6
Besov space, 66
Borel σ-algebra, 6
Borel set, 6

C

Caccioppoli set, 52
Cameron–Martin
  – space, 25
  – formula, 26
Chebyshev–Hermite polynomial, 29
Choquet capacity, 266
canonical triangular mapping, 359
capacity, 265
  – Choquet, 265
  – Gaussian, 272
  – tight, 266
centered Gaussian measure, 24
closable Dirichlet form, 416
closable norm, 58, 229
compact embedding, 4
completely asymmetric measure, 108
conditional expectation, 22
conditional measure, 19
continuously embedded space, 4
convergence of measures weak, 17
convex function, 112
convex hull, 2
convex set, 2
convolution, 16
covariance operator, 24, 142
cubic measure, 343
cylindrical function, 5
cylindrical set, 15

Subject Index

D

Dirac's measure, 23
Dirichlet form, 416
  – closable, 416
  – irreducible, 417
decomposition Hahn–Jordan, 6
densely continuous measure, 142, 346
densely differentiable measure, 142, 346
derivative, 5
  – Fomin, 69, 70
  – Fréchet, 5
  – Hadamard, 5
  – Gâteaux, 5
  – Malliavin, 257
  – Skorohod, 71
  – Sobolev, 39
  – along a subspace, 5
  – approximate, 52, 53
  – generalized, 39
  – logarithmic, 79, 181, 371, 383
  – of a measure, 69, 71
    – – along a vector field, 181
    – – local, 183, 184
    – – vector, 190
    – – with respect to a class, 184
determinant
  of Fredholm–Carleman, 331
differentiable family of measures, 380
differentiability, 5
  – L^p-, 70, 386
  – Albeverio–Høegh Krohn, 70
  – Fomin, 69
  – Fréchet, 5
  – Gâteaux, 5
  – Hadamard, 5
  – Malliavin, 258
  – Skorohod, 71
diffusion, 34
  – symmetrizable, 414
diffusion process, 34
dispersion, 23
divergence of vector field, 181, 254, 256

E

embedding
  – compact, 4
  – continuous, 4
equilibrium potential, 268
equivalent measures, 8
exceptional set, 343

F

Fernique theorem, 28
Fisher information number, 67, 82, 386
Fomin
  derivative, 69, 70
  differentiability, 69
Fourier transform, 16
Fréchet
  – derivative, 5
  – space, 2
formula
  – Cameron–Martin, 26
  – Itô, 31
  – change of variables, 9
  – integration by parts, 158, 159
fractional Sobolev class, 65, 236
function
  – c-convex, 180
  – convex, 112
  – cylindrical, 5
  – of bounded variation, 51
  – smooth cylindrical, 4

G

γ-measurable linear
  – operator, 27
  – functional, 24
Galiardo–Nirenberg
  inequality, 47
Gâteaux derivative, 5
Gaussian capacity, 272
Gaussian measure, 23
  – Radon, 23
  – centered, 24
  – nondegenerate, 27
  – standard, 23, 29
Gaussian null-set, 343
Gibbs measure, 127, 422
generalized partial derivative, 233
generator of a semigroup, 236
group
  – of diffeomorphisms, 394
  – loop, 390
  – current, 390

H

Hadamard derivative, 5
Hahn–Jordan decomposition, 6
Hajek–Feldman theorem, 28
Hamza condition, 61
Hellinger integral, 8

Hermite polynomial, 29
Hilbert–Schmidt
 – norm, 3
 – operator, 3
hull
 – absolutely convex, 2
 – convex, 2
hypoelliptic operator, 401

I

Itô's integral, 30
 – multiple, 33
image of a measure, 9
index of asymmetry, 108
indicator of a set, 6
inequality
 – Galiardo–Nirenberg, 47
 – Krugova, 163
 – Poincaré, 48, 252, 277
 – Sobolev, 48
 – – logarithmic, 250
information number
   of Fisher, 67, 82, 386
integral
 – Hellinger, 8
 – Itô, 30
 – – multiple, 33
 – Ogawa, 262
 – Skorohod, 261
 – Stratonovich, 32
 – Wiener, 30

J

Jacobian approximate, 52

K

Kakutani alternative, 107
Krugova inequality, 163

L

L^p-differentiability, 70, 386
Lipschitz constant, 4
Lipschitz mapping, 4
Lusin property (N), 54, 330
locally convex space, 1
logarithmic Sobolev inequality, 250
logarithmic gradient, 190, 371
logarithmic derivative
 – of a measure, 79, 181, 371, 383
 – vector, 190
 – with respect to a class, 184

logarithmically concave measure, 112

M

Malliavin
 – matrix, 280, 373
 – operator , 258, 372
Markovian semigroup, 235
Meyer equivalence, 239
Minkowsky functional, 3
Monge–Ampère equation, 363
manifold measurable, 370
mapping
 – μ-measurable, 7
 – Lipschitzian, 4
 – absolutely continuous, 4
 – absolutely ray continuous, 230
 – absolutely summing, 4
 – approximately differentiable, 52
 – canonical triangular, 359
 – measurable, 7
 – optimal, 363
 – stochastically Gâteaux
   differentiable, 230
 – triangular, 358, 359
mean, 23, 24
measurable manifold, 370
measurable mapping, 7
measurable linear functional, 24
measurable linear operator, 27
measurable polynomial, 240, 251
measure, 6
 – H-ergodic, 249
 – H-spherically invariant, 219
 – L^p-differentiable, 70
 – S-differentiable, 71
 – Dirac, 23
 – Gaussian, 23
 – – Radon, 23
 – – centered, 24
 – – nondegenerate, 27
 – – standard, 23, 29
 – Gibbs, 127, 422
 – Poisson, 393
 – Shavgulidze, 394
 – Wiener, 23
 – analytic along h, 71
 – atomless, 7
 – convex, 112,
 – – uniformly, 362
 – completely asymmetric, 108
 – conditional, 19
 – continuous along a vector, 70

– cubic, 343
– densely continuous, 142, 346
– densely differentiable, 142, 346
– differentiable
 – – Albeverio–Høegh-Krohn, 70
 – – Fomin, 69
 – – Skorohod, 71
 – – along a vector field, 371
– exceptional in the Aronszajn sense, 346
– infinitesimally invariant, 410
– invariant, 408, 410
– logarithmically concave, 112
– of finite energy, 271
– pre-Gaussian, 143
– quasi-invariant along a vector, 70
– spectral, 108
– stable of order α, 107
– stationary, 408
– strictly stable, 108
– surface, 299
– symmetric, 108, 217
– uniformly convex, 362
– vector, 195
– with Gaussian covariance, 143
measures equivalent, 8
measures mutually singular, 8
mixture of measures, 128
modification of a function, 7
moment of a measure, 10
multiple Itô integral, 33

N

negligible set, 343
nondegenerate Gaussian measure, 27
norm
 Hilbert–Schmidt, 3
 Sobolev, 40, 228
nuclear operator, 3

O

Ogawa integral, 262
Ornstein–Uhlenbeck
– operator , 237
– semigroup, 237, 251, 257
– process, 31, 412
operator
– Hilbert–Schmidt, 3
– Malliavin, 258, 372
– Ornstein–Uhlenbeck , 237
– absolutely summing, 4
– covariance, 24, 142

– hypoelliptic, 401
– measurable linear, 27
– nuclear, 3
– trace class, 3
optimal mapping, 363
oscillatory integral, 297

P

Poincaré inequality, 48, 252, 277
Poisson measure, 393
Poisson space, 393
Prohorov theorem, 18
polar decomposition, 4
polynomial
– Chebyshev–Hermite, 29
– Hermite, 29
– measurable, 240, 251
pre-Gaussian measure, 143
process
– Ornstein–Uhlenbeck , 31, 412
– Wiener, 29
– diffusion, 34
– nonanticipating, 30
– adapted, 30
product of Radon measures, 16
product measure, 105
proper linear version, 25, 27
property (E), 330
property (N) of Lusin, 54, 330

Q

quasieverywhere, 267
quasicontinuity, 267
quasi-invariant measure, 70

R

Radon measure, 9
– Gaussian, 23
Radon–Nikodym
– density, 8
– property, 5
– theorem, 8
regular approximate derivative, 53
regularity of a measure, 9
rigged Hilbert space, 189

S

σ-algebra
– Baire, 6
– Borel, 6

Shavgulidze measure, 394
Skorohod
 – derivative, 71
 – integral , 261
Sobolev
 – class, 39, 228, 230, 233
 – – fractional, 65, 236
 – – weighted, 58
 – inequality, 48
 – – logarithmic, 250
 – norm, 40, 228
 – space, 39, 228, 230, 233
 – weighted class, 58
Souslin set, 11
Stratonovich integral, 32
semigroup
 – Markovian, 235
 – Ornstein–Uhlenbeck, 237, 251, 257
 – hypercontractive, 251
 – strongly continuous, 235
 – sub-Markovian, 235
 – symmetric diffusion, 237
semivariation of a vector measure, 195
sequentially complete space, 1
set
 – Borel, 6
 – Caccioppoli, 52
 – Gaussian null-, 343
 – Souslin, 11
 – absolutely convex, 2
 – convex, 2
 – negligible, 343
 – of full measure, 7
 – slim, 273
shift of a measure, 26, 69
slim set, 273
smooth cylindrical function, 4
solution of a stochastic equation
 – strong, 30
 – weak, 30
spectral measure, 108, 217
space
 – Besov, 66
 – Cameron–Martin, 25
 – Fréchet, 2
 – Poisson, 393
 – Sobolev, 39
 – compactly embedded, 4
 – continuously embedded, 4
 – locally convex, 1
 – rigged Hilbert, 189
 – sequentially complete, 1

stable measure, 107
standard Gaussian measure, 23, 29
stationary measure, 408
stochastic differential equation, 30
strictly stable measure, 107
strongly continuous semigroup, 235
sub-Markovian semigroup, 235
subspace of differentiability, 134
support of a measure topological, 9
surface measure, 299
symmetric measure, 108, 217

T

τ-differentiability, 380
theorem
 – A.D. Alexandroff, 52
 – Fernique, 28
 – Hajek–Feldman, 28
 – Girsanov, 34
 – Nikodym, 18
 – Radon–Nikodym, 8
 – Sobolev embedding, 46
tight capacity, 266
tight measure, 10
topological support of a measure, 9
triangular mapping, 358
 – canonical, 359

U

uniformly convex measure, 362

V

variation of vector measure, 195
vector logarithmic derivative, 190
vector measure, 195
version of a function, 7

W

Wiener
 – integral, 30
 – measure, 23
 – process, 29, 412
weak convergence of measures, 17
weighted Sobolev class, 58

Z

zero-one law, 28

DATE DUE

GAYLORD PRINTED IN U.S.A.

QA 312 .B638 2010

Bogachev, V. I. 1961-

Differentiable measures and
the Malliavin calculus